Lecture Notes in Computer Science 2565

Edited by G. Goos, J. Hartmanis, and J. van Leeuwen

Springer
*Berlin*
*Heidelberg*
*New York*
*Hong Kong*
*London*
*Milan*
*Paris*
*Tokyo*

José M.L.M. Palma Jack Dongarra
Vicente Hernández A. Augusto Sousa (Eds.)

# High Performance Computing for Computational Science – VECPAR 2002

5th International Conference
Porto, Portugal, June 26-28, 2002
Selected Papers and Invited Talks

Springer

Volume Editors

José M.L.M. Palma
A. Augusto Sousa
Faculdade de Engenharia da Universidade do Porto
Rua Dr. Roberto Frias, 4200-465 Porto, Portugal
E-mail: {jpalma,augusto.sousa}@fe.up.pt

Jack Dongarra
University of Tennessee, Department of Computer Science
Knoxville, TN 37996-1301, USA
E-mail: dongarra@cs.utk.edu

Vicente Hernández
Universidad Politécnica de Valencia
Departamento de Sistemas Informáticos y Computación
Camino de Vera, s/n, Apartado 22012, 46020 Valencia, Spain
E-mail: vhernand@dsic.upv.es

Cataloging-in-Publication Data applied for

A catalog record for this book is available from the Library of Congress

Bibliographic information published by Die Deutsche Bibliothek
Die Deutsche Bibliothek lists this publication in the Deutsche Nationalbibliographie;
detailed bibliographic data is available in the Internet at <http://dnb.ddb.de>.

CR Subject Classification (1998): D, F, C.2, G, J.2, J.3

ISSN 0302-9743
ISBN 3-540-00852-7 Springer-Verlag Berlin Heidelberg New York



Springer-Verlag Berlin Heidelberg New York
a member of BertelsmannSpringer Science+Business Media GmbH

http://www.springer.de


Printed in Germany

Typesetting: Camera-ready by author, data conversion by DA-TeX Gerd Blumenstein
Printed on acid-free paper SPIN: 10871851 06/3142 5 4 3 2 1 0

# Preface

The 5th edition of the VECPAR series of conferences marked a change of the conference title. The full conference title now reads *VECPAR 2002 — 5th International Conference on High Performance Computing for Computational Science*. This reflects more accurately what has been the main emphasis of the conference since its early days in 1993 – the use of computers for solving problems in science and engineering. The present postconference book includes the best papers and invited talks presented during the three days of the conference, held at the Faculty of Engineering of the University of Porto (Portugal), June 26–28 2002.

The book is organized into 8 chapters, which as a whole appeal to a wide research community, from those involved in the engineering applications to those interested in the actual details of the hardware or software implementation, in line with what, in these days, tends to be considered as Computational Science and Engineering (CSE).

The book comprises a total of 49 papers, with a prominent position reserved for the four invited talks and the two first prizes of the best student paper competition.

Chapter 1 deals with the application of discretization techniques to engineering problems. The invited lecture delivered by Rainald Löhner sets the tone in this chapter, comprised of six more papers covering areas such as, for instance, combustion and geology. The talks by Vipin Kumar and Leif Eriksson open Chaps. 2 and 3 on data-mining techniques and computing in chemistry and biology. Problem-solving environments is the major topic in Chap. 4, initiated by the invited talk on the cactus software by Ed Seidel. Chapter 5 has nine papers on linear and nonlinear algebra. The papers in Chap. 6, though under the title of cluster computing, may be of general interest. Cluster computing is living up to its expectations and enabling high-performance computing at an affordable price. The last chapter, Chap. 8, has 12 papers under the general title of software tools and environments, and is thus close to computer science.

### Student Papers

Awarding two first prizes for the best student paper was a consequence of the quality of and the difficulty in distinguishing between two good pieces of work originating from quite different fields. This was also the way found by the jury to honor the multidisciplinarity of the conference. At this point, we would like to wish the greatest success to the authors and in particular to the students Marc Lange and Arturo Escribano.

- *Investigation of Turbulent Flame Kernels Using DNS on Clusters*
  by Marc Lange (High-Performance Computing Center, Stuttgart, Germany)

– *Mapping Unstructured Applications into Nested Parallelism*
by Arturo González-Escribano (Universidad de Valladolid, Spain), Arjan J.C. van Gemund (Faculty of Information Technology and Systems, The Netherlands) and Valentín Cardeñoso-Payo (Universidad de Valladolid, Spain)

To conclude, we would like to state in writing our gratitude to all members of the Scientific Committee and additional referees. Their opinions and comments were essential in the preparation of this book and the conference programme. We hope that the knowledge and experience of many contributors to this book can be useful to a large number of readers.

October 2002

José M.L.M. Palma
Jack Dongarra
Vicente Hernández
A. Augusto Sousa

VECPAR is a series of conferences organized by the Faculty of Engineering of Porto (FEUP) since 1993.

# Acknowledgments

Apart from those acknowledged in the following pages, we would like to thank Vítor Carvalho, who created and maintained the conference website, and Alice Silva for the secretarial work.

A very special word of gratitude is deserved to the members of the Organizing Committee João Correia Lopes and José Magalhães Cruz. The web-based submission and refereeing and the management of the VECPAR databases would have been impossible without their precious help.

VECPAR 2002 was a landmark in many aspects. It was the first edition of the conference held at the new buildings of the Faculty of Engineering. It was the end of a series of conferences held in Porto; the next edition of the conference will be held in Valencia (Spain). We believe that now is the time to acknowledge the collaboration of many people, always behind the limelight, who have been essential to the success of the VECPAR series of conferences. To mention only a few, and extending our gratitude to all personnel under their supervision, it is our pleasure to thank Mrs. Branca Gonçalves, from the External Relations Service of FEUP and Mr. Luís Machado, from Abreu Travel Agency.

# Committees

## Organizing Committee

A. Augusto de Sousa (Chair)
José Couto Marques (Co-chair)
João Correia Lopes
José Magalhães Cruz
Fernando Silva

## Steering Committee

José Laginha Palma, Universidade do Porto, Portugal
Jack Dongarra, University of Tennessee, USA
Vicente Hernandez, Universidad Politécnica de Valencia, Spain
Michel Daydé, ENSEEIHT-IRIT, France
Satoshi Matsuoka, Tokyo Institute of Technology, Japan
José Fortes, University of Florida, USA

## Best Student Paper Award (Jury)

Álvaro Coutinho, Universidade Federal do Rio de Janeiro, Brazil
José Fortes, University of Purdue, USA
Ronan Guivarch, ENSEEIHT-IRIT, France
Rainald Löhner, George Mason University, USA
Vipin Kumar, University of Minnesota, USA
José Palma, Universidade do Porto, Portugal
Fernando Silva, Universidade do Porto, Portugal
Aad van der Steen, Utrecht University, The Netherlands

## Scientific Committee

| | |
|---|---|
| R. Benzi | Univ. de Roma, Italy |
| L. Biferale | Univ. de Roma, Italy |
| A. Coutinho | Univ. Federal do Rio de Janeiro, Brazil |
| J.C. Cunha | Univ. Nova de Lisboa, Portugal |
| F. d'Almeida | Univ. do Porto, Portugal |
| I. Duff | Rutherford Appleton Lab., UK |
| N. Ebecken | Univ. Federal do Rio de Janeiro, Brazil |
| S. Gama | Univ. do Porto, Portugal |
| J.M. Geib | Univ. Sciences et Technologies de Lille, France |
| W. Gentzsch | SUN, USA |

| | |
|---|---|
| A. George | Univ. of Florida, USA |
| L. Giraud | CERFACS, France |
| D. Knight | Rutgers Univ., USA |
| V. Kumar | Univ. Minnesota, USA |
| R. Löhner | George Mason Univ., USA |
| E. Luque | Univ. Autònoma de Barcelona, Spain |
| A. Nachbin | Inst. Matemática Pura e Aplicada, Brazil |
| A. Padilha | Univ. do Porto, Portugal |
| M. Peric | CD Star, Germany |
| T. Priol | IRISA/INRIA, France |
| I. Procaccia | Weizmann Institute of Science, Israel |
| R. Ralha | Univ. do Minho, Portugal |
| D. Ruiz | ENSEEIHT-IRIT, France |
| H. Ruskin | Dublin City Univ., Ireland |
| M. Stadtherr | Univ. Notre Dame, USA |
| F. Tirado | Univ. Complutense de Madrid, Spain |
| A. Trefethen | NAG, UK |
| M. Valero | Univ. Politécnica de Catalunya, Spain |
| A. van der Steen | Utrecht Univ., The Netherlands |
| B. Tourancheau | École Normale Supérieure de Lyon, France |
| E. Zapata | Univ. de Malaga, Spain |

## Additional Referees

A. Augusto de Sousa
A. Pimenta Monteiro
A. Teixeira Puga
Aleksander Lazarevic
Alexandre Denis
Alfredo Bautista
Ananth Grama
Christian Pérez
Christophe Calvin
Cláudio L. Amorim
Cyril Gavoille
Didier Jamet
El Mostafa Daoudi
Enrique Fernandez-Garcia
Fernando Alvarruiz
Fernando Silva
Ignacio Blanquer
J.B. Caillau
Jesus Corbal
Jorge Barbosa
José Diaz
José E. Roman
José Gonzalez
José L. Larriba-Pey
José M. Alonso
José M. Cela
Josep Vilaplana
José C. Marques
José Nuno FIdalgo
João Gama
Julien Langou
Kirk Schloegel
Laurent Lefevre
Loic Prylli
Luis B. Lopes
Luis Torgo
M. Eduardo Correia
M. Prieto Matias
M. Próspero dos Santos
M. Radlinski
M. Velhote Correia
Martin van Gijzen

Michael Steinbach
Mário Oldemiro
Nuno Correia
Paulo B. Vasconcelos
Pedro D. Medeiros
Pierre Manneback
R.E. Goodman
Ramdev Kanapady
Renaud Lottiaux
Ricardo Lopes
Ricardo Rocha
Ronan Guivarch
S. Oral
Serge Miguet
Vivek Sarin
Xavier Pueyo
Yvon Jegou

# Invited Lecturers

- Yutaka Akiyama
  Computational Biology Research Center, Japan
- Leif Eriksson
  Uppsala University, Sweden
- Vipin Kumar
  University of Minnesota, USA
- Rainald Löhner
  George Mason University, USA
- Edward Seidel
  Max-Planck-Institut für Gravitationsphysik,
  Albert-Einstein-Institut, Germany

# Sponsoring Organizations

The Organizing Committee is very grateful to the following organizations for their support

| | |
|---|---|
| UP | – Universidade do Porto |
| FEUP | – Faculdade de Engenharia da Universidade do Porto |
| INESC Porto | – Instituto de Engenharia de Sistemas e de Computadores do Porto |
| FCT | – Fundação para a Ciência e Tecnologia |
| POSI | – Programa Operacional Sociedade de Informação |
| FEDER | – Fundo Europeu de Desenvolvimento Regional |
| EOARD | – European Office of Aerospace Research and Development |
| FCCN | – Fundação para a Computação Científica Nacional |
| FCG | – Fundação Calouste Gulbenkian |
| FLAD | – Fundação Luso Americana para o Desenvolvimento |
| FO | – Fundação Oriente |
| OE | – Ordem dos Engenheiros |
| CVRVV | – Comissão de Viticultura da Regi ao dos Vinhos Verdes |
| SCW | – Scientific Computing World |
| OCB | – Oporto Convention Bureau |

# Table of Contents

## Chapter 1: Fluids and Structures



## Chapter 2: Data Mining

## Chapter 3: Computing in Chemistry and Biology



## Chapter 4: Problem Solving Environments



## Chapter 5: Linear and Non-linear Algebra

**Chapter 6: Cluster Computing**

## Chapter 7: Imaging



## Chapter 8: Software Tools and Environments

# Chapter 1

# Fluids and Structures

# Fluid-Structure Interaction Simulations Using Parallel Computers

## Invited Talk

Rainald Löhner[1], Joseph D. Baum[2], Charles Charman[3], and Daniele Pelessone[4]

[1] School of Computational Sciences and Informatics
M.S. 4C7, George Mason University, Fairfax, VA 22030-4444, USA
[2] Applied Physics Operations
Science Applications International Corp., McLean, VA 22102, USA
[3] General Atomics, San Diego, CA 92121, USA
[4] Engineering Software System Solutions, Solana Beach, CA 92075, USA

**Abstract.** A methodology to simulate large-scale fluid-structure interaction problems on parallel machines has been developed. Particular emphasis was placed on shock-structure interaction problems. For the fluid, a high-resolution FEM-FCT solver based on unstructured grids is used. The surface motion is handled either by moving, body fitted grids, or via surface embedding. For the structure, a Lagrangean large-deformation finite element code is employed. The coupled system is solved using a loose coupling algorithm, with position and velocity interpolation and force projection. Several examples, run on parallel machines, demonstrate the range of applicability of the proposed methodology.

## 1 Introduction

The threat of intentional (Lockerbie (1988) [2], World Trade Center (1993) [3], Khobar Towers (1996), Nairobi and Dar Es Salaam (1998), Aden (2000), New York (2001), Washington (2001), etc.) or accidental (Seveso (1976), Toulouse (2001), oilrigs, oil refineries, etc.) explosions is a reality that planners have to take seriously when designing buildings, particularly those that will have a large concentration of people (airports, embassies, etc.). Any comprehensive methodology that seeks to predict or reproduce these events must account for the relevant physical phenomena, which include:

- Shock wave propagation through fluid and solid media;
- HE ignition and detonation;
- Afterburn phenomena;
- Particle transport and burn;
- Load and position/velocity transfer between fluid and structure;
- Possible catastrophic failure of the structure with consequent changes to the flowfield domain/topology.

At the same time, the simulations required must be run as fast as possible in order to have an impact on design and/or understanding. This implies that

J.M.L.M. Palma et al. (Eds.): VECPAR 2002, LNCS 2565, pp. 3–23, 2003.

- Integrated pre- and post-processing for fluid and structural dynamics;
- Fully automatic grid generation; and
- Scalability on parallel computers

should be an integral part of any such simulation methodology.

This paper reports on recent developments towards and simulations with such a methodology.

## 2 CFD Solver

### 2.1 Equations Solved

Most of the shock phenomena present in explosions are well described by the Euler equations, given by:

$$\mathbf{u}_{,t} + \nabla \cdot \mathbf{F}^a = \mathbf{S} \ , \tag{1}$$

$$\mathbf{u} = \{\rho \ , \ \rho v_i \ , \ \rho e\} \ , \tag{2}$$

$$\mathbf{F}^a_j = \{(v_j - w_j)\rho \ , \ (v_j - w_j)\rho v_i + p\delta_{ij} \ , \ (v_j - w_j)\rho e + v_j p)\} \ , \tag{3}$$

where $\rho, v_i, w_j, e$ and $p$ denote, respectively, the density, cartesian fluid and mesh velocity component in direction $x_i$, total energy and pressure. This set of equations is closed by adding an equation of state for the pressure, given as a function of the density and internal energy $e_i$:

$$p = p(\rho, e_i) \ \ , \ \ e_i = e - \frac{1}{2} v_j v_j \ . \tag{4}$$

For ideal gases, this reduces to:

$$p = (\gamma - 1)\rho e_i \ , \tag{5}$$

where $\gamma$ denotes the ratio of specific heats. In order to solve numerically these PDEs, an explicit finite element solver on unstructured grids is employed. For the spatial discretization linear tetrahedral elements are used, and the resulting discretization is cast in an edge-based formulation to reduce indirect addressing and operation count. Monotonicity, a key requirement to propagate accurately shocks, is maintained via Flux-corrected transport (FEM-FCT) [14].

### 2.2 Moving Bodies/Surfaces

For problems that exhibit surface movement, two classic solution strategies have emerged:

a) Body-Conforming Moving Meshes: here, the PDEs describing the flow are cast in an arbitrary Lagrangean-Eulerian (ALE) frame of reference, the mesh is moved in such a way as to minimize distortion, if required the topology is reconstructed, the mesh is regenerated and the solution reinterpolated as needed. While used extensively (and successfully) [15, 4, 5, 23, 24, 35], this solution strategy exhibits some shortcomings: the topology reconstruction can sometimes fail for singular surface points, and there is no way to remove subgrid features from surfaces, leading to small elements due to geometry; reliable parallel performance beyond 16 processors has proven elusive for most general-purpose grid generators; the interpolation required between grids will invariably lead to some loss if information; and there is an extra cost associated with the recalculation of geometry, wall-distances and mesh velocities as the mesh deforms.
b) Embedded Fixed Meshes: here, the mesh is not body-conforming and does not move. At every time-step, the edges crossed by CSD faces are identified and proper boundary conditions are applied in their vicinity. While used extensively - particularly in the context of adaptive Cartesian grids - [36, 28, 33, 11, 32, 12, 1], this solution strategy also exhibits some shortcomings: the boundary, which typically has the most profound influence on the ensuing physics, is also the place where the solution accuracy is worst; no stretched elements can be introduced to resolve boundary layers; adaptivity is essential for most cases; there is an extra cost associated with the recalculation of geometry (when adapting) and the crossed edge information.

We have recently explored the second type of solution strategy. This switch was prompted by the inability of CSD codes to ensure strict no-penetration during contact. Several blast-ship interaction simulations revealed that the amount of twisted metal was so considerable that any enforcement of strict no-penetration (required for consisted topology reconstruction) was hopeless.

The embedded surface technique used is the simplest possible. Given the CSD triangulation and the CFD mesh, the CFD edges cut by CSD faces are found and deactivated. If we consider an arbitrary field point $i$, the time-advancement of the unknowns $uvec^i$ is given by:

$$\mathbf{M}^i \Delta \mathbf{u}^i = \Delta t \sum_{\Omega} C^{ij} \left( F_i + F_j \right) , \tag{6}$$

where $\mathbf{M}^i, \Delta\mathbf{u}^i, \Delta t, C^{ij}, F_i$ denote, respectively, the lumped mass-matrix, increments of unknowns, time-step, edge-coefficients and fluxes. For any edge $ij$ crossed by a CSD face, the coefficients $C^{ij}$ are set to zero. This implies that for a uniform state $\mathbf{u} = const.$ the balance of fluxes for interior points with cut edges will not vanish. This is remedied by defining a new boundary point to impose total/normal velocities, as well as adding a 'boundary contribution', resulting in:

$$\mathbf{M}^i \Delta \mathbf{u}^i = \Delta t \left[ \sum_{\Omega} C^{ij} \left( F_i + F_j \right) + C^i_\Gamma F_i \right] . \tag{7}$$

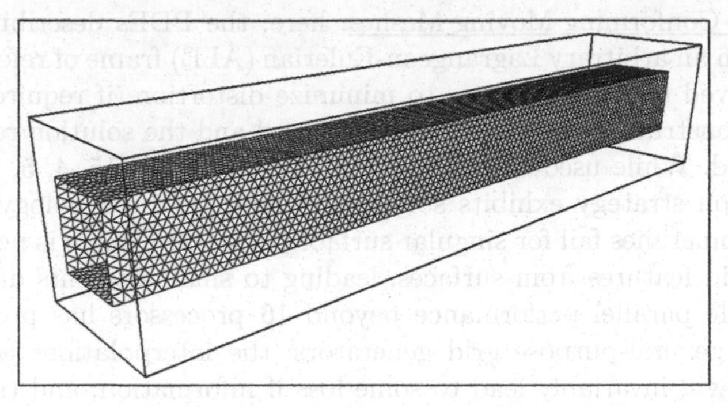

**Fig. 1a.** Shock Tube Problem: Embedded Surface

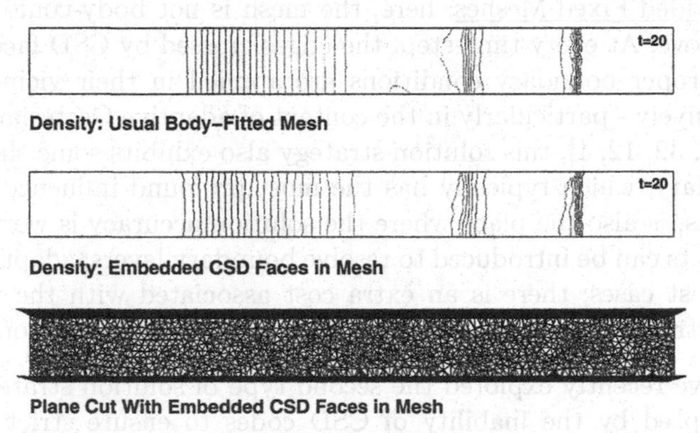


**Fig. 1b.** Shock Tube Problem: Density Contours

The point-coefficients $C_\Gamma^i$ are obtained from the condition that $\Delta\mathbf{u} = 0$ for $\mathbf{u} = const$. The mass-matrix $M^i$ of points surrounded by cut edges must be modified to reflect the reduced volume due to cut elements. Again, the simplest possible modification of $M^i$ is used. In a pass over the edges, the smallest cut edge fraction' $\xi$ for all the edges surrounding a point is found. The modified mass-matrix is then given by:

$$\mathbf{M}_*^i = \frac{1+\xi_{min}}{2}\mathbf{M}^i \; . \tag{8}$$

Note that the value of the modified mass-matrix can never fall below half its original value, implying that time-step sizes will always be acceptable. The embedded CSD technique is demonstrated by comparing the results on the Sod shock-tube problem for a 'clean-wall', body fitted mesh and an equivalent embedded CSD mesh. Initially, a diaphragm placed halfway down a tube separates two states $\rho_1 = p_1 = 1.0$ and $\rho_2 = p_2 = 0.1$.

The embedded geometry can be discerned from Figure 1a. Figure 1b shows the results for the two techniques. Although the embedded technique is rather primitive, the results are surprisingly good. The main difference is little bit more noise in the contact discontinuity region, which may be expected, as this is a linear discontinuity.

The motion of the surfaces attached to deforming fluids is imposed from the structural dynamics solver. If body-fitted grids and mesh movement are used, the mesh motion is smoothed using a nonlinear Laplacian [19] in order to reduce the amount of global and local remeshing required. For the embedded formulation, all the face-edge intersections checks are carried out incrementally, and points that 'cross states' due to CSD face movement need to be extrapolated appropriately.

At the end of each fluid time-step, the loads exerted on the 'wetted surfaces' belonging to the structural dynamics solver are computed and transferred back.

## 2.3 HE Ignition and Detonation

Most high explosives are well modeled by the Jones-Wilkins-Lee (JWL) equation of state, given by:

$$p = A\left(1-\frac{\omega}{R_1 v}\right)e^{-R_1 v} + B\left(1-\frac{\omega}{R_2 v}\right)e^{-R_2 v} + \omega\rho e_i \ , \tag{9}$$

where $v$ denotes the relative volume of the gas:

$$v = \frac{V}{V_0} = \frac{\rho_0}{\rho} \ . \tag{10}$$

Observe that in the expanded state ($v \to \infty$), the JWL equation of state reduces to

$$p = \omega\rho e_i = (\gamma - 1)\rho e_i \ , \tag{11}$$

where the correlation of $\omega$ and $\gamma$ becomes apparent. The transition to air is made by comparing the density of air to the density of the high explosive. Given that $A >> B$ while $R_1 = O(R_2)$, the decay of the first term in Eqn.(1) with increasing $v$ (decreasing $\rho$) is much faster. This implies that as $v$ increases, we have

$$p \to Be^{-R_2 v} + \omega\rho e \ . \tag{12}$$

The mixture of high explosive and air is considered as air when the effect of the $B$-term may be neglected, i.e.

$$\frac{p}{p_{cj}} = \epsilon = Be^{-R_2 v} \ , \tag{13}$$

where $p_{cj}$ denotes the Chapman-Jouget pressure and $\epsilon = O(10^{-3})$. Afterburning is modeled by adding energy via a burn coefficient $\lambda$ that is obtained from

$$\lambda_{,t} = a\left(\frac{p}{p_0}\right)^{\frac{1}{6}}\sqrt{1-\lambda} \ , \tag{14}$$

where $\lambda = 0$ for the unburned state, and $\lambda = 1$ for the fully burned material. After updating $\lambda$, the energy released is added as follows:

$$(\rho e)|^{n+1} = (\rho e)|^{n} + \rho Q(\lambda^{n+1} - \lambda^{n}) \ , \tag{15}$$

where $Q$ is the afterburn energy. Compared to the five unknowns required for the Euler equations with an ideal air equation of state, we require an additional two: the burn fraction $b$ to determine which part of the material has ignited, and the afterburn coefficient $\lambda$. Both of these scalars have to be transported through the mesh.

For a typical explosion simulation, the major portion of CPU time is required to simulate the burning material. This is because the pressures are very high, and so are the velocities of the fluid particles. Once the material has burned out, one observes a drastic reduction of pressures and velocities, which implies a dramatic increase in the allowable time-step. Even though shocks travel much larger distances, this post-burn 'diffraction phase' takes less CPU-time than the burn phase. In order to speed up the simulation, the portions of the grid outside the detonation region are deactivated. The detonation velocity provides a natural speed beyond which no information can travel. Given that the major loops in an unstructured-grid flow solver are processed in groups (elements, faces, edges, etc.) for vectorization, it seemed natural to deactivate not the individual edge, but the edge-group. In this way, all inner loops can be left untouched, and the test for deactivation is carried out at the group-level. The number of elements in each edge-, face-, or element-group is kept reasonably small (O(128)) in order to obtain the highest percentage of deactivated edges without compromising performance. The points are renumbered according to their ignition time assuming a constant detonation velocity. In this way, the work required for point-loops is minimized as much as possible. The edges and points are checked every 5-10 time-steps and activated accordingly. This deactivation technique leads to considerable savings in CPU at the beginning of a run, where the time-step is very small and the zone affected by the explosion only comprises a small percentage of the mesh.

## 3 CSD Solver

For the solid, a Lagrangian finite element code based on the large displacement formulation is used. Truss-, beam-, triangular and quadrilateral shells, as well as hexahedral solid elements have been incorporated. Innovative features of this structural dynamics code include adaptive refinement [29, 30], advanced contact algorithms [31] and the ability to fail elements or open cracks based on physics (i.e. not predetermined). At the beginning of each structural dynamics time-step, the loads from the fluid are updated. At the end of each structural dynamics time-step, the positions and velocities of the 'wetted surfaces' are extracted and transferred back to the flow solver.

# 4 Load and Position/Velocity Transfer

Optimal discretizations for the structure and the fluid are, in all probability, not going to be the same. As an example, consider an explosion in a room. For an accurate fluid solution, element sizes below 1cm have proven reliable. For the structure, shell or brick elements of 10cm are more than sufficient. There may be situations where fluid and structural surfaces are represented by abstractions of different dimensionality. A typical example here would be a commercial aircraft wing undergoing aeroelastic loads. For an accurate fluid solution using the Euler equations, a very precise surface representation with 60-120 points in the chordwise direction will be required. For the structure, a $20 \times 40$ mesh of plate elements may be more than sufficient. Any general fluid-structure coupling strategy must be able to handle efficiently the information transfer between different surface representations. This is not only a matter of fast interpolation techniques [17, 27], but also of accuracy, load conservation [6, 7], geometrical fidelity [7], and temporal synchronization [13, 7].

## 4.1 Load Transfer and Conservation

When considering different mesh sizes for the CSD and CFD surface representation, the enforcement of accuracy in the sense of:

$$\sigma_s(\mathbf{x}) \approx \sigma_f(\mathbf{x}) \tag{16}$$

and conservation in the sense of:

$$\mathbf{f} = \int \sigma_s \mathbf{n} d\Gamma = \int \sigma_f \mathbf{n} d\Gamma \tag{17}$$

proves to be non-trivial. The best way to date to handle this problem for similar surface representations is via an adaptive Gaussian quadrature [6]. In some cases, the conservation of forces may not be as important as conservation of other physically meaningful quantities, e.g. work or torque. It is important to note that conservation of only one of these: force, torque or work, can be guaranteed. The reliable and accurate treatment different levels of abstraction or dimensionality between fluid and structures represents an open question to date.

## 4.2 Geometrical Fidelity

In many instances, the structural model will either be coarse as compared to the fluid model, or may even be on a different modeling or abstraction level. On the other hand, it is the structural model that dictates the deformation of the fluid surface. It is not difficult to see that an improper transfer of structural deformation to the fluid surface mesh can quickly lead to loss of geometrical fidelity. For the cases considered here, the fluid surface was 'glued' to the surface of the structure. For 'non-glued' fluid and structural surfaces a number of recovery techniques have been proposed to date [9, 18, 7]. However, we consider

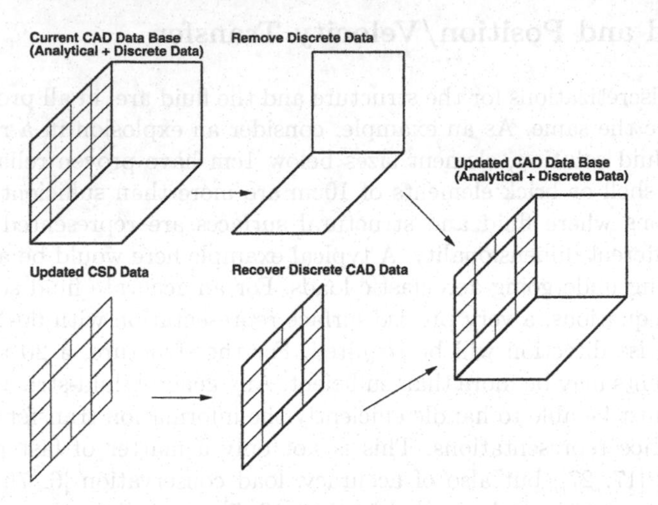


**Fig. 2.** Automatic Surface Reconstruction

the proper treatment of surface deformation for 'non-glued' fluid and structural surfaces, as well as some form of error indicator to warn the unsuspecting user, to be areas deserving further study.

## 5 Topology Change

Suppose that due to cracking, failure, spalation, etc., the 'wetted surface' of the structure has been changed. This new surface, given by a list of points and faces, has to be matched with a corresponding fluid surface. The fluid surface data typically consists of surface segments defined by analytical functions that do not change in time (such as exterior walls, farfield boundaries, etc.), and surface segments defined by triangulations (i.e. discrete data) that change in time. These triangulations are obtained from the 'wetted structure surface' at every time-step. When a change in topology is detected, the new surface definition is recovered from the discrete data, and joined to the surfaces defined analytically, as indicated in Figure 2.

The discrete surface is defined by a support triangulation, with lines and end-points to delimit its boundaries. In this sense, the only difference with analytically defined surfaces is the (discrete) support triangulation. The patches, lines and end-points of the 'wetted structure surface' are identified by comparing the unit surface normals of adjacent faces. If the scalar product of them lies below a certain tolerance, a ridge is defined. **Corners** are defined as points that are attached to:

- Only one ridge;
- More than two ridges; or
- Two ridges with considerable deviation of unit side-vector.

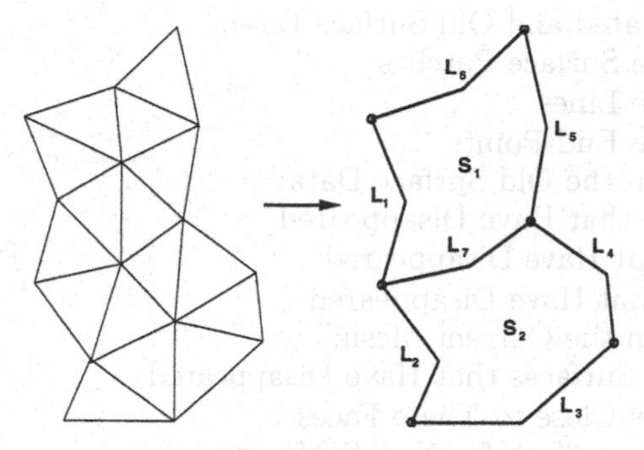


**Fig. 3.** Discrete Surface Recovery

Between corners, the ridges form **discrete lines**. These discrete lines either separate or are embedded completely (i.e. used twice) in **discrete surface patches**. Figure 3 sketches the recovery of surface features and the definition of discrete surface patches for a simple configuration. For more information, see [20].

For the old surface definition data set, the surface patches attached to wetted structure surfaces are identified and all information associated with them is discarded. The remaining data is then joined to the new wetted structure surface data, producing the updated surface definition data set. This data set is then used to generate the new surface and volume grids.

The surface reconstruction procedure may be summarized as follows:

- For the Updated Discrete Data, Obtain:
  - Surface Patches + B.C.
  - Lines
  - End-Points
  - Sources
- For the Old Analytical+Discrete Data:
  - Remove Discrete Data
  - Reorder Arrays
- Merge:
  - Old Analytical Data
  - Updated Discrete Data

A topology change as described above will imply a complete surface mesh regeneration, with subsequent flowfield mesh regeneration. If topology change occurs frequently, this can represent a considerable portion of the total computing cost. A way that has proven effective in reducing this cost is to identify which surfaces have actually changed, and to regenerate locally only the regions where changes have occurred. This local mesh generation procedure may be summarized as follows:

– For the Updated and Old Surface Data:
  • Compare Surface Patches
  • Compare Lines
  • Compare End-Points
– Remove from the Old Surface Data:
  • Surfaces that Have Disappeared
  • Lines that Have Disappeared
  • Points that Have Disappeared
– Remove from the Current Mesh:
  • Faces of Surfaces that Have Disappeared
  • Elements Close to These Faces
– Obtain Surface Mesh for New Surfaces
– Locally Remesh Close to New Surfaces

The procedure is shown schematically in Figure 4.

## 6 Scalability on Parallel Computers

From an operational point of view, the fluid-structure simulations considered here are characterized by:

– Accent on complex physics: Accuracy in the modeling of the important physical phenomena is much more important than raw speed. While some operational runs require very fast turnaround, most of the runs can be performed in a few days without affecting negatively the analysis process. This, speed by itself is not the critical factor. For complex geometry and complex physics, data assimilation (geometry, boundary conditions, materials, etc.) and input, as well as the evaluation of results, requires many more man-hours than the actual run.

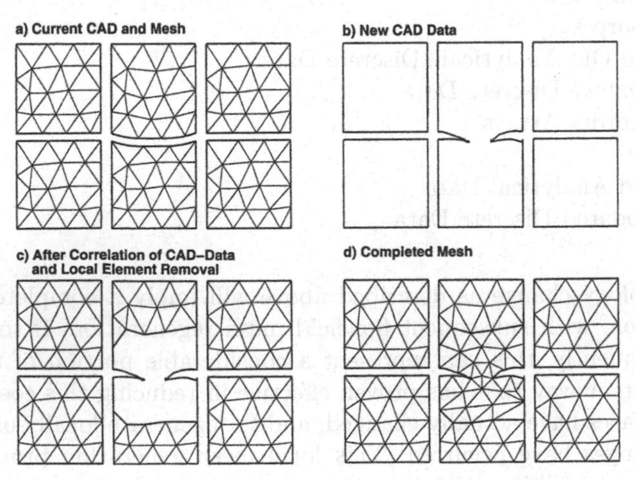


**Fig. 4.** Topology Change With Local Remeshing

- Interaction of several field solvers: The simulations are, by their nature, multidisciplinary (CFD, CSD, Transfer Library); at present, CFD consumes the vast majority of CPU time; this, however, may change if larger and more sophisticated CSD models are employed;
- Potentially large load imbalances: Remeshing (for regions with distorted elements and/or topology change [15, 23]), adaptive mesh refinement (for accuracy [16]) particle-mesh techniques (for afterburn) and contact detection and force evaluation [10] are but four of many potential sources of severe load imbalances. In order to obtain sustainable efficiency for distributed memory machines, dynamic load balancing is indispensable.
- Limited size: Experience indicates that the majority of the simulations can be performed with less than $10^8$ elements. This implies that the useful number of parallel processors is limited ($< 128$). With the appearance of sufficiently powerful departmental servers that allow shared memory parallel processing (e.g. SGI-O3K, IBM-SP4), blast-induced fluid-structure simulation has become a very tractable problem.

If we add to this list the fact that:

- Some parts (e.g. remeshing) can be quite tedious to code, debug and maintain on distributed memory parallel machine;
- Distributed memory parallel computing, with its many domain decompositions (CFD mesh, CFD particles, CSD mesh, CSD contact, CFD/CSD transfer), multiple file structures, represents an extra burden or 'pain factor' for the user; and
- Distributed memory parallel computing implies extra development effort and debug time (that are purely artificial and complicate most algorithms),

it is not difficult to see why the shared memory paradigm has remained highly popular.

From cost considerations alone, parallelizing 99% of all operations required by these codes is highly optimistic. This implies that the maximum achievable gain will be restricted to $O(1:100)$. There exist a number of platforms already that allow for efficient shared memory processing with $O(100)$ processors.

Instead of the spatial or logical decompositions encountered in typical distributed memory applications, for shared memory machines **renumbering techiques** [22, 26] play an important role in achieving acceptable scaling. Figure 5 shows the performance of FEFLO, the fluid code used in the present work, on a variety of common US-DOD-HPC platforms. One can see that the speedup obtained using shared and distributed memory approaches is similar.

## 7 Examples

We show three cases to illustrate the range of problems covered by the capabilities developed. The coupling complexity increases with each example. For the first example, the bodies move freely, based on the loads of the structure.

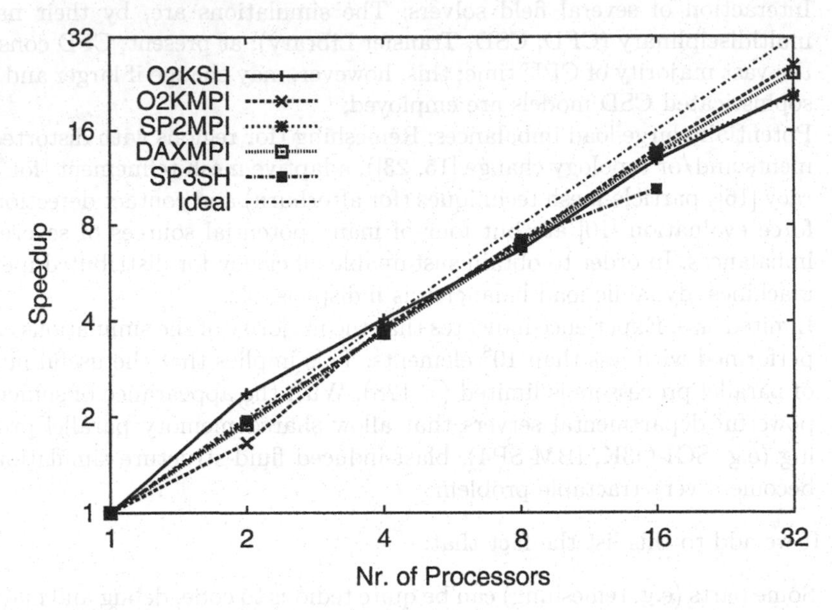


**Fig. 5.** Performance of FEFLO on Different Platforms

The last two examples show structural failure, rupture and topology change. All cases were run on SGI Origin 2000 machines in shared memory mode, using up to 64 processors.

### 7.1 Weapon Fragmentation

As the first example, we consider a weapon fragmentation experiment. The weapon used consisted of a thick wall cylinder, serrated into 32 rows of fragments, 16 fragments per row, for a total of 512 'small' fragments, and thick nose and tail plates. While the cutting pattern was identical for all rows, the pattern was rotated between rows by angles that produced no symmetric pattern. Thus, the whole weapon had to be modeled. Each fragment weighed about 380 grams.

The serrated weapon had tabs that kept the minimum distance between the fragments to 0.5 *mm*, while the average size of each fragment was a $1 - 4$ *cm* per side. In contrast, the room size was several meters. This huge disparity in dimensions required the use of sources attached to each flying body in order to ensure a uniform, high-resolution mesh about each fragment. Figure 6a shows the computational domain as well as snapshots of the surface grid employed. Each of the fragments is treated as a separate, freely flying rigid body whose velocity and trajectory were determined by integrating the six ordinary differential equations describing the balance of forces and moments. The mesh used for the CFD domain is body conforming and moves along with the fragments. Figures 6b-6i show a sequence of snapshots of pressure, detonation products velocity, and

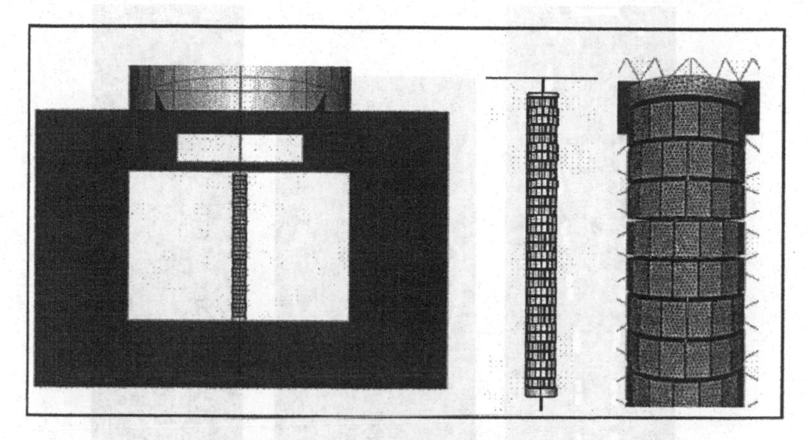

**Fig. 6a.** Computational Domain

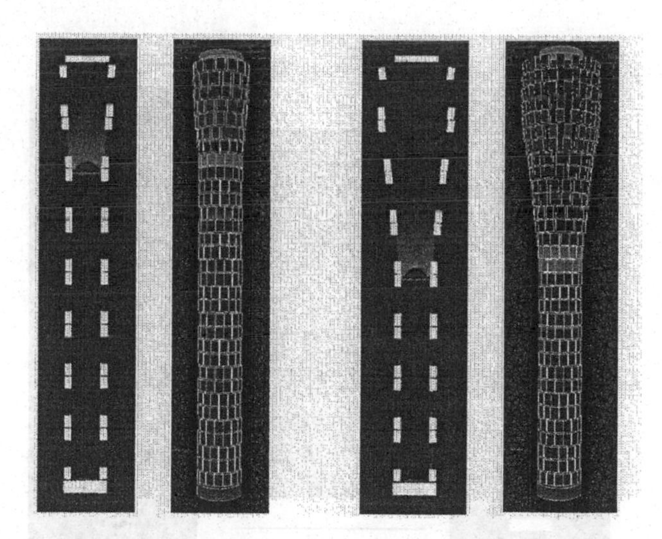

**Fig. 6b,c.** Pressure Contours and Fragment Velocities at $t = 72\ \mu sec$ and $t = 139.7\ \mu sec$

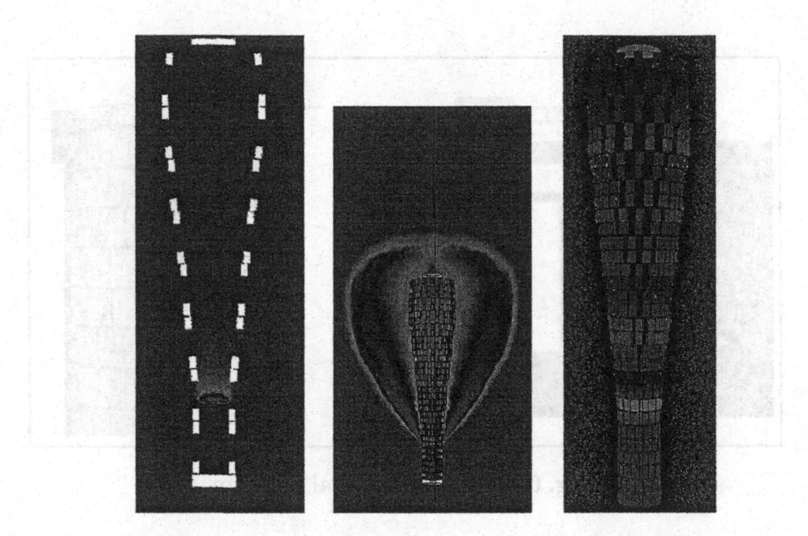

**Fig. 6d,e.** Pressure Contours, Detonation Products, and Fragment Velocity at $t = 225.7$ $\mu sec$

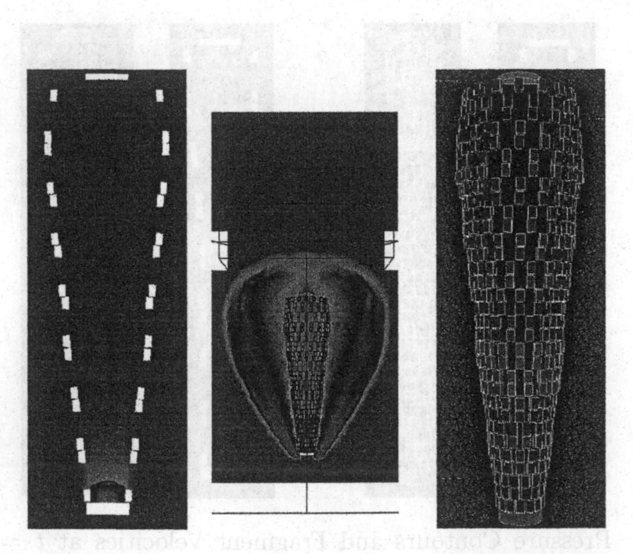

**Fig. 6f,g.** Pressure Contours, Detonation Products, and Fragment Velocity at $t = 297.3$ $\mu sec$

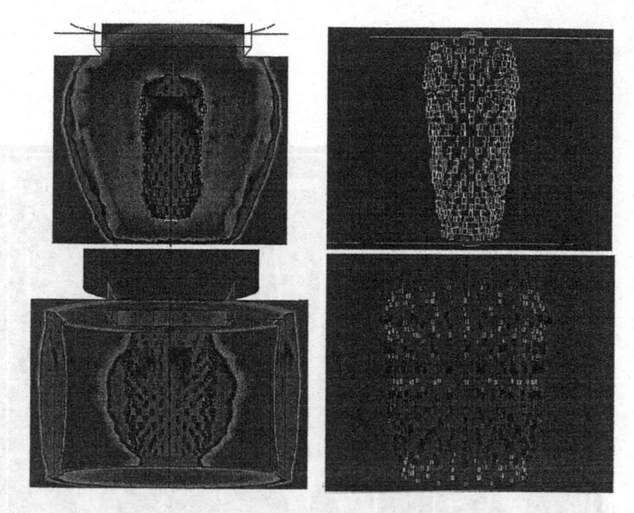

**Fig. 6h,i.** Pressure Contours and Fragment Velocities at $t = 501$ $\mu sec$ and $t = 900$ $\mu sec$

fragment and mesh velocity, at several times. The pressure results are shown for a planar cut. Note that this is not a plane of symmetry. The figures show the detonation wave, the accelerating fragments, and the sharp capturing of the shock escaping through the opening gaps. The results indicate that acceleration to the final velocity takes about 120 microseconds. The predicted velocities fall within the 10% error band of the experimental measurements. Given the complexity of the physical phenomena being modeled, such a correlation with experimental data is surprisingly good. Typical meshes for this simulation were of the order of 3-4 Mtets. The simulation was carried out on an 8 processor SGI Origin 2000, and took approximately 2 weeks. For more details, see [5].

### 7.2 Generic Weapon Fragmentation

The second example shows a fully coupled CFD/CSD run. The structural elements were assumed to fail once the average strain in an element exceeded 60%. At the beginning, the fluid domain consisted of two separate regions. These regions connected as soon as fragmentation started. In order to handle narrow gaps during the break-up process, the failed structural elements were shrunk by a fraction of their size. This alleviated the time-step constraints imposed by small elements without affecting the overall accuracy. The final breakup led to approximately 1200 objects in the flowfield. Figure 7 shows the fluid pressure and surface velocity of the structure at three different times during the simulation. As before, the CFD is body conforming and moves according to the fragments. Typical meshes for this simulation were of the order of 8.0 Mtets, and the simulations required of the order of 50 hours on the SGI Origin 2000 running on 32 processors.

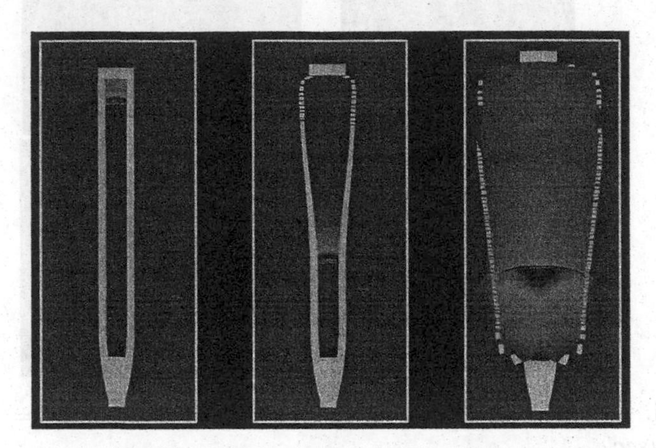

**Fig. 7a.** Pressure at Different Times

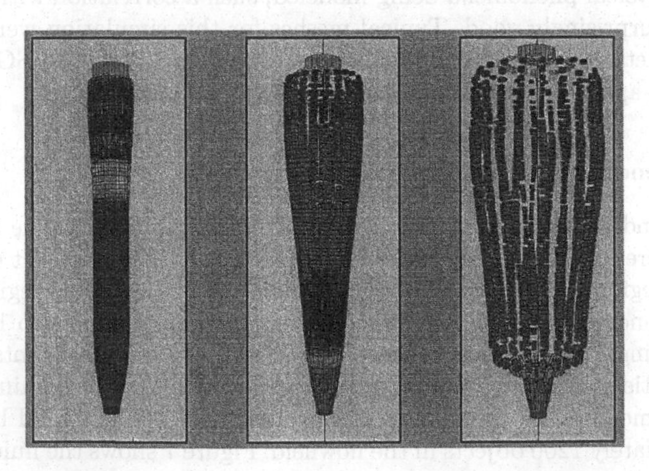

**Fig. 7b.** Surface Velocity at Different Times

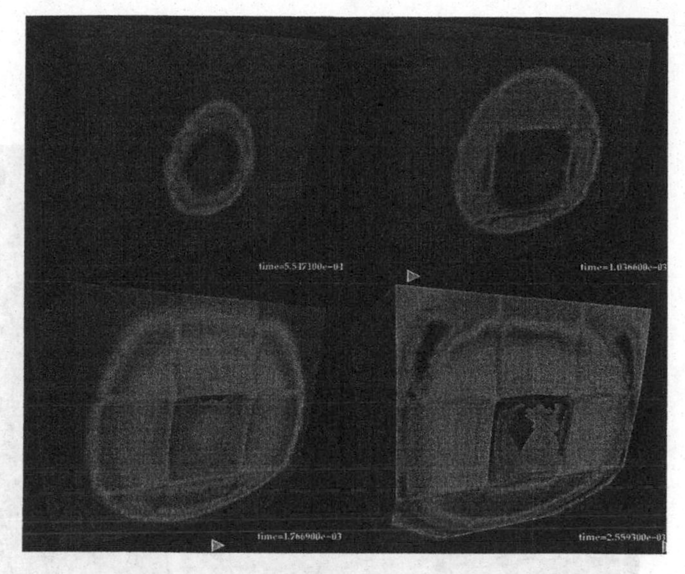

**Fig. 8a.** Surface Pressure at Different Times

### 7.3 Shock Interaction with a Ship Hull

The third example shows another fully coupled CFD/CSD run. In this case, an explosion close to a generic ship hull was considered. The structure was modeled with quadrilateral shell elements. The structural elements were assumed to fail once the average strain in an element exceeded 60%. As the shell elements fail, the fluid domain underwent topological changes. The damage is rather severe, and strict contact could not be maintained by the CSD solver.

After several unsuccessful attempts using the body conforming, moving mesh option, the case was finally run using the embedded surface technique. The CFD mesh had approximately 16.0 Mtets, and was refined in the regions where the initial shock impact occurs. Figure 8 shows the pressure contours on the surface, as well as the velocity of the structure, at several times during the run. The influence of walls can clearly be discerned for the velocity. Note also the failure of the structure, and the invasion of high pressure into the chamber.

## 8 Conclusions

A methodology to simulate large-scale fluid-structure interaction problems on parallel machines has been developed. Particular emphasis was placed on shock-structure interaction problems. For the fluid, a high-resolution FEM-FCT solver based on unstructured, moving, body fitted grids is used. For the structure, a Lagrangian large-deformation finite element code is employed. The coupled system is solved using a loose coupling algorithm, with position and velocity interpolation and force projection. The described methodology has been maturing

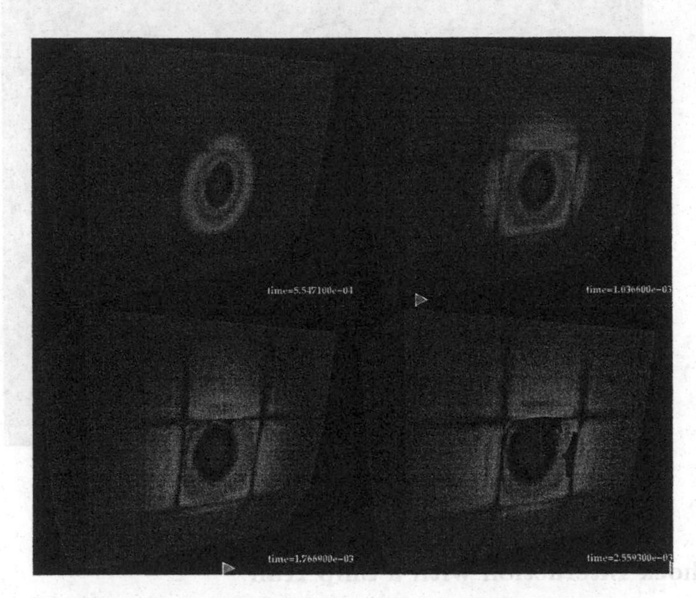


**Fig. 8b.** Surface Velocity at Different Times

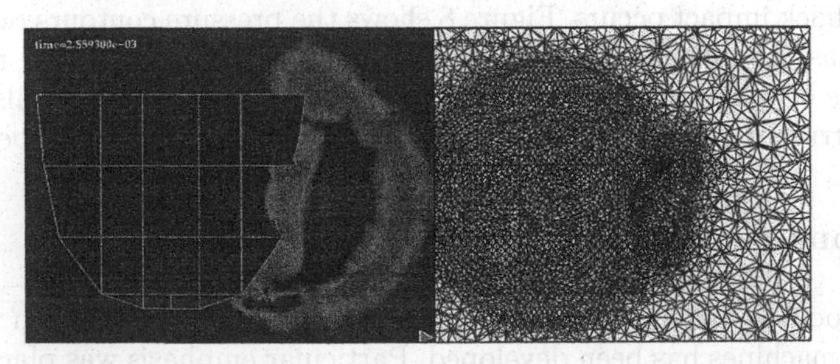


**Fig. 8c.** Plane Cut at T=0.000255 (Pressure, Grid)

rapidly over the last years, and is used on a routine basis to predict the complex phenomena encountered in shock-structure interaction problems.

Areas of current research include:

– Time-accurate implicit solvers for the drag phase;
– RANS griding for arbitrary body shapes;
– Automatic element size reduction for gaps and narrow openings during topology change;
– Parallel grid generation for large number of processors;
– Failure criteria for concrete based in first principle physics;
– 'Hard' contact algorithms that guarantee no penetration; and
– Further parallelization of fluid, structure and coupling libraries.

## Acknowledgments

This work was partially supported by DTRA, with Dr. Michael Giltrud as the technical monitor, as well as AFOSR, with Dr. Leonidas Sakell as the technical monitor.

## References

[1] M. J. Aftosmis, M. J. Berger and G. Adomavicius - A Parallel Multilevel Method for Adaptively Refined Cartesian Grids with Embedded Boundaries; *AIAA*-00-0808 (2000).

[2] J. D. Baum. H. Luo and R. Löhner - Numerical Simulation of a Blast Inside a Boeing 747; *AIAA*-93-3091 (1993).

[3] J. D. Baum, H. Luo and R. Löhner - Numerical Simulation of Blast in the World Trade Center; *AIAA*-95-0085 (1995).

[4] J. D. Baum, H. Luo, R. Löhner, C. Yang, D. Pelessone and C. Charman - A Coupled Fluid/Structure Modeling of Shock Interaction with a Truck; *AIAA*-96-0795 (1996).

[5] J. D. Baum, H. Luo and R. Löhner - The Numerical Simulation of Strongly Unsteady Flows With Hundreds of Moving Bodies; *AIAA*-98-0788 (1998).

[6] J. R. Cebral and R. Löhner - Conservative Load Projection and Tracking for Fluid-Structure Problems; *AIAA J.* 35, 4, 687-692 (1997a).

[7] J. R. Cebral and R. Löhner - Fluid-Structure Coupling: Extensions and Improvements; *AIAA*-97-0858 (1997b).

[8] J. R. Cebral and R. Löhner - Interactive On-Line Visualization and Collaboration for Parallel Unstructured Multidisciplinary Applications; *AIAA*-98-0077 (1998).

[9] G.P. Guruswamy and C. Byun - Fluid-Structural Interactions Using Navier-Stokes Flow Equations Coupled with Shell Finite Element Structures; *AIAA*-93-3087 (1993).

[10] E. Haug, H. Charlier, J. Clinckemaillie, E. DiPasquale, O. Fort, D. Lasry, G. Milcent, X. Ni, A. K. Pickett and R. Hoffmann - Recent Trends and Developments of Crashworthiness Simulation Methodologies and their Integration into the Industrial Vehicle Design Cycle; *Proc. Third European Cars/Trucks Simulation Symposium (ASIMUTH)*, Oct. 28-30 (1991).

[11] S.L. Karman - SPLITFLOW: A 3-D Unstructured Cartesian/ Prismatic Grid CFD Code for Complex Geometries; *AIAA*-95-0343 (1995).

[12] A.M. Landsberg and J.P. Boris - The Virtual Cell Embedding Method: A Simple Approach for Gridding Complex Geometries; *AIAA*-97-1982 (1997).

[13] M. Lesoinne and Ch. Farhat - Geometric Conservation Laws for Flow Problems With Moving Boundaries and Deformable Meshes, and Their Impact on Aeroelastic Computations; *Comp. Meth. Appl. Mech. Eng.* 134, 71-90 (1996).

[14] R. Löhner, K. Morgan, J. Peraire and M. Vahdati - Finite Element Flux-Corrected Transport (FEM-FCT) for the Euler and Navier-Stokes Equations; ICASE Rep. 87-4, *Int. J. Num. Meth. Fluids* 7, 1093-1109 (1987).

[15] R. Löhner - Three-Dimensional Fluid-Structure Interaction Using a Finite Element Solver and Adaptive Remeshing; *Computer Systems in Engineering* 1, 2-4, 257-272 (1990).

[16] R. Löhner and J.D. Baum - Adaptive H-Refinement on 3-D Unstructured Grids for Transient Problems; *Int. J. Num. Meth. Fluids* 14, 1407-1419 (1992).

[17] R. Löhner - Robust, Vectorized Search Algorithms for Interpolation on Unstructured Grids; *J. Comp. Phys.* 118, 380-387 (1995a).

[18] R. Löhner, C. Yang, J. Cebral, J.D. Baum, H. Luo, D. Pelessone and C. Charman - Fluid-Structure Interaction Using a Loose Coupling Algorithm and Adaptive Unstructured Grids; *AIAA*-95-2259 [Invited] (1995b).

[19] R. Löhner and Chi Yang - Improved ALE Mesh Velocities for Moving Bodies; *Comm. Num. Meth. Eng.* 12, 599-608 (1996a).

[20] R. Löhner - Regridding Surface Triangulations; *J. Comp. Phys.* 126, 1-10 (1996b).

[21] R. Löhner - Extensions and Improvements of the Advancing Front Grid Generation Technique; *Comm. Num. Meth. Eng.* 12, 683-702 (1996c).

[22] R. Löhner - Renumbering Strategies for Unstructured-Grid Solvers Operating on Shared-Memory, Cache-Based Parallel Machines; *Comp. Meth. Appl. Mech. Eng.* 163, 95-109 (1998).

[23] R. Löhner, Chi Yang, J.D. Baum, H. Luo, D. Pelessone and C. Charman - The Numerical Simulation of Strongly Unsteady Flows With Hundreds of Moving Bodies; *Int. J. Num. Meth. Fluids* 31, 113-120 (1999a).

[24] R. Löhner, C. Yang, J. Cebral, J.D. Baum, H. Luo, E. Mestreau, D. Pelessone and C. Charman - Fluid-Structure Interaction Algorithms for Rupture and Topology Change; *Proc. 1999 JSME Computational Mechanics Division Meeting*, Matsuyama, Japan, November (1999b).

[25] R. Löhner - A Parallel Advancing Front Grid Generation Scheme; *AIAA*-00-1005 (2000).

[26] R. Löhner and M. Galle - Minimization of Indirect Addressing for Edge-Based Field Solvers; *AIAA*-02-0967 (2002).

[27] N. Maman and C. Farhat - Matching Fluid and Structure Meshes for Aeroelastic Computations: A Parallel Approach; *Computers and Structures* 54, 4, 779-785 (1995).

[28] J.E. Melton, M.J. Berger and M.J. Aftosmis - 3-D Applications of a Cartesian Grid Euler Method; *AIAA*-93-0853-CP (1993).

[29] D. Pelessone and C.M. Charman - Adaptive Finite Element Procedure for Non-Linear Structural Analysis; *1995 ASME/JSME Pressure Vessels and Piping Conference*, Honolulu, Hawaii, July (1995).

[30] D. Pelessone and C.M. Charman - An Adaptive Finite Element Procedure for Structural Analysis of Solids; *1997 ASME Pressure Vessels and Piping Conference*, Orlando, Florida, July (1997).

[31] D. Pelessone and C. M. Charman - A General Formulation of a Contact Algorithm with Node/Face and Edge/Edge Contacts; *1998 ASME Pressure Vessels and Piping Conference*, San Diego, California, July (1998).
[32] R. B. Pember, J. B. Bell, P. Colella, W. Y. Crutchfield and M. L. Welcome - An Adaptive Cartesian Grid Method for Unsteady Compressible Flow in Irregular Regions - sl J. Comp. Phys. 120, 278 (1995).
[33] J. J. Quirk - A Cartesian Grid Approach with Hierarchical Refinement for Compressible Flows; *NASA CR-194938, ICASE Report No.* 94-51, (1994).
[34] R. Ramamurti and R. Löhner - Simulation of Flow Past Complex Geometries Using a Parallel Implicit Incompressible Flow Solver; pp. 1049,1050 in *Proc. 11th AIAA CFD Conf.* , Orlando, FL, July (1993).
[35] D. Sharov, H. Luo, J. D. Baum and R. Löhner - Time-Accurate Implicit ALE Algorithm for Shared-Memory Parallel Computers; *First International Conference on Computational Fluid Dynamics*, Kyoto, Japan, July 10-14 (2000).
[36] D. de Zeeuw and K. Powell - An Adaptively-Refined Cartesian Mesh Solver for the Euler Equations; *AIAA*-91-1542 (1991).

# Investigation of Turbulent Flame Kernels Using DNS on Clusters

## Best Student Paper Award: First Prize

Marc Lange

High-Performance Computing-Center Stuttgart (HLRS)
Allmandring 30, D-70550 Stuttgart, Germany
Phone: +49 711 685 2507
Fax: +49 711 685 7626
lange@hlrs.de

**Abstract.** A parallel code for the direct numerical simulation (DNS) of reactive flows using detailed chemical kinetics and multicomponent molecular transport is presented. This code has been used as a benchmark on several commodity hardware clusters including two large systems with current state-of-the-art processor and network technologies, i. e. up to 180 Alpha EV68 833 MHz and up to 400 AthlonMP 1.4 GHz CPUs have been used, which are arranged in dual-nodes with Myrinet-2000 interconnect in both cases. The other part of the paper presents the results of an investigation of flames evolving after induced ignition of a premixed mixture under turbulent conditions.

## 1 DNS of Turbulent Reactive Flows

Energy conversion in numerous industrial power devices like automotive engines or gas turbines is still based on the combustion of fossil fuels. In most applications, the reactive system is turbulent and the reaction progress is influenced by turbulent fluctuations and mixing in the flow. The understanding and modeling of turbulent combustion is thus vital in the conception and optimization of these systems in order to achieve higher performance levels while decreasing the amount of pollutant emission.

During the last few years, direct numerical simulations (DNS), i. e. the computation of time-dependent solutions of the Navier-Stokes equations for reacting ideal gas mixtures (as given in Sect. 2), have become an important tool to study turbulent combustion at a detailed level. As DNS does not make use of any turbulence or turbulent combustion models, this technique may be interpreted as high-resolution (numerical) experiments, enabling new investigations on fundamental mechanisms in turbulence-chemistry-interaction and aiding in the refinement of turbulent combustion models.

However, many of the DNS carried out so far have used simple one-step chemistry, although important effects cannot be captured by simulations with such oversimplified chemistry models [1, 2]. By making efficient use of the computational power provided by massively parallel systems, it is possible to perform

J.M.L.M. Palma et al. (Eds.): VECPAR 2002, LNCS 2565, pp. 24–38, 2003.

DNS of reactive flows using detailed chemical reaction mechanisms at least in two spatial dimensions. Nevertheless, computation time is still the main limiting factor for the DNS of reacting flows, especially when applying detailed chemical schemes.

## 2 Governing Equations

In the context of DNS, chemically reacting flows are described by a set of coupled partial differential equations expressing the conservation of total mass, chemical species masses, momentum and energy [3]. Using tensor notation[1] these equations can be written in the form

$$\frac{\partial \varrho}{\partial t} + \frac{\partial(\varrho u_j)}{\partial x_j} = 0 \quad , \tag{1}$$

$$\frac{\partial(\varrho Y_\alpha)}{\partial t} + \frac{\partial(\varrho Y_\alpha u_j)}{\partial x_j} = -\frac{\partial(\varrho Y_\alpha V_{\alpha,j})}{\partial x_j} + M_\alpha \dot{\omega}_\alpha \qquad (\alpha = 1, \ldots, N_\mathrm{S}) \quad , \tag{2}$$

$$\frac{\partial(\varrho u_i)}{\partial t} + \frac{\partial(\varrho u_i u_j)}{\partial x_j} = \frac{\partial \tau_{ij}}{\partial x_j} - \frac{\partial p}{\partial x_i} \, , \tag{3}$$

$$\frac{\partial e}{\partial t} + \frac{\partial((e+p)u_j)}{\partial x_j} = \frac{\partial(u_j \tau_{kj})}{\partial x_k} - \frac{\partial q_j}{\partial x_j} \quad , \tag{4}$$

wherein $\varrho$ is the density and $u_i$ the velocity component in $i$th coordinate direction. $Y_\alpha$, $\boldsymbol{V}_\alpha$ and $M_\alpha$ are the mass fraction, diffusion velocity and molar mass of the species $\alpha$. $N_\mathrm{S}$ is the number of chemical species occurring in the flow, $\tau_{ij}$ denotes the viscous stress tensor and $p$ the pressure, $\boldsymbol{q}$ is the heat flux and $e$ is the total energy given by

$$e = \varrho \left( \frac{u_i u_i}{2} + \sum_{\alpha=1}^{N_\mathrm{S}} h_\alpha Y_\alpha \right) - p \tag{5}$$

with $h_\alpha$ being the specific enthalpy of the species $\alpha$.

The chemical production rate $\dot{\omega}_\alpha$ of the species $\alpha$, which appears in the source-term on the right-hand sides of the $N_\mathrm{S}$ species mass equations (2), is given as the sum over the formation rate equations for all $N_\mathrm{R}$ elementary reactions

$$\dot{\omega}_\alpha = \sum_{\lambda=1}^{N_\mathrm{R}} k_\lambda (\nu_{\alpha\lambda}^{(p)} - \nu_{\alpha\lambda}^{(r)}) \prod_{\alpha=1}^{N_\mathrm{S}} c_\alpha^{\nu_{\alpha\lambda}^{(r)}} \quad . \tag{6}$$

The rate coefficients $k_\lambda$ of the elementary reactions are given by a modified Arrhenius law

$$k_\lambda = A_\lambda T^{b_\lambda} \exp\left( -\frac{E_{a\lambda}}{RT} \right) \quad , \tag{7}$$

[1] Summation convention shall be applied to roman italic indices $(i, j, k)$ only, i. e. no summation is carried out over greek indices $(\alpha, \lambda)$.

the parameters $A_\lambda$ and $b_\lambda$ of the pre-exponential factor and the activation energy $E_{a\lambda}$ are determined by a comparison with experimental data [4]. The chemical reaction mechanism for the $H_2/O_2/N_2$ system which has been used in all simulations presented in this paper contains $N_S = 9$ species and $N_R = 37$ elementary reactions [5].

The viscosity and multicomponent diffusion velocities are computed using standard formulae [6, 7], thermodynamical properties are computed using fifth-order fits of experimental measurements [4]. The set of equations is complemented by the ideal-gas state-equation

$$p = \frac{\varrho}{\overline{M}} RT \quad , \tag{8}$$

wherein $R$ is the gas constant and $\overline{M}$ the mean molar mass of the mixture.

## 3 Structure of the Parallel DNS Code

Due to the broad spectrum of length and time scales apparent in turbulent reactive flows, a very high resolution in space and time is needed to solve the system of equations given in the prvious section. The computation of the chemical source-terms and the multicomponent diffusion velocities are the most time-consuming parts in such DNS. Therefore, almost all DNS carried out so far has been (at least) restricted to the use of simplified models (e. g. one global reaction and equal diffusivities) or to two-dimensional simulations. Even with these restrictions it is crucial to make efficient use of HPC-systems, to be able to carry out DNS of reactive flows in acceptable time.

A code has been developed for the DNS of reactive flows on parallel computers with distributed memory using message-passing communication [8, 9, 2]. In favor of being able to include detailed models for the chemical reaction kinetics and the molecular transport as outlined in Sect. 2, only two-dimensional simulations are performed.

The spatial discretization in DNS codes is typically done by spectral methods or high-order finite-difference schemes. The main advantages of the latter ones are a greater flexibility with respect to boundary-conditions and the possibility of a very efficient parallelization, as in a fully explicit formulation no global data-dependencies occur. We chose a finite-difference scheme with sixth-order central-derivatives, avoiding numerical dissipation and leading to very high accuracy. The integration in time is carried out using a fourth-order fully explicit Runge-Kutta method with adaptive time-stepping. The control of the timestep is based on (up to) three independent criteria: A Courant-Friedrichs-Lewy (CFL) criterion and a Fourier criterion for the diffusion terms are checked to ensure the stability of the integration and an additional accuracy-control of the result for one or more selectable variables is obtained through timestep doubling.

The fully explicit formulation leads to a parallelization strategy, which is based on a regular two-dimensional domain-decomposition. For a given computational grid and number of processors it is tried to minimize the length of

**Table 1.** Scaling behavior on a Cray T3E-900 for a simulation with 9 species and 37 reactions on a $544^2$ points grid

| # PEs | 1 | 4 | 8 | 16 | 32 | 64 | 128 | 256 | 512 |
|---|---|---|---|---|---|---|---|---|---|
| speedup | 1.0 | 4.3 | 8.1 | 15.9 | 30.5 | 57.9 | 108.7 | 189.0 | 293.6 |
| efficiency / % | 100.0 | 106.6 | 100.7 | 99.2 | 95.3 | 90.4 | 84.9 | 73.8 | 57.4 |

the subdomain boundaries and thus the amount of communication. After the initial decomposition, each processor node controls a rectangular subdomain of the global computational domain. By using halo-boundaries, an integration step on the subdomain can be carried out independently from the other nodes. After each integration step, the values in the three points wide halo-regions are updated by point-to-point communications with the neighboring nodes. MPI is used for the communication.

Besides the computations on clusters discussed in this paper, our main platform for production runs have been Cray T3E systems. Table 1 lists the speedups and corresponding parallel efficiencies achieved on a Cray T3E-900 for a test case with a constant global size of $544 \times 544$ grid points [10]. The average performance per PE within this simulation (including I/O) for the computation using 64 processors was 86.3 MFLOP/s. For a problem with a scaled size of $32 \times 32$ grid points per processor, a speedup of 197 has been obtained using 256 PEs of a Cray T3E-900 corresponding to an efficiency of 79 % .

## 4 Performance on Clusters

A big trend in parallel computing today is the use of clustered systems. By using commodity hardware for the nodes as well as for the network, clusters of PCs can provide an excellent performance/price ratio – at least for problems which they are well suited for. Therefore, there is a broad interest in the performance behavior of real world applications on such clusters. We have studied the performance of several clusters with different types of nodes and network hardware for our application.

As a starting point of this investigation, benchmarks for different problem sizes using one and 16 processors of the following clusters have been performed:

**Alpha-Myrinet (Alpha-M)** Nodes: 64 Compaq DS10 workstations, each having one 466 MHz Alpha 21264 (EV6) processor (2 MB cache) and 256 MB RAM. Interconnect: Myrinet.

**Dual-Pentium-Myrinet (DualP3-M)** Nodes: 8 IBM dual-processor Netfinity servers, each having two 600 MHz Pentium III processors and 896 MB RAM. Interconnect: Myrinet.

**Pentium-Ethernet (P3-E)** Nodes: 16 PCs with 660 MHz Pentium III processors and 512 MB RAM. Interconnect: 100 MBit Ethernet.

**Table 2.** Monoprocessor performance for different problem sizes related to the performance of the Cray T3E-1200 solving the 50 × 50 problem

| Problem Size | 50 × 50 | 100 × 100 | 200 × 200 | 400 × 400 |
|---|---|---|---|---|
| T3E-1200 | 1.00 | 1.11 | 1.19 | 1.23 |
| Alpha-M | 2.39 | 2.40 | 2.33 | (not enough memory) |
| DualP3-M | 1.21 | 1.24 | 1.20 | 1.19 |
| P3-E | 1.28 | 1.25 | 1.22 | 1.26 |

Table 2 lists the relative performance of one processor of each of these systems for a computation on a grid with $50^2$, $100^2$, $200^2$, $400^2$ points, respectively. The relative performance is obtained from normalizing the inverse of the CPU-time needed to carry out a fixed number of timesteps with the problem size and dividing this value by the corresponding value of a Cray T3E-1200 PE for the 50 × 50 points case. The relative performance of a T3E-900 PE for the $50^2$ grid points computation is 0.82. The PE of the T3E performs better for the larger grids whereas the differences of the other systems with respect to the differently sized problems remain within a few percent. The behavior of the T3E may be attributed to the stream buffers, but does not hold for even larger grid sizes as can be seen from the results given in the last section: the superlinear speedup for 4 PEs (see Table 1) is possible only due to the relative performance per PE being better for a problem size of $272^2$ than of $544^2$ grid points. The Pentium III systems are slightly faster in these tests than the Alpha 21164, and the Alpha 21264 is almost twice as fast as the Pentium III nodes. Of course the exact relations between the different nodes strongly depend on the levels of optimization performed by the compilers. In this case, the default compilers and switches called from the `mpif90` command were used on the Linux-clusters, i. e. PGI V3.1-3 with `-o` on the Pentium III systems and Compaq Fortran CFAL 1.0 with `-O4` on the Alpha-M. On the T3Es CF90 3.5 with `-O3` has been used.

In Table 3, the measured parallel efficiencies $E_s$ for 16 processors of each of the clusters are given for two problems with a fixed load per processor of $50^2$ and $100^2$ grid points respectively. For the parallelization on the DP3-M cluster, at least from the application programmers point of view, no differences between intra-node communication and inter-node communication have been made. The

**Table 3.** Scaled parallel efficiency (in percent) using 16 processors for two different loads per processor

| Problem Size per Processor | 50 × 50 | 100 × 100 |
|---|---|---|
| T3E-1200 | 91.0 | 92.0 |
| Alpha-M | 80.5 | 90.4 |
| DualP3-M | 83.0 | 81.1 |
| P3-E | 77.5 | 85.7 |

**Table 4.** Time for communication and synchronization $t_c$/s for 10 timesteps using 16 processors

| Problem Size per Processor | 50 × 50 | 100 × 100 |
|---|---|---|
| T3E-1200 | 0.80 | 2.52 |
| Alpha-M | 0.81 | 1.43 |
| DP3-M (-o) | 1.36 | 6.02 |
| P3-E | 1.82 | 4.30 |

loads per processor chosen here correspond to real production runs, for which global grids with between $400^2$ and $1000^2$ grid points are typical. The main result is, that all the systems tested deliver reasonable parallel performance – at least for our application and up to 16 processors.

For our application, it is possible to calculate the overhead for the parallelization, i. e. the time spent for communication and synchronization, from the run-times of the serial run and the scaled parallel run [11]. Table 4 lists the time spent for communication and synchronization $t_c$ for a fixed number of timesteps using 16 processors. The number of messages in these tests is independent from the load per processor, and the length of the messages between neighboring nodes scales with the length of the domain-boundaries which is in these benchmarks of the same size as in typical production runs. From the times achieved on the Alpha-Myrinet cluster it can be seen, that for our application and 16 nodes, the Myrinet is only slightly slower than the T3E-network in the case with the smaller messages and 76 % faster in the case with the larger messages. As expected, the network performance of the Ethernet cluster is clearly lower than that of these two systems. The Dual-Pentium-Myrinet cluster suffers obviously from the very low parallel efficiency inside the nodes. The memory conflicts in the bus-based shared memory architecture also cause the performance of the dual-nodes to be more sensitive to the problem size.

The main questions arising from these benchmarks concern the efficiency of using dual-nodes and the scaling behavior for large configurations. These are adressed by measurements performed on two large clusters with current state-of-the-art processor and interconnect technology:

**Dual-Alpha-Myrinet2000 (API-M2K)** Nodes: 96 API NetWorks CS20 1U servers, each having two 833 MHz Alpha 21264B (EV68) processors (4 MB L2 cache) and 2 GB RAM. Interconnect: Myrinet 2000.

**Dual-Athlon-Myrinet2000 (HELICS)** Nodes: 256 PCs with two 1.4 GHz AMD AthlonMP processors and 1 GB RAM. Interconnect: Myrinet 2000.

Access to the first system has been granted by Cray Inc., the latter system has been installed at the Interdisciplinary Center for Scientific Computing (IWR) of Heidelberg University together with the University Computing Centers on Heidelberg (URZ) and Mannheim (RUM) in March 2002 (HEidelberger LInux Cluster System, see http://helics.iwr.uni-heidelberg.de/) and is currently

**Table 5.** Monoprocessor performance for different problem sizes related to the performance of the Cray T3E-1200 solving the 50 × 50 problem

| Problem Size | 50 × 50 | 100 × 100 | 200 × 200 | 400 × 400 |
|---|---|---|---|---|
| API-M2K | 4.40 | 4.18 | 3.67 | 2.69 |
| HELICS | 4.22 | 3.87 | 3.84 | 3.69 |

one of the most powerful PC-clusters worldwide. Table 5 lists the relative performance of one processor of each of these two systems analogous to Table 2. Both processors perform at a similar level, but the decrease of the performance with increasing problem size is much stronger on the Alpha than on the Athlon system. Compaq Fortran 1.0-920 has been used on the API-M2K. As the HELICS system had just been installed and was still in its testing phase, another IA32-Linux-cluster has been used as a development platform (http://www.hlrs.de/hw-access/platforms/volvox/). The Intel Fortran Compiler ifc 6.0$\beta$ has been used on this system to produce the binaries for the HELICS.

To be able to distinguish the influence of the intra-node parallelization from the influence of the network performance, several test cases have been run using both processors of $p/2$ nodes as well as using only one processor on each of the $p$ nodes used, where $p$ is the number of processors to be used in the considered test case. In the following, this is denoted by the suffixes -dual and -single. The parallelization overhead times $t_c$ are listed in Table 6. In our benchmark, symmetric boundary conditions are used in the $x$-direction and periodic boundary conditions in the $y$-direction. In the cases using two processors with a load of $N \times N$ grid points per processor, the global grid has $2N$ points in the y-direction and the boundary between the subdomains is parallel to the $x$-axis. Due to the periodicity each of the processors has to exchange the values at two sides of its subdomain-boundary with the other processor. In the cases with 16 and more processors of course all processors, except for those at the left and right boundaries of the global domain, have to exchange the values on all four sides of their four subdomain-boundaries with the neighboring processors. Thus, the typical number of messages to be exchanged per processor in the two-processor case is half of that of the runs with 16 and 64 processors. The communication per-

**Table 6.** Parallelization overhead $t_c$/s for 10 timesteps

| Number of Processors | 2 | 2 | 16 | 16 | 64 |
|---|---|---|---|---|---|
| Problem Size per Processor | 50 × 50 | 400 × 400 | 50 × 50 | 100 × 100 | 50 × 50 |
| T3E-1200 | - | - | 0.80 | 2.52 | 0.80 |
| API-M2K-single | - | - | - | 1.33 | 0.81 |
| API-M2K-dual | 0.35 | 21.67 | 0.56 | 1.68 | 0.58 |
| HELICS-single | 0.19 | 3.91 | 0.33 | 0.85 | 0.36 |
| HELICS-dual | 0.57 | 36.55 | 0.75 | 4.36 | 0.80 |

**Table 7.** Overall time (in seconds) for computing ten timesteps on a 400 × 400 grid

| Number of Processors | 1 | 16 | 64 |
|---|---|---|---|
| T3E-1200 | 419.64 | 31.55 | 8.88 |
| API-M2K-single | 191.33 | 9.02 | 2.43 |
| API-M2K-dual | 191.33 | 9.38 | 2.41 |
| HELICS-single | 139.54 | 9.16 | 2.27 |
| HELICS-dual | 139.54 | 12.67 | 2.71 |

formance in the single-processor per node configurations for the Myrinet 2000 on the HELICS is clearly better than the communication performance of the T3E. The tests with two processors in the dual-case again show, that (for our test case) the parallelization overhead for the intra-node parallelization strongly depends on the problem size. The efficiency of the intra-node parallelization is higher on the API-M2K than on the HELICS. In addition, the strong decrease of this efficiency occurs at a larger problem size on the API-M2K as can be seen from the value of $t_c$ in the case with $100^2$ grid points on each of 16 processors which is still good for the API-M2K but quite poor for the HELICS.

The computation times for computing ten timesteps in the case with 400×400 grid points (global) are given in Table 7. From these times, speedups for this problem of fixed size can easily be computed, but the resulting values are dominated by the dependence of the single processor performance from the problem size. The computing times however give an overview of the overall performance of these systems for our application.

From the presented results it is already expected that these clusters provide an excellent platform for our application also for very large configurations. This is confirmed by the parallel efficiencies for a scaled problem with 50 × 50 grid points per processor as given in Table 8. As the time spent for global communication in our application during the normal integration in time is negligible, the scaled parallel efficiency should stay almost constant for runs with 16 and more processors as it is the case here. There is some global communication involved when gathering data for I/O, but this is not taken into account here as the accumulated time for gathering data and writing it to disk is only about one percent of the total time of a typical production run.

From the benchmarks presented, it is clear that the performance of a configuration with $p$ single-processor nodes is superior to a configuration with $p/2$ dual-nodes. However, the latter configuration is typically much less expensive and the price/performance ratio is often better for a system with dual-nodes. The main question in this context is therefore often, if it is worthwhile for a system with a given number of nodes to invest into dual-nodes instead of single-nodes. If we compare the computation time for a fixed problem size of 400 × 400 grid points when using both processors of the dual nodes with that when only one processor per node is used, we find a performance gain of 55.3 % from the additional CPUs

**Table 8.** Scaled parallel efficiency (in %) for a load of 50 × 50 grid points per processor

| Number of Processors | 16 | 64 | 180 | 400 |
|---|---|---|---|---|
| T3E-1200 | 91.0 | 90.6 | - | - |
| API-M2K-single | - | 75.3 | - | - |
| API-M2K-dual | 76.7 | 75.9 | 74.6 | - |
| HELICS-single | 85.2 | 84.0 | 83.5 | - |
| HELICS-dual | 71.7 | 70.5 | 69.9 | 70.1 |

for 64 nodes of HELICS (i. e. the run time is reduced by a factor of 1/1.553). When using 256 processors on 128 nodes instead of only one processor on each of the 128 nodes for the same case with $400^2$ grid points the performance gain is 69.6 %.

In the following section, results of DNS of turbulent flame kernels are presented. Two simulations with 704 × 704 grid points each have been performed, one using 256 PEs of a Cray T3E-900 and one using 256 processors of HELICS (128 nodes). The computation time for advancing the solution over a physical time of 0.1 ms (from $t = 0.45$ ms to $t = 0.55$ ms) was 14857 seconds on the T3E-900 and 3639 seconds on HELICS. These times include writing four full solutions to disk, i. e. all local values of the main physical variables were saved after every 0.025 ms. About 13600 timesteps were computed for each of these parts of the integrations in time with the difference of the number of timesteps in both simulations (caused by the slightly different initial conditions) being less than 0.3 %.

# 5 Investigation of Turbulent Flame Kernels

Induced ignition and the following evolution of premixed turbulent flames is a phenomenon of large practical importance, e. g. in Otto engine combustion and safety considerations. DNS studies of this process have been performed using simple one-step chemistry in a model configuration of an initially uniform premixed gas under turbulent conditions which is ignited by an energy source in a small region at the center of the domain [12]. Our investigation extends these studies by using a detailed reaction mechanism (9 species, 37 elementary reactions) and a detailed transport model including thermodiffusion. The Soret effect (i. e. diffusion caused by temperature gradients) has often been neglected in simulations of reacting flows with multicomponent transport, but we found it to be of importance in onedimensional simulations of induced ignition of laminar $H_2/O_2/N_2$ mixtures.

In the simulations carried out in this study, a cold ($T = 298$ K) uniform mixture consisting of 23.3 % $H_2$, 11.7 % $O_2$, and 65 % $N_2$ (mole fractions) has been superimposed with a turbulent flow field computed by inverse FFT from a von-Kármán-Pao-spectrum with randomly chosen phases. The mixture is ignited by

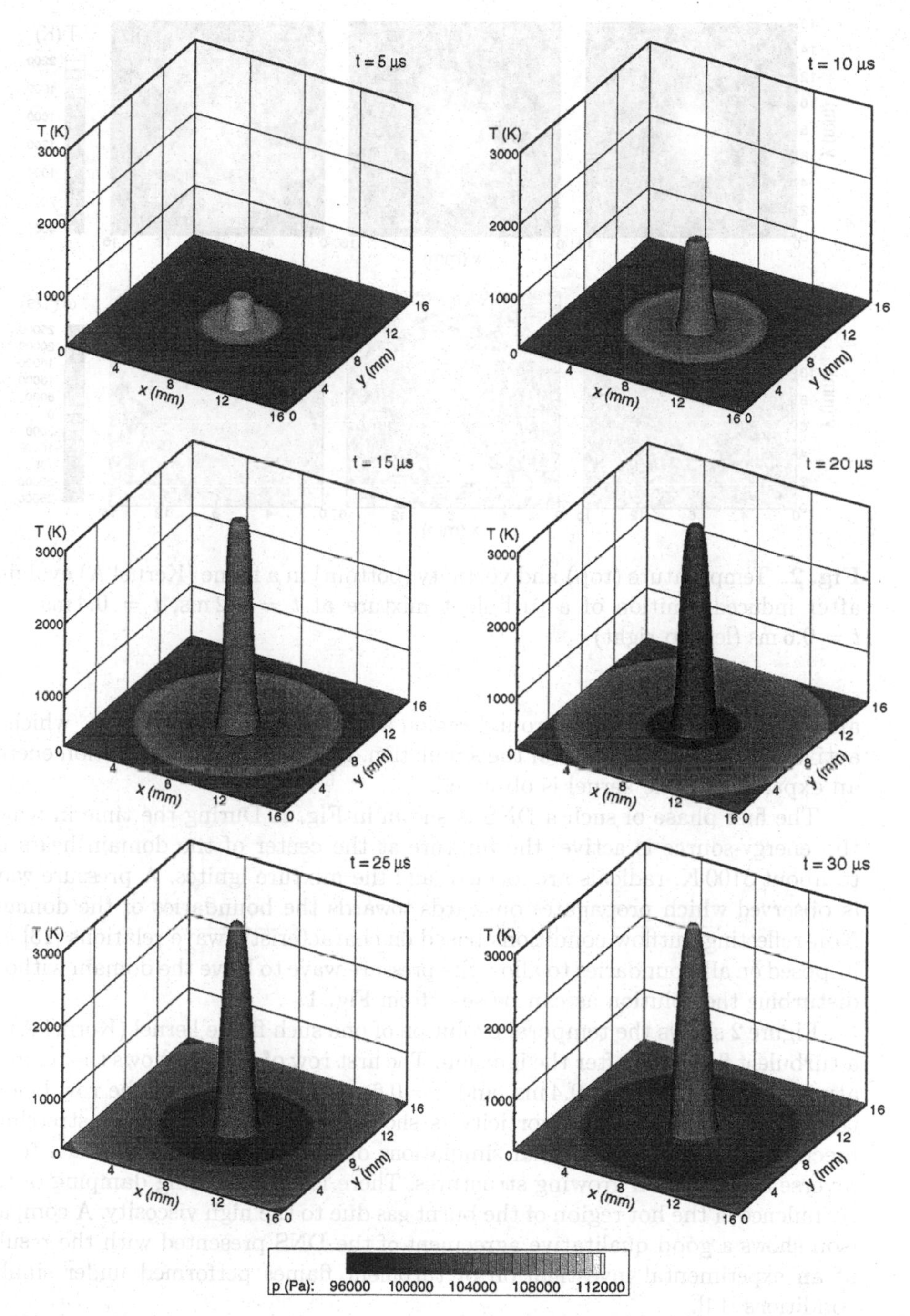


**Fig. 1.** Temporal evolution of pressure and temperature during the ignition of a turbulent hydrogen-air mixture

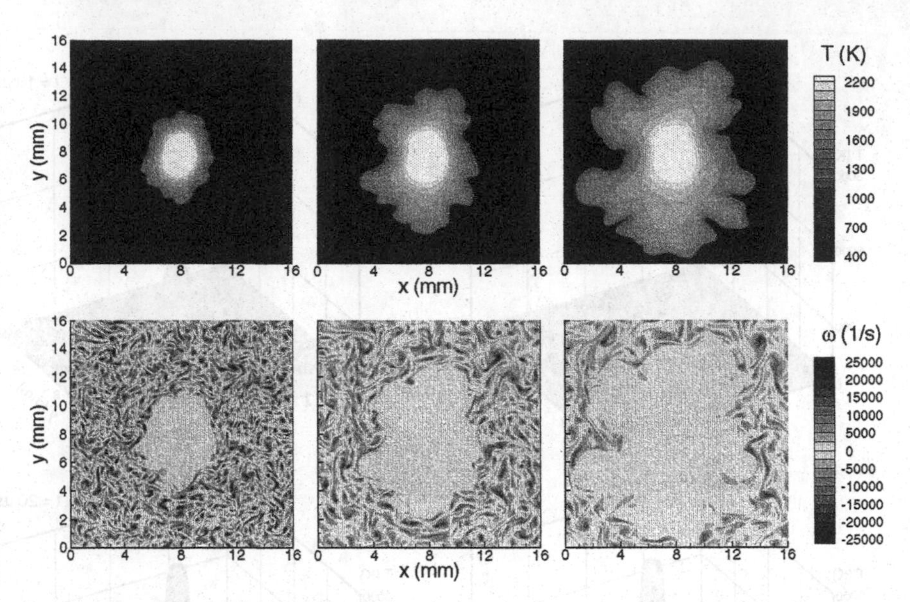


**Fig. 2.** Temperature (top) and vorticity (bottom) in a flame (Kernel A) evolving after induced ignition of a turbulent mixture at $t = 0.2\,\text{ms}$, $t = 0.4\,\text{ms}$, and $t = 0.6\,\text{ms}$ (left to right)

an energy-source in a small round region at the center of the domain, which is active during the first 15 μs of the simulation. Above a minimum ignition energy an expanding flame kernel is observed.

The first phase of such a DNS is shown in Fig. 1. During the time in which the energy-source is active, the mixture at the center of the domain heats up to about 3100 K, radicals are formed, and the mixture ignites. A pressure wave is observed which propagates outwards towards the boundaries of the domain. Non-reflecting outflow conditions based on characteristic wave relations [13] are imposed on all boundaries to allow the pressure wave to leave the domain without disturbing the solution as can be seen from Fig. 1.

Figure 2 shows the temporal evolution of one such flame kernel (Kernel A) in a turbulent flow field after the ignition. The first row of images shows the temperature at $t = 0.2\,\text{ms}$, $t = 0.4\,\text{ms}$, and $t = 0.6\,\text{ms}$, respectively. In the row below, the temporal evolution of vorticity is shown. The lack of a vortex-stretching mechanism in two-dimensional simulations of decaying turbulence leads to an inverse cascade with growing structures. There is a very strong damping of the turbulence in the hot region of the burnt gas due to the high viscosity. A comparison shows a good qualitative agreement of the DNS presented with the results of an experimental investigation of turbulent flames performed under similar conditions [14].

A DNS of a second flame kernel (Kernel B) has been carried out with the following two changes: 1. The energy-source causing the ignition has a smaller

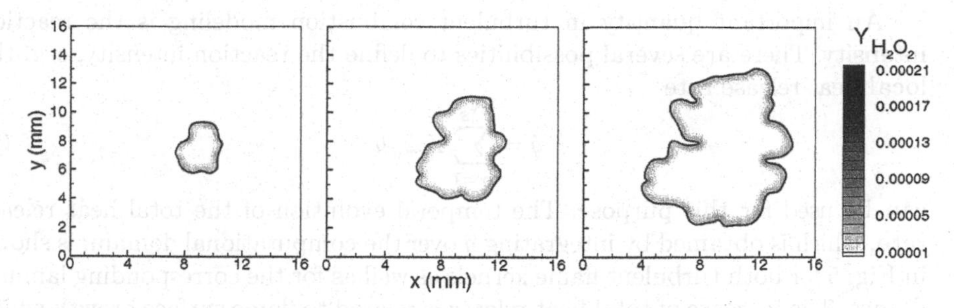


**Fig. 3.** $H_2O_2$ mass fraction in a turbulent flame kernel (Kernel B) at $t = 0.2$ ms, $t = 0.4$ ms, and $t = 0.6$ ms

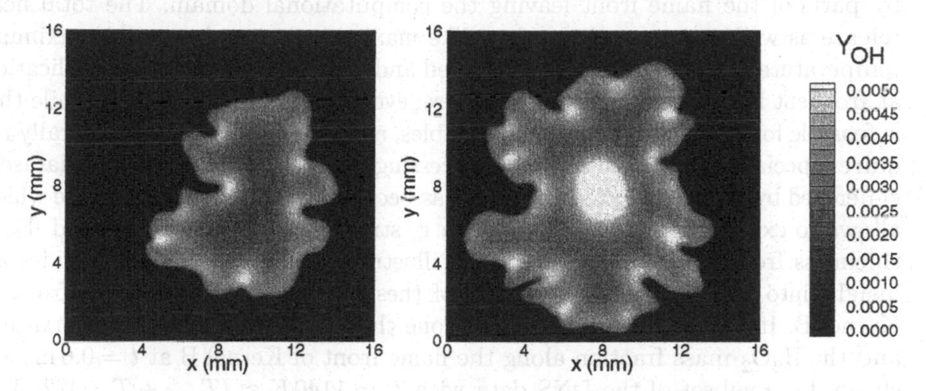


**Fig. 4.** OH mass fraction in two turbulent flame kernels (left: Kernel B, right: Kernel A) 0.6 ms after ignition caused by differently sized energy sources

diameter of $d_{\text{ign}} = 0.4$ mm instead of 1.6 mm while the amount of energy released per unit time and area is the same. 2. The individual realization of the turbulent flow field is different whereas it features the same statistical properties, i. e. the same input parameters – a root mean square of the velocity fluctuations of 3 m/s and an integral length scale of 2 mm – have been chosen for the initialization. Figure 3 shows the spatial distribution of the mass fraction of $H_2O_2$, a radical which is confined to a very thin layer in the flame front, at a physical time of 0.2 ms, 0.4 ms, and 0.6 ms. (The variations of the $H_2O_2$ concentration in the reaction zone along the flame are more clearly visible in Fig. 6.) The hydroxy radical (OH) is a very important intermediate species which is often measured in experiments. The spatial distributions of this radical in the two flame kernels at $t = 0.6$ ms are shown in Fig. 4. High concentrations of OH are found behind the flame front in regions with negative curvature (convex towards the side with the burnt gas). In the case of the larger ignition source (Kernel A), OH has been formed in excess during the ignition phase leading to the peak of OH mass fraction near the center of the domain.

An important quantity in turbulent combustion modeling is the reaction intensity. There are several possibilities to define the reaction intensity, e. g. the local heat release rate

$$\dot{q} = \sum_{\alpha=1}^{N_S} Y_\alpha \dot{\omega}_\alpha h_\alpha \tag{9}$$

can be used for this purpose. The temporal evolution of the total heat release rate, which is obtained by integrating $\dot{q}$ over the computational domain, is shown in Fig. 5 for both turbulent flame kernels as well as for the corresponding laminar flames. The increase of total heat release is related to flame surface growth which is enhanced by the wrinkling of the flame front by the turbulence. The decrease of total heat release starting at $t \approx 0.7\,\mathrm{ms}$ in the case of Kernel A is caused by parts of the flame front leaving the computational domain. The total heat release as well as some other values like maximum heat release and maximum temperature in the domain are computed and written to disk by the application at frequent intervals during the run (here: every twentieth timestep). While this is feasible for some global physical variables, a quantitative analysis generally requires specialized tools for the postprocessing of the large and complex datasets generated by DNS. An extensible tool has been developed for this purpose which allows to extract several features like e. g. strain, stretch, curvature, and flame thickness from these datasets. This is illustrated in Fig. 6 which provides an insight into the correlations of some of these properties in the flame front of Kernel B. In this figure, the reaction zone thickness, the flame front curvature and the $H_2O_2$ mass fraction along the flame front of Kernel B at $t = 0.6\,\mathrm{ms}$ are shown for a subset of the DNS data with $T = 1140\,\mathrm{K} \approx (T_{\mathrm{max}} + T_{\mathrm{min}})/2$. The values of the reaction zone thickness are based on the local temperature gradient and given relative to the reaction zone thickness of the corresponding circular flame propagating in a laminar mixture with the same chemical composition. High concentrations of the $H_2O_2$ radical in the reaction zone evidently occur in regions with negative curvature, i. e. convex towards the burnt gas. (The maxima of the $H_2O_2$ mass fraction in flame normal direction are however located at temperatures around 625 K.) Such dependencies of the local chemical composition on other variables like e. g. strainrate or curvature constitute a central point in many combustion models. As the changes of the chemical composition in curved flame fronts are caused by the combined effects of curvature and preferential diffusion, these dependencies can only be found from the analysis of data obtained by high-resolution experiments or simulations with a sufficiently accurate description of molecular diffusion and chemical kinetics.

## Acknowledgements

The author would like to thank Cray Inc. for providing access to the Dual-Alpha(EV68)/Myrinet-2000 cluster and M. Wierse for carrying out the benchmark runs on this cluster. The author gratefully acknowledges the Interdisciplinary Center for Scientific Computing (IWR) at Heidelberg for granting him access to the Dual-AthlonMP/Myrinet-2000 cluster (HELICS).

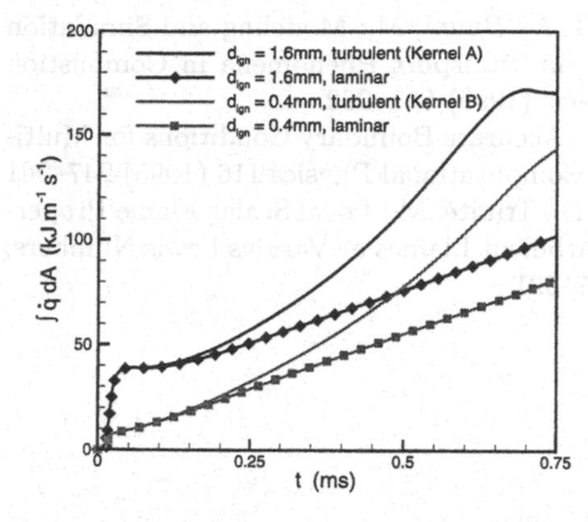


Fig. 5. Temporal evolution of overall heat-release in flame kernels evolving in turbulent and laminar mixtures after induced ignition

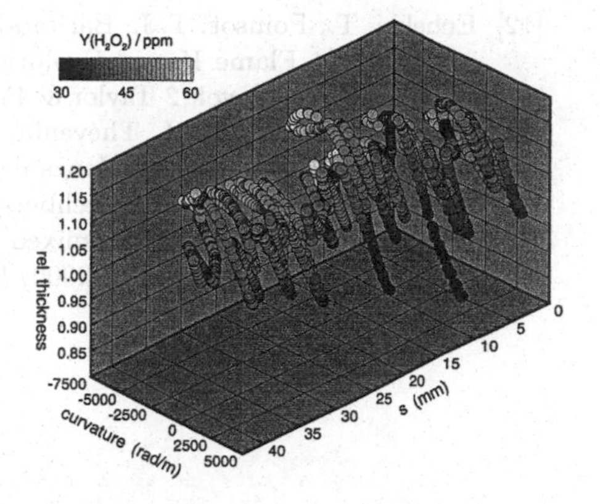


Fig. 6. Reaction zone thickness, curvature, and$H_2O_2$ mass fraction along the flame front of Kernel B at $t = 0.6\,\mathrm{ms}$

## References


[1] Mantel, T., Samaniego, J.-M.: Fundamental Mechanisms in Premixed Turbulent Flame Propagation via Vortex-Flame Interactions, Part II: Numerical Simulation, Combustion and Flame 118 (1999) 557–582

[2] Lange, M., Warnatz, J.: Investigation of Chemistry-Turbulence Interactions Using DNS on the Cray T3E, in High Performance Computing in Science and Engineering '99 (Krause, E., Jäger, W., eds.) Springer (2000) 333–343

[3] Bird, R. B., Stewart, W. E., Lightfoot, E. N.: Transport Phenomena, Wiley (1960)

[4] Warnatz, J., Maas, U., Dibble, R. W.: Combustion, 2nd ed., Springer (1999)

[5] Maas, U., Warnatz, J.: Ignition Processes in Hydrogen-Oxygen Mixtures, Combustion and Flame 74 (1988) 53–69

[6] Hirschfelder, J. O., Curtiss, C. F., Bird, R. B.: Molecular Theory of Gases and Liquids, Wiley (1954)

[7] Kee, R. J., Dixon-Lewis, G., Warnatz, J., Coltrin, M. E., Miller, J. A.: A Fortran Computer Code Package for the Evaluation of Gas-Phase Multicomponent Transport Properties, Tech. Rep. SAND86-8246 (1986)

[8] Thévenin, D., Behrendt, F., Maas, U., Przywara, B., Warnatz, J.: Development of a Parallel Direct Simulation Code to Investigate Reactive Flows, Computers and Fluids 25(5) (1996) 485–496

[9] Lange, M., Thévenin, D., Riedel, U., Warnatz, J.: Direct Numerical Simulation of Turbulent Reactive Flows Using Massively Parallel Computers, in Parallel Computing: Fundamentals, Applications and New Directions (D'Hollander, E., Joubert, G., Peters, F., Trottenberg, U., eds.) Elsevier (1998) 287–296

[10] Lange, M., Warnatz, J.: Direct Simulation of Turbulent Reacting Flows on the Cray T3E, in Proceedings of the 14th Supercomputer Conference '99 in Mannheim (Meuer, H.-W., ed.) (1999)

[11] Lange, M.: Direct Numerical Simulation of Turbulent Flame Kernels Using HPC, in High Performance Computing in Science and Engineering '01 (Krause, E., Jäger, W., eds.) Springer (2002) 418–432

[12] Echekki, T., Poinsot, T. J., Baritaud, T. A., Baum, M.: Modeling and Simulation of Turbulent Flame Kernel Evolution, in Transport Phenomena in Combustion (Chan, S. H., ed.) vol. 2 Taylor & Francis (1995) 951–962
[13] Baum, M., Poinsot, T. J., Thévenin, D.: Accurate Boundary Conditions for Multicomponent Reactive Flows, Journal of Computational Physics 116 (1995) 247–261
[14] Renou, B., Boukhalfa, A., Puechberty, D., Trinité, M.: Local Scalar Flame Properties of Freely Propagating Premixed Turbulent Flames at Various Lewis Numbers, Combustion and Flame 123 (2000) 507–521

# A Parallel, State-of-the-Art, Least-Squares Spectral Element Solver for Incompressible Flow Problems*

Margreet Nool[1] and Michael M. J. Proot[2]

[1] CWI, P.O. Box 94079, 1090 GB Amsterdam, The Netherlands
Margreet.Nool@cwi.nl
[2] Delft University of Technology, Kluyverweg 1, Delft, The Netherlands
m.m.j.proot@lr.tudelft.nl

**Abstract.** The paper deals with the efficient parallelization of least-squares spectral element methods for incompressible flows. The parallelization of this sort of problems requires two different strategies. On the one hand, the spectral element discretization benefits from an element-by-element parallelization strategy. On the other hand, an efficient strategy to solve the large sparse global systems benefits from a row-wise distribution of data. This requires two different kinds of data distributions and the conversion between them is rather complicated. In the present paper, the different strategies together with its conversion are discussed. Moreover, some results obtained on a distributed memory machine (Cray T3E) and on a virtual shared memory machine (SGI Origin 3800) are presented.

## 1 Introduction

Least-squares spectral element methods are based on two important and successful numerical methods: spectral/*hp* element methods and least-squares finite element methods. Least-squares methods lead to symmetric and positive definite algebraic systems which circumvent the Ladyzhenskaya-Babuška-Brezzi stability condition and consequently allow the use of equal order interpolation polynomials for all variables. The accuracy of a least-squares spectral element discretization of the Stokes problem (cast in velocity-vorticity-pressure form) has been reported in [5, 6] for different boundary conditions. The present paper deals with the efficient parallelization of this solver.

Parallelization of the least-squares spectral element method (LSQSEM) requires two different kinds of distribution of data and the conversion is rather complicated. The spectral element structure enables one to calculate the local

* Funding for this work was provided by the National Computing Facilities Foundation (NCF), under project numbers NRG-2000.07 and MP-068. Computing time was also provided by HP$\alpha$C, Centre for High Performance Applied Computing at the Delft University of Technology.

J.M.L.M. Palma et al. (Eds.): VECPAR 2002, LNCS 2565, pp. 39–52, 2003.

matrices corresponding to each spectral element (also called cells), simultaneously. After the (parallel) calculation of the local systems, we have to switch from a local numbering to a global numbering to complete the gathering procedure. When this has been completed, one obtains a global Compressed Sparse Row (CSR) formatted matrix which can be easily distributed along an arbitrary number of processors. Each processor has to send data from one cell to a *few* other processors or possibly to itself, a very unbalanced task due to the chosen numbering. However, if this task is completed, each processor contains a part of the global assembled matrix, and the data per processor will be balanced again.

Since the global system is symmetric and positive definite, the robust preconditioned conjugate gradient method (PCG) can be applied directly. In this paper, we describe results obtained with the simple, easy to parallelize, Jacobi or diagonal preconditioning. In the future, we plan to test the efficiency of other preconditioning methods: block-Jacobi, FEM-matrix and Additive Schwarz methods [3, 4, 7]. The results will be discussed in forthcoming work.

The parallelization of the complete algorithm, including the important conversion of a spectral element-by-element distribution to a row-wise CSR-format distribution, and numerical results of large-scale problems arising in scientific computing are also discussed in the present paper.

## 2 The Least-Squares Spectral Element Formulation of the Stokes Problem

In order to obtain a *bona fide* least-squares formulation, the Stokes problem is *first* transformed into a system of first order partial differential equations by introducing the vorticity as an auxiliary variable. By using the identity

$$\nabla \times \nabla \times \mathbf{u} = -\Delta \mathbf{u} + \nabla(\nabla \cdot \mathbf{u})$$

and by using the incompressibility constraint $\nabla \cdot \mathbf{u} = 0$, the governing equations subsequently read

$$\nabla \mathbf{p} + \nu \nabla \times \omega = \mathbf{f} \quad \text{in } \Omega, \tag{1}$$

$$\omega - \nabla \times \mathbf{u} = \mathbf{0} \quad \text{in } \Omega, \tag{2}$$

$$\nabla \cdot \mathbf{u} = 0 \quad \text{in } \Omega, \tag{3}$$

where, in the particular case of the two-dimensional problem, $\mathbf{u}^{\mathrm{T}} = [u_1, u_2]^{\mathrm{T}}$ represents the velocity vector, $\mathbf{p}$ the pressure, $\omega$ the vorticity, $\mathbf{f}^{\mathrm{T}} = [f_1, f_2]$ the forcing term (if applicable) and $\nu$ the kinematic viscosity. For simplicity it is further assumed that the density equals $\rho = 1$. In two dimensions, system (1)-(3), consists of four equations and four unknowns and is uniformly elliptic of order four. The velocity boundary condition ($\mathbf{u}$ given) is used to supplement the governing equations (1)-(3). The linear Stokes operator and its right-hand-side read:

$$\mathcal{L}(U) = F \Longleftrightarrow \begin{bmatrix} 0 & 0 & \nu \frac{\partial}{\partial x_2} & \frac{\partial}{\partial x_1} \\ 0 & 0 & -\nu \frac{\partial}{\partial x_1} & \frac{\partial}{\partial x_2} \\ \frac{\partial}{\partial x_2} & -\frac{\partial}{\partial x_1} & 1 & 0 \\ \frac{\partial}{\partial x_1} & \frac{\partial}{\partial x_2} & 0 & 0 \end{bmatrix} \begin{bmatrix} u_1 \\ u_2 \\ \omega \\ p \end{bmatrix} = \begin{bmatrix} f_1 \\ f_2 \\ 0 \\ 0 \end{bmatrix} \quad \text{in } \Omega. \tag{4}$$

## 3 General Implementation Aspects

### 3.1 Discretization

The domain is discretized with a mesh of $\mathcal{K}$ non-overlapping conforming quadrilateral spectral elements of the same order $N$. Each quadrilateral spectral element is mapped on the parent spectral element $\Omega^e$ by using an iso-parametric mapping to the bi-unitsquare $[-1,1] \times [-1,1]$ with local coordinates $\xi_1$ and $\xi_2$. In the parent element all variables, located at the Gauss-Legendre-Lobatto collocation (GLL) points, can be approximated by the same Lagrangian interpolant, since the least-squares formulation is not constrained by the Ladyzhenskaya-Babuška-Brezzi stability condition. Most spectral element methods are based on the GLL numerical integration for reason of accuracy. For the two-dimensional Stokes problem, the discrete spectral element approximation yields

$$\mathbf{U}^h = \sum_{j=0}^{N_2} \sum_{i=0}^{N_1} h_i(\xi_1)\, h_j(\xi_2) \begin{bmatrix} \hat{u}_1 \\ \hat{u}_2 \\ \hat{\omega} \\ \hat{p} \end{bmatrix}_{i,j}, \tag{5}$$

where $h_i(\xi_1)$ with $0 \le i \le N_1$ and $h_j(\xi_2)$ with $0 \le j \le N_2$ represent the Lagrange interpolants in the $\xi_1$ and $\xi_2$-direction through the GLL points, respectively. The vector $[\hat{u}_1, \hat{u}_2, \hat{\omega}, \hat{p}]^{\mathrm{T}}$ is the vector of unknown coefficients, evaluated at the GLL collocation point. Hence, each spectral element gives rise to a local system of the form:

$$A_i\, z_i = f_i, \qquad \text{with } i = 1, \cdots, \mathcal{K}, \tag{6}$$

where the matrices $A_i$ and right-hand side vectors $f_i$ are given by

$$A_i = \int_{\Omega_e} \left[\mathcal{L}(\psi_{0,0}), \cdots, \mathcal{L}(\psi_{N,N})\right]^T \left[\mathcal{L}(\psi_{0,0}), \cdots, \mathcal{L}(\psi_{N,N})\right] d\Omega, \tag{7}$$

and

$$f_i = \int_{\Omega_e} \left[\mathcal{L}(\psi_{0,0}), \cdots, \mathcal{L}(\psi_{N,N})\right]^T \mathcal{F}\, d\Omega, \tag{8}$$

respectively and where $\psi_{i,j} = h_i(\xi_1)\, h_j(\xi_2)$.

### 3.2 Local Numbering versus Global Numbering

Due to the continuity requirements between the $C^0$-spectral elements, some of the variables $z_i$ corresponding to an internal edge will belong to more than one local system which necessitates the introduction of a global numbering.

In Fig. 1 an example is given of a domain discretized with a mesh of four spectral elements. Each spectral element contains nine local nodes, numbered from 1 to 9 (small-size digits). In the same figure, also a global numbering (normal-size digits) is shown. First, the internal nodes or variables are numbered $(1, \cdots, 9)$, then the *knowns* $(10, \cdots, 25)$ given by the boundary conditions.

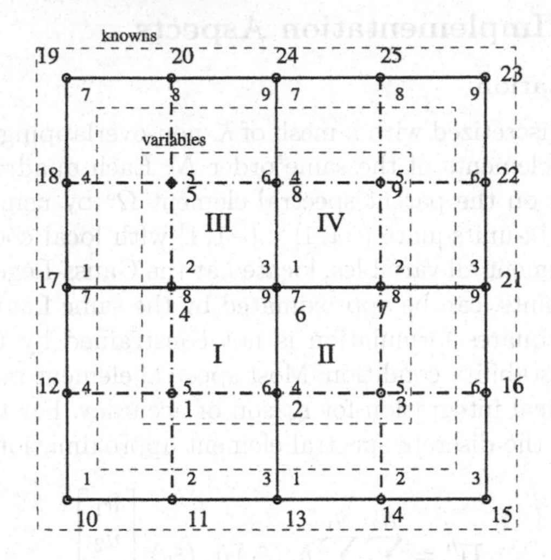


**Fig. 1.** Example of local and global numbering. The domain has been divided into four cells: I, II,III, IV. Each cell contains 9 nodes, denoted by a ∘

Since each local variable corresponds to a global variable, one can establish the local-global mapping operator $gm$ for each spectral element. For the given example, we have

$$\begin{aligned} gm_{\mathrm{I}} &= [\,10, 11, 13, 12, 1, \; 2, 17, \; 4, \; 6\,], \\ gm_{\mathrm{II}} &= [\,13, 14, 15, \; 2, 3, 16, \; 6, \; 7, 21\,], \\ gm_{\mathrm{III}} &= [\,17, \; 4, \; 6, 18, 5, \; 8, 19, 20, 24\,], \\ gm_{\mathrm{IV}} &= [\; 6, \; 7, 21, \; 8, 9, 22, 24, 25, 23\,]. \end{aligned} \tag{9}$$

The local-global mapping operator $gm_{\mathcal{I}}$ can also be expressed by the sparse *gathering matrix* $\mathcal{G}_i$ which has nonzero entries according to $\mathcal{G}_i(i, gm_{\mathcal{I}}(i)) = 1, \mathcal{I} = \mathtt{I}, \cdots, \mathtt{IV}$. The global assembly of the $\mathcal{K}$ local systems (6) can now readily be obtained with:

$$KU = F \Leftrightarrow \left[ \sum_{i=1}^{\mathcal{K}} \mathcal{G}_i^T A_i \mathcal{G}_i \right] U = \sum_{i=1}^{\mathcal{K}} \mathcal{G}_i^T f_i, \tag{10}$$

where the matrix $K$ represents the symmetrical, globally gathered matrix of full bandwidth and the vectors $U$ and $F$ represent the global nodes (e.g., degrees of freedom and *knowns*) and the global right-hand side function, respectively.

Since the known values are numbered last, one can subdivide the vector $U$ into an unknown component $U_1$ and a known component $U_2$. Consequently, the matrix $K$ can be factored into submatrices $K_{1,1}$, $K_{1,2}$, $K_{1,2}^T$ and $K_{2,2}$. Also the right-hand side vector $F$ can be factored into the vectors $F_1$ and $F_2$. Hence, system (10) has the following matrix structure

$$\begin{bmatrix} K_{1,1} & K_{1,2} \\ K_{1,2}^T & K_{2,2} \end{bmatrix} \begin{bmatrix} U_1 \\ U_2 \end{bmatrix} = \begin{bmatrix} F_1 \\ F_2 \end{bmatrix}, \tag{11}$$

which readily allows "*static condensation*" of the knowns, leading to the following sparse symmetric and positive definite system

$$K_{1,1}U_1 = F_1 - K_{1,2}U_2 \, , \tag{12}$$

which can be solved efficiently in parallel with the Jacobi preconditioned conjugate gradient method. The results of the iterative solver in [2] revealed that it is necessary to construct the global system (12) in parallel to obtain a good scalable solver. To this end, the matrices $A_i$, the vectors $f_i$, the local-global mapping operator $gm_{\mathcal{I}}$ and the gathering procedure, must be performed in parallel.

### 3.3 Investigation of the Grid

In order to construct the matrices $A_i$ and vectors $f_i$, one only needs the coordinates of the corners of the cells and its GLL orders in $\xi_1$- and $\xi_2$ direction, $N_1$ and $N_2$, respectively. For the calculation of the local-global mapping operator $gm_{\mathcal{I}}$, detailed grid information regarding the neighbouring cells is required to obtain the global numbering in parallel. Hence, the coordinates of the corners, the neighbouring cells and the GLL order together with a unique global number (see Section 3.4) are the minimal information needed to construct the global system (12) in parallel. These data will be collected on one processor when the grid is investigated and broadcasted from the root processor to $\min((\mathcal{K}, \mathcal{P}))$ processors, where $\mathcal{P}$ is the number of available processing elements.

This information can be collected the following way. The order of the spectral elements $p = N_1 = N_2$ is an input variable. When the grid file is read, the nodes, the nodal coordinates and the vertices of the spectral element mesh become available. The neighbors of the spectral elements can be obtained by investigating the mesh. This can be done the following way. The vertices of a spectral element are given in fixed order, see Fig. 2. In case two elements share the same node, their position with respect to each other is known. E.g., if node 0 of element $S_5$ equals node 3 of element $S_2$ then $S_2$ must be the *South*-neighbor of $S_5$. If node 2 of $S_5$ equals node 3 of $S_6$ then $S_6$ will be the *East*-neighbor of $S_5$, and so on. If a spectral element has less than four neighbors it must be a boundary element.

For the parallel global numbering of the GLL collocation points, it is necessary to assign the points to one particular cell. For the inner points this is trivial, but the situation is more complicated for points located at the interfaces. Therefore, we must define to which spectral element the vertices and edges belong. There are several possibilities, with the same complexity, and we choose one of them:

- a spectral element consists of all its internal GLL points, the *lower-left* vertex, the collocation points at the *South* and *West* edge, without the corner nodes.
- if a spectral element has an external boundary, the set is extended with the boundary points and possibly by vertices that do not belong to other spectral elements.

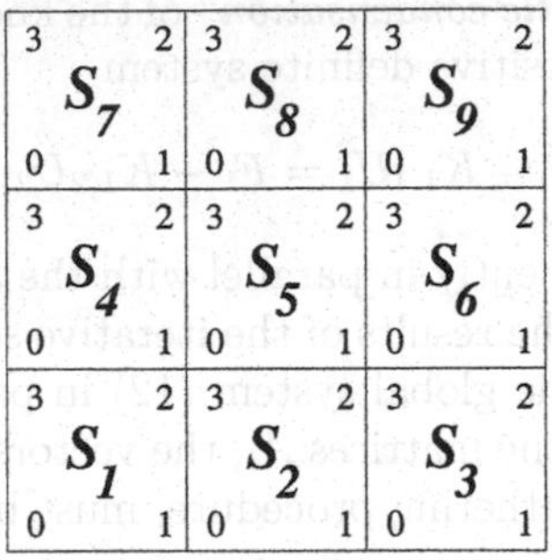

| | | |
|---|---|---|
| 3 2 $S_7$ 0 1 | 3 2 $S_8$ 0 1 | 3 2 $S_9$ 0 1 |
| 3 2 $S_4$ 0 1 | 3 2 $S_5$ 0 1 | 3 2 $S_6$ 0 1 |
| 3 2 $S_1$ 0 1 | 3 2 $S_2$ 0 1 | 3 2 $S_3$ 0 1 |

**Fig. 2.** Ordering of vertices: the vertices are numbered in fixed order, such that it is easy to recognize which spectral elements are neighbors

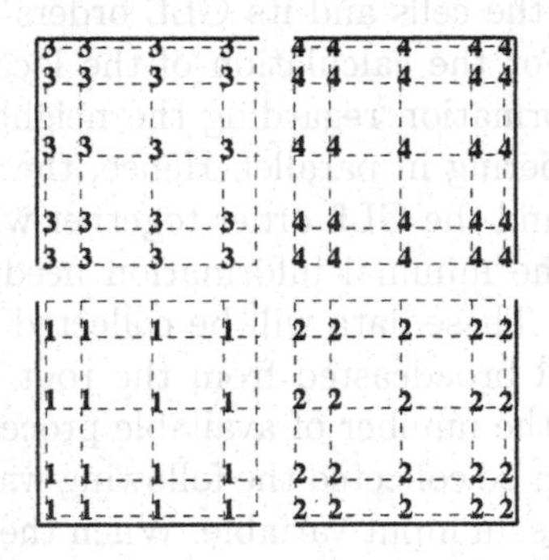

**Fig. 3.** An example of a grid discretized into four spectral elements. The nodes numbered 1 belong to spectral element 1, those numbered 2 belong to spectral element 2, and so on. The solid line corresponds to the physical boundary

Fig. 3 displays a mesh of four spectral elements and the numbers 1 to 4 elucidate to which spectral element the nodes belong.

Communication is required to obtain the missing values on the internal *North* and *East* interfaces: first data on the edge are sent from *North* to *South*. Subsequently, data on the edge are sent from *East* to *West*. After these steps, also the right-upper vertex has got its correct component values.

### 3.4 Unique Global Numbering

To obtain a unique global numbering, first the interface points are numbered which belong to a single spectral element: this local numbering starts with 1. Then we know exactly how much *degrees of freedom* $N^k_{dof}$ are present per element and transfer this information along the processors, such that we can compute $M_i$

$$M_i = \sum_{k=1}^{i-1} N^k_{dof}, \tag{13}$$

being the sum of the degrees of freedom on preceding spectral elements $k, k < i$. The value $M_i$ will be added to the local number. The *knowns* are numbered analogously. The numbering starts with $M_{\mathcal{K}} + N_{dof}^{\mathcal{K}} + 1$. Hence, one obtains the global node numbering discussed in Section 3.2.

### 3.5 The Construction of Matrix $K$

The parallel global assembly of the $\mathcal{K}$ local systems can now be performed resulting in the sparse symmetric positive definite global matrix $K$. Since the solution of eq. (12) is by far the most time consuming part, it must be optimally performed. Therefore, $K$ will be distributed along the processors into balanced parts of rows. The sparsity of $K$ asks for a CSR-formatted storage approach. The symmetry is not exploited: all nonzero elements of $K$ are stored. However, the CG method can only be applied successfully if $K$ is symmetric and positive definite.

The conversion of the local systems $A_i$ into the global system $K$ is complicated and requires communication. Beforehand, it is difficult to say which processors will communicate with each other and how long the messages will be. The global number determines to which processor the data will be sent. Therefore, a type **gambit** has been defined which couples the global number with the processor number and the row number on that particular processor:

```
TYPE :: gambit
   INTEGER  :: globalnr
   INTEGER  :: procnr
   INTEGER  :: rownr
END TYPE
```

Let $N_{dof}$ denote the degrees of freedom. Because only the first $N_{dof}$ rows of matrix $K$ are desired, the value **procnr** is not important in case the global number is larger than $N_{dof}$: to avoid confusion a negative value is added to **procnr**. For each nonzero element of $A_i$ it is then easy to determine whether it contributes to $K_{1,1}$, to $K_{1,2}$, or whether it can be neglected (row number $> N_{dof}$).

All nonzero elements of $A_i$ are considered and on account of its global number it is known to which processor it has to be sent. Before such an element and corresponding information about column number in $K$ is sent, all data are gathered and put into one message intended for a particular processor. Three different cases can occur:

1. The message must be sent to the processor on which it is already present. In that case, a send instruction is superfluous and the data can be processed immediately, or after the other messages have been sent.
2. The message is empty: no communication will take place, except that a message will be sent to the receiving processor that no contribution of the send processor can be expected.

3. The message is not empty and the send and receive processors are unequal. In that case communication is necessary.

The order in which the messages will be received is not known, and does not matter either. If nonzero elements are already present (see case 1 above) the first *very sparse* matrix $L_j$ can be built up. This matrix $L_j$ contains the nonzero values of an $A_i$ descendent of a spectral element stored in CSR-format. Contributions of other processors will be added to $L_j$ - i.e. each step the sum of two sparse matrices, stored in CSR-format, is calculated with a SPARSKIT [9] subroutine. In case no data are present, e.g., the rank of the processor is larger than the number of cells, the first data received will be used to build up the basis matrix $L_j$. At the end, $L_j$ will be the $j$-th `strip` of the matrix $K$. In the worst case, to create $L_j$, data from $\mathcal{K}$ spectral elements are needed. In practice, this number will be less. We do not investigate to find for all $(\mathcal{K}, \mathcal{P})$ combinations a favorite global numbering.

We observe that two steps are required to construct the matrix $K$ of (11) completely: one to construct $K_{1,1}$ and one to construct $K_{1,2}$ of eq. (12). The distribution of $K_{1,1}$ and $K_{1,2}$ along the processors is similar.

# 4 Parallel Platform and Implementation Tools

The calculations have been performed on

– Cray T3E system Vermeer (named after the Dutch painter) at HP$\alpha$C with 128 user processing elements (PE's) interconnected by the fast 3D torus network with a peak performance of 76.8 Gigaflop/s. The T3E uses the DEC Alpha 21164 for its computational tasks. Each node in the system contains one PE which in turn contains a CPU, memory, and a communication engine that takes care of communication between PE's. The bandwidth between nodes is quite high: 325 MB/s, bi-directional. Each PE is configured with

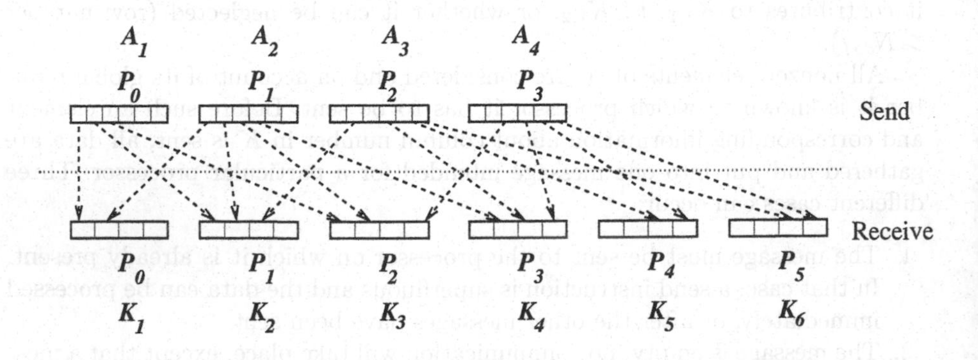


**Fig. 4.** Example of mapping of local matrices $A_i$ into the global matrix $K$, partitioned into matrix parts $K_j$ of consecutive rows

128 Mbytes of local memory, providing more than 16 Gbytes of globally addressable distributed memory.
- The SGI Origin 3800 Teras with 1024 500 MHz RI 14000 processors, subdivided into six partitions, two (interactive) 32-CPU partitions and four batch partitions of 64, 128, 256 and 512 CPU's, respectively. The theoretical peak performance is 1 Teraflop/s. Parallelization is done either automatically or explicitly by the user. A message passing model is allowed on the Origin using optimized SGI versions of e.g. `MPI`.

For more information about the Cray T3E used, we refer to [10].

To get good portable programs which may run on distributed-memory multiprocessors, networks of workstations as well as shared-memory machines we use `MPI`, Message Passing Interface. **Standard** or blocking communication mode is used: a send call does not return until the message data have been safely stored away so that the sender is free to access and overwrite the send buffer. Besides the standard communication mode, it appears to be necessary to use the buffered-mode send operation `MPI_BSEND` (see e.g., [8]). This operation can be started whether or not a matching receive has been posted. Its completion does not depend on the occurrence of a matching receive. In this way a deadlock situation, in which all processes are blocked is prevented. The execution of the send-receive process demonstrated in Fig. 4 profits of the buffered-mode send operation.

All routines have been implemented in `FORTRAN 90`, making frequently use of dynamic allocation of memory and derived types. As described in Sect. 3.5 the conversion of the cell-wise distribution into the row-wise CSR-format storage is difficult to implement and asks for a high level of flexibility. On fore hand it is not clear which processing elements will communicate and how much data will actually transferred. An upper bound for the number of processors which may need data of cell $i$ can be $\mathcal{K}$, the number of cells involved, whereas an upperbound of data from cell $i$ to PE $j$ can be the whole contents of that particular cell $i$. Apparently, multiplication of these two upperbounds leads to a storage demand per processor which grows with the number of processors used in the execution. The solution can be found in the usage of data structures with pointers to variable array sizes.

## 5 Numerical Results

### 5.1 The $h$- and $p$-Refinement Approach and Its Accuracy

In (least-squares) spectral element applications, two different kinds of refinement strategies are commonly used: $h$-refinement and $p$-refinement. The purpose of the numerical simulations is to check the accuracy performance for both refinement strategies. To this end, the least-squares spectral element formulation of the velocity-vorticity-pressure formulation of the Stokes problem is demonstrated by means of the smooth model problem of Gerritsma-Phillips [1] with $v = 1$. This model problem involves an exact periodic solution of the Stokes problem defined on the unit-square $[0,1] \times [0,1]$. The velocity boundary condition is used for all the numerical simulations. The pressure constant is set at the point $(0,0)$.

Six different grids are used to check the accuracy of the $h$-refinement. As can be observed in Table 1, the polynomial order of all the spectral elements equals 4, which means that each direction has four Gauss-Legendre-Lobatto(GLL) collocation points, and the number of spectral elements is varied from 4 to 144. In the middle column of Tables 1 and 2 the order of the large sparse global system is given together with the number of iterations required to solve this system using CG. The right column in the tables lists the $L_2$ norm of the different components, like the velocity ($L_2$ norm of $x-$ and $y-$components agree), the vorticity and pressure.

Only four different grids have been used to check the accuracy in case of the $p$-refinement (see Table 2). Each grid contains four spectral elements. The order of the approximating polynomial varies from 4 to 10 and is the same in all the variables. A growth of the polynomial order in the $p$-refinement case will increase the number of nodes per cell and likewise the amount of computational effort per cell.

### 5.2 The Parallel Performance

The least-squares spectral element solver (LSQSEM) code allows for the following situations:

$\mathcal{K} \geq \mathcal{P}$ : At least several processors contain more than one spectral element. It is not required that $\mathcal{P}$ is a true divisor of $\mathcal{K}$, or put it another way, the number of spectral elements per PE may differ.

**Table 1.** The different grids used for the investigation of the $h-$refinements

| Spectral elements | GLL-order | size of global system | # iterations | $L_2$ norm Velocity | Vorticity | Pressure |
|---|---|---|---|---|---|---|
| $2 \times 2$ | 4 | 259 | 132 | $9.2\ 10^{-4}$ | $4.8\ 10^{-2}$ | $1.8\ 10^{-2}$ |
| $4 \times 4$ | 4 | 1027 | 232 | $5.0\ 10^{-5}$ | $1.6\ 10^{-3}$ | $7.1\ 10^{-4}$ |
| $6 \times 6$ | 4 | 2307 | 326 | $5.2\ 10^{-6}$ | $2.8\ 10^{-4}$ | $6.9\ 10^{-5}$ |
| $8 \times 8$ | 4 | 4099 | 431 | $1.1\ 10^{-6}$ | $8.7\ 10^{-5}$ | $1.3\ 10^{-5}$ |
| $10 \times 10$ | 4 | 6403 | 569 | $3.2\ 10^{-7}$ | $3.5\ 10^{-5}$ | $3.6\ 10^{-6}$ |
| $12 \times 12$ | 4 | 9219 | 707 | $1.2\ 10^{-7}$ | $1.7\ 10^{-5}$ | $1.3\ 10^{-6}$ |

**Table 2.** The different grids used for the investigation of the $p$-refinements

| Spectral elements | GLL-order | size of global system | # iterations | $L_2$ norm Velocity | Vorticity | Pressure |
|---|---|---|---|---|---|---|
| $2 \times 2$ | 4 | 259 | 132 | $9.2\ 10^{-4}$ | $4.8\ 10^{-2}$ | $1.8\ 10^{-2}$ |
| $2 \times 2$ | 6 | 579 | 224 | $8.7\ 10^{-6}$ | $7.5\ 10^{-4}$ | $1.9\ 10^{-3}$ |
| $2 \times 2$ | 8 | 1027 | 305 | $6.5\ 10^{-8}$ | $7.1\ 10^{-6}$ | $1.6\ 10^{-6}$ |
| $2 \times 2$ | 10 | 1603 | 388 | $4.4\ 10^{-10}$ | $4.5\ 10^{-8}$ | $7.6\ 10^{-9}$ |

$\mathcal{K} < \mathcal{P}$ : A less plausible possibility. As we may conclude from Tables 1 and 2 that in order to increase the accuracy of the solution it is to preferred to enlarge the GLL order rather than to add more spectral elements. The latter gives rise to much larger global systems and accordingly more execution time. Indeed, during the construction of the local systems $\mathcal{P} - \mathcal{K}$ processors are idle, but for the same accuracy, the wall-clock time can be less. We remark, that MPI gives the opportunity to define an intracommunicator, which is used for communicating within a group, for instance within the group of PE's which correspond to a spectral element.

In the code we may consider several phases in the parallelization. The phases described correspond to the parts shown in Tables 3 and 4:

- The decomposition of the domain, or the investigation of the grid (cf. Sect. 3.3). In this phase, information is read from file: the number of spectral elements, the coordinates of the corners of the spectral elements, and of each spectral element its vertices are listed on file. Further the position of the spectral elements with respect to its neighbors is derived. This phase will be performed on a single processor.
- the second phase called `GAMBIT`, performs the computation on the internal GLL nodes and on the edge and vertices. Also the local-global mapping (cf. eq. (10)) is calculated. Communication is required for internal edges and vertices.
- The third phase computes the local systems and the right hand side values. This part is the most time-consuming part of the calculations on spectral elements and as can be conclude from Table 3 and 4, it is well scalable.
- The fourth and fifth phases consider the conversion of the local systems into the global systems. In `SUM` the contribution of the local systems to the construction of matrix $K$ is gathered. `CSR` performs the actual construction of $K_{1,1}$ and $K_{1,2}$ of eq. (11) by means of computing the sum of sparse matrices. Also the communication as described by Fig. 4 is included in `CSR`.
- In the last phase of the program PCG is used to solve the global system. In [2] it is described how the matrix-vector product in PCG are computed in parallel and gives pictures of speed-ups of this part of the code achieved on Teras and Vermeer.

The last column of the tables gives the wall-clock time for a complete run. We emphasize that the conversion part may not be neglected, especially the man power that was needed to give the code the flexibility for allowing an arbitrary number of spectral elements with respect to the number of processors involved, but for 32 processors it takes about 1.2 % of the total wall clock time. Notice that super linear speed up is achieved in the CSR-phase.

The wall-clock times of a cylinder grid as shown in Fig. 5 are listed in Table 4. It concerns a grid of 86 spectral elements including 66 boundary elements. The no-slip condition holds at the cylinder boundary, the velocity boundary condition on the outside boundary and the forcing term is zero. On Vermeer, the GLL order

**Table 3.** Wall-clock timings in seconds for the different parts in the parallel LSQSEM code obtained for the grid of 12 × 12 spectral elements and GLL order=6

| $\mathcal{P}$ | Spectral elements | | | Conversion | | PCG | Execution |
|---|---|---|---|---|---|---|---|
| | read | GAMBIT | Stokes | SUM | CSR | time | time |
| 4 | 0.19 | 0.38 | 118.95 | 3.01 | 21.94 | 584.57 | 729.19 |
| 8 | 0.20 | 0.30 | 59.44 | 1.56 | 7.41 | 308.55 | 377.65 |
| 16 | 0.18 | 0.23 | 29.70 | 0.85 | 2.65 | 176.34 | 210.15 |
| 32 | 0.19 | 0.21 | 16.54 | 0.59 | 1.11 | 121.67 | 140.77 |

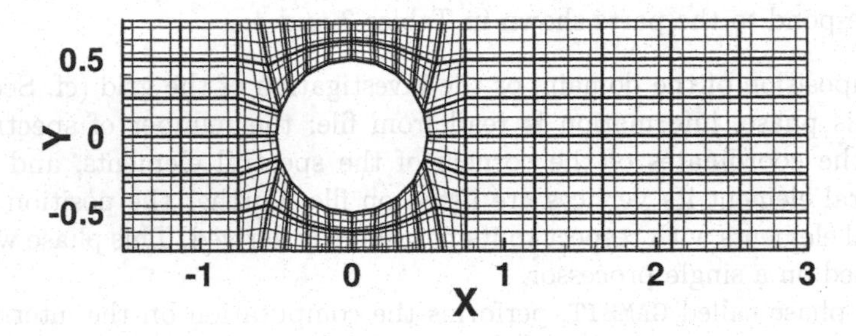


**Fig. 5.** Cylinder grid existing of 86 spectral elements with $p = 4$

**Table 4.** Wall-clock timings in seconds for the different parts in the parallel LSQSEM code obtained for the cylinder grid

| $\mathcal{P}$ | Spectral elements | | | Conversion | | PCG | Execution |
|---|---|---|---|---|---|---|---|
| | read | GAMBIT | Stokes | SUM | CSR | time | time |
| Teras, $p = 8$ | | | | | | | |
| 1 | 0.04 | 0.15 | 61.30 | 3.45 | 44.85 | 1522.1 | 1633.3 |
| 2 | 0.06 | 0.06 | 30.60 | 2.05 | 18.66 | 966.6 | 1018.2 |
| 4 | 0.04 | 0.04 | 15.65 | 1.05 | 10.53 | 475.1 | 502.5 |
| 8 | 0.06 | 0.03 | 7.93 | 0.59 | 2.83 | 232.0 | 243.6 |
| 16 | 0.11 | 0.08 | 4.40 | 0.33 | 1.24 | 142.0 | 150.3 |
| Vermeer, $p = 6$ | | | | | | | |
| 2 | 0.24 | 0.15 | 142.43 | 4.59 | 48.90 | 853.2 | 1049.0 |
| 4 | 0.24 | 0.14 | 72.58 | 2.42 | 26.97 | 444.0 | 546.7 |
| 8 | 0.19 | 0.08 | 36.24 | 1.26 | 7.63 | 228.0 | 273.6 |
| 16 | 0.19 | 0.14 | 19.80 | 0.74 | 2.77 | 120.4 | 144.1 |
| 32 | 0.20 | 0.05 | 9.88 | 0.54 | 1.11 | 68.7 | 80.7 |

is six in both directions, on TERAS the order is eight. The number of iterations is 1261 and 1904, respectively.

Finally, Fig. 6 shows the overall performance of the code on on a square domain partioned into decomposed into 8 × 8, 10 × 10 and 12 × 12 spectral elements. On Vermeer, it is not possible to run the largest problem on less than

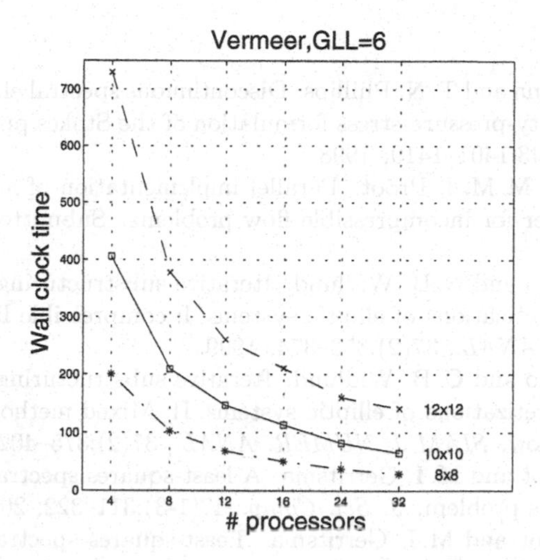


**Fig. 6.** Wall-clock timings achieved for a grid decomposed into $8 \times 8$, $10 \times 10$ and $12 \times 12$ spectral elements

four processors. We remark, that not in all cases the number of cells is a multiple of $\mathcal{P}$, e.g., the $10 \times 10$ grid has only $\mathcal{P} = 4$ as a true divisor. The speedup obtained when eight times more processors are used, (viz. from $\mathcal{P} = 4$ to $\mathcal{P} = 32$) is more than 5.2 and for the $8 \times 8$ is nearly 7.

## 6 Conclusions and Outlook

Least-squares spectral element methods result in symmetric and positive definite systems of linear equations which can be efficiently solved in parallel by PCG. The parallelization of this kind of problems requires two different strategies. Indeed, the spectral element discretization benefits from an element-by-element parallelization strategy whereas an efficient strategy to solve the large sparse global systems benefits from a row-wise distribution of data.

The numerical results, obtained with a simple model problem, confirm the good parallel properties of the element-by-element parallelization strategy. The combination of this strategy with the parallel PCG solver resulted in a good parallelizable code to solve incompressible flow problems. In the near future, the code will be extended to allow the usage of unstructured grids. More effective preconditioning methods for least-squares spectral element methods will be developed, too.

## References

[1] M. I. Gerritsma and T. N. Phillips. Discontinuous spectral element approximations for the velocity-pressure-stress formulation of the Stokes problem. *Int. J. Numer. Meth. Eng.*, 43:1401–1419, 1998.

[2] M. Nool and M. M. J. Proot. Parallel implementation of a least-squares spectral element solver for incompressible flow problems. Submitted to *J. Supercomput.*, 2002.

[3] L. F. Pavarino and O. B. Widlund. Iterative substructuring methods for spectral element discretizations of elliptic systems. I: compressible linear elasticity. *SIAM J. NUMER. ANAL.*, 37(2):353–374, 1999.

[4] L. F. Pavarino and O. B. Widlund. Iterative substructuring methods for spectral element discretizations of elliptic systems. II: Mixed methods for linear elasticity and Stokes flow. *SIAM J. NUMER. ANAL.*, 37(2):375–402, 1999.

[5] M. M. J. Proot and M. I. Gerritsma. A least-squares spectral element formulation for the Stokes problem. *J. Sci. Comp.*, 17(1-3):311–322, 2002.

[6] M. M. J. Proot and M. I. Gerritsma. Least-squares spectral elements applied to the Stokes problem. *submitted.*

[7] A. Quateroni and A. Valli. *Domain Decomposition Methods for Partial Differential equations.* Oxford University Press, 1999.

[8] Marc Snir, Steve W. Otto, Steven Huss-Lederman, David W. Walker, and Jack Dongarra. *MPI: The Complete Reference.* MIT Press, Cambridge, Massachusetts, 1996.

[9] Yousef Saad. SPARSKIT: a basic tool-kit for sparse matrix computations (Version 2) `http://www.cs.umn.edu/research/arpa/SPARSKIT/sparskit.html`

[10] Aad J. van der Steen. Overview of recent supercomputers, Issue 2001 *National Computing Facilities Foundation*, Den Haag, 2001.

# Edge-Based Interface Elements for Solution of Three-Dimensional Geomechanical Problems

Rubens M. Sydenstricker, Marcos A. D. Martins, Alvaro L. G. A. Coutinho and José L. D. Alves

Center for Parallel Computing and Department of Civil Engineering
COPPE/Federal University of Rio de Janeiro
PO Box 68506, Rio de Janeiro, RJ 21945-970, Brazil
rubens, marcos, alvaro}@nacad.ufrj.br
jalves@lamce.ufrj.br
http://www.nacad.ufrj.br

**Abstract.** An edge-based three-dimensional interface element for simulation of joints, faults and other discontinuities present in several geomechanical applications is proposed. Edge-based data structures are used to improve computational efficiency of Inexact Newton methods for solving finite element nonlinear problems on unstructured meshes. Numerical experiments in the solution of three-dimensional problems in cache based and vector processors have shown that memory and computer time are reduced.

## 1 Introduction

Solving three-dimensional large-scale geomechanical problems undergoing plastic deformations is of fundamental importance in several science and engineering applications. Particularly, interface finite elements have been widely used to model joints in rocks, faults in geologic media, as well as soil/structure interaction. Essentially, these elements must allow slip and separation between two bodies in contact, or separated by a thin material layer. Stresses are constrained so that traction in the interface element does not occur, and shear stresses are limited by a yield criterion. Although the required mechanical behavior of interface elements is simple, many analysts have reported numerical problems, such as oscillatory response.

Several interface elements have been proposed in the literature, as those proposed by Herrmann [4] and Goodman *et al.* [3] for two-dimensional analysis. Although they can be extended to the three-dimensional case, it can be shown that the resulting elements have the same features of the two-dimensional corresponding elements, including the kinematic inconsistency pointed by Kaliakin and Li [5] for the element proposed by Goodman *et al.* [3]. In this work, we develop a simple triangular interface element for three dimensional analysis, which may be viewed as an extension of the element proposed Goodman *et al.* [3]. A similar approach was used by Schellekens and De Borst [10] to derive quadrilateral interface elements.

This element has been implemented into an edge-based nonlinear quasi-static solid mechanics code to solve three-dimensional problems on unstructured finite element meshes [1, 9]. Edge-based data structures are used to improve the computational efficiency of the sparse matrix-vector products needed in the inner iterative driver of the Inexact Newton method. The interface element can be regarded as a prism with triangular base. Thus, we developed an edge-based decomposition of its stiffness matrix to preserve the good computational properties of the existing edge-based code. To minimize indirect addressing effects, we also introduced new *superedge* arrangements for interface elements. The remainder of this work is organized as follows. In the next section we briefly review the governing nonlinear finite element equations and the Inexact Newton method. Section 3 describes the interface element and its edge-based implementation. Section 4 shows numerical examples on the Athlon AMD 1.4 GHz, Intel Pentium IV 1.8 GHz, Cray SV1 and Cray T90 processors. The paper ends with a summary of the main conclusions.

## 2 Incremental Equilibrium Equations and the Inexact Newton Method

The governing equations for the quasi-static deformation of a body occupying a volume $\Omega$, in indicial notation, is:

$$\frac{\partial \sigma_{ij}}{\partial x_j} + \rho b_i = 0 \qquad in\ \Omega \tag{1}$$

where $\sigma_{ij}$ is the Cauchy stress tensor, $x_j$ is the position vector, $\rho$ is the weight per unit volume and $b_i$ is a specified body force vector. Equation (1) is subjected to the kinematic and traction boundary conditions,

$$u_i(x,t) = \overline{u}_i(x,t) \ \ in\ \Gamma_u; \ \ \sigma_{ij} n_j = h_i(x,t) \ \ in\ \Gamma_h \tag{2}$$

where $\Gamma_u$ represents the portion of the boundary where displacements are prescribed ($\overline{u}_i$) and $\Gamma_h$ represents the portion of the boundary on which tractions are specified ($h_i$). The boundary of the body is given by $G = G_u \cup G_h$, and $t$ represents a pseudo-time (or increment). Discretizing the above equations by a displacement-based finite element method we arrive to the discrete equilibrium equation,

$$\{F_{int}\} + \{F_{ext}\} = \{0\} \tag{3}$$

where $\{F_{int}\}$ is the internal force vector and $\{F_{ext}\}$ is the external force vector, accounting for applied forces and boundary conditions. Assuming that external forces are applied incrementally and restricting ourselves to material nonlinearities only, we arrive, after a standard linearization procedure, to the nonlinear finite element system of equations to be solved at each load increment,

$$[K_T]\{\Delta u\} = \{R\} \tag{4}$$

where $[K_T]$ is the tangent stiffness matrix, function of the current displacements, $\{\Delta u\}$ is the displacement increments vector and $\{R\}$ is the unbalanced residual vector, that is, the difference between internal and external forces.

*Remark.* We consider here that solid elements are perfect-plastic materials described by Mohr-Coulomb yield criterion. Stress updating is performed by an explicit, Euler-forward subincremental technique [2].

When solving iteratively the finite element system of linear equations, it is straightforward to employ inexact versions of the standard Newton-like methods [6]. In this case, tolerances for the inner iterative driver may be adaptively selected to minimize computational effort towards the solution. The iterative driver is the preconditioned conjugate gradient method, implemented employing edge-based strategies [1], that is, computing matrix-vector products as,

$$[K^e]=\sum_{s=1}^{m}[K_s^e] \tag{5}$$

where $[K_s^e]$ is the contribution of edge $s$ to $[K^e]$ and $m$ is the number of element edges.

## 3 Edge-Based Interface Elements

The interface element, which will be named GTB_3D, is composed by two triangular faces, and contains six nodes with 3 degrees of freedom per node as shown in Fig. 1.

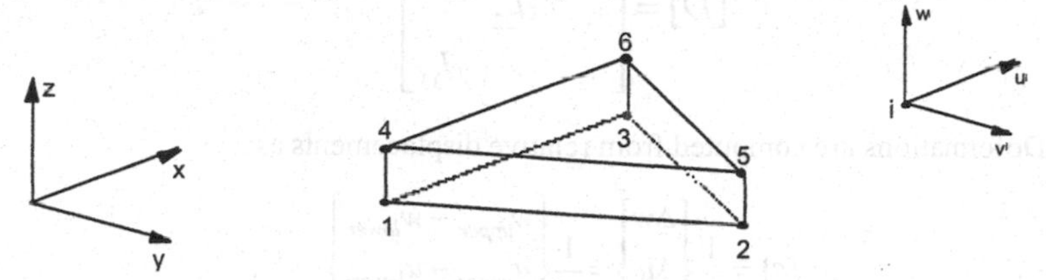


**Fig. 1.** Interface element

The element formulation is an extension of the original element proposed by Goodman *et al.* [3] for two-dimensional analysis. Displacements at upper and lower faces are independently interpolated as:

$$\{d_{upper}\}=[N]\{\hat{d}_{up}\} \quad ; \quad \{d_{lower}\}=[N]\{\hat{d}_{lw}\} \tag{6}$$

where $\{d_{upper}\}$ and $\{d_{lower}\}$ contain the displacement components as:

$$\{d_{upper}\}=\begin{Bmatrix} w_{upper} \\ u_{upper} \\ v_{upper} \end{Bmatrix} \quad ; \quad \{d_{lower}\}=\begin{Bmatrix} w_{lower} \\ u_{lower} \\ v_{lower} \end{Bmatrix} \tag{7}$$

$\{\hat{d}_{lw}\}$ and $\{\hat{d}_{up}\}$ contain the nodal displacements organized as:

$$\{\hat{d}_{lw}\}^T=[\{\hat{d}_1\}^T \ \ \{\hat{d}_2\}^T \ \ \{\hat{d}_3\}^T] \quad ; \quad \{\hat{d}_{up}\}^T=[\{\hat{d}_4\}^T \ \ \{\hat{d}_5\}^T \ \ \{\hat{d}_6\}^T] \tag{8}$$

where $\{\hat{d}_i\}^T=[\hat{u}_i \ \ \hat{v}_i \ \ \hat{w}_i]$ and $[N]$ is the interpolation function matrix given by:

$$[N]=\begin{bmatrix} 0 & 0 & L1 & 0 & 0 & L2 & 0 & 0 & L3 \\ L1 & 0 & 0 & L2 & 0 & 0 & L3 & 0 & 0 \\ 0 & L1 & 0 & 0 & L2 & 0 & 0 & L3 & 0 \end{bmatrix} \quad (9)$$

where $L1$, $L2$ and $L3$ are area coordinates.

The material behavior may be viewed as being elastic orthotropic with $E_x=E_y=0$, $G_{xy}=0$ and $\nu_{xy}=\nu_{xz}=\nu_{yz}=0$. As a consequence, normal stresses $\sigma_x$, $\sigma_y$ and the in-plane shear stress $\tau_{xy}$ are null, and stress-strain relations may be stated as:

$$\{\sigma\}=[D]\{\varepsilon\} \quad (10)$$

where:

$$\{\sigma\}^T=[\sigma_z \;\; \tau_{xz} \;\; \tau_{yz}] \;\; ; \;\; \{\varepsilon\}^T=[\varepsilon_z \;\; \gamma_{xz} \;\; \gamma_{yz}] \quad (11)$$

and $[D]$ is the constitutive matrix:

$$[D]=\begin{bmatrix} d_{11} & & \\ & d_{22} & \\ & & d_{33} \end{bmatrix} \quad (12)$$

Deformations are computed from relative displacements as:

$$\{\varepsilon\}=\frac{1}{h}\begin{Bmatrix} \Delta w \\ \Delta u \\ \Delta v \end{Bmatrix}=\frac{1}{h}\begin{Bmatrix} w_{upper}-w_{lower} \\ u_{upper}-u_{lower} \\ v_{upper}-v_{lower} \end{Bmatrix} \quad (13)$$

where h is the thickness of the element. From equations (6) and (13) we write:

$$\{\varepsilon\}=[B]\{\hat{d}\} \quad (14)$$

where:

$$[B]=\frac{1}{h}[B^*]=\frac{1}{h}[-[N] \;\; [N]] \quad (15)$$

and:

$$\{\hat{d}\}=\begin{Bmatrix} \{\hat{d}_{lw}\} \\ \{\hat{d}_{up}\} \end{Bmatrix} \quad (16)$$

Strain energy is evaluated as:

$$U=\frac{1}{2}\int_V \{\varepsilon\}^T\{\sigma\}\,dV=\frac{1}{2}\{\hat{\mathrm{d}}\}^{\mathrm{T}}\int_0^h\int_A [B]^T[D][B]\,\mathrm{dA}\{\hat{\mathrm{d}}\}=$$

$$=\frac{1}{2}\{\hat{\mathrm{d}}\}^{\mathrm{T}}\int_A [B^*]^T[DI][B^*]\,\mathrm{dA}\{\hat{\mathrm{d}}\} \quad (17)$$

where:

$$[DI]=\begin{bmatrix} DI_{11} & 0 & 0 \\ 0 & DI_{22} & 0 \\ 0 & 0 & DI_{33} \end{bmatrix}=\frac{1}{h}\begin{bmatrix} d_{11} & 0 & 0 \\ 0 & d_{22} & 0 \\ 0 & 0 & d_{33} \end{bmatrix}. \tag{18}$$

The stiffness matrix is obtained from the last integral of equation (17), and is explicit given by:

$$[K_{GTB3D}]=\frac{A}{12}\begin{bmatrix} [KK] & -[KK] \\ -[KK] & [KK] \end{bmatrix} \tag{19}$$

where:

$$[KK]=\begin{bmatrix} 2[DD] & [DD] & [DD] \\ [DD] & 2[DD] & [DD] \\ [DD] & [DD] & 2[DD] \end{bmatrix} \tag{20}$$

$$[DD]=\begin{bmatrix} DI_{22} & 0 & 0 \\ 0 & DI_{33} & 0 \\ 0 & 0 & DI_{11} \end{bmatrix} \tag{21}$$

and $A$ is the area of the triangular surface.

In order to implement the interface element into our edge-based solver, we disassemble the element matrix into its edge contributions as shown in Fig. 2.

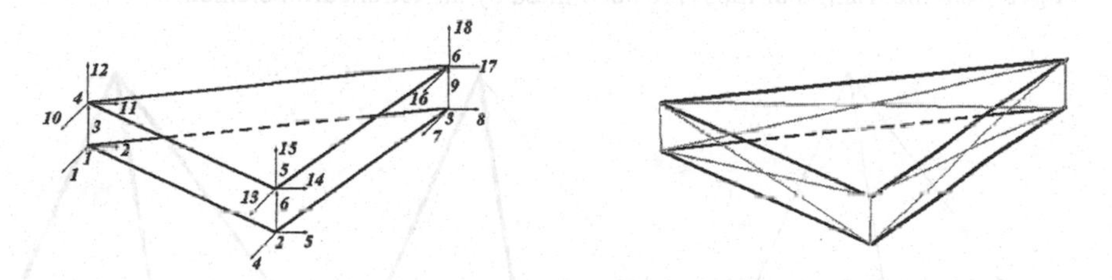


**Fig. 2.** Schematic representation of interface element (left) and its edge disassembling (right)

Observe in Fig. 2 that the interface element generates 6 geometrical edges and 9 computational edges. The resulting interface element geometrical edges are shared with the solid elements. Table 1 presents the number of floating point operations (flops) and indirect addressing (ia) for the matrix-vector multiplication needed in the preconditioned conjugate gradient method for element-by-element and edge-by-edge implementations. We show in column (a) the estimates for a tetrahedron and in column (c) for the interface element, while in columns (b) and (d) are shown the estimates for the edge implementation for both element types.

**Table 1.** Computational parameters for matrix-vector multiplication

| Parameter | Tetrahedron (a) | Edge tet × 1.5 (b) | Interface (c) | Edge int × 6 (d) |
|---|---|---|---|---|
| Flops | 252 | 54 | 108 | 36 |
| Ia | 36 | 27 | 54 | 108 |

We may observe in Table 1 the gain in flops when the edge based scheme is used for both elements. In columns (b) and (d) estimates are affected respectively by factors 1.5 and 6. These factors represent the ratio of edges to number of elements for each element type in a typical unstructured mesh. However, the interface element has more ia operations. This burden is alleviated by the small number of interface elements with respect to solid elements and the fact that algorithmic complexity is maintained, that is, instead of having two element types, we have just one, the edge. Table 2 shows the memory demand for storing stiffness coefficients element-by-element and edge-by-edge for both element types.

**Table 2.** Memory demand for storing stiffness coefficients

| Data Structure | Tetrahedron | Interface |
|---|---|---|
| Element | 78 | 171 |
| Edge | 9 | 6 |

Besides the improvements over the element based structure, there are more advantages in ia and flops parameters by grouping edges as depicted in Fig. 3, in *superedges* [1, 8]. The *superedges* assembled from interface elements are indicated as *superedge4* and *superedge9*, resulting from crossing edges and vertical edges, as indicated by bold lines in Fig. 3. The edges from the triangular faces are accounted by the tetrahedron elements.

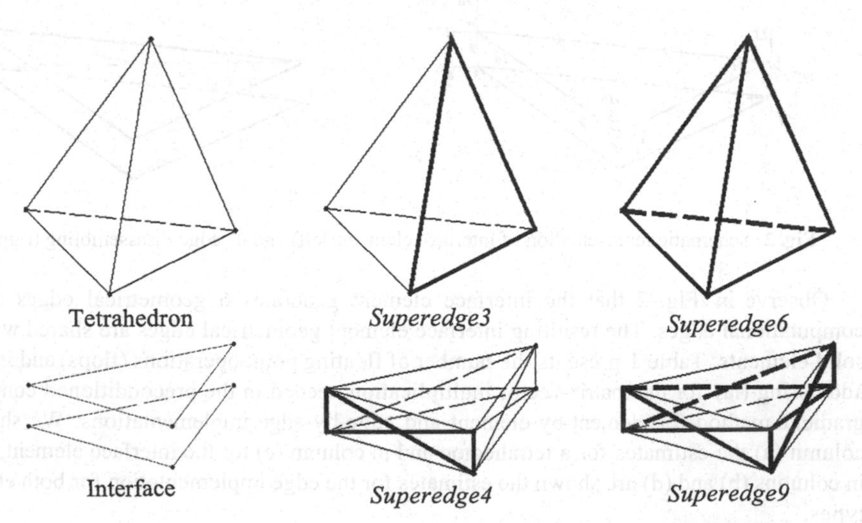


**Fig. 3.** Schematic representation of *superedges* for tetrahedron and interface elements

The idea behind *superedges* is, when loading data from memory to cache or vector registers, keep it there at the most, in order to operate them completely. This concept is completed considering that one edge carry out all contributions from its neighborhood elements, avoiding unnecessary floating points operations and several memory accesses for each element, when the edge is loaded to the processor and operated at once, so maximizing the flops/ia balance. Indeed, an ordered access to memory increases the processor performance for any kind of modern processors (cisc, risc) or shared or distributed memory parallel computers. One alternative to optimize the balance of flops/ia is to reordering the nodal points in such way that memory contention is minimized [7].

Improvements in flops and ia parameters are shown in Table 3 for several *superedges* groupings. However, observe that these gains are not for the whole mesh, since in several unstructured meshes just part of the edges can be grouped into these *superedge* arrangements [1, 9].

**Table 3.** Computational parameters for matrix-vector multiplication with *superedges*

| Parameter | Simple Edge | Superedge3 | Superedge4 | Superedge6 | Superedge9 |
|---|---|---|---|---|---|
| Flops | 36 | 130 | 190 | 268 | 436 |
| Ia | 18 | 27 | 36 | 36 | 54 |
| Flops/ia | 2.0 | 4.8 | 5.3 | 7.4 | 8.1 |

As one can see in Table 3, the balance of flops/ia increases with the edge grouping indicating great advantages over the simple edge. However, cache size and register availability may be observed in order to best fit flops and ia operations in the architecture, taking into account code simplicity and maintenance.

# 4 Numerical Examples

## 4.1 Footing-Soil Interaction

In this example we simulate a footing resting on a soil mass and submitted to a vertical and a skew horizontal force, as shown in Fig. 4. The angle between the force and $x$-axis is 30° as indicated. The vertical force was maintained constant and equal to one, and the skew horizontal force was monotonically increased from zero to 0.4 in 5 load steps. For the soil mass, all displacements at the base are prescribed, as well as all normal to the lateral bounding displacements. Soil mass and footing are separated by interface elements. Both soil and footing are linear elastic, and a Mohr-Coulomb criterion with a friction coefficient equal to 0.5 and null cohesion for the interfaces was adopted. Both soil mass and footing were represented by linear tetrahedra, and the interface between soil and block was represented by GTB_3D elements. Material properties for soil and footing are:

$$E_{block}=10^5 \quad ; \quad E_{soil}=10^3 \quad ; \quad \nu_{block} = \nu_{soil} = 0.0$$

and for the interface elements we considered:

$$DI_{11}= 10^{10} \quad ; \quad DI_{22}= DI_{33}=10^4$$

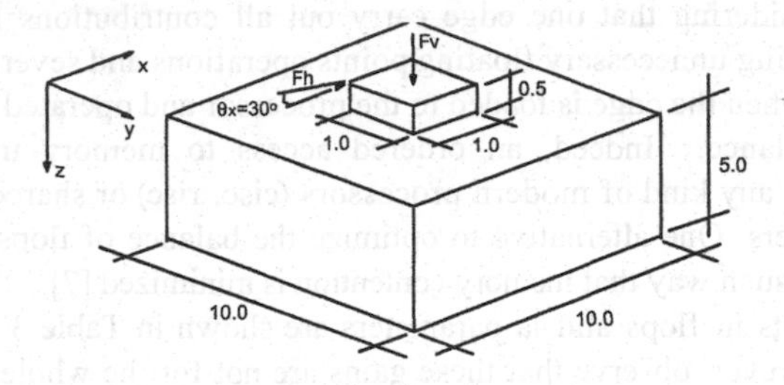


**Fig. 4.** Soil-footing interaction

In Table 4 we present the mesh properties and, in Fig. 5, we present details of the mesh. The problem was solved by using a tangent inexact Newton method [1], with PCG tolerances varying between $10^{-2}$ and $10^{-6}$.

**Table 4.** Mesh properties for soil-footing interaction problem

| Number of tetrahedra | Number of interfaces | Number of nodes | Number of equations |
|---|---|---|---|
| 143,110 | 5,000 | 27,890 | 83,238 |

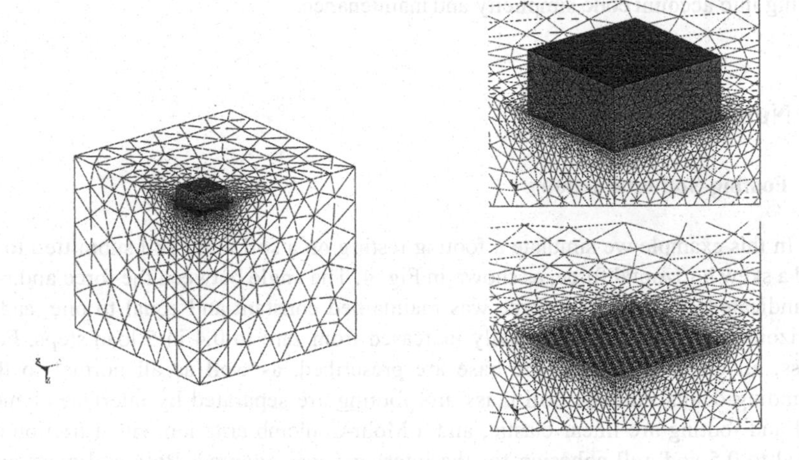

**Fig. 5.** Soil-footing interaction - details the mesh

In the first load step, almost all elements remain closed without slipping. For the remaining load steps, the interface path is shown in Fig. 6, in which interface elements that experienced opening and slipping are respectively indicated in white and gray, whereas the elements that remained closed without slipping are indicated in black. We note that, at load step 5, there is an opening region bellow the point where the skew horizontal force was applied. We see that, at this stage, almost the whole interface path is slipping.

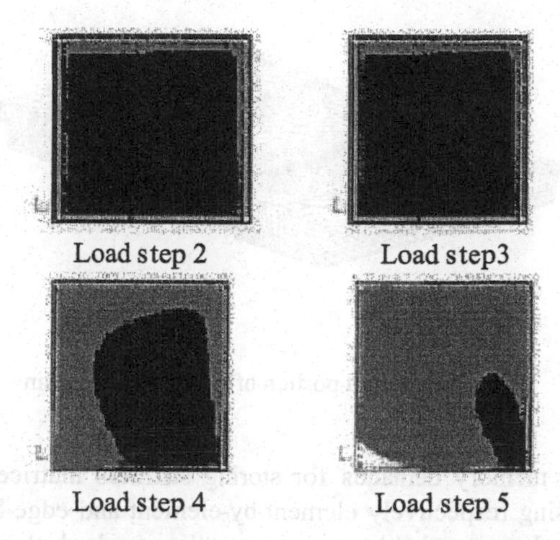


**Fig. 6.** Interface bonding (black), slipping (gray) and separation (white).

Table 5 shows the computational performance on the Athlon AMD 1.4 GHz and Intel Pentium IV 1.8 GHz for the solution of this problem. We may observe that the super-edge solutions were faster than the simple edge solutions in both machines.

**Table 5.** Computational performance for Athlon 1.4 GHz and Intel Pentium IV 1.8 GHz processors.

| Processor | Simple Edge (h) | Superedge (h) | Ratio Super/simp. |
|---|---|---|---|
| AMD Athlon 1,4 | 9.1 | 7.0 | 0.77 |
| Intel Pentium IV | 7.9 | 6.1 | 0.77 |

## 4.2 Block of a Sedimentary Basin

In this example we simulate the interaction of four blocks of a sedimentary basin separated by three geological faults, as shown in Fig. 7. Geometry and material properties are similar to a portion of a sedimentary basin in north-east Brazil. The blocks were submitted to a longitudinal compression applied at the left lateral surface. All displacements on the top are free and, at the basis and lateral surfaces, displacements normal to the mesh are prescribed. The mesh comprises 26,860 nodes, 94,058 linear tetrahedral elements, 12,960 interface elements and 76,380 equations. The resulting number of edges is 187,266 and 40.08% were grouped in *superedge6*, 16.99% in *superedge3*, 8.94% in *superedge9* and 0.88% in *superedge4*; the percentage of remaining simple edges is 33.12%. To improve data locality we reordered nodes to reduce wave front.

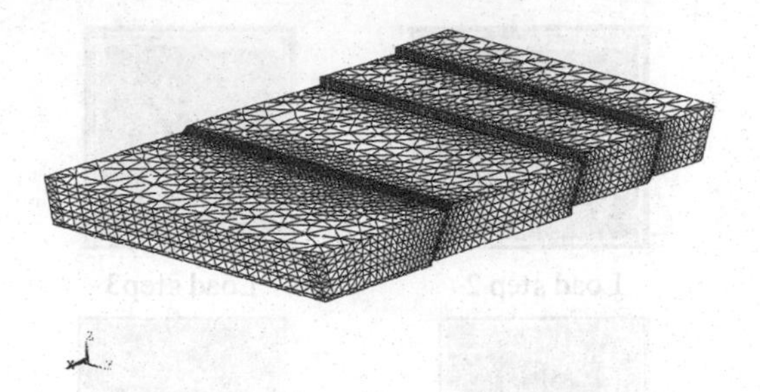

**Fig 7.** Mesh for a portion of a sedimentary basin

Table 6 presents memory demands for storing stiffness matrices for tetrahedral and interface elements using respectively element-by-element and edge-based data structures. Note that we achieved a remarkable saving over the standard element by element data structure, particularly for interface elements.

We applied prescribed displacements in six increments, stopping before the full slipping of the first inner fault. On average two nonlinear iterations were performed per displacement increment, using a tangent inexact Newton method, with PCG tolerances varying between $10^{-2}$ and $10^{-6}$. An average number of 4,000 PCG iterations per increment were needed to solve the linearized systems.

**Table 6.** Memory demand for storing stiffness matrices for element and edge data structures.

| Data type | Element (Mw) | Edge (Mw) | Ratio El/Ed |
|---|---|---|---|
| Tetrahedra | 7.0 | 1.6 | 4.4 |
| Interface | 2.1 | 0.27 | 7.8 |

In Fig. 8 we present the vertical displacements at the 6[th] displacement increment. The maximum displacement occurs at the top of the fault. Fig. 9 shows the evolution of slipping interface elements at three inner faults for 2[nd] and 6[th] steps. In this figure, the slipping elements are indicated as black and represent about 40% for 2[nd] step and almost 50% for 6[th] step of the total number of interface elements. After the 6[th] step we observed a full slipping of the first fault, producing a structural instability.

**Fig 8.** Vertical displacements for 6[th] step.

Fig 9. Slipping interface elements at inner faults for (a) 2nd and (b) 6th steps.

Table 7 shows the total CPU time obtained on the Athlon AMD 1.4 GHz, Intel Pentium IV 1.8, Cray SV1 and Cray T90 vector processor using only edges and all *superedges* arrangements for tetrahedral and interface elements. Note that performance improvements for *superedges* were greater in the scalars processors. We may associate this behavior to their cache based architectures.

**Table 7.** CPU times for symmetric edge-based matrix-vector product for different processors.

| Processor | Simple Edge (s) | Superedge (s) | Ratio Super/Simp |
|---|---|---|---|
| Athlon AMD 1.4 | 10,702.0 | 8,257.0 | 0.77 |
| Intel Pentium IV 1.8 | 7,722.0 | 5.868.0 | 0.76 |
| Cray SV1 | 3,230.0 | 2,954.0 | 0.89 |
| Cray T90 | 1,076.0 | 1.049.0 | 0.97 |

## 5. Conclusions

In this work we have introduced a new implementation of a three-dimensional interface element integrated to a highly optimized edge-based solver for geomechanical problems. Results indicate that this element represents correctly the mechanical behavior of geologic faults and geomechanical interfaces. To maximize individual processor performance we introduced new *superedge* groupings. These were particularly effective in cache-based processors. Memory costs for element data structure is almost eight times the observed for edge-based structure. However, since the number of interface elements was much smaller than the number of solid elements, the overall memory reduction is more modest.

## Acknowledgments

This work is supported by CNPq grant 522692/95-8 and by MCT/FINEP/PADCT project on Modeling of Sedimentary Basins. We are indebted to the Brazilian Petroleum Agency Program on Civil Engineering Human Resources for Oil and Gas (PRH02-UFRJ-COPPE/PEC). Computer time was provided by the Center of Parallel Computation of COPPE/UFRJ, the National Center for Supercomputing, CESUP/UFRGS and Cray Inc.

## References

1. Coutinho, A.L.G.A., Martins, M.A.D., Alves, J.L.D., Landau, L., Moraes, A.: Edge-Based Finite Element Techniques for Nonlinear Solid Mechanics Problems. Int. J. Num. Meth. Engng., 50 (2001) 2053-2058.
2. Crisfield, M.A., Non-Linear Finite Element Analysis of Solids and Structures. Advanced Topics, v. 2, John Wiley & Sons, London, UK, 1997.
3. Goodman, R.E., Taylor, R.L., Brekke, T.L.: A Model for the Mechanics of Jointed Rock. J. Soil Mech. Fdns. Div., ASCE, 99 (1968), 637-659.
4. Herrmann, L.R.: Finite Element Analysis of Contact Problems. J. Eng. Mech., ASCE, 104 (1978) 1043–1059.
5. Kaliakin, V.N., Li, J.: Insight into Deficiencies Associated with Commonly Used Zero-Thickness Interface Elements. Computers and Geotechnics, 17 (1995) 225-252.
6. Kelley, C.T. : Iterative Methods for Linear and Nonlinear Equations. Frontiers in applied mathematics, SIAM, Philadelphia, 1995.
7. Löhner, R., Galle, M., Minimization of Indirect addressing for edge-based field solvers, Comm. Num. Meth. Engng. ,18 (2002) 335-343
8. Löhner, R., Edges, Stars, Superedges and Chains, Comp. Meth. Appl. Mech. Engng. (1994), 255-263.
9. Martins, M.A.D., Alves, J.L.D., Coutinho, A.L.G.A.: Parallel Edged-Based Finite Techniques for Nonlinear Solid Mechanics. In: Palma, J.M.L.M., Dongarra, J., Hernández, V. (eds.): Vector and Parallel Processing – VECPAR 2000. Lecture Notes on Computer Science, Vol. 1981, Springer-Verlag, Berlin Heidelberg (2001), pp 506-518.
10. Schellekens, J.C.J., De Borst, R.: On the Numerical Integration of Interface Elements, Int. J. Num. Meth. Engng., 36 (1993) 43-66.

# Finite Element Simulation of Seismic Elastic Two Dimensional Wave Propagation: Development and Assessment of Performance in a Cluster of PCs with Software DSM

Francisco Quaranta[1], Maria C. S. de Castro[2,3], José L. D. Alves[1], and Claudio L. de Amorim[2]

[1] Laboratório de Métodos Computacionais em Engenharia
LAMCE/PEC/COPPE/UFRJ
CP 68552 CEP 21949-900 Rio de Janeiro-RJ Brasil
{quaranta,jalves}@lamce.ufrj.br
http://www.lamce.ufrj.br

[2] Laboratório de Computação Paralela - LCP/PESC/COPPE/UFRJ
{clicia,amorim}@cos.ufrj.br

[3] Departamento de Informática e Ciência da Computação - UERJ

**Abstract.** In this work we approach the seismic wave propagation problem in two dimensions using the finite element method (FEM). This kind of problem is essential to study the structure of the earth's interior and exploring petroleum reservoirs. Using a representative FEM-based application, we propose and evaluate two parallel algorithms based on the inverse mapping and on the mesh coloring, respectively. The distinguishing feature of our parallel versions is that they were implemented in a distributed shared-memory system (SDSMs), which offers the intuitive shared-memory programming model on a cluster of low-cost high-performance PCs. Our results for several workloads show that the inverse mapping scheme achieved the best speedup at 7.11 out of 8 processors, though the mesh coloring algorithm scales well. Overall, these preliminary results we obtained indicate that cluster-based SDSMs represent a cost-effective friendly-programming platform for developing parallel FEM-based applications.

## 1 Introduction

In solid mechanics, solving transient-state problems and the study of wave propagation effects are common research activities that are very important to several fields of engineering. In particular, a major application of interest in petroleum engineering is the analysis of seismic wave propagation in the interior of earth coupled with the study of seismic methods for determining the geological subsurface structure. Both the analysis and knowledge of sub-surface geology are essential requirements to find out petroleum reservoirs. However, applying analytical methods to such class of problems become a challenging task due to

J.M.L.M. Palma et al. (Eds.): VECPAR 2002, LNCS 2565, pp. 65–78, 2003.

the problem's complexity. Therefore, an increasingly cost-effective alternative is to develop computational environments based on numerical methods such as the highly successful finite element method (FEM) to deal with the problem's complexity. Nevertheless, given the large size of typical input data, even current high-performance microprocessors are not able to execute FEM-based wave propagation applications satisfactorily because either they are not fast enough or they do not provide sufficient RAM to efficiently process such large input data, or both. Usually, large-scale FEM-based applications run on high-performance vector computers or more recently on clusters of high-performance PCs using traditional message-passing libraries such as MPI and PVM. We note that past works [1, 2, 7, 12] examined performance of FEM-based applications on high-cost hardware shared-memory processors and described experiments using applications similar to the ones we use in this paper. In contrast to these works, we obtained a competitive solution using a low-cost cluster platform.

On the other hand, recent research results show that software distributed shared-memory systems (SDSMs) offer a cost-effective platform for parallel computing that combines the intuitive shared-memory programming model with the low-cost of distributed memory systems, as offered by clusters of PCs.

With this in mind, in this work, we design and implement parallel wave propagation applications in TreadMarks, a popular SDSM, running on a cluster of 8 high-performance PCs and evaluate performance of our applications for several workloads. We introduce two parallel versions of well-known methods to handle element assembling in FEM codes: 1) the inverse mapping scheme [12] and 2) mesh colouring[7].

Our preliminary results show that our parallel implementation of wave propagation applications attains speed-ups as much as 7.11 out 8 processors. These results suggest that cluster-based software DSMs represent a cost-effective platform for running FEM-based wave propagation applications.

The remainder of this paper is organized as follows. Next section overviews the linear elasticity problem. Section 3 summarizes the TreadMarks software distributed shared-memory system we use. Section 4 describes our parallel versions of wave propagation application. In Section 5 we present the experimental methodology. Section 6 analyzes performance of our parallel applications on top of TreadMarks. Finally, in Section 7 we draw our conclusions and outline future work.

## 2 The Linear Elasticity Problem

In this section we treat the elasticity problem in solid mechanics by presenting a formal definition of the problem and its associated finite element method (FEM) semi-discrete form. In addition, we comment on relevant computational aspects of FEM.

### 2.1 The Elasticity Problem in Solid Mechanics

A solid body, can be submitted to tractions and stresses over a period of time, as the deformed body in the Figure 1 shows. The figure represents a physical system that consists of a body with domain $\Omega$ and boundary $\Gamma$ in a reference time $t$.

The body can be submitted to prescribed displacements, surface tractions, and volume forces, at any given time, so that the body will be displaced, deformed, and it will occupy its new domain in the next time step. For instance, the body occupies the domain ${}^0\Omega$ in time $t = 0$ (left side) and the new domain ${}^t\Omega$ in a time $t > 0$ (right side), after forces are applied to it.

The system behavior is governed by four equation sets, namely, momentum conservation equation, constitutive equation, initial conditions, and boundary conditions. Both the equations and formulations for linear analysis in solid mechanics are described in detail in [3, 4, 6, 8, 13].

### 2.2 Formal Definition

A formal definition of the problem can be described as follows. Consider the surface tractions ${}^t\overline{\mathbf{q}}$ in ${}^0\Gamma_F$ and body forces ${}^t\mathbf{b}$ in ${}^0\Omega$. The problem consists in finding the displacement fields ${}^t\mathbf{u}$ in all times of the interval $t \in [0, T]$, that satisfy the problem conditions. The problem's equations are shown in Figure 2.

In Figure 2 ${}^t\sigma_{ij}$ is the Cauchy stress tensor, ${}^tx_j$ are material coordinates, ${}^tb_i$ is the body force, ${}^t\rho$ is the mass density, ${}^ta_i$ are the accelerations, $C_{ijkl}$ is the elastic constitutive tensor, ${}^t\varepsilon_{kl}$ is the strain tensor and ${}^t\overline{\mathbf{u}}$ is a prescribed displacement. Displacement and velocity at the initial time are represented by $\mathbf{u}(\mathbf{x}, \mathbf{0})$ and $\mathbf{v}(\mathbf{x}, \mathbf{0})$, respectively.

The FEM-based numerical solution for the problem can be found using a discretization of the domain. For instance, we used the bilinear quadrilateral element to obtain a discrete domain. The next section focus on the spatial and temporal finite element discretization.

### 2.3 Spatial and Temporal Discretization of the Problem

The spatial discretization of a solid body is achieved by subdividing the body domain $\Omega$ into finite elements with domain $\Omega^e$, as shown in Figure 3. In these

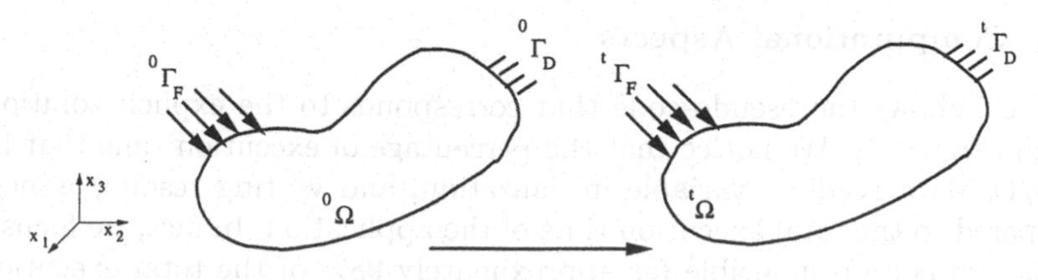


**Fig. 1.** System with a deformed solid body

1. Momentum conservation equation

$$\frac{\partial\, {}^t\sigma_{ij}}{\partial\, {}^t x_j} + {}^t b_i = {}^t\rho\, {}^t a_i \quad in \quad {}^t\Omega \tag{1}$$

2. Constitutive equation

$${}^t\sigma_{ij} = C_{ijkl}\, {}^t\varepsilon_{kl} \tag{2}$$

3. Initial conditions

$$\begin{aligned} \mathbf{u}(\mathbf{x}, \mathbf{0}) &= {}^0\mathbf{u} \quad in \quad {}^0\Omega \\ \mathbf{v}(\mathbf{x}, \mathbf{0}) &= {}^0\mathbf{v} \quad in \quad {}^0\Omega \end{aligned} \tag{3}$$

4. Boundary conditions

$${}^t\mathbf{u} = {}^t\overline{\mathbf{u}} \quad in \quad {}^t\Gamma_D \tag{4}$$

$${}^t\sigma_{ij}\, {}^t n_{ij} = {}^t\overline{q}_i \quad in \quad {}^t\Gamma_F \tag{5}$$

**Fig. 2.** Formal definition of the problem

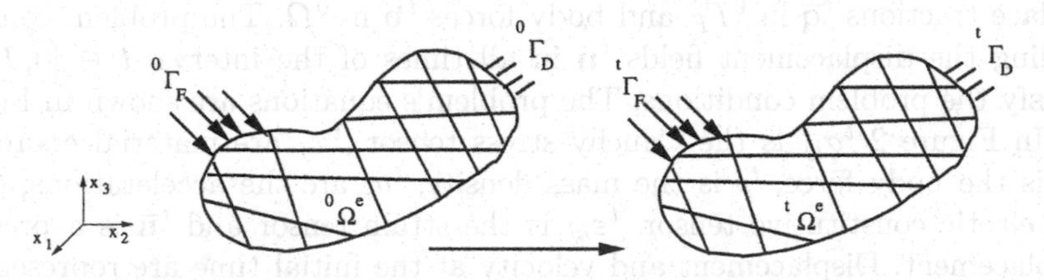

**Fig. 3.** Finite element discretization of the system

elements, the displacement field is approximated by shape functions that interpolate the displacement of nodal points [3, 4, 8, 13].

In addition, we use the central difference operator to discretize the interval time of interest $[0, T]$, yielding an explicit method of solution [4]. This method is used in the finite element solution of short transient problems in linear elasticity and is summarized by the algorithm in the Figure 4.

In the next section, we show the routine that represents the algorithm of Figure 4 and comment on relevant computational aspects.

### 2.4 Computational Aspects

Figure 5 shows the pseudo-code that corresponds to the explicit solution algorithm (Figure 4). We notice that the percentage of execution time that is spent on I/O (data reading, variable initialization, and writing results) is negligible compared to the total execution time of the application. In fact, we focus on the portion of code responsible for approximately 98% of the total execution time of the application for all the input workloads we tested.

1. Initialization of variables $t = 0$, $n = 0$. Compute $\Delta t$
2. Apply the initial values $\mathbf{u^0}$, $\mathbf{v^0}$, $\mathbf{F_p^0}$ and $\mathbf{F_{int}^0}$
3. Compute the lumped mass matrix $\mathbf{M}$
4. Compute initial acceleration $\mathbf{a^0} = \mathbf{M^{-1}}(\mathbf{F_p^0} - \mathbf{F_{int}^0})$
5. Compute $\mathbf{v^{n+\frac{1}{2}}} = \begin{cases} \mathbf{v^{n-\frac{1}{2}}} + \mathbf{a^n}\mathbf{\Delta t^n}, \text{ if } n > 0 \\ \mathbf{v^n} + \frac{1}{2}\mathbf{a^n}\mathbf{\Delta t^n}, \quad \text{if } n = 0 \end{cases}$
6. Compute displacements $\mathbf{u^{n+1}} = \mathbf{u^n} + \mathbf{v^{n+\frac{1}{2}}}\mathbf{\Delta t}$
7. Update $\mathbf{u^{n+1}}$, with boundary conditions ${}^t\overline{\mathbf{u}}$
8. Compute stress $\sigma^{n+1}$, internal forces $\mathbf{F_{int}^{n+1}}$ and external forces $\mathbf{F_p^{n+1}}$
9. Compute acceleration $\mathbf{a^{n+1}} = \mathbf{M^{-1}}(\mathbf{F_p^{n+1}} - \mathbf{F_{int}^{n+1}})$
10. Update $t = t + \Delta t$
11. If $t > T$ stop, otherwise update $n = n + 1$ and start again from item 5.

**Fig. 4.** Explicit solution algorithm

```
ExpConv Routine
Vector and matrix declarations
Initialization of variables
For all number of steps do          !TIME INTEGRATION
    Call DISASSEMBLING vector of displacements
    Call DISASSEMBLING vector of velocity
    Call EVALUATE internal forces
    Call ASSEMBLING element contribution
    For all unknowns do             !UPDATE GLOBAL UNKNOWNS
        Update accelerations, velocities, and displacements
```

**Fig. 5.** Explicit solution routine

The explicit-solution algorithm divides the problem into the following four main tasks: gather operation, evaluation of internal forces, scatter operation, and update of global unknowns. The evaluation of internal forces and update of unknowns are relatively simple arithmetic operations that divide the loops of elements and equations among processors, since there is no data dependence between the components within the same loop.

The scatter and gather tasks perform write and read accesses to memory. The scatter (disassembling) operation is immediate because array components are independently computed. However, scatter (assembling) operation requires special attention, since different processors can read and modify the same value thus requiring controlled access. Indeed, with an inadequate treatment, assembling operations can consume up to 40% of total computing time [1, 2, 12].

We consider and evaluate two well-known algorithms in the design of the assembling process. More specifically, the inverse mapping scheme introduced by Torigaki [12] uses the concept of equation connectivity while mesh colouring, presented in [7] uses a reordering of the finite element mesh in blocks of disjoint elements. Each algorithm has advantages and disadvantages that will be discussed in Section 6. A similar performance study of these algorithms was performed in [1, 2] using a shared-memory vector/parallel machine. Before describing our two parallel versions of the assembling task, we introduce next basic concepts related to software DSMs.

# 3 TreadMarks

Software distributed shared-memory systems or simply SDSM protocols implement the abstraction of shared-memory on a distributed memory system. Therefore, SDSM offer an attractive platform for parallel computing by combining the low-cost of distributed memory systems with the intuitive shared-memory programming model. However, software DSM protocols incur substantial overheads to ensure memory coherence over high-latency communication networks, that degrade SDSMs performance significantly[9].

In this work, we use TreadMarks (Tmk), a well-known page-based multiple-writer protocol SDSM as the reference SDSM protocol. Tmk protocol implements the lazy release consistency model [10](RLC) in which accesses to shared data are protected by acquire (e.g., lock) and release (e.g., unlock) operations on locks. Lock acquire is used to gain access to shared data and lock release to give such access away. In the LRC model, modifications made to shared data by a processor need to be seen only by the next processor that acquires the lock. To know which modifications precede the lock acquire Tmk divides the program execution into intervals in such a way that the intervals can be partially ordered by synchronization accesses. So, on lock acquire, the acquiring processor asks the last releaser of the lock for the information about the modifications made to shared data during the intervals that precede the acquiring processor's current interval. A barrier operation can be seen as if each processor executes a lock release followed by a lock acquire. To alleviate the problem of false-sharing as Tmk use the large page as the coherence unit, Tmk implements a multiple-writer protocol that allows multiple processors to write to the same page concurrently. Further details on Tmk can be found in [11].

Overall, SDSM systems achieve respectable performance gains on regular applications in which memory accesses are coarse-grained [5]. Since our applications belong to this class, they are likely to achieve good speedups.

# 4 The Parallel Explicit-Solution Algorithm

In this section, we describe briefly the parallel versions of the assembling phase in the explicit-solution algorithm. In this application, we implemented the assembling phase using both the inverse mapping scheme and mesh colouring. We adapted these schemes and their data structures to run on a software distributed-shared memory platform. As a result of the simplicity of the shared-memory model our parallel implementations were straightforward. More specifically, we use lock/unlock operations when pairs of processors need to synchronize and barriers to synchronize all processors.

```
Assembly_inverse mapping Routine
Distribute elements among processors
For all private equations of processor do
    Assembly element contributions
For all processors do
    If a shared equation is assigned to processor then
        Request exclusive access !LOCK acquire
            Assembly elements contributions
        Release exclusive access !LOCK release
Synchronize all processors !BARRIER
```

**Fig. 6.** Parallel inverse mapping scheme assembly

**Parallel Inverse Mapping scheme**

Figure 6 shows the parallel inverse mapping scheme. The algorithm starts by dividing the total number of elements among the processors. Each processor first assembles the contributions of elements that are in its private data area. Next, the algorithm collects the contributions of the elements that have common nodes between two processors. To build these shared equations, it is necessary to use a lock operation to ensure exclusive access to shared data as well as to enforce data coherence. Lastly, a barrier is issued to synchronize all processors before initializing the next integration time step.

**Parallel Mesh Colouring**

Figure 7 shows the parallel mesh colouring algorithm. This algorithm executes a preliminary operation, not shown, that consists in reordering the elements in sets of disjoints elements. These sets are blocks of elements that do not have nodes in common so that elements in a block have the same *colour*. For each block, we divide the elements of the block among the processors. Next, we compute their contributions for all private and shared equations. At the end of block processing, a barrier primitive synchronizes all the processors. In the mesh colouring algorithm, we do not need to use a lock primitive, because of the characteristic of the elements that belong to the same block. As the elements within the same block do not have nodes in common, the processors never contribute to the same equation at the same time.

## 5 Experimental Methodology

In order to assess the performance of the two parallel versions of the explicit solution algorithm we used a cluster of 8 PCs. Each PC consists of a 650 MHz Dual-Pentium III, 512 MBytes of RAM, 256 KBytes of cache (L2), connected by a Fast-Ethernet switch with 100 Mbits/s. Our cluster runs Linux Red Hat 6.2 and our TreadMarks version uses only one processor per node. The application was implemented in Fortran. We ran each experiment 10 times and presented the average execution time in seconds. The standard deviation was less than 2%.

```
Assembly_Mesh_Colouring Routine
For all element blocks do
    Divide elements of each block among processors
    For all elements assigned to processor do
        If is a privative element then
            Assembly elements contributions on private area
        Else
            Assembly elements contributions on shared area
    Synchronize all processors      !BARRIER
Synchronize all processors      !BARRIER
```

**Fig. 7.** Parallel mesh colouring assembly

### 5.1 Numerical Example

A bulk of layered soil is modeled for the numerical synthesis of seismic data for VSP (Vertical Seismic Profiling). This is a typical application within the scope of Geophysics with great importance for finding petroleum reservoirs [2].

In this kind of analysis, the propagation of seismic waves, generated from explosive charges detonation on the surface, through various subsoil layers is measured and interpreted to obtain geological data.

The analyzed model considers three horizontal layers, each of which presents different speeds of propagation for the dilatational wave. Figure 8 illustrates the problem.

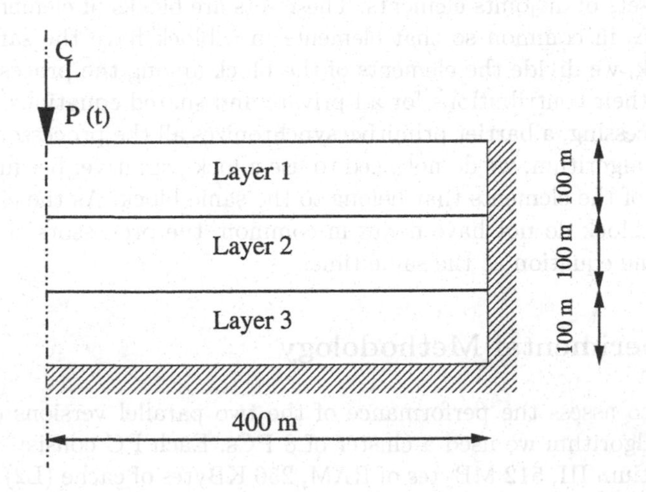


**Fig. 8.** Problem description

**Table 1.** Discretization characteristics

| Mesh | Domain | Elements | Nodes | Equations |
|---|---|---|---|---|
| GEO5 | 400m × 300m | 19200 | 19481 | 38280 |
| GEO6 | 800m × 300m | 38400 | 38841 | 76680 |
| GEO7 | 1600m × 300m | 76800 | 77561 | 153480 |
| GEO8 | 3200m × 300m | 153600 | 155001 | 307080 |

**Table 2.** The physical characteristics

| Layer | Young modulus ($KN/m^2$) | Poisson ratio | Density ($KN/m^3$) |
|---|---|---|---|
| 1 | 1.598917E+03 | 0.33 | 2.55E-01 |
| 2 | 3.197840E+02 | 0.33 | 2.04E-01 |
| 3 | 7.185440E+02 | 0.33 | 2.34E-01 |

For simulation of the physical behavior model and analysis of application performance, four different finite element meshes were used. In all cases mesh resolution was kept constant at a 2.5 square meters. In the vertical direction, the number of elements was kept always constant for each case. In the horizontal direction, the number of elements was twice the previous one. Table 1 shows the main features of each finite element mesh we used. It shows for each set of input data its label, the domain size, the number of elements, nodes, and equations. For all cases, due to the sáme dimension of the elements, an identical time step $\Delta t = 1.0E - 04s$ was used as well as 7000 steps for time integration.

Table 2 presents relevant physical characteristics of the model. They are identical to those found in [2].

Figure 9 illustrates the results for the propagation of seismic waves in the addressed problem. The figures show distribution of nodal velocities in the mesh at analysis times t = 0.0700 s, t = 0.1575 s, t = 0.2275 s and t = 0.3150 s. Due the boundary conditions and physical characteristics adopted in this example, we observe at time t = 0.1575 s the effect of different wave speed and refraction of the waves between layers 1 and 2. At times t = 0.2275 s and t = 0.3150 s we observe the propagation of the waves after the reflection at the end of layer 3.

# 6 Performance Analysis

In this section, we present and discuss the performance results obtained from the analysis of the numerical example described in the previous section. For each parallel application, whether based on the inverse mapping scheme or the mesh colouring algorithm using the explicit solution algorithm we show the total execution time in seconds and speedups. We ran the application for 1, 2, 4, and 8 processors.

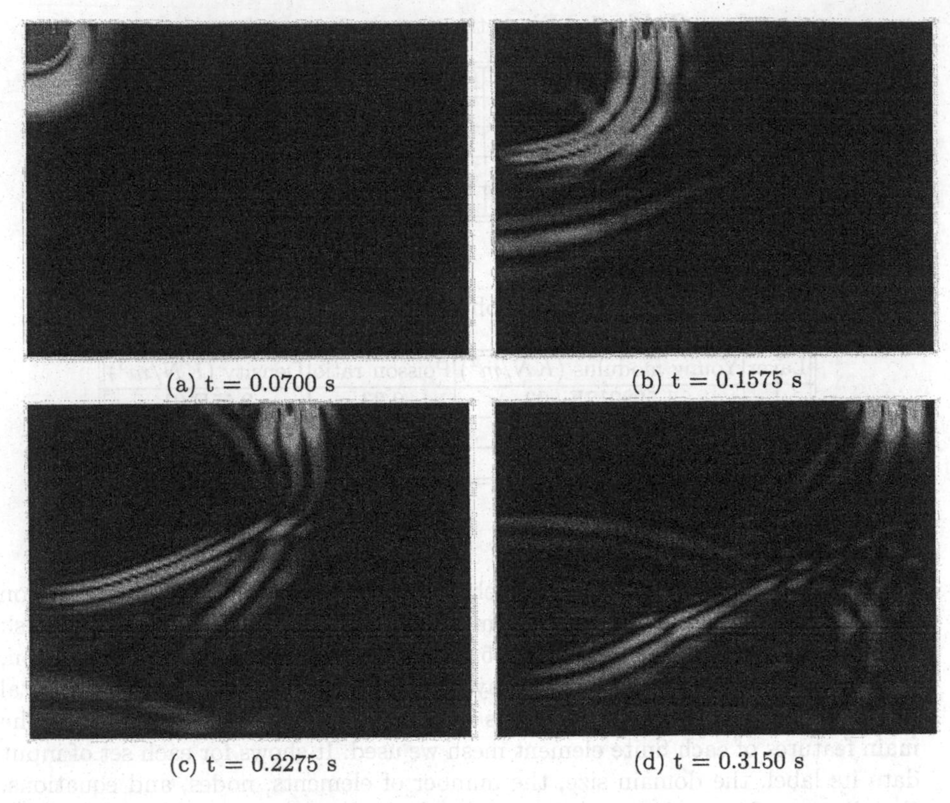

(a) t = 0.0700 s

(b) t = 0.1575 s

(c) t = 0.2275 s

(d) t = 0.3150 s

**Fig. 9.** Seismic Wave Propagation

### 6.1 Inverse Mapping Scheme

The inverse mapping application achieved the best performance. Table 3 shows the execution time (ET) and speedups (Sp) for all the input data workloads. Figure 10 shows the speedup curve related to that table.

The total execution times of the inverse mapping scheme were approximately proportional to the input data size. The speedups we obtained were very good for all input data and numbers of processors, revealing that the application is potentially scalable. The main reason is that it exhibits a coarse-grained access pattern. Also, the application presented superlinear speedups when we ran input data GEO7 for 2 processors. This is explained by the fact that as we move from 1 to 2 processors the larger memory size reduces cache pollution. Notice that the application achieved speed-up greater than 7 for GEO7. From 4 to 8 processors we obtained a slight decrease in application performance due to TreadMarks cache-coherence invalidate-based protocol. More specifically, as we increase the number of processors, so does the number of shared equations, which in turn causes the effects of page invalidations to get worse as we added more processors.

**Table 3.** Results using the inverse mapping scheme

| # Procs. | Input Data | | | | | | | |
|---|---|---|---|---|---|---|---|---|
| | GEO5 | | GEO6 | | GEO7 | | GEO8 | |
| | ET | Sp | ET | Sp | ET | Sp | ET | Sp |
| 1 | 766.6 | 1.00 | 1561.5 | 1.00 | 3231.7 | 1.00 | 6640.5 | 1.00 |
| 2 | 393.9 | 1.95 | 789.3 | 1.98 | 1596.6 | **2.02** | 3342.1 | 1.99 |
| 4 | 212.6 | 3.61 | 423.8 | 3.69 | 853.1 | 3.79 | 1770.1 | 3.75 |
| 8 | 119.1 | 6.44 | 235.9 | 6.62 | 454.5 | **7.11** | 1068.8 | 6.21 |

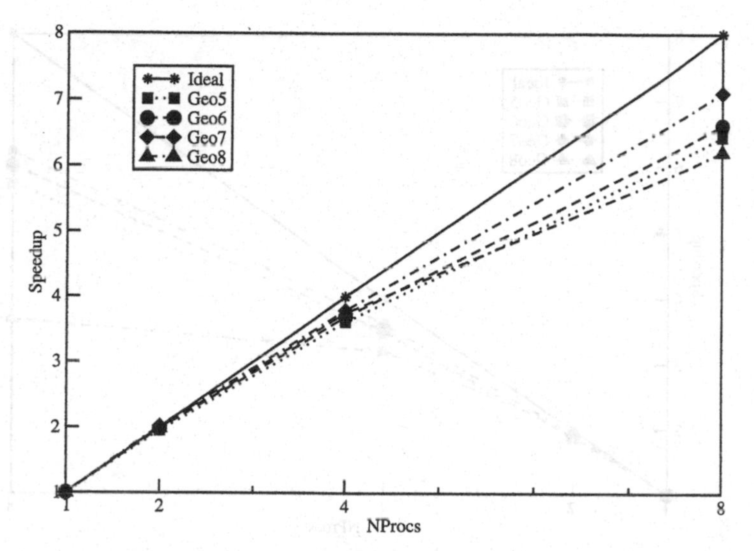


**Fig. 10.** Results using the inverse mapping scheme

## 6.2 Mesh Colouring

The mesh colouring application attained a lower performance than the inverse mapping-based one. Table 4 shows the execution time (ET) and speedups (Sp) for all input data workloads. Figure 11 shows the speedup curve related to that table.

For all input data sizes the speedups were inferior to that of the inverse mapping scheme, and total execution times were significantly greater for 4 and 8 processors. In general, the mesh colouring scheme did not show to be as scalable as the inverse mapping scheme, although it also has a coarse-grained access pattern. Note that for 8 processors and input data GEO8, it even slowed down. The reason is the same as in the inverse mapping scheme, TreadMarks' cache-coherence protocol and memory consistency model. In addition, the mesh colouring algorithm deals with a different finite element mesh data structure that aggravates coherence-related overheads.

**Table 4.** Results using mesh colouring

| # Procs. | Input Data | | | | | | | |
|---|---|---|---|---|---|---|---|---|
| | GEO5 | | GEO6 | | GEO7 | | GEO8 | |
| | ET | Sp | ET | Sp | ET | Sp | ET | Sp |
| 1 | 804.8 | 1.00 | 1641.6 | 1.00 | 3334.1 | 1.00 | 6852.5 | 1.00 |
| 2 | 432.7 | 1.86 | 869.8 | 1.89 | 1720.9 | 1.94 | 3465.9 | 1.98 |
| 4 | 232.2 | 3.47 | 468.2 | 3.51 | 921.4 | 3.62 | 2145.2 | 3.20 |
| 8 | 140.3 | 5.73 | 273.0 | 6.01 | 536.0 | 6.22 | 1866.2 | 3.68 |

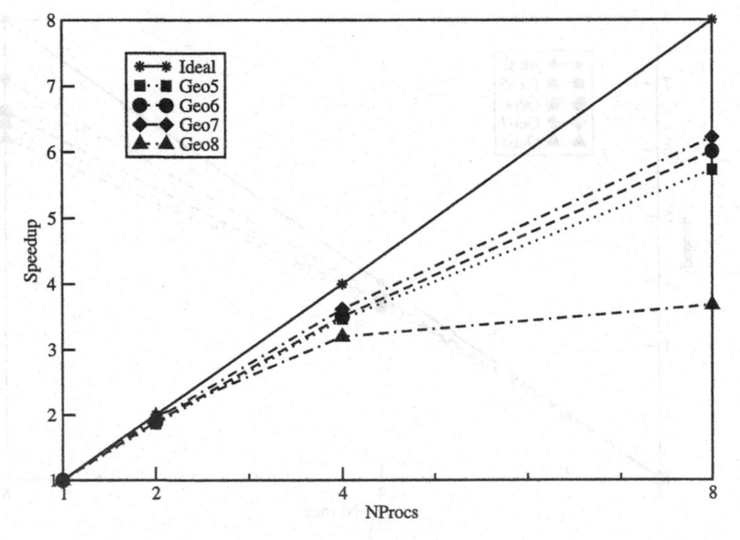


**Fig. 11.** Results using mesh colouring

### 6.3 Discussion

So far, the results we obtained confirm the potential parallelism of both algorithms. In the inverse mapping scheme, it is necessary to attribute the maximum number of elements that can be connected to every node within the finite element mesh, which limits the data structures that are suitable for a specific implementation. For instance, due to the two dimensional and regular characteristics of the meshes we used, our choice was 4 elements. In case of the data structure varies and the maximum value increases, a new assembling routine will be required. Although the mesh colouring algorithm performed worse than the inverse mapping scheme, it is more general and thus must be also taken into consideration.

## 7 Conclusions and Future Works

In this work, we proposed and evaluated the performance of two new parallel implementations of inverse mapping and mesh colouring algorithms that have been proposed for solving the elastic wave propagation problem in solid mechanics using the finite element method. We observed that this class of problem has great impact on several areas of engineering, and direct applications to some industrial sectors. In particular, it is afundamental problem in subsurface geology analysis to find reservoirs of petroleum.

In contrast with previous works, our parallel algorithm versions are unique since they use the intuitive shared-memory model implemented in a cluster of PCs as provided by current software DSM systems. We evaluate our parallel algorithms for several workloads running on a cluster of 8 PCs under TreadMarks SDSMs. Our results show that the best speedup was approximately 7 out of 8 processors for the inverse mapping scheme, but almost all experiments we performed presented speedups as much as 6. The reason for the inferior mesh colouring performance was the larger communication overheads produced by TreadMarks cache-coherence protocol. We noticed that, TreadMarks executed relatively more synchronization operations for the mesh colouring implementation than it performed for the inverse mapping one, as we added more processors and increased the input data size. Overall, these preliminary performance results indicate that cluster-based SDSMs offer a cost-effective friendly-programming platform for developing parallel FEM-based applications.

We still have opportunities for performance improvements. For instance, SDSM systems that use home-based cache coherence protocols is likely to improve performance of the parallel mesh colouring algorithm because it requires less memory than the inverse mapping scheme. This work is in progress at COPPE/UFRJ to examine related issues and also to evaluate other FEM-based applications, including experiments that use high-performance cluster networks such as Giganet and Myrinet. In this regard, preliminary results confirm performance improvements up to 20% for the mesh colouring algorithm using a Myrinet switch.

## Acknowledgements

This work was partially supported by FINEP grant "Projeto de Computação Paralela". Our experiments were performed on the cluster of PCs of Laboratório de Computação Paralela - LCP/PESC/COPPE/UFRJ. The authors also would like to acknowledge the collaboration and help-on-line of our colleagues of LCP team Ines Dutra, Lauro Whately, Raquel Pinto, Silvano Bolfoni, Thobias Salazar and Rodrigo Fonseca.

## References


[1] Alvaro L. G. A. Coutinho, Jose L. D. Alves and Nelson F. F. Ebecken: A Study of Implementation Schemes for Vectorized Sparse EBE Matrix-Vector Multiplication. Adv. Eng. Software. **Vol.13 N.3** (1991) 130–134

[2] J. L. D. Alves: Análise Transiente de Grandes Deformações em Computadores de Arquitetura Vetorial/Paralela. Tese de Doutorado em Engenharia Civil, COPPE/UFRJ. (1991)

[3] K. J. Bathe: Finite Element Procedures in Engineering Analysis. Prentice-Hall, Englewood Cliffs. (1982)

[4] T. Belytschko: An Overview of Semidiscretization and Time Integration Procedures. in T. Belytschko and T. J. R. Hughes (eds.), Computational Methods for Transient Analysis, North-Holland, Amsterdam. (1983) 1–65

[5] M. C. S. Castro and C. L. Amorim: Efficient Categorization of Memory Sharing Patterns in Software DSM Systems. in Proc. of the 15th IEEE Int'l Parallel and Distributed Processing Symposium, IPDPS'2001, San Francisco/USA, April 2001

[6] R. D. Cook, D. S. Malkus, M. E. Plesha: Concepts and Application of Finite Element Analysis. 3rd. Ed., Wiley, New York. (1989)

[7] R. M. Ferencz: Element-by-Element Preconditioning Techniques for Large-Scale, Vectorized Finite Element Analysis in Nonlinear Solid and Structural Mechanics. Ph.D. Dissertation, Stanford University. (1989)

[8] T. J. R. Hughes: The Finite Element Method: Linear Static and Dynamic Finite Element Analysis. Prentice-Hall, Englewood Cliffs (1987)

[9] Y. Zhou, L. Iftode, K. Li, J. P. Singh, B. R. Toonen, I. Schoinas, M. D. Hill, D. A. Wood: Relaxed Consistency and Coherence Granularity in DSM Systems: A Performance Evaluation. PPOPP97 (1997) 193–205

[10] P. Keleher, A. L. Cox and W. Zwaenepoel: Lazy Release Consistency for Software Distributed Shared Memory. ISCA92. (1992) 13–21

[11] P. Keleher, S. Dwarkadas and A. L. Cox and W. Zwaenepoel: TreadMarks: Distributed Shared Memory on Standard Workstations and Operating Systems. Proc. of the 1994 Winter Usenix Conference. (1994)

[12] T. Torigaki: A Structure of an Explicit Time Integration Dynamic Finite Element Program Suitable for Vector and Parallel Computers. Proceedings of the 4th International Conference on Supercomputing and 3rd World, Supercomputer Exhibition (ICS'89). Santa Clara CA (1989) 449–453

[13] O. C. Zienkiewicz, R. L. Taylor: The Finite Element Methods Vol.2: Solid and Fluid Mechanics, Dynamics and Non-Linearity. McGraw Hill, London. (1991)

# Parallel Implementation for Probabilistic Analysis of 3D Discrete Cracking in Concrete

Carmen N. M. Paz[1], Luiz F. Martha[1], José L. D. Alves,[2] Eduardo M. R. Fairbairn[2], Nelson F. F. Ebecken[2], and Alvaro L. G. A. Coutinho[2]

[1]Department of Civil Engineering
Pontifical Catholic University of Rio de Janeiro, (PUC-Rio)
Street Marquês de São Vicente 225, 22453-900 - Brazil
{mena,lfm}@tecgraf.puc-rio.br
[2]COPPE/UFRJ - Program of Civil Engineering
PO Box 68506, 21945-970
Federal University of Rio de Janeiro - Brazil
jalves@lamce.ufrj.br
eduardo@labest.coc.ufrj.br
nelson@ntt.ufrj.br
alvaro@nacad.ufrj.br

**Abstract.** This work presents computational strategies used in an implementation of the probabilistic discrete cracking model for concrete of Rossi suitable to parallel vector processor (PVP). The computational strategies used for each realization, within the framework of Monte Carlo simulation, are the inexact Newton method to solve the highly nonlinear problem and element-by-element (EBE) iterative strategies considering that nonlinear behavior is restricted to interface elements. The simulation of a direct tension test is used to illustrate the influence of adaptive inexact Newton strategy in code performance on a CRAY T90.

## 1 Introduction

Concrete, the most used construction material in the world, is a very heterogeneous material, because of its composite structure composed by aggregates, mortar, fibers, etc. and also because of the physical phenomena that take place during chemical and physical evolution: autogenous shrinkage, initial stresses and microcracking, etc. The material can also present fracture size effects that may introduce large differences for the ultimate loads and non-objectivity of the analysis. To handle with these characteristics, two basic approaches are generally used to model the mechanical behavior of the material: (i) the heterogeneous material is assimilated to a statistically equivalent homogeneous material and a deterministic approach together with a size effect law is used; (ii) homogeneity is no longer considered, at a certain scale, and the heterogeneity of the material, introduced by a stochastic approach, is the responsible for the scale effect.

Regarding concrete, the second approach is quite recent, and in this paper we refer to Rossi's *et al.* [1,2,3,4] probabilistic model that seems to take into account size effects when the statistical parameters are well determined. This method uses Monte Carlo simulations to find the average behavior of the structure, which corresponds to the converged mean values of some typical results.

The remaining of this work is organized as follows; section 2 describes the probabilistic model; section 3 describes the formulation of the interface element; section 4 describes the computational strategies employed and section 5 addresses the numerical experiments carried out. Finally, section 6 closes this work presenting some conclusions drawn from the experiments.

## 2 Probabilistic Model

Among several other relevant factors, such as water/cement ratio of the paste, casting and curing conditions, loading conditions, etc, concrete cracking depends on the random distribution of constituents and initial defects. The heterogeneity governs the overall cracking behavior and related size effects on concrete fracture. The probabilistic crack approach, based on the direct Monte Carlo method, developed by Rossi and co-workers ( Rossi *et al.* [1,2,3,4]) takes this stochastic process into account by assigning in finite element analysis, randomly distributed material properties, such as tensile strength and Young's modulus to both the solid elements and the contact elements interfacing the former, that is, a discrete crack approach. This approach considers that all nonlinearities are restricted to contact elements modeling cracks. Therefore, the stochastic process is introduced at the local scale of the material, by considering that cracks are created within the concrete with different energy dissipation depending on the spatial distribution of constituents and initial defects. The local behavior is assumed to obey a perfect elastic brittle material law. Thus, the random distribution of local cracking energies can be replaced by a random distribution of local strengths. The present probabilistic model involves a number of mechanical properties of the material to be determined, which constitutes the modeling data. From a large number of direct tensile tests, it was found that a normal law describes rather well the experimental distribution of the relevant material data (Rossi *et al.* [2]). These characteristics are the means of the tensile strength ( $f_{ct,\mu}$ ) and of the Young's modulus ( $E_\mu$ ); the standard deviations of the tensile strength ( $f_{ct,\sigma}$ ) and of the Young's modulus ( $E_\sigma$ ). The following analytical expressions were proposed (Rossi *et al.* [2]):

$$f_{ct,\mu} = 6.5 \cdot C \cdot (V_T / V_g)^{-a} \qquad f_{ct,\sigma} = 0.35 \cdot f_{ct,\mu} (V_T / V_g)^{-b} \tag{1}$$

$$E_\mu = E \qquad E_\sigma / E = 0.15\,(V_t / V_g)^{-c} \tag{2}$$

$$a = 0.25 - 3.6 \cdot 10^{-3}(f_c/C) + 1.3 \cdot 10^{-5}(f_c/C)^2$$

$$b = 4.5 \cdot 10^{-2} + 4.5 \cdot 10^{-3}(f_c/C) + 1.8 \cdot 10^{-5}(f_c/C)^2 \quad (3)$$

$$c = 0.116 + 2.7 \cdot 10^{-3}(f_c/C) - 3.4 \cdot 10^{-6}(f_c/C)^2$$

where $V_T$ is the volume of the two finite elements contiguous to an individual contact element of the mesh, $V_g$ is the volume of the coarsest grain, $C = 1MPa$, $f_c$ is the cylinder compressive strength determined by the standardized test (French standard) on a cylinder 160 mm in diameter and 320 mm high, and $E$ is the mean value (obtained from tests) of the Young's modulus which does not exhibit significant volume effects.

In these expressions, the compressive strength $f_c$ represents the quality of the concrete matrix, while the volume of the coarsest aggregate $V_g$, refers to the elementary material heterogeneity.

Equations (1) to (3) show that the smaller the scale of observation, the larger the fluctuation of the local mechanical properties, and thus the (modeled) heterogeneity of the matter. In other words, the finer the mesh, the greater the modeled heterogeneity in terms of Young's modulus and tensile strength. The empirical expressions corresponding to equations (1) to (3) were calibrated to fit a number of material tests for different types of concretes and volumes. They give a first approximation to capture volume effects on concrete fracture, but they are not expected to be universal. Furthermore, little is known about the validity of these empirical formulas for small $V_t / V_g$. These volume ratios, which may be used in finite element structural analysis, are too little to be determined by means of experimental tests.

The mesh has $nv$ volume elements and $ni$ interface elements. Each interface element follows a rigid-brittle constitutive law characterized by an individual tensile strength $f_{ct,i}$. The volume elements are elastic and have individual Young's modulus referenced by $E_v$.

Following Rossi et al. [2] findings, these individual local tensile strengths and Young's modulus are represented by normal distributions having the densities:

$$g_f(f_{ct}) = \frac{1}{f_{ct,\sigma}\sqrt{2\pi}} \exp\left[-\frac{1}{2}\left(\frac{f_{ct,\mu}}{f_{ct,\sigma}}\right)^2\right] \quad (4)$$

$$g_E(E) = \frac{1}{E_\sigma\sqrt{2\pi}} \exp\left[-\frac{1}{2}\left(\frac{E - E_\mu}{E_\sigma}\right)^2\right] \quad (5)$$

where $g_f(f_{ct})$ and $g_E(E)$ are the density function for the tensile strength $f_{ct}$ and the Young's modulus $E$, $x_\mu$ and $x_\sigma$ denote the mean and standard derivation of the distribution of quantity $x$. For the problem at hand, it is possible to find a sample of *ni* values $f_{ct,i}$, each value corresponding to an interface element, and *nv* values $E_v$, each value corresponding to a volume element, by using a standard routine for generation of random numbers for a given normal distribution (Press *et al.*, [6]).

If the heterogeneous characteristics of the material are well established and quantified by the statistical moments it is possible that the model displays the size effects related to the material heterogeneity. The problem with this approach is that these statistical moments are not known, a priori, for the characteristic volume of the finite elements used. However, some methods have been proposed to determine these parameters by means of inverse analysis using neural networks (Fairbairn *et al*, [7,8]).

The solution for this probabilistic approach is obtained by means of a Monte Carlo simulation. A number of *n* samples is generated for a given normal distribution, and some characteristic responses of the structure are computed; for example, stress crack-width $\sigma - w$ curve, or load displacement $P - \delta$ curve. Let the *j*th samples correspond to, e.g., the *j*th $\sigma - w$ curve. This *j*th $\sigma - w$ curve is composed of discrete values, $\sigma_k^j$ and $w_k^j$, where the superscript *j* indicates the sample and the subscript *k* the discrete value of the $\sigma - w$ curve. The same discrete $w-$ values are assumed and the response is defined exclusively by the values of $\sigma_k^j$. The mean curve composed by pairs $\sigma_k^{mean}$, $w_k$ then simply reads

$$\sigma_k^{mean,j} = \frac{1}{j}\sum_{l=1}^{j} \sigma_k^l \tag{6}$$

The Monte Carlo simulation is stopped when

$$\left| \sigma_k^{mean,j} - \sigma_k^{mean,j-1} \right| \le tol \tag{7}$$

where *tol* is the prescribed tolerance to check the convergence of the procedure. With the convergence of the procedure, the total number of samples is set to $n = j$. This total number of samples *n*, corresponding to a Monte Carlo converged simulation, clearly depends on *tol*, which is a measure of the precision required by the analysis. It also depends on the heterogeneity of the material represented by the standard deviation. The more heterogeneous is the material the greater is the number of samples necessary to obtain a converged solution. Our experience in this field indicates that 25 to 50 samples are sufficient to obtain a converged $\sigma - w$ curve.

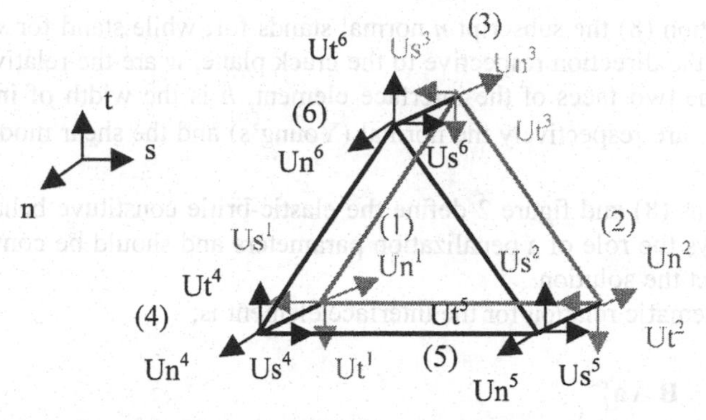


**Fig. 1.** An Interface Element and its degrees of freedom in a local system

## 3 Interface Elements for Modeling Discrete Cracking

The finite element cracking model is a discrete model for which volume elements are always elastic and cracking occurs in elastic-brittle (almost rigid brittle) contact elements placed between two volume elements. The three dimensions interface element (Paz, [5]) depicted in figure 1 can be thought as a triangular base prism connecting adjacent faces of neighboring tetrahedra. These elements are formulated to describe relative displacements between the triangular faces simulating crack opening.

The constituve law of the 3D interface element is defined by equation (8) for non cracked elastic state characterized by $\sigma_n < f_{ct,i}$. When the tensile strength is reached the elements attains the cracked stage and modulus $E_c$ and $G_c$ are set to zero (figure 2).

$$\Delta\boldsymbol{\sigma} = \mathbf{D}_{cr}\ \Delta\mathbf{w} = \begin{Bmatrix} \Delta\sigma_n \\ \Delta\sigma_s \\ \Delta\sigma_t \end{Bmatrix} = \begin{bmatrix} E_c/h & 0 & 0 \\ 0 & G_c/h & 0 \\ 0 & 0 & G_c/h \end{bmatrix} \begin{Bmatrix} \Delta w_n \\ \Delta w_s \\ \Delta w_t \end{Bmatrix} \qquad (8)$$

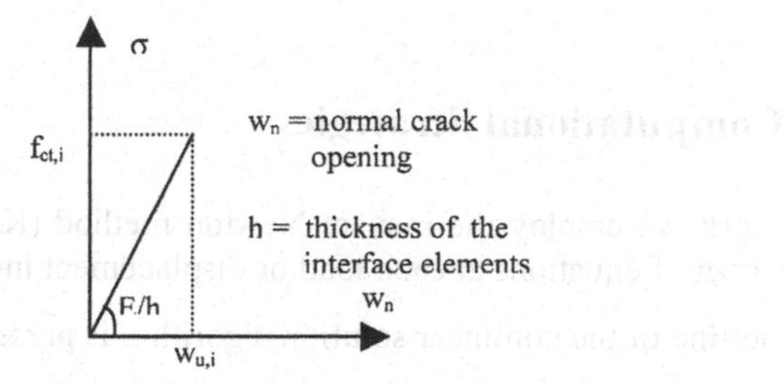


**Fig. 2.** Elastic-Brittle Contact law

In equation (8) the subscript $n$ normal stands for, while stand for $s$ and $t$ tangential indicating the direction respective to the crack plane, $w$ are the relative displacements between the two faces of the interface element, $h$ is the width of interface element, $E_c$ and $G_c$ are respectively the normal (Young's) and the shear modulus along crack plane.

Equations (8) and figure 2 define the elastic-britle constituve behavior. The thickness $h$ plays the role of a penalization parameters and should be conveniently chosen not to affect the solution.

The kinematic relation for the interface element is;

$$\Delta \mathbf{w} = \mathbf{B}\, \Delta \mathbf{a}_l^e \tag{9}$$

where

$$\mathbf{B} = \frac{1}{3}\begin{bmatrix} -1 & 0 & 0 & -1 & 0 & 0 & -1 & 0 & 0 & 1 & 0 & 0 & 1 & 0 & 0 & 1 & 0 & 0 \\ 0 & -1 & 0 & 0 & -1 & 0 & 0 & -1 & 0 & 0 & 1 & 0 & 0 & 1 & 0 & 0 & 1 & 0 \\ 0 & 0 & -1 & 0 & 0 & -1 & 0 & 0 & -1 & 0 & 0 & 1 & 0 & 0 & 1 & 0 & 0 & 1 \end{bmatrix} \tag{10}$$

$\Delta \mathbf{w}$ is the crack opening incremental and $\Delta \mathbf{a}_\mathrm{l}^\mathrm{e}$ is the vector incremental nodal displacements for the interface element

Applying a standard displacement based F. E. formulation, the resulting tangent stiffness matrix for the interface element is given is:

$$\mathbf{K}_{\mathrm{Intf}}^{\mathrm{e}} = \int_{\Omega} \mathbf{B}^T\, \mathbf{D}_{cr} \mathbf{B}\, \mathrm{d}\Omega \tag{11}$$

The interface elements are generated contiguous to the faces of selected tetrahedra elements. This selection is performed by the user, defining a 3D box inside the mesh that contains the target elements.

**Remark** Our experience indicates that to increase robustness of nonlinear solution process we have to limit only one interface element to "crack" at each nonlinear iteration.

## 4 Computational Strategies

In this work we employ the inexact Newton method (Kelley, [9]) to solve resulting nonlinear set of equations at each load or displacement increment.

The outline of the nonlinear solution algorithm is presented in Box 1.

Given $u_{tol}, r_{tol}, \eta$ relative and residual tolerance.
Compute stiffness tetrahedra matrix $\mathbf{K}_{Tetra}$
do k=1,2….., number of load increments do
Compute external forces vector

$$\mathbf{F}^k_{ext} = \mathbf{F}^K_{nodal} + \mathbf{F}_{volume} + \mathbf{F}_{\bar{\sigma}} - \left( \mathbf{K}_{tetra}\,\bar{\mathbf{U}}_k + \mathbf{K}_{Intef}\,\bar{\mathbf{U}}_k \right)$$

do i=1,2 ..., while convergence
Compute internal forces vector,

$$\mathbf{F}^i_{\text{int}} = \left(\mathbf{F}^i_{\text{int}}\right)_{Tetra} + \left(\mathbf{F}^i_{\text{int}}\right)_{Intf}$$

Compute residual vector,

$$\boldsymbol{\psi}^i = \mathbf{F}^i_{\text{int}} - \mathbf{F}^i_{ext}$$

Update stiffness interface matrix $\mathbf{K}^{\text{i}}_{\text{Intf}}$
Assembly matrix

$$\mathbf{A}^i = \mathbf{K}_{Tetra} + \mathbf{K}^i_{Intf}$$

Compute tolerance for iterative driver , $\eta_i$

Solver: $\mathbf{A}^i\ \Delta\mathbf{u} = \boldsymbol{\psi}^i$ for tolerance $\eta_i$
Update solution,

$$\mathbf{U} = \mathbf{U} + \Delta\mathbf{u}$$

if $\dfrac{\|\Delta\mathbf{u}\|}{\|\mathbf{U}\|} \leq utol$ and $\dfrac{\|\Delta\boldsymbol{\psi}^i\|}{\|\mathbf{F}^k_{ext}\|} \leq rtol$ then convergence
end while i.
end do k.

**Box 1.** Inexact Newton Algorithm

Note that in $\mathbf{F}^k_{ext}$ we account for nodal forces, body forces and prescribed displacements and stresses $\bar{\mathbf{U}}, \bar{\boldsymbol{\sigma}}$. The total internal forces vector $\mathbf{F}^i_{\text{int}}$ is the sum of the tetrahedra element vector internal forces $\left(\mathbf{F}^i_{\text{int}}\right)_{Tetra}$ plus the interface element internal forces vector $\left(\mathbf{F}^i_{\text{int}}\right)_{Intf}$. The total stiffness matrix is the sum of the continuum matrix $\mathbf{K}_{Tetra}$ plus the interface matrix $\mathbf{K}^i_{Intf}$ updated at every nonlinear iteration. An approximate solution is obtained when the Inexact Newton termination criterion is satisfied, that is

$$\left\| \mathbf{A}^i\,\Delta\mathbf{u} - \boldsymbol{\psi}^i \right\| \leq \eta_i \left\| \boldsymbol{\psi}^i \right\| \tag{12}$$

Tolerance $\eta_i$ may be kept fixed throughout the nonlinear iterations or may be adaptively selected as the iterations progress towards the solution. Our choice for $\eta_i$ follows the criteria suggested by Kelley.

The iterative driver for the solution of the linearized set of equations is the element-by-element PCG. Several preconditioner options are available ranging from the simple diagonal up to incomplete factorizations. In this work, to stress the benefits of the inexact Newton method, we restricted ourselves to simple diagonal and nodal block diagonal.

**Remarks**

1) In the present implementation matrix vector products in EBE, PCG are computed as:

$$\mathbf{A}\,\mathbf{p} = \sum_{i=1}^{N_{tetra}} \left(\mathbf{K}_{Tetra}\ \mathbf{p}_i\right) + \sum_{j=1}^{N_{Intf}} \left(\mathbf{K}_{Intf}\ \mathbf{p}_j\right) \tag{13}$$

where, $N_{tetra}$ is the number of tetrahedra, $N_{intf}$ is the number of interface elements, $\mathbf{K}_{Tetra}$ and $\mathbf{K}_{Intf}$ are respectively element matrices for the tetrahedra and interface; $\mathbf{p}_i$ and $\mathbf{p}_j$ are the components of $\mathbf{p}$ restricted to the degrees of freedom of two element type.
2) Stiffness matrices for tetrahedra are computed and stored at the beginning of the analysis since they are elastic.
3) Stiffness matrix for interface elements are updated at every nonlinear iteration.
4) The arrays of element stiffness matrices are stored taking into account their symmetry; in the case of the element tetrahedra 78 coefficients are stored and for the interface element only 18 coefficients are stored, exploring the particular structure of the discrete gradient operator given in equation (10).
5) The mesh coloring algorithm of Hughes [10] was extended to block both solid and interface elements into disjoint groups thus enabling full vectorization an parallelization of the operations involved in equation (13).

# 5 Numerical Experiments

## 5.1 Direct tension test

The experimental results of concrete uniaxial tension tests published by (Li *et al* [12]) were used to validate the developments presented in this paper. The specimens are cylinders 101.6 mm in diameter and 203.2 mm high. This specimens had 25.4 mm notches at their midheight on both sides (figure 3).

Concrete with maximum aggregate diameter of 9.525 mm was used. Its average tensile strength and Young's modulus at the age of 28 days were: $f_{ct} = 4{,}72 \quad MPa$ and $E_c = 42000 \quad MPa$.

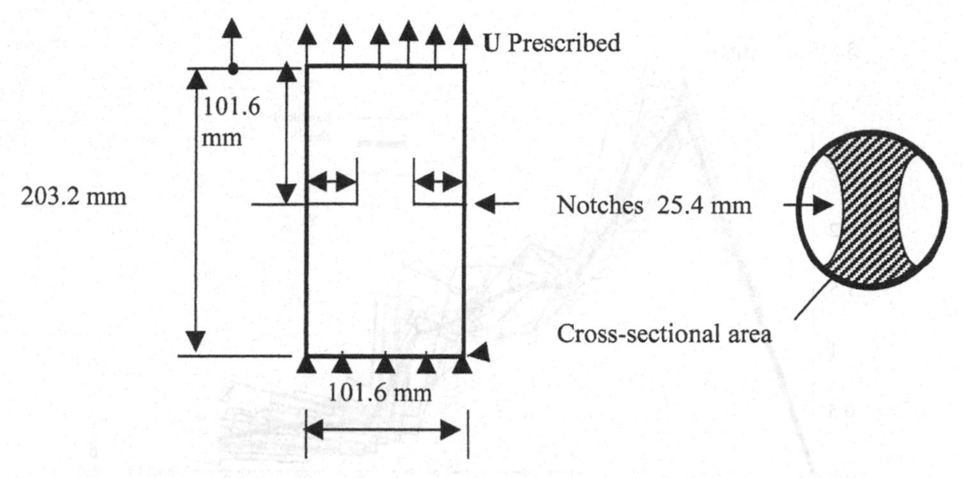


**Fig. 3.** Uniaxial tension specimen geometry, dimensions, load, boundary conditions

The numerical experiments were controlled by a field of uniform displacements applied at the top of the test specimen in 30 incremental steps $\Delta u = 5.3\,x\,10^{-6}$ . The boundary conditions restrain the degrees of freedom in the vertical at the bottom. The final mesh has 11,933 elements, where 4,775 are tetrahedra and 7,158 are interface elements (figure 4).

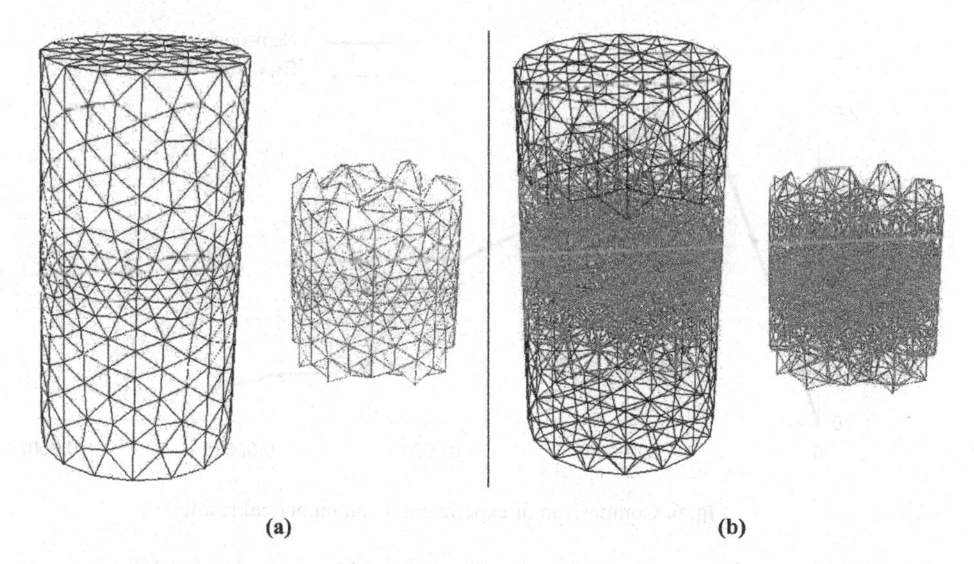


**Fig. 4.** Hidden line (a) and wire frame (b) representations of the computational mesh for the simulation of a direct tension test. It is also shown aside the resulting mesh for the interface elements considering

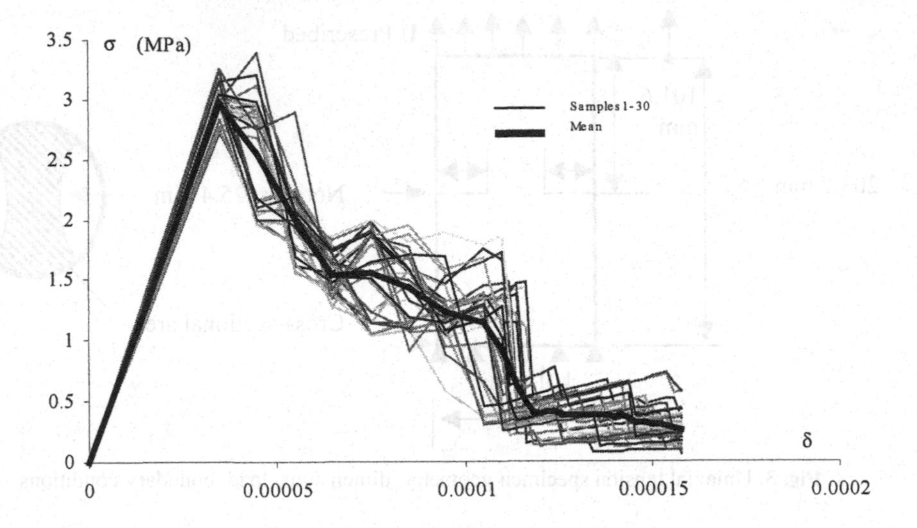


**Fig. 5.** Results for the complete Monte Carlo simulation

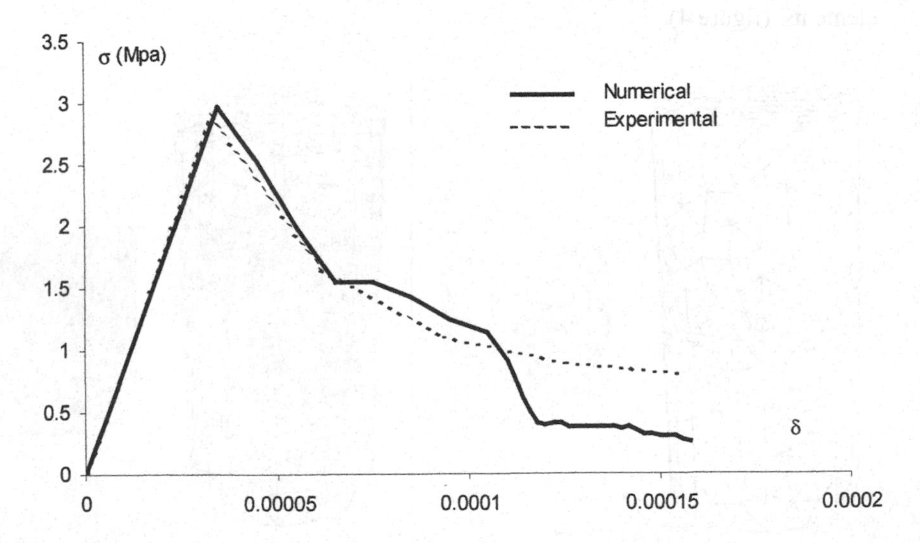


**Fig. 6.** Comparison of experimental and numerical results

Curves $\sigma - \delta$ for the several realizations of the Monte-Carlo simulation are given in figure 5. The comparison between experimental by Li *et al* [12] and the convergent Monte Carlo $\sigma - \delta$ curve is given in figure 6, figure 7 presents the crack configuration for a given sample at a stage corresponding to the softening branch of the $\sigma - \delta$ curve.

**Fig. 7.** Crack evolution for numerical simulation

### 5.2 Comparison of Inexact Newton Strategies

An assessment of the implemented inexact Newton strategies was carried out for one realization of the direct tension test simulation the selected preconditioner for this experiment was the diagonal nodal block.

A fixed tolerance $\eta = 10^{-3}$ for PCG was selected for the first analysis requiring 52 inexact Newton iterations. For the second analysis, now employing the adaptive inexact Newton method, the tolerance range for PCG was set to $10^{-3} \leq \eta \leq 10^{-6}$, requiring 54 inexact Newton iterations. Tolerances for the inexact Newton were iterations set to $u_{tol} = r_{tol} = 10^{-3}$ , in both cases.

As can be seen in Figure (8), comparing the number of nonlinear iterations for both analyses, it remained fairly the same irrespective to the selection of the inexact Newton approach, either fixed or adaptive, indicating that the overall nonlinear character of the solution was not affected the adaptive approach. Figure (9) shows a plot of the number of PCG iterations per incremental step for both analyses. It is evident in this figure the reduction of PCG iterations due to the adaptive strategy.

### 5.3 Computational Performance

A detailed vector performance analysis is obtained by the summary of the PERFVIEW's Report presented in Table 1. The CPU time of the vectorized single processor run for CRAY T90 is 10.38 hours. This table list for a single CPU run, the Mflop/s rates for the three top routines. The routines **Smatv-fint** and **Smatv-tetra** are respectively responsible for the matrix-vector products for the interface and tetrahedra elements, routines are the computations kernels the routine **PCG-block,** implements the iterative solver the nodal block diagonal Preconditioned Conjugate Gradient.

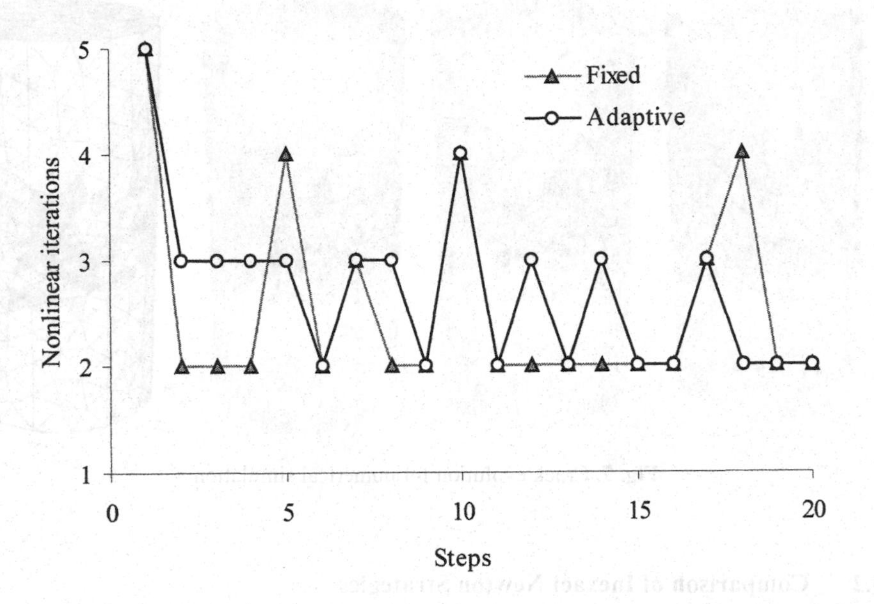


**Fig. 8.** Nonlinear iterations per incremental step for the direct tension test simulation

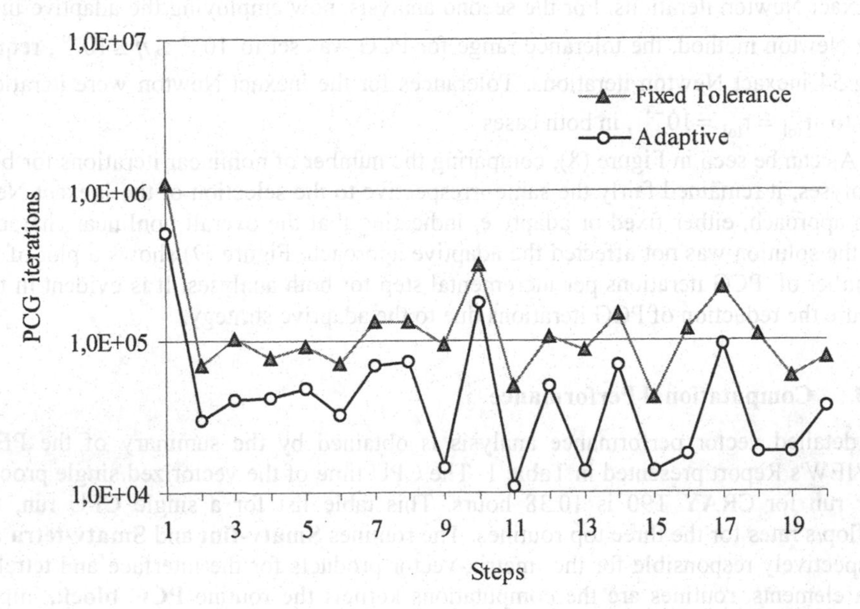


**Fig. 9.** PCG iterations per incremental step for the direct tension test simulation

**Table 1** Performance Analysis -The top 3 subroutines

| Routines | Single CPU (%) | Performance (Mflop/s) |
|---|---|---|
| smatv-intf | 52.80 | 613.7 |
| smatv-tetra | 17.32 | 554.0 |
| PCG-block | 25.90 | 82.4 |
| Others | 3.98 | - |

**Table 2** Summary of the ATEXPERT's Report for the 5 dominant loops

| Routines | % Parallel | Dedicated Speedup | Actual Speedup |
|---|---|---|---|
| Smatv-tetra | 98.9 | 3.96 | 3.8 |
| Smatv-intf | 99.8 | 3.83 | 3.8 |
| Fint-tetra | 92.5 | 3.83 | 3.2 |
| Fint-intf | 99.1 | 3.55 | 3.5 |
| A-Kintf | 85.8 | 3.53 | 2.6 |

The code achieved good vectorization on the CRAY T90 for a mesh with 11933 elements, comprising (7158 interface and 4775 tetrahedra elements). The top three subroutines consume the major CPU utilization in the whole analysis.

The parallel performance is shown in table 2 and the figure 10 (a) and (b) as obtained from a summary ATEXPERT report, autotasking performance tool. The routines **Fint-tetra** and **Fint-intf** evaluate respectively the internal force vector of the interface elements and the tetrahedra elements. Routine **A-Kintf** computer update interface stiffness.

According to ATEXPERT tool this program appears to be 99.2 percent parallel and 0.8 percent serial. Amdahl's Law predicts the program could expect to achieve a 3.9 times speedup on 4 cpus. A 3.8 speedup is predicted with 4 cpus on a dedicated system.

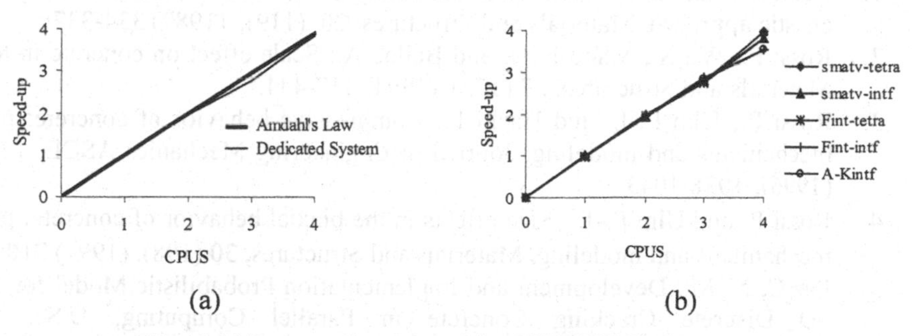


**Fig. 10** (a) Program Summary of the ATEXPERT's Report
(b) ATEXPERT's Report -The top 6 subroutines on a Dedicated Speedup

# 6 Concluding Remarks

This paper presented computational strategies used in an implementation of Rossi's probabilistic model for the simulation of cracking in concrete structures.

The resulting code achieved very good level, for parallel and vetor performance single CPU achieved 614 Mflop/s while parallel speed-up reached 3.8 in a 4 CPU system. The results emphasizes the suitability of the implemented code on a PVP system the CRAY T90

Extensive use of element-by-element techniques within the computational kernels comprised in the iterative solution drivers provided a natural way for achieving high flop rates and good parallel speed-up's.

The use of an adaptive inexact Newton method showed to be effective in reducing overall linear iterations, by a factor of two in comparison to the fixed tolerance inexact Newton scheme.

# Acknowledgments

The authors are indebted to the Computer Graphics Technology Group TECGRAF/PUC-Rio, High Performance Computing Center NACAD/COPPE/UFRJ, and the Laboratory of Computational Methods in Engineering of the Program of Civil Engineering LAMCE /COPPE/UFRJ. CESUP/UFRGS is gratefully acknowledged for the computer time provided for the performance experiments.

This work was partially supported by CAPES and CNPq's grant N° 150039/01-8(NV).

# References

1. Rossi P., Richer S.: Numerical modeling of concrete cracking based on a stochastic approach. Materials and Structures, 20 (119), (1987) 334-337.
2. Rossi P., Wu X., Maou F. le, and Belloc A.: Scale effect on concrete in tension. Materials and Structures, 27 (172), (1994) 437-444.
3. Rossi P., Ulm F.-J., and Hachi F.: Compressive behavior of concrete: physical mechanisms and modeling. Journal of Engineering Mechanics ASCE, 122 (11), (1996), 1038-1043.
4. Rossi P. and Ulm F.-J.: Size effects in the biaxial behavior of concrete: physical mechanisms and modeling. Materials and Structures, 30 (198), (1997) 210-216.
5. Paz C. N. M., Development and Implementation Probabilistic Model for 2D and 3D Discrete Cracking Concrete in Parallel Computing, D.Sc. Thesis, COPPE/UFRJ, the Graduate Institute of Federal University of Rio de Janeiro, Brazil (2000) [in Portuguese].
6. Press W. Teukolski H., S., Vetterling W.T. and Flannery B.: Numerical Recipes, Cambridge University Press (1992).

7. Fairbairn E. M. R., Paz C. N. M., Ebecken N. F. F., and Ulm F-J., Use of neural network for fitting of probabilistic scaling model parameter, Int. J. Fracture, 95, (1999), 315-324.
8. Fairbairn E. M. R., Ebecken N. F. F., Paz C. N. M., and Ulm F-J.: Determination of probabilistic parameters of concrete: solving the inverse problem by using artificial neural networks, Computers and Structures, 78, (2000) 497-503.
9. Kelley C. T.: Iterative Methods for Linear and Nonlinear Equations. Frontiers in applied mathematics, SIAM Society for Industrial and Applied Mathematics, Philadelphia, (1995).
10. Hughes T. J. R., Ferenez R. M. , Hallquist J. O.: Large-Scale Vectorized Implicit Calculation in Solid Mechanics on a CRAY X-MP/48 Utilizing EBE Preconditionated Conjugate Gradients Computer Methods in applied Mechanics and Engineering, 61, (1987) 2115-248.
11. Papadrakakis M.: Solving Large-Scale Problems in Mechanics: The Development and Application of Computational Solution, Editor, M. Papadrakakis, John Wiley and Sons, (1993).
12. Li Q. and Ansari F.: High Concrete in Uniaxial Tension, ACI Material J. 97-(1), (2000) 49- 57.
13. Coutinho A. L. G. A., Martins M. A. D., Alves J. L. D., Landau L., and Moraes A.: Edge-based finite element techniques for nonlinear solid mechanics problems, Int. J. for Numerical Methods in Engineering, 50 (9), (2001) 2050-2068.
14. Fairbairn E. M. R., Debeux V.J.C., Paz C. N. M., and Ebecken N. F. F.: Application of probabilistic Approach to the Analysis of gravity Dam Centrifuge Test, $8^{th}$ ASCE Specialty Conference on Probabilistic Mechanics and Structural Reliability (2000) PMC 2000-261.
15. Kelley C. T.: Iterative methods for optimization. Frontiers in applied mathematics, SIAM Society for Industrial and Applied Mathematics, Philadelphia, (1999).

# An *A Posteriori* Error Estimator for Adaptive Mesh Refinement Using Parallel In-Element Particle Tracking Methods*

Jing-Ru C. Cheng[1] and Paul E. Plassmann[2]

[1] Information Technology Laboratory
US Army Engineer and Research Development Center
Vicksburg, Mississippi 39180
ruth.c.cheng@erdc.usace.army.mil
[2] Department of Computer Science and Engineering
The Pennsylvania State University, University Park, PA 16802, USA
plassmann@cse.psu.edu

**Abstract.** Particle tracking methods are a versatile computational technique central to the simulation of a wide range of scientific applications. In this paper we present an *a posteriori* error estimator for adaptive mesh refinement (AMR) using particle tracking methods. The approach uses a parallel computing framework, the "in-element" particle tracking method, based on the assumption that particle trajectories are computed by problem data localized to individual elements. Adaptive mesh refinement is used to control the mesh discretization errors along computed characteristics of the particle trajectories. Traditional *a posteriori* error estimators for AMR methods inherit flaws from the discrete solution of time-marching partial differential equations (PDEs)—particularly for advection/convection-dominated transport applications. To address this problem we introduce a new *a posteriori* error estimator based on particle tracking methods. We present experimental results that detail the performance of a parallel implementation of this particle method approach for a two-dimensional, time-marching convection-diffusion benchmark problem on an unstructured, adaptive mesh.

## 1 Introduction

### 1.1 Adaptive Mesh Refinement

A variety of numerical errors (e.g., spurious oscillation, numerical spreading, grid orientation, and peak clipping/valley elevating) have been described in the solution of advection/convection-dominated problems [3, 31, 30]. These errors can be partially mitigated through the implementation of higher-order finite element or finite difference discretizations, for example TVD and ENO methods [12, 29], because of the increased spatial and/or temporal resolution. However, instead of

* This work was supported by NSF grants DGE–9987589 and ACI–9908057, DOE grant DG-FG02-99ER25373, and the Alfred P. Sloan Foundation.

J.M.L.M. Palma et al. (Eds.): VECPAR 2002, LNCS 2565, pp. 94–107, 2003.

uniformly refining the computational mesh throughout the domain, an adaptive mesh refinement approach can be more effective in its use of computer resources for large-scale applications. Adaptive mesh refinement (AMR) improves the mesh resolutions in small areas where the numerical error is large, while remaining areas retain a coarse discretization because the computed solution in this part of domain is relatively accurate. A wide variety of AMR approaches have been explored in the literature [6, 24, 27, 8, 20, 21, 10, 13].

Jones and Plassmann [13] summarized a typical parallel finite-element AMR approach. Within this framework the mesh is adaptively refined when the maximum of computed local error on an element is greater than a prescribed error tolerance. The error estimator or indicator used is, therefore, crucial to the accuracy of the computed solution. *A priori* error estimates are, however, often only sufficient to guarantee the regularity of the solution [25]. To estimate the global error and to provide information for the adaptive mesh refinement algorithms discussed above we require *a posteriori* error estimates.

Among all the partial differential equations, the mathematical analysis and numerical solution of convection-diffusion (or transport) equations has been intensively studied during the last decade [18, 23]. One often encounters numerical difficulties when solving the transport equations because of their parabolic and hyperbolic nature. Kay and Silvester [14] proposed a simple *a posteriori* error estimator based on the solution of local Poisson problems for the steady-state advection-dominated transport equations. They conclude that the method works effectively as long as streamline-diffusion stabilization (e.g., upstream winding) is incorporated. However, the solution of the unsteady transport equations are required in the majority of real-world applications for which this method would not be applicable.

## 1.2 Particle Tracking Applications

Particle tracking methods are used in many diverse applications. For example, in computer graphics the term "particle system" is used to describe the particle-based techniques for modeling, rendering, and animation. The Particle System Application Programmer Interface (API) was developed to provide a set of functions allowing applications to simulate the dynamics of particles [16]. This API design is aimed to add a variety of particle-based effects to interactive or non-interactive graphics applications. However, it was not intended for scientific simulation—although it has been implemented on both uniprocessor and multiprocessor shared memory systems.

Parallel particle tracking methods have been employed to solve a variety of scientific and engineering simulation problems. Usually particle-based simulations are used in the solution of N-body problems. For these systems, the particles' motion is determined from a force such as that resulting from gravitation, the Coulomb force, or the strong nuclear force in plasma models (such as those used in modeling fusion reactors). The bottleneck in these simulations is the computation of force on each particle, not the movement of particles. For these particle-particle (PP) methods the computational domain is not discretized—

the particle motion is solved for by using a forward Euler method with a small time step chosen to ensure stability.

On the other hand, the particle-mesh (PM) method uses not only a discretization in time but also a discretization of the spatial domain. For PM methods an algorithm is required to handle the assignment of "virtual" mass/charge values to the computational mesh. The PM approximation scheme can result in a computational efficient methods, but the accuracy is often not adequate for most applications.

A third method used to handle particle-based models is the Tree Code (TC) method (such as Barnes-Hut [1]). This approach is again aimed at reducing the overall computation cost, but while retaining a specified level of accuracy. A quadtree in two dimensions or octree in three dimensions is built to count the total mass of all particles in a tree box. At the tree box center will be placed a virtual particle with the same amount of the the counted total mass. This method is similar to the PP method, except the tree implementation can reduce the overall complexity from $O(N^2)$ in the PP approach to $O(N \log N)$.

All the above methods can be parallelized on distributed memory MIMD (Multiple Instruction Multiple Data) machines. However, the main focus of these methods is the calculation of the forces on the particles because computing the particle motion is straightforward when the restriction of a small time-step imposed. Another application is the Direct Simulation Monte Carlo (DSMC) method used to model the dynamics of dilute gas. Wong and Long [28] implemented a parallel DSMC algorithm using a forward Euler, explicit time marching scheme in the particle movement phase of the algorithm. They pointed out that two different levels of data parallelism (i.e., molecules and cells) cause parallel processing difficulties. Nance et al. [19] parallelized DSMC using the runtime library CHAOS [17] on a three-dimensional, uniformly discretized mesh. Robinson and Harvey [22] parallelized DSMC by use of a spatial mesh decomposition over the processors. This domain decomposition is based on a localized "load table" computed on each processor for its neighbors. Based on this load table cells are donated to these neighboring processors if they have lesser load. This approach was demonstrated for a two-dimensional driven cavity with approximately 10,000 uniform cells.

Another field of application is the calculation of streamlines for a given vector field, for which the most common technique is the integration of particle paths numerically or analytically. Often, for the numerical approach, a high-order ordinary differential equation (ODE) solver such as a Runge-Kutta method or Adam's method is required to maintain a high order of accuracy [15]. However, in two and three dimensions, the evaluation of velocity between grid points across processors required by the high-order ODE solver becomes too computationally expensive, and alternative methods have to be considered. Analytic solutions for streamlines within tetrahedra have been derived based on linear interpolation and, therefore, produces exact results for linear velocity fields [9]. However, these analytic solutions apply to only steady flow and are not applicable to unsteady, or time-dependent, flows.

### 1.3 The Parallel In-Element Particle Tracking Method

Cheng et al. [2] developed the sequential in-element particle tracking technique to accurately and efficiently trace fictitious particles in the velocity fields of real-world systems. Since in-element particle tracking methods are implemented with respect to the element basis, a natural approach is to develop a parallel framework based on specifying local, element-based operations. The key to the parallel algorithm is the correlated partitioning of the particle system and the unstructured element mesh. We partition particles to processors based on their element location—this approach ensures that the data required for the computation of the particle movement phase by the in-element method involves only data local to a processor. The correspondence between particles and elements is maintained through explicit references in the element and particle data structures. This correspondence is essential to ensure the correct reassignment of particles between processors. The reassignment occurs when particles move between elements owned by a processor and the ghost elements (elements with shared faces, edges, or vertices that are owned by another processor) [5].

In this paper, we present a new *a posteriori* error estimator to control adaptive mesh refinement of time-marching transport equations using parallel particle tracking methods. We present this research as follows. In Sect. 2, we define measure metrics of the error estimator to assess the quality of an *a posteriori* error estimator. In Sect. 3, we briefly describe the implementation of parallel in-element particle tracking technology. In Sect. 4, we present the new *a posteriori* error estimator for AMR methods after reviewing the state-of-the-art error estimators. A detailed illustration is provided to clarify how it can achieve our goals. In Sect. 5, experimental results are presented to demonstrate the improvement of our implementation. Both accuracy and performance aspects are included. In Sect. 6, we summarize these results and discuss planned future work.

## 2 Measure Metrics of the Error Estimator

To measure the quality of an *a posteriori* error estimator, an index is defined as the ratio of the estimated error and the true error and is called efficiency index, $I_{\text{eff}}$. According to [25, 26], $I_{\text{eff}}$ is given as

$$I_{\text{eff}} = \frac{\eta}{\|u - u_h\|}, \tag{1}$$

where $\|u - u_h\|$ is the true error measured by the norm of the difference between the true solution and the discretized solution; and $\eta$ is the estimated global error as shown in (2)

$$\eta = \left( \sum_{K \in \Gamma_h} \eta_K^2 \right)^{1/2}, \tag{2}$$

where $\eta_K$ is the estimated error on the element $K$ of the computational mesh $\Gamma_h$, i.e., via the error estimator. If both $I_{\text{eff}}$ and $I_{\text{eff}}^{-1}$ are bounded for all discretizations, the *a posteriori* error estimator is called efficient. Moreover, if they are bounded independently of the coefficients of the particular problem, then the efficient *a posteriori* error estimator is called robust for that class of problems [25].

To investigate the performance of the *a posteriori* error estimator with respect to adaptive mesh refinement, in Sect. 5 we compute the computational efficiency for the adaptive refinement algorithm incorporating this error estimator. Being compared with the exact solution, the adaptive mesh can be evaluated based on whether the refinement correctly takes place—that is in regions with higher discretization error. Moreover, the accuracy of the results computed on different estimator-controlled meshes is one way to evaluate the satisfactory of *a posteriori* error estimators.

From the results obtained by [26] for steady-state solutions of convection-diffusion equations, it is shown that the residual-based error estimator in the $L^2$–norm ($\eta_{\text{res}-L^2}$) is reliable if the solution possess regular boundary layers. In addition, the $\eta_{\text{res}-L^2}$ norm works well in conjunction with adaptive mesh refinement. The gradient indicator (gradind) $\eta_{\text{gradind}} = \|\nabla u_h\|_{L^2,K}$, the $L^2$ norm of the gradient of $u_h$ on the element $K$, is widely used in software packages for adaptive mesh refinement because of its simplicity. However, with the gradient indicator it is not possible to obtain an estimate of the global error, and this approach is unsatisfactory when used in conjunction with adaptive mesh refinement for this class of problems.

# 3 A Parallel Particle Tracking Algorithm

Cheng et al. [2] developed the in-element particle tracking technique to accurately and efficiently trace fictitious particles in the velocity fields of real-world systems. Since in-element particle tracking methods are implemented with respect to the element basis, a natural approach is to develop a parallel framework based on specifying local, element-based operations. The key to the parallel algorithm is the correlated partitioning of the particle system and the unstructured element mesh [4]. We partition particles to processors based on their element location—this approach, as shown in Algorithm 3.1 ensures that the data required for the computation of the particle movement phase by the in-element method involves only data local to a processor. To balance the work load on each processor, which can change significantly at each time step because of the particle movement and adaptive mesh refinement, a load balanced parallel "in-element" particle tracking method has been developed [5].

# 4 A New *A Posteriori* Error Estimator

The use of traditional error estimators, such as the energy norm error estimator for the time-marching transport equations, could result in the selection of incorrect locations for adaptive refinement. This is because the discrete nature of the solution of the PDE causes discontinuities in time. Thus, the first iteration at each time step uses the mesh computed at the previous time step as the domain

---

**Algorithm 3.1** The Parallel In-Element Particle Tracking Algorithm

Let $P_i$ be the set of particles on processor $i$
Set the residual time $t_r$ to the time-step size
$n \leftarrow \sum_i \|P_i\|$
**While** $(n > 0)$ **do**
  $n_i \leftarrow 0$
  **Foreach** $(p \in P_i$ $)$ **do**
    **While** $(p \in P_i$ and $status(p) \neq$ finish$)$ **do**
      Refine $M$ to the prescribed number of subelements
      Track $p$ subelement by subelement until time is exhausted or
        hit the boundary of the element $M$
      Compute $t_r$, velocity, and identify the possible neighbor
        element $M'$ for continued tracking
      **If** $t_r > 0$ **do**
        **If** $owner(M') \neq i$ **do**
          Pack $p$ and remove $p$ from $P_i$
          $n_i \leftarrow n_i + 1$
        **Endif**
      **Elseif** $t_r = 0$ **do**
        Interpolate concentration
        $status(p) \leftarrow$ finish
      **Endif**
    **Endwhile**
  **Endfor**
  $n = \sum_i n_i$
  Send and receive message, unpack messages to form new $P_i$
**Endwhile**
**If** $backward\_tracking$ **do**
  Update concentration on each vertex
**Endif**

---

discretization—unfortunately, this discretization may not be appropriate for the current time step. As might be expected, results obtained at the first iteration at the current time step using this mesh can be highly inaccurate if there has been significant change (for example, a change in a source/sink strength or boundary condition) between these two time steps. Based on this incorrect solution, we can repeatedly refine the mesh around the incorrectly computed moving front until it is within our error tolerance. Within this scenario, the traditional error estimator has failed to compute an accurate solution to the PDE system.

To circumvent the failure of the traditional error estimator our aim is the development of a new *a posteriori* error estimator to address these problems. As shown in Algorithm 4.1, at time $n$, we use the particle tracking method to track all the vertices forward and backward in time. The value, $C_j^f$, of the $j$-th forward particle is compared with the value, $C_j^i$, obtained from an interpolation derived from the mesh element vertices whose values originated from backward tracking and interpolation (we denote these values by $C^*$). Once computed, we examine each element's estimated error based on the first and second relative errors, which are obtained via a comparison with $C_j^f$ and with the maximum value ($C_M^f$) throughout the entire domain, respectively. These two errors represent the local and global nature of the error. Only when both relative error criteria (3)

$$\begin{aligned} \mid C_j^f - C_j^i \mid / C_j^f &\leq \epsilon_1 \\ \mid C_j^f - C_j^i \mid / C_M^f &\leq \epsilon_2 \end{aligned} \tag{3}$$

are violated, will the associated element containing this vertex be marked for refinement and be added to the set $S^n$ of elements to be refined. The elements in $S^n$ are refined based on the underlying AMR algorithm and generate the mesh $T^n$ which is used to solve the system using a traditional finite element method.

We employ the following iterative procedure to find an optimal mesh for the current time step. The mesh is refined or coarsened based on whether the computed local error on an element is larger than the prescribed upper error tolerance or smaller than the lower error tolerance, respectively. Whenever the mesh is changed, we need to assemble the appropriate matrices and vectors and then solve a linear system based on this new mesh. After this new solution has been obtained, the local error estimate on each element according to the traditional error estimator,

$$\eta_{gradind,K} = \|\nabla u_h\|_{L^2,K} \ , \tag{4}$$

is computed. If the errors at any element are unacceptable, this set of elements, $S_k$, are refined to generate a new mesh $T_{k+1}$. The values for the new vertices added to refine the elements in $S_k$ are determined by interpolation. The problem is iteratively solved on the new discretized problem $T_{k+1}$ until neither refinement nor coarsening has been performed.

## 5 Experimental Results

In this section we present experimental results that demonstrate the improved accuracy of our proposed algorithm. We also discuss the computational overhead associated with this new method and demonstrate its parallel efficiency. Our experimental results are based on a representative two-dimensional transport problem. This problem is solved using the parallel particle tracking software incorporated within the SUMAA3d [11] programming environment. The parallel particle tracking incorporated within this software framework is used for the forward and backward tracking of each vertex. The particle tracking is required for the evaluation of the new local error estimates for the AMR method discussed in the previous section.

**Algorithm 4.1** A parallel framework for the adaptive solution of PDEs

Track each vertex backward and record the value $C^*$ on each vertex
Track each vertex (represented by a particle) forward and update the particle location
Estimate the error on each element using the proposed new estimator
Based on these error estimates, determine a set of elements, $S^n$, to refine to form $T^n$, including nonconforming elements
$k = 0$
**while** $T_k$ is different than $T_{k-1}$ **do**
  Assemble matrix and vector on $T_k$
  Solve the associated matrix equation
  Estimate the error on each element using a traditional estimator
  Based on these error estimates, determine a set of elements, $S_k$, to be refined to form $T_{k+1}$ including nonconforming elements
  $k = k + 1$
**endwhile**

### 5.1 Problem Description

The test problem we have used is one of benchmark problems from the Convection-Diffusion Forum [7]. This is an initial value problem with a small source transported by a shear flow with diffusion. In order to have an analytical solution, a two-dimensional uni-directional flow with velocity $u_0$ in the x-direction with shear $\lambda = du/dy$ is used. This advection inherent in this flow transports the initial source while diffusion, whose magnitude is determined by a coefficient $D$, disperses the source distribution. The governing equation for this problem is given by

$$\frac{\partial C}{\partial t} + (u_0 + \lambda y)\frac{\partial C}{\partial x} = D\left(\frac{\partial^2 C}{\partial x^2} + \frac{\partial^2 C}{\partial y^2}\right) \quad (-\infty < x < \infty) . \tag{5}$$

The initial and boundary conditions are specified as

$$\begin{aligned} C(x,y,0) &= C_0(x,y) = M\delta(x - x_0, y - y_0) \\ C(x,y,t) &\to 0 \quad \text{as } |x| \to \infty \quad or \quad |y| \to \infty . \end{aligned}$$

The analytical solution of the system is

$$C(x,y,t) = \frac{M}{4\pi Dt(1 + \lambda^2 t^2/12)^{\frac{1}{2}}} \exp - \left\{ \frac{(x - \overline{x} - 0.5\lambda yt)^2}{4\pi Dt(1 + \lambda^2 t^2/12)} + \frac{y^2}{4Dt} \right\} , \tag{6}$$

where

$$\overline{x} = x_0 + u_0 t . \tag{7}$$

Let the initial distribution be finite with the peak magnitude of 1 at location $(\overline{x}, y_0)$, i.e., $M = 4\pi D t_0 (1 + \lambda^2 t^2/12)^{\frac{1}{2}}$. The simulation starts at time $t = t_0$.

**Table 1.** Parameters of the problem

| Parameter | Value |
|---|---|
| $x_0$ | 7,200 |
| $y_0$ | 0 |
| $t_0$ | 2,400 |
| $u_0$ | 0.5 |
| $D$ | 10 |
| $\lambda$ | 0.0005 |

The domain is defined to be the region ($x \in [0, 24000], y \in [-3400, 3400]$). The problem parameters used are specified in Table 1. This domain is initially discretized into 7,992 vertices and 15,957 triangular elements as shown in Fig. 1. The initial conditions are specified by (6). The traditional error estimator used in Algorithm 4.1 is the gradient indicator $\eta_{gradind}$, which is the $L^2$-norm of the gradient of $u_h$ as shown in (4) for element $K$. Results are shown at times 4800, 7200, and 9600, with the implicit time scheme using a time-step of size 50. The time-step size is adjusted to 50 because the mesh is generated based on the simple $\eta_{gradind}$ error estimator—and the problem cannot be solved on this mesh by the traditional finite element method without this shorter (than the benchmark specified of 96) time step size. This restriction is crucial for the comparison of the accuracies of the two methods. The error tolerances used for the new error estimator given in (3) are $\epsilon_1 = 0.01$ and $\epsilon_2 = 0.0001$. This problem was run in parallel on an IBM SP2 at Argonne National Laboratory.

### 5.2 Computational Results

Fig. 2 and Fig. 3 illustrate the difference in the accuracy of these two methods. We show the computed solution using the finite element method for solving this time-marching transport problem on the mesh generated by the same AMR technique with and without incorporation of the proposed new error estimator. These results clearly demonstrate that the capture of the moving front and the profile similarity are improved with the use of our refinement criteria when compared to the analytical solution. In Fig. 3 we demonstrate this improvement by comparing the pointwise absolute error for these two approaches. From Fig. 4, we observe a significant speedup on the total running time and the time spent in particle tracking for the implementation of the presented *a posteriori* error estimator.

Although the computation of the estimator does impose a non-negligible overhead, the tradeoff in using this method is that we obtain significantly more accurate results. In addition, the performance of this algorithm can also be shown to be scalable. This relative overhead drops dramatically when more processors are used. Fig. 4 also shows that a finer mesh is generated to model the critical region when using one iteration of the new error estimator. This approach causes more inner iterations for the AMR method in solving the transient problem using the same number of time steps. The plot for the efficiency index, on the

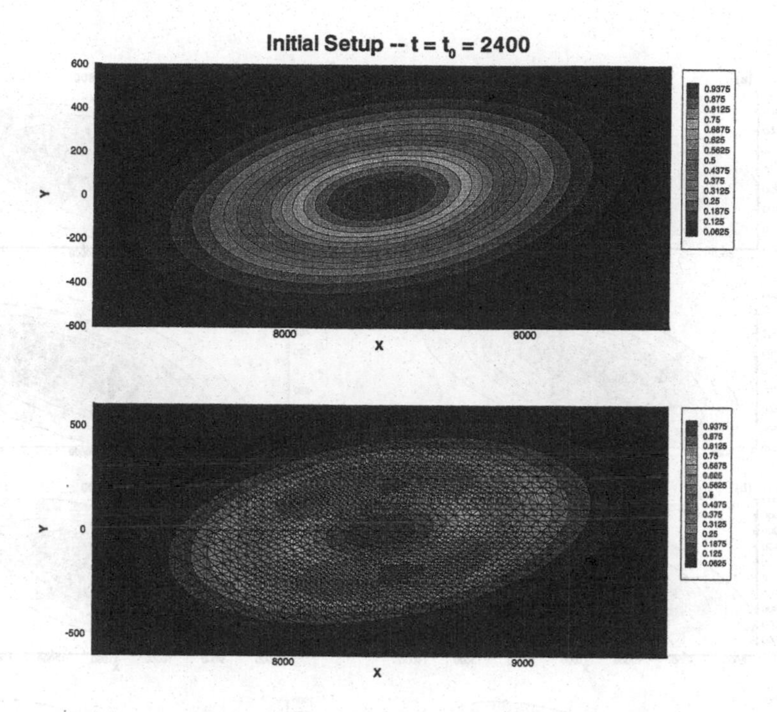


**Fig. 1.** The initial distribution in the source area (top) and the discretization of the domain (bottom)

right of Fig. 4, shows that the proposed error estimator is efficient based on the characterization given in Sect. 2.

## 6 Summary and Future Plans

In this paper, we have presented a new *a posteriori* error estimator based on a particle tracking method. With the incorporation of this error estimator, we have demonstrated that the accuracy of the computed solution of a representative convection-diffusion benchmark problem can be significantly improved. Although the overhead that the required particle tracking incurs cannot be neglected, this improvement in accuracy could not be achieved via the traditional *a posteriori* error estimator developed for steady state convection-diffusion equations. The intuitive reason for improvement is that the particle tracking method can accurately trace particle trajectories; thus, the new error estimator can correctly identify and refine the mesh at the critical region. Finally, our parallel implementation of the particle tracking method shows a significant speedup and computational efficiency. However, a more detailed study of the parallel performance and accuracy of the proposed error estimator under different flow fields and applications from other application areas is required. In particular, speedup and scalability studies on larger numbers of processors would be of great interest.

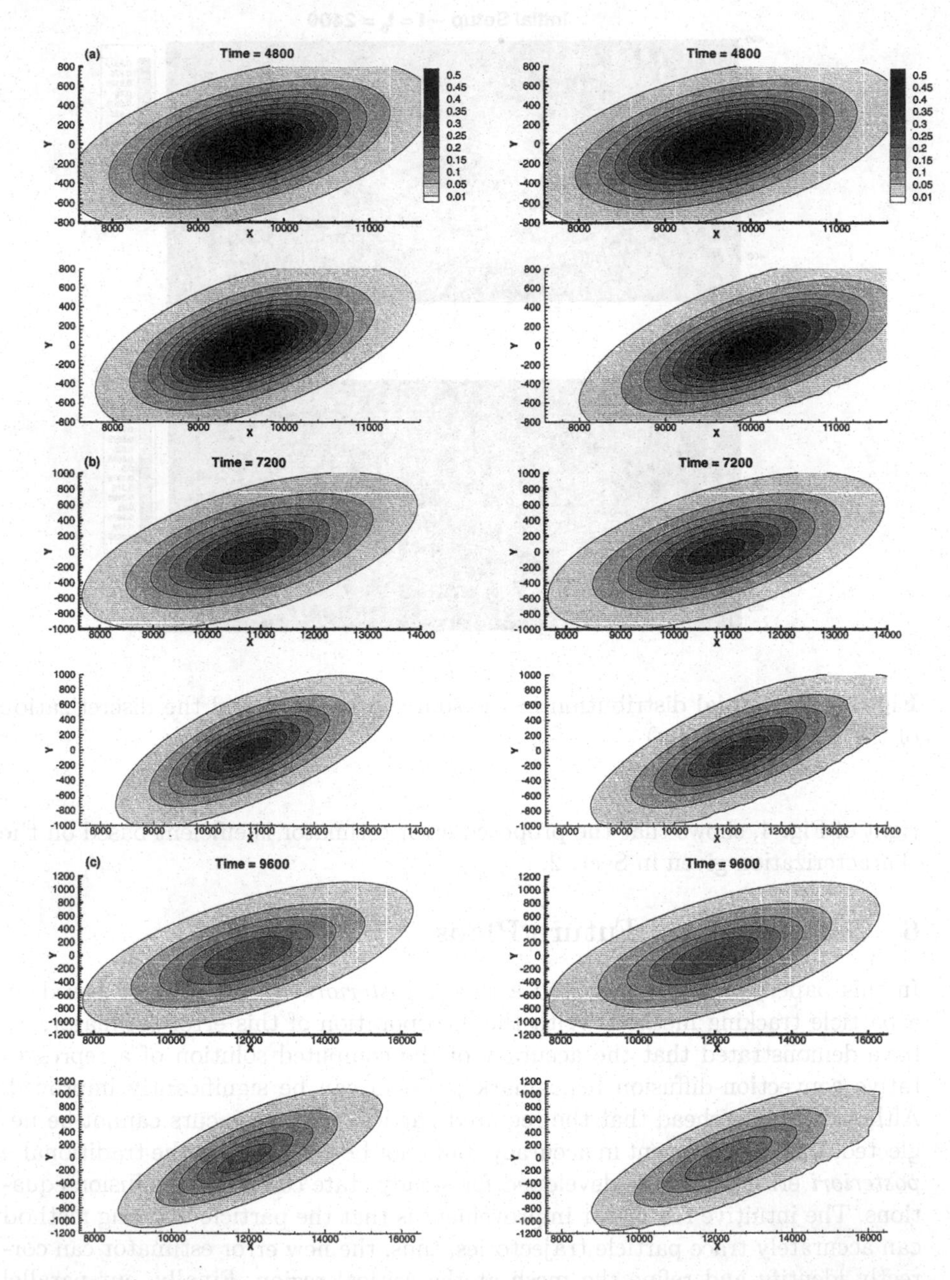


**Fig. 2.** (a) At time 4800, (b) At time 7200, and (c) At time 9600, Left: the analytical solution (top) and the simulation result (bottom) with the incorporation of the new error estimator. Right: the analytical solution (top) and the simulation result (bottom) without the incorporation of the new error estimator

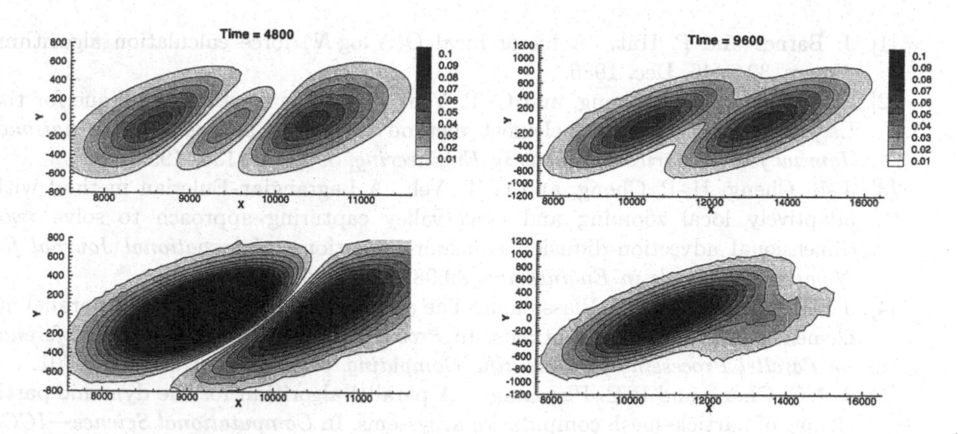


**Fig. 3.** The pointwise absolute error $| u - u_h |$ of the numerical results at time 4800 (left) and at time 9600 (right) using the numerical method with (top) and without (bottom) the incorporation of the new error estimator

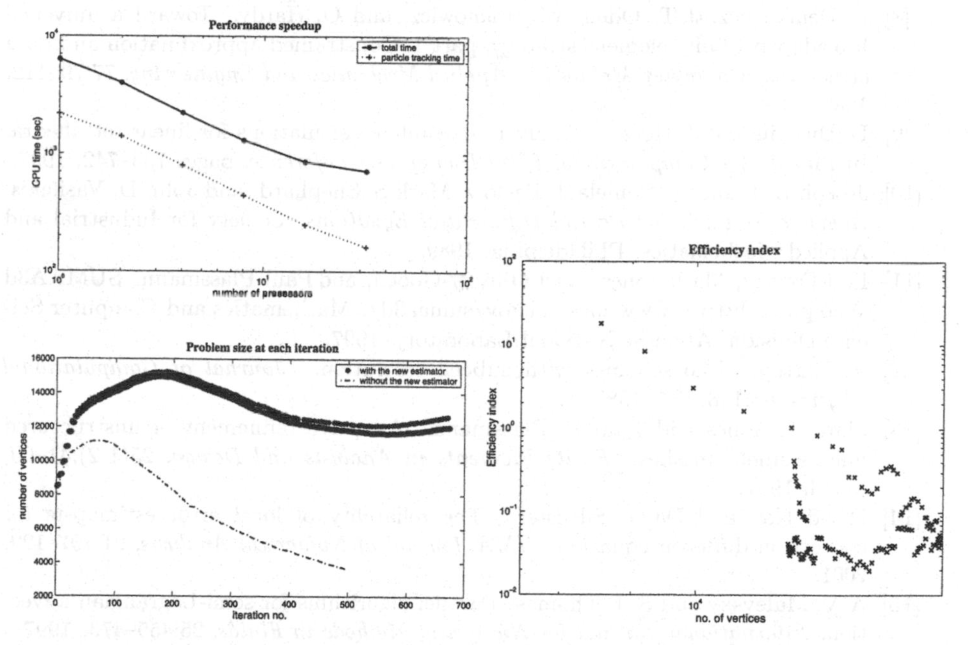


**Fig. 4.** Left: The performance speedup (top) and total number of vertices generated by AMR with and without the incorporation of the new error estimator (bottom), Right: Efficiency index with respect to different discretized mesh

## References

[1] J. Barnes and P. Hut. A hierarchical $O(N \log N)$ force calculation algorithm. *Nature*, 324:446, Dec. 1986.

[2] H.-P. Cheng, J.-R. Cheng, and G. T. Yeh. A particle tracking technique for the Lagrangian-Eulerian finite-element method in multi-dimensions. *International Journal for Numerical Methods in Engineering*, 39:1115–1136, 1996.

[3] J.-R. Cheng, H.-P. Cheng, and G. T. Yeh. A Lagrangian-Eulerian method with adaptively local zooming and peak/valley capturing approach to solve two-dimensional advection-diffusion transport equations. *International Journal for Numerical Methods in Engineering*, 39:987–1016, 1996.

[4] J-R. C. Cheng and P. E. Plassmann. The accuracy and performance of parallel in-element particle tracking methods. In *Proceedings of the Tenth SIAM Conference on Parallel Processing for Scientific Computing*, pages 252–261. SIAM, 2001.

[5] J-R. C. Cheng and P. E. Plassmann. A parallel algorithm for the dynamic partitioning of particle-mesh computational systems. In *Computational Science—ICCS 2002, The Springer Verlag Lecture Notes in Computer Science (LNCS 2331) series, Part III*, pages 1020–1029. Springer Verlag, 2002.

[6] L. P. Chew. Guaranteed-quality triangular meshes. Technical Report TR-89–983, Computer Science Department, Cornell University, 1989.

[7] Convection-Diffusion Forum. Specification of five convection-diffusion benchmark problems. 7th International Conference on Computational Methods in Water Resources, Massachusetts Institute of Technology, June 1988.

[8] L. Demkowicz, J. T. Oden, W. Rachowicz, and O. Hardy. Toward a universal h-p adaptive finite element strategy, part 1. Constrained approximation and data structures. *Computer Methods in Applied Mechanics and Engineering*, 77:79–112, 1989.

[9] D. Diachin and J. Herzog. Analytic streamline calculations for linear tetrahedra. In *13th AIAA Computational Fluid Dynamics Conference*, pages 733–742, 1997.

[10] Joseph E. Flaherty, Pamela J. Paslow, Mark S. Shephard, and John D. Vasilakis. *Adaptive Methods for Partial Differential Equations.* Society for Industrial and Applied Mathematics, Philadelphia, 1989.

[11] Lori Freitag, Mark Jones, Carl Ollivier-Gooch, and Paul Plassmann. SUMAA3d Web page. http://www.mcs.anl.gov/sumaa3d/, Mathematics and Computer Science Division, Argonne National Laboratory, 1997.

[12] A. Harten. Eno schemes with subcell resolution. *Journal of Computational Physics*, 83:148–184, 1989.

[13] Mark T. Jones and Paul E. Plassmann. Adaptive refinement of unstructured finite-element meshes. *Finite Elements in Analysis and Design*, 25(1-2):41–60, March 1997.

[14] David Kay and David Silvester. The reliability of local error estimators for convection-diffusion equations. *IMA Journal of Numerical Analysis*, 21:107–122, 2001.

[15] A. V. Malevsky and S. J. Thomas. Parallel algorithms for semi-Lagrangian advection. *International Journal for Numerical Methods in Fluids*, 25:455–473, 1997.

[16] David K. McAllister. The design of an API for particle systems. Technical Report UNC-CH TR00-007, Department of Computer Science, University of North Carolina, 1999.

[17] Bongki Moon, Mustafa Uysal, and Joel Saltz. Index translation schemes for adaptive computations on distributed memory multicomputers. In *Proceedings of the 9-th International Parallel Processing Symposium*, pages 812–819, Santa Barbara, California, April 24-28, 1995. IEEE Computer Society Press.

[18] K. W. Morton. *Numerical Solution of Convection-Diffusion Problems.* Klumer Academic Publishers, 1995.

[19] Robert P. Nance, Richard G. Wilmoth, Bongki Moon, H. A. Hassan, and Joel Saltz. Parallel DSMC solution of three-dimensional flow over a finite flat plate. In *Proceedings of the ASME 6-th Joint Thermophysics and Heat Transfer Conference*, Colorado Springs, Colorado, June 20-23, 1994. AIAA.

[20] J. T. Oden, L. Demkowicz, W. Rachowicz, and T. A. Westermann. Toward a universal h-p adaptive finite element strategy, part 2. A posterior error estimation. *Computer Methods in Applied Mechanics and Engineering*, 77:113–180, 1989.

[21] W. Rachowicz, J. T. Oden, and L. Demkowicz. Toward a universal h-p adaptive finite element strategy, part 3. Design of h-p meshes. *Computer Methods in Applied Mechanics and Engineering*, 77:181–212, 1989.

[22] C. D. Robinsion and J. K. Harvey. A fully concurrent DSMC implementation with adaptive domain decomposition. In *Proceedings of the Parallel CFD Meeting, 1997*, 1997.

[23] H.-G. Roos, M. Stynes, and L. Tobiska. *Numerical Methods for Singularity Perturbed Differential Equations.* Springer, 1996.

[24] J. Ruppert. A new and simple algorithm for quality 2-dimensional mesh generation. In *Proc. 4th ACM-SIAM Symp. on Disc. Algorithms*, pages 83–92, 1993.

[25] Rüdiger Verfürth. *A Review of* A Posteriori *Error Estimation and Adaptive Mesh-Refinement Techniques.* John Wiley & Sons Ltd and B. G. Teubner, 1996.

[26] John Volker. A numerical study of a posteriori error estimators for convection–diffusion equations. *Computer Methods in Applied Mechanics and Engineering*, 190:757–781, 2000.

[27] N. P. Weatherill, O. Hassan, D. L. Marcum, and M. J. Marchant. Grid generation by the Delaunay triangulation. Von Karman Institute for Fluid Dynamics, 1993-1994 Lecture Series, NASA Ames Research Center, January 1994.

[28] B. C. Wong and L. N. Long. A data-parallel implementation of the DSMC method on the connection machine. *Computing Systems in Engineering Journal*, 3(1-4), Dec. 1992.

[29] H. C. Yee, R. F. Warming, and A. Harten. Implicit total variation diminishing (TVD) schemes for steady-state calculations. *Journal of Computational Physics*, 57:327–360, 1985.

[30] G. T. Yeh, J.-R. Cheng, H.-P. Cheng, and C. H. Sung. An adaptive local grid refinement based on the exact peak capturing and oscillation free scheme to solve transport equations. *International Journal of Computational Fluids Dynamics*, 24:293–332, 1995.

[31] G. T. Yeh, J.-R. Cheng, and T. E. Short. An exact peak capturing and essentially oscillation-free scheme to so lve advection-dispersion-reactive transport equations. *Water Resources Research*, 28(11):2937–2951, 1992.

# Chapter 2

# Data Mining

# High Performance Data Mining*

Vipin Kumar, Mahesh V. Joshi, Eui-Hong (Sam) Han,
Pang-Ning Tan, and Michael Steinbach

University of Minnesota
4-192 EE/CSci Building, 200 Union Street SE
Minneapolis, MN 55455, USA
{kumar,mjoshi,han,ptan,steinbac}@cs.umn.edu

**Abstract.** Recent times have seen an explosive growth in the availability of various kinds of data. It has resulted in an unprecedented opportunity to develop automated data-driven techniques of extracting useful knowledge. Data mining, an important step in this process of knowledge discovery, consists of methods that discover interesting, non-trivial, and useful patterns hidden in the data [SAD+93, CHY96]. The field of data mining builds upon the ideas from diverse fields such as machine learning, pattern recognition, statistics, database systems, and data visualization. But, techniques developed in these traditional disciplines are often unsuitable due to some unique characteristics of today's data-sets, such as their enormous sizes, high-dimensionality, and heterogeneity. There is a necessity to develop effective parallel algorithms for various data mining techniques. However, designing such algorithms is challenging, and the main focus of the paper is a description of the parallel formulations of two important data mining algorithms: discovery of association rules, and induction of decision trees for classification. We also briefly discuss an application of data mining to the analysis of large data sets collected by Earth observing satellites that need to be processed to better understand global scale changes in biosphere processes and patterns.

## 1 Introduction

Recent times have seen an explosive growth in the availability of various kinds of data. It has resulted in an unprecedented opportunity to develop automated data-driven techniques of extracting useful knowledge. Data mining, an important step in this process of knowledge discovery, consists of methods that discover interesting, non-trivial, and useful patterns hidden in the data [26, 6]. The field of data mining builds upon the ideas from diverse fields such as machine learning, pattern recognition, statistics, database systems, and data visualization. But,

* This work was supported by NSF CCR-9972519, by NASA grant # NCC 2 1231, by Army Research Office contract DA/DAAG55-98-1-0441, by the DOE grant LLNL/DOE B347714, and by Army High Performance Computing Research Center cooperative agreement number DAAD19-01-2-0014. Access to computing facilities was provided by AHPCRC and the Minnesota Supercomputer Institute. Related papers are available via WWW at URL: http://www.cs.umn.edu/~kumar.

J.M.L.M. Palma et al. (Eds.): VECPAR 2002, LNCS 2565, pp. 111–125, 2003.

techniques developed in these traditional disciplines are often unsuitable due to some unique characteristics of today's data-sets, such as their enormous sizes, high-dimensionality, and heterogeneity.

To date, the primary driving force behind the research in data mining has been the development of algorithms for data-sets arising in various business, information retrieval, and financial applications. Businesses can use data mining to gain significant advantages in today's competitive global marketplace. For example, retail industry is using data mining techniques to analyze buying patterns of customers, mail order business is using them for targeted marketing, telecommunication industry is using them for churn prediction and network alarm analysis, and credit card industry is using them for fraud detection. Also, recent growth of electronic commerce is generating wealths of online web data, which needs sophisticated data mining techniques.

Due to the latest technological advances, very large data-sets are becoming available in many scientific disciplines as well. The rate of production of such data-sets far outstrips the ability to analyze them manually. For example, a computational simulation running on the state-of-the-art high performance computers can generate tera-bytes of data within a few hours, whereas human analyst may take several weeks or longer to analyze and discover useful information from these data-sets. Data mining techniques hold great promises for developing new sets of tools that can be used to automatically analyze the massive data-sets resulting from such simulations, and thus help engineers and scientists unravel the causal relationships in the underlying mechanisms of the dynamic physical processes.

The huge size of the available data-sets and their high-dimensionality make large-scale data mining applications computationally very demanding, to an extent that high-performance parallel computing is fast becoming an essential component of the solution. Moreover, the quality of the data mining results often depends directly on the amount of computing resources available. In fact, data mining applications are poised to become the dominant consumers of supercomputing in the near future. There is a necessity to develop effective parallel algorithms for various data mining techniques. However, designing such algorithms is challenging.

In the rest of this chapter, we present an overview of the parallel formulations of two important data mining algorithms: discovery of association rules, and induction of decision trees for classification. We also briefly discuss an application of data mining [13, 14, 10] to the analysis of large data sets collected by Earth observing satellites that need to be processed to better understand global scale changes in biosphere processes and patterns.

## 2 Parallel Algorithms for Discovering Associations

An important problem in data mining [6] is discovery of associations present in the data. Such problems arise in the data collected from scientific experiments, or monitoring of physical systems such as telecommunications networks, or from transactions at a supermarket. The problem was formulated originally in the

**Table 1.** Transactions from supermarket

| TID | Items |
|---|---|
| 1 | Bread, Coke, Milk |
| 2 | Beer, Bread |
| 3 | Beer, Coke, Diaper, Milk |
| 4 | Beer, Bread, Diaper, Milk |
| 5 | Coke, Diaper, Milk |

context of the transaction data at supermarket. This *market basket* data, as it is popularly known, consists of transactions made by each customer. Each transaction contains items bought by the customer (see Table 1). The goal is to see if occurrence of certain items in a transaction can be used to deduce occurrence of other items, or in other words, to find associative relationships between items. If indeed such interesting relationships are found, then they can be put to various profitable uses such as shelf management, inventory management, etc. Thus, *association rules* were born [2]. Simply put, given a set of items, association rules predict the occurrence of some other set of items with certain degree of confidence. The goal is to discover *all* such *interesting* rules. This problem is far from trivial because of the exponential number of ways in which items can be grouped together and different ways in which one can define interestingness of a rule. Hence, much research effort has been put into formulating efficient solutions to the problem.

Let $T$ be the set of transactions where each transaction is a subset of the itemset $I$. Let $C$ be a subset of $I$, then we define the *support count* of $C$ with respect to $T$ to be:

$$\sigma(C) = |\{t | t \in T, C \subseteq t\}|.$$

Thus $\sigma(C)$ is the number of transactions that contain $C$. An *association rule* is an expression of the form $X \stackrel{s,\alpha}{\Longrightarrow} Y$, where $X \subset I$ and $Y \subseteq I$. The *support* $s$ of the rule $X \stackrel{s,\alpha}{\Longrightarrow} Y$ is defined as $\sigma(X \cup Y)/|T|$, and the confidence $\alpha$ is defined as $\sigma(X \cup Y)/\sigma(X)$. For example, for transactions in Table 1, the support of rule $\{\text{Diaper, Milk}\} \Longrightarrow \{\text{Beer}\}$ is $\sigma(Diaper, Milk, Beer)/5 = 2/5 = 40\%$, whereas its confidence is. $\sigma(Diaper, Milk, Beer)/\sigma(Diaper, Milk) = 2/3 = 66\%$.

The task of discovering an association rule is to find all rules $X \stackrel{s,\alpha}{\Longrightarrow} Y$, such that $s$ is greater than or equal to a given minimum support threshold and $\alpha$ is greater than or equal to a given minimum confidence threshold. The association rule discovery is usually done in two phases. First phase finds all the *frequent* itemsets; i.e., sets satisfying the support threshold, and then they are post-processed in the second phase to find the high confidence rules. The former phase is computationally most expensive, and much research has been done in developing efficient algorithms for it. A comparative survey of all the existing techniques is given in [15]. A key feature of these algorithms lies in their method of controlling the exponential complexity of the total number of itemsets ($2^{|I|}$). Briefly, they all use the anti-monotone property of an itemset support, which states that an itemset is frequent only if all of its sub-itemsets are frequent. *Apriori* algorithm [4] pioneered the use of this property to systematically search the exponential space of itemsets. In an iteration $k$, it generates all the *candidate* $k$-itemsets (of length $k$) such that all their $(k-1)$-subsets are frequent. The number of occurrences of these candidates are then counted in the transaction database, to determine frequent $k$-itemsets. Efficient data structures are used to perform fast counting.

Overall, the serial algorithms such as *Apriori* have been successful on a wide variety of transaction databases. However, many practical applications of association rules involve huge transaction databases which contain a large number of distinct items. In such situations, these algorithms running on single-processor machines may take unacceptably large times. As an example, in the Apriori algorithm, if the number of candidate itemsets becomes too large, then they might not all fit in the main memory, and multiple database passes would be required within each iteration, incurring expensive I/O cost. This implies that, even with the highly effective pruning method of Apriori, the task of finding all association rules can require a lot of computational and memory resources. This is true of most of the other serial algorithms as well, and it motivates the development of parallel formulations.

Various parallel formulations have been developed so far. A comprehensive survey can be found in [15, 28]. These formulations are designed to effectively parallelize either or both of the computation phases: candidate generation and candidate counting. The candidate counting phase can be parallelized relatively easily by distributing the transaction database, and gathering local counts for the entire set of candidates stored on all the processors. The CD algorithm [3] is an example of this simple approach. It scales linearly with respect to the number of transactions; however, generation and storage of huge number of candidates on all the processors becomes a bottleneck, especially when high-dimensional problems are solved for low support thresholds using large number of processors. Other parallel formulations, such as IDD [11], have been developed to solve these problems. Their key feature is to distribute the candidate itemsets to processors so as to extract the concurrency in candidate generation as well as counting phases. Various ways are employed in IDD to reduce the communication overhead, to exploit the total available memory, and to achieve reasonable load balance. IDD algorithm exhibits better scalability with respect to the number of candidates. Moreover, reduction of redundant work and ability to overlap counting computation with communication of transactions, improves its scalability with respect to number of transactions. However, it still faces problems when one desires to use large number of processors to solve the problem. As more processors are used, the number of candidates assigned to each processor decreases. This has two implications on IDD. First, with fewer number of candidates per processor, it is much more difficult to achieve load balance. Second, it results in less computation work per transaction at each processor. This reduces the overall efficiency. Further lack of asynchronous communication ability may worsen the situation.

Formulations that combine the approaches of replicating and distributing candidates so as to reduce the problems of each one, have been developed. An example is the HD algorithm of [11]. Briefly, it works as follows. Consider a $P$-processor system in which the processors are split into $G$ equal size groups, each containing $P/G$ processors. In the *HD* algorithm, we execute the *CD* algorithm as if there were only $P/G$ processors. That is, we partition the transactions of the database into $P/G$ parts each of size $N/(P/G)$, and assign the task of computing the counts of the candidate set $C_k$ for each subset of the transactions to each one of these groups of processors. Within each group, these counts are computed using the *IDD* algorithm. The *HD* algorithm inherits all the good features of the *IDD* algorithm. It also provides good load balance and enough computation work by maintaining minimum number of candidates per processor. At the same time, the amount of data movement in this algorithm is cut down to $1/G$ of that of *IDD*. A detailed parallel runtime analysis of HD is given in [12]. It shows that HD is scalable with respect to both number of transactions and number of candidates. The analysis also proves the necessary conditions under which HD can outperform CD.

*Sequential Associations* The concept of association rules can be generalized and made more useful by observing another fact about transactions. All transactions have a timestamp associated with them; i.e. the time at which the transaction occurred. If this information can be put to use, one can find relationships such as if a customer bought [The C Programming Language] book today, then he/she is likely to buy a [Using Perl] book in a few days time. The usefulness of this kind of rules gave birth to the problem of discovering *sequential patterns* or *sequential associations*. In general, a sequential pattern is a sequence of item-sets with various timing constraints imposed on the occurrences of items appearing in the pattern. For example, (A) (C,B) (D) encodes a relationship that event D occurs after an event-set (C,B), which in turn occurs after event A. Prediction of events or identification of sequential rules that characterize different parts of the data, are some example applications of sequential patterns. Such patterns are not only important because they represent more powerful and predictive relationships, but they are also important from the algorithmic point of view. Bringing in the sequential relationships increases the combinatorial complexity of the problem enormously. The reason is that, the maximum number of sequences having $k$ events is $O(m^k 2^{k-1})$, where $m$ is the total number of distinct events in the input data. In contrast, there are only $\binom{m}{k}$ size-$k$ item-sets possible while discovering non-sequential associations from $m$ distinct items. Designing parallel algorithms for discovering sequential associations is equally important and challenging. In many situations, the techniques used in parallel algorithms for discovering standard non-sequential associations can be extended easily. However, different issues and challenges arise specifically due to the sequential nature and various ways in which interesting sequential associations can be defined. Details of various serial and parallel formulations and algorithms for finding such associations can be found in [17, 15].

## 3 Parallel Algorithms for Induction of Decision Tree Classifiers

Classification is an important data mining problem. The input to the problem is a data-set called the training set, which consists of a number of examples each having a number of attributes. The attributes are either *continuous*, when the attribute values are ordered, or *categorical*, when the attribute values are unordered. One of the categorical attributes is called the *class label* or the *classifying attribute*. The objective is to use the training set to build a model of the class label based on the other attributes such that the model can be used to classify new data not from the training data-set. Application domains include retail target marketing, fraud detection, and design of telecommunication service plans. Several classification models like neural networks [19], genetic algorithms [9], and decision trees [22] have been proposed. Decision trees are probably the most popular since they obtain reasonable accuracy [7] and they are relatively inexpensive to compute.

Most of the existing induction–based algorithms like *C4.5* [22], *CDP* [1], *SLIQ* [20], and *SPRINT* [23] use Hunt's method [22] as the basic algorithm. Here is its recursive description for constructing a decision tree from a set $T$ of training cases with classes denoted $\{C_1, C_2, \ldots, C_k\}$.

**Case 1** $T$ contains cases all belonging to a single class $C_j$. The decision tree for $T$ is a leaf identifying class $C_j$.

**Case 2** $T$ contains cases that belong to a mixture of classes. A test is chosen, based on a single attribute, that has one or more mutually exclusive outcomes $\{O_1, O_2, \ldots, O_n\}$. Note that in many implementations, $n$ is chosen to be 2 and this leads to a binary decision tree. $T$ is partitioned into subsets $T_1, T_2, \ldots, T_n$, where $T_i$ contains all the cases in $T$ that have outcome $O_i$ of the chosen test. The decision tree for $T$ consists of a decision node identifying the test, and one branch for each possible outcome. The same tree building machinery is applied recursively to each subset of training cases.

**Case 3** $T$ contains no cases. The decision tree for $T$ is a leaf, but the class to be associated with the leaf must be determined from information other than $T$. For example, *C4.5* chooses this to be the most frequent class at the parent of this node.

Figure 1 shows a training data set with four data attributes and two classes and its classification decision tree constructed using the Hunt's method. In the case 2 of Hunt's method, a test based on a single attribute is chosen for expanding the current node. The choice of an attribute is normally based on the entropy gains [22] of the attributes. The entropy of an attribute, calculated from class distribution information, depicts the classification power of the attribute by itself. The best attribute is selected as a test for the node expansion.

Highly parallel algorithms for constructing classification decision trees are desirable for dealing with large data sets in reasonable amount of time. Classification decision tree construction algorithms have natural concurrency, as once a node is generated, all of its children in the classification tree can be generated

| Outlook | Temp(F) | Humidity(%) | Windy? | Class |
|---|---|---|---|---|
| sunny | 75 | 70 | true | Play |
| sunny | 80 | 90 | true | Don't Play |
| sunny | 85 | 85 | false | Don't Play |
| sunny | 72 | 95 | false | Don't Play |
| sunny | 69 | 70 | false | Play |
| overcast | 72 | 90 | true | Play |
| overcast | 83 | 78 | false | Play |
| overcast | 64 | 65 | true | Play |
| overcast | 81 | 75 | false | Play |
| rain | 71 | 80 | true | Don't Play |
| rain | 65 | 70 | true | Don't Play |
| rain | 75 | 80 | false | Play |
| rain | 68 | 80 | false | Play |
| rain | 70 | 96 | false | Play |

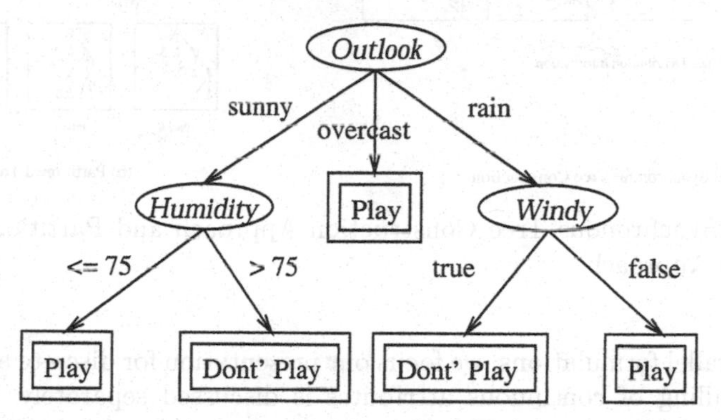


**Fig. 1.** A small training data set [Qui93] and its final classification decision tree

concurrently. Furthermore, the computation for generating successors of a classification tree node can also be decomposed by performing data decomposition on the training data. Nevertheless, parallelization of the algorithms for construction the classification tree is challenging for the following reasons. First, the shape of the tree is highly irregular and is determined only at runtime. Furthermore, the amount of work associated with each node also varies, and is data dependent. Hence any static allocation scheme is likely to suffer from major load imbalance. Second, even though the successors of a node can be processed concurrently, they all use the training data associated with the parent node. If this data is dynamically partitioned and allocated to different processors that perform computation for different nodes, then there is a high cost for data movements. If the data is not partitioned appropriately, then performance can be bad due to the loss of locality.

Several parallel formulations of classification decision tree have been proposed recently [21, 8, 23, 5, 18, 16, 24]. In this section, we present two basic parallel formulations for the classification decision tree construction and a hybrid scheme that combines good features of both of these approaches described in [24]. Most of other parallel algorithms are similar in nature to these two basic algorithms, and their characteristics can be explained using these two basic algorithms. For

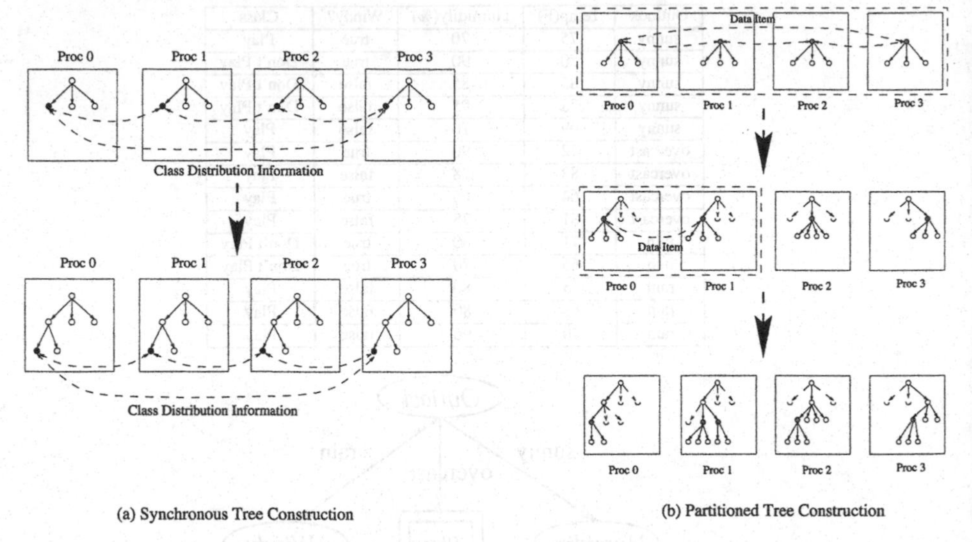


**Fig. 2.** Synchronous Tree Construction Approach and Partitioned Tree Construction Approach

these parallel formulations, we focus our presentation for discrete attributes only. The handling of continuous attributes is discussed separately. In all parallel formulations, we assume that $N$ training cases are randomly distributed to $P$ processors initially such that each processor has $N/P$ cases.

*Synchronous Tree Construction Approach* In this approach, all processors construct a decision tree synchronously by sending and receiving class distribution information of local data. Figure 2 (a) shows the overall picture. The root node has already been expanded and the current node is the leftmost child of the root (as shown in the top part of the figure). All the four processors cooperate to expand this node to have two child nodes. Next, the leftmost node of these child nodes is selected as the current node (in the bottom of the figure) and all four processors again cooperate to expand the node.

*Partitioned Tree Construction Approach* In this approach, whenever feasible, different processors work on different parts of the classification tree. In particular, if more than one processors cooperate to expand a node, then these processors are partitioned to expand the successors of this node. Figure 2 (b) shows an example. First (at the top of the figure), all four processors cooperate to expand the root node just like they do in the synchronous tree construction approach. Next (in the middle of the figure), the set of four processors is partitioned in three parts. The leftmost child is assigned to processors 0 and 1, while the other nodes are assigned to processors 2 and 3, respectively. Now these sets of processors proceed independently to expand these assigned nodes. In particular, processors 2 and processor 3 proceed to expand their part of the tree using the

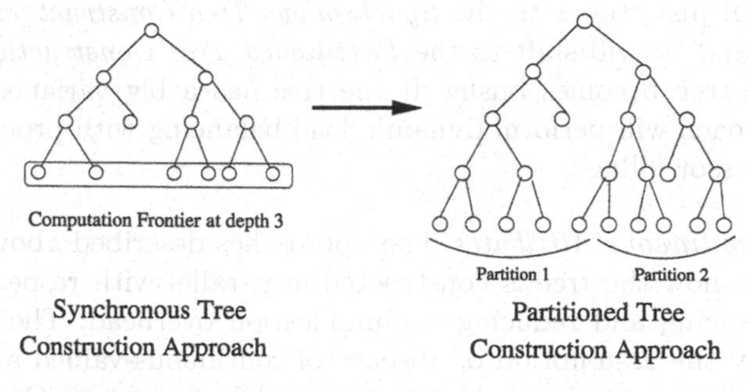


**Fig. 3.** Hybrid Tree Construction Approach

serial algorithm. The group containing processors 0 and 1 splits the leftmost child node into three nodes. These three new nodes are partitioned in two parts (shown in the bottom of the figure); the leftmost node is assigned to processor 0, while the other two are assigned to processor 1. From now on, processors 0 and 1 also independently work on their respective subtrees.

*Hybrid Parallel Formulation* The hybrid parallel formulation has elements of both schemes. The *Synchronous Tree Construction Approach* incurs high communication overhead as the frontier gets larger. The *Partitioned Tree Construction Approach* incurs cost of load balancing after each step. The hybrid scheme keeps continuing with the first approach as long as the communication cost incurred by the first formulation is not too high. Once this cost becomes high, the processors as well as the current frontier of the classification tree are partitioned into two parts. Figure 3 shows one example of this parallel formulation. At the classification tree frontier at depth 3, no partitioning has been done and all processors are working cooperatively on each node of the frontier. At the next frontier at depth 4, partitioning is triggered, and the nodes and processors are partitioned into two partitions.

A key element of the algorithm is the criterion that triggers the partitioning of the current set of processors (and the corresponding frontier of the classification tree ). If partitioning is done too frequently, then the hybrid scheme will approximate the partitioned tree construction approach, and thus will incur too much data movement cost. If the partitioning is done too late, then it will suffer from high cost for communicating statistics generated for each node of the frontier, like the synchronized tree construction approach. In the hybrid algorithm, the splitting is performed when the accumulated cost of communication becomes equal to the cost of moving records and load balancing in the splitting phase.

The size and shape of the classification tree varies a lot depending on the application domain and training data set. Some classification trees might be shallow and the others might be deep. Some classification trees could be skinny others could be bushy. Some classification trees might be uniform in depth while other trees might be skewed in one part of the tree. The hybrid approach adapts well to all types of classification trees. If the decision tree is skinny, the hybrid approach will just stay with the *Synchronous Tree Construction Approach.* On the other hand, it will shift to the *Partitioned Tree Construction Approach* as soon as the tree becomes bushy. If the tree has a big variance in depth, the hybrid approach will perform dynamic load balancing with processor groups to reduce processor idling.

*Handling Continuous Attributes* The approaches described above concentrated primarily on how the tree is constructed in parallel with respect to the issues of load balancing and reducing communication overhead. The discussion was simplified by the assumption of absence of continuous-valued attributes. Presence of continuous attributes can be handled in two ways. One is to perform intelligent discretization, either once in the beginning or at each node as the tree is being induced, and treat them as categorical attributes. Another, more popular approach is to use decisions of the form $A < x$ and $A \geq x$, directly on the values $x$ of continuous attribute $A$. The decision value of $x$ needs to be determined at each node. For efficient search of $x$, most algorithms require the attributes to be sorted on values, such that one linear scan can be done over all the values to evaluate the best decision. Among various different algorithms, the approach taken by SPRINT algorithm[23], which sorts each continuous attribute only once in the beginning, is proven to be efficient for large datasets. The sorted order is maintained throughout the induction process, thus avoiding the possibly excessive costs of re-sorting at each node. A separate list is kept for each of the attributes, in which the record identifier is associated with each sorted value. The key step in handling continuous attributes is the proper assignment of records to the children node after a splitting decision is made. Implementation of this offers the design challenge. SPRINT builds a mapping between a record identifier and the node to which it goes to based on the splitting decision. The mapping is implemented as a hash table and is probed to split the attribute lists in a consistent manner.

Parallel formulation of the SPRINT algorithm falls under the category of synchronous tree construction design. The multiple sorted lists of continuous attributes are split in parallel by building the entire hash table on all the processors. However, with this simple-minded way of achieving a consistent split, the algorithm incurs a communication overhead of $O(N)$ per processor. Since, the serial runtime of the induction process is $O(N)$, SPRINT becomes unscalable with respect to runtime. It is unscalable in memory requirements also, because the total memory requirement per processor is $O(N)$, as the size of the hash table is of the same order as the size of the training dataset for the upper levels of the decision tree, and it resides on every processor. Another parallel algorithm, ScalParC [16], solves this scalability problem. It employs a distributed hash table to achieve a consistent split. The communication structure, used to construct and access this hash table, is motivated by the parallel sparse matrix-vector multiplication algorithms. It is shown in [16] that with the proper implementation of the parallel hashing, the overall communication overhead does not exceed $O(N)$, and the memory required does not exceed $O(N/p)$ per processor. Thus, ScalParC is scalable in runtime as well as memory requirements.

## 4 Example Application: Data Mining for Earth Science Data

Data mining techniques have recently been used to find interesting spatio-temporal patterns from Earth Science data. This data consists of time series measurements for various Earth science and climate variables (e.g. soil moisture, temperature, and precipitation), along with additional data from existing ecosystem models (e.g., Net Primary Production). See figures 4 and 5. The ecological patterns of interest include associations, clusters, predictive models, and trends.

To find association patterns we transformed these time series into transactions and then applied existing algorithms traditionally used for market-basket data. We found that association rules can uncover interesting patterns for Earth Scientists to investigate. For example, we found that high temperature was well correlated with high plant growth in the forest and cropland regions in the northern hemisphere. However, significant challenges for association analysis arise due to the spatio-temporal nature of the data and the need to incorporate domain knowledge to prune out uninteresting patterns. For further detail on this work, see [27].

To predict the effect of the oceans on land climate, Earth Scientists have developed ocean climate indices (OCIs), which are time series that summarize the behavior of selected areas of the Earth's oceans. For example, the Southern Oscillation Index (SOI) is an OCI that is associated with El Nino. In the past, Earth scientists have used observation and, more recently, eigenvalue analysis

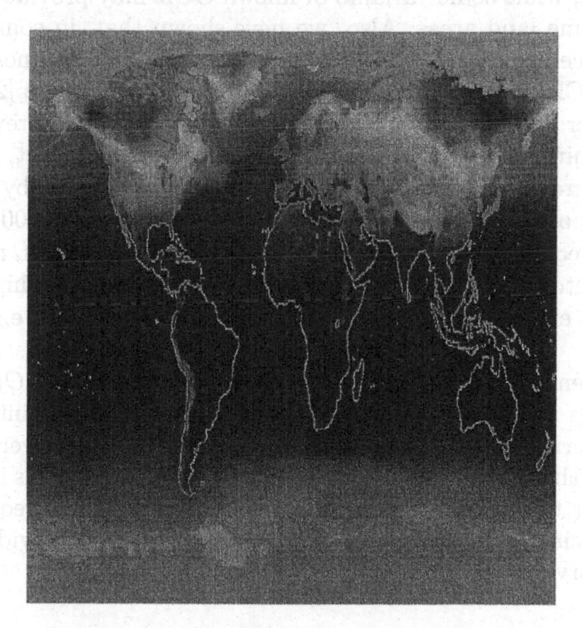

**Fig. 4.** Land and sea temperature

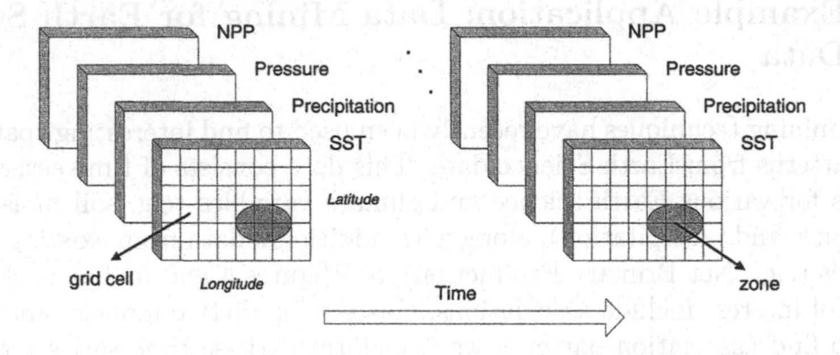


**Fig. 5.** Illustration of Earth Science Data

techniques, such as principal components analysis (PCA) and singular value decomposition (SVD), to discover ocean climate indices. However, these techniques are only useful for finding a few of the strongest signals and, furthermore, impose a condition that all discovered signals must be orthogonal to each other. We have developed an alternative methodology for the discovery of OCIs that overcomes these limitations and is based on clusters that represent ocean regions with relatively homogeneous behavior [25]. The centroids of these clusters are time series that summarize the behavior of these ocean areas. We divide the cluster centroids into several categories: those that correspond to known OCIs, those that are variants of known OCIs, and those that represent potentially new OCIs. The centroids that correspond to known OCIs provide a validation of our methodology, while some variants of known OCIs may provide better predictive power for some land areas. Also, we have shown that, in some sense, our current cluster centroids are relatively complete, i.e., capture most of the possible candidate OCIs. For further details, the reader is referred to [25].

A number of aspects of Earth Science data and the previously described analyses require the use of high-performance computing. First, satellites are providing measurements of finer granularity. For instance, a 1° by 1° grid produces 64,800 data points, while a 0.1° by 0.1° grid produces 6,480,000 data points. Second, more frequent measurements, e.g., daily measurements, multiply monthly data by a factor of 30. Also, looking at weather instead of climate requires finer resolution to enable the detection of fast changing patterns, e.g., the movement of fronts

Our current clustering analysis, while effective, requires $O(n^2)$ comparisons since it needs to evaluate the correlation of every ocean point with every land point. Furthermore, association rule algorithms can also be very compute intensive. Indeed, the computational complexity of these algorithms is potentially very much greater than $O(n^2)$. Finally, the amount of memory required for clustering and association rule algorithms can exceed the 4GB provided by traditional sequential servers.

## 5 Conclusion

This paper presented an overview of parallel algorithms for two of the commonly used data mining techniques: classification and associations. Key issues such as load balancing, attention to locality, extracting maximal concurrency, avoiding hot spots in contention, and minimizing parallelization overhead are just as central to these parallel formulations as they are to the traditional scientific parallel algorithms. In fact, in many cases, the underlying kernels are identical to well known algorithms, such as sparse matrix-vector product.

To date, the parallel formulations of many decision-tree induction and association rule discovery algorithms are reasonably well-understood. Relatively less work has been done on the parallel algorithms for other data mining techniques such as clustering, rule-based classification algorithms, deviation detection, and regression. Some possible areas of further research include parallelization of many emerging new and improved serial data mining algorithms, further analysis and refinements of existing algorithms for scalability and efficiency, designs targeted for shared memory and distributed shared memory machines equipped with symmetric multiprocessors, and efficient integration of parallel algorithms with parallel database systems.

High-performance data mining algorithms and tools are needed for mining large-scale data sets that arise in a variety of applications. This paper presented a possible application, i.e., large data sets collected by Earth observing satellites that need to be processed to better understand global scale changes in biosphere processes and patterns. Other examples of important applications of data mining include understanding gene functions in the field of genomics, the categorization of stars and galaxies in the field of astrophysics, and using data obtained through monitoring network traffic to detect illegal network activities. The key technical challenges in mining these data sets include (i) high volume, dimensionality and heterogeneity; (ii) the spatio-temporal aspects of the data; (iii) possibly skewed class distributions; (iv) the distributed nature of the data; (v) the complexity in converting raw collected data into high level features. High performance data mining is essential to analyze the growing data and provide analysts with automated tools that facilitate some of the steps needed for hypothesis generation and evaluation.

## References

[1] R. Agrawal, T. Imielinski, and A. Swami. Database mining: A performance perspective. *IEEE Transactions on Knowledge and Data Eng.*, 5(6):914–925, December 1993.

[2] R. Agrawal, T. Imielinski, and A. Swami. Mining association rules between sets of items in large databases. In *Proc. of 1993 ACM-SIGMOD Int. Conf. on Management of Data*, Washington, D. C., 1993.

[3] R. Agrawal and J. C. Shafer. Parallel mining of association rules. *IEEE Transactions on Knowledge and Data Eng.*, 8(6):962–969, December 1996.

[4] R. Agrawal and R. Srikant. Fast algorithms for mining association rules. In *Proc. of the 20th VLDB Conference*, pages 487–499, Santiago, Chile, 1994.

[5] J. Chattratichat, J. Darlington, M. Ghanem, Y. Guo, H. Huning, M. Kohler, J. Sutiwaraphun, H. W. To, and D. Yang. Large scale data mining: Challenges and responses. In *Proc. of the Third Int'l Conference on Knowledge Discovery and Data Mining*, 1997.

[6] M. S. Chen, J. Han, and P. S. Yu. Data mining: An overview from database perspective. *IEEE Transactions on Knowledge and Data Eng.*, 8(6):866–883, December 1996.

[7] D. J. Spiegelhalter D. Michie and C. C. Taylor. *Machine Learning, Neural and Statistical Classification.* Ellis Horwood, 1994.

[8] S. Goil, S. Aluru, and S. Ranka. Concatenated parallelism: A technique for efficient parallel divide and conquer. In *Proc. of the Symposium of Parallel and Distributed Computing (SPDP'96)*, 1996.

[9] D. E. Goldberg. *Genetic Algorithms in Search, Optimizations and Machine Learning.* Morgan-Kaufman, 1989.

[10] R. Grossman, C. Kamath, P. Kegelmeyer, V. Kumar, and R. Namburu. *Data Mining for Scientific and Engineering Applications.* Kluwer Academic Publishers, 2001.

[11] E.-H. Han, G. Karypis, and V. Kumar. Scalable parallel data mining for association rules. In *Proc. of 1997 ACM-SIGMOD Int. Conf. on Management of Data*, Tucson, Arizona, 1997.

[12] E. H. Han, G. Karypis, and V. Kumar. Scalable parallel data mining for association rules. *IEEE Transactions on Knowledge and Data Eng.*, 12(3), May/June 2000.

[13] J. Han and M. Kamber. *Data Mining: Concepts and Techniques.* Morgan-Kaufman, 2000.

[14] D. Hand, H. Mannila, and P. Smyth. *Principles of Data Mining.* MIT Press, 2001.

[15] M. V. Joshi, E.-H. Han, G. Karypis, and V. Kumar. Efficient parallel algorithms for mining associations. In M. J. Zaki and C.-T. Ho, editors, *Lecture Notes in Computer Science: Lecture Notes in Artificial Intelligence (LNCS/LNAI)*, volume 1759. Springer-Verlag, 2000.

[16] M. V. Joshi, G. Karypis, and V. Kumar. ScalParC: A new scalable and efficient parallel classification algorithm for mining large datasets. In *Proc. of the International Parallel Processing Symposium*, 1998.

[17] M. V. Joshi, G. Karypis, and V. Kumar. Universal formulation of sequential patterns. Technical Report TR 99-021, Department of Computer Science, University of Minnesota, Minneapolis, 1999.

[18] R. Kufrin. Decision trees on parallel processors. In J. Geller, H. Kitano, and C. B. Suttner, editors, *Parallel Processing for Artificial Intelligence 3.* Elsevier Science, 1997.

[19] R. Lippmann. An introduction to computing with neural nets. *IEEE ASSP Magazine*, 4(22), April 1987.

[20] M. Mehta, R. Agrawal, and J. Rissanen. SLIQ: A fast scalable classifier for data mining. In *Proc. of the Fifth Int'l Conference on Extending Database Technology*, Avignon, France, 1996.

[21] R. A. Pearson. A coarse grained parallel induction heuristic. In H. Kitano, V. Kumar, and C. B. Suttner, editors, *Parallel Processing for Artificial Intelligence 2*, pages 207–226. Elsevier Science, 1994.

[22] J. Ross Quinlan. *C4.5: Programs for Machine Learning.* Morgan Kaufmann, San Mateo, CA, 1993.

[23] J. Shafer, R. Agrawal, and M. Mehta. SPRINT: A scalable parallel classifier for data mining. In *Proc. of the 22nd VLDB Conference*, 1996.

[24] A. Srivastava, E.-H. Han, V. Kumar, and V. Singh. Parallel formulations of decision-tree classification algorithms. *Data Mining and Knowledge Discovery: An International Journal*, 3(3):237–261, September 1999.

[25] M. Steinbach, P. Tan, V. Kumar, S. Klooster, and C. Potter. Temporal data mining for the discovery and analysis of ocean climate indices. In *KDD Workshop on Temporal Data Mining(KDD'2002)*, Edmonton, Alberta, Canada, 2001.

[26] M. Stonebraker, R. Agrawal, U. Dayal, E. J. Neuhold, and A. Reuter. DBMS research at a crossroads: The vienna update. In *Proc. of the 19th VLDB Conference*, pages 688–692, Dublin, Ireland, 1993.

[27] P. Tan, M. Steinbach, V. Kumar, S. Klooster, C. Potter, and A. Torregrosa. Finding spatio-temporal patterns in earth science data. In *KDD Workshop on Temporal Data Mining(KDD'2001)*, San Francisco, California, 2001.

[28] M. J. Zaki. Parallel and distributed association mining: A survey. *IEEE Concurrency (Special Issue on Data Mining)*, December 1999.

# Data Mining for Data Classification Based on the KNN-Fuzzy Method Supported by Genetic Algorithm

José L. A. Rosa and Nelson F. F. Ebecken

COPPE, Universidade Federal do Rio de Janeiro, Brazil

**Abstract.** This paper presents a classification method based on the KNN-Fuzzy classification algorithm, supported by Genetic Algorithm. It discusses how to consider data clustering according to the Fuzzy logic and its consequences in the area of Data Mining. Analyses are made upon the results obtained in the classification of several data bases in order to demonstrate the proposed theory.

## 1 Introduction

Data classification is one of the most used data mining tasks.

ANDERBERGS [1] defines classification as being the process or act to associate a new item or comment to a category. As example, a person can be classified, according some attributes: sex (female or male), nationality (country where it was born), naturalness (state where it was born), instruction degree (illiterate or not), height (low, high).

Most of the knowledge discovery techniques are based on mathematical statistics and Machine Learning fundamentals [2]. Many classification methods are used and some of them are described in [3].

A memory based reasoning strategy was adopted in this paper. Basically it needs: the set of stored cases, a distance metric to compute distances between cases and the number of nearest neighbors to retrieve. This corresponds to a classical non-parametric architecture.

The basic idea is very straight-forward. For training, all input-to-output pairs in the training set are stored into a database. When an estimation or a classification is needed on a new input pattern, the answer is based on the nearest training patterns in the database.

The method requires no training time other than the time required to preprocess and store the entire training set. It is very memory intensive since the entire training set is stored. Classification/estimation is slow since the distance between input pattern and all patterns in the training set must be computed.

This method typically performs better for lower dimensional problems (less than 10), as theoretical and empirical results suggest that the amount of training data needed in higher dimensions is greater than that required in other models.

One of the difficulties that arises when utilizing this technique is that each of the labeled samples is given equal importance in deciding the class memberships of the pattern to be classified and there is no indication of its "strength" of memberships in that class. This problem can be addressed by incorporating fuzzy set theory.

The basis of the fuzzy algorithm is to assign membership as a function of the vector's distance from its nearest neighbors and those neighbors' memberships in the possible classes.

The main questions now are:

- How many nearest neighbors (K) must be considered? (empirical formulas, K should be less than the square root of the total number of training patterns)
- How weight the distance to calculate each neighbor's contribution to the membership value?

In this paper those questions are answered by using a genetic algorithm. A parallel version was implemented, including all the commented improvements to build a more precise and efficient classification methodology.

Section 2 presents the KNN-Fuzzy Method and the genetic algorithm approach used to optimize it. The parallelization schema and its consequences, and the available computational environment are discussed in section 3. The studied case is analyzed with the Amdahl's law point of view in section 4. Some conclusions and comments are presented in section 5.

## 2 KNN-Fuzzy Method

KNN (K-Nearest Neighbors) consists of the identification of groups of individuals with similar features and posterior grouping.

The calculation of the distance of the unknown sample in relation to the known samples is made by:

$$\text{dist}_i = \sum_{j=1}^{j=m} (\text{vu}_{ij} - \text{vk}_{ij})^2 \tag{1}$$

Assuming i known samples with two classes, 0 and 1, the KNN algorithm can be summarized as indicated below.

```
BEGIN
   READ i known samples (ks)
   READ unknown sample (us)
   Compute each distance from us to ksi
   ASSIGN both distance and class to a distance vector
   SORT distance vector by distance
   SELECT K vector distance samples
   Compare each class from the K selected samples
   IF (class is equal 0)
         Increment class0
   ELSE
         Increment class1
   ENDIF
END
```

Each distance is related to the corresponding class in a vector of distances. After the calculation of all the distances, the K smaller distances are selected; the unknown sample then is classified as being of the same present class.

As shown in [2], the value of K must be that one that satisfies the function max that defines the maximum value of neighbors (samples) Kmax, who are near the unknown sample, in relation to all the samples of the considered set, K tot:

$$K_{máx} = \max (K_{tot}) \tag{2}$$

In the classification task, it must be stressed that a variable can have more contribution in the definition of the class than others. A possible solution may be to consider the weight of the variables in accordance with its degree of importance for the classification process. According to [1], the balance of the variables transforms the original space in a representation space that can be wider or with reduced number of dimensions, which gives better classification accuracy.

In accordance with the work presented in [4], the basis of the KNN-Fuzzy classification algorithm is to associate the relevancy of a vector to its next K-neighbors and those members in possible classes.

Let W = { Z1, Z2..., Zc } be the set of c prototype representing the c classes. Let μ (x) the membership (to be computed) associated to vector x. As seen in (3):

$$\mu_i (x) = \frac{1 / \|x - Z_i\|^{2/(m-1)}}{\sum_{i=1}^{c} (1 / \|x - Z_i\|^{2/(m-1)})} \tag{3}$$

the associate memberships of x, are influenced by inverse of the distances of the neighbors and its c membership's classes. The inverse of the distance serves as weight, taking in consideration how close or not the vector to be classified is from the classified vector.

The variable m determines how heavily is the weight attributed at a distance of each neighbor to the value of membership. If the value of m is equal two, then the contribution of each point of the neighborhood is considered reciprocal to the long-distance of the point to be classified. As m increases, it diminishes the contribution of each neighbor; however, the reduction of m (m greater than one), implies more strong contribution. Assuming i known samples with two classes, 0 and 1 the KNN-Fuzzy algorithm can be summarized as follow.

```
BEGIN
    READ i known samples (ks)
    READ unknown sample (us)
    Compute each distance from us to ks_i
    ASSIGN both distance and class to a distance vector
    SORT distance vector by distance (column 1)
    SELECT K vector distance samples
    Compute μ(x) using (3)
END
```

### 2.1 The Proposed Methodology

In this paper, a standard genetic algorithm scheme is employed to evolve the KNN-Fuzzy method in order to achieve the best classification precision

The binary coding [5] was used to generate the chromosome (gene). The representation corresponds to 2n chromosomes, where n is the amount of genes that each chromosome possesses. The set of all the patterns that the chromosome can assume forms its space of search. In the present study the K, m and weight variables are taken into account.

To evolve the population the genetic operators promote modifications that improve the average performance of the population to each generation, using information on the previous generations. The new population starts with the selection of the individuals of the ascending population with bigger probability to compose the following generation. In a population of n individuals, each one with fitness f, a chromosome c has it probability p of selection of this individual given by:

$$p_i = \sum_{i=1}^{N} f_i \tag{4}$$

There are many selection methods, as described in [6]. In this work the method of stochastic selection via tournament was used.

Among these chromosomes, the one with bigger aptitude is selected for the following population. The process then is repeated until it fills the descending population. Once selected the individuals of the descending population, the operators of crossover and mutation are applied. The crossover operator makes swap on fragments between pairs of chromosomes, whose strings are cut in a random position producing, for each individual, two amounts of alelos that will have to be reallocated.

Mutation inverts one or more bits of an individual in a probability generally low, improves the diversity of the chromosomes in the population and destroys the information contained in the chromosome.

In this way the values of K, m and the weights of the variables were obtained to achieve the best classification precision.

## 3 Parallel Implementation

The main issues under consideration in the study of parallel implementation of conceived sequential algorithms are the partitioning algorithm schema and the target machine.

### 3.1 Partitioning Algorithm Schema

For the partitioning algorithm schema the study considers the use of the data parallelism approach that keeps a copy of the entire labeled samples, with the internal variables and functions, in each processing node partitioning training data set among nodes. This approach ensures that all values needed during the training phase like K, m and variable weights, are locally available, reducing the number of messages passing among nodes and the algorithm synchronization. In fact, all nodes perform only one communication after each complete presentation of a training data subset. In distributed memory machines this approach seems to be very efficient leading to great gain of performance. This implementation of data parallelism uses a control node that performs genetic operations, realizes chromosome encoding and decoding, broadcasts the values of K, m and weights of the variables, and summarize the interested value (total of right classes). All nodes start with same internal parameters but with different subsets of training data. During training phase of KNN-Fuzzy method, each node passes its training data subset producing partial errors that must be combined with partial errors produced by other nodes, generating an overall error. This error is used to update the weights of the variables in each node, until the procedure reaches the minimal acceptable overall error. It can be pointed out that each node only broadcasts its partial error to the control node at this time. All other calculations involve local data and can be made without synchronization. This drastically reduces the communications between nodes.

The application uses few and simple MPI standard primitives and can be portable for several machines.

### 3.2 Target Machine

The machine available in this work was an academic pc-cluster of the NACAD/COPPE/UFRJ High Performance Computing Laboratory. This machine contains 8 nodes (Intel Pentium III with 256 MB memory) each connected by a Fast Ethernet network). The Linux operating system and LAM MPI [7] programming environment supports the parallel execution of jobs for that environment. LAM features a full implementation of the MPI communication standard among parallel process. The access obtained for the execution of the application in this environment was shared, i.e., all programs share all nodes during execution. It could be emphasized here that time tests were made with unbalanced work on nodes.

## 4 The Studied Case

The data set has been extracted from a real world insurance database that contains relationships among clients, contracts and tariffs. The former objective of this application is modeling classification criteria for two predefined classes and performance assessment. Some characteristics and values have been protected by non-disclosure policies. Despite of the relation's complexity, the attributes and its values were grouped in one single table for this study. This table contains 80 attributes and 147478 registers for the predefined classes. The data set had been pre-processed, clearing some aspects not suitable for data mining methods. The resultant set contains 64 attributes and 130143 lines of registers.

In table 1 are presented the elapsed time spent in a limited execution of the application with different number of nodes for each job submission, the ideal (IdSp) and obtained speedup (Sp) and efficiency (Ep).

The application in each node reads the data files (known , training and test sets), according to the "fagknnp.cfg" configuration file, which can be altered depending on desired data set partitioning. The application of Amdahl's Law for performance evaluation shows a 99,6% of parallelization coefficient for the developed algorithm. This is explained by the large time expended by KNN-Fuzzy procedure in the overall application time (in a serial evaluation the total computation time is equal 105057 s and the KNN-Fuzzy procedure evaluation time is equal 104634 s).

**Table 1.** Jobs submission report

| nodes | Time (s) | IdSp | Sp | Ep |
|---|---|---|---|---|
| 2 | 57804 | 1,991980 | 1,817469 | 0,995990 |
| 3 | 38287 | 2,976035 | 2,743934 | 0,992012 |
| 4 | 30173 | 3,952260 | 3,481821 | 0,988065 |
| 5 | 23945 | 4,920749 | 4,387430 | 0,984150 |
| 6 | 19905 | 5,881592 | 5,277920 | 0,980265 |
| 7 | 17775 | 6,834881 | 5,910380 | 0,976412 |
| 8 | 15880 | 7,780703 | 6,615680 | 0,972588 |

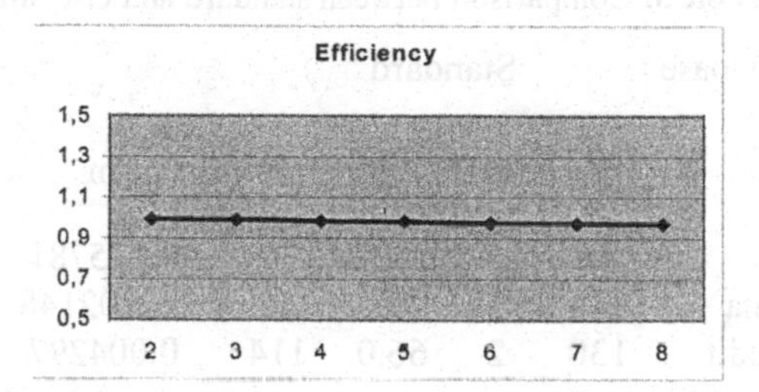


**Fig. 1.** Representation of the ideal and obtained speedup, according to Amdahl's Law

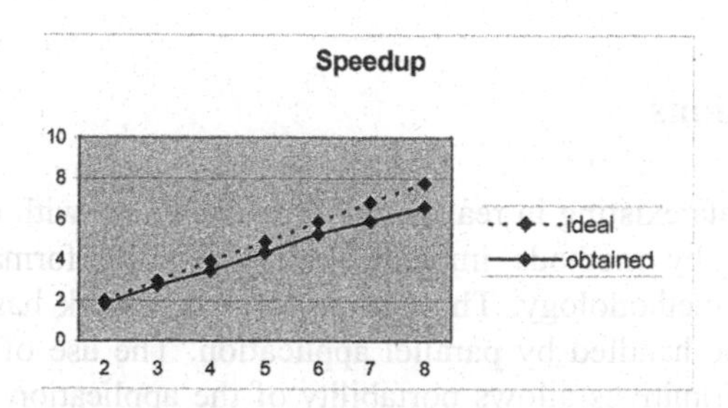


**Fig. 2.** Obtained efficiency

## 5 Classification Task

To demonstrate the efficiency of the genetic algorithm in the optimization of database classification, were used three small well known databases: Iris, Pima and Erudit. Table 2 shows the descriptions of each one of the analyzed databases:

**Table 2.** Databases descriptions

| Database | Samples | Variables | Classes |
|---|---|---|---|
| Iris Plant Database | 150 | 4 | 3 |
| Pima Indians Diabetes | 768 | 8 | 2 |
| Erudit | | 27 | 2 |

In a first step, the databases had been analyzed without the aid of the genetic algorithm, getting itself, therefore, results on this side of the waited one. In a second step, the method KNN-Fuzzy was optimized by a genetic algorithm to find the ideal values of the classification parameters: K, m and weights of the variable. Table 3 presents the obtained classification results.

**Table 3.** Comparison between standard and GA strategies

| Database | Standard | | | GA | | |
|---|---|---|---|---|---|---|
| | k | m | % | k | m | % |
| Iris | 90 | 2 | 97.0 | 30 | 0.025781 | 100 |
| Pima | 384 | 3 | 79.0 | 56 | 0.0202148 | 89.0 |
| Erudit | 130 | 2 | 66.0 | 114 | 0.004297 | 94.4 |

As it can be observed, a significant improvement in the correct classification rate was achieved in all studied cases.

## 6 Conclusions

The volume of data existing in real world data warehouse with complex relations can only be handled by methods implemented in high performance computers with efficient parallel methodology. The case used in this work has these characteristics and could only be handled by parallel application. The use of a standard library of communication primitives allows portability of the application for several machines, bringing great vantages for the methodology. The KNN-Fuzzy method, optimized by a genetic algorithm, has the ability of produce a very precise classification model comparing with other paradigms used in classification problems. The obtained performance of the parallel implementation has shown very good results.

For classification tasks, the KNN-fuzzy method showed to be very useful. It improves the accuracy and results in high classification precision, not obtained with crisp implementations.

The genetic algorithm optimization has great importance in the improvement of the accuracy, finding optimal values of K and m.

The generation of the population of chromosomes in a random form showed, in practical, to be more attractive than the exploitation of previous populations.

Depending on the generation of the population, values close or not the global minimum can occur, resulting in a variation in the processing efforts.

## References


1. ANDERBERG, MICHAEL R. Cluster Analysis for Applications. Academic Press, 1973.
2. MITCHIE, D., SPIEGELHALTER, D. J., TAYLOR, C. C. Machine Learning, Neural and Statistical Classification. Ellis Horwood, 1994.
3. MANLY, B. F. J. Multivariate Statistical Methods. A Primer , Chapman & Hall, London, 1986.
4. KELLER, J. M. , GRAY, M. R. , GIVENS JR. , J. A. , A Fuzzy K-Nearest Neighbor Algorithm, IEEE Transactions on Systems, Man, and Cybernetics, v. SMC-15, n. 4, pp. 258-260, July/August 1985.
5. HEATH, F. G. , Origins of Binary Code, Scientific American, v.227, n.2, pp.76-83, August 1972.
6. ESPÍNDOLA, ROGÉRIO P. Optimizing Classification Fuzzy Rules by Genetic Algorithms, M.Sc thesys, COPPE/PEC, May, 1999.
7. LAM / MPI Parallel Computing home page. Internet: http://www.lam-mpi.org , 2002.
8. Databases from UCI Machine Learning Repository obtained in http://www.ics.uci.edu/~mlearn/ MLRepository.html
9. Databases from European Network in Uncertainty Techniques Developments for Use in Information Technology - ERUDIT, obtained in http://www.erudit.de/erudit/index.htm
10. KDD-Sisyphus. KDD Sisyphus Workshop – Data Preparation, Preprocessing and Reasoning for Real-World Data Mining Applications. Internet: http://research.swisslife.ch/kdd-sisyphus, 1998.

# Chapter 3

# Computing in Chemistry and Biology

# Lignin Biosynthesis and Degradation – a Major Challenge for Computational Chemistry

Bo Durbeej[1,2], Yan-Ni Wang[1,3], and Leif A. Eriksson[1]

[1] Division of Structural and Computational Biophysics
Department of Biochemistry, Box 576, Uppsala University
751 23 Uppsala, Sweden
[2] Department of Quantum Chemistry, Box 518, Uppsala University
751 20 Uppsala, Sweden
[3] currently at: Advanced Biomedical Computing Center
SAIC Frederick, P.O. Box B, National Cancer Institute at Frederick
Frederick, MD 21702-1201, USA

**Abstract.** In the present chapter we review a series of computational studies related to lignin biosynthesis and degradation, with the aim to understand at a molecular level processes crucial to paper and pulp industries. Due to the complexity of the problem, a wide variety of computational approaches are employed, each with its own merits. From the theoretical studies we are able to draw conclusions regarding the behavior of lignol monomers and their corresponding dehydrogenated radicals in aqueous solution and in lipid bilayers, and reaction mechanisms, conformations and relative stabilities of lignol dimers.

## 1 Introduction

Despite the global economic significance of pulp and paper industries, computational chemistry techniques have to date been very little employed in order to increase our understanding of processes pertaining to biosynthesis and degradation of wood fibers and their constituents. The main reason for this is the complex constitution of wood fibers, putting very high demands on the computational method to be used both with respect to applicability and size of the systems that in the end can be explored at high accuracy. To this should also be added the limited amount of available high-resolution (*i.e.* atomic level) structural data, which makes construction of computational models a difficult task.

A wood fiber cell wall is composed of bundles of cellulose polymers, so-called microrfibrils, which are synthesized by trans-membrane enzymes arranged in a 'rosette-like' fashion, Fig. 1 [1]. Each microfibril normally contains 20-40 cellulose polymers, has a cross-section diameter of 20-30 μm, and is coated with a layer of branched chains of hemicelluloses [2, 3]. The vacant regions between the microfibrils are filled with very low-structured lignin polymers, providing rigidity (*i.e.* mechanical strength) to the plant. In addition, lignin reduces the water permeability (by increasing the hydrophobicity) of the cell walls, and participates in the plant's defense as a physicochemical barrier against pathogens [4].

The wood cell wall as such also has an overall complex structure. In the simplest possible picture, the cell walls formed in higher plants can be thought of as comprising four major morphological regions, differing in the relative concentration of cellulose, hemicellulose and lignin as well as the relative orientation of the cellulose microfibrils (Fig. 2) [5].

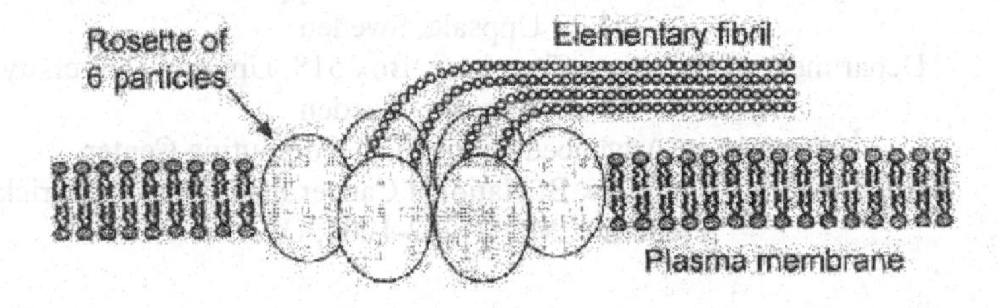


**Fig. 1.** Schematic model of cellulose synthesizing enzymes in a wood fiber cell wall (from ref [1])

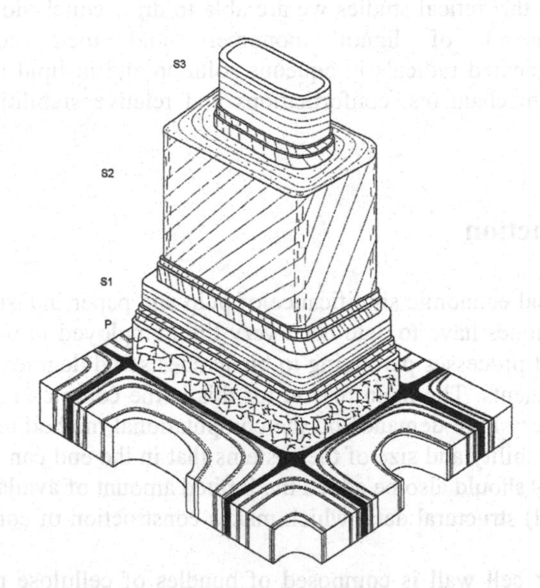


**Fig. 2.** Schematic model of the different morphological regions of a wood fiber cell wall. P stands for the primary cell wall, and S1, S2, and S3 stand for the outer, middle, and inner layers of the secondary cell wall, respectively (reprinted by courtesy of the Wood Ultrastructure Research Center (WURC) – www-wurc.slu.se)

In this review, we will try to illustrate how different computational chemistry techniques – ranging from classical molecular dynamics simulations to quantum chemical methods – can be employed to provide valuable new information on, in particular, various aspects of lignin biosynthesis. We will present results from recent studies performed in our group focusing on the initial stages of this process.

Besides playing a crucial role in wood anatomy and physiology, lignin is also important from an economical point of view. In paper making, it is considered an undesirable component – yielding paper with poor printing quality that readily fragments and turns yellow with age [6] – that has to be removed by alkaline pulping and chemical bleaching. Even though the present work, as noted above, primarily concerns lignin biosynthesis, we hope it will be evident how computational studies of this type also can benefit industry focussing on lignin degradation.

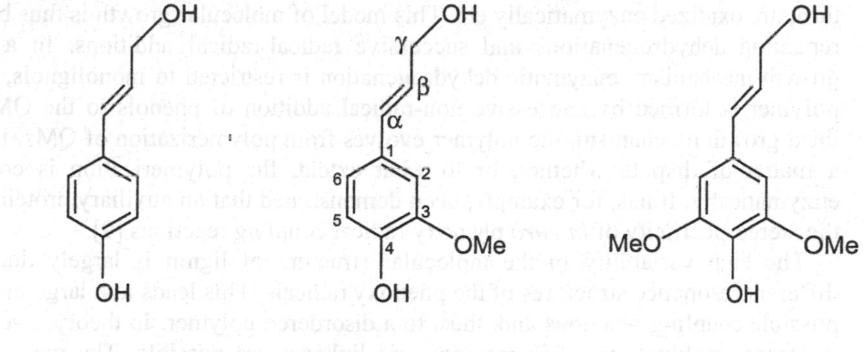

**Fig. 3.** The three different hydroxycinnamyl alcohols (*p*-coumaryl, coniferyl and sinapyl alcohol) that form the building blocks of lignin biopolymers

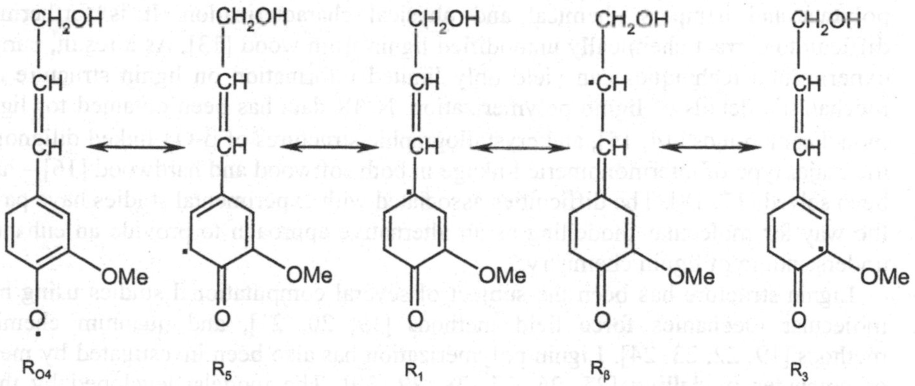

**Fig. 4.** Main mesomeric forms of a phenoxy radical produced by oxidation of a coniferyl alcohol by peroxidase

The biosynthesis of the lignin polymer is initiated by enzymatic dehydrogenation of hydroxycinnamyl alcohols (monolignols) of three different type : *p*-coumaryl alcohol, coniferyl alcohol, and sinapyl alcohol (Fig. 3) [7, 8]. These monolignols are oxidized by peroxidases and/or laccases to form phenoxy radicals that have five different resonance structures (Fig. 4). It has been proposed that the phenoxy radicals couple non-enzymatically to produce quinone methides (QMs) as reactive intermediates [7, 8]. The dimeric QMs are then converted into various dilignols by addition of water, or by intra-molecular nucleophilic attack by primary alcohols or quinone groups. The mechanism by which lignin polymerization proceeds has not yet been fully understood, and several proposals exist (ref [7], and references therein). All dilignols are phenols, and can therefore also undergo enzymatic dehydrogenation to form dilignol radicals. Hence, in one model of lignin growth, dilignol radicals are proposed to couple with phenoxy radicals to produce phenolic trilignols which, in turn, are oxidized enzymatically etc. This model of molecular growth is thus based on repeating dehydrogenations and successive radical-radical additions. In a second growth mechanism, enzymatic dehydrogenation is restricted to monolignols, and the polymer is formed by successive non-radical addition of phenols to the QMs. In a third growth mechanism, the polymer evolves from polymerization of QMs. It is also a matter of dispute whether, or to what extent, the polymerization is controlled enzymatically. It has, for example, been demonstrated that an auxiliary protein affects the stereospecificity of *in vitro* phenoxy radical coupling reactions [9].

The high variability in the molecular structure of lignin is largely due to the different resonance structures of the phenoxy radicals. This leads to a large number of possible coupling reactions and, thus, to a disordered polymer. In theory, twenty-five different combinations of intermonomeric linkages are possible. The most frequent linkages in both softwood and hardwood are β-O4, β-5, β-β, β-1, 5-5, and 5-O4 (ref [10], and references therein; cf. Fig. 5). The relative distribution of the linkages is however not clearly established, since different analytical methods give different results [11, 12].

The complex constitution of lignin makes it difficult to solubilize the whole polymer and hampers chemical and physical characterization. It is furthermore difficult to extract chemically unmodified lignin from wood [13]. As a result, current experimental techniques can yield only limited information on lignin structure and mechanistic details of lignin polymerization. NMR data has been obtained for lignin model compounds [14, 15], and crystallographic structures of β-O4 linked dilignols – the major type of intermonomeric linkage in both softwood and hardwood [16] – have been solved [17, 18]. The difficulties associated with experimental studies have paved the way for molecular modelling as an alternative approach to provide an enhanced understanding of lignin chemistry.

Lignin structure has been the subject of several computational studies using both molecular mechanics force field methods [19, 20, 21], and quantum chemical methods [19, 22, 23, 24]. Lignin polymerization has also been investigated by means of computer modelling [25, 26, 27, 28, 29, 30]. The models developed by these researchers account for many experimental findings – the Monte Carlo simulation strategy of Roussel and Lim [29, 30], *e.g.*, reproduce both the distribution of intermonomeric couplings and the number of bonds per monomer rather well – and

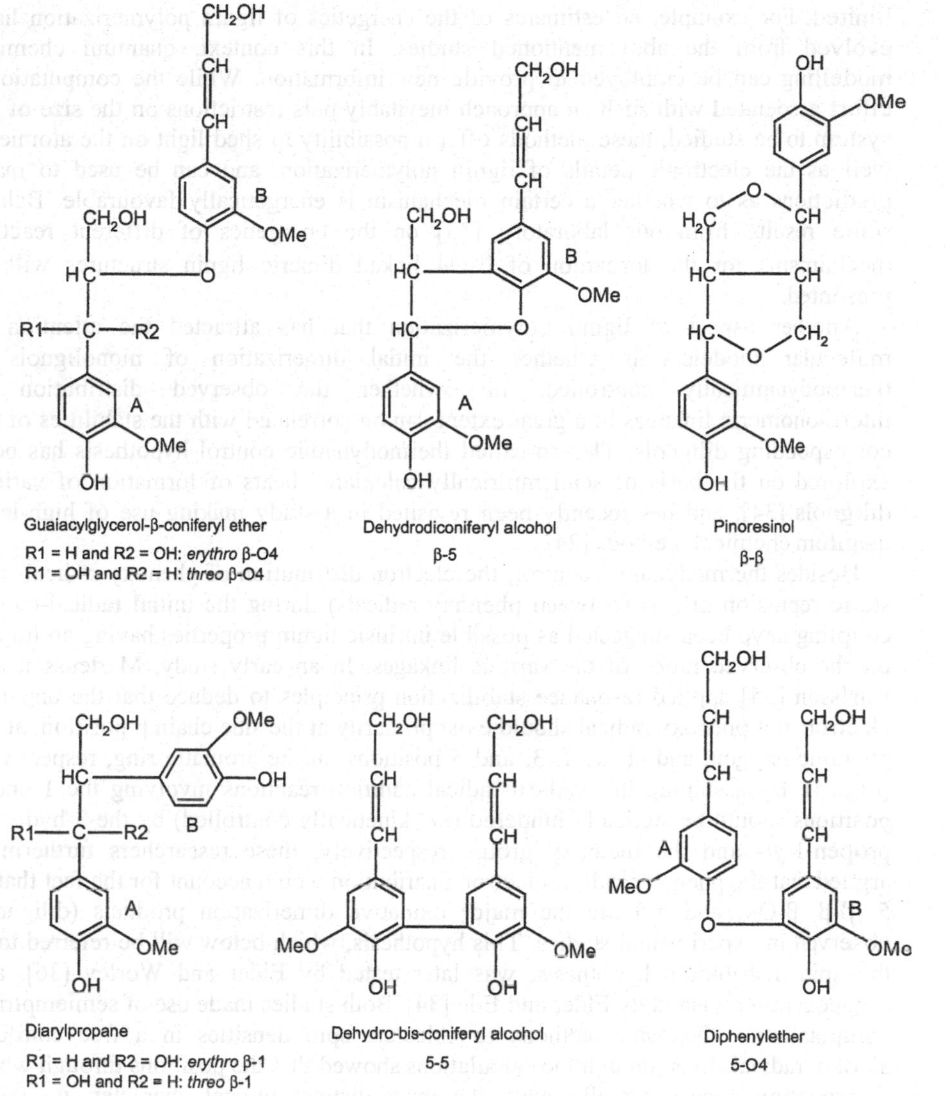

CH₂OH CH CH B CH₂OH HC O R1 C R2 A OMe OH OMe

Guaiacylglycerol-β-coniferyl ether

R1 = H and R2 = OH: *erythro* β-O4
R1 = OH and R2 = H: *threo* β-O4

Dehydrodiconiferyl alcohol

β-5

Pinoresinol

β-β

Diarylpropane

R1 = H and R2 = OH: *erythro* β-1
R1 = OH and R2 = H: *threo* β-1

Dehydro-bis-coniferyl alcohol

5-5

Diphenylether

5-O4

**Fig. 5.** Dimeric lignin structures containing the principal linkage modes between coniferyl alcohol monolignols. The acronyms reflect the sites of initial radical-radical coupling, *e.g.* the β-O4 dilignol is formed by coupling of the mesomeric forms $R_{\beta}$ and $R_{O4}$ of Figure 4

enable investigations of comparatively large systems. None of these models, however, have included a three-dimensional representation of the atomic structure of lignin. Even though a subsequent model devised by Jurasek [31, 32] attempts to make use of realistic lignin structures, and indeed provides valuable new insight into the process of lignification, the knowledge of the mechanistic details at the atomic level remains limited. For example, no estimates of the energetics of lignin polymerization have evolved from the abovementioned studies. In this context, quantum chemical modelling can be employed to provide new information. While the computational effort associated with such an approach inevitably puts restrictions on the size of the system to be studied, these methods offer a possibility to shed light on the atomic as well as the electronic details of lignin polymerization, and can be used to make predictions as to whether a certain mechanism is energetically favourable. Below, some results from our laboratory [33] on the energetics of different reaction mechanisms for the formation of β-O4 linked dimeric lignin structures will be presented.

Another aspect of lignin polymerization that has attracted the attention of molecular modelers is whether the initial dimerization of monolignols is thermodynamically controlled, *i.e.* whether the observed distribution of intermonomeric linkages to a great extent can be correlated with the stabilities of the corresponding dilignols. This so-called thermodynamic control hypothesis has been explored on the basis of semiempirically calculated heats of formation of various dilignols [34], and has recently been revisited in a study making use of high-level quantum chemical methods [24].

Besides thermodynamic control, the electron distribution of phenoxy radicals and steric repulsion effects (between phenoxy radicals) during the initial radical-radical coupling have been suggested as possible intrinsic lignin properties having an impact on the observed ratios of the various linkages. In an early study, Mårtensson and Karlsson [35] applied resonance stabilization principles to deduce that the unpaired electron of a phenoxy radical should exist primarily at the side chain β position, at the phenolic oxygen, and at the 1, 3, and 5 positions on the aromatic ring, respectively (Fig. 4). By assuming that radical-radical addition reactions involving the 1 and 3 positions should be sterically hindered (*i.e.* kinetically controlled) by the 3-hydroxy-propen-1-yl- and the methoxy group, respectively, these researchers furthermore argued that the phenoxy radical electron distribution would account for the fact that β-5, β-β, β-O4, and 5-5 are the major oxidative dimerization products (dilignols) observed in experimental studies. This hypothesis, which below will be referred to as the spin distribution hypothesis, was later tested by Elder and Worley [36], and subsequently revisited by Elder and Ede [34]. Both studies made use of semiempirical computational chemistry methods to calculate spin densities in a free coniferyl alcohol radical. Even though the calculations showed that the positions through which dimerization occurs actually have the most distinct radical character, no direct correlation between calculated spin densities and linkage ratios could be pointed out. Below, the first evaluation of the spin distribution hypothesis that goes beyond a semiempirical level of approximation will be presented [37].

The effect of the biological environment on the polymerization process is another important aspect of lignin biosynthesis that needs to be further investigated. The various monolignols are synthesized intracellularly, and thus have to cross the plasma membrane to enter the cell wall. However, it still remains unclear how this transport occurs, and how the monolignols aggregate in various biological environments. As a first step towards gaining insight into these issues, we will below present some preliminary results from molecular dynamics simulations of monolignols in aqueous solution and in a lipid bilayer [38, 39].

The structural complexity of lignin makes lignin degradation in industrial processes a difficult task. Since certain intermonomeric linkages are stronger than others, a better understanding of the chemistry of the different linkages would shed light on how these may be weakened (*i.e.* more susceptible to cleavage) through chemical modification. Such information could be utilized to optimize the conditions for the pulping and bleaching processes. In this context, it should also be noted that enzymatic and photochemical bleaching have been proposed as alternatives to the traditional bleaching chemicals chlorine ($Cl_2$), chlorine dioxide ($ClO_2$), hydrogen peroxide ($H_2O_2$) and similar [40, 41, 42].

The present chapter is organized as follows. In the next section, the computational techniques used in the different studies are outlined. In sections 3 and 4, results from molecular dynamics simulations of monolignols and their corresponding dehydrogenated radicals in aqueous solution and in a lipid bilayer are presented. In section 5, structural features of the most common lignin dimerization products are discussed, and the thermodynamic control and spin distribution hypotheses are evaluated. In section 6, finally, an investigation of the energetics of different reaction mechanisms for the formation of β-O4 linked dimeric lignin structures is presented.

The systems explored range in size from those containing thirty or so atoms to those with several thousands of atoms, and the aim with this review is to illustrate that a combination of different methodologies – each with its own advantages, bottlenecks, and computational requirements – is required if one wishes to study a complex biochemical problem such as lignin biosynthesis using the tools of computational chemistry.

## 2 Computational Approaches

When embarking on a computational project, the choice of computational scheme will depend upon the nature of the chemical problem at hand, and the applicability of the method in terms of accuracy and computational cost. If, for example, we are interested in studying a biochemical system consisting of, say, 10000 atoms, the method of choice is to run a classical molecular dynamics (MD) simulation of the system. The result of such a simulation is a trajectory, which shows how the atomic positions vary in time. If the simulation is carried out long enough, the system will eventually reach an equilibrium state with respect to certain thermodynamic and structural properties. By averaging over an equilibrium trajectory, many macroscopic physical properties can subsequently be calculated. In a system consisting of an enzyme and its corresponding substrate, or a solute in a liquid, this methodology offers a possibility to calculate, *e.g.*, the binding constant of the substrate to the enzyme and the radial distribution function between specific solute/solvent atom pairs, respectively. MD simulations also provide a means to calculate non-equilibrium macroscopic properties such as diffusion constants of molecules within biological membranes.

A key problem within this branch of molecular modeling is that of choosing an adequate force field by which the different bonded and non-bonded interactions are described through potential energy functions. These functions depend on the atomic positions only, *i.e.* electrons are not considered explicitly. This means that MD simulations are applicable only to systems that remain in their electronic ground state, and furthermore that a proper treatment of chemical reactions is beyond reach.

In the work presented in section 3, we have performed MD simulations in order to investigate the behavior of the three different monolignols, and their corresponding phenoxy radicals, in aqueous solution. The force field by Cornell *et al.* [43] was applied to all atoms except to those of the unsaturated hydrocarbon tail, which were described using the MM3 force field [44]. The TIP3P water solvent model was used throughout all simulations [45]. Force field charges for the monolignol atoms were obtained from initial quantum chemical calculations (level of theory given in section 3 below).

In order to gain insight into the trans-membrane transport of lignin precursors, MD simulations of *p*-coumaryl alcohol – the smallest monolignol – in a dipalmitoyl-phosphatidylcholine (DPPC) membrane were also carried out; the results of which are presented in section 4. In this case, the monolignol was described using the AMBER force field [46], whereas a refined version thereof [47] previously employed in MD simulations of membranes was applied to the DPPC model [48]. Again, the TIP3P water model was used and monolignol charges were obtained from initial quantum chemical calculations.

All MD simulations described here were performed using the AMBER 5.0.1 simulation package [49], and further computational details thereof are given in the respective sections below.

In cases where we are interested in calculating electronic and spectroscopic properties of a certain chemical compound, or would like to study processes that involve the breaking and formation of covalent bonds (*i.e.* chemical reactions), a quantum mechanical description in which electrons are treated explicitly becomes mandatory. Including electronic interactions dramatically increases the complexity of the equations that in the end need to be solved. As a result, the number of atoms that can be included in a quantum chemical calculation typically lies in the 10-100 atoms range, depending on the methodology chosen. The more accurate the method, the more demanding the calculation will be. This implies that quantum chemical calculations generally are performed on small *model* systems, and that solvation effects are either neglected or taken into account through the use of so-called dielectric continuum models [50, 51, 52]. Based on existing benchmark studies, the quantum chemical method best-suited for the system under study can be selected and subsequently applied to provide information on explicit orbital interactions, estimates of bond strengths, charge distributions, spectroscopic properties, and similar; information that might be difficult to obtain experimentally. By using the highly robust geometry optimization algorithms of today, which have been implemented in several commercially available quantum chemical computer codes, it is furthermore possible to locate stationary points (*i.e.* minima and transition states) along the reaction coordinate of a chemical reaction. Thereby, the effects of different substituents on energy barriers and reaction energies can be calculated, and short-lived intermediates can be identified; a cumbersome, if not impossible, task from an experimental point of view.

The success of quantum chemical modeling (in terms of providing results in good agreement with experimental data) is partly a result of many chemical phenomena being highly localized. It is thereby possible to construct a simplified computational model that both captures the important features of the macroscopic system using only a small number of relevant chemical groups, and enables the calculations to be performed at a necessary high level of theory.

Of the many quantum chemical methods available today, those based on density functional theory (DFT) [53, 54] have over the past decade become increasingly more popular, and have been extensively used in fields as diverse as solid state physics, atmospheric chemistry, and biochemistry. A key theorem in DFT says that all ground-state properties of a many-body system, such as the total electronic energy $E_{tot}$, can be obtained from the ground-state electron density $\rho(\mathbf{r})$ of that particular system. We hence wish to find an approach where we can use use $\rho(\mathbf{r})$, which depends on 3 spatial coordinates (the value of the density at point **r**), rather than the wave function, which depends on 3N spatial coordinates (with N being the number of electrons and assuming fixed nuclear positions), as the basic variable for the description of the many-body system under study, and express the total electronic energy as a functional of the electron density; $E_{tot} = E[\rho(\mathbf{r})]$. In combination with a variational principle for the electron density, and a computational methodology known as the Kohn-Sham self-consistent field approach, a computational scheme has been developed that is considerably less complicated than those arising within wave-function formalism. In addition to being applicable to larger systems, DFT methods already from the start take electron correlation effects into account through the so-called *exchange-correlation potential*, whereas conventional *ab initio* wave-function methods include these effects through perturbation series expansions or increasing orders of excited state configurations added to the 'zeroth-order' Hartree-Fock solution. Thus, DFT methods can be used to perform accurate calculations at a comparatively low computational cost. For more details on DFT we refer to any of the recent textbooks on the subject [55, 56, 57].

In the quantum chemical studies presented in sections 5 and 6, we have exclusively employed the B3LYP hybrid density functional [58, 59, 60] as implemented in the Gaussian 98 program [61]. This exchange-correlation functional consists of a mixture of local and gradient corrected density functionals, and also includes a fraction of exact (Hartree-Fock) exchange. Bulk solvation effects of the surrounding medium are accounted for using the self-consistent polarized continuum model (PCM) by Tomasi and coworkers [50]. In this model, the surrounding medium is treated as a macroscopic continuum (with a dielectric constant $\varepsilon$), and the molecule under study is thought of as filling a cavity in this continuous medium. In order to mimic an aliphatic environment, $\varepsilon$ was chosen to be equal to 4. Further computational details are given in the respective sections below.

## 3 MD Simulations of Monolignols and the Corresponding Phenoxy Radicals in Aqueous Solution

In order to better understand the hydrogen bonding and other interactions between the monolignols and their local surrounding, we undertook a detailed study of the three different hydroxycinnamyl alcohols and their corresponding radical forms in aqueous solution. The geometries of the different monolignols were first optimized using the

B3LYP functional in conjunction with the 6-31G(d,p) basis set. The optimized structures were then placed into rectangular water boxes with periodic boundary conditions, followed by initial energy minimization to remove the strain between the solute and surrounding water molecules. After equilibration simulations, the temperature was raised stepwise, and finally kept at 300 K using the Berendsen coupling algorithm [62]. A single temperature scaling factor was used for all atoms. The pressure was set to 1 bar (ca. 1 atm.) with isotropic position scaling and 500 ps relaxation time. The density of the equilibrated system was about 1.0 g/cm$^3$. A 4 ns NPT ensemble production run was finally performed with 2 fs step size. A 9 Å atom based cut-off was used to separate short-range and long-range interactions, and Ewald summation [63] was employed to calculate long-range electrostatic interactions. The non-bonded pair list was updated every 10th time step.

For the corresponding phenoxy radicals, similar energy minimization and equilibration simulations were carried out, except that the computationally faster Particle Mesh Ewald (PME) method [64, 65] was used to calculate long range electrostatic interactions. Finally 1 ns periodic boundary NPT production simulations were performed to analyze the behaviour of the radicals in solution.

Lignin biosynthesis is a process involving dehydrogenated precursors, and thus the interaction between the hydroxyl groups and its environment is of key importance. In the current simulations information about the intermolecular solute-solvent atom pairs were analysed in terms of radial distributions functions (RDF) – data that provide information about the solvation shells and hydrogen bonding interaction.

In Fig. 6 and 7 we show the RDF between the phenolic hydrogen and the oxygens of the surrounding water molecules, and between the phenolic oxygen of the corresponding dehydrogenated radicals and hydrogen of water, respectively. The simulations of the parent compounds (Fig. 6) reveal that for p-coumaryl alcohol there is a strong hydrogen bond formed to the hydroxyl hydrogen, ca 1.8 Å long and with an average number of coordinated solvation oxygen atoms of 1.2. For both coniferyl and sinapyl alcohol much weaker hydrogen bonds are found (at 1.9 and 2.0 Å average distance, respectively). These results clearly reflect that the neighbouring methoxy groups influence the interactions between lignin and the surrounding significantly by hindering the water molecules to approach the phenolic end, hence reducing the possibility of hydrogen bonding. The same results were found when analysing the O4 - - $O_{water}$ RDF (not shown). The second solvation shell is seen as a broad shoulder around 3.4 Å, and is similar for all three monolignols.

Turning to the dehydrogenated radicals (Fig. 7), the first peak of the O4 - - $H_{water}$ RDF display the same trends as for the parent compounds, albeit not quite as pronounced. For p-coumaryl alcohol the first hydrogen bonding shell appears at 1.8 Å (0.8 hydrogens), whereas for coniferyl and sinapyl the corresponding data are 1.9 Å (0.7 hydrogens) and 2.0 Å (0.6 hydrogens). Again steric hindrance is excerted by the increasing number of rotating methoxy groups in the series p-coumaryl (no OMe group), coniferyl (1 OMe) and sinapyl (2 OMe) alcohol.

Similar analyses were made for the γ-hydroxy groups (of the 'tail' of the compounds). In this case all three systems, radical as well as non-radical, display identical RDF. The first peak in the $H_\gamma$ - - $O_{water}$ RDF appears at 1.9 Å, with magnitude 1.0, and with the second solvation shell at 3.4 Å. Common for all systems investigated is that all the hydrogen bonds are linear.

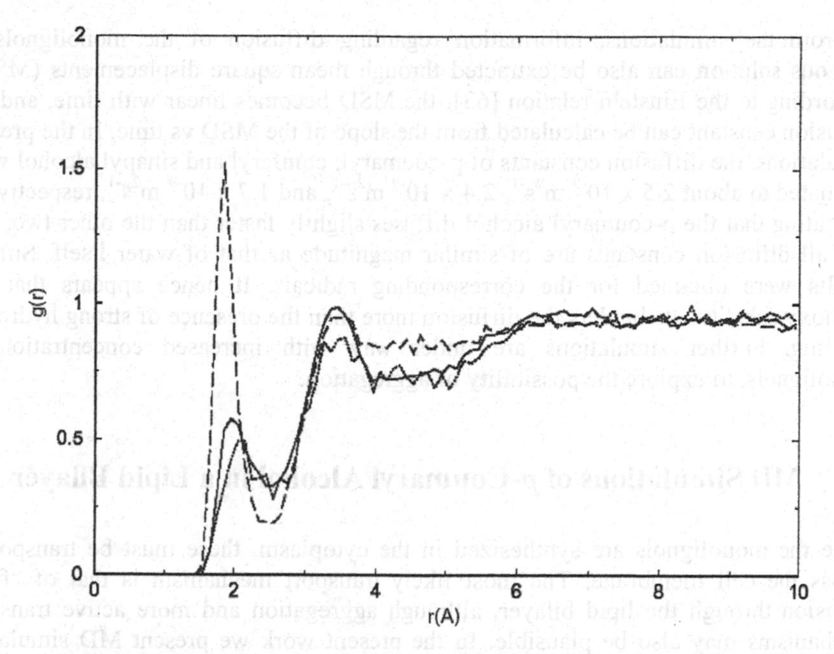


**Fig. 6.** Radial distribution functions between cinnamyl alcohol hydroxyl group H4 and oxygen of surrounding water molecules. Dashed line: p-coumaryl alcohol; solid line: coniferyl alcohol; dot-dashed line: sinapyl alcohol

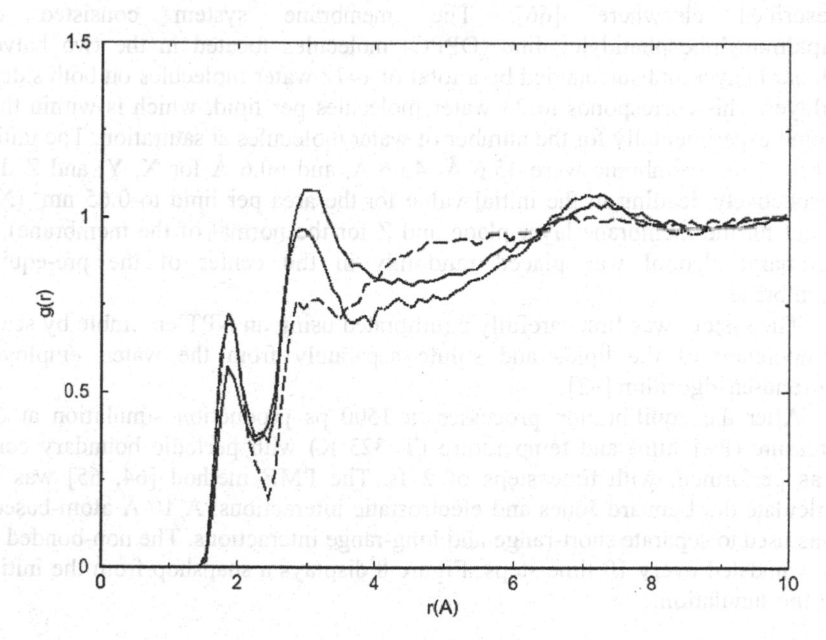


**Fig. 7.** Radial distribution functions between cinnamyl alcoholyl radical O4 and hydrogen of surrounding water molecules. Dashed line: p-coumaryl alcohol; solid line: coniferyl alcohol; dot-dashed line: sinapyl alcohol

From the simulations, information regarding diffusion of the monolignols in aqueous solution can also be extracted through mean square displacements (MSD). According to the Einstein relation [63], the MSD becomes linear with time, and the diffusion constant can be calculated from the slope of the MSD vs time. In the present simulations, the diffusion constants of p-coumaryl, coniferyl and sinapyl alcohol were estimated to about 2.5 x $10^{-9}$ $m^2s^{-1}$, 2.4 x $10^{-9}$ $m^2s^{-1}$, and 1.7 x $10^{-9}$ $m^2s^{-1}$, respectively, indicating that the p-coumaryl alcohol diffuses slightly faster than the other two, and that all diffusion constants are of similar magnitude as that of water itself. Similar results were obtained for the corresponding radicals. It hence appears that the methoxy substituents hinders the diffusion more than the presence of strong hydrogen bonding. Further simulations are under way with increased concentration of monolignols, to explore the possibility of aggregation.

# 4 MD Simulations of *p*-Coumaryl Alcohol in a Lipid Bilayer

Once the monolignols are synthesized in the cytoplasm, these must be transported across the cell membrane. The most likely transport mechanism is that of "free" diffusion through the lipid bilayer, although aggregation and more active transport mechanisms may also be plausible. In the present work we present MD simulation data for passive motion of *p*-coumaryl alcohol in a lipid bilayer. The initial lipid model was obtained by extracting a single configuration from a previous well-equilibrated simulation [48]. The details of the model and the simulations have been described elsewhere [66]. The membrane system consisted of 64 dipalmitoylphosphatidylcholine (DPPC) molecules located in the two halves of a planar bilayer and surrounded by a total of 1472 water molecules on both sides of the bilayer. This corresponds to 23 water molecules per lipid, which is within the range found experimentally for the number of water molecules at saturation. The initial edge sizes of the membrane were 45.6 Å, 45.6 Å, and 60.6 Å for X, Y, and Z direction respectively, leading to the initial value for the area per lipid to 0.65 $nm^2$ (X and Y stand for the membrane layer plane and Z for the normal of the membrane). The *p*-coumaryl alcohol was placed randomly in the center of the pre-equilibrated membrane.

The system was first carefully equilibrated using an NPT ensemble by scaling the temperature of the lipids and solute separately from the water, employing the Berendsen algorithm [62].

After the equilibration processes, a 1500 ps production simulation at constant pressure (P=1 atm) and temperature (T=323 K) with periodic boundary conditions was performed, with time steps of 2 fs. The PME method [64, 65] was used to calculate the Lennard-Jones and electrostatic interactions. A 10 Å atom-based cutoff was used to separate short-range and long-range interactions. The non-bonded pair list was updated every 10 time steps. Figure 8 displays a snapshop from the initial stage of the simulation.

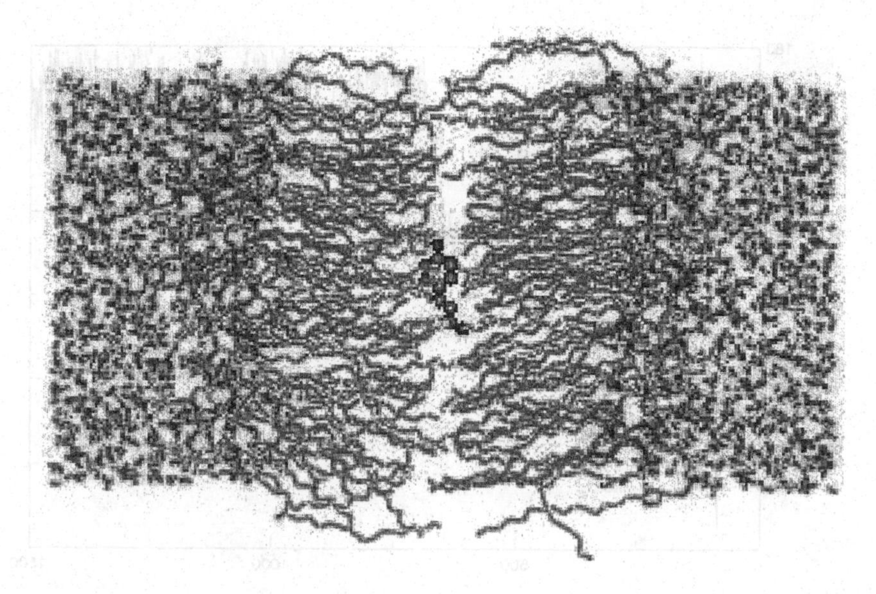

**Fig. 8.** Snapshop from the initial stage of the p-coumaryl – DPPC simulation. The monolignol is seen in 'ball-and-stick' image at the center of the bilayer

The reorientation of the lignol was analysed through the relative orientation of the C1-C4 axis and the axis formed by the mass centers of the two lipid bilayers. The angle between the two axes as a function of time is shown in Fig. 9. From the analysis it is clearly seen that the lignol very rapidly rotates/reorients to a situation where it is parallel to the norm of the bilayer (angle $\phi \approx 180°$), with the phenoxyl head pointing towards the lipid-water interface. Rotation of the entire molecule and its constituent groups were also investigated. It was concluded that the lignol rotates very slowly with time, that the molecule as such is rigid but that the two OH groups rotate freely.

Diffusion of the *p*-coumaryl alcohol inside the lipid bilayer was analysed in terms of the Einstein relation [63], and showed that the rate of diffusion was essentially constant at $3 - 4 \times 10^{-7}$ $cm^2/s$ during the entire 1500 ps simulation. The only exception was a segment in the time interval 250-500 ps from the start, when the monolignol rapidly diffused in the direction of the norm of the bilayer at a rate ca 10 times higher than during the rest of the simulation. According to previous experimental and theoretical studies the diffusion constants in lipid bilayers of solutes of similar size as here are on the order of $10^{-6}$ $cm^2/s$, whereas larger, more complex solutes diffuse in lipid bilayers at rates of the order $10^{-7} - 10^{-8}$ $cm^2/s$ [67, 68, 69]. The present data is hence consistent with previous observations.

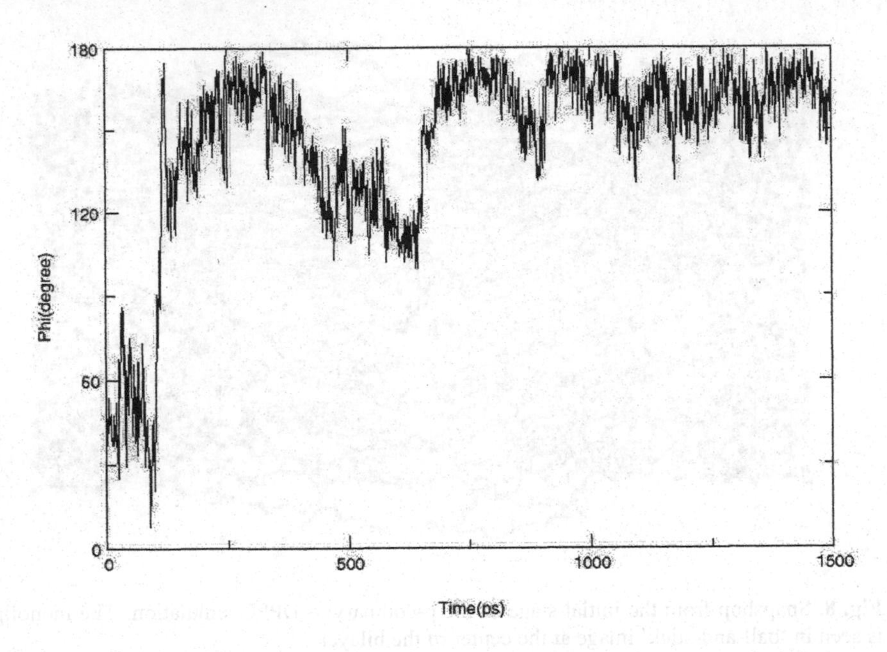


**Fig. 9.** Angle between the monolignol C1-C4 axis and the norm of the bilayer as function of time

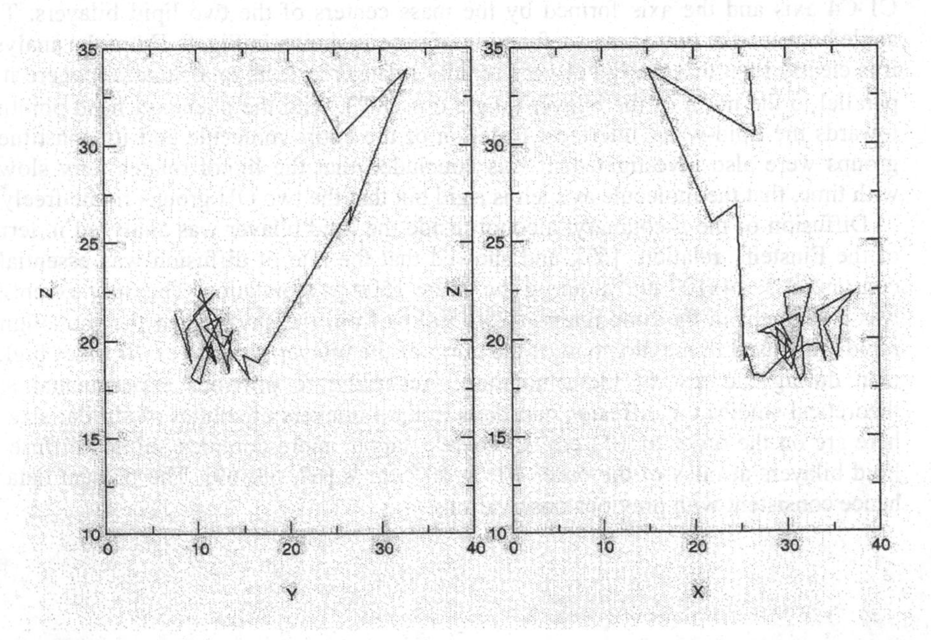


**Fig. 10.** Diffusion movement of the *p*-coumaryl alcohol in DPPC bilayer, in (left) Y-Z direction and (right) X-Z direction, in 50ps intervals. Simulations start at the top of the figures. X and Y denote the plane and Z the norm of the bilayer

The diffusion was also mapped in terms of the relative position of O4 (the phenoxylic oxygen) in the XZ and YZ planes every 50 ps of the simulation, Fig. 10. These maps display that the rapid diffusion movement is correlated with a fast motion in the Z-direction, i.e. along the norm of the bilayer. The movement corresponds to ca 15 Å distance, which places the monolignol in the polar head-group region of the phospholipids. The present data shows that the lignin monomers move much faster in the hydrophobic (inner) region of the bilayer than in the head group region with $D_{center}/D_{head} \approx 10$. The barrier to trans-membrane diffusion of monolignols thus appears to be the polar head group region – again consistently with other simulations of small solutes in lipid bilayers. Further studies of larger (di-, tri-) lignols, and of a higher concentration monolignols within the bilayer are currently under way.

### 4.1 Structural Features of Lignin Dimerization Products

Once the monolignols have entered the cell wall region, polymerization will take place. We have here investigated two different hypotheses for the initial stages of dimerisation, the thermodynamic and spin control hypetheses, and begin with an outline of the possible dimers that can be formed from coniferyl alcohol. Six different coniferyl alcohol dimerization products were considered (β-O4, β-5, β-β, β-1, 5-5, and 5-O4, see Fig. 5) in the study of lignin structure [24]. Standard bond lengths and angles were used to construct an initial conformational data set consisting of one unique starting geometry for 5-5 and 5-O4, *threo* and *erythro* β-O4 and β-1 diastereomers (Fig. 5), *cis* and *trans* β-5 stereoisomers, and several β-β stereoisomers with the $C_\beta$ hydrogens arranged either *cis* or *trans*. These geometries were then subjected to a systematic conformational search with respect to key dihedral angles, using the semiempirical PM3 method [70, 71]. The lowest-energy conformers from each search were identified and used as input in subsequent B3LYP-DFT calculations.

The geometries were first optimized with the small 3-21G basis set, whereby only a minority of the conformers obtained at the PM3 level survived. The remaining structures were refined through geometry optimizations using the larger 6-31G(d,p) basis set. In order to reduce the computational effort, only those conformers within each dimerization product having a relative energy within 10 kcal $mol^{-1}$ of the most stable one at the B3LYP/3-21G level were considered in these calculations. Altogether, geometry optimizations of thirty-two conformers (eight β-O4 conformers, five β-5, six β-β, five β-1, four 5-5, and four 5-O4 conformers, respectively) were carried out. Zero-point vibrational energy (ZPE) corrections were calculated with the 6-31G(d,p) basis set, also providing a control that the stationary points obtained were in fact true minima. Finally, better relative energies were evaluated using the larger 6-311G(2df,p) basis set and the PCM solvent approximation.

Figures 11 and 12 display the lowest-energy conformer of each dilignol obtained from the above calculations, and Tables 1 and 2 list key geometric parameters. A common feature of the lowest-energy conformers of β-O4, β-5, β-1, and 5-5 is the presence of intra-molecular hydrogen bonding that significantly stabilizes each of these structures. In the preferred *erythro* β-O4 conformer, the A-ring (see fig. 5) γ-hydroxyl hydrogen bonds to the B-ring 3-methoxy, and the A-ring α-hydroxyl is

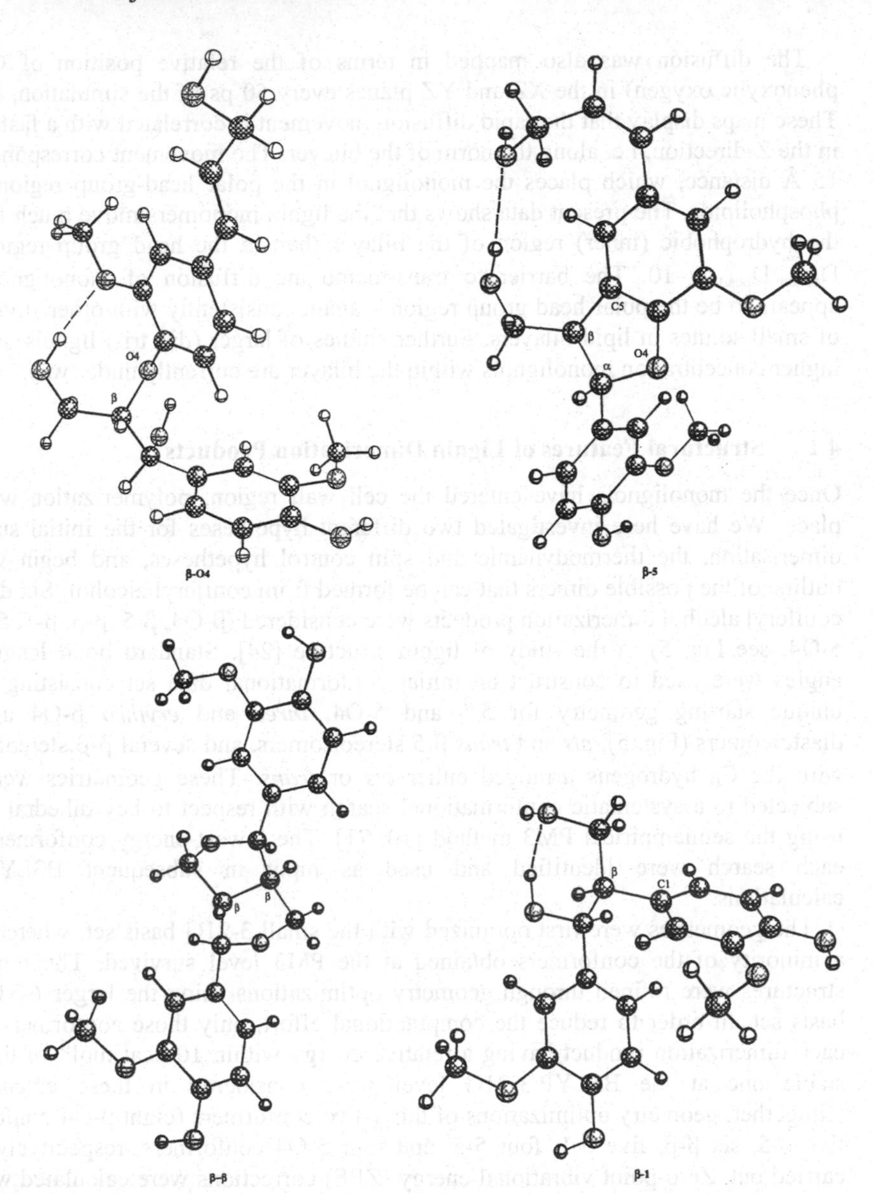


**Fig. 11.** Geometry-optimized structures of β-O4, β-5, β-β, and β-1 dilignols. Numerical values are listed in Tables 1 and 2

directed towards the B-ring ether oxygen. The most stable *threo* β-O4 conformer (not shown), in turn, has its A-ring α-hydroxyl hydrogen-bonded to the B-ring 3-methoxy and its A-ring γ-hydroxyl directed towards the B-ring ether oxygen. This conformer lies 6.5 kcal/mol higher in energy than the preferred *erythro* conformer, a result that confirms earlier experimental work showing that the *erythro* β-O4 is more stable than *threo* β-O4 [72].

In the lowest-energy conformer of the β-5 (*trans* β-5) and β-1 dilignols, the intra-molecular hydrogen bonding pattern is such that the A-ring γ-hydroxyl either acts as a hydrogen-bond donor to the B-ring γ-hydroxyl (β-5), or as a hydrogen-bond acceptor from the A-ring α-hydroxyl (β-1). The most stable 5-5 conformer, in turn, has the 4-hydroxyl of one of the aromtic rings hydrogen-bonded to the 4-hydroxyl of the other ring.

The lowest-energy β-β conformer has its $C_β$ hydrogens arranged in a *cis* conformation, and the methoxy and hydroxyl of the two 3-methoxy-4-hydroxyphenyl groups have similar orientations, being aligned in the plane of the corresponding aromatic ring (the C3-C4-O4-H4 and C2-C3-O3-CMe dihedrals are close to zero) with the hydroxyl directed towards and interacting with the lone electron pairs of the neighbouring methoxy oxygen. The A-ring 3-methoxy and 4-hydroxyl groups of the most stable conformer of β-O4, β-5, and β-1, as well as the B-ring 3-methoxy and 4-hydroxyl groups of β-1 and 5-04, are aligned in a similar fashion. Given that the calculations yielded a couple of 5-O4 conformers with the B-ring 4-hydroxyl hydrogen-bonded to the A-ring 3-methoxy, it is somewhat surprising that the most stable 5-O4 conformer actually lacks an intra-molecular hydrogen bond.

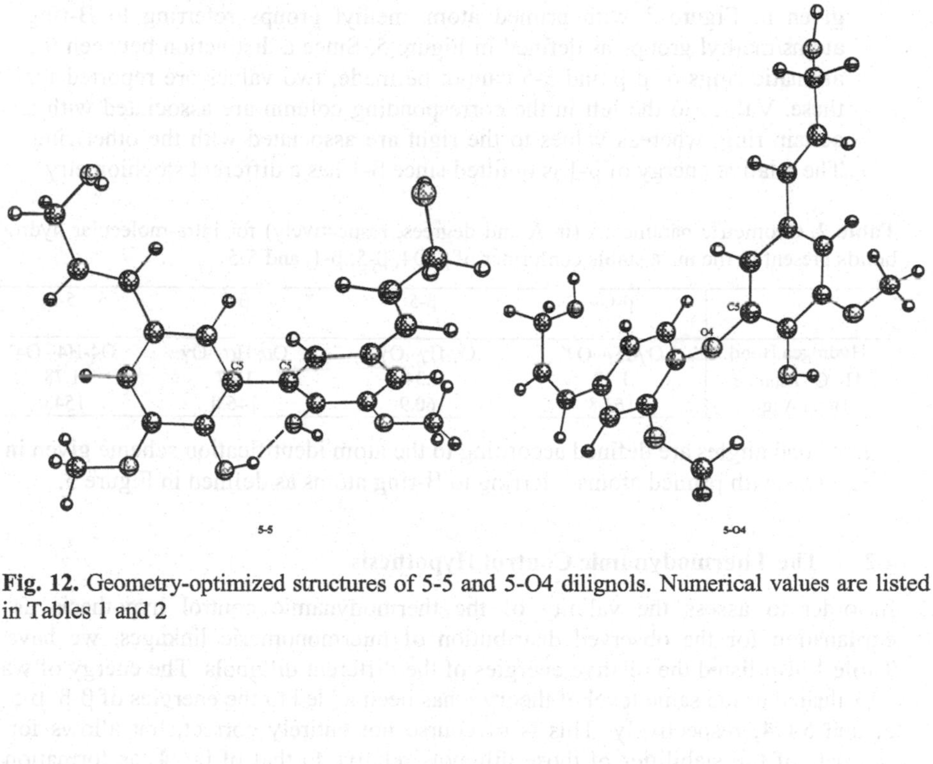

**Fig. 12.** Geometry-optimized structures of 5-5 and 5-O4 dilignols. Numerical values are listed in Tables 1 and 2

**Table 1.** Selected dihedral angles (in degrees) and zero-point corrected relative energies (in kcal mol$^{-1}$) for the most stable conformer of β-O4, β-5, β-β, β-1, 5-5, and 5-O4

| | β-O4 | β-5 | β-β | β-1 | 5-5 | 5-O4 |
|---|---|---|---|---|---|---|
| Relative Energy | (11.6) | 2.6 | 0.0 | - | 6.6 | 22.6 |
| Dihedral Angles | | | | | | |
| Cα-Cβ-C1′-C2′ | | | | −105.5 | | |
| Cα-Cβ-O4′-C4′ | −126.1 | | | | | |
| Cβ-Cα-C1-C2 | −115.1 | 77.5 | −132.8, 56.8 | 86.8 | | |
| Cβ-O4′-C4′- C5′ | 66.3 | | | | | |
| C3-C4-O4-H4 | −0.7 | 0.1 | −0.4, 0.5 | 2.0 | 144.2, 0.6 | |
| C3′-C4′-O4′-H4′ | | | | 1.5 | | −1.3 |
| C2-C3-O3-CMe | −2.0 | 0.3 | −2.0, 1.7 | 4.1 | −0.7, −0.1 | 110.2 |
| C2′-C3′-O3′-CMe′ | 0.3 | 0.5 | | 7.6 | | −1.0 |
| C3-C4-O4-C5′ | | | | | | 165.0 |
| C4-C5-C5′-C4′ | | | | | 48.0 | |
| C4-O4-C5′-C4′ | | | | | | −78.2 |
| Hα-Cα-Cβ-Hβ | | −145.4 | | −180.0 | | |
| Hβ-Cβ-Cβ′-Hβ′ | | | −19.0 | | | |

The dihedral angles are defined according to the atom identification scheme given in Figure 3 with primed atoms/methyl groups referring to B-ring atoms/methyl groups as defined in Figure 5. Since a distinction between the aromatic rings of β-β and 5-5 cannot be made, two values are reported for these. Values to the left in the corresponding column are associated with a certain ring, whereas values to the right are associated with the other ring. The relative energy of β-1 is omitted since β-1 has a different stochiometry.

**Table 2.** Geometric parameters (in Å and degrees, respectively) for intra-molecular hydrogen bonds present in the most stable conformer of β-O4, β-5, β-1, and 5-5

| | β-O4 | β-5 | β-1 | 5-5 |
|---|---|---|---|---|
| Hydrogen Bond | Oγ-Hγ···O3′ | Oγ-Hγ···Oγ′ | Oα-Hα···Oγ | O4-H4···O4′ |
| H···O Distance | 1.99 | 2.34 | 1.87 | 1.78 |
| Bond Angle | 163.8 | 160.9 | 146.0 | 154.0 |

The bond angles are defined according to the atom identification scheme given in Figure 3 with primed atoms referring to B-ring atoms as defined in Figure 5.

## 4.2 The Thermodynamic Control Hypothesis

In order to assess the validity of the thermodynamic control hypothesis as an explanation for the observed distribution of intermonomeric linkages, we have in Table 1 also listed the relative energies of the different dilignols. The energy of water – evaluated at the same level of theory – has been added to the energies of β-β, β-5, 5-5, and 5-O4, respectively. This is of course not entirely correct, but allows for an estimate of the stabilities of those dilignols relative to that of β-O4, as formation of the latter involves addition of a water molecule to the corresponding quinone methide.

Even though factors such as type of solvent [73, 74, 75] and pH [73] certainly come into play, the major oxidative dimerization products in aqueous solution are β-5

(50%), β-β (30%), β-O4 (20%), and, in small amounts, 5-5 [**76**]. This is in sharp contrast to the occurrence in natural lignin, in which β-O4 couplings are predominant [10]. From the data presented in Table 2, it is thus interesting to note that 5-O4 is the least stable dilignol, and that β-5 and β-β are the two most stable structures. The data is, however, somewhat inconclusive in that 5-5 lies 5.0 kcal $mol^{-1}$ lower in energy than β-O4, which clearly contradicts the thermodynamic control hypothesis. The calculated energies furthermore suggest that β-β rather than β-5 is the predominant dimerization product. If the relative concentrations of the various dilignols follow a strict Boltzmann distribution, i.e., that it is only the relative product energies that control the formation, then the computed data listed in Table 2 would yield β-β (98.8%) and β-5 (1.2%) as the only products observed at room temperature. It is hence obvious that our results do not support the thermodynamic control hypothesis from a quantitative point of view. We nevertheless claim that one should not entirely dismiss this hypothesis on the basis of a comparison between our theoretically estimated relative concentrations and experimental data since small errors (~1-2 kcal $mol^{-1}$) in the calculated relative energies of β-5 and β-β could have a dramatic effect on the estimated relative concentrations.

Elder and Ede have on the basis of calculated heats of formation of various dilignols argued that their results suggest dimerization of monolignols to be thermodynamically controlled [34]. Their conclusion was later questioned by Houtman, who pointed at the fact that the 12 kcal $mol^{-1}$ difference in heats of formation between the most stable (β-β) and the second most stable (β-5) dilignol would predict the relative concentration of β-β to be 100% at room temperature [75]. The semiempirical calculations performed by Elder and Ede [34] provided relative energies that differ quite substantially from those reported here, but the proposed order of thermodynamic stability – which at a qualitative level correlates with experimental data – agrees with our results.

### 4.3 The Spin Distribution Hypothesis

As pointed out in the introduction, all dilignols are phenols that – as the polymerization of lignin proceeds – also might undergo enzymatic dehydrogenation. The evaluation of the spin distribution hypothesis was therefore extended to include computations of spin densities in β-O4, β-5, β-β, β-1, 5-5, and 5-O4 dehydrogenated dilignols (DHDLs) [33]. Starting from the B3LYP/6-31G(d,p) geometry-optimized structures shown in Figures 11 and 12, an initial set of DHDL geometries was created by removal of a phenolic hydrogen from the corresponding closed-shell conformers. The resulting open-shell structures were then optimized at the same level, followed by computation of spin densities using the extended 6-311++G(2df,p) basis set and standard Mulliken population analysis.

The β-β, 5-5, and β-1 dilignols all have two phenolic hydrogens. Therefore, abstraction of either of the two phenolic hydrogens was taken into account by initial geometry optimizations at the B3LYP/3-21G level. Only the most stable B3LYP/3-21G hydrogen abstraction product was then considered at the higher level of theory.

The calculated spin densities are presented in Table 3. We note that our calculations on the coniferyl alcohol radical assign negative spins to the C2, C4, and C6 positions. The odd-alternant pattern with negative spins at those particular positions is a well-characterized feature of phenyl radicals (cf. Refs [77, 78]). The calculations by Elder and Worley [36] showed that the atomic sites having most of the unpaired spin actually are those through which polymerization occurs, and that the spin densities of the atomic sites within the aromatic ring are higher than that of the phenolic oxygen. Our study verifies that the unpaired spin of the coniferyl alcohol radical exists primarily on atoms involved in bond formation during polymerization, but assigns the highest spin to the O4 position (*i.e.* the phenolic oxygen). This result agrees with previous findings for the tyrosyl radical, and is supported by experimental studies of similar systems (*e.g.* Ref [79], and references therein).

The formation of dilignols in aqueous solution at room temperature typically yields 50% β-5, 30% β-β, 20% β-O4, and small amounts of 5-5 couplings [76]. The calculations on the coniferyl alcohol radical provide spin densities that, apart from distinguishing the atomic sites through which polymerization occur, correlate rather weakly with this distribution. Given the predominance of β-O4 couplings in natural lignin [10], it is however noteworthy that dehydrogenation of dilignols, in turn, also seems to produce radicals with most of the spin residing at the phenolic oxygen. For all DHDLs except 5-5, the calculations assign most of the spin to the O4 position. This result might partly explain the large percentage of β-O4 couplings in natural lignin.

**Table 3.** Calculated spin distributions for coniferyl alcohol radical and DHDLs

| Atom | Coniferyl | β-O4 | β-5 | β-β | β-1 | 5-5 | O4-5 |
|---|---|---|---|---|---|---|---|
| C1 | 0.25 | 0.22 | 0.29 | 0.29 | 0.30 | 0.30 | 0.28 |
| C2 | −0.10 | −0.11 | −0.11 | −0.12 | −0.11 | −0.10 | −0.11 |
| C3 | 0.22 | 0.25 | 0.23 | 0.24 | 0.26 | 0.23 | 0.24 |
| C4 | −0.00 | 0.01 | −0.02 | −0.03 | −0.04 | −0.06 | −0.09 |
| O4 | 0.32 | 0.37 | 0.36 | 0.36 | 0.36 | 0.29 | 0.30 |
| C5 | 0.18 | 0.16 | 0.22 | 0.24 | 0.22 | 0.24 | 0.23 |
| C6 | −0.08 | −0.04 | −0.07 | −0.09 | −0.08 | −0.10 | −0.10 |
| Cα | −0.10 | 0.01 | −0.00 | 0.01 | 0.00 | −0.12 | −0.11 |
| Cβ | 0.21 | 0.01 | 0.00 | 0.02 | 0.01 | 0.23 | 0.23 |
| Cγ | −0.01 | 0.01 | −0.00 | 0.01 | 0.01 | −0.01 | 0.00 |

The atoms are numbered according to the scheme given in Figure 3. The spin densities of the DHDLs are those of the atomic sites at the monolignol subunit from which the phenolic hydrogen has been abstracted. Without exception, the spin densities of the atomic sites on the second subunit are negligible.

## 5 Formation of the β-O4 Dimeric Linkage

As mentioned above, the distribution of dimeric linkages found in natural lignin differs considerably from that observed when letting lignin react in aqueous solution. The large fraction of unpaired spin at the O4 position of the dehydrogenated radicals may be one explanation for this effect, but also other mechanistic and environmental features can come into play. Our investigation of the mechanistic details of the formation of β-O4 linked dimeric lignin structures involves two alternative chemical reactions, as shown in Figure 13 [37]. The first reaction corresponds to the formation of a β-O4 linked QM from two coniferyl alcohol radicals. As such, this scheme represents a prototype lignin polymerization mechanism based on successive radical-radical additions. The second reaction corresponds to the addition of a coniferyl alcohol to its phenoxy radical, yielding a radical hereafter denoted QMR. This particular reaction constitutes an alternative 'competing' pathway to the formation of β-O4 linked dimeric lignin structures, and will below be assessed through a comparison between the calculated energetics of the two pathways.

In order to reduce the computational effort, the 3-hydroxy-propen-1-yl tail of the $R_{O4}$ radical was replaced by a hydrogen atom in the modeling of the reactions described in Figure 13. Given that the structural and electronic properties of the full β-O4 dilignols are similar to those of models lacking this particular side chain [24], we believe that this simplification is justified.

In order to locate a transition state structure for a radical-radical addition reaction, whereby an energy barrier can be obtained, one should ideally employ a multi-configurational *ab initio* method such as CASSCF [**80**] to account for non-dynamical electron correlation effects. It should therefore be noted that the B3LYP hybrid density functional method, while requiring considerably less of computational resources and thus being applicable to larger systems, analogously with most *ab initio* methods such as HF, MP2, QCISD and similar is a single-reference method.

An approximation to the energy barrier for the formation of a β-O4 linked QM was obtained by constrained triplet state geometry optimizations in which the Cβ-O4 distance between the two radicals was held fixed at different values ranging from 2.52 to 1.72 Å. All other geometrical parameters were allowed to relax during the optimizations. For each of these geometries, the corresponding B3LYP/6-31G(d,p) singlet state energy – which should be higher than the triplet energy at large Cβ-O4 distances and lower at small distances – was subsequently evaluated, and the geometry at which the singlet and triplet surfaces intersect was identified (Cβ-O4 distance 2.25 Å). In addition, singlet and triplet B3LYP/6-311G(2df,p) and B3LYP/6-311++G(2df,p) energy curves were computed using the B3LYP/6-31G(d,p) triplet state geometries. From the corresponding intersection points, an approximate energy barrier with respect to the isolated reactants could be obtained.

The results from the above calculations are listed in Table 4 together with the corresponding reaction energies. Since we have not explicitly located a transition state structure, ZPE corrections have not been added to the presented energies. We observe that the energy barrier for the formation of QM is low (~2-5 kcal/mol), and that the reaction energy is negative by more than 20 kcal/mol. These results show that the process is both fast and thermodynamically favourable, and in agreement with results from studies of other radical-radical additions.

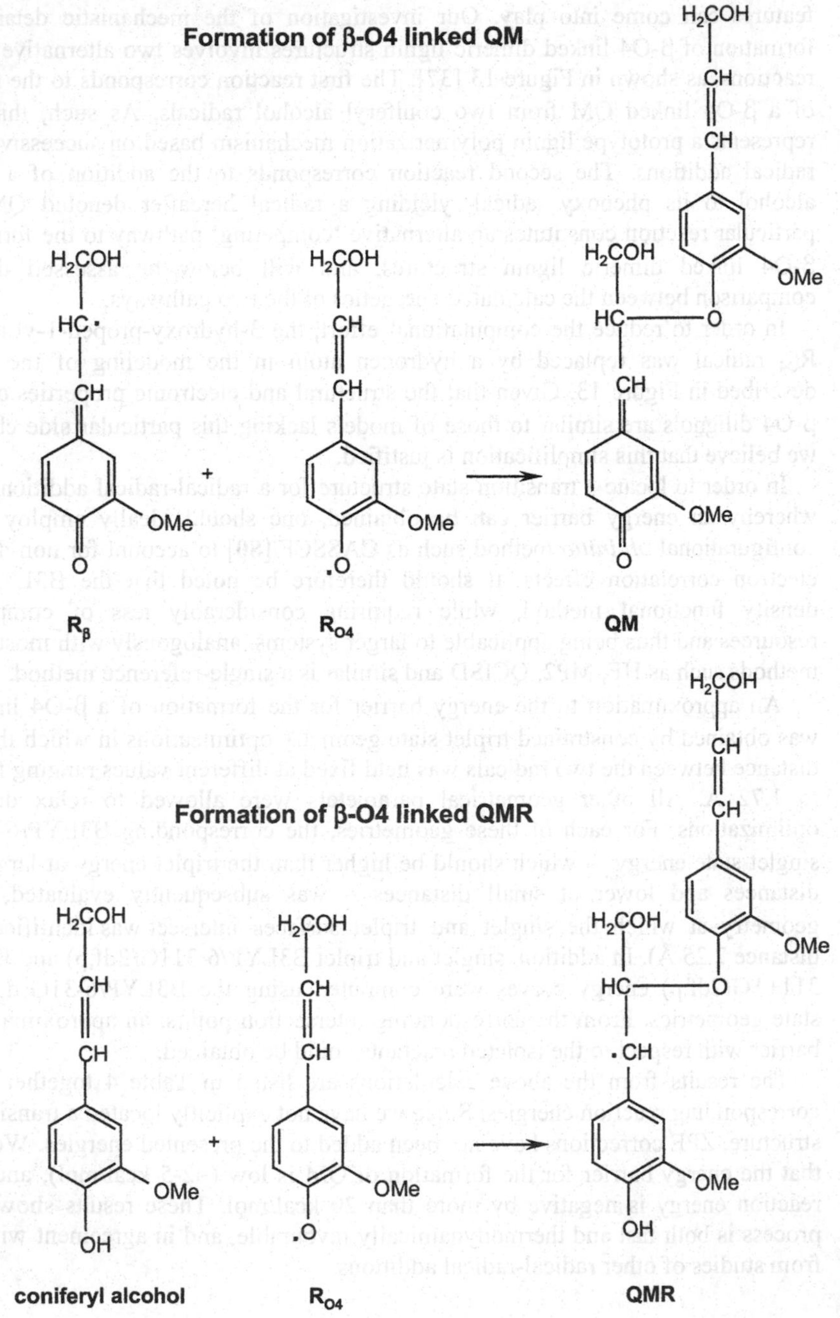

**Formation of β-O4 linked QM**

$R_{\beta}$ + $R_{O4}$ → QM

**Formation of β-O4 linked QMR**

coniferyl alcohol + $R_{O4}$ → QMR

**Fig. 13.** Formation of β-O4 linked dimeric lignin structures

**Table 4.** Energy barriers ($\Delta E^{\dagger}$) and reaction energies ($\Delta E$) for the formation of β-O4 linked QM (kcal/mol)

| Level | $\Delta E^{\dagger}$ | $\Delta E$ |
|---|---|---|
| B3LYP/6-311G(2df,p) | 1.6-3.1 | –24.9 |
| B3LYP/6-311G(2df,p)/PCM | 3.3-5.2 | –23.0 |
| B3LYP/6-311++G(2df,p) | 3.6-5.2 | –23.0 |

All calculations carried out using B3LYP/6-31G(d,p) optimized geometries. The energy barrier is calculated as $\Delta E^{\dagger} = E(S/T) - E(R_{\beta}) - E(R_{O4})$, where E(S/T) is the energy at which the singlet and triplet energy curves intersect; the reaction energy is calculated as $\Delta E = E(QM) - E(R_{\beta}) - E(R_{O4})$. The uncertainties in the calculated energy barriers reflect the uncertainties in E(S/T). PCM indicates that the polarized continuum model has been used.

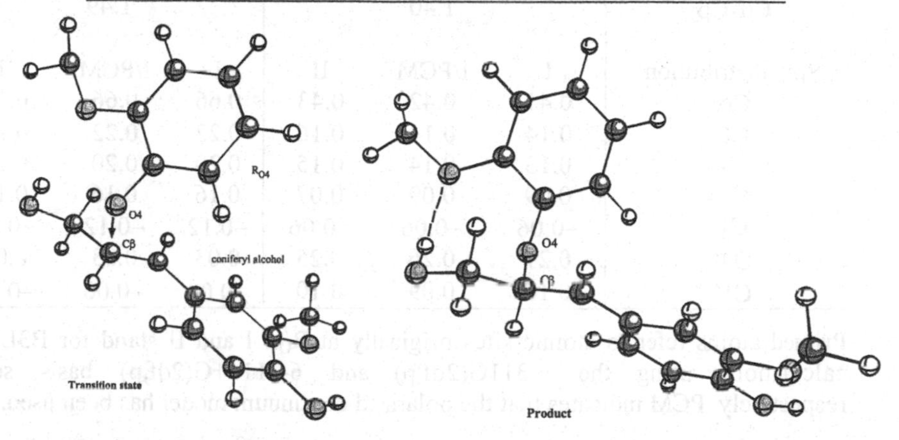


**Fig. 14.** Geometry-optimized structures of transition state and product for the formation of β-O4 linked QMR. Numerical values are listed in Table 5

The geometry-optimized structure of the transition state for the formation of a β-O4 linked QMR is shown in Figure 14 together with the corresponding product structure (both obtained at the B3LYP/6-31G(d,p) level). Relevant geometric parameters are given in Table 5, which also lists total atomic spin densities calculated from Mulliken population analysis at different levels of theory using the B3LYP/6-31G(d,p) geometries. Energy barriers and reaction energies are presented in Table 6.

We observe that the Cβ-O4′ and C4′-O4′ bond lengths of the transition state equal 1.95 and 1.32 Å, respectively (cf. 1.48 and 1.38 Å in the QMR product). The C4′-O4′ bond thus displays partial double-bond character. This feature is the result of $R_{O4}$ not being the only possible resonance structure of a coniferyl alcohol radical (Fig. 4). As the reaction proceeds, however, the unpaired spin is transferred from the $R_{O4}$ moiety to the coniferyl alcohol moiety. The different resonance structures thereby lose significance, which provides an explanation for the fact that the C4′-O4′ bond length of the product is 0.06 Å larger than that of the transition state. As expected, the Cα-Cβ bond is stretched during the course of the reaction. The two structures furthermore differ in that the product is stabilized by an intra-molecular hydrogen bond between the γ-hydroxyl and the 3′-methoxy. This particular hydrogen bond is also present in the lowest-energy *erythro* conformer of β-O4 dilignols [24].

**Table 5.** Geometric parameters (in Å and degrees, respectively) for transition state (TS) and product (PROD) for the formation of β-O4 linked QMR

| Geometric parameter | TS | | | PROD | | |
|---|---|---|---|---|---|---|
| Cβ-O4′ | 1.95 | | | 1.48 | | |
| C4′-O4′ | 1.32 | | | 1.38 | | |
| Cβ-C4′ | 2.87 | | | 2.47 | | |
| Oγ-O3′ | 2.93 | | | 2.86 | | |
| Hγ-O3′ | 2.20 | | | 1.97 | | |
| Oγ-Hγ-O3′ | 131.0 | | | 151.6 | | |
| Cα-Cβ | 1.40 | | | 1.49 | | |
| Spin distribution | I | I/PCM | II | I | I/PCM | II |
| Cα | 0.41 | 0.42 | 0.43 | 0.66 | 0.66 | 0.66 |
| C6 | 0.14 | 0.14 | 0.16 | 0.22 | 0.22 | 0.24 |
| C4 | 0.13 | 0.14 | 0.15 | 0.20 | 0.20 | 0.21 |
| C2 | 0.09 | 0.09 | 0.07 | 0.16 | 0.16 | 0.12 |
| C1 | −0.06 | −0.06 | −0.06 | −0.12 | −0.12 | −0.11 |
| O4′ | 0.27 | 0.26 | 0.25 | 0.03 | 0.03 | 0.02 |
| C1′ | 0.10 | 0.09 | 0.10 | −0.00 | −0.00 | −0.00 |

Primed atoms refer to atomic sites originally at $R_{O4}$. I and II stand for B3LYP calculations using the 6-311G(2df,p) and 6-311++G(2df,p) basis sets, respectively. PCM indicates that the polarized continuum model has been used.

**Table 6.** ZPE-corrected (and uncorrected) energy barriers ($\Delta E^{\dagger}$) and reaction energies ($\Delta E$) for the formation of β-O4 linked QMR (kcal/mol)

| Level | $\Delta E^{\dagger}$ | $\Delta E$ |
|---|---|---|
| B3LYP/6-311G(2df,p) | 7.6 (6.7) | −1.1 (−3.5) |
| B3LYP/6-311G(2df,p)/PCM | 11.5 (10.6) | 2.8 (0.4) |
| B3LYP/6-311++G(2df,p) | 9.7 (8.8) | 0.9 (−1.5) |

All calculations carried out using B3LYP/6-31G(d,p) optimized geometries.

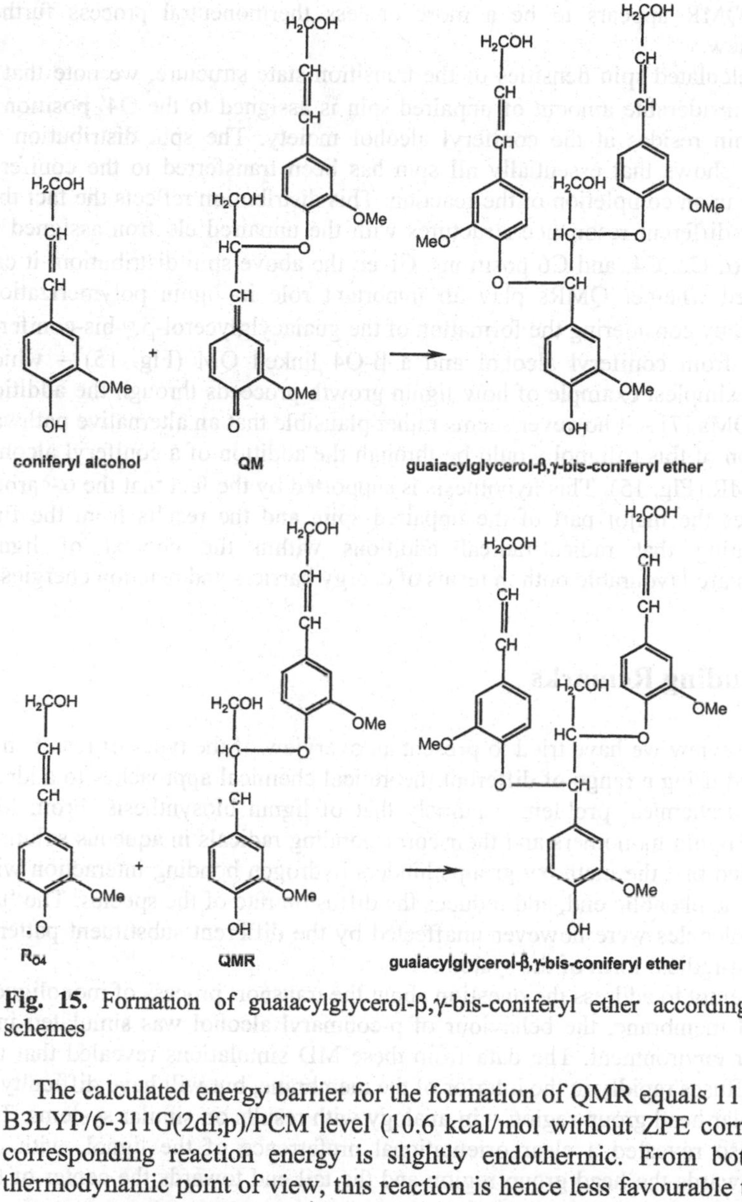

coniferyl alcohol + QM → guaiacylglycerol-β,γ-bis-coniferyl ether

$R_{O4}$ + QMR → guaiacylglycerol-β,γ-bis-coniferyl ether

**Fig. 15.** Formation of guaiacylglycerol-β,γ-bis-coniferyl ether according to two different schemes

The calculated energy barrier for the formation of QMR equals 11.5 kcal/mol at the B3LYP/6-311G(2df,p)/PCM level (10.6 kcal/mol without ZPE corrections), and the corresponding reaction energy is slightly endothermic. From both a kinetic and thermodynamic point of view, this reaction is hence less favourable than the previous radical-radical addition ($\Delta E^{\dagger}$ = 3.3-5.2 and $\Delta E$ = –23.0 kcal/mol at the B3LYP/6-311G(2df,p)/ PCM level). This holds true also at the other levels of theory employed. Even though the difference in energy barrier between the two reactions according to transition state theory translates into reaction rates differing by several orders of magnitude, we believe that the possibility that QMRs are actually formed should not be discarded since ~8-12 kcal/mol clearly also constitutes a surmountable energy barrier under standard conditions of temperature and pressure. The fact that the formation of QMR appears to be a more or less thermoneutral process further supports this view.

From the calculated spin densities of the transition state structure, we note that – even though considerable amount of unpaired spin is assigned to the O4′ position – most of the spin resides at the coniferyl alcohol moiety. The spin distribution of QMR, in turn, shows that essentially all spin has been transferred to the coniferyl alcohol moiety upon completion of the reaction. This distribution reflects the fact that QMR has four different resonance structures with the unpaired electron assigned to either of the Cα, C2, C4, and C6 positions. Given the above spin distribution, it can thus be debated whether QMRs play an important role in lignin polymerization. Hypothetically, by considering the formation of the guaiacylglycerol-β,γ-bis-coniferyl ether *tri*lignol from coniferyl alcohol and a β-O4 linked QM (Fig. 15) – which constitutes the simplest example of how lignin growth proceeds through the addition of phenols to QMs [7] – it however seems rather plausible that an alternative pathway to the formation of this trilignol would be through the addition of a coniferyl alcohol radical to a QMR (Fig. 15). This hypothesis is supported by the fact that the α-carbon of QMR carries the major part of the unpaired spin, and the results from the first section indicating that radical-radical additions within the context of lignin polymerization are favourable both in terms of energy barriers and reaction energies.

## 6 Concluding Remarks

In the present review we have tried to present an overview of the types of results that can be obtained using a range of different theoretical chemical approaches to address a complex (bio)chemical problem – namely that of lignin biosynthesis. From MD simulations of lignin monomers and their corresponding radicals in aqueous solution, it was concluded that the methoxy groups hinders hydrogen bonding interaction with the solvent at the phenolic end, and reduces the diffusion rate of the species. The 'tail end' of the molecules were however unaffected by the different substituent patterns and radical/non-radical form of the system.

In a first attempt to address the question about the transport process of monolignols across the cell membrane, the behaviour of p-coumaryl alcohol was simulated in a DPPC + water environment. The data from these MD simulations revealed that the monolignol diffuses rapidly in the interior of the membrane, but will have difficulty in passing the polar head group region – in analogy with results on similar systems. The simulations also revealed a clear orientational preferrence of the lignol, with the phenol ring towards the head group region and the tail end towards the center of the bilayer.

Using quantum chemical methods we then explored the relative stabilities of various dilignols, and tested the two dominating hypotheses regarding the observed product distribution in aqueous solution. It is concluded that, albeit the relative stabilities provide a good qualitative explanation for the observed distribution, the thermodynamic control hypothesis does not explain all details. Looking at the spin distributions of the dehydrogenated radicals, on the other hand, the odd-alternant pattern again yields increased radical character (and, hence, reactivity) at those positions known to be the sites of cross-link formation. The main fraction of the unpaired spin is however located on the phenolic oxygen (O4), and the spin control hypothesis can also not alone explain the observed product ratio in aqueous solution. Instead it appears that the elevated level of unpaired spin at O4 may be a contributing factor for the predominance of β-O4 cross-links in natural lignin.

In the final section, we investigated two different mechanisms for the formation of the β-O4 dimers; a radical – radical and a radical – non-radical addition mechanism. The radical – radical addition mechanism was found to have a barrier of only a few kcal/mol, and be exothermic by 23 kcal/mol, whereas the radical – non-radical mechanism excerts a ca 10 kcal/mol high barrier to reaction and is essentially thermoneutral. Based on this data we conclude that the first mechanism is the most likely, albeit the second is also plausible under standard conditions. We ended the review with an outline of how the corresponding mechanisms may come into play in the formation of tri- and higher lignols.

## Acknowledgements

The Swedish Science Research Council (VR), the Carl Trygger Foundation and the Wood Ultrastructure Research Center (WURC) are gratefully acknowledged for financial support. We also acknowledge Dr. David van der Spoel and Prof. Aatto Laaksonen for valuable discussions.

## References

[1] Newman, R. in *Microfibril Angle In Wood.* IAWA/IUFRO International Workshop on the *Significance of Microfibril Angle to Wood Quality.* Ed. B.G. Butterfield, Westport, New Zealand, p. 81, **1998**.

[2] Heyn, A.N.J., *J. Ultrastruct. Res.* **1969**, *26*, 52.

[3] Higuchi, T., *Wood Sci. Technol.* **1990**, *24*, 23.

[4] Baucher, M. B. Monties, M. Van Montagu and W. Boerjan, *Critical Reviews in Plant Sciences.* **1998**, *17*, 125.

[5] Cotê, W.A., *Cellular Ultrastructure of Woody Plants.* Syracuse Univ. Press, New York, Syracuse, **1965**.

[6] Weir, N.A., J. Arct and A. Ceccarelli, *Polymer Degradation and Stability* **1995**, *47*, 289, and references therein.

[7] Freudenberg, K., *Science* **1965**, *148*, 595.

[8] Higuchi, T., in *Biosynthesis and Biodegradation of Wood Components.* Ed. T. Higuchi. Academic Press, Orlando. p 141, **1985**.

[9] Davin, L.B., H.-B. Wang, A.L. Crowell, D.L. Bedgar, D.M. Martin, S. Sarkanen and N.G. Lewis, *Science* **1997**, *275*, 362.

[10] Monties, B., in *Methods in Plant Biochemistry: Plant Phenolics, Vol. 1.* Eds. P.M. Dey, J.B. Harborne. Academic Press, New York. p 113, **1989**.

[11] Monties, B., in *The Biochemistry of Plant Phenolics (Ann. Proc. Phytochem. Soc. Eur. Vol 25)*, p 161, **1985**.

[12] Monties, B., *Polym. Degrad. Stabil.,* **1998,** *59,* 53.
[13] Sjöström, E., *Wood Chemistry – Fundamentals and Applications, 2nd ed.* Acad. Press, San Diego, CA, **1993.**
[14] Lundquist, K., in *Methods in Lignin Chemistry*. Eds. S.Y. Lin, C.W. Dence. Springer-Verlag, Berlin. p 242, **1992.**
[15] Robert, D., in *Methods in Lignin Chemistry*. Eds. S.Y. Lin, C.W. Dence. Springer-Verlag, Berlin. p 250, **1992.**
[16] Sakakibara, A., *Wood Sci. Technol.* **1980,** *14,* 89.
[17] Stomberg, R. and K. Lundquist, *Acta Chem. Scand.* **1986,** *A40,* 705.
[18] Stomberg, R., M. Hauteville and K. Lundqvist, *Acta Chem. Scand.* **1988,** *B42,* 697.
[19] Elder, T.J., M.L. McKee and S.D. Worley, *Holzforschung* **1988,** *42,* 233.
[20] Faulon, J.-L. and P.G. Hatcher, *Energy and Fuels* **1994,** *8,* 402.
[21] Simon, J.P. and K.-E.L. Eriksson, *Holzforschung* **1995,** *49,* 429.
[22] Simon, J.P. and K.-E.L. Eriksson, *J. Mol. Struct.* **1996,** *384,* 1.
[23] Simon, J.P. and K.-E.L. Eriksson, *Holzforschung* **1998,** *52,* 287.
[24] Durbeej, B. and L.A. Eriksson, *Holzforschung,* in press, **2002.**
[25] Glasser, W.G. and H.R. Glasser, *Macromolecules* **1974,** *7,* 17.
[26] Glasser, W.G. and H.R. Glasser , *Paperi Ja Puu* **1981,** *2,* 71.
[27] Glasser, W.G., H.R. Glasser and N. Morohoshi. *Macromolecules* **1981,** *14,* 253.
[28] Lange, H., B. Wagner, J.F. Yan, E.W. Kaler and J.L. McCarthy, in *Seventh Intl. Symposium on Wood and Pulping Chemistry, Vol 1* .Beijing. p 111, **1993.**
[29] Roussel, M.R. and C. Lim, *Macromolecules* **1995,** *28,* 370.
[30] Roussel, M.R. and C. Lim, *J. Comp. Chem.* **1995,** *16,* 1181.
[31] Jurasek, L., *J. Pulp Paper Sci.* **1995,** *21,* J274.
[32] Jurasek, L., *J. Pulp Paper Sci.* **1996,** *22,* J376.
[33] Durbeej, B. and L.A. Eriksson, in preparation.
[34] Elder, T.J. and R.M. Ede, in *Proceedings of the 8th International Symposium on Wood and Pulping Chemistry, Vol. 1.* Gummerus Kirjapaino Oy, Helsinki. p 115, **1995.**
[35] Mårtensson, O. and G. Karlsson, *Arkiv för Kemi* **1968,** *31,* 5.
[36] Elder, T.J. and S.D. Worley, *Wood Sci. Technol.* **1984,** *18,* 307.
[37] Durbeej, B. and L.A. Eriksson, *Holzforschung,* in press, **2002.**.
[38] Wang, Y.-N. and L.A. Eriksson, submitted for publication.
[39] Wang, Y.-N. And L. A. Eriksson, in preparation.
[40] Onysko, K.A., *Biotech. Adv.* **1993,** *11,* 179.
[41] Reid, I.D. and M.G. Paice, *FEMS Microbiol. Rev.* **1994,** *13,* 369.
[42] Viikari, L., A. Kantelinen, J. Sundqvist and M. Linko, M., *FEMS Microbiol. Rev.* **1994,** *13,* 335.
[43] Cornell, W.D., P. Cieplak, C.I. Bayly, I.R. Gould, K.M. Merz, D.M. Ferguson, D.C. Spellmeyer, T. Fox, J.W. Caldwell and P.A. Kollman, *J. Am. Chem. Soc.,* **1995,** *117,* 5179.
[44] Allinger, N.L., Y.H. Yuh, J.-H. Lii, *J. Am. Chem. Soc.* **1989,** *111,* 855.
[45] Jorgensen, W.L., J. Chandrasekhar, J.D. Madura, R.W. Impey and M.L. Kelwin, *J. Chem. Phys.* **1983,** *79,* 926.
[46] Weiner, S.J., P.A. Kollman, D.A. Case, U.C. Singh, C. Ghio, G. Alagona, S. Profeta and P. Weiner, *J. Am. Chem. Soc.* **1984,** *106,* 765.

[47] Smondyrev, A.M. and M.L. Berkowitz, *J. Comp. Chem.* **1999**, *20*, 531.
[48] Söderhäll, J.A. and A. Laaksonen, *J. Phys. Chem. B* **2001**, *105*, 9308.
[49] AMBER 5. Case, D.A., *et al.*, University of California, San Francisco, **1987**.
[50] Miertus, S., E. Scrocco and J. Tomasi, *J. Chem Phys.* **1981,** *55,* 117.
[51] Miertus, S. and J. Tomasi, *Chem Phys.* **1982,** *65,* 239.
[52] Cossi, M., V. Barone, R. Cammi, R. and J. Tomasi, *Chem. Phys. Lett.* **1996,** *255*, 327.
[53] Hohenberg, P. and W. Kohn, *Phys. Rev. B* **1964,** *136,* 864.
[54] Kohn, W. and L.J. Sham, *Phys. Rev. A* **1965,** *140,* 1133.
[55] Parr, R.G. and W. Yang, *Density Functional Theory of Atoms and Molecules,* Oxford Univ. Press: N.Y., **1989**.
[56] Gross, E.K.U. and R.M. Dreizler, *Density Functional Theory*, Springer, **1992**.
[57] Koch, W. and M.C. Holthausen, *A Chemist's Guide to Density Functional Theory,* Wiley-VCH: Weinheim, **2001**.
[58] Becke, A.D., *J. Chem. Phys.* **1993,** *98,* 5648.
[59] Lee, C., W. Yang and R.G. Parr, *Phys. Rev.* **1988,** *B37*, 785.
[60] Stevens, P.J., F.J. Devlin, C.F. Chabalowski and M.J. Frisch, *J. Phys. Chem.* **1994**, *98,* 11623.
[61] Gaussian 98, rev A7-A11. Frisch, M.J. *et al.* Gaussian Inc., Pittsburgh, PA **1998**.
[62] Berendsen, H.J.C., J.P.M. Postma, W.F. Gunsteren, A. DiNola and J.R. Haak, *J. Chem. Phys.* **1984**, *81,* 3684.
[63] Allen, M.P. and D.J. Tildesley, *Computer Simulation of Liquids.* Oxford Science, Oxford, **1987**.
[64] Darden, T., D. York and L.G. Pedersen, *J. Chem. Phys.* **1993**, *98*, 10089.
[65] Pedersen, L.G., *J. Chem. Phys.* **1995**, *103*, 3668.
[66] Berger, O., O. Edholm and F. Jähnig, *Biophys. J.* **1997**, *72*, 2002.
[67] Alper, H.E. and T.R. Stouch, *J. Phys. Chem.* **1995**, *99*, 5724.
[68] Mason, R.P. and D.W. Chester, *Biophys. J.* **1989**, *56*, 1193.
[69] Wu, E.-S., K. Jacobson and D. Papahadjopoulos, *Biochemistry.* **1977**, *16*, 3936.
[70] Stewart, J.J.P, *J. Comp. Chem.* **1989**, *10*, 209.
[71] Stewart, J.J.P, *J. Comp. Chem.* **1989**, *10*, 221.
[72] Miksche, G., *Acta Chem. Scand.* **1972**, *26*, 4137.
[73] Chioccara, F., S. Poli, B. Rindone, T. Pilati, G. Brunow, P. Pietikäinen and H. Setälä, *Acta Chem. Scand.* **1993**, *47*, 610.
[74] Terashima, N. and R.H. Atalla, in *Proceedings of the 8th International Symposium on Wood and Pulping Chemistry, Vol. 1.* Gummerus Kirjapaino Oy, Helsinki. p 69, **1995**.
[75] Houtman, C.J., *Holzforschung* **1999**, *53*, 585.
[76] Tanahashi, M., H. Takeuchi and T. Higuchi, *Wood Res.* **1976**, *61*, 44.
[77] Himo, F., A. Gräslund and L.A. Eriksson, *Biophys. J.* **1997**, *72*, 1556.
[78] Himo, F., G.T. Babcock and L.A. Eriksson, *Chem. Phys. Lett.* **1999**, *313*, 374.
[79] Gräslund, A. and M. Sahlin, *Annu. Rev. Biophys. Biomol. Struct.* **1996**, *25*, 259.
[80] Roos, B., in *Advanced Chemistry and Physics Vol. 69 (Ab Initio Methods in Quantum Chemistry, Part II).* Ed. K.P. Lawley. Wiley, New York. p 399, **1987**.

# High Performance Computing in Electron Microscope Tomography of Complex Biological Structures

José J. Fernández[1], Albert F. Lawrence[2], Javier Roca[1], Inmaculada García[1], Mark H. Ellisman[2], and José M. Carazo[3]

[1] Dpt. Arquitectura de Computadores, Universidad de Almería
04120 Almería, Spain
[2] National Center for Microscopy and Imaging, University of California
San Diego, La Jolla, CA 92093-0608, USA
[3] Biocomputing Unit. Centro Nacional de Biotecnología
Universidad Autónoma, 28049 Madrid, Spain
jose@ace.ual.es, lawrence@sdsc.edu, jroca@ace.ual.es
inma@ace.ual.es, mark@ncmir.ucsd.edu, carazo@cnb.uam.es

**Abstract.** Series expansion reconstruction methods using smooth basis functions are evaluated in the framework of electron tomography of complex biological specimens, with a special emphasis upon the computational perspective. Efficient iterative algorithms that are characterized by a fast convergence rate have been used to tackle the image reconstruction problem. The use of smooth basis functions provides the reconstruction algorithms with an implicit regularization mechanism, very appropriate for noisy conditions. High Performance Computing (HPC) techniques have been applied so as to face the computational requirements demanded by the reconstruction of large volumes. An efficient domain decomposition scheme has been devised that leads to a parallel approach with capabilities of interprocessor communication latency hiding. Comparisons with Weighted BackProjection (WBP), the standard method in the field, are presented in terms of computational demands as well as in reconstruction quality under highly noisy conditions. The combination of efficient iterative algorithms and HPC techniques have proved to be well suited for the reconstruction of large biological specimens in electron tomography, yielding solutions in reasonable computational times.

**Keywords:** Electron Tomography; High Performance Computing; Iterative Reconstruction Algorithms; Parallel Computing

## 1 Introduction

Electron microscopy is central to the study of many structural problems in disciplines related to Biosciences. In concrete, electron microscopy together with sophisticated image processing and three-dimensional (3D) reconstruction techniques allow to obtain quantitative structural information about the 3D struc-

J.M.L.M. Palma et al. (Eds.): VECPAR 2002, LNCS 2565, pp. 166–180, 2003.

ture of biological specimens [1, 2]. Obtaining this information is essential for interpreting the biological function of the specimens.

In contrast to the conventional instrument, high-voltage electron microscopes (HVEMs) are able to image relatively thick specimens that contain substantial 3D structure. Electron Tomography makes it possible the determination and the analysis of their complex 3D structure from a set of projection HVEM images acquired from different orientations by tilting the specimen, typically following single-tilt axis [2] or, lately, double-tilt axis [3], geometries.

Rigorous structural analysis require that the tomographic image acquisition and the subsequent reconstructions should have sufficient resolution for a proper interpretation and measurement of the structural features. As a consequence, Electron Tomography of large complex biological specimens requires large projection HVEM images (typically from 256×256 up to 1024×1024 pixels) and provides large volumes, which involves an intensive use of computational resources and considerable processing time. High Performance Computing (HPC) then arises as the appropriate tool to face those high computational demands by using parallel computing on supercomputers or networks of workstations.

As far as the reconstruction methods are concerned, Weighted Back-Projection (WBP) [1] may be considered as the standard method in electron tomography of large specimens. The relevance of WBP in this field mainly stems from the computational simplicity of the method ($O(N^3)$, where $N^3$ is the number of voxels of the volume). However, WBP exhibits the strong disadvantages that (i) the results may be heavily affected by the limited tilt capabilities of the HVEM, and (ii) WBP does not implicitly take into account the noise conditions. As a consequence, *a posteriori* regularization techniques (such as low-pass filtering) may be needed to attenuate the effects of the noise.

Series expansion reconstruction methods constitute one of the main alternatives to WBP to image reconstruction. Despite their potential advantages [4, 5], these methods still have not been extensively used in the field of electron tomography due to their high computational costs ($O(2KN^3V^2)$, where $K$ is the number of iterations, $V$ is the extension of the volume elements, and $N$ is the same as above). One the main advantages of these methods is the flexibility to represent the volume elements by means of basis functions different from the traditional voxels. During the nineties [6, 7], spherically symmetric volume elements (blobs) have been thoroughly investigated and, as a consequence, the conclusion that blobs yield better reconstructions than voxels has been drawn in Medicine [8, 9] and Electron Microscopy [4, 5].

This work addresses the application of series expansion methods using blobs as basis functions in electron tomography of complex biological specimens from a computational throughput perspective. The aim is to show that blob-based series expansion methods may be so efficient that may constitute real alternatives to WBP as far as the computation time is concerned. In that sense, this work contributes (i) the use of very recent series expansion methods [10, 11] (component averaging methods) characterized by a really fast convergence, achieving least-squares solutions in a very few iterations, and (ii) the use of HPC techniques, which allows to determine the 3D structure of large volumes in reasonable computation time with either reconstruction method, taking advantage of parallel computer architectures. On the other hand, the use of blob basis functions provides the series expansion methods with an implicit regularization mechanism which makes them better suited for noisy conditions than WBP.

This work has been structured into the following: Section 2 reviews iterative reconstruction methods. Section 3 is devoted to the HPC approach devised for this application. Section 4 then presents the experimental results, in terms of computation time and speedup rate, that have been obtained. Also the experimental results from the application of the approach to real biological data are shown, for quality comparison purposes. Finally, the results are discussed in Section 5.

# 2 Iterative Image Reconstruction Methods

## 2.1 Series Expansion Methods

Series expansion reconstruction methods assume that the 3D object or function $f$ to be reconstructed can be approximated by a linear combination of a finite set of known and fixed basis functions $b_j$

$$f(r, \phi_1, \phi_2) \approx \sum_{j=1}^{J} x_j b_j(r, \phi_1, \phi_2) \tag{1}$$

(where $(r, \phi_1, \phi_2)$ are spherical coordinates)

and that the aim is to estimate the unknowns $x_j$. These methods also assume an image formation model where the measurements depend linearly on the object in such a way that:

$$y_i \approx \sum_{j=1}^{J} l_{i,j} x_j \tag{2}$$

where $y_i$ denotes the $i$th measurement of $f$ and $l_{i,j}$ the value of the $i$th projection of the $j$th basis function.

Under those assumptions, the image reconstruction problem can be modeled as the inverse problem of estimating the $x_j$'s from the $y_i$'s by solving the system of approximate equations. Algebraic Reconstruction Techniques (ART) constitute a well known family of iterative algorithms to solve such systems [12].

Series expansion methods exhibit some advantages over WBP: First, the flexibility in the spatial relationships between the object to be reconstructed and the measurements taken, and second, the possibility of incorporate spatial constraints. The main drawbacks of series expansion methods are (i) their high computational demands and (ii) the need of parameter optimization to properly tune the model.

### 2.2 Component Averaging Methods

Component averaging methods have recently arisen [10, 11] as efficient iterative algorithms for solving large and sparse systems of linear equations. In essence, these methods have been derived from ART methods [12], with the important novelty of a weighting related to the sparsity of the system which allows a convergence rate that may be far superior to the ART methods, specially at the early iteration steps. Assuming that the whole set of equations in the linear system (Eq. (2)) may be subdivided into $B$ blocks each of size $S$, a generalized version of component averaging methods can be described via its iterative step from the $k$-th estimate to the $(k+1)$-th estimate by:

$$x_j^{k+1} = x_j^k + \lambda_k \sum_{s=1}^{S} \frac{y_i - \sum_{v=1}^{J} l_{i,v} x_v^k}{\sum_{v=1}^{J} s_v^b (l_{i,v})^2} l_{i,j} \tag{3}$$

where:

$\lambda_k$ denotes the relaxation parameter.
$b = (k \bmod B)$, and denotes the index of the block.
$i = bS + s$, and represents the $i$-th equation of the whole system, and
$s_v^b$ denotes the number of times that the component $x_v$ of the volume contributes with nonzero value to the equations in the $b$-th block.

The processing of all the equations in one of the blocks produces a new estimate. And one iteration of the algorithm involves processing all the blocks. This technique produces iterates which converge to a weighted least squares solution of the system of equations provided that the relaxation parameters are within a certain range and the system is consistent [11]. The efficiency of component averaging methods stems from the explicit use of the sparsity of the system, represented by the $s_v^b$ term in Eq. (3).

Component averaging methods, as ART methods, can be classified into the following categories as a function of the number of blocks involved:

**Sequential.** This method cycles through the equation one-by-one producing consecutive estimates ($S = 1$). This method exactly matches the well-known *row-action* ART method [12] and is characterized by a fast convergence as long as relaxation factors are optimized.

**Simultaneous.** This method, known simply as "component averaging" (CAV) [10], uses only one block ($B = 1$), considering all equations in the system in every iterative step. This method is inherently parallel, in the sense that every equation can be processed independently from the others. However, it is characterized, in general, by a slow convergence rate.

**Block-Iterative.** Block-Iterative Component Averaging methods (BICAV) represent the general case. In essence, these methods sequentially cycle block-by-block, and every block is processed in a simultaneous way. BICAV exhibits an initial convergence rate significantly superior to CAV and on a par with row-action ART methods, provided that the block size and relaxation parameter are optimized. BICAV methods (with $S > 1$) methods also exhibit an inherent nature for parallelization.

### 2.3 Basis Functions

In the field of image reconstruction, the choice of the set of basis functions to represent the object to be reconstructed greatly influences the result of the algorithm [6]. Spherically symmetric volume elements (blobs) with smooth transition to zero have been thoroughly investigated [6, 7] as alternatives to voxels for image representation, concluding that the properties of blobs make them well suited for representing natural structures of all physical sizes. The use of blobs basis functions provides the reconstruction algorithm with an implicit regularization mechanism. In that sense, blobs are specially suited for working under noisy conditions, yielding smoother reconstructions where artifacts and noise are reduced with relatively unimpaired resolution. Specifically in electron tomography, the potential of blob-based iterative reconstruction algorithms with respect to WBP were already highlighted by means of an objective task-oriented comparison methodology [4].

Blobs are a generalization of a well-known class of window functions in digital signal processing called *Kaiser-Bessel* [6]. Blobs are spatially limited and, in practice, can also be considered band-limited. The shape of the blob is controlled by three parameters: the radius, the differentiability order and the density drop-off. The appropriate selection of them is highly important. The blob full width at half maximum (FWHM), determined by the parameters, is chosen on the basis of a compromise between the resolution desired and the data noise suppression: narrower blobs provide better resolution, wider blobs allow better noise suppression. For a detailed description of blobs, refer to [6].

The basis functions in Eq. (1) are shifted versions of the selected blob arranged on a *simple cubic* grid, which is the same grid as the one used with voxels, with the important difference that blobs have an overlapping nature. Due to this overlapping property, the arrangement of blobs covers the space with a pseudo-continuous density distribution (see Fig. 1) very suitable for image representation. On the contrary, voxels are only capable of modeling the structure by non-overlapping density cubes which involve certain spatial discontinuities.

## 3 High Performance Computing in Electron Tomography

Parallel Computing has been widely investigated for many years as a means of providing high-performance computational facilities for large-scale and grand-challenge applications. In the field of electron tomography of large specimens, parallel computing has already been applied to face the computational demands required to reconstruct huge volumes by means of WBP as well as voxel-based iterative methods [2]. The use of voxels as basis functions makes the single-tilt axis reconstruction algorithms relatively straightforward to implement on massively parallel supercomputers. The reconstruction of each of the one-voxel-thick slices orthogonal to the tilt axis of the volume is assigned to an individual node on the parallel computer. In that sense, this is an example of an *embarrassingly parallel* application (see [13] for a revision of the most important concepts

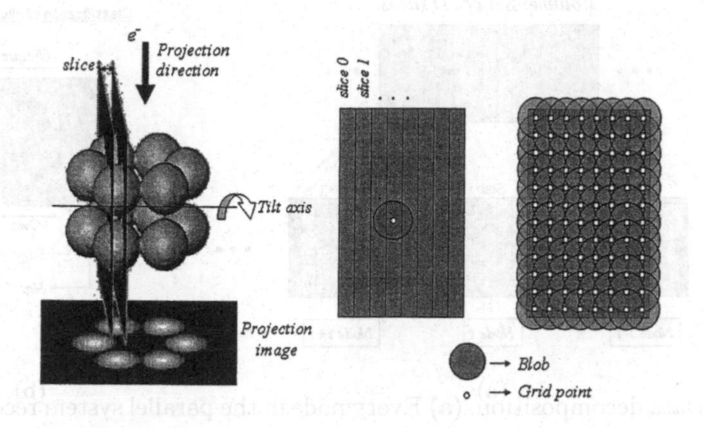


**Fig. 1.** Image representation with blobs. Left: the single-tilt axis acquisition geometry is sketched. The slices are those one-voxel-thick planes orthogonal to the tilt axis. Center: depicts the slices by means of columns, and the circle represents a generic blob which, in general, extends beyond the slice where it is located. Right: Blobs create a pseudo-continuous 3D density distribution representing the structure

in parallel computing) that is no longer valid for blob-based methodologies, as will be shown below. For the blob case, more effort is needed for a proper data decomposition and distribution across the nodes.

### 3.1 Data Decomposition

The single-tilt axis data acquisition geometries typically used in electron tomography allow the application of the *Single-Program Multiple-Data* (SPMD) computational model for parallel computing. In the SPMD model, all the nodes in the parallel computer execute the same program, but for a data subdomain. Single-tilt axis geometries allow a data decomposition consisting of dividing the whole volume to be reconstructed into slabs of slices orthogonal to the tilt axis. The SPMD model then involves that the task to be carried out by every node is to reconstruct its own slab.

Those slabs of slices would be independent if voxel basis functions were used. However, due to their overlapping nature, the use of blobs as basis functions makes the slices, and consequently the slabs, inter-dependent (see Fig. 1). Therefore, the nodes in the parallel computer have to receive a slab composed of its corresponding subdomain together with additional redundant slices from the neighbor nodes. The number of redundant slices depends on the blob extension. Fig. 2(a) shows an scheme of the data decomposition.

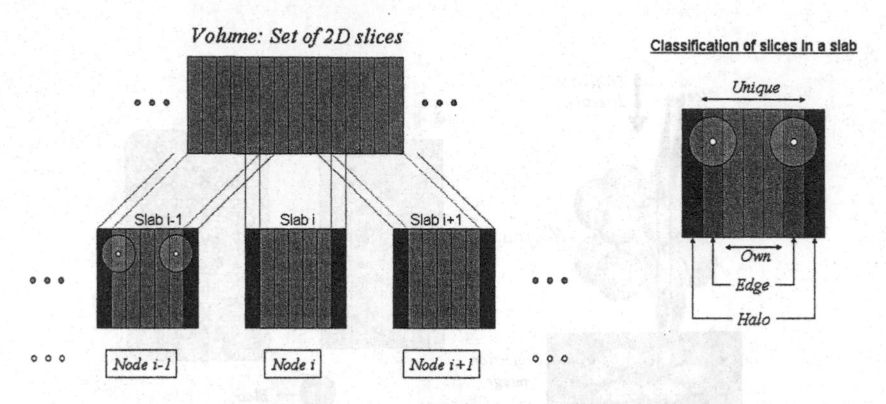


**Fig. 2.** Data decomposition. (a) Every node in the parallel system receives a slab including unique slices (light-gray), and additional redundant slices (dark-gray) according to the blob extension. (b) Classification of the slices in the slab

The slices in the slab received by a given node are classified into the following categories (see scheme in Fig. 2(b)):

**Halo** These slices are only needed by the node to reconstruct some of its other slices. They are the redundant slices mentioned above, and are located at the extremes of the slab. Halo slices are coming from neighbor nodes, where they are reconstructed.

**Unique** These slices are to be reconstructed by the node. In the reconstruction process, information from neighbor slices is used. These slices are further divided into the following subcategories:

- **Edge** Slices that require information from the halo slices coming from the neighbor node.
- **Own** Slices that do not require any information from halo slices. As a result, these slices are independent from those in the neighbor nodes.

It should be noted that edge slices in a slab are halo slices in a neighbor slab.

## 3.2 The Parallel Iterative Reconstruction Method

In this work, the BICAV (with $S > 1$) and CAV methods have been parallelized following the SPMD model and the data decomposition just described. The row-action version of component averaging methods has been discarded since the BICAV method yields a convergence rate on-a-par with it but with better warranties of speedup due to its inherent parallel nature.

Conceptually, BICAV and CAV reconstruction algorithms may be decomposed into three subsequent stages: (i) Computation of the forward-projection of the model; (ii) Computation of the error between the experimental and the calculated projections; and (iii) refinement of the model by means of backprojection

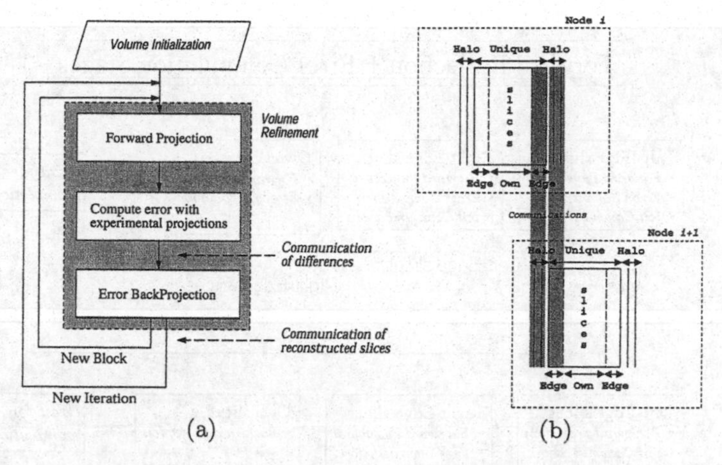


**Fig. 3.** The Parallel Iterative Reconstruction Method. (a) Flow chart of the iterative reconstruction algorithm, including communication/synchronization points. (b) Communications in the parallel algorithm

of the error. Those stages can be easily identified in Eq. (3). Those reconstruction algorithms iteratively pass through those stages for every block of equations and for every iteration, as sketched in Fig. 3(a). Initially, the model may be set up to zero, constant value or even from the result of another reconstruction method.

The interdependence among neighbor slices due to the blob extension implies that, in order to compute either the forward-projection or the error backprojection for a given slab, there has to be a proper exchange of information between neighbor nodes. Specifically, updated halo slices are required for a correct forward-projection of the edge slices. On the other hand, updated halo error differences are needed for a proper error backprojection of the edge slices. The need of communication between neighbor nodes for a mutual update of halo slices is clear. The flow chart in Fig. 3(a) shows a scheme of the iterative algorithm, pointing out the communication points: Just before and after the error backprojection. Fig. 3(b) also shows another scheme depicting the slices involved in the communications: halo slices are updated with edge slices from the neighbor node.

Our parallel SPMD approach then allows all the nodes in the parallel computer to independently progress with their own slab of slices. Notwithstanding this fact, there are two implicit synchronization points in every pass of the algorithm in which the nodes have to wait for the neighbors. Those implicit synchronization points are the communication points just described.

In any parallelization project where communication between nodes is involved, *latency hiding* becomes an issue. That term stands for overlapping communication and computation so as to keep the processor busy while waiting the communications to be completed. In this work, an approach that further exploits the data decomposition has been devised for latency hiding. In essence,

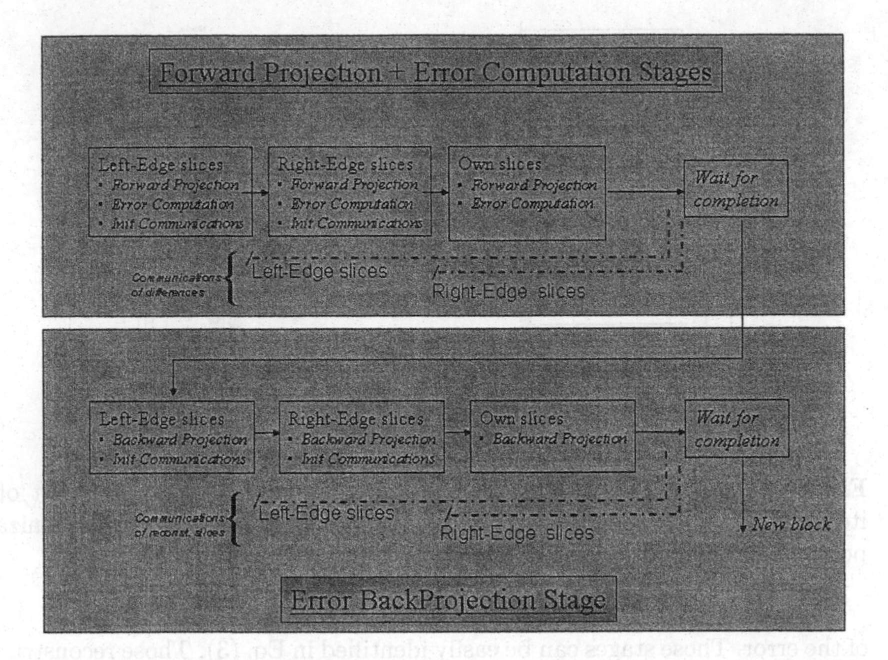


**Fig. 4.** Latency hiding: Overlapping communication and computation. The boxes represent stages of the iterative algorithm, and the discontinuous lines denote the transmission of the data already processed. The latency hiding is applied for both communication points in the parallel algorithm: Just before and after the error backprojection stage

the approach is based on ordering the way the slices are processed between communication points. Fig. 4 sketches this approach. First of all, the left edge slices are processed, and they are sent as soon as they are ready. The communication of left edge slices is then overlapped with the processing of right slices. Similarly, the communication of right edge slices is overlapped with the processing of own slices. This strategy is applied to both communication points, just before and after the error backprojection stage. On the other hand, the ordered way the nodes communicate with each other also makes this parallel approach deadlock-free.

## 4 Results

The experiments that have been carried out had a two-fold aim. First, the effective speedup and the computation times that our parallel approach yields were computed so as to evaluate its efficiency. Second, we intended to do a comparison between WBP and the component averaging methods in terms of the quality of the reconstructions when applied to real data under extremely noisy conditions.

For both purposes, we tested WBP, CAV and BICAV algorithms under different parameters that proved to be representative [6, 7, 9, 10, 11]:

- block sizes for component averaging methods: 1, 10, ..., 70. We have considered that the block size should be multiple of the size of the projection images so that all the pixels of the same image belong to the same block.
- blob parameters. We have tested different blob radii: 1.5, 2.0, 2.25, 2.795. Those values represent a spectrum of different resolution properties. Those four blobs have different FWHMs: 1.20, 1.33, 2.15 and 2.45, respectively, relative to the sampling interval. Consequently, those blobs have a direct influence on the maximum resolution attainable in the reconstructions.
- different values for the relaxation factor.

The software developed for this application is based on C programming language and the MPI (Message-Passing Interface) library. The portability of the software has been tested in a number of computer platforms.

### 4.1 Evaluation of the Parallel Approach

The performance results that will be described and analyzed here were measured in the cluster of workstations at the NCMIR (National Center for Microscopy and Imaging Research). Such a cluster consists of 30 computers (uni- and dual-Pentium III) which involve 50 processors, and a total of 17 GBytes of RAM memory distributed across them. The computers in the cluster are switched by a fast ethernet network.

The efficiency of any parallelization approach is usually evaluated in terms of the speedup [13], defined as the ratio between the computation time required for the application to be executed in a single-processor environment and the corresponding time in the parallel system. Fig. 5(a) shows the speedup obtained for the component averaging methods in reconstructing a 512×512×512 volume from 70 projection images using blob radius of 2.0 (standard blob) and 2.795 (wide blob). Specifically, we have computed the speedup for the extreme cases of BICAV: BICAV using 70 blocks and CAV (equivalent to BICAV using only one block). In Fig. 5(a) only one curve for CAV is presented because the speedup curves that have been obtained for both the standard and the wide blob present negligible differences (lower than 0.01%).

In general, the speedup curves in Fig. 5(a) exhibit a global behavior that approaches the linear speedup. Still, CAV yields better speedup rates than BICAV. The difference between CAV and BICAV in this experiment is that the amount of communications in BICAV is scaled, with respect to CAV, by the number of blocks (70 in this case). In theory, the approach for overlapping communication and computation that we have developed here should provide latency hiding. Consequently, in principle, there should be no significant differences in the speedup rates. We speculate that the causes underlying those differences are related to the communication startup time involved in any communication step and which is not avoidable for any latency hiding approach. As the number of processors in the system increases, the time needed for all the processors to start the communication is longer, possibly due to the network overhead. This hypothesis is justified by the fact that the reconstruction algorithm implemented here works synchronously.

Finally, it is clearly appreciated in Fig. 5(a) that there seems to be an inflexion point in all the curves when using around 20 processors. From 20 processors on, all the speedup curves follows a slope lower than the one exhibited for less processors, progressively moving away from the linear speedup. This behavior still has to be further investigated.

On the other hand, in order to compare component averaging methods with WBP in terms of the computational burden, we have measured the time required for every iteration, using different numbers of blocks for the BICAV methods. We have taken such measures for different reconstruction sizes (256×256×256, 512×512×512, 1024×1024×200), different number of tilt angles (61 or 70) and different blob sizes, obtaining similar relative behaviors. Fig. 5(b) shows the results obtained for a 512×512×512 reconstruction from 70 projections, using the standard blob (radius 2.0, bars in light-gray) and wide blob (radius 2.795, bars in dark-gray). In that figure, it is clearly appreciated that WBP is the lowest resources-consuming method, since it requires 235 seconds for the whole reconstruction. However, one iteration of the component averaging methods only requires a little bit more time (324-358 seconds and 438-496 seconds for blob with radius, respectively, 2.0 and 2.795). Taking into account that these iterative methods are really efficient and yield suitable solutions after a few iterations (in a range usually between 1 and 10), these computation times make component averaging methods real alternatives to WBP, as opposed to the traditional iterative ART methods.

Fig. 5(b) shows that, in iterative methods, the time per iteration is slightly larger as the number of blocks increases (from 324/438 seconds in CAV to 358/496 seconds in BICAV with 70 blocks, using a blob with radius 2.0/2.795, respectively). As happened with the speedup, we speculate that this behavior may be caused by the communication startup times.

### 4.2 Application to Electron Tomography of Real Mitochondria Data

In this work, we have applied the BICAV and WBP reconstruction methods to real mitochondria data obtained from HVEM and prepared using photooxidation procedures, which are characterized by the low contrast and the extremely low signal to noise ratio (approaching SNR = 1.0) exhibited by the images. Seventy projection images (with tilt angles in the range $[-70, +68]$) were combined to obtain the reconstructions. The projection images were 1024×480, and the volume 1024×480×256. We have tested the algorithms under different conditions, but here only the most significant results are presented.

A montage showing one z-section of the volume reconstructed with the different methods is presented in Fig. 6. Fig. 6(a) shows the result coming from WBP. Fig. 6(b) shows the reconstruction obtained from BICAV with 70 blocks

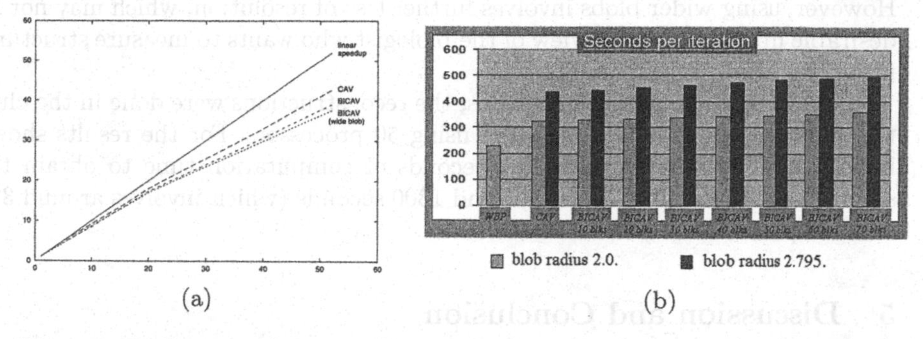


**Fig. 5.** Evaluation of the parallel approach. The measures were obtained from reconstruction processes of a 512×512×512 volume from 70 projection images using blob radius of 2.0 (standard blob) and 2.795 (wide blob). (a) Speedup exhibited by the component averaging methods (CAV and BICAV with 70 blocks). (b) Computation times for component averaging methods and WBP

after 4 iterations with a relaxation factor of 1.0 and a blob with radius 2.0. We have also made tests with CAV, but the solutions are still blurred after 30-50 iterations, due to its relatively slow convergence rate compared to BICAV.

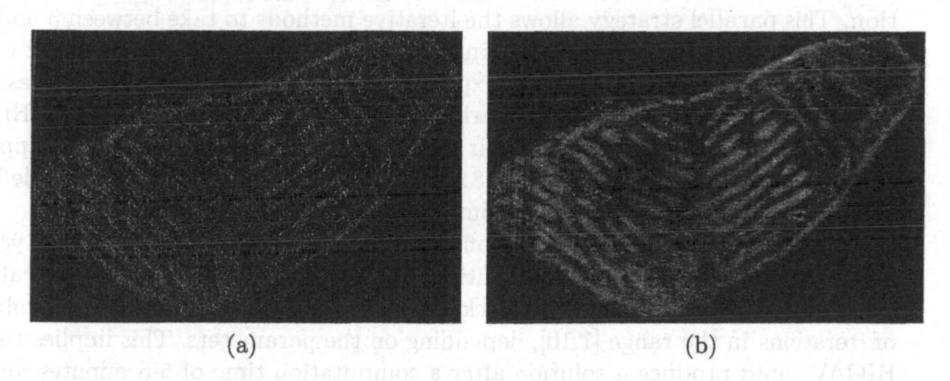


**Fig. 6.** Comparison of WBP and component averaging methods in an experimental application. One of the slices along the Z axis is shown. (a) Result from WBP. (b) Result from BICAV with 70 blocks after 4 iterations

Fig. 6 clearly shows that blob-based BICAV yields a solution much cleaner than WBP and, moreover, at the same resolution level. The excellent behavior exhibited by BICAV under so noisy situations comes from the regularization nature of blobs. However, WBP provides a "noisy" solution due to the high noise level in the experimental projection images. We have also tested wider blobs in the BICAV methods, and the results prove to be smoother than those presented.

However, using wider blobs involves further loss of resolution, which may not be desirable from the point of view of the biologist who wants to measure structural features in the reconstructions.

Regarding the computation times, the reconstructions were done in the cluster of workstations at the NCMIR, using 50 processors. For the results shown in Fig. 6, WBP took around 200 seconds of computation time to obtain the solution, whereas BICAV took around 1300 seconds (which involves around 325 seconds per iteration).

## 5 Discussion and Conclusion

In this work we have analyzed the application of blob-based series expansion methods in electron tomography of complex biological specimens, with a special emphasis on the computational perspective. First, we have made use of efficient iterative methods to tackle the problem of image reconstruction. On the other hand, high performance computing (HPC) techniques have been applied so as to face the high computational demands and take advantage of parallel systems.

A parallel approach for the iterative algorithms has been devised, exhibiting a speedup nearly linear with the number of processors in the system. This approach has also proved to be very efficient to deal with the communications among the processors. The results indicate that the latency due to communications is almost completely hidden by overlapping computation and communication. This parallel strategy allows the iterative methods to take between 5 and 8 computation minutes per iteration in the reconstruction of a $512 \times 512 \times 512$ volume. The parallel approach is also exploited for WBP in such a way that a result of the same size is obtained after nearly 4 computation minutes. In this way, High Performance Computing is making it possible to afford "grand challenge" applications (e.g., reconstructions of $2048 \times 2048 \times 2048$) currently unapproachable by uni-processor systems due to the computational resources requirements.

The new iterative reconstruction methods we have applied here are very efficient, providing least squares solutions in a very few of iterations. Specifically, BICAV with a large number of blocks yields good reconstructions in a number of iterations in the range [1,10], depending on the parameters. This implies that BICAV could produce a solution after a computation time of 5-8 minutes for a $512 \times 512 \times 512$ volume, compared to 4 minutes needed by WBP.

On the other hand, the use of blobs has a two-fold benefit. First, blobs are better than voxels from the computational point of view, since they allow the use of look-up tables to speed up the forward- and backward- projection stages. Second, blobs provide the reconstruction algorithms with an implicit regularization mechanism which makes them well suited for noisy environments. As a consequence, the solutions yielded by blob-based iterative methods are smoother than those by WBP, but with relatively unimpaired resolution. In particular, under extremely noisy conditions, this type of algorithms clearly outperform WBP. In those situations, WBP would require a strong low-pass filtering post-reconstruction stage which would limit the maximum resolution attainable.

Regularized iterative reconstruction methods have additional and interesting implications in data post-processing, particularly the segmentation of the structural components from the reconstruction. The results yielded by regularized methods usually present smoothly varying densities that are specially propitious for automatic segmentation techniques.

Finally, this work has shown that the combination of efficient iterative methods and HPC is well suited to tackle the reconstruction of large biological specimens in electron tomography, yielding solutions in reasonable computation times.

## Acknowledgments

The authors wish to thank Dr. G. Perkins who kindly provided the real mitochondria data and contributed with valuable comments on the reconstructions. The authors also wish to thank C.O.S Sorzano and J.G. Donaire for fruitful discussions during this work. This work has been partially supported through grants from the Spanish CICYT TIC99-0361, TIC2002-00228, BIO98-0761 and BIO2001-1237, from the Spanish MECD PR2001-0110, from the Commission for Cultural, Educational and Scientific Exchange between USA and Spain, grant #99109. This work was also supported by the NIH/National Center for Research Resources through grants P41-RR08605 and P41-RR04050. Some of this work was performed using NSF-NPACI facilities under grant CISE NSF-ASC 97-5249.

## References

[1] J. Frank, editor. *Electron Tomography. Three-Dimensional Imaging with the Transmission Electron Microscope.* Plenum Press, 1992.

[2] G. A. Perkins, C. W. Renken, J. Y. Song, T. G. Frey, S. J. Young, S. Lamont, M. E. Martone, S. Lindsey, and M. H. Ellisman. Electron tomography of large, multicomponent biological structures. *J. Struc. Biol.*, 120:219–227, 1997.

[3] D. N. Mastronarde. Dual-axis tomography: An approach with alignment methods that preserve resolution. *J. Struct. Biol.*, 120:343–352, 1997.

[4] R. Marabini, E. Rietzel, E. Schroeder, G. T. Herman, and J. M. Carazo. Three-dimensional reconstruction from reduced sets of very noisy images acquired following a single-axis tilt schema. *J. Struct. Biol.*, 120:363–371, 1997.

[5] R. Marabini, G. T. Herman, and J. M. Carazo. 3D reconstruction in electron microscopy using ART with smooth spherically symmetric volume elements (blobs). *Ultramicroscopy*, 72:53–56, 1998.

[6] R. M. Lewitt. Alternatives to voxels for image representation in iterative reconstruction algorithms. *Phys. Med. Biol.*, 37:705–716, 1992.

[7] S. Matej, R. M. Lewitt, and G. T. Herman. Practical considerations for 3-D image reconstruction using spherically symmetric volume elements. *IEEE Trans. Med. Imag.*, 15:68–78, 1996.

[8] S. Matej, G. T. Herman, T. K. Narayan, S. S. Furuie, R. M. Lewitt, and P. E. Kinahan. Evaluation of task-oriented performance of several fully 3D PET reconstruction algorithms. *Phys. Med. Biol.*, 39:355–367, 1994.

[9] P. E. Kinahan, S. Matej, J. S. Karp, G. T. Herman, and R. M. Lewitt. A comparison of transform and iterative reconstruction techniques for a volume-imaging PET scanner with a large axial acceptance angle. *IEEE Trans. Nucl. Sci.*, 42:2281–2287, 1995.

[10] Y. Censor, D. Gordon, and R. Gordon. Component averaging: An efficient iterative parallel algorithm for large and sparse unstructured problems. *Parallel Computing*, 27:777–808, 2001.

[11] Y. Censor, D. Gordon, and R. Gordon. BICAV: A block-iterative, parallel algorithm for sparse systems with pixel-related weighting. *IEEE Trans. Med. Imag.*, 20:1050–1060, 2001.

[12] G. T. Herman. *Medical Imaging. Systems, Techniques and Applications. C. Leondes (Ed.)*, chapter Algebraic Reconstruction Techniques in Medical Imaging, pages 1–42. Gordon and Breach Science, 1998.

[13] B. Wilkinson and M. Allen. *Parallel Programming.* Prentice Hall, 1999.

# Visualization of RNA Pseudoknot Structures

Wootaek Kim[1], Yujin Lee[2], and Kyungsook Han[2]

[1] Department of Automation Engineering, Inha University
[2] Department of Computer Science and Engineering, Inha University
Inchon 402-751, South Korea
khan@inha.ac.kr

**Abstract.** RNA pseudoknots are not only important structural elements for forming the tertiary structure, but also responsible for several functions of RNA molecules such as frame-shifting, read-through, and the initiation of translation. There exists no automatic method for drawing RNA pseudoknot structures, and thus representing RNA pseudoknots currently relies on significant amount of manual work. This paper describes the first algorithm for automatically generating a drawing of H-type pseudoknots with RNA secondary structures. Two basic criteria were adopted when designing the algorithm: (1) overlapping of structural elements should be minimized to increase the readability of the drawing, and (2) the whole RNA structure containing pseudoknots as well as the pseudoknots themselves should be recognized quickly and clearly. Experimental results show that this algorithm generates uniform and clear drawings with no edge crossing for all H-type RNA pseudoknots. The algorithm is currently being extended to handle other types of pseudoknots.

## 1 Introduction

Visualization of complex molecular structures helps people understand the structures, and is a key component of support tools for many applications in biosciences. This paper describes a new algorithm for automatically producing a clear and aesthetically appealing drawing of RNA pseudoknot structures.

A pseudoknot is a tertiary structural element of RNA formed when bases of a single-stranded loop pair with complementary bases outside the loop. Pseudoknots are not only widely occurring structural motifs in all kinds of viral RNA molecules, but also responsible for several important functions of RNA. For example, pseudoknot structures present in coding regions can stimulate ribosomal frame-shifting and translational read-through during elongation. Pseudoknots in non-coding regions can initiate translation either by being part of the so-called internal ribosomal entry site (IRES) in the 5′ non-coding region or by forming translational enhancers in the 3′ non-coding region [1]. Although 14 types of topologically distinct pseudoknots are possible according to this broad definition [2], most of the pseudoknots are of the so-called H-type, where H stands for hairpin loop, in which bases in a hairpin loop pair with complementary bases outside the hairpin loop (see Figure 1).

J.M.L.M. Palma et al. (Eds.): VECPAR 2002, LNCS 2565, pp. 181–194, 2003.

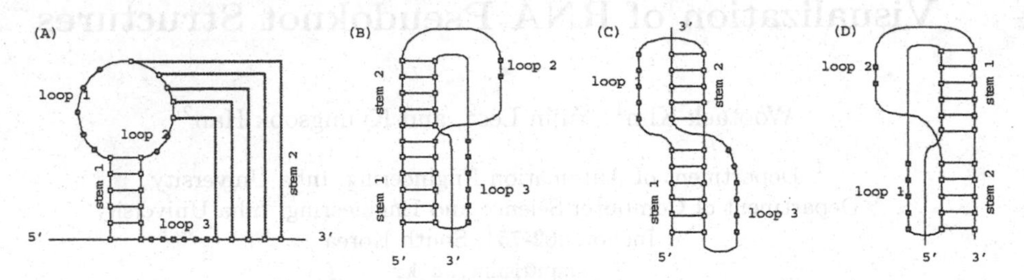


**Fig. 1.** Schematic representation of H-type pseudoknots, adapted from [3]. (A) General configuration. (B) Loop 1 is eliminated. (C) Loop 2 is eliminated. (D) Loop 3 is eliminated. The pseudoknot shown in (C) is the one most abundant in natural RNA

There currently exists no automatic method for drawing RNA pseudoknot structures. Several computer programs have been developed for drawing RNA structures (for example, [4, 5, 6, 7], but all of them are intended for drawing RNA secondary structures only. Therefore, representing RNA pseudoknots heavily relies on manual work. In a sense of graph theory, a drawing of RNA pseudoknots is a graph (and possibly non-planar graph) while that of RNA secondary structures is a tree. Thus drawing RNA pseudoknot structures is computationally much harder than RNA secondary structures. One of the difficulties in drawing RNA structures is overlapping of structural elements, which reduces the readability of the drawing. In most drawing programs of RNA secondary structures, computational load is increased because of the work associated with removing overlap of structural elements, performed either by an iterative process of the programs or with user intervention.

We have developed an algorithm for generating overlap-free drawings of H-type pseudoknot structures, and implemented the algorithm in a working program called PSEUDOVIEWER. This is the first algorithm for automatically drawing RNA pseudoknot structures. We adopted two basic criteria when designing the algorithm: (1) Overlapping of structural elements should be minimized to increase the readability of the drawing, and (2) Not only pseudoknots themselves but also the whole RNA structure containing pseudoknots should be recognized quickly and clearly. The rest of this paper describes the algorithm and its experimental results.

## 2 Preliminaries

An RNA molecule is made up of ribonucleotides linked together in a chain. Each ribonucleotide comprises one of four different bases, adenine (A), cytosine (C), guanine (G), and uracil (U). Single-stranded RNA often folds back onto itself in structures stabilized by hydrogen bonds between A and U and between G and C. The so-called G-U wobble pair is also frequently found although it may have a weaker bond than the A-U and G-C pairs.

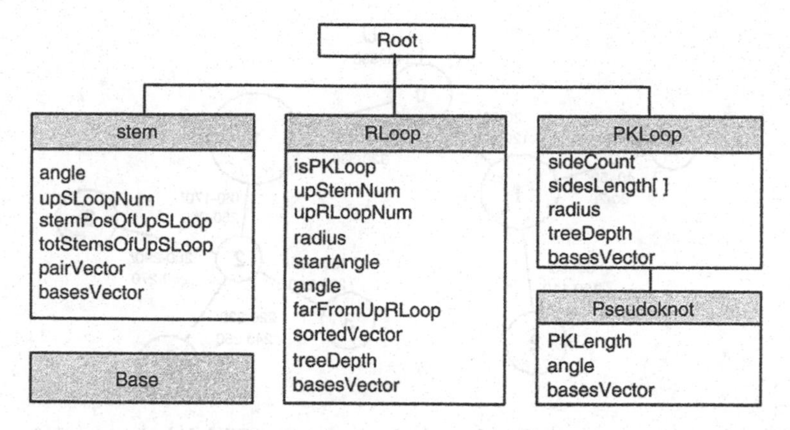


**Fig. 2.** The relationship of the classes representing the structural elements. RLoop stands for a regular loop, and PKLoop stands for a pseudoknot loop

In RNA secondary structures, there are two types of structural elements:

- Stem (also called helix): double-stranded part, which is a contiguous region of base pairs.
- Regular loop: single-stranded part such as a hairpin loop, internal loop, bulge loop, multiple loop, or dangling end.

Since our algorithm draws pseudoknot structures, we consider additional structural elements as well:

- Pseudoknot: structural element formed by pairing of bases in a regular loop with complementary bases outside the loop (see Figure 1 for an example).
- Pseudoknot loop: loop that contains a pseudoknot as well as single-stranded part (see Figure 9 for an example).

Figure 2 shows the relationship of the classes representing the above structural elements. Every class has a variable called basesVector, which contains the bases of the class.

# 3 Drawing RNA Pseudoknots

As mentioned earlier, a drawing of RNA pseudoknots is a graph with inner cycles within a pseudoknot as well as possible outer cycles formed between a pseudoknot and other structural elements. What we call a "PK loop" represents the outer cycle. Given RNA pseudoknots and secondary structures, we represent the whole structure as a tree rather than as a graph (see Figure 3 for an example of the abstract tree). This is possible by representing both regular loops and PK loops as nodes of the tree. Edges of the tree represent stems of the secondary structure. The root node of the tree is the loop with the smallest starting base number. If

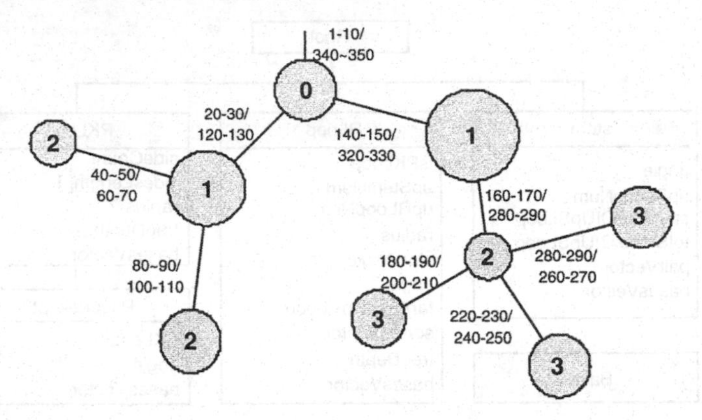


**Fig. 3.** An abstract tree for the whole structure. The label in each loop represents the level (i.e., depth) of the loop in the abstract tree. The labels in each stem represent the starting and ending base numbers of the stem. For example, the stem with label 1-10/340-350 indicates the stem is formed by pairing bases 1-10 with bases 340-350

there is a dangling end, we add artificial bases to pair the first and last bases. The artificial bases introduced in this step are not actually shown in the final drawing. A pseudoknot itself is not represented in the abstract tree; it is part of a node of the tree.

By hiding the inner cycles as well as the outer cycles in the nodes of the abstract tree, we can represent the whole, top-level RNA structure as a tree, making the drawing process simple. Loops of the tree are placed and drawn in increasing order of their depth values (the root node has the smallest depth value). The member treeDepth of classes RLoop and PKLoop represents the depth of the loop in the abstract tree. The outline of both a regular loop and a PK loop is drawn in a circle shape. The algorithm of PSEUDOVIEWER is outlined as follows:

1. Stems, regular loops, pseudoknots and PK loops are identified from the input structure data.
2. An abstract tree is constructed for representing the entire structure.
3. For each node and edge of the tree, its size and shape is determined.
4. Starting with the root node, each node and edge of the tree is positioned level by level by both translation and rotation.

Following principles are used in drawing RNA pseudoknot structures.

– All angles are measured with respect to the positive y-axis.
– Units of angles are radians rather than degrees, unless specified otherwise.
– Modulus operator (%) is used for computing angle values since angle values should be in the range $[0, 2\pi]$.
– In the drawing, unfilled circles represent the bases of a stem while filled circles represent the bases in the single-stranded part.

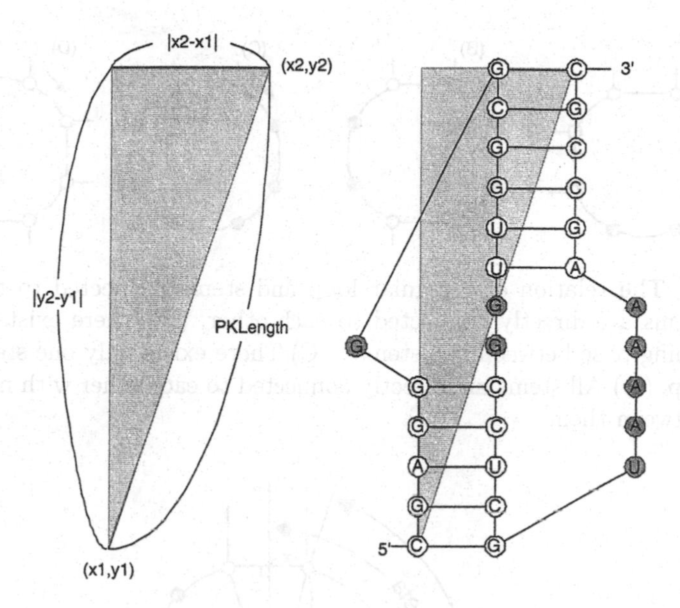


**Fig. 4.** Drawing of a pseudoknot

### 3.1 Pseudoknots

PSEUDOVIEWER takes as input an ASCII file in pairing format, which is widely used for representing pseudoknots [9]. The paring format describes pseudoknots as well as secondary structures in the following style. In this example, a pseudoknot is formed by base pairing $G_{56}CGGUU_{61}$ with $A_{74}GCCGC_{79}$.

```
  50        60        70
$ CGAGGGGCGGUUGGCCUCGUAAAAAGCCGC
% (((((:[[[[[[::))))):::::]]]]]]
```

Given this input, PSEUDOVIEWER generates a drawing of a pseudoknot as shown in Figure 4. The data member PKLength of class Pseudoknot represents the diagonal length of a bounding box of the pseudoknot and is computed by

$$PKLength = \sqrt{(x_{max} - x_{min})^2 + (y_{max} - y_{min})^2} \tag{1}$$

$x_{min}, y_{min}$: minimum x and y coordinates of the bounding box
$x_{max}, y_{max}$: maximum x and y coordinates of the bounding box

### 3.2 Regular Loops

The relation of a regular loop and stems connected to the loop can be classified into 4 types, as shown in Figure 5.

The outline of a regular loop is drawn in a circle shape. In Figure 6, the distance between the centers of adjacent bases of a regular loop is $2x$. Since

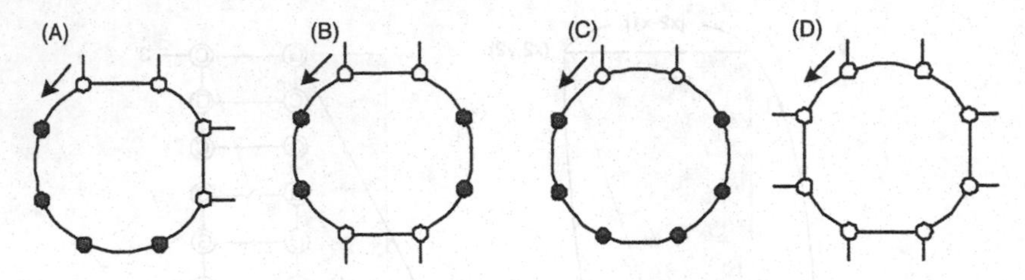


**Fig. 5.** The relation of a regular loop and stems connected to the loop. (A) Two stems are directly connected to each other. (B) There exists at least one intervening base between two stems. ( C) There exists only one stem connected to a loop. (D) All stems are directly connected to each other with no intervening base between them

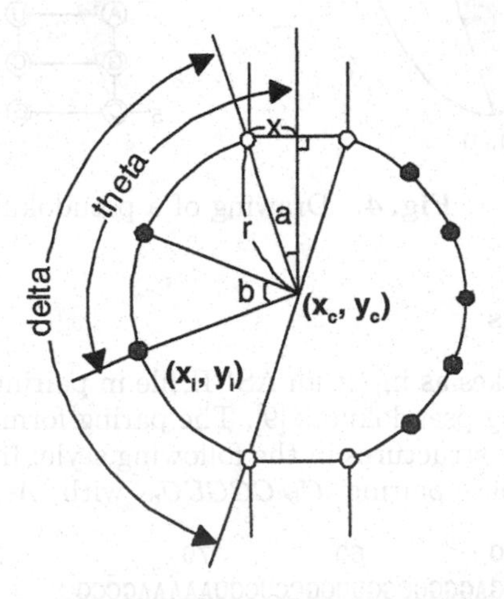


**Fig. 6.** The radius of a regular loop and the angle between adjacent stems

a regular loop is considered to consist of several isosceles triangles, the vertical angle can be computed using equation 2, where $n_{lb}$ is the number of bases on the regular loop.

$$a = \frac{1}{2} \cdot \frac{2\pi}{n_{lb}} = \frac{\pi}{n_{lb}} \tag{2}$$

The radius of a regular loop can be easily computed using equation 3.

$$r = \frac{x}{\sin a} \tag{3}$$

Once we determine the radius of a regular loop, we can calculate the angle between adjacent stems and the angle $\theta_i$ between the i-th base of a regular loop and the positive y-axis. Let $n_{lb}$ be the number of bases on a regular loop and $n_b$ be the number of intervening bases between adjacent stems. Then,

$$\delta = 2a(n_b + 2) \tag{4}$$

$$\theta_i = (2i + 1)a, \ \ i = 0, 1, 2, \ldots, n_{lb} - 1 \tag{5}$$

Using the angles computed by equations 2-5, the bases of a regular loop can be positioned as follows.

$$x_i = -r \sin \theta_i + x_c \tag{6}$$

$$y_i = r \cos \theta_i + y_c \tag{7}$$

$x_i, y_i$: x and y coordinates of the i-th base of a regular loop
$x_c, y_c$: x and y coordinates of the center of a regular loop
$\theta_i$: angle between the i-th base of a regular loop and the positive y-axis
$r$: radius of a regular loop

In Figure 7, an arrow points to the base with the smallest number of a stem. In the first regular loop, both the stem angle $\phi$ of the first stem and the startAngle

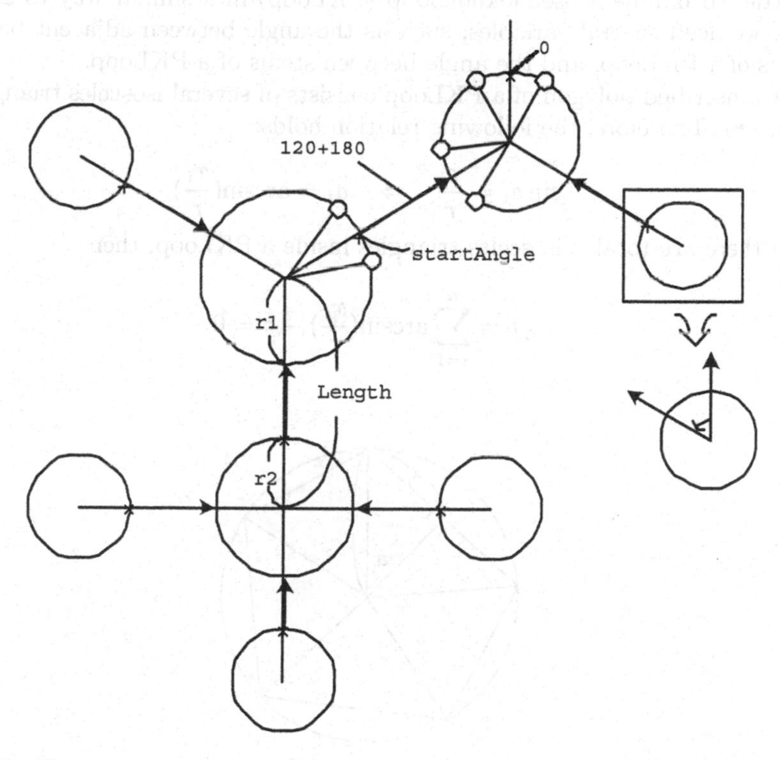


**Fig. 7.** The stem angle, startAngle, and the distance between the centers of regular loops

of the loop are zeros. The stem angles $\phi$ of other stems are measured with respect to the positive y-axis, and are calculated using equation 8.

$$\phi = D_{loop} + \pi + \delta \tag{8}$$

where $D_{loop}$ is the startAngle of the upper regular loop. To make the stem point to the upper regular loop, $\pi$ is added in computing the stem angle $\phi$. The StartAngle of the lower regular loop is equal to the stem angle of the current stem. The distance $D_{rl}$ between the centers of regular loops is computed using equation 9.

$$D_{rl} = r_c + (n_{ub} - 1) \times h_s + r_u \tag{9}$$

$r_c$: radius of the current regular loop
$r_u$: radius of the upper regular loop
$n_{ub}$: number of bases of the upper stem
$h_s$: distance between adjacent base pairs of a stem

### 3.3 Pseudoknot Loops

In order to handle a pseudoknot loop (PKLoop) in a similar way to a regular loop, we need several variables, such as the angle between adjacent bases, the radius of a PKLoop, and the angle between stems of a PKLoop.

An inscribed polygon of a PKLoop consists of several isosceles triangles (see Figure 8). Therefore, the following relation holds:

$$\sin a_i = \frac{x_i}{r} \quad \Rightarrow \quad a_i = \arcsin(\frac{x_i}{r}) \tag{10}$$

If there are total $n$ isosceles triangles inside a PKLoop, then

$$f = \sum_{i=1}^{n} \arcsin(\frac{x_i}{r}) - \pi = 0 \tag{11}$$

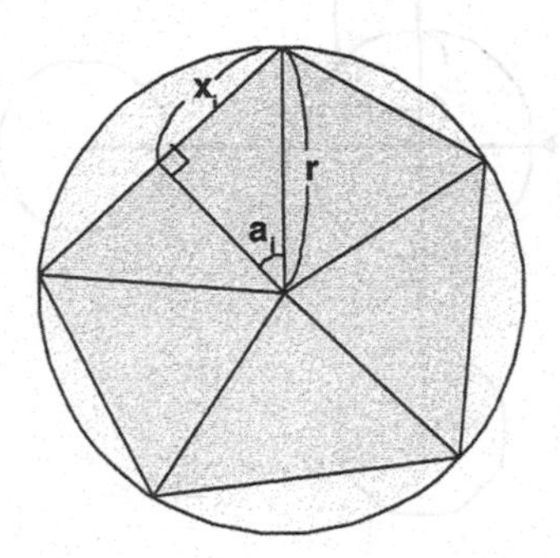


**Fig. 8.** An inscribed polygon of a PKLoop consists of several isosceles triangles

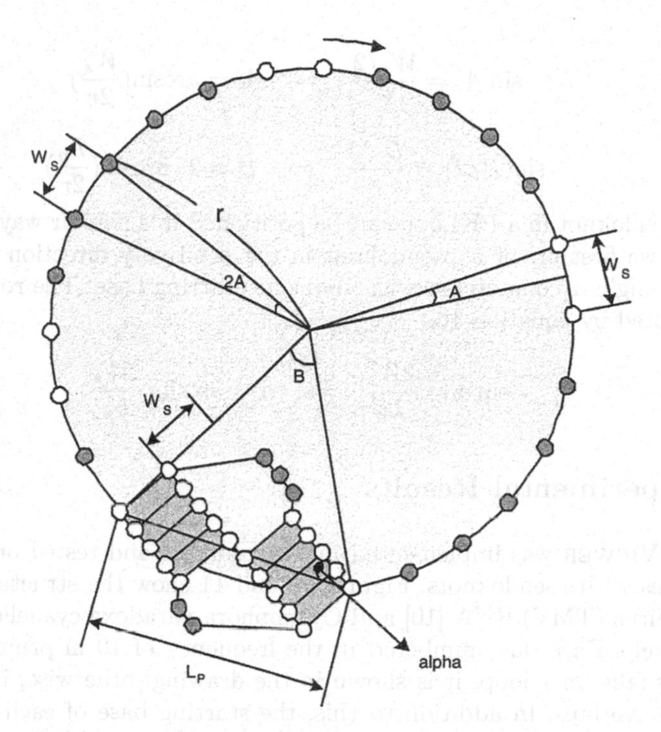


**Fig. 9.** Components of a PKLoop. $W_s$: width of a stem, which equals the distance between adjacent bases of a loop, $L_p$: diagonal length of a pseudoknot, $r$: radius of a PKLoop, $\alpha$: angle between the diagonal direction and the stem direction of a pseudoknot

Function $f$ is monotonically decreasing after a certain point, and Newton's method can be used to determine $r$ that satisfies $f = 0$. However Newton's method requires differentiation of function $f$, which takes a significant amount of computation as shown in equation 12.

$$\frac{df}{dr} = \sum_{i=1}^{n} \frac{-x_i}{r^2\sqrt{1 - \frac{x_i^2}{r^2}}} = \sum_{i=1}^{n} \frac{-x_i}{r\sqrt{r^2 - x_i^2}} \tag{12}$$

Instead of using Newton's method, we determine the value of $r$ by incrementing $r$ by a small step. Once we determine the radius $r$ of a PKLoop, several angles associated with a PKLoop can be computed using equations 13-16 (see Figure 9). Let $L_p$ be the diagonal length of the pseudoknot region, $x_e$ and $y_e$ be the coordinates of the ending base of a pseudoknot, and $x_s$ and $y_s$ be the coordinates of the starting base of a pseudoknot. Then,

$$L_p = \sqrt{(x_e - x_s)^2 + (y_e - y_s)^2} \tag{13}$$

$$\sin A = \frac{W_s/2}{r} \quad \Rightarrow \quad A = \arcsin(\frac{W_s}{2r}) \tag{14}$$

$$\sin(B/2) = \frac{L_p/2}{r} \quad \Rightarrow \quad B = 2 \cdot \arcsin(\frac{L_p}{2r}) \tag{15}$$

A pseudoknot in a PKLoop can be positioned in a similar way to a stem. To do this, we first orient a pseudoknot in the positive y direction by rotating it through angle $\alpha$ counterclockwise about its starting base. The rotation angle $\alpha$ is computed by equation 16.

$$\sin \alpha = \frac{2W_s}{L_p} \quad \Rightarrow \quad \alpha = \arcsin(\frac{2W_s}{L_p}) \tag{16}$$

## 4 Experimental Results

PSEUDOVIEWER was implemented in Microsoft C♯ and tested on several RNA structures with pseudoknots. Figures 10 and 11 show the structures of tobacco mosaic virus (TMV) RNA [10] and Cyanophora paradoxa cyanelle tmRNA [11], respectively. Bases are numbered in the frequency of 10 in principle. If a base numbers falls on a loop, it is shown in the drawing; otherwise, it is not shown to avoid overlaps. In addition to this, the starting base of each pseudoknot is shown in green background color with its base number. Pseudoknots are shown in yellow background color, and thus are easily distinguished from other structural elements. Filled circles represent the canonical base pairs (A-U or G-C), and open circles represent the G-U wobble pairs. There is no overlapping of structural elements, and the PK loop with several pseudoknots takes on a circle shape similar to a regular loop. When drawing a pseudoknot by first generating a secondary structure with an ordinary secondary drawing program and then modifying the secondary structure, pseudoknots as well as other structural elements are often distorted, and thus the readability of the drawing is much reduced.

In addition to the standard drawings, as shown in Figures 10 and 11, PSEUDOVIEWER produces outline drawings as well. The outline drawings display the structure in the form of a backbone in which loops are replaced by circles and helices by line segments. PSEUDOVIEWER also provides an interactive editing facility for manually fixing drawings when overlaps occur. In the editing mode, the user can drag a regular loop to any position.

## 5 Conclusion and Future Work

Generating a clear representation of RNA pseudoknot structures is computationally harder than RNA secondary structures since a drawing of pseudoknots is a graph (and possibly nonplanar) rather than a tree and contains more overlapping of structural elements. We have developed a new algorithm for visualizing H-type RNA pseudoknots as a planar graph and implemented the algorithm in a

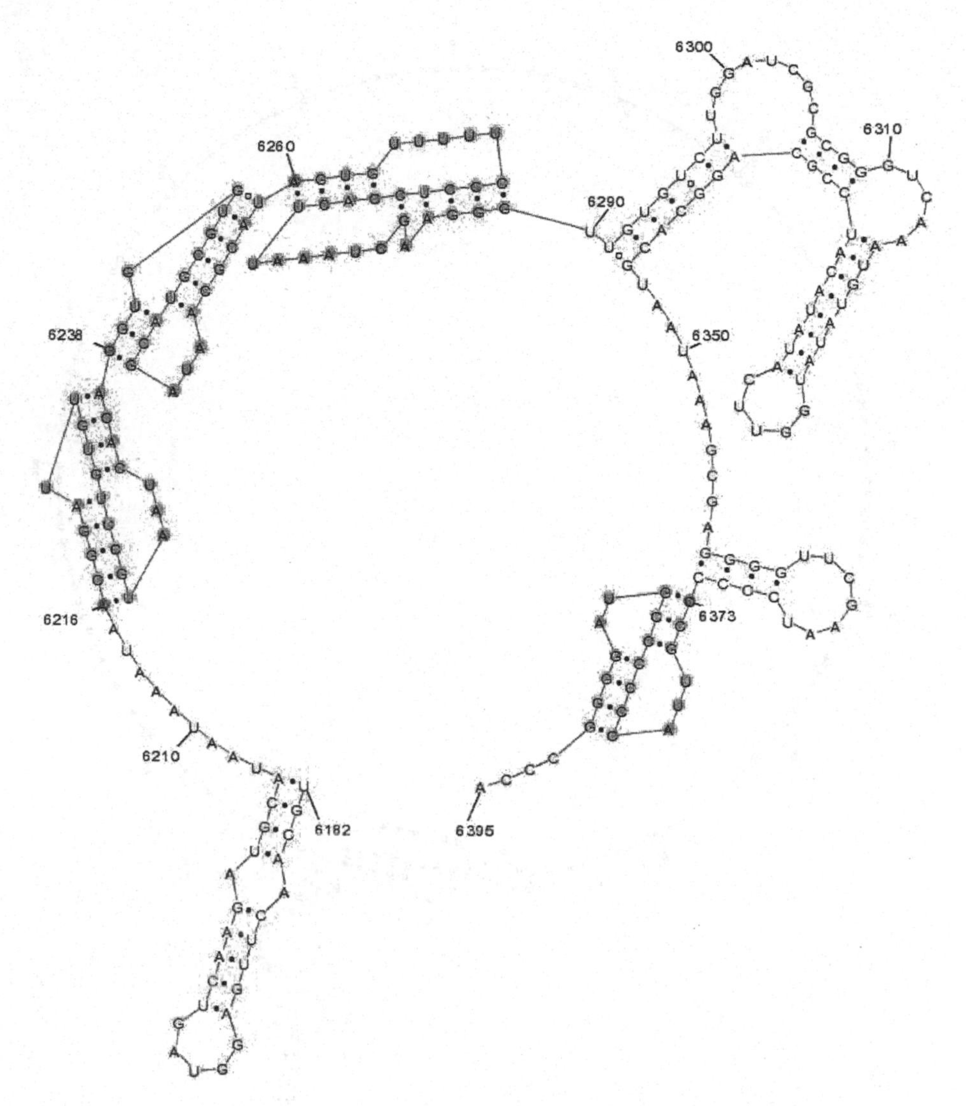


**Fig. 10.** Structure of tobacco mosaic virus, visualized by PSEUDOVIEWER

working program called PSEUDOVIEWER. We believe PSEUDOVIEWER is the first program for automatically drawing RNA secondary structures containing H-type pseudoknots. We adopted two basic criteria when designing PSEUDOVIEWER: (1) overlapping of structural elements should be minimized to increase the readability of the drawing, and (2) the whole RNA structure containing pseudoknots as well as the pseudoknots themselves should be recognized quickly and clearly. Experimental results showed that PSEUDOVIEWER is capable of generating a clear and aesthetically pleasing drawing of all H-type RNA pseudoknots.

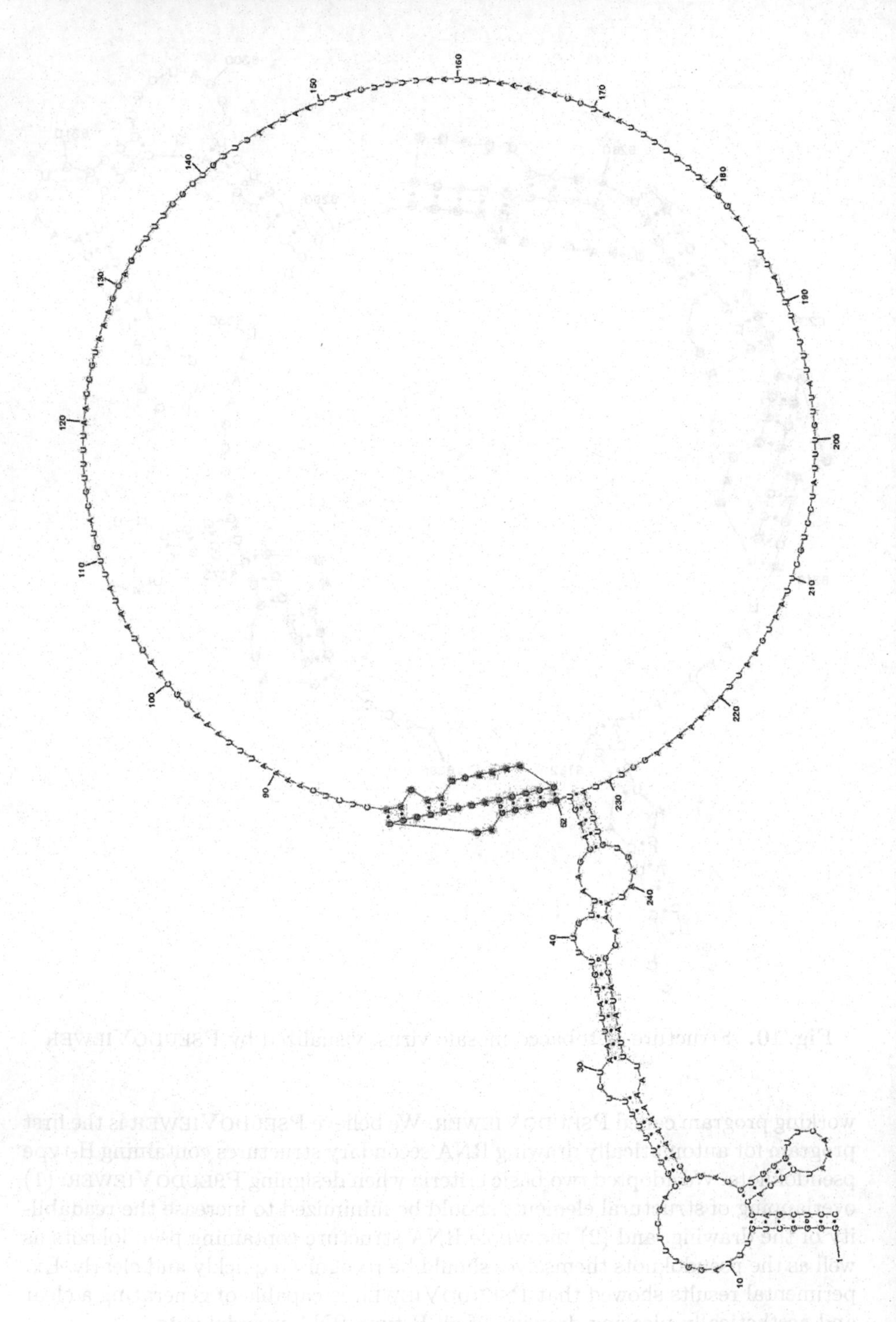

**Fig. 11.** Structure of Cyanophora paradoxa cyanelle tmRNA

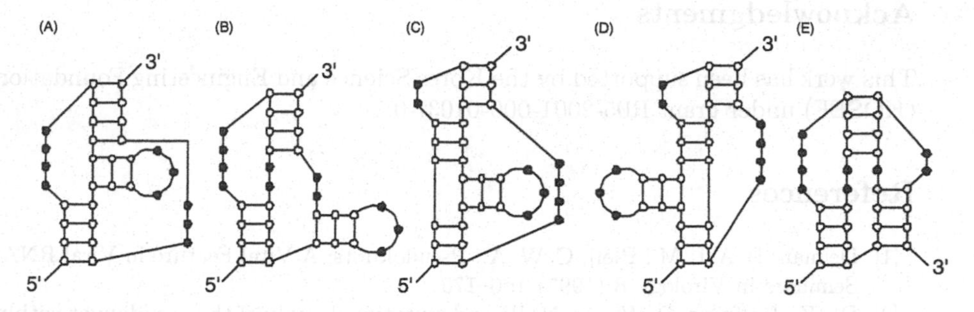


**Fig. 12.** Pseudoknots of other types. (A) LL type. (B) HL_OUT type. (C) HL_IN type. (D) HH type. (E) HHH type. H: hairpin loop, L: bulge loop, internal loop, or multiple loop

**Table 1.** Classification of 236 pseudoknots in PseudoBase

| Pseudoknot type | Number of occurrences | Ratio of occurrences |
|---|---|---|
| H | 180 | 76.3% |
| LL | 12 | 5.1% |
| HL_OUT | 24 | 10.2% |
| HL_IN | 11 | 4.7% |
| HH | 1 | 0.4% |
| HHH | 6 | 2.5% |
| Unclassified | 2 | 0.8% |
| total | 236 | 100.0% |

The development of PSEUDOVIEWER is not complete yet, and is being extended in several directions. First, it can currently visualize typical H-type pseudoknots only. Although the most commonly occurring pseudoknots are of the H-type, PSEUDOVIEWER is being extended to handle other types of pseudoknots as well. Basepairing of a hairpin loop with a single stranded part outside the hairpin loop forms the H-type pseudoknot, which has been discussed in this paper. Basepairing of a hairpin loop with another hairpin loop forms a HH type pseudoknot, while basepairing of a hairpin loop with a single stranded part of a bulge, internal, or multiple loop forms a HL type (Figure 12). We have analyzed all the pseudoknots of PseudoBase [9] for the purpose of drawing them in combination of basic pseudoknot types and concluded that there are 6 basic pseudoknot types in total – H-type and 5 other types displayed in Figure 12 (see also Table 1).

Second, PSEUDOVIEWER will be made available as a web-based application program so that it can be executed anywhere using a general web browser.

Finally, PSEUDOVIEWER currently represents RNA pseudoknots as a two-dimensional drawing. We would like to display RNA pseudoknots as a three-dimensional drawing. Perhaps this would be the ultimate goal of our visualization work of RNA pseudoknots.

## Acknowledgments

This work has been supported by the Korea Science and Engineering Foundation (KOSEF) under grant R05-2001-000-01037-0.

## References

[1] Deiman, B. A. L. M., Pleij, C. W. A.: Pseudoknots: A Vital Feature in Viral RNA. Seminars in Virology **8** (1997) 166–175

[2] Du, Z., Hoffman, D. W.: An NMR and mutational study of the pseudoknot within the gene 32 mRNA of bacteriophage T2: insights into a family of structurally related RNA pseudoknots. Nucleic Acids Research **25** (1997) 1130–1135

[3] Hilbers, C. W., Michiels, P. J. A., Heus, H. A.: New Developments in Structure Determination of Pseudoknots. Biopolymers **48** (1998) 137–153

[4] Han, K., Kim, D., Kim, H.-J.: A vector-based method for drawing RNA secondary structure. Bioinformatics **15** (1999) 286–197

[5] Chetouani, F., Monestie, P., Thebault, P., Gaspin, C., Michot, B.: ESSA: an integrated and interactive computer tool for analyzing RNA secondary structure. Nucleic Acids Research **25** (1997) 3514–3522

[6] Shapiro, B. A., Maizel, J., Lipkin, L. E., Currey, K., Whitney, C.: Generating non-overlapping displays of nucleic acid secondary structure. Nucleic Acids Research **12** (1984) 75–88

[7] Winnepenninckx, B., Van de Peer, Y., Backeljau, T., De Wachter, R.: CARD: A Drawing Tool for RNA Secondary Structure Models. BioTechniques **16** (1995) 1060–1063

[8] De Rijk, P., De Wachter, R.: RnaVIz, a program for the visualization of RNA secondary structure. Nucleic Acids Research **25** (1997) 4679–4684

[9] van Batenburg, F. H. D., Gultyaev, A. P., Pleij, C. W. A., Ng, J., Olihoek, J.: PseudoBase: a database with RNA pseudoknots. Nucleic Acids Research **28** (2000) 201–204

[10] Gultyaev, A. P., van Batenburg, E. Pleij, C. W. A.: Similarities between the secondary structure of satellite tobacco mosaic virus and tobamovirus RNAs. Journal of General Virology **75** (1994) 2851–2856

[11] Knudsen, B., Wower, J., Zwieb, C., Gorodkin, J.: tmRDB (tmRNA database). Nucleic Acids Research **29** (2001) 171–172

# Chapter 4

# Problem Solving Environments

# The Cactus Framework and Toolkit: Design and Applications
## Invited Talk

Tom Goodale[1], Gabrielle Allen[1], Gerd Lanfermann[1], Joan Massó[2], Thomas Radke[1], Edward Seidel[1], and John Shalf[3]

[1] Max-Planck-Institut für Gravitationsphysik, Albert-Einstein-Institut
Am Mühlenberg 1, 14476 Golm, Germany
{goodale,allen,lanfer,tradke,eseidel}@aei.mpg.de
[2] Departament de Física, Universitat de les Illes Balears
E-07071 Palma de Mallorca, Spain
jmasso@gridsystems.com
[3] Lawrence Berkeley National Laboratory
Berkeley, CA
jshalf@lbl.gov

**Abstract.** We describe Cactus, a framework for building a variety of computing applications in science and engineering, including astrophysics, relativity and chemical engineering. We first motivate by example the need for such frameworks to support multi-platform, high performance applications across diverse communities. We then describe the design of the latest release of Cactus (Version 4.0) a complete rewrite of earlier versions, which enables highly modular, multi-language, parallel applications to be developed by single researchers and large collaborations alike. Making extensive use of abstractions, we detail how we are able to provide the latest advances in computational science, such as interchangeable parallel data distribution and high performance IO layers, while hiding most details of the underlying computational libraries from the application developer. We survey how Cactus 4.0 is being used by various application communities, and describe how it will also enable these applications to run on the computational *Grids* of the near future.

## 1 Application Frameworks in Scientific Computing

Virtually all areas of science and engineering, as well as an increasing number of other fields, are turning to computational science to provide crucial tools to further their disciplines. The increasing power of computers offers unprecedented ability to solve complex equations, simulate natural and man-made complex processes, and visualise data, as well as providing novel possibilities such as new forms of art and entertainment. As computational power advances rapidly, computational tools, libraries, and computing paradigms themselves also advance. In such an environment, even experienced computational scientists and engineers can easily find themselves falling behind the pace of change, while they

J.M.L.M. Palma et al. (Eds.): VECPAR 2002, LNCS 2565, pp. 197–227, 2003.

redesign and rework their codes to support the next computer architecture. This rapid pace of change makes the introduction of computational tools even more difficult in non-traditional areas, e.g., social arts and sciences, where they may potentially have the most dramatic impact.

On top of this rapidly changing background of computation, research and engineering communities find themselves more and more dependent on each other as they struggle to pull together expertise from diverse disciplines to solve problems of increasing complexity, e.g., simulating a supernova or the Earth's climate. This creates a newfound need for *collaborative* computational science, where different groups and communities are able to co-develop code and tools, share or interchange modules and expertise, with the confidence that everything will work. As these multidisciplinary communities develop ever more complex applications, with potentially hundreds or thousands of modules and parameters, the potential for introducing errors, or for incorrectly using modules or setting parameters, also increases. Hence a software architecture that supports a wide range of automated consistency checking and multi-architecture verification becomes essential infrastructure to support these efforts.

Application developers and users also need to exploit the latest computational technologies without being exposed to the raw details of their implementation. This is not only because of the time needed to learn new technologies, or because their primary interest is in what the application itself can do, and not how it does it, but also because these technologies change rapidly; e.g. time invested in learning one technology for message passing is largely wasted when a developer has to learn the next. What the developer really needs is an *abstracted* view of the operations needed for data distribution, message passing, parallel IO, scheduling, or operations on the Grid [1] (see section 16.5), and not the particular implementation, e.g., a particular flavour of MPI, PVM, or some future replacement. The abstract concept of passing data between processors does not change over time, even though a specific library to do it will. A properly designed *application framework* with suitable abstractions could allow new or alternate technologies, providing similar operations, to be swapped in under the application itself. There is a great need for developing *future-proof* applications, that will run transparently on today's laptops as well as tomorrow's Grid.

At the same time, many application developers, while recognising the need for such frameworks, are reluctant to use "black box", sealed packages over which they have little control, and into which they cannot peer. Not only do some application developers actually want to see (perhaps some) details of different layers of the framework, in some cases they would like to be able to extend them to add functionality needed for their particular application. Such transparency is nearly impossible to provide through closed or proprietary frameworks without resorting to excessively complex plug-in SDKs. For this reason, freely available, *open source* tools are preferred by many communities (see, for example, Linux!). Source code is available for all to see, to improve, and to extend; these improvements propagate throughout the communities that share open source tools.

The idea of open source tools, once accepted by one segment of a community, tends to be contagious. Seeing the benefits, application developers making use of such open tools are often inclined to make their own modules freely available as well, not only in the computational science areas, but also for modules specific to research disciplines. In the computational science disciplines, modular components that carry out specific tasks are increasingly able to inter-operate. Developing application components for, or in, one framework makes it easier to use them with another framework. This is a particular goal of the "Common Component Architecture" [2], for example. As another example, we describe below application modules developed for numerical relativity and astrophysics; a growing number of scientists in this domain make use of the Cactus framework [3], and many of them make their application specific modules available to their community. Not only does this encourage others to share their codes, it raises the overall quality; knowing modules will become public increases the robustness of the code, the accompanying documentation, and the modularity.

For all of these reasons, and more that we touch on below, open, modular, application frameworks that can both hide computational details, and enable communities to work together to develop portable applications needed to solve the complex problems in their disciplines, are becoming more and more important in modern computational science. This will become particularly true as the Grid, described below, becomes a reality [1]. "Grid-enabling frameworks" will be crucial if application scientists and engineers are to make use of new computational paradigms promised by the Grid [4].

## 2 Cactus as an Application Framework

We have motivated various general reasons why application frameworks are needed, and in the rest of this paper we describe a specific example. The programming framework *Cactus* [5, 6], developed and used over a number of years, was designed and written specifically to enable scientists and engineers to perform the large scale simulations needed for their science. From the outset, Cactus has followed two fundamental tenets: respecting user needs and embracing new technologies. The framework and its associated modules must be driven from the beginning by user requirements. This has been achieved by developing, supporting, and listening to a large user base. Among these needs are ease of use, portability, the abilities to support large and geographically diverse collaborations, and to handle enormous computing resources, visualisation, file IO, and data management. It must also support the inclusion of legacy code, as well as a range of programming languages. It is essential that any living framework be able to incorporate new and developing, cutting edge computation technologies and infrastructure, with minimal or no disruption to its user base. Cactus is now associated with many computational science research projects, particularly in visualisation, data management, and the emerging field of Grid computing [7].

These varied needs for frameworks of some kind have long been recognised, and about a decade ago the US NSF created a funding program for "Grand Challenges" in computational science. The idea was to bring computer and computational scientists together with scientists in specific disciplines to create the software needed to exploit large scale computing facilities to solve complex problems. This turned out to be a rather difficult task; in order to appreciate the evolution of attempts to tackle this problem, and the maturity of current solutions, it is instructive to review a bit of history before plunging directly into the present design of Cactus 4.0.

Like many of these NSF Grand Challenge projects, a particular project called the "Black Hole Grand Challenge Alliance", aimed to develop a "Supertoolkit" for the numerical relativity community to solve Einstein's equations for the study collisions of black holes. The thought was that, as these equations are so complex, with equations of many types (e.g., hyperbolic, elliptic, and even parabolic), a properly designed framework developed to solve them would also be a powerful tool in many fields of science and engineering. (This has turned out to be true!) A centrepiece of this toolkit was, in it simplest description, a parallelisation software known as DAGH [8] (which has now evolved into a package called GrACE [9]), that would allow the physicist to write serial Fortran code that could be easily parallelised by using the DAGH library.

At the same time, a series of independent parallel codes were developed to solve Einstein's equations, with the expectation that they would later make use of DAGH for parallelism. However, some of these codes, notably the "G-code" [10], took on a life of their own, and became mini-frameworks (in Fortran!) themselves, evolving into workhorses for different collaborations within the project. As the codes and collaborations grew, it became clear that the original designs of these codes needed major revision. A prototype framework, naturally called "The Framework" [11], was designed and developed by Paul Walker at the Max Planck Institute for Gravitational Physics (AEI), with the idea that different physics codes could be plugged in, like the G-code and others, and that it would be an interface to DAGH to provide parallelism.

Learning from this experiment, Paul Walker and Joan Masso began to develop Cactus in January 1997. Cactus 1.0 was a re-working of the Framework for uni-grid applications using a simple parallel uni-grid driver as a temporary replacement for DAGH. In Cactus there were two types of object, the "flesh", a core component providing parameter parsing, parallelisation, and IO, and "thorns" which are optional modules compiled in. In October 1997 Cactus 2.0 was released, providing for the first time the ability to dynamically pick which routines were run at run-time, and some ability to order them, this ability removed the necessity of modifying the flesh to support new thorns. In these early versions of Cactus all thorns received all variables defined for a simulation; Cactus 3.0 removed this necessity, further enhancing the modularity.

Cactus became the major workhorse for a collaboration of numerical relativists spanning several large projects in Europe and the US. This community requires very large scale, parallel computational power, and makes use of computational resources of all types, from laptops for development to heavy-duty production simulations on virtually every supercomputing architecture, including Linux clusters, "conventional" supercomputers (e.g., SGI Origins or IBM SP systems), and Japanese vector parallel machines. Hence complete portability and very efficient parallel computation and IO are crucial for its users.

Lessons learnt through its heavy use in this community led to the development of a still more modular and refined version of Cactus, Cactus 4.0, now the flesh contains almost no functionality; anything can be implemented as a thorn, not just physics routines, but all computational science layers. These computational science thorns implement a certain function, say message passing, hydrodynamics evolution, or parallel IO, and are interchangeable; any thorn that implementing a given function can be interchanged with any other one. For example, different "driver layer" thorns (see section 10) that provide message passing, can be implemented and interchanged, and can be written using any message passing library; properly written applications that make use of Cactus can use any of them without modification. Entire software packages, such as GrACE for parallelism, PETSc [12] for linear solvers, and others, can be made available for Cactus users through thorns that are designed to connect them. Hence, Cactus should be thought of as a Framework for conveniently connecting modules or packages with different functionality, that may have been developed completely outside of Cactus. In principle, Cactus can accommodate virtually any packages developed for many purposes, and it is very powerful. Moreover, while earlier versions of Cactus were limited in the number of architectures they supported (T3E, IRIX, OSF, Linux, NT), the community's need for portability lead to a complete redesign of the make system to enhance portability and reduce compilation times, and now Cactus compiles on practically any architecture used for scientific computing, and the make system, which is described in section 6 is designed to make porting to new architectures as easy as possible. These points and others will be developed in depth in the sections below.

Cactus 4.0 has been developed as a very general programming framework, written to support virtually any discipline. It has been developed through a long history of increasingly modular and powerful frameworks, always aimed at supporting existing communities. Hence it is by now a very mature, powerful, and most importantly, widely *used* framework; its design has been completely derived from the needs of its users, through both careful planning and much trial, error, and redesign. The first official release of the Cactus Computational Toolkit [13], which builds on the Cactus flesh to provide standard computational infrastructure is released along with Cactus 4.0, as well as application specific toolkits. The success of the Cactus approach is borne out by the wide use of Cactus not only by the numerical relativity community in five continents, but also its emerging adoption in other fields such as climate modelling, chemical engineering, fusion modelling, and astrophysics. In the next sections, we detail the design and capabilities of Cactus 4.0.

Note that Cactus is not an application in itself, but a development environment in which an application can be developed and run. A frequent misinterpretation is that one "runs Cactus", this is broadly equivalent to stating that one "runs perl"; it is more correct to say that an application is run within the Cactus framework.

## 3 Design Criteria for Cactus 4.0

In reviewing previous versions of Cactus and other codes while designing the new Cactus framework, it became clear that the new framework had to meet some wide ranging design criteria.

It must be able to run on a wide range of different machines, so requires a flexible and modular make system. Moreover people must be able to use the same source tree to build executables for different architectures or with different options on the same architecture without having to make copies of the source tree. The make system must be able to detect the specific features of the machine and compilers in use, but must also allow people to add configuration information that cannot otherwise be auto-detected in a transparent and concise way – without having to learn new programming languages.

Similarly it should be easy to add new modules, and as much functionality as possible should be delegated to modules rather than the core. Data associated with individual modules should exist in separate name spaces to allow independently developed modules to co-exist without conflict. Functionally equivalent modules should be inter-changeable without other modules which use or depend on them noticing the exchange; many such modules should be able to be compiled into the executable at once and the desired one picked at program startup.

A core requirement is to provide transparent support for parallelism. The framework should provide abstractions for distributed arrays, data-parallelism, domain decomposition and synchronisation primitives. Furthermore, the primitives used to manage the parallelism must be minimal and posed at a high enough level of abstraction so as not to be tedious to use or to expose system architecture or implementation dependencies. The framework must be able to trivially build applications that work in uniprocessor, SMP, or MPP environments without code modification or placement of explicit conditional statements in their code to account for each of these execution environments. If the programmer wishes to build a serial implementation on a multiprocessor platform for testing purposes, the framework must support that build-time choice.

Another important class of modules is those responsible for input and output. There are many different file formats in use, and the framework itself should not dictate which is used. Nor should other modules be unnecessarily tied to any particular IO module – as new IO modules are developed people should be able to access their functionality without modification to existing code.

In past versions of Cactus it was possible to set input parameters to values which were meaningless or, in the case of keyword parameters, to values not recognised by any module. This is obviously undesirable and should be picked up when the parameter file is read.

There are many frameworks (for a review, see [14]), however most of them require a new code to be written to fit within the framework, and often restrict the languages which can be used. One design criteria was to be able to support legacy codes, and, in particular, to support the large number of codes and programmers who use FORTRAN 77.

Another criterion was to be able to allow applications using the framework to utilise features such as parallelism, adaptive mesh refinement (AMR) and out of core computation without having to modify their code, as long as their code is structured appropriately.

All the above requirements can be summed up in saying that Cactus must be portable and modular, should abstract the interfaces to lower-level infrastructure as much as possible, but be easy to use and capable of supporting legacy codes. By themselves these requirements would lead to something with cleaner interfaces than older versions of Cactus, however we also chose to design the framework to be as flexible and future proof as possible. The framework should allow the modules to provide access to new technologies, such as those to provide collaborative working, or facilities emerging from the Grid community, while at the same time the framework should not come to depend upon such technologies. This list is by no means a complete list of requirements, however these are the more important ones coming out of our experiences with earlier versions of cactus and our desire for the framework to be as future-proof as possible.

## 4 Structure

The code is structured as a core – the 'flesh' – and modules which are referred to as 'thorns'.

The flesh is independent of all thorns and provides the main program, which parses the parameters and activates the appropriate thorns, passing control to thorns as required. It contains utility routines which may be used by thorns to determine information about variables, which thorns are compiled in or active, or perform non-thorn-specific tasks. By itself the flesh does very little apart from move memory around, to do any computational task the user must compile in thorns and activate them at run-time.

Thorns are organised into logical units referred to as 'arrangements'. Once the code is built these have no further meaning – they are used to group thorns into collections on disk according to function or developer or source.

A thorn is the basic working module within Cactus. All user-supplied code goes into thorns, which are, by and large, independent of each other. Thorns communicate with each other via calls to the flesh API, plus, more rarely, custom APIs of other thorns.

The connection from a thorn to the flesh or to other thorns is specified in configuration files which are parsed at compile time and used to generate glue code which encapsulates the external appearance of a thorn.

When the code is built a separate build tree, referred to as a 'configuration' is created for each distinct combination of architecture and configuration options,

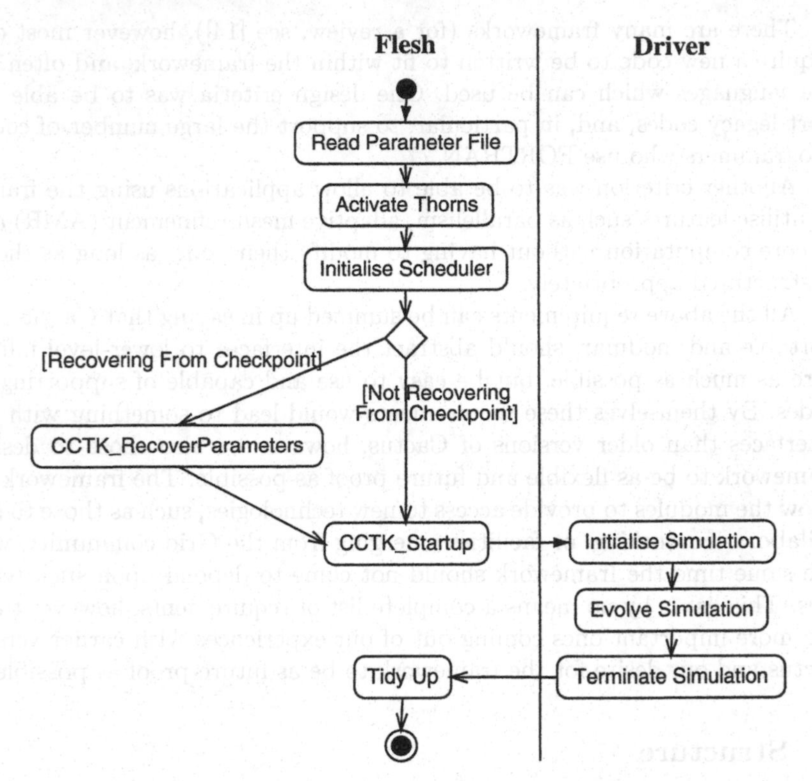


**Fig. 1.** The main flow of control in the Cactus Framework. The flesh initialises the code, then hands control to the driver thorn (see section 10). The actions in the driver swimlane are detailed in figures 2, 3, and 4.

e.g. optimisation or debugging flags, compilation with MPI and selection of MPI implementation, etc. Associated with each configuration is a list of thorns which are actually to be compiled into the resulting executable, this is referred to as a 'thornlist'.

At run time the executable reads a parameter file which details which thorns are to be active, rather than merely compiled in, and specifies values for the control parameters for these thorns. Non-active thorns have no effect on the code execution. The main program flow is shown in figure 1.

# 5 Code Management and Distribution

Since the code is developed and maintained by a geographically dispersed team of people, a well-defined code management system is necessary. The Cactus development follows clearly defined coding guidelines and procedures which are set out in the Cactus Maintainers' Guide [15].

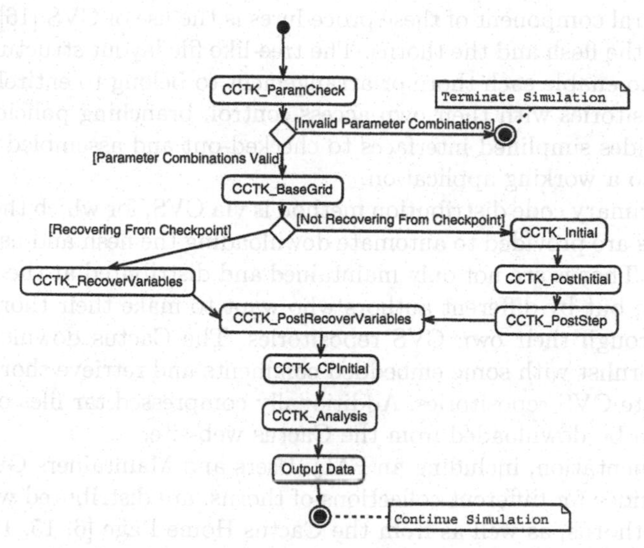


**Fig. 2.** Initialisation action diagram (corresponding to the *initialise simulation* action in figure 1). All activities prefixed with "CCTK_" are schedule bins (see section 9)

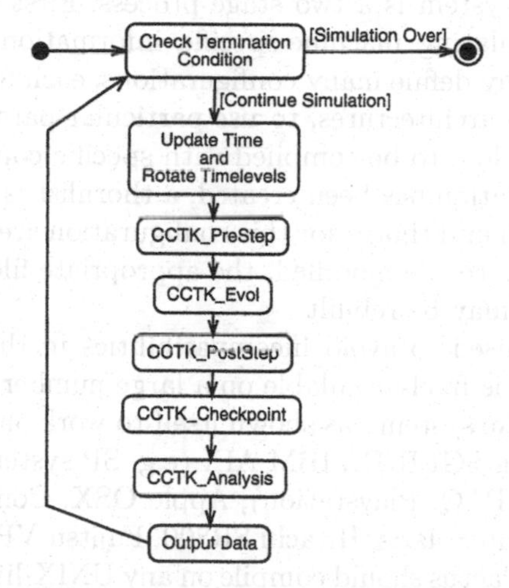


**Fig. 3.** Evolution action diagram (corresponding to the *evolve simulation* action in figure 1). All activities prefixed with "CCTK_" are schedule bins (see section 9)

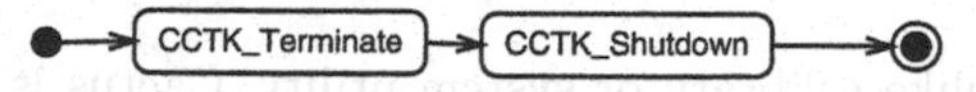


**Fig. 4.** Termination action diagram (corresponding to the *terminate simulation* action in figure 1). All activities prefixed with "CCTK_" are schedule bins (see section 9)

A central component of these procedures is the use of CVS [16] to manage revisions in the flesh and the thorns. The tree-like file layout structure is purposely designed to enable each thorn or arrangement to belong to entirely independent CVS repositories with their own access control, branching policies. The Cactus flesh provides simplified interfaces to checked-out and assembled together these thorns into a working application.

The primary code distribution method is via CVS, for which the several wrapper scripts are provided to automate downloading the flesh and user-defined lists of thorns. Thorns are not only maintained and distributed at the central Cactus repository, but by different authors who want to make their thorns available to others through their own CVS repositories. The Cactus download scripts can take a thornlist with some embedded comments and retrieve thorns from all the appropriate CVS repositories. Additionally compressed tar files of the flesh and thorns can be downloaded from the Cactus web-site.

Documentation, including an FAQ, Users and Maintainers Guide, as well as Thorn Guides for different collections of thorns, are distributed with the Cactus flesh and thorns, as well as from the Cactus Home Page [6, 15, 13].

## 6 A Portable and Modular Make System

The Cactus make system is a two stage process. First of all a "configuration" is created, encapsulating machine-specific information and also user defined choices; the user may define many configurations, each for specific purposes, e.g. for use on different architectures, to use particular parallel libraries, to include debugging information, to be compiled with specific compilers, etc.

Once a configuration has been created, a thornlist is created for the configuration, and the flesh and thorns for this configuration are built. If the thornlist or source files for a thorn are modified, the appropriate files for this configuration and its executable may be rebuilt.

GNU Make is used to avoid incompatibilities in the make program across architectures. This is freely available on a large number of platforms.

The Cactus make system has shown itself to work on a large number of Unix platforms, including SGI IRIX, IBM AIX (e.g. SP systems), Linux (IA32, IA64, Alpha, PowerPC, IPAQ, Playstation), Apple OSX, Compaq OSF, Cray Unicos (T3E, SV1, etc), Sun Solaris, Hitachi SR800, Fujitsu VPP, OpenBSD, and NEC SX-5; in principle Cactus should compile on any UNIX-like platform with at most minimal effort. To date the only non-Unix platform used has been Windows, where the Cygwin [17] environment is used to provide a Unix-like environment; the resulting code is, however, independent of Cygwin if non-Cygwin compilers are used.

Note that unlike a library or system utility, Cactus is not "installed". Currently each user has a copy of the sources and generates their own configurations based upon their individual needs and preferences.

## 6.1 Creating a Configuration

In order to support as wide a range of computers as possible, GNU Autoconf [18] is used to detect specific features of compilers and libraries. However not all such things can be easily detected, and it was deemed important that in those cases a user compiling on a previously unsupported architecture should not need to learn detailed shell programming and M4 to extend the detection system. To this end, architecture specific information, such as flags necessary for full optimisation, or debugging, or warnings, or which libraries need to be linked against for the Fortran compiler, is stored in 'known-architecture' files, which are simple shell scripts which are sourced by the master configuration script once it has detected the operating system. These files are simple in format, and new architectures can be supported merely by placing a new file in the directory holding these files, with a name corresponding to the operating system name returned by the 'config.sub' script invoked by Autoconf.

A similar system is used to configure for optional packages such as MPI, pthreads or HDF5. To add a new optional package with its own configuration options, a user may just create a directory in a specific place, and add a script file in this place which may add extra definitions to the master include file or add extra information to the make system. This is heavily used to support the various different flavours of available MPI – native MPI on SGI, Cray and IBM AIX machines, or MPICH, LAM, HPVM, MPIPro on these or other systems.

When creating a configuration the user may pass instructions to the configuration scripts to override detection or defaults. For example the user may choose to use a specific C or Fortran compiler, or choose to configure the code for parallel operation by configuring with MPI, or specific the user may want this configuration have different debugging or optimisation flags. The user may also pass a specific architecture name to the configuration system in order to cross-compile for a different architecture; in this case the known-architecture file for the target platform must contain certain details which the configure script would otherwise detect.

## 6.2 Building a Configuration

Once a configuration has been created the user must pick a set of thorns to build. If the user doesn't provide one, the make system will search for thorns and generate a thornlist and give the user the option to edit it. The thornlist is a plain text file and may be edited with any text editor.

Once there is a thornlist associated with a configuration, the CST, which is described in detail in section 8.2, is run to validate the thorns and provide bindings between the flesh and the thorns, and then the code is built based upon the flesh source and the sources of all thorns in the thornlist. The flesh and each thorn are compiled into libraries which are linked to form the final executable.

In individual thorns the thorn writer has the option of using their own Makefile or of using a simplified system whereby they list the files and subdirectories which should be compiled, and a standard make file is used for that thorn. The vast majority of thorn writers use the latter approach, which enables them to develop with minimal knowledge of make file syntax or operation. More advanced users may also similarly write make rules specifying dependencies between source files in their thorn.

Since the object files for each configuration are kept in different directories from the original source files, and a user may be building several configurations simultaneously from one source tree, care has to be taken with the working directory of the compilation, especially when Fortran 90 modules or C++ templates are used. This is currently solved by setting the working directory of any compilation as a specific scratch directory in the configuration-specific build area.

Currently each thorn's object files are packaged into libraries which are then linked together to form the final executable. In future versions of Cactus it is planned that these libraries may be dynamically loadable, on platforms which allow this, thus allowing new thorns providing new capabilities, in particular data analysis, to be loaded at run-time in long running simulations.

At build time the user may pass options to the make system to echo all commands to the screen or to switch on compiler warnings. However no option which would change the final executable is allowed at this stage, on the principle that the files stored with the configuration should completely define the way the executable is produced.

### 6.3 Managing Configurations

The configuration is the basic unit with which an end-user operates. The Cactus framework provides many options to manage a configuration, some examples are: building an executable (this option checks the thornlist and dependencies of the executable and its sources, rebuilds any object files which are out of date, and re-links the final executable if necessary as described above); running a test suite (see section 13); removing object files or stored dependency information to force a clean rebuild or save disk space; or producing a document describing all the thorns in the configuration's thornlist.

## 7 Code Portability

The previous section has described how the make system is able to configure and make the code on all platforms. However this would be of no use if the code itself were not portable. In order to ensure code portability, the flesh is written in ANSI C. Thorns, however, may be written in C, C++, FORTRAN 77 or Fortran 90. In order to ensure that thorns are platform independent, the configuration script determines various machine-specific details, such as the presence or absence of certain non-ANSI C functions (thorns are encouraged to use equivalent functions in the Cactus utility libraries (see section 12) which are guaranteed to be there), or the sizes of variables, which vary from platform to platform – in order to avoid problems with this, e.g. in cross-language calls or parallel jobs distributed across heterogeneous platforms, Cactus defines a set of types, such as CCTK_REAL,

CCTK_INT, which are guaranteed to have the same size on all platforms and in all languages. This also means that runs can produce exactly the same results on different machines, and checkpoint files created on one machine can be restarted on any other.

# 8 Modularity

This section describes how thorns identify themselves to the flesh and the make system, and the way in which functionality is encapsulated and name spaces are realised. Two important pieces of terminology must be introduced before preceding. *Grid variables* are variables which can be passed between thorns or routines belonging to the same thorn through the defined flesh interface; this implies it is related to the computational grid rather than being an internal variable of the thorn or one of its routines. An *implementation* is an abstract name for the functionality offered by a thorn; all thorns providing the same implementation expose the same interfaces to the outside world, and only one such thorn may be active at a time.

## 8.1 CCL

Each thorn provides three specification files written in the Cactus Configuration Language (CCL). These files designate which implementation the thorn provides and the variables that the thorn expects to be available, describe the parameters it uses, and specify routines from the thorn which must be called.

Variable scope may be either private, i.e. visible only to routines from this thorn; protected, i.e. visible to all thorns in the group of friends of this thorn's implementation; or public, i.e. visible to all thorns which inherit from this thorn's implementation.

Grid Variables fall into three categories: Grid Scalars (GSs), which are single numbers (per processor); Grid Functions (GFs), which are distributed arrays with a size fixed to the problem size (all GFs have the same size); and Grid Arrays (GAs) which are similar to GFs but may have any size. Grid Variables must be one of the defined CCTK types (e.g. CCTK_REAL, CCTK_INT, etc), whose size is determined when a configuration is created (see section 6.1). Grid Functions are by default vertex-centred, however they may be staggered along any axis.

Parameter scope may similarly be private, or may be declared to be 'restricted', which means that other thorns may choose to share these parameters. Parameters may be numeric, i.e. integer or floating point, boolean, keywords, i.e. the value is a distinct token from a defined list, or arbitrary strings. The specification of a parameter includes a specification of the allowed values, i.e. a range of numbers for numeric, a list of keywords for keyword parameters, or a set of regular expressions for string parameters.

Parameters may also be declared to be *steerable*. This means that they are not fixed for the duration of the simulation, but may have their value changed. By default parameters are non-steerable, however a thorn author can declare that a parameter may be steered at all times, or when the code is recovering from a checkpoint. This is an important distinction as, while parameters are provided as local variables to routines for speed of access, a lot of parameters are used to setup data structures or to derive other values, which would thus be unaffected if the parameter value was subsequently changed; by declaring a parameter "steerable" the thorn author is certifying that changing the value of the parameter will make a difference. (Parameters values may only be changed by calling a routine in the flesh which validates the change.)

Routines from the thorn may be scheduled to run at various time bins and relative to the times when other routines from this thorn or other thorns are run. The thorn author may also schedule groups within which routines may be scheduled; these groups are then themselves scheduled at time bins or within schedule groups analogously to routines. The scheduling mechanism is described in more detail in section 9. The scheduling of routines may be made conditional on parameters having certain values by use of `if` statements in the CCL file.

Memory for variables may be allocated throughout the course of the simulation, or allocated just during the execution of a particular scheduled routine or schedule group.

Additionally thorn authors may define specific functions which they provide to other thorns or expect other thorns to provide. This provides an aliasing mechanism whereby many thorns may provide a function which may be called by another thorn with a particular name, with the choice of which one is actually called being deferred until run-time. In the absence of any active thorn providing such a function an error is returned.

## 8.2 CST

When the executable for any particular configuration is built, the Cactus Specification Tool (CST) is invoked to parse the thornlist and then the individual thorns' CCL files. When the CST has parsed the CCL files it performs some consistency checks. These include: checking that all definitions of a particular implementation are consistent; checking that default values for parameters are within the allowed ranges; and checking that parameters shared from other implementations exist and have the correct types.

Once the CCL files have been verified the CST uses the specifications from these files to create small C files which register information with the flesh about the thorns. This mechanism allows the flesh library to be completely independent of any thorns, while at the same time minimising the amount of machine-generated code. Along with the registration routines the CST generates various macros which are used to provide argument lists to routines called from the scheduler, as well as wrapper routines to pass arguments to Fortran routines. The parameter specifications are turned into macros which place the parameters as local variables in any routine needing access to parameters. All the macros are generated on a thorn-by-thorn basis and contain only information relevant to that thorn. The schedule specifications are turned into registration routines which register relevant thorn routines with the scheduler.

# 9 The Scheduling Mechanism

Routines (or schedule groups – for scheduling purposes they are the same) scheduled from thorns may be scheduled *before* or *after* other routines from the same or other thorns, and *while* some condition is true. In order to keep the modularity, routines may be given an alias when they are scheduled, thus allowing all thorns providing the same implementation to schedule their own routine with a common name. Routines may also be scheduled with respect to routines which do not exist, thus allowing scheduling against routines from thorns or implementations which may not be active in all simulations. Additionally the **schedule.ccl** file may include 'if' statements which only register routines with the scheduler if some condition involving parameters is true.

Once all the routines have been registered with the scheduler, the before and after specifications form a directed acyclic graph, and a topological sort is carried out. Currently this is only done once, after all the thorns for this simulation have been activated and their parameters parsed.

The *while* specification allows for a degree of dynamic control for the scheduling, based upon situations in the simulation, and allows looping. A routine may be scheduled to run while a particular integer grid scalar is non-zero. On exit from the routine this variable is checked, and if still true, the routine is run again. This is particularly useful for multi-stage time integration methods, such as the method of lines, which may schedule a schedule group in this manner.

This scheduling mechanism is rule-based as opposed to script-based. There are plans to allow scripting as well; see section 17 for further discussion of this.

# 10 Memory Management and Parallelisation

In order to allow the use of the same application codes in uni-grid, parallel and adapted-grid modes, each routine called from the scheduler is assigned an n-dimensional block of data to work on; this is the classical distributed computing paradigm for finite difference codes. In addition to the block of data and its size, information about each boundary is given, which specifies if this boundary is a boundary of the actual computational domain, or an internal grid boundary. The routine is passed one block of data for each variable which has storage assigned during this routine (see section 8.2 for argument list generation). The memory layout of arrays is that of Fortran, i.e. fastest-changing index first; in C these appear as one-dimensional arrays and macros are provided to convert an n-dimensional index into a linear offset withing the array.

In order to isolate the flesh from decisions about block sizes, a particular class of thorns, which provide the DRIVER implementation, is used. Drivers are responsible for memory management for grid variables, and for all parallel operations, in response to requests from the scheduler.

The driver is free to allocate memory for variables in whatever way is most appropriate, some possibilities for uni-grid are: having memory for the whole domain in one block, with this block being passed to routines; all the memory in one block, but passing sub-blocks to routines; splitting the domain into sub-domains and processing these one by one, e.g. for out of core computation; or splitting into sub-domains and then processing sub-domains in parallel.

Since the application routines just get such a block of data, data layout may change with time, and indeed will do with adaptive mesh refinement or dynamic load balancing.

As the driver is responsible for memory management, it is also the place which holds the parallelisation logic. Cactus 4.0 supports three basic parallelisation operations: ghost-zone synchronisation between sub-domains; generalised reduction operators (operations which take one or more arrays of data and return one or more scalar values from data across the whole domain); and generalised interpolation operators (operations which take one or more arrays and a set of coordinates and return one or more scalars for each coordinate value).

Synchronisation is called from the scheduler – when a routine or schedule group is scheduled it may list a set of grid variables which should be synchronised upon exit. Routines may also call synchronisation internally, however such routines would fail when more than one block of data is processed by a processor or thread, such as with adaptive mesh refinement or out of core computation, as the routine would not yet have been invoked on the remaining blocks so ghost-zone exchange would be impossible.

Similarly the other operations – reduction and interpolation – being global in nature, should only be performed one per grid, however many sub-grids there are; routines which call these routines must be scheduled as "global". This information is then passed to the driver which should then only invoke such routines once per processor, rather than once per sub-block per processor.

The flesh guarantees the presence of these parallel operations with a function registration mechanism. A thorn may always call for synchronisation, reduction or interpolation, and these are then delegated to functions provided by the driver, or return appropriate error codes if no such function has been provided.

These parallel operations are in principle all that an application programmer needs to think about as far as parallelisation is concerned. Such a programmer is insulated by the scheduler and by these functions of any specific details of the parallelisation layer; it should be immaterial at the application level whether the parallelisation is performed with MPI, PVM, CORBA, threads, SUN-RPC, or any other technology.

Since all driver thorns provide the DRIVER implementation, they are interchangeable, no other thorn should be affected by exchanging one driver for another. Which driver is used is determined by which is activated at run-time.

At the time of writing there are four driver thorns known to the authors: a simple example uni-grid, non-parallel one (`SimpleDriver`); a parallel uni-grid one (`PUGH`); a parallel fixed mesh refinement driver (`Carpet`) ; and a parallel AMR one (`PAGH`). `PUGH` is the most frequently used one, and is contained in the computational toolkit available from the Cactus web site; `Carpet` and `PAGH` have been developed independently.

### 10.1 PUGH

PUGH is the only driver currently distributed with the Cactus Computation Toolkit. This driver is MPI based, although it may also be compiled without MPI for single processor operation. In single-processor mode PUGH currently creates one block of data per variable and passes that through to routines; there has also been discussion of enhancing this to generate smaller sub-blocks tuned to the cache size to increase performance. In parallel mode PUGH decomposes the processors in the n-dimensions such that the smallest dimension has the fewest number of processors, which allows greater cache-efficiency when constructing messages to be passed to other processors, however this may be over-ridden by parameter setting which allow any or all of the dimensions to be set by hand. There are plans to add features to allow the processor topology to be optimised for node-based machines such as the IBM SP or Hitachi SR8000 which consist of small (typically 8 or 16 processor) nodes connected together to produce a larger machine. The load-balancing algorithm in PUGH is also customisable, and a modified version of PUGH has been used in large scale distributed computing [19].

Along with PUGH, the Computational Toolkit provides the auxiliary thorns PUGHReduce and PUGHInterp to provide reduction and interpolation when PUGH is used as a driver, and PUGHSlab, which provides the hyperslabbing capability used by the Toolkit's IO thorns.

## 11 IO

Input and Output are vital to any simulation, and indeed to any framework. In keeping with the overall design of Cactus, the flesh has minimal knowledge of the IO, but provides a mechanism so that thorn authors can call for output of their variables and the appropriate IO routines will be called.

All thorns providing IO routines register themselves with the flesh, saying they provide an *IO method*, which is a unique string identifier. Associated with an IO method are three functions: one to output all variables which need output; one to output a specific variable; and one to determine if a variable requires output. The first two of these routines then have analogues in the flesh which traverses all IO methods calling the appropriate routines, or call the routine corresponding to a specific IO method. Once per iteration the master evolution loop calls the routine to output all variables by all IO methods.

Thorns providing IO methods typically have string parameters which list the variables which should be output, how frequently (i.e. how many iterations between output), and where the output should go. In order to simplify such thorns, and to provide standards for parameter names and meanings, the computational toolkit contains the IOUtil thorn. This thorn provides various default parameters such as the output directory, the number of iterations between output of variables, various down-sampling parameters and other parameters which may need to be used by thorns providing output of the same data but in different formats. It also provides common utility routines for IO.

The computational toolkit contains thorns which output data to screen, to ASCII output files in either xgraph/ygraph [20], or GNUPlot [21] format, binary output in IEEEIO [22] or HDF5 [23] formats, or as jpegs of 2d slices of datasets. This data may also be streamed to external programs such as visualisation clients instead of or in addition to being written to disk. Streamed visualisation modules exist for Amira [24] and the widely available OpenDX [25] visualisation toolkits.

Since all IO thorns must operate on m-dimensional slabs of the n-d data, which is (section 10) laid out and distributed in a driver dependent manner, there is also a defined interface provided by thorns providing the HYPERSLAB implementation. This interface is driver independent, and allows an IO thorn to get or modify a particular slab of data without being tied to a particular driver.

## 12 Utilities

The flesh provides sets of utility functions which thorns developers may use. These include common data storage utilities such as hash tables and binary trees; an arithmetical expression parser which may be used to perform calculations on data based upon an expression given as a string; access to regular expression functions (using the GNU regex library if the configure script does not find one); and an interface to create and manipulate string-keyed tables of data.

Apart from these general utilities, the flesh also provides an infrastructure to time the execution of code blocks. This consists of a way to create and manipulate timer objects, each of which has an associated set of clocks, one for each time-gathering function, such as the Unix getrusage call, the wall-clock time or the MPI_Wtime function. These timers and their clocks may be switched on or off or reset as necessary. The scheduler in the flesh uses these functions to provide timing information for all scheduled functions.

## 13 Test Suites

An essential component of any code is a well-defined way to perform regression tests. This is especially true for a framework, where the final result of a set of operations may depend upon many independently written and maintained modules. Cactus includes an automated test-suite system which allows thorn authors to package parameter files and the results they are expected to produce with their thorns, and then allows the current code to be tested against these results. The test-suite compares numbers in the output files to a specified tolerance, which is generally set to the accuracy which can be expected given different machines' round-off error and machine precision, but can be over-ridden by the thorn author if this measure is not appropriate.

## 14 Thorn Computational Infrastructure

The overriding design criterion for Cactus has been to put as little in the flesh as possible, which allows the maximum flexibility for users to develop new functionality or modify existing functionality. The corollary to this is that the flesh by itself is not very useful. Sections 10 and 11 described two sets of infrastructure thorns, the driver and the IO thorns respectively. This section describes some of the other infrastructure which is available in the computational toolkit.

### 14.1 Interpolation and Reduction

While the flesh provides standard interfaces for generalised interpolation and reduction operations, and guarantees that these functions may always be called, the actual operations need to be done by the driver thorn (see section 10 , or by a closely associated thorn. A thorn providing such an operation registers a function with the flesh, with a unique name associated with the operator, such as "maximum". A thorn which needs to interpolate data in a grid variable, or to perform a reduction on it, then calls the appropriate flesh function, with the name of the operation, and the flesh then delegates the call to the thorn providing the operation. The Computational Toolkit provides several thorns for operations such as maximum, L1 and L2 norms, parallel interpolation, etc.

### 14.2 Coordinates

A particular grid has a single coordinate system associated with it, out of the many possibilities, and each grid array may be associated with a different coordinate system. To facilitate the use of different coordinate systems we have a thorn providing the COORDBASE implementation, which defines an interface to register and retrieve coordinate information.

This interface allows thorns providing coordinate systems to store information such as the coordinates of the boundaries of the computational grid in physical space, and the coordinates of the interior of the grid – i.e. that part of the grid which is not related to another by a symmetry operation or by a physical boundary condition. The thorn also associates details of how to calculate the coordinate value of a point with the coordinate system; this data may be calculated by origin and delta information, by associating a one-dimensional array with the coordinate direction to hold the coordinate data, or, for full generality, with an n-dimensional GA (or GF) where each point in this GA (or GF) holds the coordinate value for this coordinate for this point.

Based upon this coordinate system information, IO methods may output appropriate coordinate information to their output files or streams.

### 14.3 Boundary Conditions

Boundary conditions are another basic feature of any simulation. They fall into two broad classes – symmetry and physical boundary conditions. Additionally boundary conditions in both these classes may be either local or global. For example periodic boundary conditions are global symmetry conditions, the Cartoon [26] boundary condition is a symmetry condition but is local, and radiative boundary conditions are physical and local.

If all boundary conditions were local, there would be no problem calling them from within an application routine, however as soon as global conditions are used, calling from a routine suffers the same problems as any parallel operation. Additionally it is undesirable that routines should have to know anything about the symmetries of the grid – they should just know about the physical boundary conditions, and when these are applied the appropriate symmetries should be performed.

In order to address these needs, we have defined a scheme whereby instead of calling the boundary conditions directly, calculation routines merely call a function to indicate that a particular variable requires a particular physical boundary condition, and then the thorn providing this routine schedules a schedule group called ApplyBCs after the calculation routine (schedule groups may be scheduled multiple times, so there is no problem with conflicts between multiple thorns scheduling this group). All symmetry boundary condition and global physical boundary condition thorns schedule a routine in the ApplyBCs group, and the Boundary thorn schedules a routine to apply local physical boundary conditions in the group as well. Thus when the ApplyBCs group is active, each routine scheduled in it examines the list of variables which have been marked as needing boundary conditions applied, and, if it is a symmetry routine applies the symmetry, or if it is a physical boundary condition and the variable requires that physical boundary condition, applies the appropriate physical boundary condition – local physical boundary conditions are registered with the Boundary thorn, which then uses the routine it has scheduled in the group to dispatch variables to the appropriate routine.

This scheme allows new symmetry conditions to be added with no modification of any calculation routine, and allows these routines to work appropriately in the presence of multiple blocks of data per process or thread.

## 14.4 Elliptic Solvers

The computational toolkit provides a thorn Ell_Base which provides an experimental interface for elliptic solvers. This thorn provides a set of registration functions to allow thorns which solve certain sets of elliptic equations to be accessed via a specified interface; e.g. a thorn wanting to solve Helmholtz's equation makes a call to the function in Ell_Base passing as one argument a well-known name of a particular solver, such as PETSc [12] , which is then invoked to perform the calculation. Thus the actual elliptic solver used may be decided at run-time by a parameter choice, and new elliptic solvers may be used with no change to the application thorn's code.

## 14.5 Utility Thorns

In addition to necessities such as drivers, IO, coordinates and boundary conditions, there are various general utility thorns. For example we have a thorn which takes datasets and checks for NaNs; this can be either done periodically as set by a parameter, or called from other thorns. On finding a NaN the thorn can issue a warning, report the location of the NaN, stop the code, or any combination of these.

Another thorn, which is under development at the moment, interacts with the Performance API (PAPI) [27] to allow profiling of a simulation. PAPI allows access to hardware performance counters in a platform-independent manner.

# 15 External Interaction

The most basic way to run a simulation is to start it and then examine the data written to disk during or after the run. A framework, however, is free to provide modules to allow interaction with a running simulation. This may range from just querying the status of the simulation, through visualising data remotely, to steering the control parameters of a simulation based upon user input or predictions and analysis performed by other programs.

## 15.1 HTTPD

The Cactus Computational Toolkit contains a thorn, HTTPD (based on an original idea and implementation by Werner Benger), which acts as a web-server. The basic thorn allows users to connect to the running simulation with a web-browser and to examine and, if authenticated, to modify, the values of parameters. An authenticated user may also choose to terminate or pause the simulation; the user may also tell the code to pause at a specific iteration or after a certain amount of simulation time. The ubiquitous nature of web-browsers makes this an almost ideal way to interact with the simulation.

Additional thorns may be used to provide further capabilities to the web server. For example, application thorns can also add their own information pages and provide custom interfaces to their parameters. The toolkit contains the thorn HTTPDExtra which provides access to the IO methods within the framework by providing a view port by which users may view two-dimensional slices through their data using IOJpeg, and through which any file known by the simulation may be downloaded. This allows users to examine the data produced by a remotely running simulation on the fly by streaming data to visualisation tools such as OpenDX [25], Amira [24] or GNUPlot [21], see section 15.2

The combination of the ability to pause the simulation, to examine data remotely, and to steer simulation parameters provides a powerful method for scientists to work with their simulations. This combination could be enhanced further to provide a debugging facility.

Currently HTTPD only supports un-encrypted HTTP connections, however there are plans to enhance it to use HTTP over TLS [28] [29]/SSL [30] or over GSI [31], providing capabilities for secure authentication and encrypted sessions.

## 15.2 Remote Visualisation and Data Analysis

As mentioned above, the Cactus Computational Toolkit provides mechanisms to query and analyse data from a running simulation. While it is true that because the simulation may be scheduled in the batch queue at unpredictable times, the scientist will not necessarily be on hand to connect to the simulation at the moment that it is launched, the simulations that create the most egregious difficulties with turnaround times typically run for hours or days. So while the interactive analysis may not necessarily capture the entire run, it will intercept at least part of it. Any on-line analysis of the running job before the data gets archived to tertiary storage offers an opportunity to dramatically reduce the turn-around time for the analysis of the simulation outcome.

Keeping up with the data production rate of these simulations is a tall order, whether the data is streamed directly to a visualisation client or pre-digested data is sent. Earlier pre-Cactus experiments with a circa 1992 CAVE application called the Cosmic Worm demonstrated the benefits of using a parallel isosurfacer located on another host to match this throughput. However, it also pointed out the fact that it doesn't take too many processors limits primarily from the I/O rate are reached. It does no good to perform the isosurface any faster if its performance is dominated by the rate at which data can be delivered to it.

For the first remote visualisation capability in Cactus, it was decided to implement a parallel isosurfacer in-situ with the simulation so that it would use the same domain decomposition and number of processors as the simulation itself. The isosurfacer would only send polygons to a very simple remote visualisation client that runs locally on the user's desktop; offering sometimes orders of magnitude in data reduction with throughput that exactly matched the production rate of the simulation code. This paradigm was further enhanced to include geodesics and other geometric visualisation techniques.

Other ways to reduce the throughput to data clients is to send only sub-sets of the data. The Cactus IO methods, in particular the `HDF5` method, allow the data to be down-sampled before being written to disk or streamed to a visualisation client. Another way to stream less data is to pick a data-set of a lower dimension from the full data-set, and we have IO methods which will produce 1 or 2-dimensional subsets of three-dimensional data-sets.

These and other remote technologies are being actively developed through a number of projects for example [32, 33, 34].

# 16 Applications

## 16.1 Numerical Methods

Although Cactus was designed to support, or to be extensible to support, different numerical methods, most of the existing infrastructure has been developed around regular structured meshes with a single physical domain and coordinate system, and more specifically for finite difference methods with spatial dimensions of three or less. This is well-suited to many calculations, particularly in the problem domains for which Cactus was first developed. However there are many other problem domains for which these restrictions pose problems.

Fundamental support for other numerical methods typically involves the development of a standard driver and associated thorns (with possible extensions of CCL features). It is hoped that many of the existing infrastructure thorns, such as those for IO and coordinates, can be developed to support additional methods. For example, the addition of a hyperslabbing thorn for a particular driver should allow for IO thorns to provide for different underlying methods.

With the addition of appropriate drivers (see section 17), and associated driver infrastructure thorns, the Cactus framework can be used with structured, unstructured, regular and irregular meshes, and can implement finite differencing, finite volumes or finite elements, particle and spectral methods, as well as ray tracing, Monte Carlo etc.

## 16.2 Scalar Wave Equation

The solution of the scalar wave equation using 3D Cartesian coordinates provides a prototypical example of how a wide class of initial value problems can be implemented with finite differencing methods in Cactus. The `CactusWave` arrangement contains thorns which implement this solution in each of the currently supported programming languages, along with additional thorns providing initial data and source terms [13].

## 16.3 General Relativity

The introduction describes the historical relationship between Cactus and numerical relativity. The requirements of this diverse community for collaboratively developing, running, analysing and visualising large-scale simulations continue to drive the development of Cactus, motivating computational science research into advanced visualisation and parallel IO as well as Grid computing.

The Cactus framework and Computational Toolkit is used as the base environment for numerical relativity codes and projects in nearly two dozen distinct groups worldwide, including the 10-institution EU Network collaboration and others in Europe, as well as groups in the US, Mexico, South Africa, and Japan. In addition, the Cactus Einstein Toolkit provides an infrastructure and common tools for numerical relativity, and groups which choose to follow the same conventions can easily inter-operate and collaborate, sharing any of their own thorns, and testing and verifying thorns from other groups (and many do).

The numerical relativity community's use of Cactus provides an example of how a computational framework can become a focal point for a field. With many different relativity arrangements now freely available, both large groups and individuals can work in this field, concentrating on their physics rather than on computational science. In this environment, smaller groups are more easily able to become involved, as they can build on top of open and existing work of other groups, while concentrating on their particular expertise. For example, an individual of group with expertise in mathematical analysis of gravitational waves can easily implement thorns to do this, using other groups' initial data and evolution thorns. Such an approach is increasingly becoming a point of entry into this field for groups around the world.

In addition to the leverage and community building aspect of an open framework like Cactus, the abstractions it provides make it possible for a new technology, such as fixed or adaptive mesh refinement, improved IO, new data types, etc, to be added and made available to an entire community with minimal changes to their thorns. Not only does this have obvious benefits to the user communities, it has the added effect of attracting computational science groups to work with Cactus in developing their technologies! They find not only a rich application oriented environment that helps guide development of their tools, but they also find satisfaction in knowing their tools find actual use in other communities.

## 16.4 Other Fields

Although Cactus began as a project to support the relativity community it is an open framework for many applications in different disciplines. Here we simply list a few of a growing number of active projects of which we are aware at present. As part of an NSF project in the US, the Zeus suite of codes for Newtonian magnetohydrodynamics has been developed into a set of thorns (in the `Zeus` arrangement, which use AMR, and is publicly available. In another discipline, Cactus has been used as a framework to write parallel chemical reactor flow applications, where it was found to be an effective tool for speeding up simulations across multiple processors [35]. Climate modellers at NASA and in the Netherlands have also taken interest in Cactus, and are actively developing thorns for both shallow water models and a coupling of two ocean models [36]. Other application communities prototyping applications in Cactus, or investigating its use as a framework for their applications, include the fusion simulation community, avalanche simulators, and geophysics groups, to name a few.

## 16.5 Grid Computing

Last but not least, we turn to Cactus as a framework, not only for traditional applications, but also for developing applications that may run in a Grid environment. The Grid is an exciting new development in computing, promising to seamlessly connect computational resources across an organisation or across the world, and to enable new kinds of computational processes of unprecedented scales. As network speeds increase, the effective "distance" between computing devices decreases. By harnessing PCs, compute servers, file servers, handhelds, etc, distributed across many locations, but connected by ever better networks, dynamically configurable virtual computers will be created for many uses [1].

However, even if the networks, resources, and infrastructure are all functioning perfectly, the Grid presents both new challenges and new possibilities for applications. Although the applications of today may (or may well not) actually run on a Grid, in order to run efficiently, or more importantly, to take advantage of *new* classes of computational processes that will be possible in a Grid world, applications must be retooled, or built anew. And yet, as we stressed in the introduction, most application developers and users want to focus on their scientific or engineering discipline, and not on the underlying computational technology.

For the same reasons that Cactus has been an effective tool for building applications that take advantage of advanced computational science capabilities on many different computer architectures, while hiding many details from the user, it has also been one of the leading sources of early Grid applications. Its flexible and portable architecture make it a good match for the varying needs of Grid applications. In particular, it has traditionally been used for many remote and distributed computing demonstrations [37, 38, 39, 19]. Because of the abstractions available in Cactus, it has been possible to modify the various data distribution, message passing, and IO layers so they function efficiently in a distributed environment, without modifying the applications themselves [19]. For example, it has been possible to take very complex, production code for solving Einstein's equations which is coupled to GR hydrodynamics to simulate colliding neutron stars, and distributed across multiple supercomputers on different continents, while it is remote visualised and controlled from yet another location.

This is just the beginning of what can be done on future Grids. We imagine a world where applications are able autonomously to seek out resources, and respond to both their own changing needs and the changing characteristics of the Grid itself, as networks and computational resources change with time. Applications will be able not only to parallelise across different machines on the Grid, but will also be able spawn tasks, migrate from site to site to find more appropriate resources for their current task, acquire additional resources, move files, notify users of new events, etc [4]. Early prototypes of this kind of capability have already been developed in Cactus; the "Cactus Worm" demonstration of Supercomputing 2000 was a Cactus application (any application inserted in the Cactus framework would do) that was able to: run on any given site, make use of the existing Grid infrastructure to move itself to a new site on a Grid, register itself with a central tracker so its users could monitor or control it, and then move again if desired [40]. This demonstration of new types of computation in a Grid environment foreshadows a much more complex world of the future.

However, the underlying Grid technology is quite complex, and varies from site to site. Even if an application programmer learns all the details of one Grid technology, and implement specific code to take advantage of it, the application would likely lose some degree of portability, crucial in the Grid world. Not only does one wish to have applications that run on any major computing site, with different versions of Grid infrastructure, but also one wants them to run on local laptops or mobile devices, which may have no Grid infrastructure at all!

Learning from the Cactus Framework, one solution is to develop a Grid Application Toolkit (GAT), that abstracts various Grid services for the application developer, while inter-operating with any particular implementation. The GridLab project [41], is currently underway to do just this. The goal is to develop a toolkit for *all applications*, whether they are in the Cactus framework or not. The GAT will provide an abstracted view of Grid services of various kinds to the user/developer, and will dynamically discover which services are actually available that provide the desired functionality (e.g., "find resource", "move file", "spawn job", etc). It is far beyond the scope of this article to go into further detail, but we refer the reader [4, 42, 43] for more detail.

## 17 Future Plans

Cactus 4.0 is a great step from previous versions of Cactus, providing much more modularity and flexibility. However there are numerous shortcomings, and many things are planned for future versions.

Support for unstructured meshes is intended, which opens up the use of Cactus for finite volume and finite element calculations. We have had many discussions with communities, e.g. aerospace and earthquake simulation, which use these methods, and have well-developed plans on how to support such meshes.

Another plan we have is to enable more complicated scenarios, consider the following: (*i*) Many CFD calculations also use multi-block grids – grids which are themselves made up of many sub-grids, some being structured and some unstructured, e.g. simulation of flow past an aircraft wing may use an overall structured mesh, with a small unstructured grid block around the wing and its stores; (*ii*) Climate modelling simulations have many distinct physical domains, such as sea, icebergs, land and air, each of which has its own physical model, and then interaction on the boundaries; (*iii*) Some astrophysical simulations require several coordinate patches to cover a surface or volume.

These are all sub-classes of what we refer to as "multi-model". While all can be done currently within Cactus using Grid Arrays, this is not the most natural way to provide this functionality, and requires more communication and coordination between individual module writers than is desirable. We have defined a specification to enable any combination of these types of scenarios and intend to enable this functionality in a future release.

Currently all Grid Functions or Grid Arrays are distributed according to their dimension across all processors. This is the most common need, however there are many situations where one would like to define a variable which only lives on the boundary of a grid. This can be faked at the moment by defining a GA with one lower dimension than GFs and ignoring it on internal processors, however even then the distribution of this GA may not correspond to the distribution of a GF on that face. To solve this problem we intend to introduce a new class of grid objects, sub-GFs, which will have the appropriate distribution.

Another class of methods not currently supported are particle methods, such as Particle-In-Cell (PIC) and Smoothed-Particle-Hydrodynamics (SPH). Both these methods require a small amount of additional communication infrastructure to enable exchange of particle information between different processors.

Currently thorns can only be activated when the parameter file is read, and not during the course of the simulation. An obvious enhancement is to enable thorn activation, and even de-activation, at any point in the program execution.

This would allow the code to act as a compute server, receiving requests for simulations from external sources, and, if thorns were dynamically loadable libraries as opposed to statically linked, would allow new functionality to be incorporated at will into long-running simulations.

Another enhancement would be to allow scripting as an alternative to the current scheduling mechanism. The current mechanism allows thorns to interoperate and for simulations to be performed with the logic of when things happen encapsulated in the schedule CCL file; other frameworks do the same thing by providing a scripting interface, which gives more complete control of the flow of execution, at the expense of the user needing to know more of the internals. Both schemes have advantages and disadvantages. In the future we would like to allow users to script operations using Perl, Python, or other scripting languages.

Currently Cactus only allows thorns in C, C++, and Fortran. Addition of other languages is fairly straightforward as long as there exists a method to map between the language's data and C data. We plan to add support for thorns written in Perl, Python and Java in the future; this will be done by defining an interface which would allow a thorn to inform the make system and the CST about the mappings and how to produce object files from a language's source files, thus allowing support for any language to be added by writing an appropriate thorn.

In section 6 it was stated that Cactus is not installed. In the future we would like to support the installation of the Cactus flesh and the Computational Toolkit arrangements in a central location, with other arrangements installed elsewhere, for example in users' home directories, thus providing a consistent checked out version of Cactus across all users on a machine. In principle it would also be possible for configurations to be centralised to some extent – the object files and libraries associated with the Computational Toolkit for a particular configuration could be provided by the system administrator, and only the object files and libraries of user-local arrangements would then need to be compiled by the user. However, given the dynamic nature of code development it is not clear that such centralised configurations would be practicable.

All the above require modifications to the flesh. However the primary method to add functionality is by adding new thorns.

One basic feature intended for the future is to develop a specification for a set of parallel operation, the Cactus Communication Infrastructure (CCI) which will abstract the most commonly used parallel operations, and provide a set of thorns which implement this infrastructure using common communication libraries such as MPI, PVM, CORBA, SUN-RPC, etc. This would then allow driver thorns themselves to be built independently of the underlying parallel infrastructure, and greatly increase the number of parallel operations available to other thorns.

Another set of useful functionality would be to develop remote debuggers and profilers based upon the current `HTTPD` thorn and the expression parser. Indeed HTTP is far from being the only possible way to interact with a remote code, and thorns could be written to present allow interaction with a simulation via a SQL or LDAP interface, or as a web-service; there is already work on steering directly from visualisation clients using the streamed HDF5 interface [44].

Multi-grid methods are commonly used to solve elliptic equations, however all current methods of putting elliptic solvers into Cactus require knowledge of the driver or at least of the parallelisation technology. It should be possible to develop a set of thorns which use the scheduler in a driver and parallel-layer independent manner to solve elliptic problems.

The current set of thorns allows IO in many formats, however it is straightforward to add other formats, and there is at least one group with plans to write a NetCDF IO thorn. Additionally the infrastructure for IO makes it plausible to write IO methods that do transformations on the data first, e.g. spectral decomposition, before writing them out to disk or a stream; such methods would need to call routines from lower-level IO methods, and we have designed an interface which presents lower-level IO methods as stream-equivalents, thus insulating the higher-level IO methods from knowledge of specific lower-level IO methods.

## Acknowledgements

Cactus has evolved and been developed over many years, with contributions from a great many people. Originally an outgrowth of work in the NCSA numerical relativity group in the early 1990's, the first framework with the name Cactus was developed at the Max Planck Institute for Gravitational Physics by Paul Walker and Joan Massó. Its creation and further development was influenced greatly, and directly contributed to, by many friends, especially Werner Benger, Bernd Brügmann, Thomas Dramlitsch, Mark Miller, Manish Parashar, David Rideout, Matei Ripeanu, Erik Schnetter, Jonathan Thornburg, Malcolm Tobias and the numerical relativity groups at the Albert Einstein Institute and Washington University. It has also been generously supported by the MPG, NCSA, Intel, SGI, Sun, and Compaq, and by grants from the DFN-Verein (TiKSL, GriKSL), Microsoft, NASA, and the US NSF grant PHY-9979985 (ASC).

## References

[1] Foster, I.: (2002). The Grid: A New Infrastructure for 21st Century Science. *Physics Today*, **Febrary**.

[2] Component Component Architecture (CCA) Home Page `http://www.cca-forum.org`.

[3] Cactus Numerical Relativity Community `http://www.cactuscode.org/Community/Relativity.html`.

[4] Allen, G., Seidel, E., and Shalf, J.: (2002). Scientific Computing on the Grid. *Byte*, **Spring**.

[5] Cactus: http://www.cactuscode.org.

[6] *Cactus Users Guide* `http://www.cactuscode.org/Guides/Stable/UsersGuide/UsersGuideStable.pdf`.

[7] Allen, G., Benger, W., Dramlitsch, T., Goodale, T., Hege, H., Lanfermann, G., Merzky, A., Radke, T., and Seidel, E.: (2001). Cactus Grid Computing: Review of Current Development. In R. Sakellariou, J. Keane, J. Gurd, and L. Freeman, editors, *Europar 2001: Parallel Processing, Proceedings of 7th International Conference Manchester, UK August 28-31, 2001*. Springer.

[8] Parashar, M. and Browne, J. C.: (2000). *IMA Volume on Structured Adaptive Mesh Refinement (SAMR) Grid Methods*, chapter System Engineering for High Performance Computing Software: The HDDA/DAGH Infrastructure for Implementation of Parallel Structured Adaptive Mesh Refinement, pages 1–18. Springer-Verlag.

[9] Grid Adaptive Computational Engine (GrACE) `http://www.caip.rutgers.edu/~parashar/TASSL/Projects/GrACE/`.

[10] Anninos, P., Camarda, K., Massó, J., Seidel, E., Suen, W.-M., and Towns, J.: (1995). Three-dimensional numerical relativity: The evolution of black holes. *Phys. Rev. D*, **52**(4):2059–2082.

[11] Walker, P.: (1998). *Horizons, Hyperbolic Systems, and Inner Boundary Conditions in Numerical Relativity*. Ph.D. thesis, University of Illinois at Urbana-Champaign, Urbana, Illinois.

[12] Balay, S., Gropp, W., McInnes, L. C., and Smith, B.: (1998). PETSc - The Portable, Extensible Toolkit for Scientific Compuation. `http://www.mcs.anl.gov/petsc/`.

[13] *Cactus Computational Toolkit Thorn Guide* `http://www.cactuscode.org/Guides/Stable/ThornGuide/ThornGuideStable.pdf`.

[14] TASC: Software Engineering Support of the Third Round of Scientific Grand Challenge Investigations: Task 4, Review of Current Frameworks `http://sdcd.gsfc.nasa.gov/ESS/esmf_tasc/t400.html`.

[15] *Cactus Maintainers Guide* `http://www.cactuscode.org/Guides/Stable/MaintGuide/MaintGuideStable.pdf`.

[16] Concurrent Versions System (CVS) Home Page `http://www.cvshome.org/`.

[17] Cygnus Solutions (Cygwin) Home Page: `http://www.cygwin.com`.

[18] The Autoconf Home Page `http://www.gnu.org/software/autoconf/autoconf.html`.

[19] Allen, G., Dramlitsch, T., Foster, I., Karonis, N., Ripeanu, M., Seidel, E., and Toonen, B.: (2001). Supporting Efficicient Execution in Heterogeneous Distributed Computing Environments with Cactus and Globus. In *Proceedings of Supercomputing 2001, Denver*. http://www.cactuscode.org/Papers/GordonBell_2001.ps.gz.

[20] The xgraph and ygraph Home Pages `http://jean-luc.aei-potsdam.mpg.de/Codes/xgraph`, `http://www.aei.mpg.de/~pollney/ygraph`.

[21] GNUPlot Home Page `http://www.gnuplot.org`.

[22] IEEEIO Home Page `http://zeus.ncsa.uiuc.edu/~jshalf/FlexIO/`.

[23] Hierarchical Data Format Version 5 (HDF5) Home Page `http://hdf.ncsa.uiuc.edu/HDF5`.

[24] *Amira - Users Guide and Reference Manual* and *AmiraDev - Programmers Guide*, Konrad-Zuse-Zentrum für Informationstechnik Berlin (ZIB) and Indeed - Visual Concepts, Berlin, `http://amira.zib.de`.

[25] Open DX Home Page `http://www.opendx.org`.

[26] Alcubierre, M., Brandt, S., Brügmann, B., Holz, D., Seidel, E., Takahashi, R., and Thornburg, J.: (2001). Symmetry without Symmetry: Numerical Simulation of Axisymmetric Systems using Cartesian Grids. *Int. J. Mod. Phys. D*, **10**:273–289. Gr-qc/9908012.
[27] The Performance API (PAPI) Home Page `http://icl.cs.utk.edu/projects/papi/`.
[28] Dierks, T. and Allen, C.: The TLS Protocol Version 1.0: RFC 2246 January 1999.
[29] Rescorla, E.: HTTP Over TLS: RFC 2818 May 2000.
[30] Frier, A., Karlton, P., and Kocher, P.: The SSL 3.0 Protocol. Netscape Communications Corp. November 18, 1996.
[31] Foster, I., Kesselman, C., Tsudik, G., and Tuecke, S.: (1998). A Security Architecture for Computational Grids.
In *Proceedings of 5th ACM Conference on Computer and Communications Security Conference*, pages 83–92.
[32] DFN Gigabit Project *"Tele-Immersion: Collision of Black Holes"* (TIKSL) Home Page `http://www.zib.de/Visual/projects/TIKSL`.
[33] DFN-Verein Project *"Development of Grid Based Simulation and Visualization Techniques"* (GRIKSL) Home Page `http://www.griksl.org`.
[34] Shalf, J. and Bethel, E.W.: Cactus and Visapult: An Ultra-High Performance Grid-Distributed Visualization Architecture Using Connectionless Protocols. *Submitted to IEEE Computer Graphics and Animation.* Submitted.
[35] Camarda, K., He, Y., and Bishop, K.A.: (2001). A Parallel Chemical Reactor Simulation Using Cactus.
In *Proceedings of Linux Clusters: The HPC Revolution, NCSA 2001*.
[36] Dijkstra, H.A., Oksuzoglu, H., Wubs, F.W., and Botta, F.F.: (2001). A Fully Implicit Model of the three-dimensional thermohaline ocean circulation. *Journal of Computational Physics*, **173**:685715.
[37] Allen, G., Goodale, T., Massó, J., and Seidel, E.: (1999). The Cactus Computational Toolkit and Using Distributed Computing to Collide Neutron Stars. In *Proceedings of Eighth IEEE International Symposium on High Performance Distributed Computing, HPDC-8, Redondo Beach, 1999*. IEEE Computer Society.
[38] Allen, G., Benger, W., Goodale, T., Hege, H., Lanfermann, G., Merzky, A., Radke, T., and Seidel, E.: (2000).
The Cactus Code: A Problem Solving Environment for the Grid.
In *Proceedings of Ninth IEEE International Symposium on High Performance Distributed Computing, HPDC-9, Pittsburgh*.
[39] Benger, W., Foster, I., Novotny, J., Seidel, E., Shalf, J., Smith, W., and Walker, P.: (March 1999).
Numerical Relativity in a Distributed Environment. In *Proceedings of the Ninth SIAM Conference on Parallel Processing for Scientific Computing*.
[40] Allen, G., Dramlitsch, T., Goodale, T., Lanfermann, G., Radke, T., Seidel, E., Kielmann, T., Verstoep, K., Balaton, Z., Kacsuk, P., Szalai, F., Gehring, J., Keller, A., Streit, A., Matyska, L., Ruda, M., Krenek, A., Frese, H., Knipp, H., Merzky, A., Reinefeld, A., Schintke, F., Ludwiczak, B., Nabrzyski, J., Pukacki, J., Kersken, H.-P., and Russell, M.: (2001). Early Experiences with the EGrid Testbed. In *IEEE International Symposium on Cluster Computing and the Grid*. Available at `http://www.cactuscode.org/Papers/CCGrid_2001.pdf.gz`.
[41] GridLab: A Grid Application Toolkit and Testbed Project Home Page `http://www.gridlab.org`.

[42] Allen, G., Goodale, T., Russell, M., Seidel, E., and Shalf, J.: (2003). *Grid Computing: Making the Global Infrastructure a Reality*, chapter Classifying and Enabling Grid Applications. Wiley.

[43] Seidel, E., Allen, G., Merzky, A., and Nabrzyski, J.: (2002). GridLab — A Grid Application Toolkit and Testbed. *Future Generation Computer Systems*, **18**:1143–1153.

[44] Allen, G., Benger, W., Goodale, T., Hege, H., Lanfermann, G., Merzky, A., Radke, T., Seidel, E., and Shalf, J.: (2001). Cactus Tools for Grid Applications. *Cluster Computing*, **4**:179–188. http://www.cactuscode.org/Papers/CactusTools.ps.gz.

# Performance of Message-Passing MATLAB Toolboxes

Javier Fernández*, Antonio Cañas*, Antonio F. Díaz*, Jesús González, Julio Ortega*, and Alberto Prieto*

Dept. of Computer Technology and Architecture
ETSII, University of Granada, 18071-Granada, Spain
{javier,acanas,afdiaz,jesus,julio,aprieto}@atc.ugr.es
Tele: +34 958 248994
Fax: +34 958 248993

**Abstract.** In this work we compare some of the freely available parallel Toolboxes for MATLAB, which differ in purpose and implementation details: while DP-Toolbox and MultiMATLAB offer a higher-level parallel environment, the goals of PVMTB and MPITB, developed by us [7], are to closely adhere to the PVM system and MPI standard, respectively. DP-Toolbox is also based on PVM, and MultiMATLAB on MPI. These Toolboxes allow the user to build a parallel application under the rapid-prototyping MATLAB environment. The differences between them are illustrated by means of a performance test and a simple case study frequently found in the literature. Thus, depending on the preferred message-passing software and the performance requirements of the application, the user can either choose a higher-level Toolbox and benefit from easier coding, or directly interface the message-passing routines and benefit from greater control and performance.

**Topics:** Problem Solving Environments, Parallel and Distributed Computing, Cluster and Grid Computing.

## 1 Introduction

MATLAB has established itself as a de-facto standard for engineering and scientific computing. The underlying high-quality state-of-the-art numerical routines, the interactive environment, matrix-oriented language and graphical and visualization tools are some of the most frequently cited reasons for its widespread use [18]. Over 400 MATLAB-based books covering virtually any topic [16] and a user community of more than 500.000 [17] account for two more reasons attracting new users to this scientific computing environment.

In recent years a number of parallel Toolboxes for MATLAB have been made available by researchers and institutions (see [7, Table 1], and [3] for a survey). The timeliness of such Toolboxes is evident, taking into account the rapid popularization of clusters of computers as a valid and cost-effective Parallel Computing platform, and the popularity of MATLAB as a software prototyping tool.

* Supported in part by the Spanish MCYT TIC2000-1348 project.

J.M.L.M. Palma et al. (Eds.): VECPAR 2002, LNCS 2565, pp. 228–241, 2003.

Following the cluster computing trend, many standard Linux distributions (RedHat, Suse) now include the PVM [11] and LAM/MPI [15, 14] systems as regular packages, making it yet easier for cluster administrators to provide message-passing environments to their users. Together with some of the Toolboxes considered here, the MATLAB environment could then be used as a fast learning tool for the corresponding message-passing library, due to its interactive nature. In this sense, these Toolboxes make it easier for the extended MATLAB user base to turn to Parallel Computing.

On the other hand, previous knowledge of the message-passing library is of course reusable. Even the highest-level Toolbox commands can be conceptually related to the underlying library calls. For instance, the `Shift` command in MultiMATLAB can be related to the `MPI_Cart_Shift()` and `MPI_Sendrecv()` routines. An experienced C/MPI developer might consider prototyping the desired application under MATLAB, taking advantage of its sophisticated numeric and graphic routines, and of its interactive nature as an additional debugging facility, in order to translate (perhaps compile) the final debugged prototype back into C language.

Finally, these Toolboxes can be used merely to reduce the execution time of sequential code in the interpreted MATLAB environment by explicit (message-passing) parallelization. The medium-coarse granularity of many MATLAB prototypes makes them suitable for the so-called *outer loop* [19] parallelization, in agreement with the manufacturer's opinion on the possible uses of parallelism under MATLAB. Computer and interconnect performances, as well as per-node memory size, have all increased since [19] was written, but it remains true that SMP parallelism is better managed by the OS (and possibly a machine-tuned BLAS, see [20]) than by MATLAB, and that the (explicit) parallelization effort should address the coarsest granularity available in the prototype, which is usually well suited to the message-passing paradigm in clusters of computers.

In Section 2 the four Toolboxes considered here are briefly described. Section 3 shows the results of a ping-pong test programmed and run under each Toolbox, and Section 4 evaluates their performance on a case study. Conclusions are summarized in Section 5.

## 2 Message-Passing Matlab Toolboxes

Table 1 lists the Toolboxes studied in this work. All of them use the MATLAB ability of calling externally compiled C or FORTRAN programs (MEX-files) in order to gain access to the PVM or MPI library routines, offer a (growing with date) number of commands and interfaced library calls, and provide transparent support for most MATLAB data-types. The user is expected to issue a `spawn` command (`Start` in MultiMATLAB) to start remote MATLAB processes which cooperate in a parallel application by explicit message-passing.

Additional information on these Toolboxes is related below:

- The Distributed Parallel Toolbox [21] is a multi-layered project whose low-level layer (`dplow`, 56 commands) interfaces around 47 PVM routines. It

**Table 1.** Message-Passing MATLAB Toolboxes

| Toolbox | library | cmds | date | URL |
|---|---|---|---|---|
| DP-TB | PVM | 56 | Mar–99 | http://www-at.e-technik.uni-rostock.de/dp |
| MultiMATLAB | MPI | 27 | Nov–97 | http://www.cs.cornell.edu/Info/People/lnt/ /multimatlab.html |
| PVMTB | PVM | 93 | Jun–99 | http://atc.ugr.es/javier-bin/pvmtb_eng |
| MPITB | MPI | 153 | Feb–00 | http://atc.ugr.es/javier-bin/mpitb_eng |

was developed at Rostock University. It also includes an intermediate layer (`dphigh`) and a high level environment (DP-MM), consisting of 15 commands, which means the user does not need to learn PVM.
The group library has not been interfaced, so collective functionality (including `pvm_barrier` and `pvm_reduce`) is missing, but could be implemented in terms of the basic DP-TB routines if required. Coverage of the PVM library is greater than 50%. DP-TB works under both Windows and Linux.

- MultiMATLAB [23, 2, 18] is a high level Toolbox which frees the user from learning MPICH. Around 20 of its 27 commands correspond to MPI bindings. It was developed at Cornell University, with cooperation from The MathWorks. CTC is now working on CMTM [4], which runs under Windows and MPI/Pro, thus covering MPI-based MATLAB demand on this platform.
- PVMTB [7, 9] is a complete interface to PVM, including library calls such as `pvm_spawn` and `pvm_tickle`. It also interfaces the `TEV_MASK` macros and allows MATLAB programmed hooks to be registered as hoster, tasker, resource manager or receive functions.
- MPITB [7, 8] is an almost complete interface to the MPI 1.2 standard, excluding the `MPI_Type_*`, `MPI_Op_*` and `MPI_Pcontrol` bindings. It also includes several MPI 2.0 bindings required to spawn instances and communicate with them —`MPI_Info_*`, `MPI_Comm_spawn_*`, `MPI_Comm_get_parent`, `MPI_Comm_accept`, `MPI_Comm_[dis]connect`, `MPI_*_port` and `MPI_*_name`.

Figure 1 shows the role of the MEX interface in these Toolboxes. Both DP-TB and MultiMATLAB utilize the *directory* technique (called *indirection table* in [18]), in which a centralized MEX-file is linked to the message-passing library and provides entry points (table offsets) which are either library call wrappers (e.g. `pvm_config`, `Nproc`) or more highly elaborated routines (`pvme_upkarray` and `Recv` do support MATLAB data-types, for instance). Strictly speaking, only the directory MEX-file in DP-TB and MultiMATLAB sits on top of the message-passing libraries in Fig. 1.

Data is comfortably received in the left-hand side of the MATLAB call, as in DP-TB's `[data,info]=pvm_upkdouble(nitem,stride)` or MultiMATLAB's `[data,src,tag]=Recv(rank)`, allowing for compact MATLAB expressions such as `Sum=Sum+Recv(i)`.

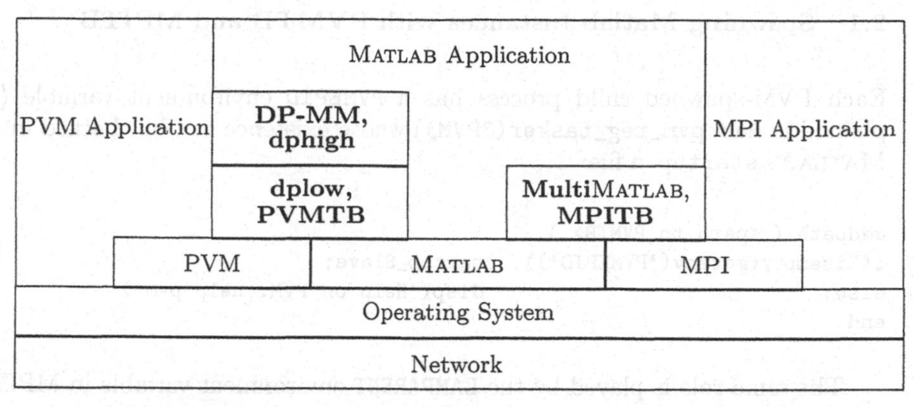


**Fig. 1.** Overview diagram of Toolbox role. The Toolbox calls PVM/MPI to provide message-passing functionality, like any other PVM/MPI application. It also calls the MATLAB API to pass arguments from/to the MATLAB process. Several MATLAB processes run on distinct cluster computers, each one being an instance of this diagram

Clearly, commands such as MultiMATLAB's `Eval` (or DP-TB's `dpmmeval`) imply coordination with slave processes, which must in some way be enforced at slave process startup. DP-TB uses a MEX *starter code* (`dpmatlab.mex`) for this purpose, equivalent to the MultiMATLAB's "separate top-level MEX routine" [18].

In PVMTB and MPITB designs [7] each message-passing routine has its corresponding MEX wrapper. The libraries have been dynamically linked to avoid disk waste. Source code for the MEX wrappers is extensively reutilized.

Commitment to the original routine parameter list means that PVMTB has commands such as `info=pvm_upkdouble(var,nitem,stride)`, and MPITB has `[info,stat]=MPI_Recv(var,src,tag,comm)`. This requires a more elaborate user code, since `var` is not returned on the left-hand side. Compact MATLAB expressions such as the accumulation shown above must be broken down into two steps (receive-add). However, using this syntax no new storage is allocated for `var`, avoiding additional memory operations in applications where the same storage can be reutilized many times.

With our Toolboxes, MATLAB users can rely on the provided startup M-file for slave MATLAB processes, or design their own M-file *starter codes* in the interpreted MATLAB language (not MEX-files), suited to their own applications. PVM hoster, tasker, resource manager, message handlers and receive functions, as well as MPI error handlers and attribute copy/delete functions, can also be programmed in MATLAB language and registered using the corresponding PVMTB or MPITB commands.

### 2.1 Spawning Matlab Instances with PVMTB and MPITB

Each PVM-spawned child process has a PVMEPID environment variable (explained in man pvm_reg_tasker(3PVM)) whose existence can be checked in the MATLAB's startup.m file:

```
addpath ( <path to PVMTB> );
if~isempty(getenv('PVMEPID')), startup_Slave;
else,                          disp('Help on PVM: help pvm')
end
```

The same role is played by the LAMPARENT environment variable in MPITB. Spawned MATLAB instances (Fig. 2(c)) can thus be discriminated from the parent instance launched from the user shell (Fig. 2(a)), and forced to perform an additional startup script —startup_Slave in the example above. This script can implement any desired protocol, and is written in MATLAB language.

Using the PVMTB commands pvm_start_pvmd (Fig. 2(b)) and pvm_spawn (Fig. 2(c)), the MATLAB user can start the Parallel Virtual Machine and children MATLAB processes from within MATLAB. MPITB users should invoke lamboot from the shell prior to issuing MPI_Init and MPI_Comm_spawn from within MATLAB—since the LAM daemons are not part of the MPI standard, there are no MPI commands to start them.

Most applications only need to specify one or more commands to be executed in the slave instance (Fig. 2(d)). The commands would be sent as strings that could be run on slave computers by using the MATLAB eval built-in. Thus, startup_Slave could be written as:

```
global TAG DEFAULT RAW PLACE,       TAG=7; DEFAULT=0; RAW=1; PLACE=2;
parent=pvm_parent; hostname         % show hostname if "xterm -e matlab"
pvm_recv(parent,TAG); pvm_unpack;   % parent should send NUMCMDS & QUIT
for PVMLoopIdx=1:NUMCMDS
  pvm_recv(parent,TAG); pvm_unpack; % and NUMCMDS strings called "cmd"
  eval(cmd);                        % containing desired MATLAB commands
end
if QUIT, quit, end
```

A message tag number is reserved for PVMTB use. The parent MATLAB should send two variables to each spawned MATLAB instance: NUMCMDS, indicating the desired number of commands to execute, and QUIT, to finish MATLAB children after the commands are executed. This flag can be cleared while developing or debugging and set when the prototype is finished. It is also customary to spawn xterm -e matlab during the debugging phase, in order to use MATLAB's built-in debugger on children instances.

The pvm_unpack Toolbox command, when called without arguments, will unpack all variables packed in the current receive buffer using their original

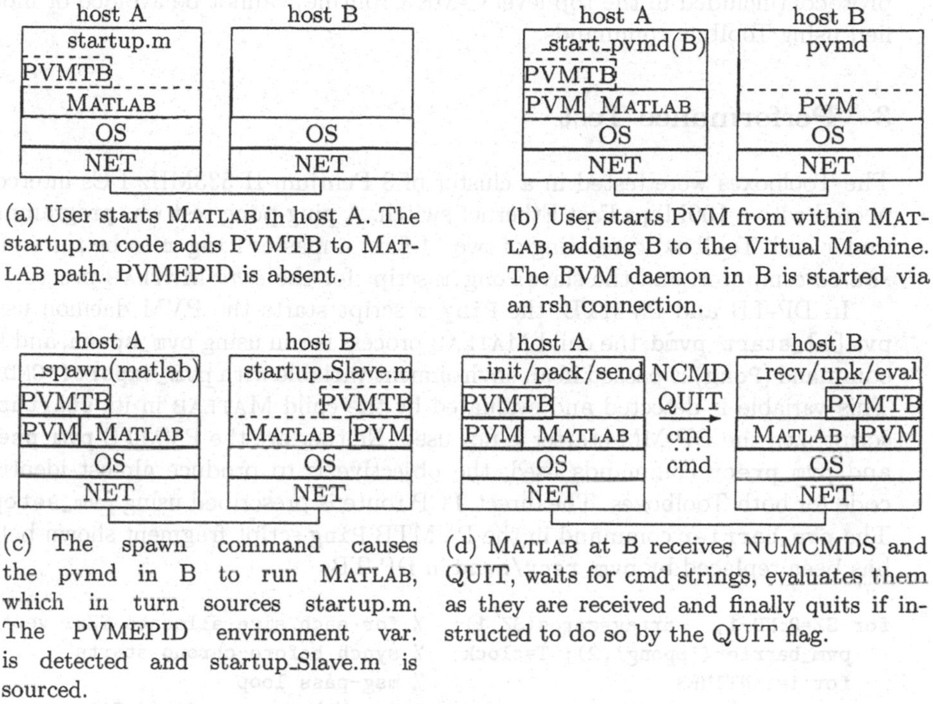

(a) User starts MATLAB in host A. The startup.m code adds PVMTB to MATLAB path. PVMEPID is absent.

(b) User starts PVM from within MATLAB, adding B to the Virtual Machine. The PVM daemon in B is started via an rsh connection.

(c) The spawn command causes the pvmd in B to run MATLAB, which in turn sources startup.m. The PVMEPID environment var. is detected and startup_Slave.m is sourced.

(d) MATLAB at B receives NUMCMDS and QUIT, waits for cmd strings, evaluates them as they are received and finally quits if instructed to do so by the QUIT flag.

**Fig. 2.** Sample control protocol for spawned MATLAB instances

names. So the master is expected to send `NUMCMDS` times a string called `cmd` which will be evaluated at the child instance.

The default `startup_Slave` provided with PVMTB is more elaborate, using a 5s-timed `pvm_trecv` to avoid blocking interactively spawned children (useful for the tutorial), and a conditional branch for `NUMCMDS==0`, so that the loop only ends when `cmd=='break'|'quit'|'exit'` (useful for applications where the total number of commands is not known beforehand).

The `startup_Slave` for MPITB is even more elaborate, building a new intercommunicator which includes the parent and spawned MATLAB processes. The 5s-timed receive is implemented using `MPI_Wtime` and `MPI_Iprobe`.

PVMTB or MPITB users might implement any other conceivable protocol in their startup files, since any desired PVM or MPI call has been interfaced and is available. DP-TB users can reproduce the protocols described here, since all the required PVM calls are interfaced in `dplow`.

A different protocol is used in the following sections, since the child(ren) will run just one (and the same) command. The startup code can be simplified, exporting the command string from the parent as an environment variable under PVMTB/DP-TB, or broadcasting it under MPITB. MultiMATLAB's startup protocol (included in the top-level C-MEX routine) cannot be avoided or modified using Toolbox commands.

## 3 Performance Test

The Toolboxes were tested in a cluster of 8 Pentium II 333MHz PCs interconnected with a 100Mbps Fast Ethernet switch. A ping-pong test was programmed under each Toolbox, consisting of two M-files: a master `Ping.m` to be run in the *source* computer, and the slave `Pong.m` script for the *echo* MATLAB process.

In DP-TB and PVMTB, the `Ping.m` script starts the PVM daemon using `pvm[e]_start_pvmd`, the child MATLAB process is run using `pvm_spawn`, and the command (`Pong`) is passed in an environment variable with `pvm_export('CMD')`. This variable is detected and evaluated by the child MATLAB in its `startup.m` script, i.e., the DP-MM starter is not used. Neither are the PVMTB `pvm_psend` and `pvm_precv` commands used; the objective is to produce almost identical code for both Toolboxes. The direct TCP route is prescribed using `pvm_setopt`. The `pvm_barrier` command in the PVMTB `Ping` script fragment shown below has been replaced by `pvm_recv/send` in DP-TB.

```
for SZ=0:ND-1,    array=zeros(SZ,1);  % for each size allocate SZx1 vector
   pvm_barrier('ppong',2); T=clock;   % synch before chrono starts
   for i=1:NTIMES                      % msg-pass loop
      pvm_initsend(PLACE); pvm_pkdouble(array); pvm_send(tid,TAG);
      pvm_recv(tid,TAG);  pvm_upkdouble(array);
   end
   T=etime(clock,T);        Data(SZ+1).time=T/2/NTIMES;
   S=whos('array');         Data(SZ+1).size=S.bytes;
   clear array
end                                    % for SZ
```

The `pvm_initsend` command specifies in-place data encoding. The `Pong` script swaps the order of the two innermost lines (`send/recv`). The `for` loop is repeated `NTIMES=10` times to reduce loop initialization and clock resolution influence on the results. The round-trip time `T/NTIMES` is halved to compute the one-way (transmit, ping) time.

According to [13] (and common practice), the test should have been repeated a number of times in order to obtain reproducible figures (minimum of averages). However, the peaks shown in Fig. 3 can be easily identified and suggest the typical distribution and magnitude of the perturbations that can be expected.

The MPITB code is very similar, using `MPI_Spawn` to start the MATLAB child and `MPI_Intercomm_merge` to create a common intracommunicator. The command is explicitly passed as a string using `MPI_Bcast` in the new communicator. The child's `startup.m` script cooperates in the collective merge and broadcast operations, and then the command string (`Pong`) is evaluated. The `Ping` loop shown below is a direct translation of the former PVM fragment to MPI:

```
MPI_Barrier(NEWORLD); T=clock;       % synch before chrono starts
for i=1:NTIMES                       % msg-pass loop
    MPI_Send(array,1,TAG,NEWORLD);
    MPI_Recv(array,1,TAG,NEWORLD);
end
```

LAM *client-to-client* mode and homogeneous operation is prescribed adjusting the LAM TROLLIUSRTF environment variable before MPI_Init invocation.

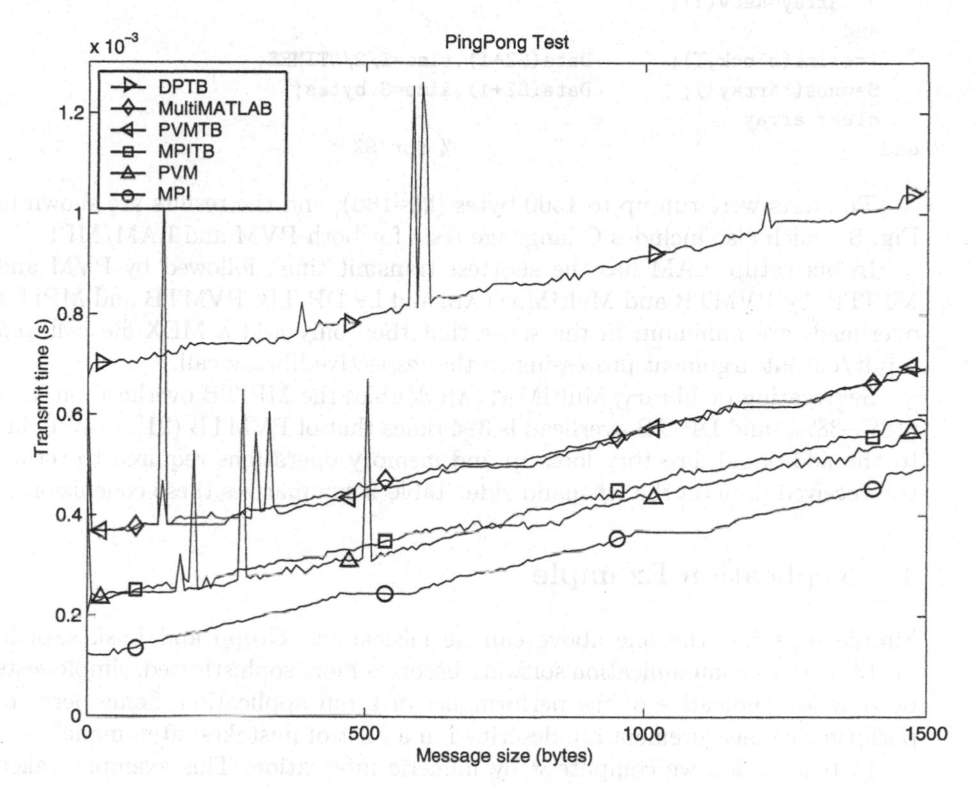


**Fig. 3.** Results of the ping-pong test

**Table 2.** Toolbox overhead in the 0–1500 bytes range, segregated by library

| Toolbox/library | 0B | overhead | 1500B | overhead |
|---|---|---|---|---|
| DP-TB | 676 $\mu$s | 3.350 | 1047 $\mu$s | 1.837 |
| MultiMATLAB | 369 $\mu$s | 3.037 | 691 $\mu$s | 1.384 |
| PVMTB | 369 $\mu$s | 1.827 | 692 $\mu$s | 1.214 |
| MPITB | 227 $\mu$s | 1.871 | 588 $\mu$s | 1.179 |
| PVM | 202 $\mu$s | 1.000 | 570 $\mu$s | 1.000 |
| MPI | 121 $\mu$s | 1.000 | 499 $\mu$s | 1.000 |

The MultiMATLAB code is more compact, requiring only a `Start` command to launch the *echo* MATLAB process, which becomes the target for the `Eval` command without any additional `startup.m` programming.

```
Eval(1,['Pong(' int2str(ND) ',' int2str(NTIMES) ')']);
for SZ=0:ND-1, array=zeros(SZ,1);     % for each size allocate SZx1 vector
  Barrier;   T=clock;                 % synch before chrono starts
  for i=1:NTIMES                      % msg-pass loop
    Send(1,array);
    array=Recv(1);
  end
  T=etime(clock,T);        Data(SZ+1).time=T/2/NTIMES;
  S=whos('array');         Data(SZ+1).size=S.bytes;
  clear array
end                                   % for SZ
```

The tests were run up to 1500 bytes (`ND=188`), and the results are shown in Fig. 3, which also includes C language tests for both PVM and LAM/MPI.

In our setup, LAM has the shortest transmit time, followed by PVM and MPITB, by PVMTB and MultiMATLAB, and by DP-TB. PVMTB and MPITB overheads are minimum in the sense that they only add a MEX-file call and input/output argument processing to the respective library call.

Segregating by library, MultiMATLAB doubles the MPITB overhead on MPI (18%–38%), and DP-TB overhead is 3–4 times that of PVMTB (21%–84%), due to the additional directory look-up and memory operations required to return the received data on the left-hand side. Table 2 summarizes these conclusions.

## 4 Application Example

Simple tests like the one above can be misleading. Gropp and Lusk explain in [13] how, as communication software becomes more sophisticated, simple tests become less indicative of the performance of a full application. Some perils of performance measurement are described in a "list of mistakes often made".

In this section we compute $\pi$ by numeric integration. This example, taken from [22] and cited in [5], is proposed as an exercise in many textbooks. Even this simple, embarrassingly parallel application brings into discussion some of the *common pitfalls* described in [13], such as:

- Ignore overlap of computation and communication.
- Confuse total bandwidth with point-to-point bandwidth.
- Ignore cache effects.
- Use a communication pattern different from the application, measure with just two processors, or measure with a single communication pattern.

Another advantage of this example is that we can show all the relevant MATLAB code without getting involved in complex code explanations. Interested readers can find more complex applications in [6] (Control Problems), [7] (Wavelet

Analysis), [10] (Chemical Manufacturing) and [12] (NanoElectronics). In particular, the latter paper includes a detailed scattering model with efficiencies ranging from 98% (2 CPUs, 1.96x speedup) to 84% (120 CPUs, 101x speedup). The Parallel NANOMOS simulator used to implement this model was built on top of PVMTB and is available from the CELAB CVS repository [12].

Returning to the $\pi$ example, the C sequential code proposed in [5] applies the mid-point integration rule to arctan′ in $[0, 1]$ (there are faster, more accurate methods for approximating $\pi$ [1]):

```
width = 1.0 / N;          sum = 0;           % subdivision width
for (i=0; i<N; ++i) {                        % subdivision index
  register double x = (i + 0.5) * width;     % abscissa mid-points
  sum += 4.0 / (1.0 + x * x);                % area = width * height
}
sum *= width;                      % interchanged width* and 4* for accuracy
```

In this example we chose `N=408800` subdivisions of the $[0, 1]$ interval, accounting for 13 decimal digits of precision and around 0.5s computation time under MATLAB. The loop can be trivially staggered among `C` computers, making each of them compute a subset `for (i=inum-1; i<N; i+=C)`, which only requires an additional *rank* or *instance* number `inum=1...C`. In order to produce intuitive scalability results, time was measured from transmission of arguments to accumulation of results, not including master process setup and spawning of slaves. In the master PVMTB code shown below, only the *inum* argument needs to be transmitted separately (and included in the measurement).

```
cmd=['Work(' int2str(C) ',' int2str(N) ')'];
putenv(['CMD=' cmd]); pvm_export('CMD');
[numt tids]=pvm_spawn('matlab',{'-nosplash'},33,'.',C);
...                                          % Direct TCP route
tic
for numt=1:C                                 % inum a-la DP-TB
pvm_initsend(PLACE); pvm_pkdouble(numt); pvm_send(tids(numt),TAG);
end

Psum=0; Sum=0;
for numt=1:C                                 % reduce a-la DP-TB
  pvm_recv(-1,TAG); pvm_upkdouble(Sum); Psum=Psum+Sum;
end
Psum = Psum/N;

T=toc;
```

DP-TB allows a more compact `Psum=Psum+pvm_upkdouble(1,1)` expression. As in the ping-pong test, group functionality (`pvm_joingroup` / `pvm_reduce` in this case) was not used under PVMTB to avoid an "apples to oranges" comparison with DP-TB, resulting in very similar code for both Toolboxes.

In this way, not only were the tests run under the same OS and the same library (in each case, the same PVM or LAM/MPI version) compiled with the same compiler and optimization switches, but also the test codes were kept as similar as possible to each other, avoiding the use of library functionality not available under all Toolboxes.

More complex applications could conceivably benefit from more sophisticated library calls, such as PVM message handlers or mailboxes, elaborate collective MPI operations — or even an `MPI_Sendrecv_replace` call, as in [7], functionality that is not present in all the Toolboxes here considered. For this simple example we can restrict ourselves to the common functionality provided by all Toolboxes without being concerned about hypothetical performance improvements due to the use of different library calls.

The same criterion was followed with MPITB, giving the sentence:

```
for numt=1:C
  MPI_Recv(Sum,MPI_ANY_SOURCE,TAG,NEWORLD); Psum=Psum+Sum;
end
```

The most compact master code is MultiMATLAB `Mast.m`:

```
tic
            Psum=0;
Eval([1:C],['Work(' int2str(C) ',' int2str(N) ')'])
for i=1:C ,  Psum=Psum+Recv; end
            Psum=Psum/N;
T=toc;
```

The slave computation loop can be vectorized to avoid its interpretation cost under MATLAB. Again, the most compact version is MultiMATLAB `Work.m`:

```
rank=ID;
  width=1/N; lsum=0;
  i=rank-1:C:N-1;                 % index subset for this rank
  x=(i+0.5)*width;                % vector of abscissa mid-points
  lsum=sum(4./(1+x.^2));          % vectorized - no for loops
Send(0,lsum)
```

MPITB uses `[info rank]=MPI_Comm_rank(NEWORLD)` instead of `ID`, and the more verbose `MPI_Send(lsum,0,TAG,NEWORLD)`, and requires the already mentioned **`startup.m`** support for merging the communicators, while PVMTB and DP-TB `Work.m` scripts explicitly receive the **`inum`** index:

```
parent=pvm_parent;
...                                               % Direct TCP route
pvm_recv(parent,TAG); pvm_upkdouble(inum);
  width=1/N; lsum=0;
  ...
pvm_initsend(PLACE); pvm_pkdouble(lsum); pvm_send(parent,TAG);
```

Figure 4 shows the measured time, accounting for slave computation, argument transmit and result collect times. Absolute times are almost indistinguishable, since computation time is in the order of 500ms (expected 62.5ms for `C==8`)

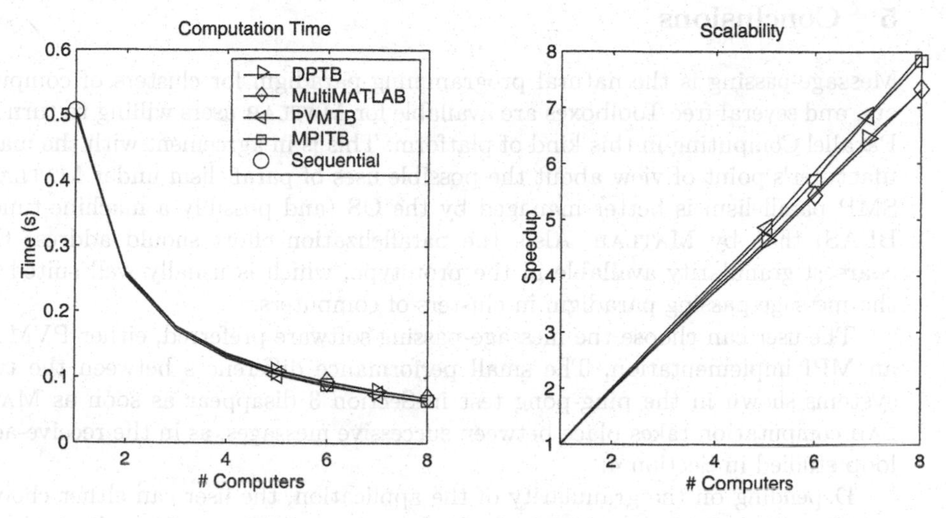


**Fig. 4.** Results of the scalability test

**Table 3.** Algorithm Time and Speedup using C=2, 4, 8 computers

| Toolbox | Time (ms) | | | Speedup | | |
|---|---|---|---|---|---|---|
| | C=2 | C=4 | C=8 | C=2 | C=4 | C=8 |
| DP-TB | 255.807 | 133.928 | 69.440 | 1.984 | 3.790 | 7.309 |
| MultiMATLAB | 259.408 | 136.134 | 69.264 | 1.957 | 3.728 | 7.328 |
| PVMTB | 254.102 | 129.118 | 63.730 | 1.997 | 3.931 | 7.964 |
| MPITB | 257.290 | 131.676 | 64.770 | 1.973 | 3.854 | 7.836 |

while expected transmit time ranges from 0.227ms (MPITB) to 0.676ms (DP-TB, see Table 2), giving computation-to-communication ratios in the order of $10^1$–$10^3$, depending on the number of computers C and the Toolbox used. The differences can be highlighted by moving these absolute times to the denominator, as in the associated speedup graph in Fig. 4.

In this application, up to C==8 messages come back to the master MATLAB process, although not in perfect synchronization, due to the initial sequential data distribution. The performance advantage of MPITB over PVMTB in Table 2 is lost in the final accumulation `for` loop in the master process. The greater advantage of MultiMATLAB over DP-TB is also lost.

In PVMTB and MPITB, the receive buffer `Sum` is reutilized, while temporal storage is allocated and discarded for each iteration in DP-TB and MultiMATLAB code. Table 3 quantifies this small performance loss.

## 5 Conclusions

Message-passing is the natural programming paradigm for clusters of computers, and several free Toolboxes are available for MATLAB users willing to turn to Parallel Computing in this kind of platform. This is in agreement with the manufacturer's point of view about the possible uses of parallelism under MATLAB: SMP parallelism is better managed by the OS (and possibly a machine-tuned BLAS) than by MATLAB. Also, the parallelization effort should address the coarsest granularity available in the prototype, which is usually well suited to the message-passing paradigm in clusters of computers.

The user can choose the message-passing software preferred, either PVM or an MPI implementation. The small performance differences between the two systems shown in the ping-pong test in Section 3 disappear as soon as MATLAB computation takes place between successive messages, as in the receive-add loop studied in Section 4.

Depending on the granularity of the application, the user can either choose a higher-level Toolbox and benefit from easier, more compact coding, or directly interface the message-passing routines and benefit from greater control and performance. For coarse-grained, embarrassingly parallel applications with high computation-to-communication ratios, the performance loss is very small. Other parallel applications with smaller granularity, complex communication patterns or repeated synchronization or *rendez-vous* steps, might require the higher performance attainable at the cost of more verbose programming.

## References

[1] J. Arndt, C. Haenel: "$\pi$ – Unleashed", Springer-Verlag, 2001, ISBN: 3540665722, **http://www.jjj.de/pibook/pibook.html**

[2] C. Chang, G. Czajkowski, X. Liu, V. Menon, C. Myers, A.E. Trefethen, L.N. Trefethen: "The Cornell MultiMATLAB Project", Proceedings of the POOMA'96 Conference, Sanfa Fe, New-Mexico, **http://www.acl.lanl.gov/Pooma96/abstracts/anne/POOMA96.html**

[3] R. Choi: Parallel MATLAB survey, **http://supertech.lcs.mit.edu/~cly/survey.html**

[4] Cornell Theory Center: "New Parallel Programming Tools for MATLAB", **http://www.tc.cornell.edu/news/releases/2000/cmtm.asp**

[5] H. Dietz: "Parallel Processing HOWTO", January 1998, Linux Documentation Project, **http://aggregate.org/PPLINUX/**

[6] S. Dormido-Canto, A.P. Madrid, S. Dormido: "Programming on Clusters for Solving Control Problems", Proceedings of the 4th Asian Control Conference ASCC2002, Suntec Singapore, Singapore, September 2002, Session WA9-11 PaperID 1343 in
**http://www.ece.nus.edu.sg/ascc2002/advprog/Prog_Summary.pdf**
**http://atc.ugr.es/~javier/investigacion/papers/Seb-ASCC.pdf**

[7] J. Fernández-Baldomero: "Message Passing under MATLAB", Proceedings of the HPC 2001 (Adrian Tentner, Ed.), pp.73–82. ASTC 2001, Seattle, Washington, **http://www.scs.org/confernc/astc/astc01/prelim-program/html/hpc-pp.html**
**http://atc.ugr.es/~javier/investigacion/papers/H010FernandezBaldomero002.pdf**

[8] J. Fernández-Baldomero: MPITB home page, **http://atc.ugr.es/javier-bin/mpitb_eng**

[9] J. Fernández-Baldomero: PVMTB home page, **http://atc.ugr.es/javier-bin/pvmtb_eng**

[10] V. García-Osorio, B. E. Ydstie: "Parallel, Distributed Modeling of Simulation of Chemical Systems", American Institute of Chemical Engineers, Annual Meeting, Reno NV, November 2001. **http://www.aiche.org/conferences/techprogram/ paperdetail.asp?PaperID=2009&DSN=annual01**

[11] A. Geist, A. Beguelin, J. Dongarra, W. Jiang, R. Mancheck, V. Sunderam: "PVM: Parallel Virtual Machine. A Users' Guide and Tutorial for Networked Parallel Computing", The MIT Press, Cambridge, Massachusetts, 1994, **http://www.netlib.org/pvm3/book/pvm-book.html**

[12] S. Goasguen, A. Butt, K. D. Colby, M. S. Lundstrom: "Parallelization of the Nanoscale Device Simulator nanoMOS 2.0 using a 100 Nodes Linux Cluster", Proceedings of the 2nd IEEE Conference on Nanotechnology IEEE-NANO'02, Arlington VA, USA, August 2002, Session WA7#0840 in
**http://www.ewh.ieee.org/tc/nanotech/nano2002/techprog.html**
**http://atc.ugr.es/~javier/investigacion/papers/SebNano02.pdf**
**http://atc.ugr.es/~javier/investigacion/papers/IEEE-nano.pdf**
**http://falcon.ecn.purdue.edu:8080/cgi-bin/cvsweb.cgi/**

[13] W. Gropp, E. Lusk: "Reproducible Measurements of MPI Performance Characteristics", Proceedings of the EuroPVM/MPI'99 Conference, Barcelona, Spain, September 1999,
**http://www-unix.mcs.anl.gov/~gropp/bib/papers/**

[14] W. Gropp, E. Lusk, A. Skjelum: "Using MPI: Portable Parallel Programming with the Message-Passing Interface", The MIT Press, 1994,
**http://www.mcs.anl.gov/mpi/usingmpi/index.html**

[15] LAM Home Page, **http://www.lam-mpi.org**

[16] The MathWorks, Inc.: "MATLAB based books" web page,
**http://www.mathworks.com/support/books**

[17] The MathWorks, Inc.: "User stories" web page,
**http://www.mathworks.com/products/user_story**
**http://www.mathworks.com/products/matlab/description/overview.shtml**

[18] V. S. Menon, A. E. Trefethen: "MultiMATLAB: Integrating MATLAB with High-Performance Parallel Computing", Proceedings of Supercomputing'97, ACM SIG ARCH and IEEE Computer Society, 1997,
**http://www.supercomp.org/sc97/proceedings/TECH/MENON/INDEX.HTM**

[19] C. Moler: "Why there isn't a parallel MATLAB", MATLAB news & notes, Cleve's Corner Spring 1995,
**http://www.mathworks.com/company/newsletter/pdf/spr95cleve.pdf**

[20] C. Moler: "MATLAB incorporates LAPACK", MATLAB news & notes, Cleve's Corner Winter 2000, **http://www.mathworks.com/company/newsletter/clevescorner/ winter2000.cleve.shtml**

[21] S. Pawletta, W. Drewelow, P. Duenow, T. Pawletta, M. Suesse: "A MATLAB Toolbox for Distributed and Parallel Processing", Proceedings of the MATLAB'95 Conference (C. Moler, S. Little, Eds.), MathWorks Inc., Cambridge, MA, October 1995, **http://citeseer.nj.nec.com/pawletta95matlab.html**

[22] M. J. Quinn: *"Parallel Computing Theory and Practice, 2nd Edition"*, McGraw Hill, New York, 1994.

[23] A. E. Trefethen, V. S. Menon, C. C. Chang, G. J. Czajkowski, C. Myers, L. N. Trefethen: "MultiMATLAB: MATLAB on Multiple Processors", Tech. Report 96-239, Cornell Theory Center, 1996, **ftp://ftp.tc.cornell.edu/pub/tech.reports/tr239.ps**

# Evaluating the Performance of Space Plasma Simulations Using FPGA's

Ben Popoola and Paul Gough

Space Science Center, University of Sussex
Falmer, Brighton, BN1 9QT, U.K
{o.m.popoola,m.p.gough}@sussex.ac.uk

**Abstract.** This paper analyses the performance of a custom compute machine, that performs electrostatic plasma simulations, using Field Programmable Gate Array's (FPGAs). Although FPGA's run at slower clock speeds than their off-the-shelf counterparts, the processing power lost in the reduced number of clock cycles per second is quickly recovered in the high degree of spatial parallelism that is achievable within the devices. We describe the development of the architecture of the machine and its support for the C-programming language via the use of a cross-compiler. Results are presented and a discussion is given on the constraints of FPGAs in particular and the hardware design process in general.

## 1 Introduction

Scientific computing is characterised by developing a mathematical model to investigate some physical phenomenon of interest, verifying that the mathematical model developed can produce reasonable results within specified input constraints and discretising the mathematical model developed for use on digital computers.

The discretized model which usually consists of a large number of differential equations, to which a numerical solution is sought, forms the basis of a computer simulation model. When expressed as a sequence of computer instructions, the physical phenomenon under interest is investigated using a computer simulation program.

Termed computer experiments [15], simulation programs consist of following the temporal evolution of a system, from time $t = 0$ in discrete time steps of $dt$ within a computional box that represents some finite region of space in the problem domain. One of the challenges in scientific computing is in achieving technically accurate and efficient simulation models, given finite and constrained computing resources.

The large amount of computing power required to solve the simulation models, had led to the proposal and/or development of varying general and special purpose computer architectures since the introduction of modern digital computing. This has been fueled further by the advent of Very Large Scale Integrated (VLSI) circuit technology that has allowed the computer architect to utilise millions of transistors on a single silicon chip.

J.M.L.M. Palma et al. (Eds.): VECPAR 2002, LNCS 2565, pp. 242–254, 2003.

One of the more successful architectures built from these proposals in the 1970's, was the CRAY-1 [22] supercomputer. It was one of only a handful of computers available at the time, that offered vector processing techniques as a means of rapidly computing replicated structures of data commonly found in simulation programs. One of the main problems posed by supercomputers, apart from their comparatively high cost, is in the slow turn-around time in the implementation of new architectures that utilise modern integrated circuit technology techniques. This arises from the long design times and the complicated nature of their custom architectures.

A more recent achievement in obtaining high performance in scientific computing, is in the use of parallel computing architectures over a cluster of workstations. The earliest approach used a Network of Workstations (NOW's) but a more modern method is in the use of off-the-shelf computing components to build Beowulf clusters [5]. Using these systems requires users to write algorthims using some sort of portable communication library [3], one of which is the Message Passing Interface(MPI) [26, 13].

Computing clusters have brought their own challenges to the scientific computing arena as users try to ensure that all processors in a computing cluster perform an equal proportion of work, a technique known as load balancing. A equally challenging problem in using computing clusters is in the granularity of the algorthims used, which is highly dependent on the communication bandwidth between participating processors.

All the techniques mention so far, use Application Specific Integrates Circuits (ASICs) devices, whose functionality cannot be altered once it has been fabricated on silicon. Recent advances in transistor and packaging technology has seen, over the last twenty years, the emergence of the Field Programmable Gate Arrays (FPGAs) [8, 1]. An FPGA is a reprogrammable chip that provides system designers with the opportunity of implementing their own custom architectures on silicon, via software, without incurring the high costs associated with fabricating ASICs.

Although functions implemented in FPGAs run at slower clock speeds than the same function implemented in an ASIC, they provide the user the opportunity of implementing application specific code speed-ups in hardware, through the use of high degrees of spatial parallelism. Algorithm improvements can also be re-implemented as FPGAs are reprogrammable. As most ASICs are fabricated as general purpose devices, clock cycles that are lost in "raw speed" in FPGA implementations of specific functions, can be easily recovered and surpassed by the use of well structured spatially and temporally parallel algorithms.

This paper describes our approach in using FPGAs in the solution of space plasma simulations and is organized as follows. In section two, we introduce the mathematical model encountered in electrostatic plasma calculations and describe the principle behind FPGAs and their use in developing custom parallel architectures. Section three describes the implementation of our custom Coprocessing Units(CUs), used to accelerate the solution of the simulation model. The

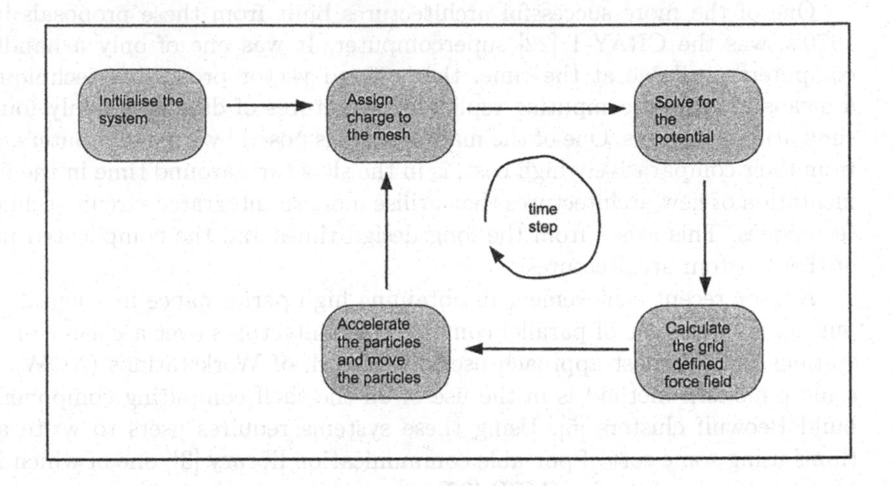


**Fig. 1.** The design flow of a typical plasma simulation program

next section describes our results and is followed by a discussion of the challenges faced in using FPGAs in scientific computing.

## 2 Background Work

Rather than develop a mathematical model from scratch for space plasma simulations, with all the technically challenging problems of energy conversvation, convergence errors and floating point rounding errors, a decision was made to use a plasma simulation code that was already established, available and had obtainable source code. A 1D simulation code that the Space Group of the University of Sussex is familiar with and has been used to investigate various phenomenon in space plasmas is the 1D electrostatic code(ES1)[4].

It can typically be used to evolve in time, hundreds of thousands of particles with position and velocity, where each iterative time step involves the following cycle of events:

- The assignment of charge onto a computational grid from the particles' positions $\mathbf{x}$ and velocities $\mathbf{v}$.
- Integration of the field equations on the grid to obtain the electric $\mathbf{E}$ and magnetic $\mathbf{B}$ fields.
- Calculating, from the electric and magnetic fields, the grid-defined force $\mathbf{F}$ field.
- An interpolation of the force field to obtain the new particles positions and velocities - particle movement.

This procedure is repeated in a continuous loop for the lifetime of the simulation with the user being offered the opportunity of viewing the results either immediately, intermittently, or after a simulation has been completed by saving the results to a disk file for post-processing.

An important calculation that lies at the heart of the ES1 code, is in obtaining the solution of the field equations for a bounded plasma by the use of a rapid elliptic solver [25] based on the Fast Fourier Transform (FFT) [6, 21]. The FFT is the method used in the ES1 code as a Poisson solver to solve the 1D Poisson partial-differential equation $-\Delta u = f(x)$.

Although FPGAs have been in use since the 1980's, only recently are they becoming available with a large amount of logic elements (LE) and internal embedded RAM. It is not uncommon today to find devices in the Altera APEX 20KE logic family which have upto 51,840 LE (APEX1500KE) which roughly translates into about 1.5 million usable gates. To put this number into per perspective, the integer unit of our MIPS I compatible processor utilises about 3 per cent of the above mentioned device (APEX1500KE).

A method used to design custom architectures for FPGAs is with traditional schematic entry software tools used for digital logic design. The software generally consists of a variety of low-level digital logic components which can be wired together to form more complicated circuits. The user also has the opportunity of obtaining higher-level functions, such as transcendential algorithms, as Intellectual Property (IP) cores from third party sources. However the method used here is to a provide a description of the functionality of the various algorithms at the Register Transfer Level (RTL) using a variant Hardware Description Language (HDL), VHDL.

An emerging standard of capturing custom architecures, which is not explored here, is in the use of the C-like hardware programming languages Handle-C and System-C. Which ever method is used, a full Electronic Design Automation (EDA) tools suite is required that allows a capturing of the design, simulation, synthesis, back simulation and the final step of place and routing. The place and routing tools' output is a disk file in the form of a serial object file. This describes the inner working of the FPGA, how each logic elements will function, how all the logic elements are interconnected and how the I/O pins are mapped to external devices. Once this file is downloaded onto an FPGA a working device with user specified functionality is obtained.

With the availability of FPGAs that are capable of potentially containing the simulation model, the choice of how to implement the design in custom hardware must be made. The method used here has been influenced by previous studies that have shown that integrating a processor with hardware function units eliminates the communication bandwidth between an off-board processor and reconfigurable logic [12, 27, 11]. Here we describe the design of the MIPS R3000 compatible processor with an accompliment of Coprocessing Units (CU's) to accelerate a modified version of the ES1 code. Each CU must conform to our coprocessor interface specification and meet the specific requirements given below.

A CU can be as small as a unit that accelerates a specific portions of a code, to special purpose units from LU decomposition matrix solvers to custom visualization units that provide standard computer graphics output. The constraints that we apply to any coprocessing unit currently is that if they use floating point arithmetic they should comply with the IEEE-754 Standard [16, 9] on floating point arithmetic, must be able to reside on the same FPGA as the MIPS 1 compatible core and should not increase the maximum CPU pipeline-stage latency. Apart from these restrictions each CU can exhibit its' own degrees of internal spatial and temporal parallelism.

The next section of this paper describes the overall architecture of the custom computing machine and the implementation of three IEEE-754 compliant coprocessors a floating point, fast fourier transform and particle mover units.

## 3 Implementation

As the intention of this project is to physically implement a prototype system, rather than produce a VHDL simulation model, a Reduced Instruction Set Computer (RISC) [20, 18, 23], with a Harvard bus architecture that is easy to implement has been chosen as our general purpose CPU. This led to the choice of a design that is compatible with the MIPS R3000 [17, 14] processor, mainly due to its widely available documentation.

Our implementation, as in the original, is a single issue CPU with temporal parallelism obtained through the use of five pipelined stages. Our current version supports a 32-bit instruction set and is characterised by a load-store architecture. A decision not to implement a superscalar multi-issue processor was made, in anticipation that the CU's will be the "work horses" of the system rather than the CPU.

Our implementation also, does not implement Level 1 (L1) instruction and data caches on chip. Instead we use ZBT SRAM [19] as off-chip Level 2 (L2) cache with access times of about 5ns. This has the disadvantage of not only limiting our system to a maximum clock frequency of 200Mhz, but also limits the number of particles available in a simulation, in our prototype model, due to the relatively high cost per bit of SRAM compared to SDRAM. It does offer however, the prototype system, the advantage of sustained memory throughput unlike SDRAM that requires a complicated cache design and non-sustainable throughputs, hence delays, due to cache misses.

An alternative approach could have been to use the internal RAM available on FPGA's as L1 cache, however as the current trend in FPGAs, is to provide an amount of internal RAM that is proportional to the LE count this would have left our design FPGA type dependent. To date we currently have 3 working CUs for plasma simulation acceleration and a visualization unit that provides output to a standard VGA monitor ( To be described in a forth-coming paper). These units and the overall architecture are shown in figure 2.

Although all units can run in parallel, each CU has a special control function that allows it to take complete control of the data buses. This prevents memory conflicts with hardware units trying to access memory simultaneously. This is especially important due to the implementation of left and right data caches.

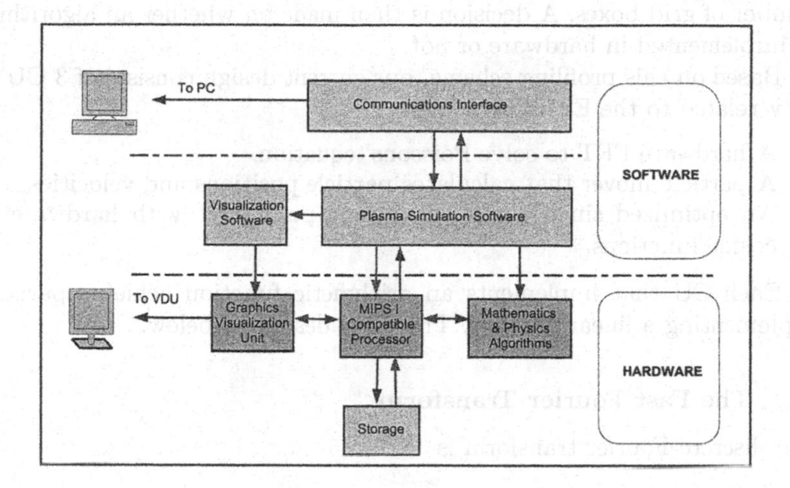


**Fig. 2.** The overall framework of an FPGA based simulation environment. A FPGA is used to implement a variety of mathematical functions used in plasma simulations. Plasma simulation codes are written in the C programming language and interface directly to the underlying hardware through an Application Programmers Interface (API)

The left data cache can be considered to be the standard data cache found in Harvard implemented architectures.

The right cache is used as a scratch area for implementing algorithms that require writing-back to the same area of memory, like the FFT. Our implementation allows a sustainable throughput of pipelined functions by reading data from memory side A, performing a function, and saving the result to side B, whilst simultaneously reading into the function new data values from side A. This allows an efficient pipelined implementation of algorithms without delays due to memory reads and writes. An even better approach would have been to use commodity dual-port RAM (DPRAM), however the very high cost per-bit of this type of RAM, excluded its use in our prototype model.

To support the monitoring, programming and data collection of the machine we connect it to our local area network using the ethernet protocol. This could have been implemented as a CU, but was considered too expensive in terms of the number of Logic Elements it would have required. Instead we interface the system to a cheaply available commodity network IC that has ethernet support. We could use this interface in the future to support simulations over the World Wide Web.

Each processing unit transparently adds functionality to the CPU's instruction set. Algorithms that should be implemented in hardware are chosen by profiling the ES1 code and calculating the relative portion of time spent. This is done for simulations containing a different number of particles and a varying number of grid boxes. A decision is then made on whether an algorithm should be implemented in hardware or not.

Based on this profiling scheme, our current design consists of 3 CU's specifically related to the ES1 code

1. A hardware FFT to solve Poissons' equation.
2. A particle mover that calculates particle positions and velocities.
3. An optimized single-precision floating point unit with hardware sine and cosine functions.

Each CU that implements an arithmetic function achieves parallelism by implementing a linear pipeline. These are described below.

### 3.1 The Fast Fourier Transform

The discrete Fourier transform is

$$F_n = \sum_{k=0}^{N-1} f_k.W^{jk} \qquad W^{jk} = e^{-j2\pi nk/N} \tag{1}$$

with $0 \leq n \leq N-1$

An N-point FFT, N being a power of 2 requires $O(Nlog_2N)$ operations and N/2 butterfly operations are computed for the $log_2$ stages. Spatial parallelism is achieved within the butterfly unit by calculating the real and imaginary parts of the complex variables concurrently. The limit to the temporal parallelism in our present FFT implementation is the rate at which the address of the variables are calculated. A higher level abstraction of the FFT is shown in figure 3.

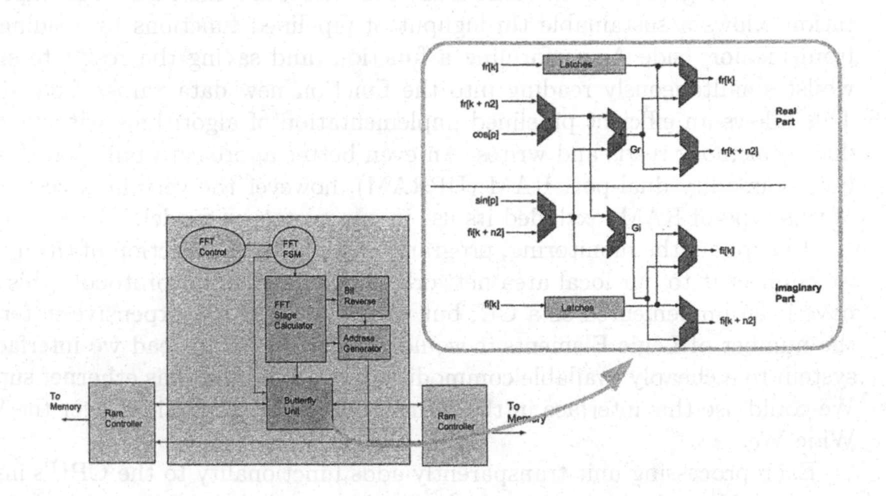


**Fig. 3.** The FFT coprocessing unit with a single butterfly unit consisting of a 15-stage arithmetic pipeline. The efficiency of the pipeline is limited by the rate at which the address of the arithmetic operators can be calculated

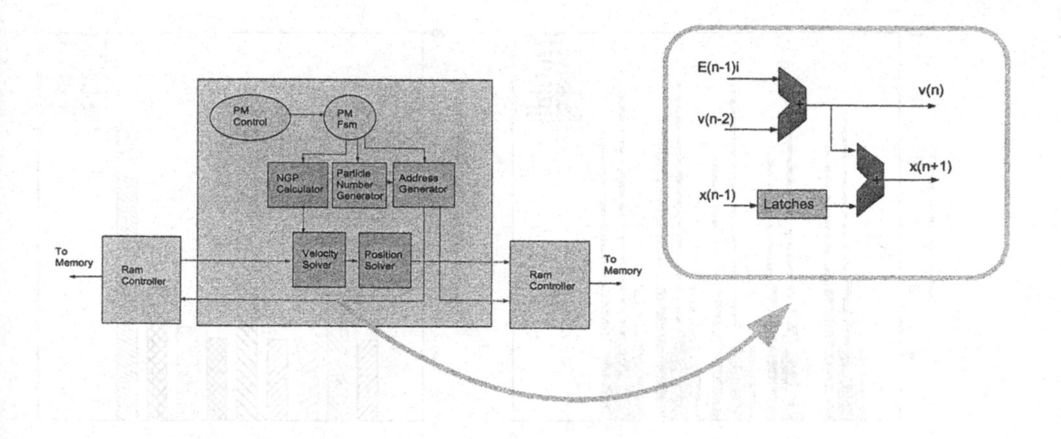


**Fig. 4.** The particle mover used to calculate the velocity and position from the electric. This particular CU users the Nearest Grid Point schema consists of an arithmetic pipeline which allows a velocity calculation to be performed in 5-stages with another 5 required for the particle position calculation

## 3.2 The Particle Mover

From the force $F_i$ on each particle i, obtained from the electric field, $E$ (in the electrostatic case) using the Lorentz force equation, we calculate the new velocity $v_i$ and position $x_i$. Of the many methods used to calculate the force within a grid our current model uses the simplest, the Nearest Grid Point (NGP) method.

In the nearest grid point schema, the new particle velocity and position are calculated in dimensionless coordinates based on the equations below.

$$v_i^n = v_i^{n-2} + E_j^{n-1} \tag{2}$$

$$x_i^{n+1} = x_i^{n-1} + v^n \tag{3}$$

where $E_j$ is electric field associated with bin $j$. The throughput to these equations is continuous with a difference of 5 clock cycles between each veloctiy and position calculation. This is shown in figure 4.

## 3.3 The C-Language Support

Software support is provided by the use of the LCC cross-compiler [10, 7]. This is run as a host on a freeBSD UNIX workstation with a backend implementation of our own processor as the target. Although GCC [24] and its associated binutils are generally regarded as a better optimizing compiler, the LCC cross compiler

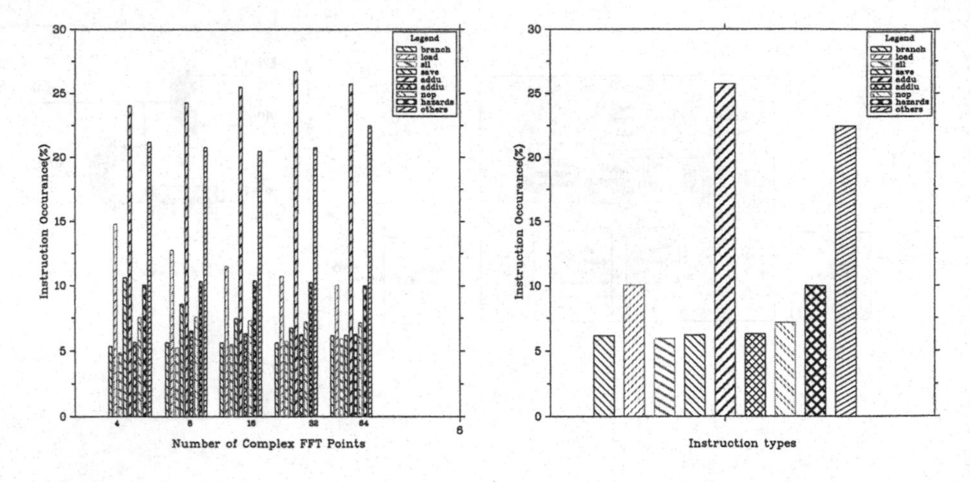


**Fig. 5.** The figures above shows the instruction set profile when using the C-Code only to implement a FFT to solve the Poisson equation using a various number of bins. The figure on the left shows the result from 4 to 64 bins. The one on the right is a magnified version for 64 bins

is smaller and has, from our point of view, more accessible documentation. Also most of our optimizing is performed by the hardware coprocessing units.

As our CPU design has a MIPS compatible base, we used the MIPS R3000 machine description (md) file as the starting point for creating our own backend to the cross-compiler. The output produced by LCC is assembly code, so a custom assembler has been written to convert the output to machine code. If we had used GCC we could have used the GCC cross-assember, GAS, but this was not looked into due to the relative ease of writing an assembler.

Using GCC would also have provided automatic support for the C++ programming language and the advantages of object-oriented programming. The CUs are integrated into the C programming language structure as a series of user callable functions that are handcoded, in assembly language, into a library. This library is linked into the main body of code by the assembler.

## 4 Results and Discussion

Although for our small sample size, of FFT and particle points, a considerable performance gain has been achieved (see figures 7 and 8) a considerable hinderance to the software model can be attributed to the compiler itself. A 15 percent drop in the software performance - 'nops' with respect to the FFT - arises from the compiler's inability to utilise the branch delay slot. A further 10 percent reduction is due to data hazards. (see figure 5).

A proportion of the add unsigned instructions - 'addu' - can also be attributed to data hazards. However bearing these in mind, a substantial increase

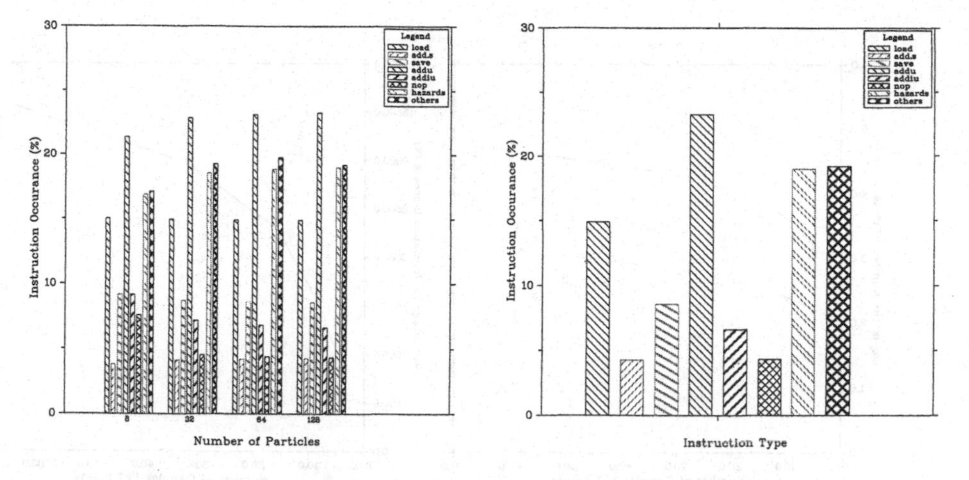


**Fig. 6.** This shows (on the left) the result of using C-Code only to implement the nearest grid point algorithim for a varying number of particles. The figure on the right is the enlarged case of 128 particles. Data hazards instructions and nops, due to control hazards, can be seen to significantly increase the software implementation time

in performance is still gained in the hardware implementation when compared to that in software only, on the same system. This can also be seen in the case of the particle mover (shown in figure 8). These results have been generated with the system running at 40Mhz.

A limiting factor of the hardware units is our restriction that they should all run at the same speed of 40 Mhz, which is the current maximum speed of our MIPS R3000 compatible core with the integrated units, although individually the units run faster. Whether each CU should be allowed to run at its full speed and an asynchronous communication protocol be set up between the units has not yet been investigated. Neither has the idea of implementing a plasma simulation code completely in hardware. However both of these methods would make the test and verification of the system more complicated.

## 5 Conclusion and Future Work

We have shown above that by implementing certain mathematical functions directly in hardware we can reduce the calculation time of an electrostatic plasma simulation. The performance of the system can be improved by writing a custom compiler rather than using a cross-compiler to produce more efficient machine code. Using a software/hardware approach still has many benefits - the network connectivity is implemented by using a software tcp library - and will be used in our future work.

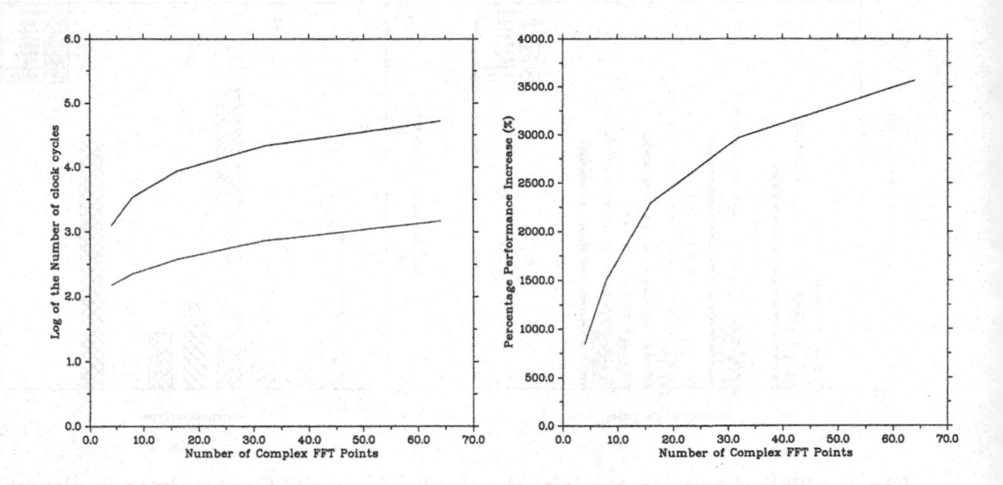


**Fig. 7.** The figure on the left shows the results of implementing the FFT purely in C-code (top blue curve) relative to the hardware implementation (lower red curve) for a varying number of bins. The corresponding percentage performance increases of the hardware version to that of the software is shown plotted on the left

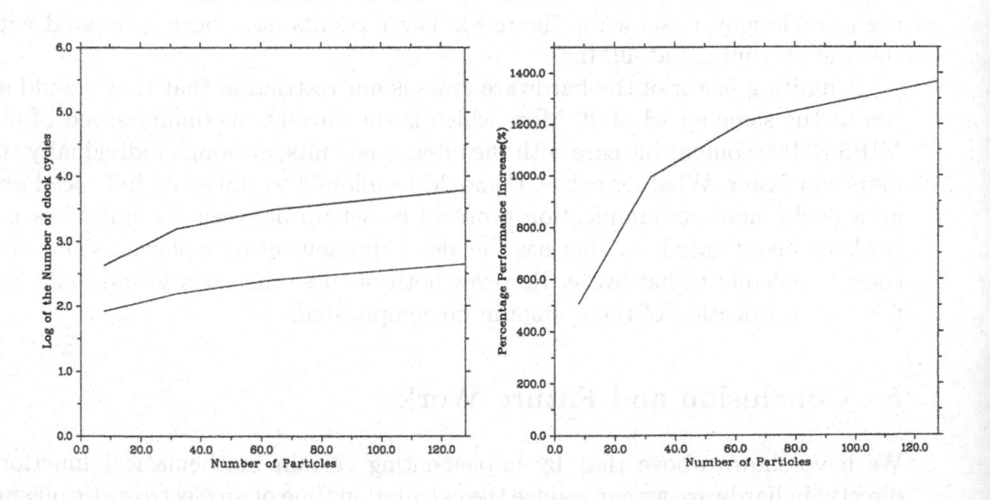


**Fig. 8.** A comparism of the C-code implementation (upper blue curve) to that of the hardware CU (lower red curve) is shown in the figure on the left. The result of the performance increases of the hardware CU relative to the software implementation is given on the right

The next stage of this project is to increase the overall speed of the system by implementing our hardware models in structural, rather than synthesizable behavioural VHDL. This should increase the system performance from around 40 Mhz to between 70 and 100 Mhz. We are also in the process of developing a set of programs to test the sysyem against desktop computers.

Our experience gained so far has also encouraged us to move away from customising the one dimensional ES1 code, to directly implementing a two dimensional electrmagnetic plasma simulation code, with a coprocessor implemented for each stage of the design flow.

The rapidly changing FPGA market has seen vendors implementing devices that not only contain different types of embedded hardware processors, but also advanced macro units - the new Altera Stratix devices [2] embed a number of dsp units containing hardwired adders and multipliers. How we will incorporate these into our work is still uncertain, but their use should also increase the overall performance of the system.

There are advantages of implementing a soft processor in hardware as the coprocessing units can communicate synchronously. A hardwire processor core embedded within a FPGA forces asynchronus communcation between the core and custom logic through the vendor defined bus standard. Also, a varying number of softcore processors with varying databus widths can be implemented on a single FPGA to suit the application at hand, in contrast to the fixed number of an embedded hardwired processor.

With the prediction that in the future designing for FPGAs will be moved away from the hardware designers domain to that of the software programmer, we will see an increase in projects that use FPGAs in scientific compututing.

## Acknowledgments

This research is funded by the Particle Physics Research Council (PPARC), U.K. The authors would like to thank Alex Benyon, Field Applications Engineer of Altera Europe, for his helpful advice and assistance on using Alteras family of FPGAs. Thanks goes also to Andy Buckley, Eduardo Bezerra and Marianne Pouchet, of the Space Science Group, University of Sussex, for their useful comments and suggestions.

## References

[1] Altera Corporation, San Jose, CA. *The APEX 20K Programmable Logic Device Family Data Book*, 2001.

[2] Altera Corporation, San Jose, CA. *The Stratix Programmable Logic Device Family Data Book*, 2002.

[3] J Bala, V Bruck. A portable and tunable collective communication library for scalable parallel computers. Technical report, IBM T. J Watson Research Center, 1993.

[4] A.B Birdsall, C.K Langdon. *Plasma Physics via Computer Simulation.* Institute of Physics Publishing Ltd, 1991.

[5] R Buyya, editor. *High Performance Cluster Computing: Architectures and Systems.* Prentice Hall, 1999.
[6] J W Cooley, J W Tukey. An algorithm for the machine calculation of complex fourier series. *Math. Comp. 19.*, pages 297–301, 1965.
[7] D R Fraser, C W Hanson. A code generation interface for ansi c. *Software-Practice and Experience*, 21(9):963–988, Sept 1991.
[8] R Freeman. User programmable gate arrays. *IEEE Spectrum*, December 1988.
[9] D Goldberg. What every computer scientist should know about floating-point arithmetic. *ACM Computing Surveys*, 23(1):5–48, 1991.
[10] C W Hanson, D R Fraser. *A Retargetable C Compiler: Design and Implementation.* Addison-Wesley Publishing Co, 1995.
[11] W T Hauck, S Fry. The chimaera reconfigurable functional unit. *IEEE Symposium on Field-Programmable Custom Computing Machines*, 1997.
[12] J Hauser, J Wawrzynek. Garp: A mips processor with a reconfigurable coprocessor. In *Proceedings of the IEEE Symposium on Field Programable Custom Computing Machines*, FCCM'97, April 1997.
[13] D Hempel, R Walker. The emergence of the mpi message passing standard for parallel computing. *Computer Standards and Interfaces*, 7:51–62, 1999.
[14] D Hennessy, J Patterson. *Computer Organisation and Design - The Hardware Software Interface.* Morgan Kaufman, 1998.
[15] J W Hockney, R W Eastwood. *Computer Simulations using Particles.* Institute of Physics Publishing Ltd, 1999.
[16] IEEE. *ANSI/IEEE Standard 754 - 1985, Standard for Binary Floating Point Arithmetic.* Institute for Electrical and Electronic Engineerings.
[17] G Kane. *MIPS RISC Architecture.* Prentice Hall, 1988.
[18] R E Kessler. The alpha 21264 microprocessor. *IEEE Micro*, March/April 1999.
[19] Micron Technology Inc. *Designing with ZBT SRAM*, 2002. Micron technology app. notes.
[20] D Patterson. Reduced instruction set computers. *Comm. ACM*, 28(1):8–21, Jan 1985.
[21] M C Pease. An adaptation of the fast fourier transform for parallel processing. *Journal of the ACM*, 15:252–264, 1968.
[22] R Russell. The cray-1 computer system. *Communications of the ACM*, 21(1), 1978.
[23] P S Song, M Denman, and J Chang. The powerpc 604 risc microprocessor. *IEEE Micro*, October 1994.
[24] R M Stallman. *Using and Porting GNU CC - Cross Compiler.* Free Software Foundation Inc, 1998.
[25] C Temperton. Direct methods for the solution of the discrete poisson equation: Some comparissons. *Journal of Computational Physics*, 31:1–20, 1979.
[26] J Walker, D Dongarra. Mpi: A standard message passing interface. *Supercomputer*, 12(1):56–68, January 1996.
[27] P Wittig, R Chow. Onechip: An fpga processor with reconfigurable logic. *IEEE Symposium on FPGA's for Custom Computing Machine*, pages 126–135, 1996.

# Remote Parallel Model Reduction of Linear Time-Invariant Systems Made Easy*

Peter Benner[1], Rafael Mayo[2],
Enrique S. Quintana-Ortí[2], and Gregorio Quintana-Ortí[2]

[1] Institut für Mathematik, Technische Universität Berlin
D-10623 Berlin, Germany
benner@math.tu-berlin.de
Tel.: +30-314-28035, Fax: +30-314-79706

[2] Depto. de Ingeniería y Ciencia de Computadores, Universidad Jaume I
12071–Castellón, Spain
{mayo,quintana,gquintan}@icc.uji.es
Tel.: +34-964-728257, Fax: +34-964-728486

**Abstract.** This paper describes a library for model reduction of large-scale, dense linear time-invariant systems on parallel distributed-memory computers. Our library is enhanced with a mail service which serves as a demonstrator of the capabilities of the library. Remote requests submitted via e-mail are executed on a cluster composed of Intel Pentium-II nodes connected via a Myrinet switch. Experimental results show the numerical and parallel performances of our model reduction routines.

**Keywords:** Model reduction, absolute error methods, matrix sign function, mail service.

## 1 Introduction

Model reduction is of fundamental importance in many modeling and control applications involving linear time-invariant (LTI) systems. In state-space form, such systems are described by the following models:

*Continuous LTI system:*

$$\begin{aligned} \dot{x}(t) &= Ax(t) + Bu(t), \quad t > 0, \quad x(0) = x^0, \\ y(t) &= Cx(t) + Du(t), \quad t \geq 0. \end{aligned} \tag{1}$$

*Discrete LTI system:*

$$\begin{aligned} x_{k+1} &= Ax_k + Bu_k, \quad x_0 = x^0, \\ y_k &= Cx_k + Du_k, \quad k = 0, 1, 2, \ldots \end{aligned} \tag{2}$$

In both models, $A \in \mathbb{R}^{n \times n}$ is the state matrix, $B \in \mathbb{R}^{n \times m}$, $C \in \mathbb{R}^{p \times n}$, $D \in \mathbb{R}^{p \times m}$, $n$ is said to be the order of the system, and $x^0 \in \mathbb{R}^n$ is the initial

* Supported by the Fundació Caixa-Castelló/Bancaixa PI-1B2001-14 and CICYT TIC2002-04400-C03-01.

J.M.L.M. Palma et al. (Eds.): VECPAR 2002, LNCS 2565, pp. 255–268, 2003.

state of the system. The associated transfer function matrix (TFM) is $G(s) = C(sI - A)^{-1}B + D$. Here, we assume that the state matrix $A$ is stable. In the continuous-time case, this implies that the spectrum of $A$ is contained in the open left half plane while in the discrete-time case, the spectrum of $A$ must lie inside the unit circle. This condition ensures that the system is stable, but we do not assume minimality of the system.

In the model reduction problem, we are interested in finding a reduced-order LTI system,

*Continuous LTI system:*

$$\begin{aligned} \dot{\hat{x}}(t) &= \hat{A}\hat{x}(t) + \hat{B}\hat{u}(t), \quad t > 0 \quad \hat{x}(0) = \hat{x}^0, \\ \hat{y}(t) &= \hat{C}\hat{x}(t) + \hat{D}\hat{u}(t), \quad t \geq 0. \end{aligned} \tag{3}$$

*Discrete LTI system:*

$$\begin{aligned} \hat{x}_{k+1} &= \hat{A}\hat{x}_k + \hat{B}\hat{u}_k, \quad \hat{x}_0 = \hat{x}^0, \\ \hat{y}_k &= \hat{C}\hat{x}_k + \hat{D}\hat{u}_k, \quad k = 0, 1, 2, \ldots, \end{aligned} \tag{4}$$

of order $r$, $r \ll n$, and associated TFM $\hat{G}(s) = \hat{C}(sI - \hat{A})^{-1}\hat{B} + \hat{D}$ which approximates $G(s)$.

There is no general technique for model reduction that can be considered as optimal in an overall sense since the reliability, performance, and adequacy of the reduced system strongly depends on the system characteristics. The model reduction methods we study here rely on truncated state-space transformations where, given $Y = [T_l^T, L_l^T]^T \in \mathbb{R}^{n \times n}$ and $Y^{-1} = [T_r, L_r]$, with $T_l \in \mathbb{R}^{r \times n}$ and $T_r \in \mathbb{R}^{n \times r}$,

$$\hat{A} = T_l A T_r, \quad \hat{B} = T_l B, \quad \hat{C} = C T_r, \quad \text{and} \quad \hat{D} = D. \tag{5}$$

Model reduction methods based on truncated state-space transformations usually differ in the measure they attempt to minimize. Balanced truncation (BT) methods [16, 20, 22, 24], singular perturbation approximation (SPA) methods [15], and Hankel-norm approximation (HNA) methods [12] all belong to the family of absolute error methods, which try to minimize $\|\Delta_a\|_\infty = \|G - \hat{G}\|_\infty$. Here, $\|G\|_\infty$ denotes the $\mathcal{L}_\infty$- or $\mathcal{H}_\infty$-norm of a stable, rational matrix function defined as

$$\|G\|_\infty = \operatorname*{ess\,sup}_{\omega \in \mathbb{R}} \sigma_{\max}(G(\jmath\omega)), \tag{6}$$

where $\jmath := \sqrt{-1}$ and $\sigma_{\max}(M)$ is the largest singular value of the matrix $M$.

The family of relative error model reduction methods attempt to minimize, on the other hand, the relative error $\|\hat{\Delta}\|_\infty$, defined implicitly by $G - \hat{G} = \hat{\Delta}G$. Among these, the balanced stochastic truncation (BST) methods [9, 13, 25] are specially interesting. Currently, BST methods are not included in the model reduction library and the associated mail service described here. Therefore, we do not include BST methods in our presentation.

Model reduction of large-scale systems arises, among others, in control of large flexible mechanical structures or large power systems as well as in circuit simulation and VLSI design; see, e.g., [7, 8, 11, 17]. LTI systems with state-space dimension $n$ of order $10^2$ to $10^4$ are common in these applications.

All absolute error model reduction methods for LTI systems with dense state matrices have a computational cost of $\mathcal{O}(n^3)$ floating-point operations (flops). Large-scale applications thus clearly benefit from using parallel computing techniques to obtain the reduced system. The usual approach a few decades ago was to solve these problems on very expensive parallel supercomputers, but the situation is changing in the last years. Clusters constructed from commodity systems (personal computers and local and system area switches) and "open" software have started to be widely used for parallel scientific computing. The enormous improvements in the performance of the commodity hardware has provided an affordable tool for dealing with large-scale scientific problems. However, many end-users (control engineers, VLSI circuit designers, etc.) are not familiar with the basics of parallel algorithms and/or parallel architectures. Thus, if a parallel library of control algorithms is to be useful to a broader spectrum of the scientific community, it should be part of our work to provide, along with the library, the necessary tools to facilitate its use.

The tool we propose to use, e-mail, has been employed from the first days of the Internet not only for personal communication but also for submission of jobs to remote systems, access to databases, document retrieval, etc. In this work we explore the use of e-mail to provide a parallel model reduction service to the scientific community. The advantage of providing this service via e-mail is that the end-user remains completely isolated from the complexities of the software and the hardware system, and the maintenance of both. A secondary goal of this service is to serve as a demonstrator of the capabilities of the parallel library, which can be then downloaded and installed on any (parallel) architecture with MPI and ScaLAPACK.

Our library provides parallel routines for model reduction using the three mentioned absolute error methods. These approaches are briefly reviewed in Section 2. The contents and structure of the library are described in Section 3. The library is completed with a mail service, presented in Section 4, which allows remote access to our parallel cluster and enables the user to apply any of the model reduction algorithms on a specific system. Finally, the numerical and parallel performances of the routines in the parallel library is reported in Section 5, and some concluding remarks are given in Section 6.

## 2 Model Reduction via Absolute Error Methods

In this section we briefly review the methods for absolute error model reduction. Serial implementations of these algorithms can be found in the *Subroutine Library in Control Theory* – SLICOT[1]. Note that we use implementations that are different from the serial versions in SLICOT but result in the same reduced-order model. Our implementation is particularly efficient on parallel computing platforms.

[1] Available from `http://www.win.tue.nl/niconet/NIC2/slicot.html`.

All these methods are strongly related to the controllability Gramian $W_c$ and the observability Gramian $W_o$ of the LTI system. In the continuous-time case, the Gramians are given by the solutions of two coupled *Lyapunov equations*

$$\begin{aligned} A\,W_c + W_cA^T + BB^T &= 0, \\ A^TW_o + W_oA\; + C^TC &= 0, \end{aligned} \tag{7}$$

while, in the discrete-time case, the Gramians are the solutions of two coupled *Stein equations*

$$\begin{aligned} A\,W_cA^T - W_c + BB^T &= 0, \\ A^TW_oA\; - W_o + C^TC &= 0. \end{aligned} \tag{8}$$

As $A$ is assumed to be stable, the Gramians $W_c$ and $W_o$ are positive semidefinite, and therefore there exist factorizations $W_c = S^TS$ and $W_o = R^TR$. Matrices $S$ and $R$ are called the *Cholesky factors* of the Gramians.

The Lyapunov equations are solved in our model reduction algorithms using the Newton iteration for the matrix sign function introduced by Roberts [19]. The Stein equations are solved using the Smith iteration proposed in [21]. Both iterations are specially adapted taking into account that, for large-scale non-minimal systems, the Cholesky factors are often of low (numerical) rank. By not allowing these factors to become rank-deficient, a large amount of workspace and computational cost can be saved. For a detailed description of this technique, see, e.g., [1].

Consider now the singular value decomposition (SVD)

$$SR^T = [U_1\; U_2] \begin{bmatrix} \Sigma_1 & 0 \\ 0 & \Sigma_2 \end{bmatrix} \begin{bmatrix} V_1^T \\ V_2^T \end{bmatrix}, \tag{9}$$

where the matrices are partitioned at a given dimension $r$ such that $\Sigma_1 = \operatorname{diag}(\sigma_1, \ldots, \sigma_r)$, $\Sigma_2 = \operatorname{diag}(\sigma_{r+1}, \ldots, \sigma_n)$, $\sigma_j \geq 0$ for all $j$, and $\sigma_r > \sigma_{r+1}$. Here, $\sigma_1, \ldots, \sigma_n$ are known as the *Hankel singular values* of the system. If $\sigma_r > \sigma_{r+1} = 0$, then $r$ is the *McMillan* degree of the system. That is, $r$ is the state-space dimension of a minimal realization of the system.

The so-called *square-root* (SR) BT algorithms determine the reduced-order model in (5) using the projection matrices

$$T_l = \Sigma_1^{-1/2} V_1^T R \quad \text{and} \quad T_r = S^T U_1 \Sigma_1^{-1/2}. \tag{10}$$

In case $\Sigma_1 > 0$ and $\Sigma_2 = 0$, the reduced-order model computed by using $T_l, T_r$ from (10) in (5) is a minimal realization of the TFM $G(s)$.

The *balancing-free square-root* (BFSR) BT algorithms often provide more accurate reduced-order models in the presence of rounding errors [24]. These algorithms share the first two stages (solving the coupled equations and computing the SVD of $SR^T$) with the SR methods, but differ in the procedure to obtain $T_l$ and $T_r$. Specifically, the following two QR factorizations are computed,

$$S^TU_1 = [P_1\; P_2] \begin{bmatrix} \hat{R} \\ 0 \end{bmatrix}, \qquad R^TV_1 = [Q_1\; Q_2] \begin{bmatrix} \bar{R} \\ 0 \end{bmatrix}, \tag{11}$$

where $P_1, Q_1 \in \mathbb{R}^{n\times r}$ have orthonormal columns, and $\hat{R}$, $\bar{R} \in \mathbb{R}^{r\times r}$ are upper triangular. The reduced system is then given by the projection matrices

$$T_l = (Q_1^T P_1)^{-1} Q_1^T, \qquad T_r = P_1, \tag{12}$$

and (5).

SR SPA or BFSR SPA reduced-order models can be obtained by first using the SR BT or BFSR BT algorithms, respectively, to obtain a minimal realization of the original system. Let the tuple $(\tilde{A}, \tilde{B}, \tilde{C}, \tilde{D})$ be a realization of this minimal system and partition

$$\tilde{A} = \begin{bmatrix} A_{11} & A_{12} \\ A_{21} & A_{22} \end{bmatrix}, \quad \tilde{B} = \begin{bmatrix} B_1 \\ B_2 \end{bmatrix}, \quad \tilde{C} = [\, C_1 \;\; C_2 ],$$

according to the desired size $r$ of the reduced-order model, i.e., $A_{11} \in \mathbb{R}^{r\times r}$, $B_1 \in \mathbb{R}^{r\times m}$, $C_1 \in \mathbb{R}^{p\times r}$, etc. Then the SPA reduced-order model is obtained by applying the following formulae

$$\begin{aligned} \hat{A} &:= A_{11} - A_{12}(\gamma I - A_{22})^{-1} A_{21}, \\ \hat{B} &:= B_1 - A_{12}(\gamma I - A_{22})^{-1} B_2, \\ \hat{C} &:= C_1 - C_2(\gamma I - A_{22})^{-1} A_{21}, \\ \hat{D} &:= \tilde{D} - C_2(\gamma I - A_{22})^{-1} B_2, \end{aligned} \tag{13}$$

where $\gamma = 0$ for continuous LTI systems and $\gamma = 1$ for discrete LTI systems [15, 23, 24].

BT and SPA model reduction methods aim at minimizing the $\mathcal{H}_\infty$-norm norm of the error system $G - \hat{G}$. However, they usually do not succeed in finding an optimal approximation. Using the *Hankel norm* of a stable rational transfer function, defined by

$$\|G\|_H = \sigma_1(G), \tag{14}$$

where $\sigma_1(G)$ is the maximum Hankel singular value of $G$, it is possible to compute an approximation minimizing $\|G - \hat{G}\|_H$ for a given order $r$ of the reduced-order system [12]. The derivation of a realization of $\hat{G}$ is quite involved. Due to space limitations, we refer the reader to [12, 27] and only describe the essential computational tools required in an implementation of the HNA method. In the first step, a balanced minimal realization of $G$ is computed. This can be done using the SR version of the BT method described above. Then a transfer function $\tilde{G}(s) = \tilde{C}(sI - \tilde{A}\,)^{-1}\tilde{B} + \tilde{D}$ is computed as follows: first, the order $r$ of the reduced-order model is chosen such that the Hankel singular values of $G$ satisfy

$$\sigma_1 \geq \sigma_2 \geq \ldots \sigma_r > \sigma_{r+1} = \ldots = \sigma_{r+k} > \sigma_{r+k+1} \geq \ldots \geq \sigma_n, \quad k \geq 1.$$

Then, by applying appropriate permutations, the balanced transformation of $G$ is re-ordered such that the Gramians become

$$\begin{bmatrix} \check{\Sigma} & \\ & \sigma_{r+1} I_k \end{bmatrix}.$$

The resulting balanced realization given by $\breve{A}, \breve{B}, \breve{C}, \breve{D}$ is partitioned according to the partitioning of the Gramians, i.e.,

$$\breve{A} = \begin{bmatrix} A_{11} & A_{12} \\ A_{21} & A_{22} \end{bmatrix}, \quad \breve{B} = \begin{bmatrix} B_1 \\ B_2 \end{bmatrix}, \quad \breve{C} = [\, C_1 \;\; C_2 ],$$

where $A_{11} \in \mathbb{R}^{n-k \times n-k}, B_1 \in \mathbb{R}^{n-k \times m}$, $C_1 \in \mathbb{R}^{p \times n-k}$, etc. Then the following formulae define a realization of $\tilde{G}$:

$$\begin{aligned}
\tilde{A} &= \Gamma^{-1}(\sigma_{r+1}^2 A_{11}^T + \breve{\Sigma} A_{11} \breve{\Sigma} + \sigma_{r+1} C_1^T U B_1^T, \\
\tilde{B} &= \Gamma^{-1}(\breve{\Sigma} B_1 - \sigma_{r+1} C_1^T U), \\
\tilde{C} &= C_1 \breve{\Sigma} - \sigma_{r+1} U B_1^T, \\
\tilde{D} &= D + \sigma_{r+1} U.
\end{aligned}$$

Here, $U := (C_2^T)^\dagger B_2$, where $M^\dagger$ denotes the pseudoinverse of $M$, and $\Gamma := \breve{\Sigma}^2 - \sigma_{r+1}^2 I_{n-k}$. This step only requires matrix scaling, matrix multiplication, and a QR decomposition. Now we can compute an additive decomposition of $\tilde{G}$ such that $\tilde{G}(s) = \hat{G}(s) + F(s)$ where $\hat{G}$ is stable and $F$ is anti-stable. Then $\hat{G}$ is an optimal $r$-th order Hankel norm approximation of $G$. We have implemented the additive decomposition of $\tilde{G}$ via a block diagonalization of $\tilde{A}$, where we first compute a block Schur form using the sign function of $\tilde{A}$ and then annihilate the off-diagonal block by solving a Sylvester equation using again a sign function-based iterative solution procedure. For further details of the implementation of the HNA method and computational aspects see [3].

## 3 Parallelization, Structure and Contents of the Library

The absolute error methods basically perform matrix operations such as the solution of linear systems and linear matrix equations, and the computation of matrix products and matrix decompositions (LU, QR, SVD, etc.). All these operations are basic matrix algebra kernels parallelized in ScaLAPACK and PBLAS. Thus, our parallel model reduction library PLICMR[2] heavily relies on the use of the available parallel infrastructure in ScaLAPACK, the serial computational libraries LAPACK and BLAS, and the communication library BLACS (see Figure 1).

Details of the contents and parallelization aspects of PLICMR are given, e.g., in [1, 2].

The functionality and naming convention of the parallel routines closely follow analogous routines from SLICOT. The contents of PLICMR are structured in four main groups:

- Model reduction computational routines.
  - `pab09ax`: BT method.
  - `pab09bx`: SPA method.
  - `pab09cx`: HNA method.

[2] `http://spine.act.uji.es/~plicmr`

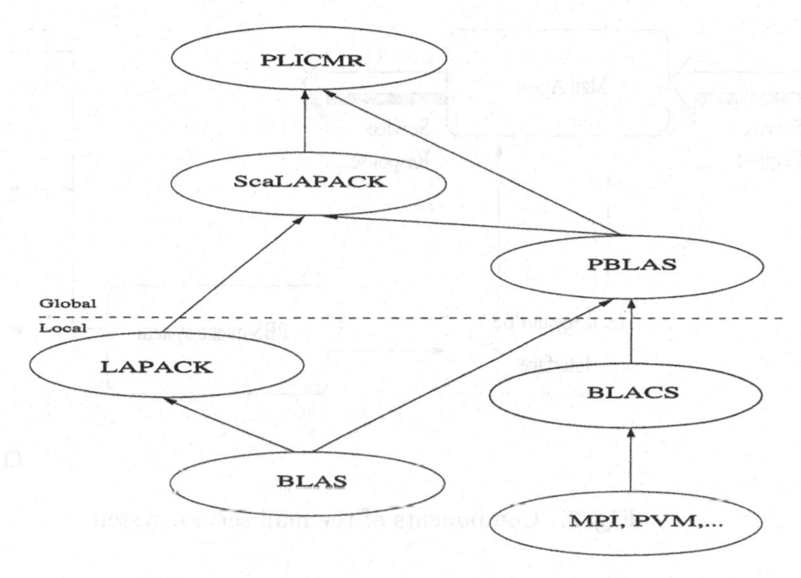


**Fig. 1.** PLICMR and the underlying libraries

- Matrix functions.
  - `pmb05rd`: Computation of the sign function of a matrix.
- Linear matrix equation solvers.
  - `psb03odc`: Solution of coupled Lyapunov equations (7) using the Newton iteration for the matrix sign function.
  - `psb03odd`: Solution of coupled Stein equations (8) using the Smith iteration.
  - `psb04md`: Solution of a Sylvester equation using the Newton iteration for the matrix sign function.
- Linear algebra.
  - `pmb03td`: Computation of the SVD of a product of two rectangular matrices.
  - `pmb03ox`: Estimation of the numerical rank of a triangular matrix.

A standardized version of the library is integrated into the subroutine library PSLICOT [5], with parallel implementations of a subset of SLICOT. It can be downloaded from `ftp://ftp.esat.kuleuven.ac.be/pub/WGS/SLICOT`. The version maintained at `http://spine.act.uji.es/~plicmr` might at some stages contain more recent updates than the version integrated into PSLICOT. The library can be installed on any parallel architecture where the above-mentioned computational and communication libraries are available. The efficiency of the parallel routines will depend on the performance of the underlying libraries for matrix computation (BLAS) and communication (usually, MPI or PVM).

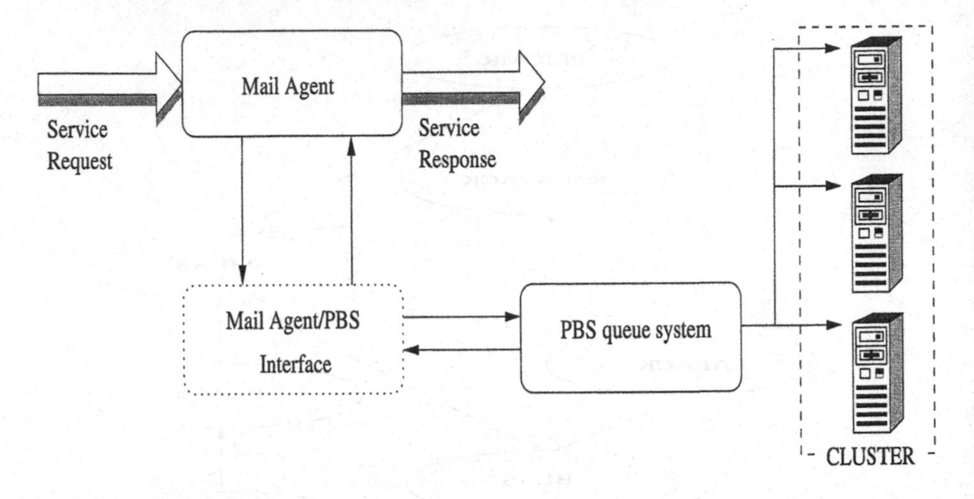


**Fig. 2.** Components of the mail service system

# 4 Mail Service

In this section we describe the access procedure and the structure of the service for parallel model reduction via e-mail. This service is provided on a parallel cluster[3], composed of 32 Intel Pentium-II personal computers, running a Linux operating system, and connected with a 1 Gbps Myrinet network.

The user can submit a job using any standard application for sending an e-mail message to `plicmr@spine.act.uji.es`. A service request specifies the parameters of the job (like the model reduction algorithm to employ, the dimensions of the system, the desired order for the reduced system, etc.) in the body of the message. Validation of access is performed via a login and password included in the e-mail. Different compression tools are supported to reduce the size of the files specifying the system matrices, provided as attachments. The results of the execution are returned to the user via e-mail, with the matrices of the reduced system as attached (compressed) files.

The service and the specific formats of the e-mail body and attachments are further described in `http://spine.act.uji.es/~plicmr`.

The model reduction mail server is structured as a Mail Agent (MA) and a set of scripts which define the interface between this agent and PBS (portable batch system), a batch queuing and workload management system (see Figure 2).

The MA is the operating system mail daemon that processes e-mails. On reception of an e-mail addressed to `plicmr@spine.act.uji.es`, the MA creates a process, PLICMR-MA, which then employs the preprocessing scripts in the MA/PBS Interface to extract the contents of the e-mail, save the system matrices in the local file system, generate the submission scripts for PBS, and submit

[3] The cluster is owned by the Parallel Scientific Research Group, at the University Jaume I; `http://spine.act.uji.es/psc.html`.

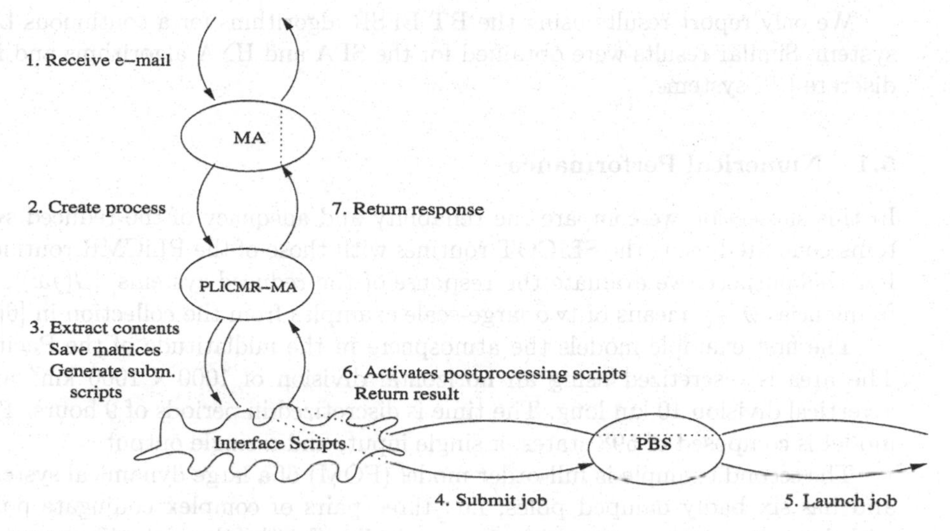


**Fig. 3.** Actors and actions involved in processing a service request for parallel model reduction

the job. PBS is in charge of launching the execution of the job in the cluster and return the result to the PLICMR-MA using the postprocessing (epilogue) scripts in the MA/PBS Interface. The PLICMR-MA then generates an appropriate e-mail with the matrices of the reduced-order system and uses the MA to send it back to the user. Figure 3 illustrates the overall process, beginning with the reception of an e-mail with a service request from a remote user, and ending when an e-mail, containing the results of the execution of the request, is returned to the user.

## 5 Experimental Results

All the experiments presented in this section were performed on Intel Pentium-II processors using IEEE double-precision floating-point arithmetic ($\varepsilon \approx 2.2204 \times 10^{-16}$). The parallel algorithms were evaluated on a parallel distributed cluster of 32 nodes. Each node consists of an Intel Pentium-II processor at 300 MHz, and 128 MBytes of RAM. We employ a BLAS library, specially tuned for the Pentium-II processor as part of the ATLAS and ITXGEMM projects [26, 14], that achieves around 180 Mflops (millions of flops per second) for the matrix product (routine DGEMM). The nodes are connected via a *Myrinet* crossbar network; the communication library BLACS is based on an implementation of the communication library MPI specially developed and tuned for this network. The performance of the interconnection network was measured by a simple loop-back message transfer resulting in a latency of 33 $\mu$sec and a bandwidth of 200 Mbit/sec. We made use of the LAPACK, PBLAS, and ScaLAPACK libraries wherever possible.

We only report results using the BT BFSR algorithms for a continuous LTI system. Similar results were obtained for the SPA and HNA algorithms and for discrete LTI systems.

## 5.1 Numerical Performance

In this subsection we compare the reliability and adequacy of the reduced systems computed using the SLICOT routines with those of the PLiCMR routines. For this purpose, we evaluate the response of the reduced systems, $|G(\jmath\omega)|$, at frequencies $\omega$, by means of two large-scale examples from the collection in [6].

The first example models the atmosphere in the midlatitude of the Pacific. The area is discretized using an horizontal division of $1000 \times 1000$ km$^2$ and a vertical division 10 km long. The time is discretized in periods of 9 hours. The model is composed of 598 states, a single input, and a single output.

The second example is full-order model (FOM) of a large dynamical system, and has six badly damped poles, i.e., three pairs of complex conjugate poles with large imaginary part which are responsible for the three significant peeks in the frequency response. Except for these peeks, the dynamics of this system are smooth as all the remaining poles are real and well separated from the imaginary axis. There is no difficulty to be expected in reducing the order of this system significantly.

Figure 4 reports the frequency responses of the original system and the reduced models of orders $r$=9 and 10 for the atmosphere and the FOM examples, respectively. In both cases, no noticiable difference is encountered between the behaviour of the reduced systems computed by means of the SLICOT (`ab09ad`) and the PLiCMR (`pab09ax`) routines. As a matter of fact, in both examples the frequency response of the reduced systems perfectly match that of the original system for all frequencies.

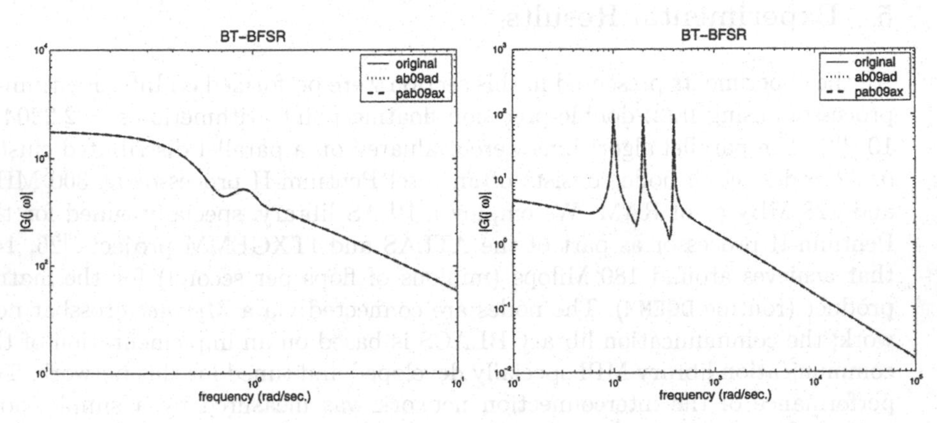


**Fig. 4.** Frequency response of the atmosphere (left) and the FOM (right) examples

## 5.2 Parallel Performance

We employ two different large-scale examples to evaluate the parallel performance of the routines in PLiCMR.

The first example comes from a finite element discretization of a steel cooling process described by a boundary control problem for a linearized 2-dimensional heat equation; see [18] and references therein. The system has 6 inputs and outputs, and $n = 821$ or 3113 states, depending on the meshsize. As there is no significant gap in the Hankel singular values of the system, in this experiment we computed a reduced system of fixed order $r = 40$.

The system with $n = 821$ was reduced using the SLICOT routines in about 4 minutes. The parallel routines computed the reduced system in half of this time, using 4 processors. However, the system with $n = 3113$ could not be solved using the serial routines as the system matrices were too large to be stored in a single node. Our parallel routines provided the solution in around 15 minutes (using 16 processors).

The second example is a random continuous LTI system constructed as follows. First, we generate a random positive semidefinite diagonal Gramian $W_c = \mathrm{diag}(\Sigma_{q_1}, \Sigma_{q_2}, 0_{q_3}, 0_{q_4})$, where $\Sigma_{q_1} \in \mathbb{R}^{q_1 \times q_1}$ contains the desired Hankel singular values for the system and $\Sigma_{q_2} \in \mathbb{R}^{q_2 \times q_2}$. Then, we construct a random positive semidefinite diagonal Gramian $W_o = \mathrm{diag}(\Sigma_{q_1}, 0_{q_2}, \Sigma_{q_3}, 0_{q_4})$, with $\Sigma_{q_3} \in \mathbb{R}^{q_3 \times q_3}$. Next, we set $A$ to a random stable diagonal matrix and compute $F = -(AW_c + W_cA^T)$ and $G = -(A^TW_o + W_oA)$. Thus,

$$
\begin{aligned}
F &= \mathrm{diag}(f_1, f_2, \dots, f_{q_1+q_2}, 0_{q_3+q_4}),\\
G &= \mathrm{diag}(g_1, g_2, \dots, g_{q_1}, 0, \dots, 0, g_{q_1+q_2+1}, \dots, g_{q_1+q_2+q_3}, 0_{q_4}).
\end{aligned}
$$

A matrix $B \in \mathbb{R}^{n \times (q_1+q_2)}$ such that $F = BB^T$ is then obtained as

$$
B = \mathrm{diag}\left(\sqrt{f_1}, \sqrt{f_2}, \dots, \sqrt{f_{q_1+q_2}}\right).
$$

The procedure for obtaining $C$ is analogous. The LTI system is finally transformed into $A := U^TAU$, $B := U^TB$, and $C := CU$ using a random orthogonal transformation $U \in \mathbb{R}^{n \times n}$. The system thus defined has a minimal realization of order $r = q_1$. The Cholesky factors satisfy $\mathrm{rank}(S) = q_1 + q_2$ and $\mathrm{rank}(R) = q_1 + q_3$.

We first evaluate the execution time of the parallel model reduction algorithms. In the example, we set $n = 1000$, $m = p = 100$, and $q_1 = q_2 = q_3 = 50$. As there is no noticeable gap in the Hankel singular value distribution of the system, we obtain a reduced-order model of order $r = 40$.

The left-hand plot in Figure 5 reports the execution time of the BFSR BT serial routine for model reduction in SLICOT and the corresponding parallel algorithm as the number of nodes, $n_p$, is increased. The results show a considerable acceleration achieved by the parallel algorithm (with even super speed-ups). This is partially due to the efficiency of the Lyapunov solvers used in our algorithms which compute factors of the solutions in compact (full-rank) form instead of

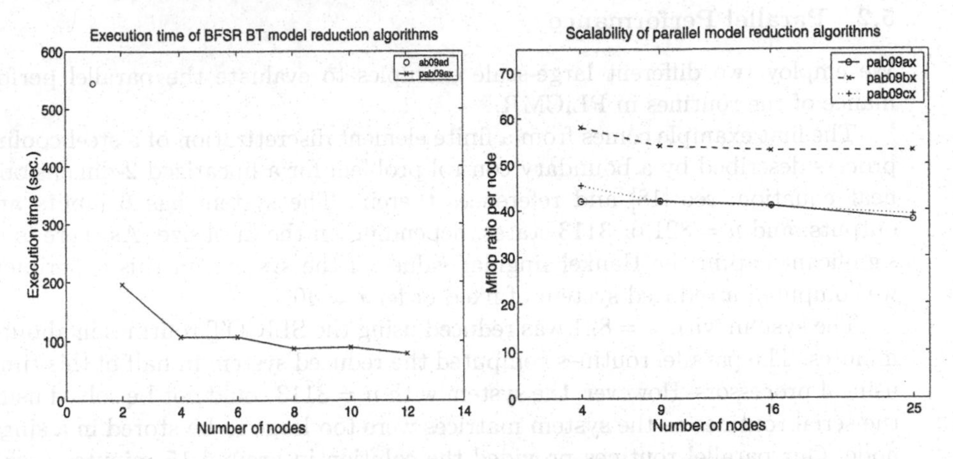


**Fig. 5.** Performance of the parallel model reduction algorithms

square matrices, thus requiring less computations. Comparison of the results on 2 and 4 nodes roughly shows the efficiency of the parallel algorithm. The execution time is reduced by a factor of almost 2 (the number of resources, that is nodes, is doubled). Using a larger number of nodes does not achieve a significant reduction of the execution time due to the small ratio $n/\sqrt{n_p}$.

We next evaluate the scalability of our parallel algorithms. As the memory of the system does not allow to test the serial algorithms on larger problems, in the experiment we fix the problem size per node using $n/\sqrt{n_p} = 800$, $m/\sqrt{n_p} = 400$, $p/\sqrt{n_p} = 400$, and $q/\sqrt{n_p} = 200$, with $q_1 = q_2 = q_3 = q$. In the right-hand plot in Figure 5 we report the Mflop ratio per node for the parallel model reduction algorithms. This is often referred as the *scaled speed-up*. The figure shows a high scalability of all three algorithms, as there is only a minor decrease in the Mflop ratio per node as the number of nodes is increased up to 25 (a problem of size $n = 4000$). The scalability confirms that a larger problem can be solved by increasing proportionally the number of nodes employed.

A more detailed analysis of the numerical accuracy and parallelism of the kernels in the parallel library for model reduction can be found in [4].

## 6 Concluding Remarks

In this paper we have described a parallel library for model reduction of large-scale, dense LTI systems using absolute error methods. Our experiments report a notable reduction in the execution time when the parallel algorithms are used. An important part of this reduction is due to the use of compact factors of the Gramian matrices.

Future extensions of the library will include parallel routines for relative error model reduction. A further extension will provide a service which allows an interaction between remote users and the parallel system through a web environment.

## References


[1] P. Benner, E. S. Quintana-Ortí, and G. Quintana-Ortí. Balanced truncation model reduction of large-scale dense systems on parallel computers. *Mathematical and Computer Modeling of Dynamical Systems*, 6(4):383–405, 2000.

[2] P. Benner, E. S. Quintana-Ortí, and G. Quintana-Ortí. PSLICOT routines for model reduction of stable large-scale systems. In *Proc. 3rd NICONET Workshop on Numerical Software in Control Engineering, Louvain-la-Neuve, Belgium, January 19, 2001*, pages 39–44. 2001.

[3] P. Benner, E. S. Quintana-Ortí, and G. Quintana-Ortí. Computing optimal Hankel norm approximations of large-scale systems. In preparation.

[4] P. Benner, E. S. Quintana-Ortí, and G. Quintana-Ortí. Experimental evaluation of the parallel model reduction routines in PSLICOT. SLICOT Working Note 2002–7, `http://www.win.tue.nl/niconet/`

[5] I. Blanquer, D. Guerrero, V. Hernández, E. S. Quintana-Ortí, and P. Ruíz. Parallel-SLICOT implementation and documentation standards. SLICOT Working Note 1998–1, `http://www.win.tue.nl/niconet/`, September 1998.

[6] Y. Chahlaoui and P. Van Dooren. A collection of benchmark examples for model reduction of linear time invariant dynamical systems. SLICOT Working Note 2002–2, February 2002. Available from `http://www.win.tue.nl/niconet/NIC2/reports.html`.

[7] J. Cheng, G. Ianculescu, C. S. Kenney, A. J. Laub, and P. M. Papadopoulos. Control-structure interaction for space station solar dynamic power module. *IEEE Control Systems*, pages 4–13, 1992.

[8] P. Y. Chu, B. Wie, B. Gretz, and C. Plescia. Approach to large space structure control system design using traditional tools. *AIAA J. Guidance, Control, and Dynamics*, 13:874–880, 1990.

[9] U. B. Desai and D. Pal. A transformation approach to stochastic model reduction. *IEEE Trans. Automat. Control*, AC–29:1097–1100, 1984.

[10] E. Elmroth, P. Johansson, B. Kågström, and D. Kressner. A Web computing environment for the SLICOT library. SLICOT Working Note 2001–2, `http://www.win.tue.nl/niconet/`, June 2001.

[11] L. Fortuna, G. Nummari, and A. Gallo. *Model Order Reduction Techniques with Applications in Electrical Engineering*. Springer-Verlag, 1992.

[12] K. Glover. All optimal Hankel-norm approximations of linear multivariable systems and their $L^\infty$ norms. *Internat. J. Control*, 39:1115–1193, 1984.

[13] M. Green. Balanced stochastic realization. *Linear Algebra Appl.*, 98:211–247, 1988.

[14] J. Gunnels, G. Henry, and R. A. van de Geijn. A family of high-performance matrix multiplication algorithms. In V. N. Alexander, J. Dongarra, B. A. Julianno, R. S. Renner, and C. J. Kenneth Tan, editors, *Computational Science - ICCS 2001, Part I, Lecture Notes in Computer Science 2073*, pages 51–60, 2001.

[15] Y. Liu and B. D. O. Anderson. Controller reduction via stable factorization and balancing. *Internat. J. Control*, 44:507–531, 1986.

[16] B. C. Moore. Principal component analysis in linear systems: Controllability, observability, and model reduction. *IEEE Trans. Automat. Control*, AC-26:17–32, 1981.

[17] C. R. Paul. *Analysis of Multiconductor Transmission Lines*. Wiley–Interscience, Singapur, 1994.

[18] T. Penzl. Algorithms for model reduction of large dynamical systems. Technical Report SFB393/99-40, Sonderforschungsbereich 393 *Numerische Simulation auf massiv parallelen Rechnern*, TU Chemnitz, Germany, 1999.

[19] J. D. Roberts. Linear model reduction and solution of the algebraic Riccati equation by use of the sign function. *Internat. J. Control*, 32:677–687, 1980. (Reprint of Technical Report No. TR-13, CUED/B-Control, Cambridge University, Engineering Department, 1971).

[20] M. G. Safonov and R. Y. Chiang. A Schur method for balanced-truncation model reduction. *IEEE Trans. Automat. Control*, AC–34:729–733, 1989.

[21] R. A. Smith. Matrix equation $XA + BX = C$. *SIAM J. Appl. Math.*, 16(1):198–201, 1968.

[22] M. S. Tombs and I. Postlethwaite. Truncated balanced realization of a stable non-minimal state-space system. *Internat. J. Control*, 46(4):1319–1330, 1987.

[23] A. Varga. Balancing-free square-root algorithm for computing singular perturbation approximations. In *Proc. of the 30th IEEE Conf. Dec. Control*, pages 1062–1065, 1991.

[24] A. Varga. Efficient minimal realization procedure based on balancing. In *Prepr. of the IMACS Symp. on Modeling and Control of Technological Systems*, volume 2, pages 42–47, 1991.

[25] A. Varga and K. H. Fasol. A new square–root balancing–free stochastic truncation model reduction algorithm. In *Prepr. 12th IFAC World Congress*, volume 7, pages 153–156, Sydney, Australia, 1993.

[26] R. C. Whaley and A. Petitet and J. Dongarrra. Automated empirical optimizations of software and the ATLAS project. *Parallel Computing*, 27(1–2):3–15, 2001.

[27] K. Zhou, J. C. Doyle, and K. Glover. *Robust and Optimal Control*. Prentice-Hall, Upper Saddle River, NJ, 1996.

# An Approach to Teaching Computer Arithmetic

Ester M. Garzón, Inmaculada García, and José Jesús Fernández

Dept. Arquitectura de Computadores y Electrónica
Universidad de Almería, 04120 Almería, Spain
{ester,inma,jose}@ace.ual.es
Tph/Fax: +34 950 015 486

**Abstract.** In this work we present an initiative to support teaching computer representation of numbers (both integer and floating point) as well as arithmetic in undergraduate courses in computer science and engineering. Our approach is based upon a set of carefully designed practical exercises which highlights the main properties and computational issues of the representation. In conjunction to the exercises, an auxiliary computer-based environment constitutes a valuable support for students to learn and understand the concepts involved. For integer representation, we have focused on the standard format, the well known 2's complement. For floating point representation, we have made use of an intermediate format as an introduction to the IEEE 754 standard. Such an approach could be included in an introductory course related to either computer structure, discrete mathematics or numerical methods.

*God made the integers, man made the rest.*
L. Kronecker. German mathematician.

## 1 Introduction

Computational science is nowadays making it possible to solve grand-challenge problems thanks to greatly improved computational techniques and powerful computers. Numerical computing is the basis underlying computational science. Virtually, any technical, medical or scientific discipline relies heavily on numerical computing. Scientists and engineers make intensive use of numerical methods and powerful computers to solve complex problems, which may range from the modeling of the microstructure of the atom to building design [1].

Numerical computing is foremost based upon floating point computation. However, computer representation of integer numbers [2] is also involved since it is always implicit and it is essential for almost any task in which the use of a computer is present. The most important computer representation for integer numbers is the well known 2's complement, which constitutes the standard used in almost every modern computer. The main issues related to that format, which any computer scientist or engineer should be familiarized with, are the numeric ranges and how the arithmetic operations are carried out. Such issues are learned

J.M.L.M. Palma et al. (Eds.): VECPAR 2002, LNCS 2565, pp. 269–283, 2003.

in first-level courses in computer science and engineering, usually making use of pencil-and-paper exercises as support.

Floating point computation [3] is undoubtedly of enormous importance in computational science. Computers have supported floating point computation since their earliest days, although using different representations developed by each computer manufacturer. However, in early 1980's, the fruitful cooperation between academic computer scientists and the most important hardware manufacturers allowed the establishment of a binary floating point standard, commonly known as the IEEE 754 Floating Point Representation [4]. In essence, the standard aimed at (1) making floating point arithmetic as accurate as possible, within the constraints of finite precision arithmetic; (2) producing consistent and sensible outcomes in exceptional situations (e.g., overflow, underflow, infinite, ...); (3) standardizing floating point operations across computer systems; and (4) providing the programmer with control over exception handling (e.g. division by zero). Nowadays, most computers offer this standard for floating point computation.

Due to the great importance of floating point computation, computer scientists and engineers should have an excellent knowledge of what a finite floating point representation and arithmetic involve and, in particular, of the ubiquitous IEEE 754 standard. First of all, from the point of view of numerical computing, computer scientists and engineers should be aware of extremely important issues such as, for instance, precision, ranges or algebraic properties related to any finite floating point representation. On the other hand, several aspects in the design of a computer system require a good knowledge on floating point. First, from the point of view of the computer architect, who has to deal with the design of instruction sets including floating point operations. Second, from the point of view of the compiler and programming language design, in the sense that the semantics of the language has to be defined precisely enough to prove statements about programs. Third, from the point of view of the operating system (as far as exception handling concerns), in the sense that trap handlers may be defined by the users/programmers to deal with the exceptions, according to the problem at hand.

The Joint IEEE Computer Society and ACM Task Force on Computing Curricula actually develops curricular guidelines for undergraduate programs in different computing disciplines [5]. Machine-level representation of data has always been a core unit within the Computer Architecture and Organization area in the introductory phase of the undergraduate curriculum, and tightly related to the Programming Languages and Computational Science areas.

In spite of its enormous importance, floating point representation still remains shrouded in mystery by the average computer science or engineering student, and only well understood by experts. Several initiatives have arisen in the nineties [6, 7, 8] and recently [10, 9] to make the floating point arithmetic accessible to the students. Some of them are focused directly on the IEEE floating point standard[7, 9, 10], explaining the representation itself and all the issues involved in it. Others [6, 8] make use of an intermediate floating point representation so as to illustrate the main concepts. The works of [6] and [10] also include exercises to clarify the concepts of precision and resolution involved in any finite floating point representation. Finally, [10] provides a set of computer programs to show clearly what the floating point computation using the IEEE standard involves.

In this work, we present our own initiative to support the teaching of the computer representation of numbers and arithmetic (both integer and floating point), intended to be included in a first level course for undergraduate computer science or engineering students. Our initiative combines (1) a set of key practical exercises for both the integer and the floating point cases, (2) the use of a supporting environment consisting of a set of auxiliary computer programs, and (3) the use of intermediate floating point representations as an introduction to the IEEE standard, with the aim of facilitating the illustration of all the computational issues involved in the 2's complement integer representation as well as in any floating point representation. Such an approach could be included in either a introductory course related to computer structure, discrete mathematics or numerical methods.

## 2 Teaching Computer Representation and Arithmetic of Integers

The most extended computer representation format for integer numbers is the well known 2's complement. The range of the 2's complement integer representation using strings of $p$ bits is $[-2^{p-1}, 2^{p-1}-1]$. This is a representation specially well suited from the point of view of the computer hardware, since it does not require additional special hardware for integer subtraction. In order to illustrate all the issues related to this representation, a complete set of carefully designed practical exercises have been designed, supported by an auxiliary computer-based environment.

### 2.1 Representation of Integer Numbers

The type of exercises in this category help the students to have experiences with the representation itself, the range of numbers that is covered, and different situations in which numbers do not fit ranges. In that sense, the exercises that are proposed include (1) conversions of integer numbers between decimal and 2's complement formats using different word lengths, and vice versa; (2) calculation of the minimum number of bits that are needed to represent given numbers; (3) determination of 2's complement representation ranges for different word lengths.

As an example of practical exercise, we ask students to do decimal-to-2's complement conversions of given numbers and then feed the results into the reverse conversion, using our support environment. As a result of such a "pipeline", only those numbers that fit the corresponding representation ranges will result in themselves. Students are then encouraged to think about the reasons underlying the different results. With this type of exercises, students experience, on their own, the limits of the representation ranges and acquire skill to realize and deal with such situations in real life.

### 2.2 Integer Arithmetic

Exercises in this category mainly include the computation of integer arithmetic operations (addition, subtraction, multiplication and division) using different word lengths. This type of exercises and the software that has been developed illustrate the algorithms used to carry out the integer arithmetic in the hardware units. In addition, the understanding of exceptional situations derived from overflow is specially facilitated. Auxiliary programs have been included in the support environment to show in detail the procedures of the integer addition and subtraction, and, specially, the Booth algorithm for integer multiplication as well as the restoring and non-restoring algorithms for integer division. Such procedures are shown by the software according to the notation used in [2]. For the particular case of multiplication, our environment also affords the chance to show the procedure in a pencil-and-paper format.

As an example of practical exercise related to integer arithmetic, we ask students to carry out different arithmetic operations with certain numbers and different word lengths, using the support environment. The numbers have been carefully chosen so that all possible situations occur, specially related to overflow. Figure 1 shows the output yielded by the support environment that describes the procedure of the integer multiplication of $-5$ and $+7$ using a word length of 4 bits through the Booth algorithm: The first column denotes the iteration of the algorithm; the second column represents the register that supposedly contains the multiplicand; the third column indicates the action to do in the corresponding step of the algorithm; finally, last column represents the double-sized register which initially contains the multiplier at its lower significant half, and into which the result of the multiplication is progressively computed and stored. The extra bit needed by the Booth algorithm is the right-most one in the last column.

## 3 Teaching Floating Point Representation and Arithmetic

Any floating point representation makes use of a exponential notation to represent real numbers, in which any number is decomposed into mantissa or significand and an exponent for a, normally, implicit base. For instance, a nonzero real number is represented by

$$\pm m \times B^E, \text{ with } 1 \leq m < B$$

where $m$ denotes the mantissa, $E$ the exponent, and $B$ the implicit base.

Any computer-based representation involves a finite number of bits to represent the main fields of the floating point representation, mantissa and exponent. Consequently, rounding techniques have to be used for the real numerical value

MULTIPLICATION $(-5) \times (+7)$

MULTIPLICAND: $-5 \equiv 1011_2$ MULTIPLIER: $+7 \equiv 0111_2$

| Iter | Multiplicand | Action | Product-Multiplier |
|---|---|---|---|
| 0 | 1011 | Initial Values | 0000 0111 0 |
| 1 | 1011 | Prod=Prod - Multiplicand | 0101 0111 0 |
| | 1011 | Right Shift | 0010 1011 1 |
| 2 | 1011 | No Operation | 0010 1011 1 |
| | 1011 | Right Shift | 0001 0101 1 |
| 3 | 1011 | No Operation | 0001 0101 1 |
| | 1011 | Right Shift | 0000 1010 1 |
| 4 | 1011 | Prod=Prod + Multiplicand | 1011 1010 1 |
| | 1011 | Right Shift | 1101 1101 0 |
| | | Final Result: -35 ==> 11011101 | |

**Fig. 1.** Integer multiplication by means of the Booth algorithm. The operands −5 and +7 are multiplied using a word length of 4 bits

to fit the number of bits of the representation format. Therefore, floating point computation is by nature inexact. In the particular case of the IEEE 754 standard, the single precision of the standard involves 32 bits, one bit for the sign, 23 bits are used for the mantissa and the exponent is represented by 8 bits, using an implicit binary base. Double precision in the standard makes use of 64 bits, 52 for the mantissa and 11 for the exponent.

Finite floating point representation involves a considerable number of issues which any computer scientist or engineer should be aware of. The most important issues derived from the use of such a representation are:

- There exists an interesting trade-off in terms of precision and range of the format. The number of bits in the exponent and the mantissa defines the range and the precision, respectively.
- The gap between successive floating point numbers of the representation varies along the real numerical intervals. The gap is smaller as the magnitudes of the numbers themselves get smaller, and bigger as the numbers get bigger.
- There is a relatively great gap between zero and the nearest non-zero floating point number. However, the use of *denormalized* numbers allows underflow to be gradual, evenly filling such a gap. Denormalized numbers are denoted by a zero exponent field and an unnormalized mantissa, assuming that the implicit bit is 0. In this way, the value of a denormalized number in the single precision IEEE standard is given by:

$$\text{Real value} = (-1)^s \times (0.m) \times 2^{-126}$$

- The machine epsilon, $\epsilon_{\text{mach}}$ refers to the gap between 1.0 and the smallest floating point number greater than 1.0. It provides an idea of the accuracy

of the floating point operations: floating point values are accurate to within a factor of about $1 + \epsilon_{\mathrm{mach}}$ [10]. Except for denormalized numbers, neighboring floating point numbers differ by one bit in the last bit of the mantissa, so $\epsilon_{\mathrm{mach}} = 2^{-n_m}$. For the single precision IEEE standard, $\epsilon_{\mathrm{mach}} = 2^{-23} \simeq 10^{-7}$.
- The zero number is represented by a string of zero bits.
- Rounding, truncating and cancellation errors as well as error accumulation are heavily involved in the floating point arithmetic.
- Special quantities (Infinite, Not-a-Number) are designed to handle exceptional situations.
- Trap handlers for exception handling are under control of the user/programmer.
- Floating point representation introduces serious anomalies with respect to the conventional algebra, in the sense that some fundamental rules of arithmetic, such as the associative or distributive properties, are no longer guaranteed.

We have developed a relatively simple floating point representation format faithfully resembling all the important issues in the IEEE 754 floating point standard, with the aim of attenuating the relatively tedious aspect of teaching these kind of issues. This format is based on a user-defined word length so that the concepts related to precision and range is easily illustrated. The base format makes use of 12 bits, 1 for sign, 5 for the exponent and 6 for the mantissa. Because the anomalies arising as a result of a finite floating point representation are amplified, the use of this base format allows to easily show them, and therefore, facilitating the understanding by the student. Table 1 summarizes the most important features of this representation. This shorter version of the IEEE standard 754 has a five-fold aim: (1) an easy identification of the relationship between the real number and its floating point representation as well as the effects of the truncation/rounding in the conversion; (2) a clear illustration of the concepts related to the range/precision trade-off; (3) the understanding of the algorithms underlying the floating point arithmetic operations; (4) a clear identification of which, when and why anomalies arise in finite precision floating point computation (such anomalies include the effects of rounding errors in arithmetic operations, the non-fulfillment of the properties of the conventional algebra, and the exceptional situations); (5) the word length may be small enough to allow calculations by-hand, if convenient.

In summary, the floating point format proposed here is specially well suited for illustrating all issues related to finite precision floating point representation, and in particular the IEEE standard 754. In that sense, a complete set of practical exercises have been thoroughly devised to highlight such aspects. Some of the exercises are based on the aforementioned format, but many others work on a general floating point format, by specifying different lengths for the exponent and mantissa fields. These practical exercises are accompanied by a computational environment that turns out to be an excellent auxiliary tool, from the pedagogical point of view, to illustrate such not-so-trivial issues.

The practical exercises that have been devised fall into different categories:

**Table 1.** The Simpler Floating Point Representation

THE MOST IMPORTANT FEATURES OF THE REPRESENTATION

| | |
|---|---|
| Length of the representation: 12 bits: | |
| sign: | 1 bit |
| exponent: | $n_e$ =5 bits |
| mantissa: | $n_m$ =6 bits |
| Exponent Bias: | $2^{n_e-1} - 1 = 15 = 01111_2$ |
| Range of the exponent: | $[-14, 15]$ |
| Machine epsilon: | $\epsilon_{\text{mach}} = 2^{-n_m} = 2^{-6}$ |
| Supported rounding modes: | *Truncation* , *Round to Nearest* |
| Denormalized numbers: | Fully supported |
| Supported exceptions: | *Overflow, Underflow, Invalid, Division by Zero* |

THE MOST IMPORTANT (POSITIVE) SPECIAL VALUES

| Special value | Representation | Value |
|---|---|---|
| Largest normalized: | 0 11110 111111 | $+1.984375 \times 2^{15}$ |
| Smallest normalized: | 0 00001 000000 | $+1.000000 \times 2^{-14}$ |
| Largest denormal.: | 0 00000 111111 | $+0.984375 \times 2^{-14}$ |
| Smallest denormal.: | 0 00000 000001 | $+0.015625 \times 2^{-14}$ |
| + Zero | 0 00000 000000 | +0.0 |
| + Infinite: | 0 11111 000000 | --- |
| Not-a-Number (NaN): | 0 11111 111111 | --- |

– The first category of exercises is intended to help students to learn the procedure to convert a real number to its floating point representation, analysing the effects of the rounding modes in terms of the relative error. In that sense, the exercises deal with the determination of the representation of different real numbers with the simpler floating point format (12 bits length), using different rounding techniques. Specially care has been taken to choose proper real values to give rise special situations in the conversions (for example, the smallest normalized floating point number, the biggest floating point number, the extreme values in the denormalized floating point range, etc).
Figure 2 shows some examples of representations, as reported by the computer program in charge of the conversion. The first example was computed using truncation as the rounding mode, resulting in that the real number 1.999 is represented as 1.984375. The second, which turns out to be a denormalized number in this format, used round-to-nearest rounding mode. The third and fourth are examples of values bringing about exceptions: 0.0000001 is too small and 65500.0 is to big to be represented in the format. The output of the program shows the input real number, the representation, and the value of the floating point representation.

– The following group of exercises aims at facilitating the understanding of the concepts of precision, range and the precision-range trade-off. Exercises in this group include the following aspects:

```
Real Number:      1.999000  = 1.999000e+00 = 0.999500 x 2^1
Representation:   0 01111 111111
Float. P. Value:  1.984375  = 1.984375e+00 = 1.984375 x 2^0
---------------------------------------------------------------
Real Number:      0.000040  = 4.000000e-05 = 0.655360 x 2^-14
Representation:   0 00000 101010
Float. P. Value:  0.000040  = 4.005432e-05 = 0.656250 x 2^-14
                  !! Denormalized Number !!
---------------------------------------------------------------
Real Number:      0.0000001 = 1.000000e-07 = 0.838861 x 2^-23
Representation:   0 00000 000000
Float. P. Value:  0.0000000 = 0.000000e+00 = 0.000000 x 2^0
                    ! Exception: Underflow !
---------------------------------------------------------------
Real Number:      65500.000000 = 6.550000e+04 = 0.999451 x 2^16
Representation:   0 11111 000000
Float. P. Value:  +Infinite
                    ! Exception: Overflow !
```

**Fig. 2.** Representation of certain values according to our floating point format, using 5 and 6 bits for exponent and mantissa, respectively

- determination of the representation of different real numbers using different formats (i.e., varying bits for exponent and mantissa).
- determination of the gap between floating point numbers in some intervals of different representation formats.
- determination of the ranges covered for different floating point formats (including the sub-range of denormalized numbers, and that of normalized numbers).
- computation of the minimum number of bits dedicated for the exponent and mantissa fields to fulfill certain given constraints, for example, to be able to distinguish between the representations of two given real numbers very close to each other.

As an example, Figure 3 summarizes the process to work out the minimum number of bits (in exponent and in mantissa fields) required to distinguish the real numbers 15.9 and 15.925 and, at the same time, to represent the real number 100000. Students start working from the simpler format, using $n_e = 5$, $n_m = 6$, and then they progressively increase the number of bits in mantissa until 15.9 and 15.925 get distinguishable representations. In that figure, it can be seen that $n_m = 8$ is the minimum number of bits required, which makes 15.9 and 15.925 get represented as 15.875 and 15.90625, respectively. Then, students try to get the representation of 100000 using the format just computed, obtaining an *Overflow* exception. They progressively increase the number of bits in the exponent field until 100000 gets representable. The final answer is that the minimum format to fulfill the requirements of the assignment is to use $n_e = 6$, $n_m = 8$.

– The following class of exercises is related to arithmetic, and is intended to help students to learn the procedures to carry out the floating point arithmetic operations. Exercises in this group are mainly focused on the computation of floating point additions, subtractions, multiplications and divisions

DETERMINATION OF THE MINIMUM NUMBER OF BITS IN MANTISSA FIELD TO DISTINGUISH 15.9 AND 15.925

| | 15.9 | 15.925 |
|---|---|---|
| Floating Point Format | Representation | Representation |
| $n_e = 5$, $n_m = 6$ | 0 10010 111111 | 0 10010 111111 |
| $n_e = 5$, $n_m = 7$ | 0 10010 1111110 | 0 10010 1111110 |
| $n_e = 5$, $n_m = 8$ | 0 10010 11111100 | 0 10010 11111101 |
| $n_e = 5$, $n_m = 8$ | 15.9→15.875 | 15.925 →15.90625 |

DETERMINATION OF THE MINIMUM NUMBER OF BITS TO DISTINGUISH 15.9 AND 15.925 AND REPRESENT 100000

| | Real number: 100000 | |
|---|---|---|
| Floating Point Format | Representation | Value |
| $n_e = 5$, $n_m = 8$ | 0 11111 00000000 | +Infinite |
| $n_e = 6$, $n_m = 8$ | 0 101111 10000110 | 99840.0 |

**Fig. 3.** Procedure to determine the minimum floating point format to fulfill the requirements that the real numbers 15.9 and 15.925 are distinguishable and that the real number 100000 is representable. Top: the minimum mantissa length to distinguish 15.9 and 15.925 is worked out to be $n_m = 8$. Bottom: $n_e = 6$ is the minimum exponent length required to represent 100000

using different pairs of operands, experiencing with the different rounding modes, and measuring the relative error of the results. The operands that are proposed are specially chosen so that different situations occur. Specially interesting is, for instance, the addition of pairs of operands that are enormously different in magnitude (for instance, our simpler format, using $n_e = 5$ and $n_m = 6$, makes $1000.0 + 2.5$ equal to 1000.0).
Figure 4 is intended to show the output of the program in charge of illustrating the process of floating point addition. As can be seen, the program specifies in detail all the stages in the procedure of such an arithmetic operation: (1) conversion of operands to their representation; (2) alignment of mantissas; (3) addition of mantissas; and finally (4) the result of the normalization and rounding. Such a program is extremely useful for students to discern all the issues concerning the round-off problems in arithmetic operations. This operation was performed using our simpler floating point representation ($n_e = 5, n_m = 6$).

- This category deals with the algebraic anomalies that arise as a result of the finite nature of the floating point representation. It is imperative that students have insights into the machinations of floating point arithmetic on computers and, in particular, into those anomalies, to succeed in real life computation problems. So, exercises in this category include:
  - analysis of the cancellation error that arises in the subtraction of two operands extremely close to each other.
  - practical exercises to show the effect of accumulation of rounding errors. For instance, multiplication of a given floating point number by a integer constant by means of accumulated sum of the former, using different rounding modes. Students have to do a graph showing the evolution of the relative error during the accumulated sum. Figure 5 shows the result

```
                    Floating Point Addition 1.0 + 0.999, using 2 guard bits.
                    Format using: 5 bits/Exp. 6 bits/Mantissa; 2 Guard bits.

1.- Representation of operands.
                           s   e     m        s  e       .m
            Operand  1 ->  0 01111 000000 -> 0 01111   1.000000  = 1.000000
            Operand  2 ->  0 01110 111111 -> 0 01110   1.111111  = 0.999000

2.- Alignment of operands.
                           s   e       .m      g
            Operand  1 ->  0 01111  1.000000 00
            Operand  2 ->  0 01111  0.111111 10

3.- Addition of mantissas.
                           s   e       .m      g
            Operand  1 ->  0 01111  1.000000 00
            Operand  2 ->  0 01111  0.111111 10
            -----------------------------------
            Addition   ->  0 01111  1.111111 10

4.- Normalization and Rounding of the result.
            Result     ->  0 10000  1.000000      = 2.000000

Result of the Addition:  0 10000 000000
Decimal Value:  2.000000 = 2.000000e+00 = 1.000000 x 2^1
```

**Fig. 4.** Procedure and result of the floating point addition of 1.0 and 0.999

of the multiplication $1.999 \times 10$ as shown by the auxiliary programs. Here, the intermediate results from the accumulated sum are shown. On the left, the sums are carried out using truncation. On the right, the round-to-nearest rounding mode is used. The last line shows the final result of the multiplication. Clearly, the superiority of the round-to-nearest rounding mode is evident. These results are rapidly generated using one of the programs of our environment. Students have to analyze the results, and generate graphs of the error evolution.

- practical exercises to show that some fundamental rules of the conventional algebra are not fulfilled in floating point computation:
  - ∗ floating point addition is not associative.
  - ∗ floating point multiplication is not associative.
  - ∗ floating point multiplication does not necessarily distribute over addition
  - ∗ ordering of operations is significant.
  - ∗ the *cancellation* property is not always valid, i.e., there exist positive floating point numbers $A$, $B$, $C$ such that $A+B = A+C$ and $B \neq C$.
  - ∗ multiplication of a floating point number by its inverse is not always equal to 1.
  - ∗ it is almost always wrong to ask whether two floating point numbers are equal.

MULTIPLICATION 1.999 × 10

| Iter | | TRUNCATION Representation | | Value | ROUND TO NEAREST Representation | | Value |
|---|---|---|---|---|---|---|---|
| 1 | -> | 0 01111 111111 | -> | 1.99900 | 0 01111 111111 | -> | 1.99900 |
| 2 | -> | 0 10000 111111 | -> | 3.96875 | 0 10000 111111 | -> | 3.96875 |
| 3 | -> | 0 10001 011111 | -> | 5.93750 | 0 10001 011111 | -> | 5.93750 |
| 4 | -> | 0 10001 111110 | -> | 7.87500 | 0 10001 111111 | -> | 7.93750 |
| 5 | -> | 0 10010 001110 | -> | 9.75000 | 0 10010 001111 | -> | 9.87500 |
| 6 | -> | 0 10010 011101 | -> | 11.62500 | 0 10010 011111 | -> | 11.87500 |
| 7 | -> | 0 10010 101100 | -> | 13.50000 | 0 10010 101111 | -> | 13.87500 |
| 8 | -> | 0 10010 111011 | -> | 15.37500 | 0 10010 111111 | -> | 15.87500 |
| 9 | -> | 0 10011 000101 | -> | 17.25000 | 0 10011 000111 | -> | 17.75000 |
| 10 | -> | 0 10011 001100 | -> | 19.00000 | 0 10011 001111 | -> | 19.75000 |

**Fig. 5.** Multiplication 1.999 × 10 by means of an accumulated sum. The simpler floating point representation has been used (5/6 bits for exponent/mantissa)

One of the most interesting examples of algebraic anomalies is related to the computation of Harmonic series:

$$H_n = 1 + 1/2 + 1/3 + \ldots + 1/n.$$

We use such a series to show how the ordering of operations may be extremely significant. We propose to compute the series, first, literally as in the formula and, second, in reverse order. The results help students to learn that summing the smallest values first, progressively increasing in magnitude, yields more accurate final results. Such an ordering avoids loosing the lowest precision bits of the smallest quantities in summing with values very different in magnitude. Figure 6 shows the results generated by our software package. This exercise aims at highlighting the fact that the ordering of floating point operations may be significant. On the left, the results for an ordering according to the original formula. On the right, the results for the reverse ordering. The results include the current term that is to be added in the series, and the intermediate value of the accumulated sum.
On the other hand, such a series is mathematically proven to diverge. However, in floating point representations, the series converge due to the round-off errors. We have also designed practical exercises on that.

– The last category of exercises comprises those related to exceptions. Since the IEEE standard 754 allows exception handling to be under control of the user/programmer, it is extremely important to familiarize students with the situations that produce exceptions, how to manage them, and the floating point values resulting from exceptions. The exercises in this category are mainly intended to come exceptions into manifest:
  - practical exercises to show that some arithmetic operations with, in principle, normal numbers may result in exceptions because the operation yields a result not representable in the format. For instance, the multiplication of extremely small floating point numbers may give rise to an

COMPUTATION OF HARMONIC SERIES

| | $\sum_{n=1}^{10} \frac{1}{n}$ | | $\sum_{n=10}^{1} \frac{1}{n}$ | |
|---|---|---|---|---|
| Index | Number | Sum 1..N | Number | Sum N..1 |
| 1 | 1/1 = 1.00000 | 1.000000 | 1/10 = 0.10000 | 0.100000 |
| 2 | 1/2 = 0.50000 | 1.500000 | 1/9 = 0.11111 | 0.210938 |
| 3 | 1/3 = 0.33333 | 1.828125 | 1/8 = 0.12500 | 0.335938 |
| 4 | 1/4 = 0.25000 | 2.062500 | 1/7 = 0.14286 | 0.476562 |
| 5 | 1/5 = 0.20000 | 2.250000 | 1/6 = 0.16667 | 0.640625 |
| 6 | 1/6 = 0.16667 | 2.406250 | 1/5 = 0.20000 | 0.843750 |
| 7 | 1/7 = 0.14286 | 2.562500 | 1/4 = 0.25000 | 1.093750 |
| 8 | 1/8 = 0.12500 | 2.687500 | 1/3 = 0.33333 | 1.421875 |
| 9 | 1/9 = 0.11111 | 2.812500 | 1/2 = 0.50000 | 1.921875 |
| 10 | 1/10 = 0.10000 | 2.906250 | 1/1 = 1.00000 | 2.937500 |

**Fig. 6.** Computation of the Harmonic series with $N = 10$ using our simpler floating point format (5 bits exp., 6 bits mantissa), and using different orderings

*Underflow* exception, or the addition/multiplication of two big numbers may result in an *Overflow* exception, etc. Two concrete examples are that $0.0009 * 0.001$ produces an *Underflow* exception and $1875 * 35$ brings about an *Overflow* one, if our simpler format (with $n_e = 5$ and $n_m = 6$) is used.

- computation of $A + \infty$, $A * \infty$, $A/0$, $A/\infty$, etc, given a floating point number $A$.
- computation of $\infty + 0$, $\infty + \infty$, $\infty - \infty$, $\infty * 0$, $\infty * (-\infty)$, $0/0$, $\infty/\infty$, $\infty/0$, etc.

Figure 7 shows the exception that is brought about and the corresponding floating point value that would be generated for different singular situations. Such a behavior should be exhibited by any floating point system guided by the IEEE standard.

To conclude this section, we should mention that all the practical exercises that we have devised are supported by the computer-based environment that will be described in the following section. The programs in the environment are intended to provide answers and their justifications to all the questions and exercises formulated in the assignments.

# 4 The Support Computational Environment

We have developed a support computer-based environment for students to practice number representation and arithmetic in computers, facilitating them the understanding of the computational issues in this field. The environment consists of computer programs, written in standard C programming language, which internally use techniques derived from multiprecision arithmetic to compute the representations and the arithmetic operations. The floating point values returned by the programs are computed using double precision data types. We have tested the programs in different computer platforms (PC, SGI, Sun, Alpha) and operating systems (Windows, Linux, IRIX, Solaris) obtaining coherent results. The programs can be executed on console, specifying the input operands and options in command line. In the particular case of the programs related to floating point, the options include the possibility of specifying the number of bits for the exponent and mantissa fields so that students can experience with representations different from the default ($n_e = 5$, $n_m = 6$). The environment is multilingual in the sense that *Spanish* and *English* languages are currently supported. An option in compilation time allows to activate the language that has been chosen.

We have also developed a friendly web-based front-end for the programs which is based on the client/server model. Students need not deal directly with the console and command lines but, instead, they access the environment through their web browser/client (e.g. netscape) and, transparently, execute the programs in the web server. This front-end is based on HTML (HyperText Markup Language) forms and CGI (Common Gateway Interface) technology [11]. Students use their web browser/client to (1) select the program to be executed, (2) specify the operands and options in the form, and (3) send the data to the server. CGI technology is in charge of obtaining the data sent by the web client, executing in the server the corresponding program with the proper parameters and, finally, sending the results back to the browser/client.

Our environment, currently available at `http://www.ace.ual.es/RAC/`, is structured into two main blocks: on the one hand, all the programs concerning integer representation and arithmetic, and those concerning floating point representation and arithmetic. There is an accompanying manual containing the theoretical concepts and an exhaustive list of assignments that covers the wide spectrum of the issues in computer number representation and arithmetic.

This environment proves to be, from the pedagogical point of view, an excellent auxiliary tool for students to learn about the finite length number repre-

| Singular operation | Exception | Returned value |
|---|---|---|
| $A - \infty$ | Overflow | -Infinite |
| $\infty + \infty$ | Overflow | +Infinite |
| $\infty - \infty$ | Invalid | NaN |
| $A * \infty$ | Overflow | ±Infinite |
| $\infty * 0$ | Invalid | NaN |
| $\infty * (-\infty)$ | Overflow | -Infinite |
| $A/0$ | Division-by-zero | ±Infinite |
| $A/\infty$ | Underflow | ±Zero |
| $0/0$ | Invalid | NaN |
| $\infty/\infty$ | Invalid | NaN |
| $-\infty/0$ | Overflow | -Infinite |

**Fig. 7.** Different situations that produce exceptions in our floating point format. The symbol ± means that the sign is dependent on the signs of the input operands

sentation and arithmetic in computers from a hands-on, practical, perspective. Thanks to this environment and the set of carefully devised exercises, students can have insights into the not-so-trivial issues in this field.

## 5 Conclusions

In this article we have described an approach to teaching computer representation of numbers and arithmetic. This initiative mainly consists of a complete set of thoroughly designed practical exercises conceived to emphasize all the important issues in finite length computer arithmetic. A computational environment that turns out to be a very valuable support tool for students to practice is also provided. Our experience after several years of lecturing and teaching computer arithmetic in an introductory course on computer organization in undergraduate computer science curricula allows us to claim the success of this initiative.

As far as the integer representation is concerned, we have experienced that the use of carefully designed practical exercises allows students to learn easily how the 2's complement represents the integer numbers, and all the issues regarding range and word lengths. The algorithms and the computer hardware involved in the integer arithmetic are also facilitated. In conjunction to the practical exercises, the computer-based environment devised in this work provides students with a valuable support for their learning.

From the pedagogical point of view, we have experienced that teaching floating point representation based upon a simple format which helps us illustrate all the issues in any floating point representation system is successful. We use this simple format as an introduction to the IEEE 754 floating point standard. Students feel more comfortable to deal with the IEEE standard once they are familiarized with the simple format and all the computational issues involved. The use of a simpler format allows all the singularities of floating point representation to be highlighted and amplified, and consequently, students are more impressed by those effects.

In ultimate instance, this approach facilitates lecturers to achieve the goal as a academic staff of making students aware of the computational issues involved in finite length number representation and arithmetic. This approach may be valuable for introductory undergraduate courses related to computer organization, programming, discrete mathematics and numerical methods in computer science or engineerings.

## Acknowledgments

This work has been partially supported by the Spanish CICYT through grants TIC99-0361 and TIC2002-00228.

## References


[1] C. W. Ueberhuber *Numerical Computation. Methods, Software, and Analysis*, Vols. 1&2. Springer-Verlag, 1997.

[2] D. A. Patterson and J. L. Hennessy. *Computer Organization and Design. The Hardware/Software Interface* Morgan Kaufmann Pub., 1998.

[3] D. E. Knuth. *The Art of Computer Programming*, 3rd ed, volume 2, Seminumerical Algorithms. Addison-Wesley, 1998.

[4] W. Kahan. IEEE Standard 754 for Binary Floating-Point Arithmetic. WWW document, 1996.
http://www.cs.berkeley.edu/ wkahan/ieee754status/ieee754.ps.

[5] Curriculum 2001 Joint IEEE Computer Society/ACM Task Force. "Year 2001 Model Curricula for Computing ((CC-2001)," Final report, December 15, 2001.

[6] T. J. Scott. "Mathematics and computer science at odds over real numbers," *ACM SIGCSE Bulletin*, 23(1):130–139, 1991.

[7] D. Goldberg. "What every computer scientist should know about floating-point arithmetic," *ACM Comp. Surveys*, 23:5–48, 1991.

[8] C. W. Steidley. "Floating point arithmetic basic exercises in mathematical reasoning for computer science majors," *Computers in Education Journal*, 2(4):1–6, 1992.

[9] W. Kahan. Ruminations on the design of floating-point arithmetic. WWW document, 2000. http://www.cs.nyu.edu/cs/faculty/overton/book/docs/).

[10] M. L. Overton. *Numerical Computing and the IEEE Floating Point Standard*. SIAM, 2001.

[11] S. Guelich, S. Gundavaram, G. Birznieks. *CGI Programming on the World Wide Web*. O'Reilly, 2000.

# Chapter 5

# Linear and Non-linear Algebra

# Fast Sparse Matrix-Vector Multiplication for TeraFlop/s Computers

Gerhard Wellein[1], Georg Hager[1], Achim Basermann[2], and Holger Fehske[3]

[1] Regionales Rechenzentrum Erlangen, D-91058 Erlangen, Germany
[2] C&C Research Laboratories, NEC Europe Ltd, D-53757 Sankt Augustin, Germany
[3] Institut für Physik, Universität Greifswald, D-17487 Greifswald, Germany

**Abstract.** Eigenvalue problems involving very large sparse matrices are common to various fields in science. In general, the numerical core of iterative eigenvalue algorithms is a matrix-vector multiplication (MVM) involving the large sparse matrix. We present three different programming approaches for parallel MVM on present day supercomputers. In addition to a pure message-passing approach, two hybrid parallel implementations are introduced based on simultaneous use of message-passing and shared-memory programming models. For a modern SMP cluster (HITACHI SR8000) performance and scalability of the hybrid implementations are discussed and compared with the pure message-passing approach on massively-parallel systems (CRAY T3E), vector computers (NEC SX5e) and distributed shared-memory systems (SGI Origin3800).

## 1 Introduction

Clusters of multiprocessor shared-memory (SMP) nodes are presently the common way to exceed the TFlop/s barrier. Although these systems outperform all other architectures with respect to peak performance there is still the question how to write portable codes which run and scale efficiently. The simplest approach is to use massively-parallel (MPP) codes totally ignoring the shared-memory within SMP nodes. A more promising approach seems to be the "hybrid programming model" which uses both message-passing model (between SMP nodes) and shared-memory model (within SMP nodes) simultaneously and maps most closely to the memory structure of the underlying target architecture. It is evident that the benefit of either programming model depends on the algorithm and the programming style used. Some studies for application programs have already been done in this context [1, 2, 3, 4, 5, 6] leading to different results. Numerical cores of many applications are iterative sparse matrix algorithms such as sparse eigenvalue solvers. In practice, such operations will have a performance bound by the memory access. Since the LINPACK benchmark [7] usually does not cover this quantity, an iterative solver benchmark (*sparsebench*) has been developed recently by Dongarra et al. [8]. The *sparsebench* benchmark solves sparse linear systems using different methods and storage schemes but is available in a non-parallel version for small problem sizes only.
The aim of this work is to discuss three different parallel implementations of a sparse matrix-vector multiplication (MVM) for SMP clusters:

J.M.L.M. Palma et al. (Eds.): VECPAR 2002, LNCS 2565, pp. 287–301, 2003.

**MPP-mode** : Each MPI process is assigned to exactly one processor of a SMP node. This implementation makes no use of the hardware structure.

**Vector-mode** : One SMP node is identified as a single process with high peak-performance and memory bandwidth similar to a single vector processor. Shared-memory parallelisation of long vector loops is done. This implementation might be beneficial for vector codes on SMP nodes providing fast collective thread operations.

**Task-mode** : Using OpenMP we perform an explicit programming of each processor of a SMP node to overlap communication and computation. An asynchronous data transfer can be simulated by defining one processor to be responsible for communication. This strategy improves performance on SMP nodes without hardware support for asynchronous data transfer and for codes where a sufficient overlap of communication and computation exists.

The different implementations have been applied to extreme sparse matrices (originating from large scale eigenvalue problems in theoretical physics) with dimensions ranging from $10^3$ to $10^8$ and have been run on a maximum of 1024 processors. For completeness a detailed performance comparison with classical supercomputers is given.

## 2 Implementation of Matrix-Vector Multiplication

Generally, the performance of iterative sparse eigenvalue solvers is determined by a matrix-vector multiplication step involving the large sparse matrix. There are two widely used methods to exploit the sparsity of the matrix to minimize the memory requirements in the MVM step which show different performance characteristics. The *Compressed Row Storage* (CRS) format [9] is generally used for RISC systems but has serious drawbacks on vector processors due to short inner loop lengths. The *Jagged Diagonal Storage* (JDS) scheme [9] provides long inner vector loops at the cost of minor performance drawbacks on RISC systems for matrices used in our work [10]. To achieve high performance on a wide range of supercomputer architectures, a parallel JDS scheme was implemented.

### 2.1 JDS Matrix-Vector Multiplication

The JDS format is a very general storage scheme for sparse matrices making no assumptions about the sparsity pattern. In a compression step, the zero elements are removed, nonzero elements are shifted to the left and the rows are made equal in length by padding them with trailing zeros. To prevent storing theses trailing zeros, the rows are rearranged according to the number of nonzero elements per row. The columns of the compressed and permuted matrix are called *Jagged Diagonals*. Obviously, the number of these *Jagged Diagonals* is equal to the maximum number `max_nz` of nonzero elements per row. The MVM is performed along the *Jagged Diagonals*, providing an inner loop length essentially equal to the matrix dimension ($D_{\mathrm{Mat}}$). For a detailed description of our JDS implementation we refer to Ref. [10].
Using permuted vectors, the numerical core of the MVM is given as follows:

```
for j= 1,...,max_nz do
  for i= 1,...,jd_ptr(j+1)-jd_ptr(j)) do
    y(i) = y(i)+ value(jd_ptr(j)+i-1)* x(col_ind(jd_ptr(j)+i-1))
  end for
end for
```

The nonzero elements of $A$ are stored in the linear array `value(:)`, stringing the *Jagged Diagonals* together one by one. An array of the same length (`col_ind(:)`) contains the corresponding column indices. The array indices belonging to the first element of each *Jagged Diagonal* are stored in the pointer array `jd_ptr(:)`. The performance of the innermost loop is clearly determined by the quality of the memory access: One **store** and **load** operations are required to perform two floating point operations (Flop). Taking into account the gather operation on the vector **x**, we do not expect MVM performance to exceed 20–25 % of the peak performance even for present-day vector processors delivering a memory bandwidth of one word per Flop. The performance on RISC systems is even worse [8].

### 2.2 Parallel Implementation

For parallel computers with distributed memory, we use a row-wise distribution of matrix and vector elements among the processes (**MPP-mode**) involved in MVM (cf. Fig. 1 (left panel)). Communication is required for all matrix elements that refer to vector data stored on remote processes. Since the matrix does not change during eigenvalue computation, a static communication scheme is predetermined beforehand. In this way, we can exchange data efficiently in anticipation of the overlapping of communication and computation in the MVM step.

Each process transforms its local rows into the JDS format as described above (see also Ref. [10]). In order to maintain a MVM core similar to that shown above, local and nonlocal vector elements of **x** are stored in the same array. It is apparent that a *Jagged Diagonal* can only be released for computation if the

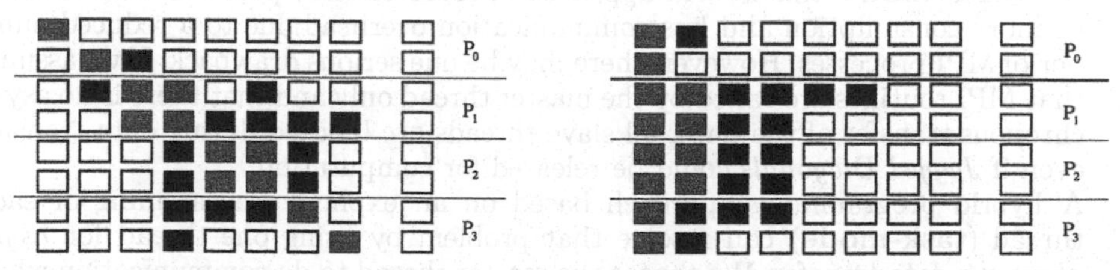


**Fig. 1.** Distribution of matrix and vector elements on 4 processes ($P_0 - P_3$). Nonzero matrix elements causing communication are marked black and local nonzero matrix elements are colored gray. The original distribution is given in the left panel. The right panel depicts the distribution after the modified JDS compression step

corresponding nonlocal vector elements were successfully received.
The parallel MVM implementation is based on MPI and uses non-blocking communication routines. After all communication calls have been initialized (pairs of `mpi_isend` and matching `mpi_irecv`), we check repeatedly for the completion of any receive call (`mpi_testany`) and release *Jagged Diagonals* for computation if possible. In principle this implementation allows overlapping of communication and computation if asynchronous data transfer is supported by the hardware and at least one local *Jagged Diagonal* exists, involving local matrix elements only.
At this point the parallel JDS scheme as described above has one serious drawback: The (process-)local compression step might cause a strong mixture of local and nonlocal matrix elements when shifting the nonzero elements to the left. For instance, process $P_3$ in Fig. 1 (left panel) requires at least one receive-operation to be completed before starting any computation when using the JDS transformation in a straightforward way. For that reason we have slightly modified the parallel compression step as follows:

1. Shift local matrix elements to the left only within the local sub-block on each process.
2. Fill each local sub-block with the nonlocal matrix elements from the right.

If the diagonal of the matrix is full this algorithm generates at least one local *Jagged Diagonal* as can be seen in Fig.1 (right panel).

## 2.3 Hybrid Parallel Implementation

The MVM implementation presented above allows the efficient use of vector processors since the long inner loop can easily be vectorised by the compiler. The same holds for vector parallel computers where the loop length decreases with increasing process number but in general remains long with respect to vector start-up times.
Of course, the inner loop can also be automatically parallelised and executed in parallel on the processors of a SMP node. Thus we can easily use our MPI based implementation on SMP clusters, assigning one MPI process to each node and running a shared memory parallelisation within each node. When compared to the **MPP-mode**, this hybrid approach (**vector-mode**) provides both a lower memory consumption and less communication overhead due to a reduced number of MPI processes. However, there may be one serious drawback: If we assume that MPI routines are called by the master thread only and that there is no asynchronous transfer of the data, all slave threads are blocked during data transfer even if *Jagged Diagonals* could be released for computation.
A hybrid programming approach based on an explicit programming of each thread (**task-mode**) can resolve that problem by using one thread for *asynchronous data transfer*. We choose the master thread to do communication while slave threads perform computations. The list of released diagonals is governed by the master thread and is protected with a lock mechanism. Each slave thread computes a fixed number of local rows to avoid the synchronisation of the slave threads at the end of the inner loop. Due to different length of the *Jagged Diagonals* an equal distribution of rows may result in a load imbalance for slave

threads. Instead we use an equal distribution with respect to the number of floating point operations to be performed on each slave thread. For portability reasons, we have chosen an OpenMP implementation for the **task-mode**:

```
!$OMP PARALLEL
thread_id = OMP_GET_THREAD_NUM()
if( thread_id .eq. 0) then
    1.) release local diagonals
    2.) start receive and send operations
    3.) wait for the completion of any receive operation
    4.) release diagonals for computation if possible
    5.) if there is still an outstanding receive operation goto 3.)
    6.) wait for the completion of all send operations
else
    1.) wait until at least one diagonal is released for computation
    2.) compute MVM for all released diagonals
    3.) if not all diagonals have been released goto 1.)
endif
!$OMP END PARALLEL
```

At this point the question arises, which scenarios favor the use of **task-mode** or **vector-mode**. The basic performance relationship between the hybrid modes can be determined qualitatively using a simple model: Assuming a perfect load balance concerning the MPI processes, a full speed up with respect to the number of threads ($N_{\mathrm{THR}}$) within a SMP node as well as no MPI communication hiding in the **vector-mode**, the total runtime per node for the **vector-mode** can be written as

$$T_{\mathbf{VECTOR}} = T_{\mathrm{COMP}}/N_{\mathrm{THR}} + T_{\mathrm{COMM}} \tag{1}$$

where $T_{\mathrm{COMM}}$ denotes the time spent for MPI communication and $T_{\mathrm{COMP}}$ is the compute time using one thread per node. Furthermore, in **task-mode** we assume that MPI communication can fully be overlapped by computation giving a total runtime which is determined solely by $T_{\mathrm{COMP}}$ with a reduced number of threads:

$$T_{\mathbf{TASK}} = T_{\mathrm{COMP}}/(N_{\mathrm{THR}} - 1) \geq T_{\mathrm{COMM}} \tag{2}$$

The performance ratio $\epsilon_{\mathrm{T,V}}$ of **task-mode** to **vector-mode** is given by the inverse ratio of the corresponding runtimes and can be written as follows:

$$\epsilon_{\mathrm{T,V}} = \frac{T_{\mathbf{VECTOR}}}{T_{\mathbf{TASK}}} = (1 - \frac{1}{\mathrm{N_{THR}}}) + (N_{\mathrm{THR}} - 1)\frac{T_{\mathrm{COMM}}}{T_{\mathrm{COMP}}} \tag{3}$$

It is evident that the minimum value of $\epsilon_{\mathrm{T,V}}$ is adopted in the limit of vanishing communication ($T_{\mathrm{COMP}} \gg T_{\mathrm{COMM}}$) where the number of compute threads determines the runtimes. The upper limit is given by Eq. (2) yielding to:

$$1 - 1/N_{\mathrm{THR}} \leq \epsilon_{\mathrm{T,V}} \leq 2 - 1/N_{\mathrm{THR}} \tag{4}$$

Obviously there is a crossover between problems with low communication intensity ($T_{\mathrm{COMM}}/T_{\mathrm{COMP}} \ll 1$) where the **vector-mode** is the proper choice and more complex problems which benefit from the use of the **task-mode**. In the limit $T_{\text{COMM}} \gg T_{\text{COMP}}$, however, Eq. (2) does not hold and the hybrid modes show approximately the same performance as the total runtime is determined mainly by communication. To sum it up, the **task-mode** benefits from larger thread counts and can provide a performance gain of up to 100% compared to the **vector-mode**. This effect is expected to be most pronounced if communication time and computation time are of similar size.

# 3 Benchmark Issues

## 3.1 Benchmark Platforms

The benchmarks have been performed on the platforms given in Table 1. While architecture and memory hierarchy of the RISC based systems (SGI/CRAY) are widely known, some important features of the NEC vector computer and the HITACHI SR8000 should be emphasized.
The NEC SX5e vector processor runs at a frequency of 250 MHz using eight-track vector pipelines. There is one load and one load/store pipeline which can either work in parallel as two load pipelines or as a single store pipeline.
The HITACHI SR8000 processor is basically an IBM PowerPC processor running at 375 MHz with important modifications in hardware and instruction set. These extensions include 160 floating point registers and the possibility of up to 16 outstanding `prefetch` or 128 outstanding `preload` instructions. While the `prefetch` loads one cache line in advance to the L1 cache, the `preload` instruction can load single data items directly from the memory to the large register set bypassing the L1 cache. Together with an extensive software pipelining done by the compiler these features are called Pseudo-Vector-Processing (PVP), as they are able to provide a continuous stream of data from the main memory to

**Table 1.** Benchmark platforms and single processor specifications. LINPACK (cf. Ref. [7]) and peak performance numbers (Peak) are given in GFlop/s, whereas the memory bandwidth (memb) per processor is given in GBytes/s. The rightmost columns contain the sizes of the L1 and L2 cache for the RISC processor based systems

| Platform | Installation site | #CPU | LINPACK | Single CPU specifications | | | |
|---|---|---|---|---|---|---|---|
| | | | | Peak | memb | L1 [kB] | L2 [MB] |
| NEC SX5e | HLR Stuttgart | 32 | 121 | 4.0 | 32.0 | — | — |
| CRAY T3E-1200 | NIC Jülich | 540 | 447 | 1.2 | 0.6 | 8 | 0.096 |
| SGI Origin3800 | TU Dresden | 128 | n.a. | 0.8 | 0.8 | 32 | 8 |
| HITACHI SR8000 | LRZ Munich | 168 × 8 | 1645 | 1.5 | 4.0 | 128 | — |

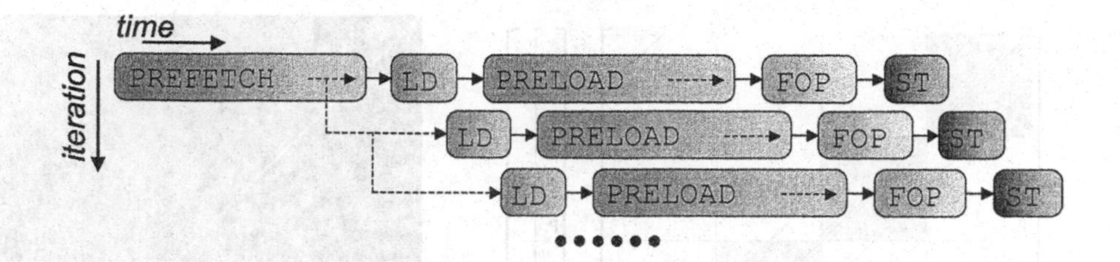


**Fig. 2.** Diagrammatic view of Pseudo-Vector-Processing for indirect memory access

the processor avoiding the penalties of memory latency. Combining `prefetch` and `preload` instructions, even vector-gather operations such as the indirect access to vector `x` in the MVM (cf. Sect. 2.1) can be executed in a vector processor like manner (cf. Fig. 2). The `prefetch` transfers a full cache-line (16 or 32 data items depending on the data type) of the index array (`col_ind`) to the L1 cache followed by `load` instructions. For each entry of the cache-line non-blocking `preload` instructions are issued. After paying only once the penalty of memory latency a continuous stream of data from memory to registers is established even for non-contiguous memory access. Eight processors form a shared memory node with a memory bandwidth of 32 GByte/s which is basically the same as for a single NEC SX5e processor. In contrast to the NEC SX5e this bandwidth holds for load and store operations. Hardware support for fast collective thread operations is provided. Therefore the system can efficiently be used either as a vector-like computer running one process comprising eight threads on one node or as an MPP assigning one process to each processor.

### 3.2 Benchmark Matrices

Two different types of matrices have been chosen for the performance evaluation. As a first example, we consider a modeling of disordered solid by a finite-dimensional Anderson Hamiltonian matrix [12]

$$H = \sum_i \epsilon_i |i\rangle\langle i| - t \sum_{\langle i,j\rangle} |i\rangle\langle j| \tag{5}$$

defined on a D-dimensional hypercubic lattice, where $|i\rangle$ denote tight-binding states situated at lattice point $i$ and $\langle i,j\rangle$ means nearest neighbours only. The on-site energies $\{\epsilon\}$ are assumed to be independent, identically distributed random variables. During the last five decades, the Anderson Hamiltonian (AH) has been intensively studied as a generic model for the localization of electronic states due to disorder effects. Working in the one-particle sector, the matrix structure of the AH is identical to the seven point discretisation of the differential operator on a three dimensional Cartesian grid (periodic boundary conditions) and will be

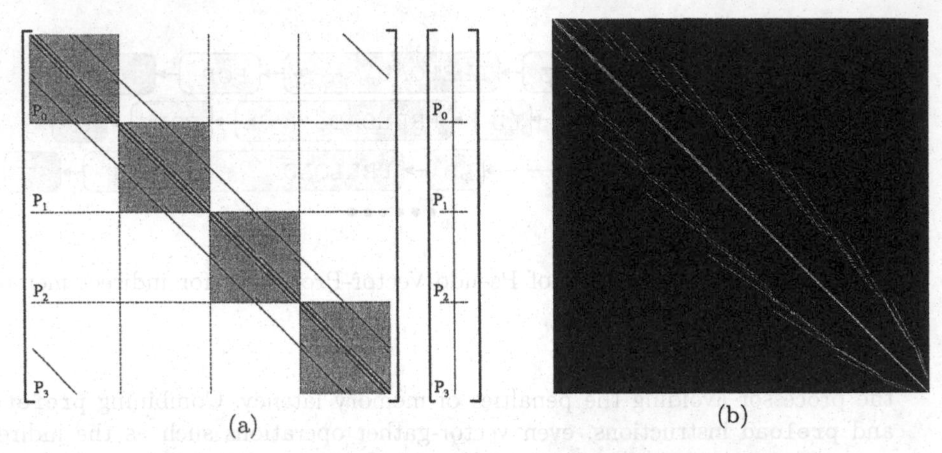


**Fig. 3.** (a) Structure and data distribution for `AH matrix` with $D_{\text{Mat}} = 8^3$. (b) Sparsity pattern of `HHM matrix` ($D_{\text{Mat}} = 0.5 \times 10^6$) with $8.6 \times 10^6$ nonzero matrix entries

denoted as `AH matrix` in the following. If the conversion of cartesian coordinates ($\{i, j, k\}$) to a linear index $l$ is defined as follows

$$\{i, j, k\} \longrightarrow l = i + (j-1) \cdot n_i + (k-1) \cdot n_i \cdot n_j \quad (6)$$
$$(i = 1, \ldots, n_i \; ; \; j = 1, \ldots, n_j \; ; \; k = 1, \ldots, n_k)$$

and parallelisation using $n_{\text{proc}}$ MPI processes is done along the $k$-direction ($n_k^{\text{loc}} = n_k / n_{\text{proc}}$), a matrix structure and data distribution is generated as depicted in Fig.3(a). It is well known that diagonal storage techniques are ideally suited for this class of matrices. Nonetheless these matrices provide a valuable case study for our JDS MVM implementation because

- problem sizes and number of non-zero matrix entries ($7 \cdot n_i \cdot n_j \cdot n_k$) can linearly be scaled,
- the communication scheme is independent of problem sizes (nearest neighbour communication only),
- the number of floating point operations ($14 \cdot n_i \cdot n_j \cdot n_k^{\text{loc}}$) and amount of data transfer ($n_i \cdot n_j$ data items per direction) per process can easily be controlled.

As a second example, we consider the Holstein Hubbard Hamiltonian

$$H = -t \sum_{i,\sigma} (c_{i\sigma}^\dagger c_{i+1\sigma} + \text{H.c.}) + U \sum_i n_{i\uparrow} n_{i\downarrow} + g\omega_0 \sum_{i,\sigma} (b_i^\dagger + b_i) n_{i\sigma} + \omega_0 \sum_i b_i^\dagger b_i, \quad (7)$$

which can be taken as a fundamental model for the interaction of electron and lattice degrees of freedom in solids. Here $c_{i\sigma}^\dagger$ creates a spin-$\sigma$ electron at Wannier site $i$ ($n_{i,\sigma} = c_{i\sigma}^\dagger c_{i\sigma}$), $b_i^\dagger$ creates a local phonon of frequency $\omega_0$, $t$ denotes the hopping integral, $U$ is the on-site Hubbard repulsion, $g$ is a measure of the electron-phonon coupling strength. Because the Holstein Hubbard model (HHM) describes two types of particles (fermions and bosons) and their interactions, a much more complicated matrix structure results (see Fig. 3(b)). Although the number of non-zero entries per row varies, we choose a simple row-wise distribution for parallelisation at the cost of some load imbalance.

## 4 Performance Analysis

### 4.1 Sequential Performance

First the sequential performance as a function of the problem size $D_{\mathrm{Mat}}$ is discussed (see Fig. 4) using a `HHM matrix` which basically reflects the qualitative effects.
For the SGI Origin we recover the well known cache effect: At small problem sizes cached data can be reused giving a performance of about 110 MFlop/s. Consistent with the large L2 cache size the performance drops around $D_{\mathrm{Mat}} \sim 10^4 - 10^5$ and an asymptotic performance of about 35 MFlop/s is sustained for local memory access. If the matrix size is further increased ($D_{\mathrm{Mat}} > 2.5 \times 10^6$) **nonlocal** memory on remote compute nodes has to be allocated. Although the network could in principle sustain the same memory bandwidth for local and remote memory access, the performance decreases to 25 MFlop/s due to higher latencies for remote memory accesses. Since the problem sizes in Fig. 4 start at $D_{\mathrm{Mat}} \sim 10^3$ there is only a minor cache-effect for CRAY T3E and HITACHI SR8000 (small cache sizes). For the HITACHI SR8000 this picture changes completely if the PVP feature (cf. discussion in Sect. 3.1) is enabled. As a consequence of the vector-like processor characteristic, performance increases with increasing problem size and saturates at long loop lengths. With an asymptotic performance of about 150 MFlop/s, the HITACHI SR8000 processor is able to maintain a reasonable portion of its high memory-bandwidth and outperforms present-day RISC processors by a factor of roughly 5. Therefore PVP has always been enabled in

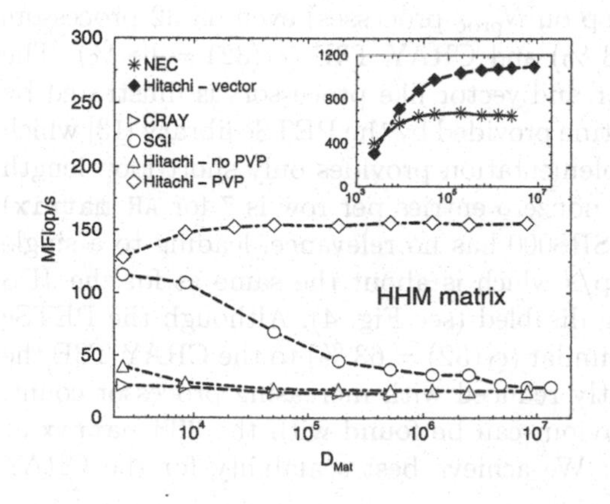


**Fig. 4.** Single processor performance as a function of the matrix dimension $D_{\mathrm{Mat}}$ (`HHM matrix`). The inset depicts a comparison of one NEC SX5e processor and one HITACHI SR8000 node running in **vector-mode**

what follows.
It is straightforward to use one HITACHI SR8000 node in **vector-mode** by enabling the automatic parallelisation compiler feature which generates a shared-memory parallelisation of the inner loop in the MVM (cf. Sect. 2.1). In the inset of Fig. 4 the MVM performance of one HITACHI SR8000 SMP node (eight processors) is shown together with the measurements for one NEC SX5e vector processor. Again the HITACHI SR8000 realizes a vector-like characteristic and saturates at a performance of about 1100 MFlop/s or 138 MFlop/s per processor. A decrease of less than 10 % compared to the single processor performance indicates only minor memory contention problems in the vector-mode. Surprisingly one SMP node exceeds the performance of the NEC SX5e at intermediate and large problem sizes although the maximum memory bandwidth is the same for both systems. Since both systems achieve approximately the same performance for vector triads (cf. Ref. [11]), this effect can not be attributed to the higher `store` bandwidth of the HITACHI SR8000 node, which is twice the `store` bandwidth of the NEC SX5e processor (cf. discussion in Sect. 3.1). Obviously the indirect load can be performed more efficient by the PVP feature than by a vector-gather instruction. However, at small problem sizes the vector processor outperforms the SMP node because of short vector pipeline start-up times.

### 4.2 MPP Programming Model

At the next step we measure the scalability of code and platforms within the **MPP-mode**. Here we assign one MPI process to each processor and thus we use eight MPI processes per node for the HITACHI SR8000. In order to minimize inter-node communication a continuous assignment of processes to nodes is chosen on the HITACHI SR8000.
As a starting point, basic questions are addressed in Fig.5(a) using `AH matrix`. Even though a rather small matrix size is considered, we find reasonable scalability measured by the parallel efficiency

$$\epsilon_1(N_{\mathrm{proc}}) = T(1)/(N_{\mathrm{proc}} \cdot T(N_{\mathrm{proc}})) \tag{8}$$

($T(N_{\mathrm{proc}})$ is the time per MVM step on $N_{\mathrm{proc}}$ processes) even on 32 processors on HITACHI SR8000 ($\epsilon_1(32) = 53$ %) and CRAY T3E ($\epsilon_1(32) = 68$ %). The advantage of JDS format for vector and vector like processors is illustrated by comparison to the sparse MVM routine provided by the PETSc library [13] which uses CRS format. As the CRS implementation provides only short loop length for the innermost loop (number of nonzero entries per row is 7 for `AH matrix`) the PVP feature of the HITACHI SR8000 has no relevance, leading to a single processor performance of 32 MFlop/s which is about the same as for the JDS format with the PVP feature being disabled (see Fig. 4). Although the PETSc routine shows scalability numbers similar ($\epsilon_1(32) = 63$ %) to the CRAY T3E the performance gap can only be slightly reduced with increasing processor count. Basically the same scalability behaviour can be found with the `HHM matrix` at larger problem size (cf. Fig. 5(b)): We achieve best scalability for the CRAY

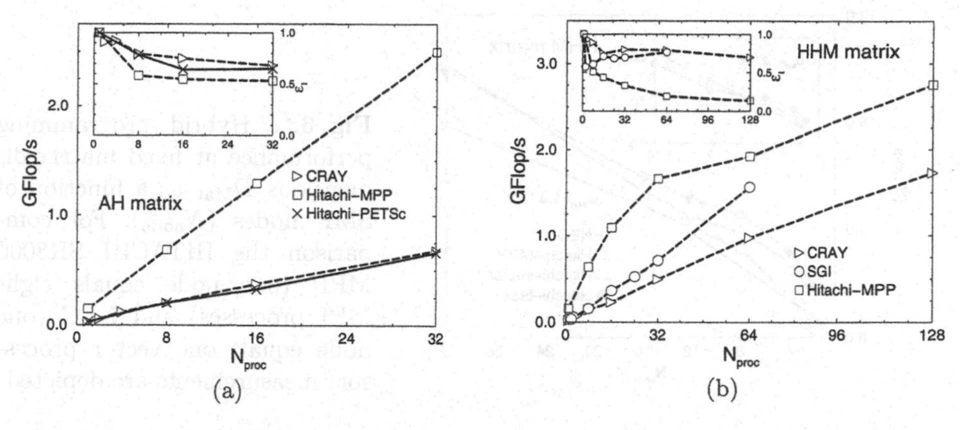


**Fig. 5.** (a) Performance of JDS implementation and PETSc for **AH matrix** with $D_{\mathrm{Mat}} = 64^3 = 0.26 \times 10^6$. (b) Scalability of the **MPP-mode** for different systems ($D_{\mathrm{Mat}} = 1.2 \times 10^6$). Both insets show the parallel efficiencies

T3E system where a parallel efficiency $\epsilon_1$ of more than 80 % can be maintained up to 64 processors and even with 256 processors 72 % is still exceeded. Due to the higher single processor performance (cf. Fig. 4) the SGI Origin outperforms the CRAY T3E with small numbers of processors. Interestingly this performance gap widens at intermediate processor counts accompanied by an increase in the parallel efficiency for the Origin system [14]. This fact can be attributed to the large amount of aggregate L2 cache size with increasing number of processors involved in the MVM, e.g. the total L2 cache of 16 processors can hold all nonzero elements of the matrix used in Fig. 5(b).
The worst scaling is again provided by the HITACHI SR8000. Even when intranode communication is used only (up to $N_{\mathrm{proc}} = 8$), a sharp drop in the efficiency occurs. Going beyond eight processors, both intra- and inter-node communication is used and a steady decrease in efficiency was measured. Mainly two reasons account for the poor scalability of the HITACHI SR8000: First the ratio between inter-node communication bandwidth (1 GByte/s) and local memory bandwidth (32 GByte/s) is not as balanced as for the CRAY T3E and SGI Origin systems. Second, the MPI performance suffers a drop if a large number of outstanding messages has to be governed [15]. Nevertheless, the HITACHI SR8000 provides the best absolute performance values due to the excellent single processor performance. At fixed problem size, however, the SGI Origin benefits from the scalability both of interconnect and code as well as the aggregate L2 cache and narrows the gap with increasing processor count.

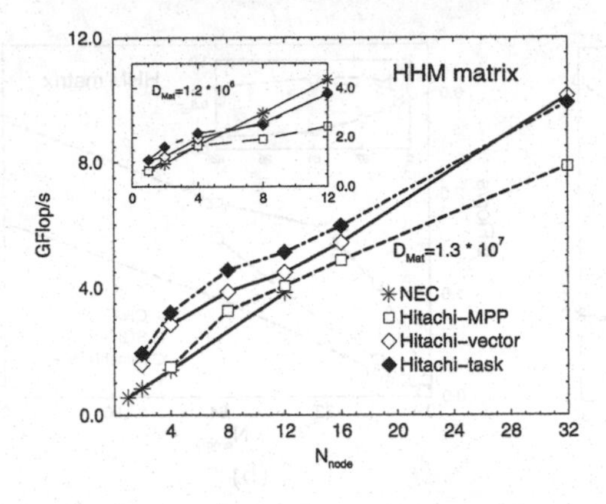


**Fig. 6.** Hybrid programming performance at fixed matrix dimensions $D_{\mathrm{Mat}}$ as a function of SMP nodes ($N_{\mathbf{node}}$). For comparison the HITACHI SR8000 MPP (one node equals eight MPI processes) and NEC (one node equals one vector processor) measurements are depicted

## 4.3 Hybrid Programming Model

It is obvious to improve the HITACHI SR8000 performance by reducing the scalability problems. The two hybrid parallel programming approaches (**vector-mode** and **task-mode**) introduced in Sect. 2.3 seem to be feasible in this context. Using the parameters of Fig. 5(b) a comparison between **MPP-mode** and hybrid modes as well as NEC SX5e is depicted in the inset of Fig. 6. We find that the hybrid modes ($\epsilon_1(12) \approx 26$ %) scale significantly better than the **MPP-mode** ($\epsilon_1(12) = 16.9$ %) resulting in a performance improvement of up to 50 % at 12 nodes. The same holds even if the problem size is increased by a factor of 10 (cf. Fig. 6). Obviously the hybrid modes benefit both from the reduction of the number of outstanding MPI messages and from an improvement in load balancing by an increase in the number of local rows per MPI process. Best scalability, however, is found for the NEC SX5e system where the MPI communication is done within a shared-memory. Achieving parallel efficiencies of more than 50 % up to 12 processors, the NEC SX5e can close or even overcome the single processor performance gap (cf. Fig. 4).

A comparison of the two hybrid modes qualitatively confirms the discussion in Sect. 2.3: The performance gain of the **task-mode** is most pronounced at low and intermediate node counts and vanishes at a large number of nodes. Increasing the problem size widens the performance gap and shifts the crossover to higher node counts.

The qualitative picture as described in the context of `HHM matrices` also holds for `AH matrices` (cf. Fig. 7), whereas the absolute performance figures for a fixed matrix size significantly increase at intermediate and large node counts due to the reduced complexity in the `AH matrix` structure. Moreover, the `AH matrices` allow a detailed analysis of the performance ratio $\epsilon_{\mathrm{T,V}}$ of **task-mode** to **vector-mode** since the dependency between between $T_{\mathrm{COMM}}/T_{\mathrm{COMP}}$ and the matrix parameters is given by the ratio of floating point operations per process to

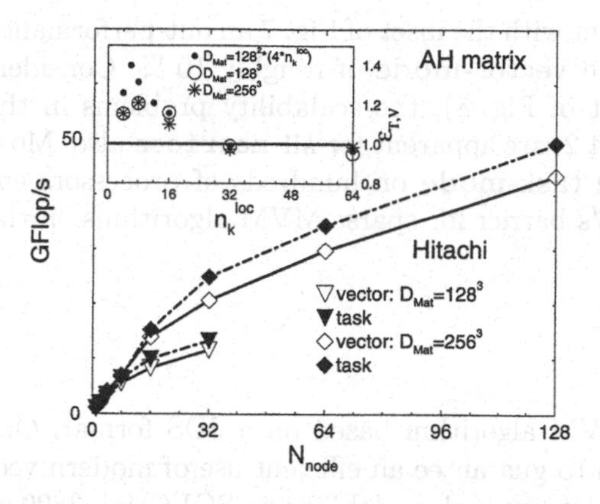


**Fig. 7.** Hybrid programming performance on HITACHI SR8000 (**AH matrix**) as a function of SMP nodes ($N_{\mathrm{node}}$). The corresponding performance ratio $\epsilon_{\mathrm{T,V}}$ is depicted in the inset together with additional data for a fixed number of nodes ($N_{\mathrm{node}}$=4) but varying $n_k^{\mathrm{loc}}$ as well as the crossover line $\epsilon_{\mathrm{T,V}} = 1$ (dotted line)

amount of data transfer per process (cf. Sect. 3.2.) and can be written as

$$\frac{T_{\mathrm{COMM}}}{T_{\mathrm{COMP}}} \propto \frac{n_i \cdot n_j}{n_i \cdot n_j \cdot n_k^{\mathrm{loc}}} = \frac{1}{n_k^{\mathrm{loc}}}. \tag{9}$$

Following Eq.(3) $\epsilon_{\mathrm{T,V}}$ should mainly be dominated by the variation of the local workload per process ($n_k^{\mathrm{loc}}$). This result is in good agreement with data shown in the inset of Fig. 7 where $\epsilon_{\mathrm{T,V}}$ is nearly independent of matrix size or number of nodes being used. Thus we are able to locate a fixed crossover point which determines when to use **task-mode** ($n_k^{\mathrm{loc}} \leq 32$) or **vector-mode** ($n_k^{\mathrm{loc}} \geq 32$). Furthermore we find that speed-up values of typically 15–30 % are achieved with the **task-mode** which is in the range given by Eq. (4). Finally, we clearly demonstrate in Fig. 8 that only the hybrid modes provide reasonable performance for large scale matrices used in our work. In this context a scalability study up to 1024 processors (128 nodes) has been performed on the HITACHI SR8000 using a fixed workload per processor with matrix dimensions ranging from $D_{\mathrm{Mat}} = 2.62 \times 10^5$ to $D_{\mathrm{Mat}} = 2.68 \times 10^8$. For the hybrid modes we

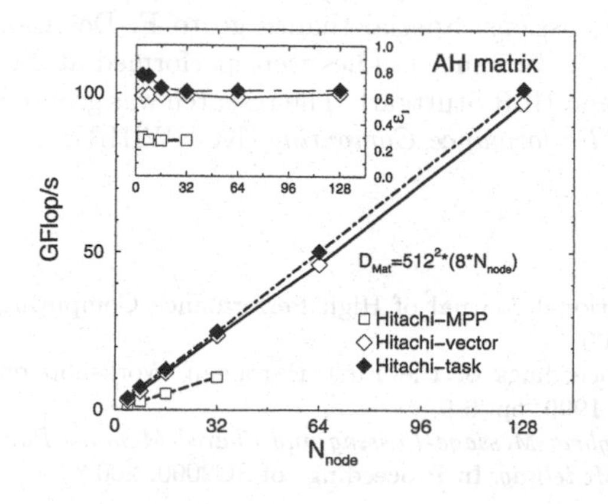


**Fig. 8.** Performance and parallel efficiency on the HITACHI SR8000 (**AH matrix**) as a function of SMP nodes ($N_{\mathrm{node}}$) using a fixed problem size per processor ($n_i = n_j = 512$; $n_k^{\mathrm{loc}} = 1(8)$ for **MPP-mode** (hybrid modes))

choose $n_k^{\rm loc} = 8$ and find in agreement with the inset of Fig. 7 an out-performance of the **task-mode** compared to the **vector-mode** of roughly 10 %. Considering the parallel efficiency (cf. inset of Fig. 8), the scalability problems in the **MPP-mode** as discussed in Sect. 4.2. are apparent for `AH matrices` also. Most notably the high scalability of the **task-mode** on hundreds of processors enables us the exceed the 100 GFlop/s barrier for sparse MVM algorithms on the HITACHI SR8000.

## 5 Conclusions

We presented a scalable sparse MVM algorithm based on a JDS format. Our parallel implementation was shown to guarantee an efficient use of modern vector computers (NEC SX5e) and to achieve high scalability on SGI Origin3800 as well as CRAY T3E systems. However, the main focus was on two "hybrid programming model" implementations optimized for SMP based supercomputers. As an example for an SMP system the HITACHI SR8000 TeraFlop/s system at LRZ Munich was chosen. We demonstrated that the "hybrid programming model" generally outperforms the MPP model for the matrices used in our work. Furthermore it was substantiated that an explicit thread programming technique simulating an asynchronous data transfer achieves best performance for MVM problems with a reasonable ratio of communication to computation. Besides the hybrid features with respect to architecture and programming model, another hybrid characteristic was clearly demonstrated for the HITACHI SR8000: Although based on RISC technology one SMP node can exceed the performance of comparable present-day vector processors (NEC SX5e) for vectorizable loops even if a vector-gather operation is involved. Combining hybrid features of the HITACHI SR8000 with "hybrid programming model" is the method of choice to achieve high performance for large scale sparse MVM algorithms.

## Acknowledgements

The authors would like to thank M. Brehm, U. Küster, T. Lanfear, R. Rabenseifner and R. Wolff for helpful discussions. Special thanks go to F. Deserno, M. Kinateder and F. Wagner. The performance studies were performed at TU Dresden, NIC Jülich, LRZ Munich and HLR Stuttgart. The research was granted by the *Bavarian Network for High Performance Computing* (KONWIHR).

## References

[1] S. W. Bova *et al.*, The International Journal of High Performance Computing Applications, **14**, pp. 49-60, 2000.
[2] L. A. Smith and P. Kent, Proceedings of the First European Workshop on OpenMP, Lund, Sweden, Sept. 1999, pp. 6-9.
[3] D. S. Henty, *Performance of Hybrid Message-Passing and Shared-Memory Parallelism for Discrete Element Modelling.* In Proceedings of SC2000, 2000.

[4] H. Shan *et al.*, *A Comparison of Three Programming Models for Adaptive Applications on the Origin2000.* In Proceedings of SC2000, 2000.
[5] W.D. Gropp *et al.*, *Performance Modeling and Tuning of an Unstructured Mesh CF Application.* In Proceedings of SC2000, 2000.
[6] R. Rabenseifner, *Communication Bandwidth of Parallel Programming Models on Hybrid Architectures.* To be published in the proceedings of WOMPEI 2002, Kansai Science City, Japan. LNCS 2327.
[7] Available at `http://www.top500.org`.
[8] J. Dongarra *et al.*, *Iterative Solver Benchmark*, available at `http://www.netlib.org/benchmark/sparsebench/`.
[9] R. Barrett *et al.*, *Templates for the Solution of Linear Systems: Building Blocks for Iterative Methods*, SIAM, Philadelphia (1994).
[10] M. Kinateder *et al.*, E. Krause and W. Jäger, eds.: High Performance Computing in Science and Engineering 2000, Springer, Berlin (2001), pp. 188–204.
[11] W. Schönauer, *Architecture and Use of Shared and Distributed Memory Parallel Computers*, eds.: W. Schönauer, ISBN 3-00-005484-7.
[12] P.W. Anderson, Phys. Rev. B **109**, 1492 (1958).
[13] Available at `http://www-fp.mcs.anl.gov/petsc`.
[14] G. Wellein *et al.*, *Exact Diagonalization of Large Sparse Matrices: A Challenge for Modern Supercomputers*, In Proceedings of CUG SUMMIT 2001, CD-ROM.
[15] M. Brehm, LRZ Munich, private communication.

# Performance Evaluation of Parallel Gram-Schmidt Re-orthogonalization Methods

Takahiro Katagiri[1,2]

[1] PRESTO, Japan Science and Technology Corporation (JST)
JST Katagiri Laboratory, Computer Centre Division, Information Technology Center
The University of Tokyo, 2-11-16 Yayoi, Bunkyo-ku, Tokyo 113-8658, JAPAN
[2] Department of Information Network Science
Graduate School of Information Systems, The University of Electro-Communications
1-5-1 Choufugaoka, Choufu-shi, Tokyo 182-8585, JAPAN
katagiri@is.uec.ac.jp
Phone: +81-424-43-5642, FAX: +81-424-43-5681

**Abstract.** In this paper, the performance of the five kinds of parallel re-orthogonalization methods by using the Gram-Schmidt (G-S) method is reported. Parallelization of the re-orthogonalization process depends on the implementation of G-S orthogonalization process, i.e. Classical G-S (CG-S) and Modified G-S (MG-S). To relax the parallelization problem, we propose a new hybrid method by using both the CG-S and MG-S. The HITACHI SR8000/MPP of 128 PEs, which is a distributed memory super-computer, is used in this performance evaluation.

## 1 Introduction

The orthogonalization process is one of the most important processes to perform several linear algebra computations, such as eigen decomposition and QR decomposition [3, 10, 11]. Many researchers have paid more attention to QR decomposition [2, 5, 6, 12, 13]. Notwithstanding, we focus on the re-orthogonalization process in this paper. This is because a lot of iterative methods for solving linear equations and eigenvector computations need the re-orthogonalization process to maintain accuracy of results. For instance, the GMRES method, which is one of the latest iterative algorithms, requires the re-orthogonalization process to obtain base-vectors according to the number of its re-start frequency [4]. The approach of parallel re-orthogonalization is the main difference with respect to conventional parallel approaches based on QR decomposition [6, 12, 13].

Recently, there are a number of parallel machines, because parallel processing technologies have been established. For this reason, many researchers try to parallelize the orthogonalization process. It is known that, however, the parallelism of the process depends on the number of orthogonalized vectors we needed [7, 8]. The orthogonalized process, therefore, is classified as the following two kinds of processes in this paper:

- QR decomposition
  The orthogonalization process to obtain the normalized and orthogonalized vectors $q_1, q_2, \cdots, q_n$ by using the normalized vectors $a_1, a_2, \cdots, a_n$.

J.M.L.M. Palma et al. (Eds.): VECPAR 2002, LNCS 2565, pp. 302–314, 2003.

– Re-orthogonalization
  The orthogonalization process to obtain the normalized and orthogonalized vector $q_i$ by using the normalized and orthogonalized vectors $q_1, q_2, \cdots, q_{i-1}$.

This paper discusses how to parallelize the re-orthogonalization process by using the Gram-Schmidt (G-S) orthogonalization method. The focus is especially the parallelization of Classical G-S (CG-S) method, since the CG-S method has high parallelism. The CG-S method, however, has a trade-off problem between accuracy and execution speed. T.Katagiri reported that more than 90% execution time is wasted by using MG-S method in a parallel eigen solver[8]. For this reason, the main aim of this paper is a proposition of a new parallel G-S algorithm to relax the trade-off problem.

This paper is organized as follows. Chapter 2 is the explanation of sequential G-S algorithms in the viewpoint of data dependency. Chapter 3 describes parallel G-S algorithms based on the sequential G-S algorithms. Chapter 4 proposes the new algorithm. Chapter 5 is the evaluation of the new algorithm by using the HITACHI's parallel super-computer. Finally, Chapter 6 summarizes the findings of this paper.

# 2 Sequential Algorithms of G-S Re-orthogonalization Method

It is widely known that there are the following two methods to perform re-orthogonalization by using the G-S method. They are known as Classical G-S (CG-S) method and Modified G-S (MG-S) method. The CG-S method is a simply implemented method in the formula of the G-S orthogonalization, and the MG-S method is a modified one in order to obtain high accuracy. In this chapter, we will explain the difference between them from the viewpoint of data dependency.

## 2.1 Classical Gram-Schmidt (CG-S) Method

The CG-S method to perform the re-orthogonalization is shown in Figure 1. The notation of $(\cdot, \cdot)$ in Figure 1 means an inner product.

Figure 1 shows that the most inner kernel $\langle 3 \rangle$ has a parallelism. The nature of this can be explained as the parallelization of the inner product $(q_j^T, a_i)$, since the inner product can be performed in parallel for $j$-loop when we obtain the initial vector of $a_i$.

$\langle 1 \rangle$ $q_i = a_i$
$\langle 2 \rangle$ **do** $j = 1, i-1$
$\langle 3 \rangle$ $\quad q_i = q_i - (q_j^T, a_i) q_j$
$\langle 4 \rangle$ **enddo**
$\langle 5 \rangle$ Normalization of $q_i$.

**Fig. 1.** The CG-S method in re-orthogonalization to the vector $q_i$

$\langle 1\rangle$ $a_i^{(0)} = a_i$
$\langle 2\rangle$ **do** $j = 1, i-1$
$\langle 3\rangle$ $a_i^{(j)} = a_i^{(j-1)} - (q_j^T, a_i^{(j-1)})q_j$
$\langle 4\rangle$ **enddo**
$\langle 5\rangle$ Normalization of $q_i = a_i^{(i-1)}$.

**Fig. 2.** The MG-S method in re-orthogonalization to the vector $q_i$

### 2.2 Modified Gram-Schmidt (MG-S) Method

The MG-S method in re-orthogonalization is shown in Figure 2.

Figure 2 shows that there is no parallelism to the inner product of $\langle 3\rangle$, since the inner product depends on the defined vector $a_i^{(j-1)}$ for $j$-loop. This is the basic difference with the CG-S method. For this reason, many researchers believe that the re-orthogonalization by the MG-S method is an unsuitable method for parallel processing.

## 3 Parallelization of G-S Orthogonalization Method

With parallel processing, the re-orthogonalization by the G-S method behaves differently according to the data distribution method for the orthogonalized vectors $q_1, q_2, ..., q_{i-1}$ [8]. We will explain this in this chapter.

### 3.1 Column-Wise Distribution

First of all, we explain a simple distribution method, named column-wise distribution (CWD). The CWD is a distribution that the whole elements of the normalized vector $a_i$ and the normalized and orthogonalized vectors $q_1, q_2, ..., q_{i-1}$ are distributed to each PE (Processing Element).

Next is a discussion on how to implement parallel algorithms based on the CWD.

**CG-S Method** Figure 3 shows a re-orthogonalization algorithm in CWD. The notation of "Local" in Figure 3 shows that the following formula should be performed by using local data (distributed data) only.

According to Figure 3, there is a parallelism for computing the kernel of $\langle 9\rangle$.

**MG-S Method** Figure 4 shows a re-orthogonalization algorithm in CWD.

Please note that there is no parallelism for computing the kernel of $\langle 4\rangle$ according to Figure 3, because PE holds $a_i^{(j-1)}$ and PE holds $a_i^{(j)}$ are located in different places in CWD.

For this reason, the re-orthogonalization by MG-S method in CWD has basically poor parallelism.

```
⟨1⟩  q_i = a_i
⟨2⟩  if (I have a_i) then
⟨3⟩    Broadcast (a_i)
⟨4⟩  else
⟨5⟩    receive (a_i)
⟨6⟩    q_i = 0
⟨7⟩  endif
⟨8⟩  do j = 1, i − 1
⟨9⟩    Local q_i = q_i − (q_j^T, a_i) q_j enddo
⟨10⟩ if (I have a_i) then
⟨11⟩   do j = 1, i − 1
⟨12⟩     receive (q_j) from PE that holds q_j
⟨13⟩     Local q_i = q_i + q_j enddo
⟨14⟩ else
⟨15⟩   send (q_i)
⟨16⟩ endif
⟨17⟩ Normalization of q_i.
```

**Fig. 3.** The CG-S method in column-wise distribution

```
⟨1⟩ a_i^(0) = a_i
⟨2⟩ do j = 1, i − 1
⟨3⟩   receive (a_i^(j−1)) from PE that holds a_i^(j−1)
⟨4⟩   Local a_i^(j) = a_i^(j−1) − (q_j^T, a_i^(j−1)) q_j
⟨5⟩   send (a_i^(j)) to PE that holds a_i^(j+1)
⟨6⟩ enddo
⟨7⟩ Normalization of q_i = a_i^(i−1).
```

**Fig. 4.** The MG-S method in column-wise distribution

### 3.2 Row-Wise Distribution

Next we will explain another well-known distribution, namely row-wise distribution (RWD). The RWD is a distribution in which the elements of vectors ($a_i$, $q_i$) and orthogonalized vectors $q_1, q_2, ..., q_{i-1}$ are distributed to different PEs. Please note that the difference between CWD and RWD is that the CWD does not distribute the elements of vectors.

In RWD, to calculate the inner products of $(q_j^T, a_i^{(j-1)})$ and $(q_j^T, a_i)$, we need a scalar reduction operation in both of CG-S and MG-S methods for the RWD, since each PE does not have whole elements to calculate the inner products.

**CG-S Method** Figure 5 shows the parallel re-orthogonalization algorithm of the CG-S method. The notation of "Global sum" in Figure 5 is the collective communication operation, which sums up distributed data and then distributes

⟨1⟩ $q_i = a_i$
⟨2⟩ **do** $j = 1, i-1$
⟨3⟩ Local $(q_j^T, a_i)$ **enddo**
⟨4⟩ Global sum of $\eta_j = (q_j^T, a_i)$.
$(j = 1, \cdots, i-1)$
⟨5⟩ **do** $j = 1, i-1$
⟨6⟩ Local $q_i = q_i - \eta_j \, q_j$
⟨7⟩ **enddo**
⟨8⟩ Normalization of $q_i$.

**Fig. 5.** The CG-S method in row-wise distribution

⟨1⟩ $a_i^{(0)} = a_i$
⟨2⟩ **do** $j = 1, i-1$
⟨3⟩ Local $(q_j^T, a_i^{(j-1)})$
⟨4⟩ Global sum of $\eta = (q_j^T, a_i^{(j-1)})$.
⟨5⟩ Local $a_i^{(j)} = a_i^{(j-1)} - \eta \, q_j$
⟨6⟩ **enddo**
⟨7⟩ Normalization of $q_i = a_i^{(i-1)}$.

**Fig. 6.** The MG-S method in row-wise distribution

the result to all PEs. The collective communication can be implemented by using the `MPI_ALLREDUCE` function on the MPI (Message Passing Interface).

For Figure 5, we find that the Global sum operation is not needed in every step because the CG-S method does not require the inner product value calculated in the most inner loop. This fact also can be found in the explanation of Chapter 2, since there is no dependency in the direction of the $j$-loop. The inner product value is easily calculated by using the value before normalization. The vector length of the Global sum operation depends on the number of orthogonalized vectors $i-1$.

**MG-S Method** Figure 6 shows the parallel re-orthogonalization algorithm for the MG-S method.

For Figure 6, we find that a scalar reduction for distributed data around PEs is essential in every steps, since the Global sum operation is located in the innermost loop. In $j+1$-th step, the MG-S method needs the inner product data calculated $j$-th step, thus we have to implement the Global sum operation in the most inner loop. The explanation in Chapter 2 also shows the fact, since there is a flow dependency in the direction of the $j$-loop.

**Comparison of the Number of Communications** Now we compare the number of execution times for the Global sum operation in the MG-S and CG-S methods. Let the number of re-orthogonalization $i$ be fixed as $k$. It is clear that the MG-S method requires $k - 1$ times of the Global sum operation while the CG-S method requires 1 times. Thus, from the view point of execution times, the CG-S method is superior to the MG-S method. The CG-S method, however, has lower accuracy than the accuracy of the MG-S method. Therefore, we can not determine the best parallel method in the RWD.

From the viewpoint of these variations, the RWD will be a good choice. The discussion on the RWD, however, is omitted in this paper. The reason is that the distributing the elements of vectors makes the implementation difficult in many iterative algorithms and poor utility for many users.

# 4 Proposal of the Hybrid Gram-Schmidt (HG-S) Method

In this chapter, we propose a new parallel re-orthogonalization method, named Hybrid Gram-Schmidt (HG-S) method.

## 4.1 Basic Idea

We have explained that there is a trade-off problem between accuracy and parallel execution speed in G-S method. The following is a summary of the problem for CWD.

- **CG-S Method** Accuracy: Low, Parallelism: Yes
- **MG-S Method** Accuracy: High, Parallelism: None

To relax the problem, using both CG-S and MG-S methods for re-orthogonalization is one of the reasonable ways. We call this "Hybrid"-ed G-S method as HG-S method in this paper.

## 4.2 The HG-S Re-orthogonalization Method

Figure 7 shows a sketch of the algorithm for the HG-S method. Please note that the code of Figure 7 shows the nature of re-orthogonalization process, since many iterative algorithms, for example the inverse iteration method for eigenvector computation, needs the re-orthogonalization like Figure 7.

Please note that we select MG-S method after CG-S method to obtain high accuracy, since the HG-S method in Figure 7 can use MG-S orthogonalized vectors in the process of CG-S method.

# 5 Experimental Results

The HG-S method using a distributed memory parallel machine is evaluated in this chapter. To evaluate the HG-S method, we implemented the HG-S method to Inverse Iteration Method (IIM) for computation of eigenvectors.

```
do i = 1,...,max_num_vectors
....
do iter = 1,...,max_iter
....
Re-orthogonalization (q_i) by using the CG-S Method.
....
if (q_i is converged) break the iter-loop.
enddo
Re-orthogonalization (q_i) by using the MG-S Method.
....
enddo
```

**Fig. 7.** The HG-S method in re-orthogonalization. The notations of *max_num_vectors* and *max_iter* mean the number of vectors to orthogonalize, and the maximal number of iterations in the iterative method, respectively

### 5.1 Experimental Environment

The HITACHI SR8000/MPP is used in this experiment. The HITACHI SR8000/MPP system is a distributed memory, message-passing parallel machine of the MIMD class. It is composed of 144 nodes, each having 8 Instruction Processors (IPs), 14.4GFLOPS of theoretical peak performance and 16 Gigabytes of main memory, interconnected via a communication network with the topology of a three-dimensional hyper-crossbar. The peak interprocessor communications bandwidths are 1.6 Gbytes/s in one-way, and 3.2 Gbytes/s in both-way.

The SR8000/MPP system has two types of parallel environments. One is intra-node parallel processing, and the other is inter-node parallel processing. The intra-node parallel processing is so-called parallel processing as a shared memory parallel machine. On the other hand, the inter-node parallel processing is similar to parallel processing as a distributed memory parallel machine, and it should perform interprocessor communications. The HITACHI Optimized Fortran90 V01-04 compiler is used. In this experiment, *-opt=ss -parallel=0* is used as a compile option in the inter-node parallel processing.

As communication library, optimized MPI (Message Passing Interface) by HITACHI is used in this experiment.

### 5.2 Details of Implemented Re-orthogonalization Methods

We implemented parallel re-orthogonalization methods in IIM for eigenvector computation [10]. The details of the IIM algorithm are summarized as follows.

- Object Matrix : Tridiagonal real symmetric matrix $T$
- Method : The Rayleigh quotient IIM
- Max Iterative Numbers : 40

- Requirement residual : $||Tx_i - \lambda_i x_i||_2 \leq ||T||_1 \times \epsilon$, $(i = 1, 2, ..., n)$, where $\epsilon$ is a machine epsilon, $x_i$ is the $i$-th eigenvector, and $\lambda_i$ is the $i$-th eigenvalue in an eigensystem.

We have developed an efficient parallel eigensolver by using a reduced communication method [9]. The IIM method is also available for the eigensolver, therefore, we measure the execution time for the routine of IIM.

The test matrices are chosen as the following.

- Matrix(1) : Tridiagonal matrix reduced from the Frank matrix
  - Dimension : 10,000
  - The length to decide the clustered eigenvalues : 58471.532
  - The number of groups for the clustered eigenvalues : 8
  - The maximum number of clustered eigenvalues : 9,993
- Matrix(2) : Glued Wilkinson $W_{21}^{+}$ Matrix, $\delta = 1.0$
  - Dimension : 10,000
  - The length to decide the clustered eigenvalues : 64.998
  - The number of groups for the clustered eigenvalues : 1
  - The maximum number of clustered eigenvalues : 10,000
- Matrix(3) : Glued Wilkinson $W_{21}^{+}$ Matrix, $\delta = 10^{-14}$
  - Dimension : 10,000
  - The length to decide the clustered eigenvalues (for a block diagonal matrix) : 0.129
  - The number of groups for the clustered eigenvalues (for a block diagonal matrix) : 13
  - The maximum number of clustered eigenvalues (for a block diagonal matrix) : 2
  - The number of block diagonal matrices : 477

Please note that the re-orthogonalization is performed to maintain accuracy of eigenvectors. The number of re-orthogonalized eigenvectors is the same as the number of clustered eigenvalues.

Next, we summarize the implemented parallel re-orthogonalization methods as follows.

- CG-S (1) : Applying MG-S as local orthogonalizations
- CG-S (2) : Applying CG-S as local orthogonalizations
- MG-S
- HG-S
- IRCG-S : Iterative Refinement CG-S Method [1, 10]
- NoOrt : Not Re-orthogonalized

There are two kinds of methods in the parallel re-orthogonalization of CG-S method. The CG-S(1) is a method in which CG-S is performed inter PEs, and MG-S is performed to intra PE data. The CG-S(2) is a method in which CG-S is performed inter PEs, and CG-S is also performed to intra PE data.

The IRCG-S method shown above is called Iterative Refinement CG-S Method, and this method performs CG-S method multiple times. In this experiment, the number of CG-S method is fixed as 2 times.

**Table 1.** The execution time of each method in the IIM. The unit is in seconds

(a) Test Matrix(1): Reduced Frank Matrix

| #PEs(IPs) | CG-S(1) | CG-S(2) | MG-S | HG-S | IRCG-S | NoOrt |
|---|---|---|---|---|---|---|
| 8 | 6604 | 6527 | 38854 | 6604 | 12883 | 23.857 |
| 16 | 3646 | 3526 | 24398 | 3574 | 6987 | 12.065 |
| 32 | 2061 | 2059 | 28050 | 2082 | 3906 | 7.044 |
| 64 | 1633 | 1559 | 27960 | 1614 | 3059 | 3.025 |
| 128 | 2091 | 2204 | >14400 | 2027 | 3978 | 1.808 |

(b) Test Matrix(2): Glued Wilkinson Matrix $\delta$ =1.0

| #PEs(IPs) | CG-S(1) | CG-S(2) | MG-S | HG-S | IRCG-S | NoOrt |
|---|---|---|---|---|---|---|
| 8 | 7779 | 7898 | 77851 | 10825 | 14673 | 30.88 |
| 16 | 4358 | 4264 | 61242 | 4348 | 7595 | 15.31 |
| 32 | 2533 | 2347 | 32131 | 2480 | 4429 | 8.99 |
| 64 | 2152 | 2110 | >28800 | 2189 | 3241 | 4.44 |
| 128 | 2696 | 2390 | >14400 | 2450 | 4699 | 2.45 |

(c) Test Matrix(3): Glued Wilkinson Matrix $\delta = 10^{-14}$

| #PEs(IPs) | CG-S(1) | CG-S(2) | MG-S | HG-S | IRCG-S | NoOrt |
|---|---|---|---|---|---|---|
| 8 | 0.558 | 0.561 | 0.510 | 0.593 | 2.494 | 0.59 |
| 16 | 0.554 | 0.696 | 1.922 | 1.697 | 0.619 | 0.33 |
| 32 | 0.364 | 0.828 | 0.970 | 0.293 | 0.858 | 0.20 |
| 64 | 0.530 | 0.451 | 0.521 | 0.431 | 0.378 | 0.38 |
| 128 | 0.277 | 0.221 | 0.315 | 0.269 | 0.259 | 0.07 |

### 5.3 The Results of Experiment

Table 1 shows the execution time for the five parallel re-orthogonalization methods in IIM. For the restriction of super-computer environment, over $\lfloor 64/\lceil p/8 \rceil \rfloor$ hours jobs are automatically killed, where $p$ is the number of PEs.

Table 2 shows the orthogonalization accuracy of the calculated eigenvectors by IIM, which is measured by the Frobenius norm. The Frobenius norm of $n \times n$ matrix $A = (a_{ij})$, $i, j = 1, ..., n$ is defined as:

$$||A\,||_F \equiv \sqrt{\sum_{i=1}^{n}\sum_{j=1}^{n} a_{ij}^2}. \qquad (1)$$

### 5.4 Discussion

**Test Matrix (1)** Table 1 (a) and Figure 8(a) shows there was poor parallelizm for MG-S. The reason is explained as no parallelizm in MG-S process in

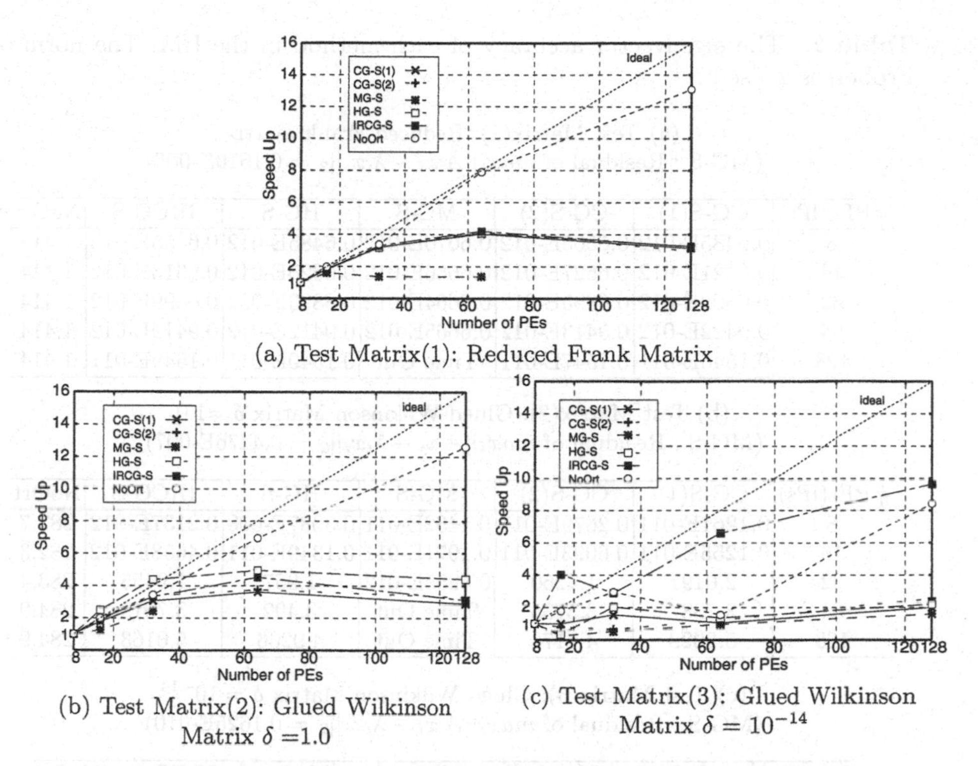


(a) Test Matrix(1): Reduced Frank Matrix

(b) Test Matrix(2): Glued Wilkinson Matrix $\delta = 1.0$

(c) Test Matrix(3): Glued Wilkinson Matrix $\delta = 10^{-14}$

**Fig. 8.** The speed-up factors of each method in the IIM. The factors are scaled on the execution time of 8 PEs

CWD. On the other hand, by using CG-S Method, we obtained speed-up factors increasing the number of PEs.

For HG-S method, the HG-S was faster than CG-S. This is a surprising result in a sense, because HG-S performs multi-times CG-S and one-time MG-S. We think the main reason is that by using the HG-S method, total number of iterations in IIM is reduced by using better accuracy eigenvalues with comparison to eigenvectors orthogonalized by CG-S only.

In the case of the Frank matrix, we could not find the difference of accuracy for eigenvectors except for not-orthogonalized cases. For this reason, HG-S is the best method in this case.

**Test Matrix (2)** First of all, we have to mention that the accuracy of eigenvectors was very poor when we use over 32 PEs except for MG-S method. For this reason, we should select the MG-S method in this case. As a result, if we use 16 PEs and 8 PEs, the methods of CG-S(1), CG-S(2) and HG-S will be considerable methods in this case.

For the execution time, we also had a surprising result for MG-S, because although the MG-S method has no parallelizm in this case, there are speed-up

**Table 2.** The eigenvector accuracy of each method in the IIM. The norm of Frobenius is used

(a) Test Matrix(1): Reduced Frank matrix
(MG-S : Residual of $max_i||A\ x_i - \lambda_i x_i||_2$ = 0.1616E-006)

| #PEs(IPs) | CG-S(1) | CG-S(2) | MG-S | HG-S | IRCG-S | NoOrt |
|---|---|---|---|---|---|---|
| 8 | 0.6485E-012 | 0.6505E-012 | 0.6670E-012 | 0.6485E-012 | 0.6455E-012 | 1.414 |
| 16 | 0.6621E-012 | 0.6627E-012 | 0.6666E-012 | 0.6620E-012 | 0.6613E-012 | 1.414 |
| 32 | 0.6892E-012 | 0.6895E-012 | 0.6664E-012 | 0.6893E-012 | 0.6899E-012 | 1.414 |
| 64 | 0.9422E-012 | 0.9413E-012 | 0.6665E-012 | 0.9412E-012 | 0.9419E-012 | 1.414 |
| 128 | 0.1546E-011 | 0.1547E-011 | Time Out | 0.1540E-011 | 0.1549E-011 | 1.414 |

(b) Test Matrix(2): Glued Wilkinson Matrix $\delta$ =1.0
(MG-S : Residual of $max_i||A\ x_i - \lambda_i x_i||_2$ = 0.4476E-007)

| #PEs(IPs) | CG-S(1) | CG-S(2) | MG-S | HG-S | IRCG-S | NoOrt |
|---|---|---|---|---|---|---|
| 8 | 0.1261E-011 | 0.2679E-011 | 0.1432E-011 | 0.3087E-008 | 0.2137E-012 | 283.7 |
| 16 | 0.1255E-011 | 0.6093E-011 | 0.1971E-011 | 0.1349E-011 | 0.4658E-012 | 282.3 |
| 32 | 2.0191 | 1.4260 | 0.5255E-012 | 1.967 | 1.9455 | 283.4 |
| 64 | 3.7387 | 3.5735 | Time Out | 3.492 | 3.7619 | 284.9 |
| 128 | 5.2028 | 4.7178 | Time Out | 4.9206 | 5.0163 | 284.9 |

(c) Test Matrix(3): Glued Wilkinson Matrix $\delta = 10^{-14}$
(MG-S : Residual of $max_i||A\ x_i - \lambda_i x_i||_2$ = 0.1625E-010)

| #PEs(IPs) | CG-S(1) | CG-S(2) | MG-S | HG-S | IRCG-S | NoOrt |
|---|---|---|---|---|---|---|
| 8 | 0.3966E-007 | 31.50 | 0.3966E-007 | 0.3966E-007 | 0.3966E-007 | 31.50 |
| 16 | 0.3966E-007 | 31.50 | 0.3966E-007 | 0.3966E-007 | 0.3966E-007 | 31.50 |
| 32 | 0.3967E-007 | 31.38 | 0.3966E-007 | 0.3967E-007 | 0.3967E-007 | 31.50 |
| 64 | 0.3966E-007 | 31.31 | 0.3966E-007 | 0.3966E-007 | 0.3966E-007 | 31.50 |
| 128 | 0.3967E-007 | 31.06 | 0.3966E-007 | 0.3967E-007 | 0.3967E-007 | 31.50 |

factors of MG-S. We think that the result is caused by reducing data amount by parallel processing, however, detailed analysis for this result is a future work.

On the other hand, we found the factors of super linear speed-up in HG-S method according to Figure 8(b) (16 and 32 PEs). Analyzing the reason is also a future work, but the acceleration of convergence by using HG-S method in IIM may be a considerable reason.

**Test Matrix (3)** In this case, the execution time of each method is quite small compared to the cases of the matrix (1) and (2). This reason is explained as the implementation of parallel IIM routine. In our IIM routine, if an element of sub-diagonals of the tridiagonal matrix $T$ is of quite small value such as machine $\epsilon$, the diagonal matrices separated by the element are treated as independent matrices of the eigensystem. This separation can dramatically reduce computational complexity and increase parallelizm in this matrix. In the case of matrix (3), the test matrix is separated by several small diagonal matrices, each has a dimension of 21. For this reason, the number of re-orthogonalizations is reduced, naturally and the execution time is also very shortened.

As for the accuracy of the results, we could find that the accuracy of CG-S(2) method was poor in this matrix with comparison to CG-S(1) method. For this result, therefore, we can say that data which holds intra PE should be re-orthogonalized by using MG-S method, even CG-S method is used for inter PEs.

**Accuracy of All Matrices** The accuracy in this experiment varies according to the number of PEs. This is also a surprising result, because the computations of each G-S method are same. We think that the result is caused by changing the order for the computation of G-S method, however, detail analysis is a part of future work.

## 6 Conclusion

In this paper, we have proposed a hybrid Gram-Schmidt re-orthogonalized method, and evaluated the five kinds of parallel re-orthogonalization methods including the new hybrid method by using IIM for eigenvector computation.

For the experimental result, the new hybrid method also has an accuracy problem for computed eigenvectors, but it has a benefit for parallel execution time compared to CG-S method. The main reason of this is using higher accuracy eigenvectors than that of CG-S orthogonalized, however, detailed analysis is needed.

The analysis and additional experiments for the test matrices are important future work.

## Acknowledgments

I would like to express my sincere thanks to staff at Computer Centre Division, Information Technology Center, the University of Tokyo, for supporting my super-computer environments. I would also like to express my sincere thanks to all members at Kanada Laboratory, Information Technology Center, for giving me useful discussions for the study.

This study is supported by PRESTO, Japan Science and Technology Corporation (JST).

## References

[1] S. Balay, W. Gropp, L. C. McInnes, and B. Smith. Petsc 2.0 users manual, 1995. ANL-95/11 - Revision 2.0.24, http://www-fp.mcs.anl.gov/petsc/.

[2] C. Bischof and C. van Loan. The wy representation for products of householder matrices. *SIAM J. Sci. Stat. Comput.*, 8(1):s2–s13, 1987.

[3] J. W. Demmel. *Applied Numerical Linear Algebra.* SIAM, 1997.
[4] J. J. Dongarra, I. S. Duff, D. C. Sorensen, and H. A. van der Vorst. *Numerical Linear Algebra for High-Performance Computers.* SIAM, 1998.
[5] J. J. Dongarra and R. A. van de Geijn. Reduction to condensed form for the eigenvalue problem on distributed memory architectures. *Parallel Computing*, 18:973–982, 1992.
[6] B. A. Hendrickson and D. E. Womble. The tours-wrap mapping for dense matrix calculation on massively parallel computers. *SIAM Sci. Comput.*, 15(5):1201–1226, 1994.
[7] T. Katagiri. A study on parallel implementation of large scale eigenproblem solver for distributed memory architecture parallel machines. *Master's Degree Thesis, the Department of Information Science, the University of Tokyo*, 1998.
[8] T. Katagiri. A study on large scale eigensolvers for distributed memory parallel machines. *Ph.D Thesis, the Department of Information Science, the University of Tokyo*, 2000.
[9] T. Katagiri and Y. Kanada. An efficient implementation of parallel eigenvalue computation for massively parallel processing. *Parallel Computing*, 27:1831–1845, 2001.
[10] B. N. Parlett. *The Symmetric Eigenvalue Problem.* SIAM, 1997.
[11] G. W. Stewart. *Matrix Algorithms Volume II:Eigensystems.* SIAM, 2001.
[12] D. Vanderstraeten. A parallel block gram-schmidt algorithm with controlled loss of orthogonality. *Proceedings of the Ninth SIAM Conference on Parallel Processing for Scientific Computing*, 1999.
[13] Y. Yamamoto, M. Igai, and K. Naono. A new algorithm for accurate computation of eigenvectors on shared-memory parallel processors. *Proceedings of Joint Symposium on Parallel Processing (JSPP)'2000*, pages 19–26, 2000. in Japanese.

# Toward Memory-Efficient Linear Solvers

Allison Baker, John Dennis, and Elizabeth R. Jessup

University of Colorado, Boulder, CO 80302, USA
{allison.baker,john.dennis,jessup}@cs.colorado.edu
http://www.cs.colorado.edu/~jessup

**Abstract.** We describe a new technique for solving a sparse linear system $Ax = b$ as a block system $AX = B$, where multiple starting vectors and right-hand sides are chosen so as to accelerate convergence. Efficiency is gained by reusing the matrix $A$ in block operations with $X$ and $B$. Techniques for reducing the cost of the extra matrix-vector operations are presented.

## 1 Introduction

The problem of solving very large sparse systems of linear equations arises, for example, in the application of implicit approximation methods for the solution of nonlinear partial differential equations (PDEs). In many cases, solving the linear systems is the most time consuming part of the overall application [1, 2]. Because PDEs play a role in evaluating physical models from such diverse areas as acoustic scattering, aerodynamics, combustion, global circulation, radiation transport, and structural analysis [3, 4, 1], fast linear system solvers are of broad importance.

In this paper, we report on our study of methods for writing efficient iterative linear solvers. While competing linear algebra algorithms have traditionally been compared in terms of floating-point operation costs, it has long been known that reducing memory access costs is essential to attaining good performance [5]. In particular, codes that can be based on the level 2 and 3 BLAS [6, 7] often achieve near peak performance because they allow substantial reuse of data at fast levels of the memory hierarchy. How to attain such performance for general applications is a more difficult question, and how to answer it is becoming increasingly difficult. That is, while the performance of microprocessors has been improving at the rate of 60% a year, the performance of DRAM (dynamic random access memory) has been improving at a rate of only 10% [8]. In addition, advances in algorithm development mean that the total number of floating-point operations required for solution of many problems is decreasing [9].

For these reasons, we focus on mechanisms for making efficient use of the memory hierarchy for computers with a single processor. Such techniques form an increasingly important area of high-performance computing. Although we will not discuss issues of parallelization per se, the overall performance in a parallel program is highly dependent on the performance of its serial components [10]. In addition, effective programming of some innovative parallel architectures may

J.M.L.M. Palma et al. (Eds.): VECPAR 2002, LNCS 2565, pp. 315–327, 2003.

require attention to some issues of memory hierarchy: on IBM's Power4 chip, two processors share a unified second level cache [11]. In addition, blocking of data to fit in the cache resembles blocking of data for distribution between processors of a parallel computer. Techniques for reduction of traffic between components of the memory hierarchy resemble techniques for reduction of data transfer between parallel processors. Thus, we expect our results to have direct applicability to parallel solvers.

We are primarily interested in the solution of the linear system $Ax = b$, where $A$ is a large, sparse, square matrix, $b$ is a single right-hand side, and $x$ is a single solution vector. Performance tuning can be carried out for such a system by means of blocking the sparse matrix operations [12, 13], but we are investigating alternatives based on solving the block linear system $AX = B$ where multiple starting vectors and right-hand sides are chosen so as to accelerate convergence. Efficiency is gained by reusing the matrix $A$ in block operations with $X$ and $B$. The cost of the extra matrix-vector operations will ultimately be reduced via innovative programming techniques.

The ultimate goal of our work is to produce a robust, memory-efficient sparse linear solver for inclusion in Argonne National Laboratory's Portable, Extensible Toolkit for Scientific Computation (PETSc) [12]. In this paper, we report on our progress toward that goal. In Section 2, we review the GMRES method for solving sparse linear systems. We also cover relevant acceleration techniques for that method. In Section 3, we describe a new acceleration method and present performance results for a Matlab prototype. In Section 4, we show how our method can be used to translate the matrix-vector system $Ax = b$ into a block system $AX = B$. In Section 5, we present the fast matrix-multivector multiply routine that will be the basis of our efficient solver. In Section 6, we briefly summarize our results.

## 2 Background

When the coefficient matrix $A$ is sparse, iterative methods are often chosen to solve the linear system $Ax = b$. A popular class of iterative methods are Krylov subspace methods. Krylov methods find an approximate solution:

$$x_i \in x_0 + K_i(A, r_0), \tag{1}$$

where $K_i(A, r_0) = span\{r_0, Ar_0, \ldots, A^{i-1}r_0\}$ denotes an $i$-dimensional Krylov subspace, $x_0$ is the initial guess, and $r_0$ is the initial residual ($r_0 = b - Ax_0$). Krylov methods are also known as polynomial methods since equation (1) implies that the residual $r_i$ can be written in terms of a polynomial of $A$: $r_i = p(A)r_0$.

When $A$ is Hermitian, the Conjugate Gradient method [14] is popular, efficient, and well-understood. However, for non-Hermitian $A$, the choice of algorithm is not as clear (e.g. [15]), though GMRES is arguably the best known method for non-Hermitian linear systems and is a popular choice. In particular, GMRES and modifications to GMRES (augmented methods, hybrid methods, and acceleration techniques, for example) are well-represented in the literature.

Currently, we are studying GMRES without preconditioners as we gain understanding of the method and its block counterpart. However, we recognize that the choice in preconditioner will play an important role in a practical implementation.

## 2.1 GMRES

GMRES is called a minimum residual method because it selects an $x_i \in x_0 + K_i(A, r_0)$ such that $\|b - A(x_0 + z)\|_2$ is a minimum over all $z \in K_i(A, r_0)$. The minimum residual requirement is equivalent to the condition $r_i \perp AK_i(A, r_0)$.

At each iteration, GMRES generates one dimension of an orthonormal basis for the Krylov subspace $K_i(A, r_0)$. The residual norm, $\|r_i\|_2$, is minimized via a least squares problem. The GMRES algorithm is efficient and stable, and is guaranteed to converge in at most $n$ ($A \in \mathcal{R}^{n \times n}$) steps. See [16] for complete details.

At each iteration of GMRES the amount of storage and computational work required increases. Therefore the standard GMRES algorithm is not practical in most cases, and a restarted version of the algorithm is used as suggested in [16]. In restarted GMRES (GMRES($m$)), the maximum dimension $m$ of the Krylov subspace is fixed. At the end of $m$ iterations, if convergence has not occurred, the algorithm restarts with $x_0 = x_m$. Parameter $m$ is generally chosen small relative to $n$ to keep storage and computation requirements reasonable. Note that GMRES($m$) is only guaranteed to converge if $A$ is positive definite (in which case it converges for any $m \geq 1$) [17].

## 2.2 Existing Modifications to GMRES

In this section, we briefly describe some existing modifications to the standard GMRES algorithm. These modifications all have the common goal of enhancing the robustness of restarted GMRES, given that full GMRES is often not computationally feasible. These modifications are of interest to us in terms of understanding various factors that affect convergence. In studying them, we gain insight into choosing appropriate additional right-hand sides to turn our single right-hand side linear system into a block system. We consider the two primary categories of modifications closest to our work: acceleration techniques and hybrid methods.

The first in the acceleration category are augmented methods. These methods seek to accelerate convergence and avoid stalling by improving information in GMRES at the time of the restart. A nearly A-invariant subspace is appended to the Krylov approximation space, resulting in an "augmented Krylov subspace" [18]. This appended subspace is typically an approximation to the invariant subspace of $A$ associated with the smallest eigenvalues. Eigenvector estimates that correspond to the smallest eigenvalues are typically used, as those eigenvalues are thought to hinder convergence. Eigenvalue estimates are easily obtained from the Arnoldi process that underlies GMRES. Recent papers that advocate including spectral information at the restart are [19], [20], [21], and [18].

Also of interest is the suggestion that convergence may be improved for $Ax = b$ by using a block method with approximate eigenvectors for the additional right-hand side vectors [18]. This is probably the first mention of using a block linear system to solve a single right-hand side system. However, the authors conclude that eigenvalue deflation on a single right-hand side system may work just as well or better than a block method. We have experimented with block techniques using approximate eigenvectors as right-hand sides and have not found them to be promising. Building a Krylov space off of an eigenvector does not make sense in exact arithmetic and results in basis vectors being determined purely by rounding error in finite precision.

Augmentation techniques are more suitable for some types of problems than others. They may have little effect on highly non-normal problems [18], or the eigenvalue problem may be too costly for the technique to be beneficial (if the approximate eigenvectors are not converging) [19]. Other approaches to deflating eigenvalues for GMRES are described in [22] and [23]. Both of these approaches construct a preconditioner to deflate the small eigenvalues from the spectrum of $A$.

Other acceleration techniques are based on the observation that the approximation space should ideally contain the correction $c$ such that $x = x_0 + c$ is the exact solution to the problem [24]. The GMRESR method [25] is one such technique that uses an outer GCR method and an inner iterative method (like GMRES for example) at step $i$ to approximate the solution to $Ac = r_i$, where $r_i$ is the current residual at step $i$. The approximate solution to $Ac = r_i$ then becomes the next direction for the outer approximation space. The goal of this method is to mimic the convergence of full GMRES with less computational cost (under certain conditions).

Another acceleration technique is the GCRO method by de Sturler [26]. His aim is twofold: to compensate for the information that is lost due to restarting as well as to overcome some of the stalling problems that GMRESR can experience in the inner iteration. GCRO is a modification to GMRESR such that the inner iterative method maintains orthogonality to the outer approximation space. Thus the approximation from the inner iteration at step $i$ takes into account both the inner and outer approximation spaces. See also [27] for more details on preserving orthogonality in the inner iteration of a nested Krylov method. Additionally, in a subsequent paper, de Sturler [28] describes GCROT, which is a truncated version of GCRO. GCROT attempts to determine which subspace of the outer approximation space should be retained for the best convergence of future iterations as well as if any portion of the inner Krylov subspace should also be kept. In [24], Eiermann, Ernst and Schneider present a thorough overview and analysis of the most common acceleration techniques.

Hybrid algorithms constitute the other class of algorithms of interest. Hybrid algorithms typically consist of 2 phases. Phase 1 consists of a method that does not require prior information about the matrix $A$ but rather provides information on the spectrum of $A$ upon completion (like GMRES). Phase 2 then typically consists of some sort of parameter-dependent polynomial iteration method that is "cheaper" than phase 1. There are many varieties of hybrid algorithms that use GMRES in phase 1; several are described in [29].

One of the simplest algorithms is that presented by Nachtigal, Reichel, and Trefethen in [29]. In phase 1, GMRES is run until the residual norm drops by a specified amount. The resulting GMRES residual polynomial is then reapplied cyclically (via a cyclic Richardson iteration) until convergence. Similar algorithms are described in [30, 31].

The main purpose of a hybrid algorithm based on GMRES is to improve the performance of GMRES($m$) (particularly when $m$ is large) by reducing the number of vector-vector operations. At the same time, hybrid methods typically increase the number of matrix-vector operations. Therefore, the performance of such methods is dependent on the machine architecture and the problem size as the number of times a large matrix is accessed from memory substantially impacts performance.

## 3 A New Algorithm

In this section, we present a new technique for accelerating GMRES($m$). We first describe the new algorithm and its implementation. We close with some performance results for a Matlab prototype.

A well-known drawback of restarted GMRES is that orthogonality to previously generated approximation spaces is neglected at each restart. In fact, GMRES($m$) can stall as a result. Stalling means that there is no decrease in the residual norm at the end of a restart cycle. Many of the algorithms in the previous section seek to overcome stalling by either appending approximations to eigenvectors corresponding to small eigenvalues or maintaining orthogonality to parts of previous approximation spaces.

We call the new algorithm LGMRES. The implementation of LGMRES requires minimal changes to the standard restarted GMRES algorithm. The new algorithm is similar in concept to both GMRESR [25] and GCRO [26] in that it includes previous approximations to the error in the new approximation space and, in this way, maintains some orthogonality to previous Krylov subspaces. While LGMRES is not mathematically equivalent to either algorithm, it is most similar to GCRO. However, the implementation is quite different.

The primary motivation for this new method is that the idea and implementation easily lend themselves to a block method for solving a single right-hand side system which is our ultimate goal.

### 3.1 LGMRES: Idea and Implementation

Recall from Section 2 that augmented methods provide a simple framework for appending (non-Krylov) vectors to the approximation space. The idea is then to append vectors to the approximation space that in some sense represent the approximation spaces from previous restart cycles. After restart cycle $i$ we have that $r_i \perp AK_m(A, r_{i-1})$. Ideally we would also have that $r_i \perp AK_m(A, r_{i-2})$, but, unfortunately, GMRES($m$) discards $K_m(A, r_{i-2})$ entirely.

Suppose that $x$ is the true solution to $Ax = b$. The error after the $i$-th restart cycle is denoted by:

$$e_i \equiv x - x_i \tag{2}$$

Because we have $x_i \in x_{i-1} + K_m(A, r_{i-1})$, we can denote

$$z_i \equiv x_i - x_{i-1} \tag{3}$$

as an approximation to the error at the $i$-th restart cycle. Note also that because $z_i \in K_m(A, r_{i-1})$, the error approximation is a natural choice of vector with which to augment our next approximation space $K_m(A, r_i)$ as it in some sense represents the Krylov subspace $K_m(A, r_{i-1})$ generated in the previous cycle. Furthermore, as explicitly pointed out in [24], if our approximation space contains the exact correction $c$ such that $x = x_i + c$, then we have solved the problem. Therefore, adding an approximation to $c$, where here $c = e_i$, to the approximation space is a reasonable strategy. Thus, the general idea of GMRES is like that of both GMRESR [25] and GCRO [26], but its implementation is like that of GMRES with eigenvectors (GMRES-E) as described in [19].

Therefore, LGMRES($m$, $k$) is a restarted augmented GMRES($m$) algorithm in which the approximation space is augmented with the $k$ previous approximations to the error. We can write LGMRES($m$, $k$) in the following way:

$$x_{i+1} = x_i + q_{i+1}^{m-1}(A) r_i + \sum_{j=i-k+1}^{i} \alpha_{ij} z_j \tag{4}$$

*Note:* $k = 0$ corresponds to standard GMRES($m$). In GMRES($m$), $q_{i+1}^{m-1}(\xi)$ is referred to as the *iteration* polynomial (degree $m-1$). The *residual* polynomial $p_{i+1}^{m}(\xi)$, on the other hand, satisfies $r_{i+1} = p_{i+1}^{m}(A) r_i$. The two polynomials are related by:

$$q_{i+1}^{m-1}(\xi) = \frac{1 - p_{i+1}^{m}(\xi)}{\xi}. \tag{5}$$

The implementation of LGMRES($m$, $k$) is as follows. At each restart cycle $i$, we generate Krylov subspace $K_m(A, r_{i-1})$ and augment with the $k$ most recent error approximations ($z_j$, $j = (i-k+1) : i$) such that the augmented approximation space is size $m + k$ ($K_m(A, r_i) + span\{z_j\}_{j=(i-k+1):i}$). We then find the approximate solution from the augmented approximation space whose corresponding residual is minimized in the Euclidean norm. This implementation requires minimal changes to the standard GMRES($m$) implementation. For similar implementation details, see Morgan's GMRES-E in [19].

We note that when implementing LGMRES($m$, $k$), only $m$ matrix-vector multiplies are required per restart cycle, irrespective of the value of $k$, provided that we form both $z_j$ and $Az_j$ at the end of cycle $j$. Note that the latter does not require an explicit multiplication by $A$ and that we need only store at most $k$ pairs of $z_j$ and $Az_j$. Typically, the number of vectors appended, $k$, is much smaller than the restart parameter, $m$.

### 3.2 Empirical Tests

As noted, LGMRES acts as an accelerator for restarted GMRES. We tested a Matlab implementation of the algorithm using 17 different problems from the Matrix Market Collection [32] (primarily Harwell-Boeing and SPARSKIT matrices) and three convection-diffusion problems. All had random right-hand sides and zero initial guesses. For each we used 3 to 4 different values of $m$, resulting in a total of 61 different test problems. No preconditioning was used.

We compared of the number of matrix-vector multiplies to convergence required for GMRES($m$) and the "best" LGMRES($m - k$, $k$) for $k \leq 5$. Notice that parameters were chosen such that $m$ is the dimension of the approximation space for both algorithms. The algorithms were implemented in MATLAB and are problems for which GMRES($m$) doesn't stall "exactly," although it may converge with excruciating slowness. Generally, $k = 1$ or $k = 2$ is observed to be optimal for LGMRES($m - k$, $k$).

Figure 1 shows how the numbers of matrix-vector multiplies compare for the two methods. The y-axis is the number of matrix-vector multiplies needed for GMRES($m$) divided by the number for LGMRES. The x-axis corresponds to the 61 test problems. The bars extending above the x-axis favor LGMRES($m-k$, $k$) (55 cases)–in these cases GMRES($m$) requires more matrix-vector multiplies than does LGMRES. The bars below the x-axis favor GMRES($m$) (two cases). The remaining four cases are equal for both methods.

## 4 Extension to a Block Method

The natural extension of LGMRES to a block method is as follows. We assume that we are solving an $n \times n$ system $Ax = b$ via a block system $AX = B$. We use a block version of LGMRES which we denote BLGMRES, for "Block" LGMRES. In BLGMRES($m$, $k$), the parameter $k$ now indicates the number of previous error approximations ($z_j$) to be used as additional right-hand sides. Therefore, matrices $X$ and $B$ are of size $n \times (k + 1)$. Now our augmented approximation space consists of the traditional Krylov portion built by repeated application of the matrix $A$ to the current residual $r_i$ together with Krylov spaces resulting from the application of $A$ to previous error approximations. In other words we are finding

$$x_{i+1} \in x_i + K_m(A, r_i) + \sum_{j=i-k+1}^{i} K_m(A, z_j) \tag{6}$$

Comparing to LGMRES($m$, $k$) as given in (4), we can write BLGMRES($m$, $k$) as

$$x_{i+1} = x_i + q_{i+1}^{m-1}(A) r_i + \sum_{j=i-k+1}^{i} \alpha_{ij}(A) z_j \tag{7}$$

Notice that the coefficients $\alpha_{ij}$ of $z_j$ are now polynomials in $A$.

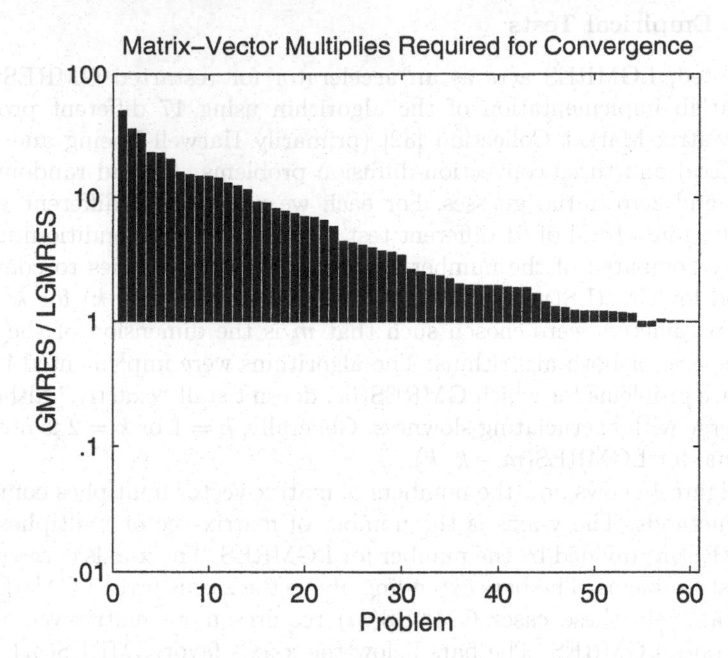


**Fig. 1.** A comparison of the matrix-vector multiplies required for convergence for LGMRES($m-k$, $k$) and GMRES($m$) in 61 test problems

As proof of concept, Table 1 lists the number of times matrix $A$ is accessed from memory (i.e., matrix accesses) for Matlab implementations of GMRES, BLGMRES($m$, 1), and LGMRES($m$, 1) for a few problems from the Matrix Market collection. For LGMRES and BLGMRES, the most recent error approximation vector $z_i$ is appended (i.e., $k = 1$). The accelerating properties of LGMRES($m$, 1) lead to a reduction in matrix accesses as compared to GMRES($m$). The block operations of BLGMRES($m$, 1) lead to further savings as compared to LGMRES($m$, 1). The effectiveness of these new algorithms depends on the matrix and on the restart parameter $m$, but the savings in matrix accesses are quite substantial in most cases. Note that the number of matrix accesses does not translate directly into time savings. In BLGMRES($m$, 1), each time matrix $A$ is accessed, it multiplies a block of vectors rather than the single vector multiplied in GMRES($m$) and LGMRES($m$, 1). In the next section, we discuss how a matrix-multivector multiply can be made efficient.

## 5 Memory-Efficient Techniques

The results of the preceding sections demonstrate that LGMRES is an effective accelerator for GMRES($m$) and that LGMRES lends itself to the translation of the matrix-vector system $Ax = b$ to the block system $AX = B$ via the

**Table 1.** Comparison of the number of accesses of $A$ required for convergence

| Matrix | size ($n$) | $m$ | GMRES($m$) | BLGMRES | LGMRES |
|---|---|---|---|---|---|
| sherman5 | 3312 | 10 | > 30000 | 8774 | > 30000 |
| | | 15 | > 30000 | 3639 | > 30000 |
| | | 20 | > 30000 | 7059 | 7031 |
| sherman4 | 1104 | 10 | 586 | 162 | 165 |
| | | 15 | 502 | 186 | 256 |
| | | 20 | 520 | 146 | 261 |
| sherman1 | 1000 | 10 | 6410 | 510 | 481 |
| | | 15 | 4060 | 459 | 526 |
| | | 20 | 3154 | 498 | 494 |
| cavity05 | 1182 | 10 | > 30000 | 2144 | 2961 |
| | | 15 | > 30000 | 2503 | 3618 |
| | | 20 | 15566 | 1215 | 4201 |
| cavity10 | 2597 | 10 | > 30000 | 3949 | 12177 |
| | | 15 | > 30000 | 2812 | 5864 |
| | | 20 | 15566 | 2675 | 4046 |
| orsirr_1 | 1030 | 10 | 9022 | 1256 | 1371 |
| | | 15 | 5855 | 1188 | 1276 |
| | | 20 | 5237 | 1032 | 1141 |
| steam_2 | 600 | 10 | > 30000 | 7926 | > 30000 |
| | | 15 | > 30000 | 788 | 3061 |
| | | 20 | 6036 | 243 | 1202 |

block implementation BLGMRES. We now visit the problem of efficient matrix-multivector multiplication. We begin with a review of the machine characteristics that drive an efficient implementation. We then describe our matrix-multivector routine and present some timing results.

### 5.1 The Memory Hierarchy

In the introduction to this paper, we review recent improvements in arithmetic and memory access costs. Advances in memory components themselves have also changed the performance picture. In particular, modern architectures all include multiple, high-speed caches to address the increasing cost of data transfer between processor and main memory. Current microprocessor designs typically have two to three levels of cache. The first level or L1 caches are typically 32 to 64 Kbytes in size and are accessible in 2 to 3 clock ticks. The L2 caches range from 96 Kbytes to 8 Mbytes in size and can have access times of 10 to 30 clock ticks. Increasingly common are L3 caches that range in size from 2 Mbytes to 32 Mbytes with latencies of 35 to 60 clock ticks. Main memories range in size from 128 Mbytes to 64 GBytes, and the memory to highest level of cache latencies are typically between 80 and 200 clock ticks. The highest level of cache is L2 or, if present, L3. (See, for example, [33, 11, 34, 35].) We confine our discussion to data movement between main memory and cache within the memory hierarchy, since this component is often the most costly in a typical computation.

Movement of data through the memory hierarchy occurs in units of cachelines. Common cacheline sizes range from 32-128 bytes. A program that accesses data items located next to each other in a cacheline in close succession is said to have good memory reference locality [36]. In a program with this property, data moved into cache are used as much as possible before they are evicted to make room for new data. The more efficiently a code uses a cacheline, the less the program's performance depends on the costs of moving data through the memory hierarchy.

Some scientific codes exhibit high degrees of memory reference locality. For example, dense linear algebra algorithms typically access all words in a cacheline in a predictable stride-1 manner. As a result, such codes have achieved excellent performance [37]. Unfortunately, not all important scientific codes have such neatly ordered memory access patterns. Applications that manipulate sparse matrices are an important class of problems in which the concept of reference locality is not always followed. These applications, which are generally based on a sparse matrix-vector multiply, typically access data structures though the use of indirect addressing. As a result, the location of the desired operand in memory is not known at compile time but rather is a function of another variable. Therefore, these codes typically don't exhibit much memory reference locality: a cacheline of four words may be read in from memory and only one word used before the line is evicted and replaced by another.

One possible solution to the poor memory behavior common to sparse linear solvers is the use of block algorithms. Block algorithms allow the multiplication of a multivector by a sparse matrix. And matrix-multivector multiplication is the routine on which the implementation of BLGMRES depends. The goal is to create an implementation of matrix-multivector multiplication in which the matrix $A$ is used as much as possible each time it is accessed. We therefore focus on how best to improve locality of reference of the multivector elements.

Kaushik [38] has developed an implementation with which the cost of multiplying a multivector of width four by a matrix is about 1.5 times the cost of calculating a single matrix-vector product on an SGI Origin 2000. We have reproduced this result by means of several common programming techniques. First, we modify the memory order of the operand vectors in order to maximize cacheline reuse. Instead of placing successive words in a single vector next to each other, the different vectors are interleaved. This interleaving places corresponding elements of each vector (i.e., $v1(j), v2(j), \ldots, vk(j)$) in the same cache line. Second, the loop across the vector elements used at one time must be hand unrolled to allow for compiler optimization. That is, the nested $ij$-loop that performs the series of scalar multiplications $a(i,j) * v1(j), a(i,j) * v2(j), \ldots, a(i,j) * vk(j)$ is converted to a single loop over the matrix rows. That loop is indexed by the loop $i$ by explicitly writing out the $k$ operations that constitute the inner $j$ loop.

We added a matrix-multivector multiply subroutine supporting both AIJ and AIJNODE matrix formats to a local installation of PETSc. The AIJ for-

**Table 2.** Comparative timings of matrix-vector and matrix-multivector multiplication

| matrix-vector multiply | matrix-multivector multiply | | matrix-vector multiply |
|---|---|---|---|
| 1 vector $Ax$ | 2 vectors $AX$ | 4 vectors $AX$ | for $i = 1 : 4$ $Ax_i$ |
| $21.81 \pm 2.77$ | $24.99 \pm 2.65$ | $29.73 \pm 3.05$ | $43.55 \pm 3.96$ |

mat is PETSc's name for the compressed sparse row format described in, for example, [17]. The AIJNODE is a variation of that format such that consecutive rows with identical non-zero blocks are grouped together. The latter format thus permits little blocks of dense matrix-vector multiplies inside the sparse system.

We timed the resulting codes on an Ultra Sparc II, which has a 400Mhz processor with an 8MB L2 cache running Solaris 8. In Table 2, the results of timing 22 matrices (14 using AIJ and 8 using AIJNODE) taken from Matrix Market are given for four different configurations. Timings are in microseconds and include mean and standard deviation. The first (leftmost) value in Table 2 represents the time to perform a single matrix-vector product. The second and third represent the time to perform matrix-multivector multiplies for two and four vectors, respectively. Finally the fourth value represents the time to perform four matrix-vector multiplies in serial. The average cost of performing a four-wide matrix-multivector product versus a single matrix-vector product is 1.36, a result that correlates well with the approximate value of 1.5 observed by Kaushik. In comparison, performing four matrix-vector multiplies in succession takes 2.00 times as long as performing a single matrix-vector multiply. Those results together show that reformulating four single matrix-vector multiplies as a single width-four matrix-multivector multiply represents a time savings of 43%.

## 6 Conclusion

In this paper, we describe the development of an accelerated variant of GMRES that lends itself to a block implementation. We further describe one important building block of a memory-efficient implementation of that block algorithm, namely a fast matrix-multivector multiplication routine. What remains is to put the pieces together as a robust code in a high-level language that will allow full performance evaluation of both algorithm and implementation.

## References


[1] Field, M.: Optimizing a parallel conjugate gradient solver. SIAM J. Sci. Stat. Comput. **19** (1998) 27–37

[2] Simon, H., Yeremin, A.: A new approach to construction of efficient iterative schemes for massively parallel applications: variable block CG and BiCG methods and variable block Arnoldi procedure. In R. Sincovec et al., ed.: Parallel Processing for Scientific Computing. (1993) 57–60

[3] Anderson, W. K., Gropp, W. D., Kaushik, D., Keyes, D. E., Smith, B. F.: Achieving high sustained performance in an unstructured mesh CFD application. In: Proceedings of Supercomputing 99. (1999) Also published as Mathematics and Computer Science Division, Argonne National Laboratory, Technical Report ANL/MCS-P776-0899

[4] Farhat, C., Macedo, A., Lesoinne, M.: A two-level domain decomposition method for the iterative solution of high frequency exterior Helmholtz problems. Numerische Mathematik **85** (2000) 283–308

[5] Dongarra, J., Hammarling, S., Sorensen, D.: Block reduction of matrices to condensed form for eigenvalue computations. J. Comp. Appl. Math. **27** (1989) 215–227

[6] Dongarra, J., , DuCroz, J., Hammarling, S., Hanson, R.: An extended set of Fortran Basic Linear Algebra Subprograms. ACM Trans. Math. Software **14** (1988) 1–17

[7] Dongarra, J., DuCroz, J., Duff, I., Hammarling, S.: A set of level 3 Basic Linear Algebra Subprograms. ACM TOMS **16** (1990) 1ff

[8] Patterson, D., Anderson, T., Cardwell, N., Fromm, R., Keeton, K., Kozyrakis, C., Thomas, R., Yelick, K.: A case for intelligent RAM. IEEE Micro **March/April** (1997) 34–44

[9] Gropp, W., Kaushik, D., Keyes, D., Smith, B.: Toward realistic performance bounds for implicit CFD codes. In the Proceedings of the International Conference on Parallel CFD (1999)

[10] Kaushik, D., Keyes, D.: Efficient parallelization of an unstructured grid solver: A memory-centric approach. In the Proceedings of the International Conference on Parallel CFD (1999)

[11] Behling, S., Bell, R., Farrell, P., Holthoff, H., O'Connell, F., Weir, W.: The POWER4 Processor Introduction and Tuning Guide. IBM Redbooks (2001)

[12] Gropp, W., et al.: PETSc 2.0 for MPI. http://www.mcs.anl.gov/petsc/ (1999)

[13] Basic Linear Algebra Subprograms Technical (BLAST) Forum: Document for the Basic Linear Algebra Subprograms (BLAS) standard. http://www.netlib.org/utk/papers/blast-forum.html (1998)

[14] Hestenes, M., Stiefel, E.: Methods of conjugate gradients for solving linear systems. J. Res. Nat. Bur. Stds. **49** (1952) 409–436

[15] Nachtigal, N. M., Reddy, S. C., Trefethen, L. N.: How fast are nonsymmetric matrix iterations? SIAM Journal on Matrix Analysis Applications **13** (1992) 778–795

[16] Saad, Y., Schultz, M.: GMRES: A generalized minimal residual algorithm for solving nonsymmetric linear systems. SIAM Journal on Scientific and Statistical Computing **7** (1986) 856–869

[17] Saad, Y.: Iterative Methods for Sparse Linear Systems. PWS Publishing Company (1996)

[18] Chapman, A., Saad, Y.: Deflated and augmented Krylov subspace techniques. Linear Algebra with Applications **4** (1997) 43–66
[19] Morgan, R. B.: A restarted GMRES method augmented with eigenvectors. SIAM Journal on Matrix Analysis and Applications **16** (1995) 1154–1171
[20] Morgan, R. B.: Implicitly restarted GMRES and Arnoldi methods for nonsymmetric systems of equations. SIAM Journal of Matrix Analysis and Applications **21** (2000) 1112–1135
[21] Saad, Y.: Analysis of augmented Krylov subspace methods. SIAM Journal on Matrix Analysis and Applications **18** (1997) 435–449
[22] Baglama, J., Calvetti, D., Golub, G., Reichel, L.: Adaptively preconditioned GMRES algorithms. SIAM Journal on Scientific Computing **20** (1998) 243–269
[23] Erhel, J., Burrage, K., Pohl, B.: Restarted GMRES preconditioned by deflation. Journal of Computational Applied Mathematics **69** (1996) 303–318
[24] Eiermann, M., Ernst, O. G., Schneider, O.: Analysis of acceleration strategies for restarted minimum residual methods. Journal of Computational and Applied Mathematics **123** (200) 261–292
[25] van der Vorst, H. A., Vuik, C.: GMRESR: a family of nested GMRES methods. Numerical Linear Algebra with Applications **1** (1994) 369–386
[26] de Sturler, E.: Nested Krylov methods based on GCR. Journal of Computational and Applied Mathematics **67** (1996) 15–41
[27] de Sturler, E., Fokkema, D.: Nested Krylov methods and preserving orthogonality. In Melson, N., Mantcuffel, T., McCormick, S., eds.: Sixth Copper Mountain Conference on Multigrid Methods. Part 1 of NASA conference Publication 3324, NASA (1993) 111–126
[28] de Sturler, E.: Truncation strategies for optimal Krylov subspace methods. SIAM Journal on Numerical Analysis **36** (1999) 864–889
[29] Nachitgal, N. M., Reichel, L., Trefethen, L. N.: A hybrid GMRES algorithm for nonsymmetric linear systems. SIAM Journal of Matrix Analysis Applications **13** (1992) 796–825
[30] Manteuffel, T. A., Starke, G.: On hybrid iterative methods for nonsymmetric systems of linear equations. Numerical Mathematics **73** (1996) 489–506
[31] Joubert, W.: A robust GMRES-base adaptive polynomial preconditioning algorithm for nonsymmetric linear systems. SIAM Journal on Scientific Computing **15** (1994) 427–439
[32] National Institute of Standards and Technology, Mathematical and Computational Sciences Division: Matrix Market. http://math.nist.gov/MatrixMarket (2002)
[33] S.Naffziger, Hammond, G.: The implementation of the next generation 64bitanium microprocessor. In: Proceedings of the IEEE International Solid-State Circuits Conference. (2002)
[34] Kessler, R. E., McLellan, E. J., Webb, D. A.: The alpha 21264 microprocessor architecture (2002) http://www.compaq.com/alphaserver/download/ev6chip.pdf.
[35] DeGelas, J.: Alphalinux: The penguin drives a Ferrari (2000) http://www.aceshardware.com/Spades
[36] Hennessey, J., Patterson, D.: Computer Architecture: A Quantitative Approach. 2nd edn. Morgan Kaufmann (1996)
[37] Dongarra, J., Bunch, J., Moler, C., Stewart, G.: LINPACK Users' Guide. SIAM Publications (1979)
[38] Gropp, W. D., Kaushik, D. K., Keyes, D. E., B. F.Smith: Toward realistic performance bounds for implicit CFD codes. In A. Ecer et al., ed.: Proceedings of Parallel CFD'99, Elsevier (1999)

# A Parallel Newton-GMRES Algorithm for Solving Large Scale Nonlinear Systems*

Jesús Peinado[1] and Antonio M. Vidal[2]

[1] Departamento de Sistemas Informáticos y Computación
[2] Universidad Politécnica de Valencia. Valencia, 46071, Spain
{jpeinado,avidal}@dsic.upv.es
Phone : +(34)-6-3877798, Fax: +(34)-6-3877359

**Abstract**. In this work we describe a portable sequential and parallel algorithm based on Newton's method, for solving nonlinear systems. We used the GMRES iterative method to solve the inner iteration. To control the inner iteration as much as possible and avoid the oversolving problem, we also parallelized several forcing term criterions. We implemented the parallel algorithms using the parallel numerical linear algebra library SCALAPACK based on the MPI environment. Experimental results have been obtained using a cluster of Pentium II PC's connected through a Myrinet network. To test our algorithms we used three different test problems, the H-Chandrasekhar problem, computing the intersection point of several hyper-surfaces, and the Extended Rosenbrock Problem. The latter requires some improvements for the method to work with structured sparse matrices and chaotic techniques. The algorithm obtained shows a good scalability in most cases. This work is included in a framework tool we are developing where, given a problem that implies solving a nonlinear system, the best nonlinear method must be chosen to solve the problem. The method we present here is one of the methods we implemented.

## 1 Introduction and Objectives

There are many engineering problems where the solution is achieved by solving a large scale nonlinear system; sometimes as a consequence of a discretization of a PDE; and sometimes as a mathematical model problem. In this work we describe a portable sequential and parallel algorithm to solve large scale nonlinear systems. The algorithm presented in this paper is based on Newton's method [9], using the GMRES [19] iterative method to solve the inner linear system. We also implemented several forcing terms criterions [5],[9] to avoid the oversolving [5],[9] problem. We used the analytic Jacobian Matrix when known, and finite difference techniques [9] to

* Work supported by Spanish CICYT. Project TIC 2000-1683-C03-03.

approximate the Jacobian Matrix otherwise. Our idea is to use as standard a method as possible. Our algorithms are designed for solving large problems.

Furthermore, our algorithms are efficient in terms of computational cost. We describe here one of the algorithms included in a framework tool we are developing where, given a nonlinear system of equations, the best parallel method to solve the system must be chosen. This framework was first introduced in [18].

In our experiments for this paper we have considered three different test problems: the H-Chandrasekhar problem detailed in [9],[14], computing the intersection point of several hyper-surfaces [17], and the Extended Rosenbrock Problem [13]. Each problem has a different way of evaluating the function and computing the Jacobian Matrix. In fact they are also different because the first one and the second one are dense problems, while the third is a sparse structured problem and requires some modification of the Newton-GMRES method to work with sparse matrices and chaotic techniques [4].

To compare our approach with other nonlinear solvers we used Powell's method, implemented in the MINPACK-1 [12] standard package. Powell's method is a robust general purpose method to solve nonlinear systems. We also compared our algorithm with a standard Parallel Newton method, which we implemented in [15],[16].

All our algorithms have been implemented using portable standard packages. In the serial algorithms we used the LAPACK [1] library, while the parallel codes make use of the SCALAPACK [3] and BLACS [21] libraries on top of the MPI [20] communication library. The exception to this is the algorithm to compute the sparse matrix-vector product in the Extended Rosenbrock problem. This algorithm has been implemented directly by us [17]. All the codes have been implemented using C++ and some routines use FORTRAN 77 algorithms.

Experimental results have been obtained using a cluster of Pentium II PC's connected through a Myrinet [11] network. However other machines could be used due to the portability of the packages and our code. We achieved good results and show that our algorithm is scalable.

We want to emphasize the behaviour of the algorithms using the cluster of PC's and the Myrinet network. This system is a good, cheap alternative to more expensive systems because the ratio performance to price is higher than when using classical MPP machines.

The rest of the paper is organized as follows: section 2 contains a brief description of the type of problem we are trying to solve, focusing on the Newton method using the GMRES method. We devote our attention on the GMRES iteration: describing a standard code we choose and giving several oversolving criterions. In section 3 we give a description of the parallel algorithm, with careful considerations on the problems that can appear as a consequence of using the SCALAPACK and PBLAS packages to parallelize algorithms that are very demanding with data distribution. In section 4 we describe three case studies. In section 5, the experimental results are presented. And finally section 6 contains our conclusions. Section 7 contains the acknowledgements and section 8 is the references section.

# 2 Solving Large Scale Nonlinear Systems. Newton and Newton-GMRES Methods

Solving large scale nonlinear systems is not a trivial task. There are several families of methods to solve this kind of problem. We cannot reference a "perfect" method to solve all the nonlinear systems, because each engineering, physics, or mathematical problem has different characteristics. Therefore we are trying to define a framework where we can analyze a given nonlinear system and give the "a priori" best method to solve the problem. As a part of that framework, we have implemented the most efficient working methods. In previous papers we presented the framework [18] and some of the methods [16],[15]. Now we want to present a Newton Iterative method as part of our work.

## 2.1 Newton's Method. Iterative Methods. The GMRES Iteration

A good method to solve the nonlinear system: $F(x) = 0, F \in \Re^n,\ x \in \Re^n$ is Newton's method, because it is powerful and has quadratic local convergence [9]. Newton's method is based on the following algorithm, where $J$ is the Jacobian Matrix and $F$ is the function whose root is to be found ($k$ is the iteration number):

$$J(x_k)s_k = -F(x_k), \quad x_{k+1} = x_k + s_k \quad J(x_k) \in \Re^{n \times n},\ s_k,\ F(x_k) \in \Re^n .$$

To solve the linear system that appears in each iteration, a method to solve a linear system must be used. We can use a direct method such as LU decomposition, which we studied in [15],[16]. Or we can use an iterative method. In this paper we focused on this latter case. The idea is that we want to approximate the solution of the Newton step using an iterative method. The nonlinear methods using this scheme are known *as Inexact Newton Methods*, [9] where each step satisfies:

$$\left\| J(x_k)s_k + F(x_k) \right\|_2 = \eta_k \left\| F(x_k) \right\|_2 . \tag{1}$$

One of the best methods to solve the iterative linear system in the nonsymmetric case is the GMRES [19] method. We will explain the method briefly. More comprehensive information can be found in [6]. This method was proposed by Saad and Schultz in 1986.

## 2.2 The Nonlinear Stopping Criterion. Forcing Terms

As stated above, an inexact Newton method is a kind of Newton's method for solving a nonlinear system, where at the $k$th iteration, the step from the current approximate solution is required to satisfy (1).

But we want to choose a strategy to select the $\eta_k$ as the outer iteration progresses. This term is known as *forcing term*. There are several possible strategies to choose a forcing term. This issue is independent of the particular linear solver. Setting to a constant for the entire iteration is often a reasonable strategy, but the choice of that constant depends on the problem. If a constant is too small much effort can be wasted in the initial stages of the iteration. This problem is known as oversolving [5],[9]. Oversolving means that the linear equation for the Newton step is solved to a precision far beyond what is needed to correct the nonlinear equation. We propose here three possible alternatives, two obtained from [5] and the third from [9]. As we will see below none of the following strategies are perfect and they can create problems such as the need to do more outer iterations. Now, we will explain briefly each strategy. The first and the second are given with the following formulae:

$$\eta_k = \frac{\left\| F(x_k) - F(x_{k-1}) - J(x_{k-1})s_{k-1} \right\|_2}{\left\| F(x_{k-1}) \right\|_2} \qquad \eta_k = \frac{\left| \left\| F(x_k) \right\|_2 - \left\| F(x_{k-1}) + J(x_{k-1})s_{k-1} \right\|_2 \right|}{\left\| F(x_{k-1}) \right\|_2}.$$

It is possible to include in these strategies some method of *safeguarding* [5], i.e. preventing the $\eta_k$ term from being too small. We will see this with the next strategy. The third strategy is a refinement of the other two strategies. A first approximation could lead us to:

$$\eta_k^A = \gamma \frac{\left\| F(x_k) \right\|_2^2}{\left\| F(x_{k-1}) \right\|_2^2}.$$

where $\gamma \in [0,1)$. Then $\eta_k$ is uniformly bounded away from 1. This guarantees q-quadratic convergence [9]. But we need to use a first term that sets $\eta_k = \eta_{\max}$ for the first iteration. And we can use a method for safeguarding criterion. The idea is that if $\eta_{k-1}$ is sufficiently large we do not let $\eta_k$ decrease by much more than a factor of $\eta_{k-1}$:

$$\eta_k^B = \begin{cases} \eta_{\max}, & n = 0 \\ \min(\eta_k^A, \eta_{\max}), & n > 0 \gamma\eta_{k-1}^2 < 0.1 \\ \min(\eta_k^A, \max(\eta_k^A, \gamma\eta_{k-1}^2)), & n > 0 \gamma\eta_{k-1}^2 > 0.1 \end{cases}.$$

The constant 0.1 is arbitrary. This safeguarding improves the performance of the iteration in some cases. This oversolving on the final step can be controlled comparing the norm of the current nonlinear residual $\left\| F(x_n) \right\|$ to the nonlinear residual norm at which the iteration would terminate and bounding from below by a constant multiple of $\tau_t / \left\| F(x_n) \right\|$. The final expression is:

$$\eta_k = \min(\eta_{\max}, \max(\eta_k^B, 0.5\tau_t / \left\| F(x_n) \right\|)).$$

Finally it is important to note that these criterions are parallelizable. We explain how this was done in section 3.

### 2.3 Newton GMRES Sequential Algorithm. Practical Implementation

In this section, we present our Newton-GMRES sequential algorithm. Our algorithm gives the user the possibility of choosing the following steps for the nonlinear system: the initial guess, the function, the Jacobian matrix, and the stopping criterion. Furthermore, the preconditioner and the oversolving criterion can be chosen for the inner iteration. Our implementation is based on using the BLAS and LAPACK numerical kernels. We chose GMRES standard code from CERFACS [6] to solve the linear system, and we will discuss later why we chose this implementation.

**Algorithm sequential Newton-GMRES**

```
Choose a starting point x_k
Compute vector F(x_k)
While the stopping criterion has not been reached
  Compute the Jacobian Matrix J(x_k)
  Solve the linear system J(x_k)s_k = -F(x_k):
    Compute the next Forcing Term η_k
    GMRES( J(x_k), F(x_k), η_k)
  Update the iterate x_k = x_k + s_k
  Compute vector F(x_k)
```

The GMRES from CERFACS implements a powerful, efficient and configurable method. And it has several advantages over other implementations: it leaves the user the possibility of using numerical kernels (we will study this matter in the next paragraph). Moreover, it can be modified to work with SIMD parallel computing. We will explain how we did this in section 3.

It is known that almost all the Krylov methods are based on Matrix-vector multiplications and dot products. This is an advantage since the GMRES code we use has a reverse communication scheme: a driver routine implements the method and the numerical operations needed by GMRES or by calling user-defined routines. The dot products are performed using DDOT from Level 1 BLAS, while the matrix-vector product is performed using DGEMV from Level 2 BLAS or using the version we have designed for the sparse structured matrices from the Extended Rosenbrock problem. The user can also change these routines for others that he desires. The preconditioning routines are left to the user. Therefore the code can be as follows:

```
Algorithm Sequential GMRES ( J(x_k), F(x_k), η_k)
Call Driver_GMRES( J(x_k), F(x_k), η_k, revcom)
Repeat
  IF revcom = MatVec (* Matrix-vector product *)
    compute MatVec          // DGEMV BLAS2 routine or a sparse user
                               defined routine
  ELSIF revcom = PrecondLeft
     Compute PrecondLet  // This is left to the user
  ELSEIF revcom = PreCondRight
     Compute PreconRight    // This is left to the user
  ELSEIF revcom = DotProd
     Compute ddot      // DDOT BLAS1 routine
Until revcomm=0
```

| | 0 | | | | 1 | | | 2 | |
|---|---|---|---|---|---|---|---|---|---|
| 0 | $j_{11}$ | $j_{12}$ | $j_{17}$ | $j_{18}$ | $j_{13}$ | $j_{14}$ | $j_{19}$ | $j_{15}$ | $j_{16}$ |
| | $j_{21}$ | $j_{22}$ | $j_{27}$ | $j_{28}$ | $j_{23}$ | $j_{24}$ | $j_{29}$ | $j_{25}$ | $j_{26}$ |
| | $j_{51}$ | $j_{52}$ | $j_{57}$ | $j_{58}$ | $j_{53}$ | $j_{54}$ | $j_{59}$ | $j_{55}$ | $j_{56}$ |
| | $j_{61}$ | $j_{62}$ | $j_{67}$ | $j_{68}$ | $j_{63}$ | $j_{64}$ | $j_{69}$ | $j_{65}$ | $j_{66}$ |
| | $j_{91}$ | $j_{92}$ | $j_{97}$ | $j_{98}$ | $j_{93}$ | $j_{94}$ | $j_{99}$ | $j_{95}$ | $j_{96}$ |
| 1 | $j_{31}$ | $j_{32}$ | $j_{37}$ | $j_{38}$ | $j_{33}$ | $j_{34}$ | $j_{39}$ | $j_{35}$ | $j_{36}$ |
| | $j_{41}$ | $j_{42}$ | $j_{47}$ | $j_{48}$ | $j_{43}$ | $j_{44}$ | $j_{49}$ | $j_{45}$ | $j_{46}$ |
| | $j_{71}$ | $j_{72}$ | $j_{77}$ | $j_{78}$ | $j_{73}$ | $j_{74}$ | $j_{79}$ | $j_{75}$ | $j_{76}$ |
| | $j_{81}$ | $j_{82}$ | $j_{87}$ | $j_{88}$ | $j_{83}$ | $j_{84}$ | $j_{89}$ | $j_{85}$ | $j_{86}$ |

| | 0 | 1 | 2 |
|---|---|---|---|
| 0 | $-F_{11}$ | | |
| | $-F_{21}$ | | |
| | $-F_{51}$ | | |
| | $-F_{61}$ | | |
| | $-F_{91}$ | | |
| 1 | $-F_{31}$ | | |
| | $-F_{41}$ | | |
| | $-F_{71}$ | | |
| | $-F_{81}$ | | |

**Fig. 1.** SCALAPACK block cyclic distribution for the Jacobian matrix and the function

With this scheme we can parallelize all the nonlinear scheme including the GMRES iteration. We will study this in the next section.

# 3 Parallel Algorithm

## 3.1 How to Parallelize the Sequential Algorithm

The parallel version we implemented uses the SCALAPACK library. Within an iteration, the function evaluation, the computation of the Jacobian Matrix, the computing of the forcing term, the solution of the linear system including (the matrix-vector products and the scalar products), and the update of the iterate are parallelized. These are all the steps of an iteration. The evaluation of the function and the computation of the Jacobian depend on the problem to be solved. We analyze this in the case study section. Below we explain how to parallelize the other steps. We must take into account the fact that the Extended Rosenbrock problem is slightly different because we used structured sparse techniques. The main differences are studied in section 4.3. The SCALAPACK library uses a 2-D block cyclic data distribution.

***Parallelizing the GMRES code***

We have two different parts to parallelize: the driver routine, and the reverse communication steps.

Driver routine: The same driver routine [6] is run now on all the mesh processors. Some data are replicated over all the processors, while others are distributed (across the mesh processors), using a cyclic block distribution.

Matrix $A$, the right hand side of the linear system $b$, the computed solution of linear system $x$, and Krylov basis $V_m$ computed by Arnoldi's process, are stored distributed.

The Hessenberg matrix $H$, the solution of the current LS problem, $y$, the sine and cosine of the Givens rotation are replied locally.

Reverse communication mechanism: this mechanism must be adapted to work in parallel. The precondition steps are problem dependant and are left to the user, but we have to modify the Matrix-vector product and the vector dots steps. To compute the Matrix-vector dot we use the SCALAPACK routine `PDGEMV`, or our sparse routine for the Extended Rosenbrock problem. `PDGEMV` uses the 2-D cyclic distribution as shown in figure 1. As already mentioned, the CERFACS GMRES code is ready [7] to work in an SIMD [10] environment, but SCALAPACK is not fully compliant with this because the right hand side vectors are not stored in all the processors. We can solve this by sending the subvector to all the processors of the same row. This can be appreciated in figure 2.

The other operation is the parallel dot product. The two data vectors are distributed as vector $F$ (see figure 1). We can compute each subvector product and then perform a global sum operation. This operation is optimized in SCALAPACK. Then the result of the dot product is broadcasted to all the processors in the column.

From the point of view of a processor that belongs to the mesh of $(P_r, P_c)$ processors and using the SCALAPACK distribution shown in figure 1., the algorithm is the following:

```
Parallel Algorithm GMRES (n, J(x_k), F(x_k), η_k)

On the processor (p_r,p_c) of the mesh:

  Call Driver_GMRES( J(x_k), F(x_k), η_k, revcom)
  Repeat
    IF revcom = MatVec        //Parallel Matrix-vector
      PMatVec                 //PDGEMV or a sparse user defined routine
      In processor(p_r,0) send the resultant subvector
      to the processors (p_r,1...P_r - 1) located at the
      same row
    ELSIF revcom = PrecondLeft
      Compute PrecondLet   // Left to the user
    ELSEIF revcom = PreCondRight
      Compute PreconRight // Left to the user
    ELSEIF revcom = DotProd
      Compute DDOT    // DDOT BLAS2 routine
      For each column c of the mesh
      Do the global_sum using processors(1...P_c - 1,p_c)
  Until revcomm=0
```

### *Forcing term*

If we take a look at the operations done to compute the forcing term criterion in the former section, we can see that almost all are norm based. Furthermore there is a Matrix-vector product. And some operations are done by using the values computed in the former iterate. In the following paragraph we explain how these operations can be carried out.

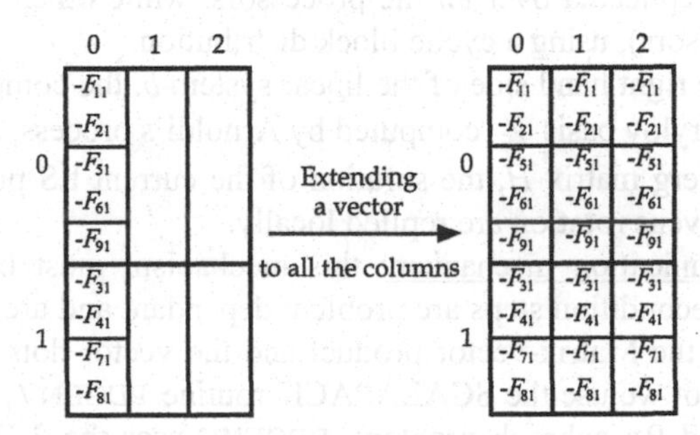


**Fig. 2.** All the column processors must have the vectors to work with the GMRES driver

The Matrix-vector dot is computed using the standard SCALAPACK routine PDGEMV or a user defined routine for the structured sparse case. The norms are computed in a similar way to that used for the parallelization of the dot products for the reverse communication mechanism: we can compute them as subvector product, then do a global sum operation, and do a square root to compute the norm. Finally the norms must be sent to all the processors. When the processors receive the norms, each of them compute $\eta_k$.

***Nonlinear iterate update***

When the linear system is solved, the solution $s$ is stored in the right hand side vector. Then we only have to broadcast $s$ to all the processors, and update the iterate. To broadcast $s$, we used two calls to the BLACS:

1. All the first column processors must have the complete $s_{(k)}$.
2. Each first column processor sends the complete vector to the processors located in its row.

With these steps the vector is located in all the processors, and we only have to update $x_{(k+1)} = x_{(k)} + s_{(k)}$.

Note that this iterate is done for each iteration. It could be interesting not to carry out this step for each iteration, only for some of them, or even for none of them. This kind of techniques are known as chaotic techniques [4]. We use these techniques with the Extended Rosenbrock Problem.

# 4 Case Studies

## 4.1 The Chandrasekhar H-Equation

The Chandrasekhar H-equation is used to solve exit distribution problems in radiative transfer. The equation can be discretized. The resulting discrete problems is:

$$F_i(x) = x_i - \left(1 - \frac{c}{2N}\sum_{j=1}^{N}\frac{\mu_i x_j}{\mu_i + \upsilon_j}\right)^{-1}.$$

It is known that there are two solutions for $c \in (0,1)$. Only one of these solutions has a physical meaning. The Jacobian can be computed by using the analytic expression or by using a finite difference formula [9]. We used both. This latter is obtained by a forward difference approximation technique. We parallelized both the computations of the function and the Jacobian matrix by using the standard SCALAPACK distribution.

## 4.2 Computing the Intersection Point of Several Hyper-Surfaces

To compute the intersection point of $n$ hyper-surfaces in an $n$-dimensional space, it is necessary to solve a system of nonlinear equations as follows:

$$F_i(x) = x^T A_i x + b_i^T x + c_i = 0 \quad \forall i: 1 \le i \le n,\ \forall\ A_i \in \Re^{n\times n}, b_i,\ x, \in \Re^{n\times 1},\ c_i \in \Re .$$

To simplify the problem we can use the particular case where $A_1 = A_2 = \ldots A_n = A$. The Jacobian matrix can be obtained by using the analytic expression or by using a finite difference formula [9]. We used the analytic expression. It can be obtained by deriving the formula of the nonlinear system.

$$J(i,j) = \frac{\partial F_i}{\partial x_j}(x) = b_i(j) - A(j,:)x + x^T A(:,j).$$

We parallelized both the computation of the function and the computation of the Jacobian matrix by using the standard SCALAPACK distribution.

### 4.3 The Extended Rosenbrock Problem

The Extended Rosenbrock function is one of the most standard problems used in the literature for nonlinear systems. We took the problem from [13]. The function and the Jacobian Matrix are as follows:
Given $n$, variable, but even:

$$f_{2i-1}(x) = 10\left(x_{2i} - x_{2i-1}^2\right), \quad f_{2i}(x) = 1 - x_{2i} \qquad \forall i: 1 \le i \le n/2.$$

The initial guess $x_0$ is :

$$x_0 = (\xi_j) \quad \text{where} \quad \xi_{2j-1} = -1.2, \xi_{2j} = 1 \qquad \forall j: 1 \le j \le n/2.$$

The Jacobian matrix $J$ is easy to obtain by deriving the expression of the function, and its expression can be written as:

$$j_{2i-1,2i-1} = -20x_{2i-1}, \quad j_{2i-1,2i} = 10, \quad j_{2i,2i-1} = -1 \qquad \forall i: 1 \le i \le n/2.$$

As we can appreciate the expression of the Jacobian leads to a structured sparse matrix. We can appreciate this by representing a 6 x 6 example:

$$J = \begin{bmatrix} x & x & 0 & 0 & 0 & 0 \\ x & 0 & 0 & 0 & 0 & 0 \\ 0 & 0 & x & x & 0 & 0 \\ 0 & 0 & x & 0 & 0 & 0 \\ 0 & 0 & 0 & 0 & x & x \\ 0 & 0 & 0 & 0 & x & 0 \end{bmatrix}$$

**Fig. 3.** Structure of the Jacobian matrix for the extended Rosenbrock problem

This kind of matrix must not be stored as a dense matrix. Moreover there are three considerations which must be taken into account for the Newton-GMRES method:

1. We cannot use the standard DGEMV/PDGEMV Matrix-vector product routines because the matrix is structured sparse and it is stored by using this kind of structure. Here we implemented a special sparse product [17] for this problem. For the parallel case (using a vertical mesh), the product is carried out on the data stored in each processor.

2. The nonlinear update cannot be carried out because the necessary communications to do this could be more expensive than the communications used for other steps of the algorithm. In fact we did not make the update and used a chaotic technique [4].
3. As the problem is sparse, we can work with very large sizes of $N$. In fact, as we see in the next section we worked with a matrix using $N = 1310400$.

## 5 Experimental Results

Below is a brief study on the performance of the parallel algorithm. The complete study can be found in [17]. We report here the most relevant results. We used the standard (non chaotic) Newton-GMRES version for the H-Chandrasekhar problem and for computing the intersection point of several hyper-surfaces. We used the chaotic version (with no nonlinear update) for the Extended Rosenbrock Problem. In all cases we used a restarted version with $m = 30$. For each problem we had the possibility of using the forward finite differences or analytic Jacobian Matrix.
Furthermore we used the techniques for avoiding the oversolving problem explained before, although these techniques showed no significant performance improvement [5] in any of the studied cases. Moreover, we compared our standard Newton-GMRES method, with the parallel Newton method we developed and used in [15],[16] and with Powell's standard method as implemented in the MINPACK-1 [12] standard sequential package. In this paper we show the comparison for the H-Chandrasekhar problem. We present here the test results using several matrices in each problem. The sizes are $N = 1200$ and $N = 1600$ for the H-Chandrasekhar problem. N=1200, N=1600, 2000 for the hyper-surfaces problem, and $N = 864000$ and $N = 1310400$ for Rosenbrock's. The hardware used was on a cluster of 16 Pentium II/300 MHz PC's with a Myrinet network [11]. Here we show the results corresponding to the computing time of Newton's and Newton-GMRES methods for H-Chandrasekhar problem.

**Table 1.** Figures corresponding to the parallel computing time in seconds of Newton's and Newton-GMRES methods for the Chandrasekhar problem

| Chandrasekhar | Newton-GMRES | | Newton | |
|---|---|---|---|---|
| Proc. | 1200 | 1600 | 1200 | 1600 |
| 1 | 1435,33 | 3402,78 | 1487,42 | 3592,15 |
| 2 | 719,14 | 1701,57 | 756,90 | 1800,26 |
| 4 | 359,91 | 850,72 | 382,47 | 904,71 |
| 6 | 239,61 | 574,80 | 256,15 | 612,86 |
| 8 | 180,37 | 425,57 | 195,75 | 456,69 |
| 9 | 160,33 | 383,08 | 181,79 | 415,56 |
| 10 | 144,44 | 341,25 | 161,66 | 366,75 |
| 12 | 120,81 | 300,02 | 145,99 | 327,57 |
| 16 | 90,41 | 213,16 | 99,52 | 231,10 |

The table shows the good behaviour obtained using the Newton-GMRES method. We improved the Newton's method we developed for other problems [15],[16]. Moreover both algorithms obtain a very good parallel performance. This is confirmed by figure 4 below, where we show the speedup for both methods. If we compare both sequential methods with Powell's method, it can be seen that our methods always outperform the MINPACK-1 implementation.

**Table 2.** Figures corresponding to the computing time in seconds of sequential Powell's, Newton's and Newton-GMRES methods for the Chandrasekhar problem

| Powell's (MINPACK-1) | | Newton-GMRES | | Newton | |
|---|---|---|---|---|---|
| 1200 | 1600 | 1200 | 1600 | 1200 | 1600 |
| 1585,35 | 3621,8 | 1435,33 | 3402,78 | 1487,42 | 3592,15 |

This behaviour is not a surprise because Powell's method has the advantage that it is a very robust but not a very fast method. Below is the behaviour of the speedup corresponding to the Newton-GMRES method. Again we compared it with Newton's method using the Chandrasekhar problem for both methods. It can be seen that the curves are very similar, and also that the Newton-GMRES speedup is slightly faster than that of the Newton's method.

Now we show the results of computing the intersection point of several hyper-surfaces. Again the behaviour for the Newton-GMRES method is better not only for computing time but also for convergence.

**Table 3.** Figures corresponding to the parallel computing time in seconds of Newton's and Newton-GMRES methods for computing the intersection of hyper-surfaces

| Hyper-surface | Newton-GMRES | | | Newton | | |
|---|---|---|---|---|---|---|
| Proc. | 1200 | 1600 | 2000 | 1200 | 1600 | 2000 |
| 1 | 40,34 | 54,99 | 81,33 | 73,22 | 229,40 | 1894,55 |
| 2 | 23,25 | 26,63 | 43,30 | 44,94 | 130,60 | 1119,46 |
| 4 | 13,41 | 14,43 | 23,48 | 23,68 | 112,90 | No Conv. |
| 6 | 10,33 | 10,84 | 17,41 | 18,16 | 60,99 | No Conv. |
| 8 | 8,87 | 8,79 | 13,72 | 15,19 | 50,33 | 274,70 |
| 9 | 7,81 | 8,18 | 12,53 | 14,88 | 38,90 | No Conv. |
| 10 | 8,21 | 8,32 | 12,52 | 13,14 | 34,45 | 354,54 |
| 12 | 6,65 | 6,70 | 10,27 | 12,13 | 32,00 | 218,62 |
| 16 | 5,67 | 5,52 | 8,28 | 9,88 | 34,26 | 182,13 |

We can appreciate that sometimes Newton's method cannot achieve convergence or it takes a long time to obtain it for some mesh configurations. This is due to the problem being ill conditioned [17].

Now we show the figure for the speedup. We can appreciate that the problem is scalable but the numerical results are worse than the Chandrasekhar case. This is due to the difficulties found in the parallelisation of the Jacobian matrix. Because to compute each element of the matrix, it is necessary [17] to use a row and column of matrix $A$ corresponding to the hyper-surface equation.

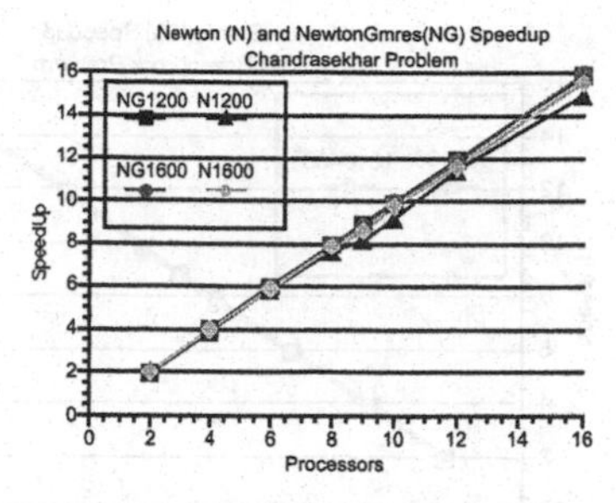


**Fig. 4.** Speedup for Newton-GMRES and Newton's methods working with the Chandrasekhar problem

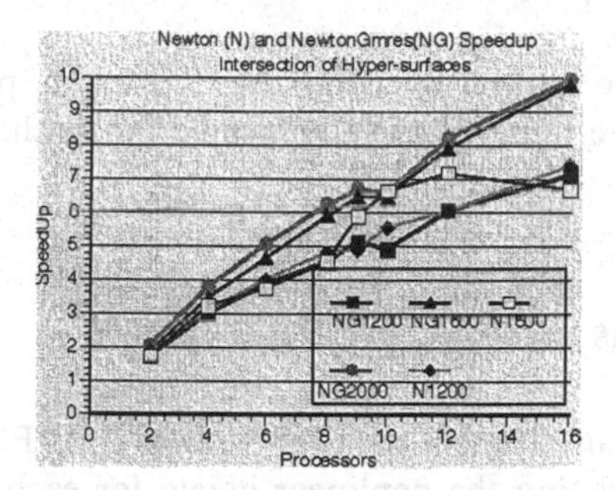


**Fig. 5.** Speedup for computing the intersection of hyper-surfaces using Newton's and Newton-GMRES methods

Now, we study the Newton-GMRES method with a totally different problem, the extended Rosenbrock Problem. The results for the Extended Rosenbrock Problem are shown in the following table:

**Table 4.** Figures corresponding to the parallel computing time in seconds of Newton's and Newton-GMRES methods for the Rosenbrock problem using several processor configurations

| Rosenbrock | Newton-GMRES | |
|---|---|---|
| Proc. | 864000 | 1310400 |
| 1 | 13,38 | 20,32 |
| 2 | 6,88 | 10,17 |
| 4 | 3,42 | 5,21 |
| 6 | 2,27 | 3,43 |
| 8 | 1,68 | 2,62 |
| 9 | 1,52 | 2,32 |
| 10 | 1,36 | 2,10 |
| 12 | 1,13 | 1,69 |
| 16 | 0,86 | 1,30 |

These results confirm the good behaviour of our methods, and the results themselves can be confirmed by studying the speedup figure for this problem.

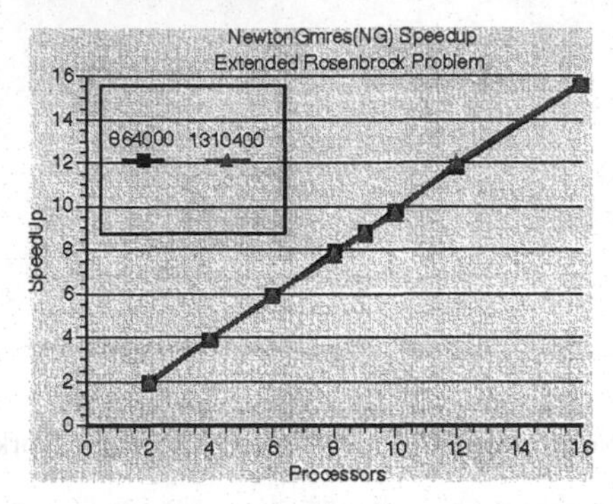


**Fig. 6.** Speedup in Newton-GMRES for the extended Rosenbrock Problem

We can appreciate similar speedup results for this problem compared with the previous problem. The figure shows the method for reaching the theoretical maximum speedup for all cases.

## 6 Conclusions

We have developed an efficient parallel Newton-GMRES method. Our method uses the possibility of updating the nonlinear iterate for each iteration or updating every several iterations (or not updating at all). We used this technique for the structured sparse case although it can be carried out for any problem. Moreover, our method works with any kind of nonlinear problem (is not problem dependant). Only the function and the Jacobian Matrix must be supplied and computed (in parallel). It is important to note that we used a sparse case to show the versatility of our method, although there are more specialized packages to solve this, for example the PETSC package [2].

In our implementation it is also important to note the portability of the code because all the software used is standard and portable, the LAPACK/SCALAPACK libraries and GMRES (from CERFACS) standard codes, and PVM/MPI parallel environments. We note the behaviour of the Myrinet network. We think it could be a good, cheap alternative to classical MPP machines.

This work is included in a framework tool we are developing where, given a problem that implies solving a nonlinear system, the best nonlinear method must be chosen to solve the problem.

## Acknowledgments

We want to thank Professor Luc Giraud from CERFACS (Parallel Algorithms Project) for his invaluable help in explaining to us the difficulties and characteristics of his GMRES code. Without his help, it would have been impossible to develop this work.

## References


[1] Anderson E., Bai Z., Bischof C., Demmel J., Dongarra J., Du Croz J., Greenbaum A., Hammarling S., Mackenney A., Ostrouchov S., Sorensen D. (1995). *LAPACK Users' Guide. Second edition.* SIAM Publications, Philadelphia.

[2] Balay S., Gropp W.D, Curfman L., and Smith B.F. (2001). *PETSc Users Manual*. T. Report ANL-95/11 – Revision 2.11. Argonne National Laboratory.

[3] Blackford L.S., Choi J., Cleary A., D'Azevedo E., Demmel J. Dhillon I., Dongarra J., Hammarling S., Henry G., Petitet A., Stanley K., Walker D., Whaley R.C. (1997). *SCALAPACK Users' Guide*. SIAM Publications, Philadelphia.

[4] Bru R, Elsner L, Neumann M. (1988). *Models of Parallel Chaotic Iteration Methods*. Linear Algebra and its Applications, 103: 175-192.

[5] Einsestat S.C., Walker S.C. (1994). Choosing the Forcing Terms in an Inexact Newton Method. Center for Research on Parallel computation. Technical Report CRPRPC-TR94463. Rice University.

[6] Frayssé V., Giraud L., Gratton S. (1997). *A Set of GMRES routines for Real and Complex Arithmetics*. CERFACS Technical Report TR/PA/97/49.

[7] Frayssé V., Giraud L., Kharraz-Aroussi. (1998). *On the influence of the othogonalization scheme on the parallel performance of GMRES*. EUROPAR 98'. Parallel Processing. pp. 751-762, Vol 1470. Springer Verlag. Eds. D. Pritchard, J. Reeve.

[8] Golub G. H. , Van Loan C. F. (1996). *Matrix Computations (third edition)*. John Hopkins University Press.

[9] Kelley C.T. (1995). *Iterative Methods for Linear and Nonlinear Equations*. Frontiers in Applied Mathematics, SIAM Publications, Philadelphia.

[10] Kumar R ., Grama A., Gupta A., Karypis G. (1994). *Introduction to Parallel Computing: Design and Analysis of Algorithms*. The Benjamin Cumimngs Publishing Company.

[11] Myrinet.*Myrinet overview*. (Online). http://www.myri.com/myrinet/overview/index. Html.

[12] Moré J.J., Garbow B.S., Hillstrom K.E, (1980). *User Guide for MINPACK-1*. Technical Report. ANL-80-74. Argonne National Laboratory.

[13] Moré J.J, Garbow B.S and Hillstrom K.E. (1981*). Testing Unconstrained Optimization Software*. ACM Transactions on Mathematical Software, Vol 7, No.1, March 1981.

[14] Moré J.J. (1990). *A Collection of Nonlinear Model problems*. Lectures in Applied Mathematics, Volume 26. ,pp 723-762.

[15] Peinado J, Vidal A.M. (2001). *A new Parallel approach to the Toeplitz Inverse Eigenproblem using the Newton's Method and Finite Difference techniques.* VecPar '2000. 4th International Meeting on Vector and Parallel Processing. Selected Papers and Invited Talks. Lecture Notes in Computer Science. Ed. Springer.

[16] Peinado J, Vidal A.M. (2001). Aproximación Paralela al Problema Inverso de los Valores Propios de Matrices Toeplitz mediante métodos tipo Newton. XII Jornadas de Paralelismo. Septiembre 2001. Valencia. pp 235-240.

[17] Peinado J, Vidal A.M. (2001). *Aproximación paralela al Método Newton-GMRES*. T.Report DSIC-II/23/01. Departamento de Sistemas Informáticos y Computación Universidad Politécnica de Valencia. December 2001. Valencia.

[18] Peinado J, Vidal A.M. (2002). *Un marco Computacional Paralelo para la Resolución de Sistemas de gran Dimension*. V Congreso de Métodos Numéricos en Ingenieria. Madrid, 3-6 de Junio de 2002. Accepted.

[19] Saad Y., Schultz M. (1986). GMRES: A generalized minimal residual algorithm for solving nonsymmetric linear systems. SIAM J. Sci. Stat, Comput. 7, 856-869.

[20] Skjellum A. (1994). Using MPI: Portable Programming with Message-Passing Interface. MIT Press Group.

[21] Whaley R.C. (1994). Basic Linear Algebra Communication Subprograms (BLACS): Analysis and Implementation Across Multiple Parallel Architectures. Computer Science Dept. Technical Report CS 94-234, University of Tennesee, Knoxville.

# Preconditioning for an Iterative Elliptic Solver on a Vector Processor

Abdessamad Qaddouri and Jean Côté

Recherche en prévision numérique, Meteorological Service of Canada
2121 Trans-Canada Highway, Dorval, QC, H9P 1J3, Canada
{abdessamad.qaddouri,jean.cote}@ec.gc.ca
http://riemann.cmc.ec.gc.ca/index.html

**Abstract.** A family of preconditioners for an iterative solver of a separable elliptic boundary value problem in spherical geometry arising in numerical weather prediction has been implemented and tested on a vector processor. The PCG-MILU(0) solver obtains practical convergence in a small number of iterations and outperforms a highly optimized direct solver at high-resolution.

## 1 Introduction

Most models in operational numerical weather prediction (NWP) use either an implicit or semi-implicit time discretization on tensor products spatial grids. This gives rise to the need to solve a separable 3D elliptic boundary value (EBV) problem at each model time step. This is the case for the Global Environmental Multiscale (GEM) model in operation at the Canadian Meteorological Center (CMC) [2]. This separable EBV problem is currently solved with a direct method which can be very efficiently implemented on Environment Canada NEC SX-4 system of vector multiprocessors. A parallel distributed-memory implementation with explicit message passing which preserves the vector performance was described recently in [5].

This direct solver, see Sect. 3 below, can be implemented either with a fast or a slow Fourier transform. In the case of the slow transform, a full matrix multiplication, the cost per grid point raises linearly with the number of grid points along the transform direction. A question that arises naturally is: could an iterative solver advantageously replace the "slow" direct solver? This is the object of the present paper in which we investigate a family of preconditioners for the Preconditioned Conjugate Gradient method, the iterative method most suited for our problem. The challenge is to preserve the vector performance using non-diagonal preconditioners, and we defer the parallelization aspect to a future study.

The paper is organized as follows: section 2 presents the problem, section 3 a quick review of the direct method, section 4 the iterative method and the family of preconditioners, section 5 the results, and finally section 6 the conclusion.

J.M.L.M. Palma et al. (Eds.): VECPAR 2002, LNCS 2565, pp. 343–353, 2003.

## 2 Elliptic Problem

The problem to solve is a 3D separable positive definite EBV problem, which after vertical separation reduces to a set of horizontal Helmholtz problems, each of the form

$$(\eta - \Delta)\,\phi = R, \eta > 0\,. \tag{1}$$

Since we will keep the vertical separation in the iterative solver, we need only to discuss the horizontal aspects of the problem. The numerical results will be obtained for the full set of size $NK = 28$, where $NK$ is the number of levels in the vertical. The domain is a sphere of radius unity, discretized in latitude $\theta$ and longitude $\lambda$ and we are looking for the solution of (1) on this grid:

$$\phi \equiv \phi(\lambda, \theta) \rightarrow \phi_{ij} = \phi(\lambda_i, \theta_j),\ \{i = 1, NI\ ;\ j = 1, NJ\}\ , \tag{2}$$

with

$$0 \leq \lambda_i < 2\pi,\ \lambda_1 = 0,\ -\frac{\pi}{2} < \theta_j < \frac{\pi}{2}\,. \tag{3}$$

There is a great freedom in the position of the grid points. The grid points do not need to be regularly spaced. At CMC, the GEM model is integrated with various grid configurations depending on the application: a uniform-resolution grid (0.9°) for medium- to long-range (5 days or more) global forecasting, a variable-resolution grid with a uniform-resolution (0.22°) window over North America and adjacent waters for short-range (2 days) regional forecasting at a continental scale, an even more focused grid for shorter-range (1 day) forecasting at the mesoscale (0.09°). The current strategy for varying the grid is to increase the grid lengths by constant factors along each coordinate direction away from the uniform-resolution region. The polar axis of the coordinate system need not coincide with the geographical one either, and we usually choose the equator of the coordinate system to bisect the region of interest. The motivation for the use of this variable-grid strategy is presented in [2].

It is convenient to introduce the grid extensions

$$\lambda_0 = \lambda_{NI} - 2\pi,\ \lambda_{NI+1} = \lambda_1 + 2\pi\,, \tag{4}$$

$$\theta_0 = -\frac{\pi}{2},\ \theta_{NJ+1} = \frac{\pi}{2}\,, \tag{5}$$

and the grid spacings

$$\Delta\lambda_i = \lambda_{i+1} - \lambda_i,\ i = 0, NI,\ (\Delta\lambda_0 \equiv \Delta\lambda_{NI})\,, \tag{6}$$

$$\Delta \sin\theta_j = \sin\theta_{j+1} - \sin\theta_j,\ j = 1, NJ\,. \tag{7}$$

For finite differencing we also introduce the following auxiliary grids

$$\tilde{\lambda}_i = \frac{(\lambda_i + \lambda_{i+1})}{2},\ i = 1, NI,\ \tilde{\lambda}_0 = \tilde{\lambda}_{NI} - 2\pi,\ \tilde{\lambda}_{NI+1} = \tilde{\lambda}_1 + 2\pi\,, \tag{8}$$

$$\tilde{\theta}_j = \frac{(\theta_j + \theta_{j+1})}{2},\ j = 1, NJ-1,\ \tilde{\theta}_0 = -\frac{\pi}{2},\ \tilde{\theta}_{NJ} = \frac{\pi}{2}\,, \tag{9}$$

and the grid spacings

$$\Delta\tilde{\lambda}_i = \tilde{\lambda}_{i+1} - \tilde{\lambda}_i,\ i = 0, NI\,, \tag{10}$$

$$\Delta \sin\tilde{\theta}_j = \sin\tilde{\theta}_{j+1} - \sin\tilde{\theta}_j,\ j = 1, NJ-1\,. \tag{11}$$

This is a tensorial grid. The discrete form of (1) is

$$\mathbf{A}\ \phi\ =\ \mathbf{r}\,, \tag{12}$$

where

$$\mathbf{A} = [\,\eta\, P(\theta) \otimes P(\lambda)\ -\ P'(\theta) \otimes P_{\lambda\lambda}\ -\ P_{\theta\theta} \otimes P(\lambda)\,]\,, \tag{13}$$

where $\otimes$ is the notation for tensor product and

$$\mathbf{r} = P(\theta)\ \otimes\ P(\lambda)\, R\,, \tag{14}$$

with

$$P(\theta) = \begin{bmatrix} \Delta\sin\tilde{\theta}_0 & & & \\ & \Delta\sin\tilde{\theta}_1 & & \\ & & \ddots & \\ & & & \Delta\sin\tilde{\theta}_{NJ-1} \end{bmatrix}, \tag{15}$$

$$P'(\theta) = \begin{bmatrix} \frac{\Delta\sin\tilde{\theta}_0}{\cos^2\theta_1} & & & \\ & \frac{\Delta\sin\tilde{\theta}_1}{\cos^2\theta_2} & & \\ & & \ddots & \\ & & & \frac{\Delta\sin\tilde{\theta}_{NJ-1}}{\cos^2\theta_{NJ}} \end{bmatrix}, \tag{16}$$

$$P_{\theta\theta} = \begin{bmatrix} -\frac{\cos^2\tilde{\theta}_1}{\Delta\sin\theta_1} & \frac{\cos^2\tilde{\theta}_1}{\Delta\sin\theta_1} & & & \\ \frac{\cos^2\tilde{\theta}_1}{\Delta\sin\theta_1} & -\frac{\cos^2\tilde{\theta}_1}{\Delta\sin\theta_1} - \frac{\cos^2\tilde{\theta}_2}{\Delta\sin\theta_2} & \frac{\cos^2\tilde{\theta}_2}{\Delta\sin\theta_2} & & \\ & \frac{\cos^2\tilde{\theta}_2}{\Delta\sin\theta_2} & \ddots & \ddots & \\ & & \ddots & -\frac{\cos^2\tilde{\theta}_{NJ-2}}{\Delta\sin\theta_{NJ-2}} - \frac{\cos^2\tilde{\theta}_{NJ-1}}{\Delta\sin\theta_{NJ-1}} & \frac{\cos^2\tilde{\theta}_{NJ-1}}{\Delta\sin\theta_{NJ-1}} \\ & & & \frac{\cos^2\tilde{\theta}_{NJ-1}}{\Delta\sin\theta_{NJ-1}} & -\frac{\cos^2\tilde{\theta}_{NJ-1}}{\Delta\sin\theta_{NJ-1}} \end{bmatrix} \tag{17}$$

$$P(\lambda) = \begin{bmatrix} \Delta\tilde{\lambda}_0 & & & \\ & \Delta\tilde{\lambda}_1 & & \\ & & \ddots & \\ & & & \Delta\tilde{\lambda}_{NI-1} \end{bmatrix}, \tag{18}$$

$$P_{\lambda\lambda} = \begin{bmatrix} -\frac{1}{\Delta\lambda_0} - \frac{1}{\Delta\lambda_1} & \frac{1}{\Delta\lambda_1} & & & \frac{1}{\Delta\lambda_0} \\ \frac{1}{\Delta\lambda_1} & -\frac{1}{\Delta\lambda_1} - \frac{1}{\Delta\lambda_2} & \frac{1}{\Delta\lambda_2} & & \\ & \frac{1}{\Delta\lambda_2} & \ddots & \ddots & \\ & & \ddots & \ddots & \frac{1}{\Delta\lambda_{NI-1}} \\ \frac{1}{\Delta\lambda_{NI}} & & & \frac{1}{\Delta\lambda_{NI-1}} & -\frac{1}{\Delta\lambda_{NI-1}} - \frac{1}{\Delta\lambda_{NI}} \end{bmatrix} . \tag{19}$$

In (12-19) the natural (dictionary) ordering of the variables is used. The matrix $\mathbf{A}$ is symmetric and positive definite provided $\eta > 0$, which is the case here. The nonzero structure of the matrix $\mathbf{A}$ is shown below for the case $NI = 4$ and $NJ = 3$. It is the classical pattern of the five-point difference scheme, except that periodicity implies that the diagonal blocks have entries in the bottom-left and top-right corners due to the fact that every grid point has right- and left-neighbors, and is connected to them.

$$\left[\begin{array}{cccc|cccc|cccc} * & * & & * & * & & & & & & & \\ * & * & * & & & * & & & & & & \\ & * & * & * & & & * & & & & & \\ * & & * & * & & & & * & & & & \\ \hline * & & & & * & * & & * & * & & & \\ & * & & & * & * & * & & & * & & \\ & & * & & & * & * & * & & & * & \\ & & & * & * & & * & * & & & & * \\ \hline & & & & * & & & & * & * & & * \\ & & & & & * & & & * & * & * & \\ & & & & & & * & & & * & * & * \\ & & & & & & & * & * & & * & * \end{array}\right] . \tag{20}$$

## 3 Direct Solution

We obtain the direct solution of (12) by exploiting the separability, and expanding $\phi$ in $\lambda$-direction eigenvectors that diagonalize $\mathbf{A}$, i.e.

$$\phi_{ij} = \sum_{I=1}^{NI} \psi_i^{[I]} \, \phi_j^{[I]} , \tag{21}$$

with

$$P_{\lambda\lambda} \, \psi^{[I]} = \varepsilon_I \, P(\lambda) \, \psi^{[I]}, \; I = 1, NI , \tag{22}$$

and the orthogonality property

$$\sum_{i=1}^{NI} \psi_i^{[I']} \, P(\lambda)_{ii} \, \psi_i^{[I]} = \delta_{II'} , \tag{23}$$

which we use to project (12) on each mode in turn

$$\sum_{i,i',I'=1}^{NI} \psi_i^{[I]} \mathbf{A}_{iji'j'} \psi_{i'}^{[I']} \phi_{j'}^{[I']} = (\mathbf{A}^{[I]} \phi^{[I]})_j = \mathbf{r}_j^{[I]} = \sum_{i=1}^{NI} \psi_i^{[I]} \mathbf{r}_{ij} . \tag{24}$$

Writing (24) in matrix form, we obtain

$$\mathbf{A}^{[I]}\, \phi^{[I]} = [\eta\, P(\theta) - \varepsilon_I\, P'(\theta) - P_{\theta\theta}]\, \phi^{[I]} = \mathbf{r}^{[I]}\;, \tag{25}$$

which in our discretization is simply a set of symmetric positive-definite tridiagonal problems.

The algorithm can then be summarized as:

1. analysis of the of right-hand side: $\mathbf{r}_j^{[I]} = \sum_{i=1}^{NI} \psi_i^{[I]}\mathbf{r}_{ij}$ ,
2. solution of the set of tridiagonal problems, $(\mathbf{A}^{[I]}\phi^{[I]})_j = \mathbf{r}_j^{[I]}$ ,
3. synthesis of the solution, $\phi_{ij} = \sum_{I=1}^{NI} \psi_i^{[I]}\phi_j^{[I]}$ .

It is well known that for a uniform $\lambda$-grid the modes $\psi^I$ are proportional to the usual Fourier modes and the analysis and synthesis steps can be implemented with a real Fourier transform, and if furthermore $NI$ factorizes properly, the Fast Fourier Transform (FFT) algorithm can be used, otherwise these steps are implemented with a full matrix multiplication. We shall refer to this case as the "slow" transform case. The operation count for each case is

$$Number\ of\ operations_{direct\ fast} \approx NI\ NJ\ (2\,C\ \ln\ NI + 5)\;, \tag{26}$$

$$Number\ of\ operations_{direct\ slow} \approx NI\ NJ\ (4\,NI + 5)\;, \tag{27}$$

where C is some constant.

The solution algorithm presented above has a great potential for vectorization since at each step we have recursive work along only one direction at a time, leaving the two others, in 3D, available for vectorization. The principles to be followed to obtain good performance on the SX-4 are the same as those to obtain good performance on vector processors in general, see [3] for an introduction to the subject. For optimal performance, one must: avoid memory bank conflicts by using odd strides, maximize vector length by vectorizing in the plane normal to the recursive work direction, avoid unnecessary data motion. We have at our disposal the multiple FFTs of Temperton [9] which vectorize very well, furthermore there exists on the SX-4 a highly optimized matrix multiply subroutine that on large problems nearly reaches the peak performance of the processors. The longitudinal part of the direct solver (1., 3.), in either the slow or fast cases, is therefore highly optimized.

For the latitudinal part of the direct solver (2.), to set-up the factorization of the tridiagonal problems ahead of time and retrieve them later would take as much time, if not more, on our system as recomputing them. The data handling is much simpler in the latter case. We use Gaussian elimination without pivoting, which is correct because the systems are positive-definite, and our elimination procedure takes into account the symmetry of the systems. In any case the tridiagonal problems solution takes only a small fraction of the execution time which is dominated by the longitudinal transforms.

## 4 Iterative Solution

Since our problem is symmetric positive definite, the Preconditioned Conjugate Gradient (PCG) method is the natural choice for an iterative method. It is more difficult to choose the preconditioner: it should make the scheme converge rapidly while requiring only a few operations per grid point, finally it should vectorize because of the large discrepancy between vector and scalar performance on the SX-4.

Preconditioning with the diagonal would have a low cost per iteration but it would require too many iterations since we have two difficulties in our problem: the metric factors reduce the diagonal dominance in the polar regions and the variable resolution induces large variations in the mesh sizes. Both difficulties would be alleviated by non-diagonal preconditioners, which are more effective at transporting global information at each iteration.

Candidates of higher complexity are the Alternating Direction Implicit (ADI) method and Incomplete LU Factorizations. The ADI method is known to work well in Cartesian geometry with uniform grid provided optimal iteration parameters are known, and since its use is more delicate it was not considered in this work. Incomplete LU Factorizations on the other hand have been developed for problems similar to ours, and they can be implemented with a simple algebraic approach. A preliminary study with P-SPARSLIB [7] was therefore undertaken which confirmed that Incomplete LU Factorization with tolerance dropping ILUT [6] was indeed an effective preconditioner. It was then decided to begin developing code for the preconditioning step (triangular solves) of the Incomplete LU Factorization method that would exploit the data regularity and yield vector performance.

The simplest Incomplete LU Factorization is ILU(0), which finds factors that have the same sparsity structure as the matrix $\mathbf{A}$. Writing $\mathbf{A}$ as

$$\mathbf{A} = \mathbf{L_A} + \mathbf{D_A} + \mathbf{L_A^T} , \tag{28}$$

where $\mathbf{L_A}$ is the strict lower triangular part of $\mathbf{A}$, and the preconditioner $\mathbf{M}$ as

$$\mathbf{M} = \left(\mathbf{I} + \mathbf{L_A}\mathbf{D}^{-1}\right)\mathbf{D}\left(\mathbf{I} + \mathbf{D}^{-1}\mathbf{L_A^T}\right) , \tag{29}$$

the ILU(0) factorization is obtained by requiring that the diagonal of $\mathbf{M}$ be the same as the diagonal of $\mathbf{A}$:

$$diag(\mathbf{M}) = \mathbf{D_A} , \tag{30}$$

from which $\mathbf{D}$ is determined recursively. Note that $\mathbf{M}$ is also symmetric and positive definite [4].

A related factorization, which can also be cast in the form (29), is the Modified Incomplete LU Factorization with zero fill-in MILU(0). Instead of (30), MILU(0) satisfies the constraint that the row sum of $\mathbf{M}$ equals the row sum of $\mathbf{A}$:

$$\mathbf{M}\, e = \mathbf{A}\, e, \; e \equiv (1, 1, ..., 1)^T , \tag{31}$$

**Table 1.** Test problems

| problem | resolution | $NI$ | $NJ$ | $NK$ |
|---|---|---|---|---|
| P1 | 0.22° | 353 | 415 | 28 |
| P2 | 0.14° | 509 | 625 | 28 |
| P3 | 0.09° | 757 | 947 | 28 |

and which can be effective for problems deriving from an elliptic operator such as (1) [8].

Another simple preconditioner that also has the form (29) is obtained by choosing $\mathbf{D} = \mathbf{D_A}$, this yields the Symmetric Gauss-Seidel (SGS) preconditioner.

We have used these three preconditioners: SGS, ILU(0) and MILU(0). The difference between them is the calculation of the diagonal $\mathbf{D}$. The set-up of each factorization is different, but thc triangular solve can proceed with the same subroutine. The operation count per iteration with these three preconditioners is

$$Number\ of\ operations_{iterative} \approx NI\ NJ\,(4 + 9 + 9 + 13)\ , \tag{32}$$

where the various constants are for the two scalar products, the three vector updates, the matrix-vector product and the preconditioning step respectively in our implementation.

## 5 Results

We present the results of numerical experiments carried out on a single processor of Environment Canada's NEC SX-4 system. Table 1 summarizes the three 3D test problems considered, they are denoted P1, P2 and P3 respectively. The operational grid is used for the first problem, the grids for the other problems are obtained by increasing the resolution in the central window to 0.14° and 0.09° respectively while keeping the stretching factors approximately constant (~1.1). We did not optimize the memory requirements, as it was not a major concern on the SX-4. We are also not bound by the fact that the dimension is a power of two, but the best performance obviously is when the vector registers are filled (size 256). General-purpose solvers are not in general vectorized and would have scalar rather than vector performances, so we do not give results for them. The factorization step of the iterative solvers is done once and for all at the beginning and is not included in the timings. The matrices corresponding to the three problems have 731769 × 28, 1589607 × 28 and 3582881 × 28 nonzero entries respectively. We use compressed diagonal storage (CDS), which makes the matrix-vector more efficient by taking advantage of vectorization [1]. We didn't use BLAS for either the dot products or the vector updates in the conjugate gradient algorithm. In order to vectorize the preconditioning step, we performed the triangular solve by moving on wavefronts or hyperplanes [10]. The right-hand side is computed from a known random solution of amplitude 1. The initial

**Table 2.** Convergence of PCG with SGS, ILU(0) and MILU(0) preconditioning

| type | SGS | | ILU(0) | | MILU(0) | |
|---|---|---|---|---|---|---|
| problem | iteration | maximum error | iteration | maximum error | iteration | maximum error |
| P1 | 151 | 9.821 E-04 | 19 | 8.692 E-04 | 8 | 4.999 E-04 |
| P2 | 185 | 9.021 E-04 | 19 | 8.743 E-04 | 8 | 5.922 E-04 |
| P3 | 201 | 9.274 E-04 | 20 | 8.100 E-04 | 9 | 3.163 E-04 |

guess for the preconditioned conjugate gradient iterations is the zero vector, and we iterate until the maximum difference between the random solution and the solution given by PCG at all grid point is less than $10^{-3}$. Reduction of the residual by 3 orders of magnitude is sufficient for all practical purposes. Note that this is the speed of convergence of the first vertical mode, for the other modes the convergence is much more rapid. We have done studies of the first vertical mode, the worst case scenario.

Table 2 shows the results of applying the PCG algorithm with the three preconditioners to the three test problems. We first note that all these methods converge. ILU(0) and MILU(0) constitute a marked improvement over the SGS preconditioner, and they seem to exhibit some kind of robustness as the grids are refined, this is probably due to the fact that the Helmholtz constants ($\eta$) are inversely proportional to the time-step squared, and the time-step varies linearly with the minimum grid size. Moreover MILU(0) converges nearly twice as fast as ILU(0), and solves the three test problems in a small number of iterations. It is worth emphasizing that this particular application requires the solutions of several linear systems with the same coefficient matrix but different right-hand sides. In this case the cost of setting up the preconditioner is amortized over many solutions, nevertheless Table 3 shows that the preprocessing cost is of the same order as the solution cost and is not prohibitive.

Table 4 compares the performance of the Direct Slow solver against that of PCG-MILU(0). We see that for problem P1 the Direct solver is faster, whereas for the bigger problems, P2 and P3, the iterative solver is three to four times faster.

Table 5 gives the execution time of the Direct Fast solver for problems of nearly the same sizes, $NI$ has to be adjusted so as to factorize properly, and we see that the fast solver is about twenty to forty times faster than the slow

**Table 3.** Cost of MILU(0) preconditioning

| problem | set-up | solution | total |
|---|---|---|---|
| P1 | 2.21 s | 3.77 s | 5.98 s |
| P2 | 4.78 s | 3.08 s | 7.86 s |
| P3 | 10.82 s | 7.56 s | 18.38 s |

**Table 4.** Comparison between PCG-MILU(0) and the Direct Slow solver

| problem | PCG-MILU(0) | Direct Slow |
|---|---|---|
| P1 | 3.77 s | 3.11 s |
| P2 | 3.08 s | 9.53 s |
| P3 | 7.56 s | 31.60 s |

**Table 5.** Cost of the Direct Fast solver for nearly identical size problems

| $NI$ | Direct Fast |
|---|---|
| 360 | 0.17 s |
| 512 | 0.33 s |
| 768 | 0.81 s |

solver. This comparison illustrates the gain due to the use of FFTs, and one should keep in mind that the problems solved by the fast solver are physically different from those solved by the slow solver, the resolution in the $\lambda$-direction being about five times finer in the slow solver case.

Figure 1 shows the execution time normalized by the number of degrees of freedom ($NI \times NJ \times NK$), a more meaningful measure than flop rate, as a function of $NI$ for the Direct Slow solver and 8 iterations of the Preconditioned Conjugate Gradient solver with MILU(0) preconditioning.

The straight line is a (near perfect) fit to the Direct Slow solver data points, and this indicates that the matrix product is dominating the cost, and (27) provides a good performance model. The PCG-MILU(0) data points should be on a constant line according to (32), and this is roughly the case for the bigger problems. That this is not the case for P1 might be due to the fact that for this problem the average vector length is smaller. In any case, the iterative code has not been fully optimized yet, and the results obtained are a very encouraging upper bound for the method.

A possible refinement would be the use of a stopping criteria, or even preconditioning, that depends on the vertical mode index, since the Helmholtz constants vary by orders of magnitude and hence the condition number of $\mathbf{A}$.

## 6 Conclusion

A family of preconditioners for an iterative solver for an EBV problem in spherical geometry arising in numerical weather prediction has been investigated. The Preconditioned Conjugate Gradient method with MILU(0) preconditioning obtains practical convergence in a small number of iterations and outperforms a highly optimized direct solver at high-resolution. We have been able to maintain the vector performance of Environment Canada SX-4 NEC. Note that the Cenju by NEC is a scalar multiprocessor.

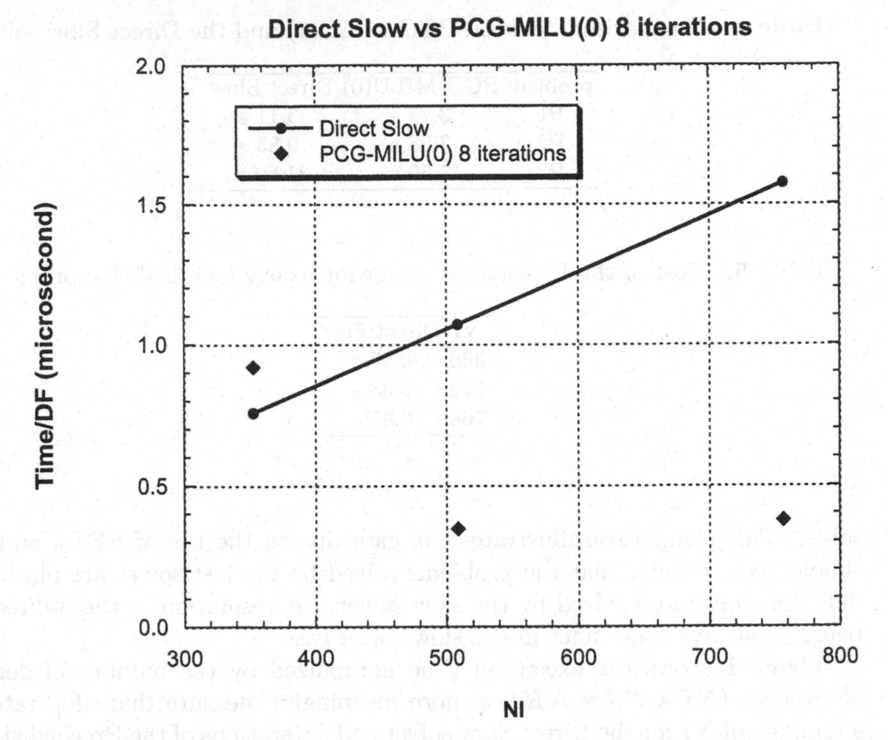


**Fig. 1.** Execution time/$(NI \times NJ \times NK)$ in microsecond as a function of $NI$ for the Direct Slow solver and 8 iterations of the Preconditioned Conjugate Gradient solver with MILU(0) preconditioning. The straight line is a (near perfect) fit to the Direct Slow solver data points

The future work will be devoted to developing a parallel distributed-memory implementation of this solver which, again will preserve the vector performance, while allowing several nodes of NEC SX-4 vector multiprocessors to work concurrently on the problem.

## Acknowledgments

This research was supported by the Office of Science (BER), U.S. Department of Energy, Grant No. DE-FG02-01ER63199. We gratefully acknowledge the help of Yves Chartier for the graphics.

## References

[1] Barrett, R., Berry, M., Chan, T. F., Demmel, J., Donato, J., Dongarra, J., Eijkhout, V., Pozo, R., Romine, C., van der Vorst, H.: Templates for the Solution of Linear Systems: Building Blocks for Iterative Methods. 2nd Edition, SIAM (1994).

[2] Côté, J., Gravel, S., Méthot, A., Patoine, A., Roch, M., Staniforth, A.: CMC-MRB Global Environmental Multiscale (GEM) model. Part I: design considerations and formulation. Monthly Weather Review 126 (1998) 1373-1395.

[3] Levesque, J. M., Williamson, J. W.: A Guidebook to FORTRAN on Supercomputers. Academic Press, San Diego, California (1989).

[4] Meijerink, J., van der Vorst, H.: An iterative solution method for linear systems of which the coefficient matrix is a symmetric M-matrix. Mathematics of Computation, 31 (1977) 148-162.

[5] Qaddouri, A., Côté, J., Valin, M.: A parallel direct 3D elliptic solver. Proceedings of the 13th Annual International Symposium on High Performance Computing Systems and Applications, Kingston, Canada, June 13-16 1999, Kluwer (1999).

[6] Saad, Y.: ILUT: A dual threshold incomplete LU Factorization. Numer. Linear Algebra Appl. 1 (1994) 387-402.

[7] Saad, Y., Malevsky, A.: P-SPARSLIB: A portable library of distributed memory sparse iterative solvers. Technical Report UMSI-95-180, Minnesota Supercomputing Institute, Minneapolis, MN (1995).

[8] Saad, Y.: Iterative methods for sparse linear systems. International Thomson Publishing (1996).

[9] Temperton, C.: Self-sorting mixed-radix fast Fourier transforms. Journal of Comp. Physics 52 (1983) 1-23.

[10] van der Vorst, H.: (M)ICCG for 2D problems on vector computers, in Supercomputing. A. Lichnewsky and C. Saquez, eds., North-Holland (1988).

# 2-D R-Matrix Propagation: A Large Scale Electron Scattering Simulation Dominated by the Multiplication of Dynamically Changing Matrices

Timothy Stitt[1], N. Stan Scott[1], M. Penny Scott[2], and Phil G. Burke[2]

[1] School of Computer Science
Queen's University Belfast, Belfast BT7 1NN, UK
{t.stitt,ns.scott}@qub.ac.uk
Phone: +44 (0)28 9027 4647/4626 Fax: +44(0)28 9068 3890
[2] Department of Applied Mathematics and Theoretical Physics
Queen's University Belfast, Belfast BT7 1NN, UK
{m.p.scott,p.burke}@qub.ac.uk
Phone: +44 (0)28 9027 3197

**Abstract.** We examine the computational aspects of propagating a global R-matrix, $\mathfrak{R}$, across sub-regions in a 2-D plane. This problem originates in the large scale simulation of electron collisions with atoms and ions at intermediate energies. The propagation is dominated by matrix multiplications which are complicated because of the dynamic nature of $\mathfrak{R}$, which changes the designations of its rows and columns and grows in size as the propagation proceeds. The use of PBLAS to solve this problem on distributed memory HPC machines is the main focus of the paper.

Topics: Large Scale Simulations in all areas of Engineering and Science; Numerical Methods; Parallel and Distributed Computing.

## 1 Introduction

Electron-atom and electron-molecule collisions drive many of the key chemical and physical processes in important environments that range from plasmas to living tissue[1]. Despite the importance of this wide range of applications, relatively little accurate data is known for many of the processes involved. This is even the case for electron collisions with the 'simplest' target, hydrogen. Here there is a dearth of accurate data once one goes beyond excitation to the few lowest lying states. One reason is that the accurate treatment of electron impact excitation and ionization of atoms and ions at intermediate energies is difficult because account must be taken of the infinite number of continuum states of the ionized target and of the infinite number of target bound states lying below the ionization threshold.

Traditional R-matrix theory[2] starts by partitioning configuration space into two regions by a sphere of radius $a$ centred on the target nucleus. The radius $a$ is

J.M.L.M. Palma et al. (Eds.): VECPAR 2002, LNCS 2565, pp. 354–367, 2003.

chosen so that the charge distribution of the target is contained within the sphere. In recent years, research has focused on the extension of this method to enable the accurate treatment of electron impact excitation and ionization of atoms/ions at intermediate energies. Two main approaches have been developed: the R-matrix with pseudo-states approach (RMPS[3]); and the intermediate energy R-matrix method (IERM[4]). Associated with the IERM method is the 2-dimensional R-matrix propagator[4]. This novel approach has the advantage of extending the boundary radius of the inner-region far beyond that which is possible using the traditional 'one-sector' technique.

During the last two years, our work has concentrated on re-engineering a prototype F77 2-D propagator program[5, 6] into a robust and flexible object-based F90 production code capable of exploiting both DM-MIMD and SMP architectures[7]. Considerable effort has been invested into abstracting from the user the need to be aware of the details of the underlying hardware and of mathematical libraries such as BLAS, LAPACK, ScaLAPACK and PBLAS [8]. Using the CSAR Cray T3E at Manchester[9], computations of increasing inner-region radius have been used to examine the disputed $^1S$ single differential cross section(sdcs) for electron impact ionization of hydrogen at 4 eV above threshold. The most recent results[10], with a radius of 600 a.u. and 110 states per angular momentum, are in excellent agreement with the highly regarded TDCC and ECS results[11].

In this paper we consider the computational aspects of 2-D R-matrix propagation on DM-MIMD machines. In §2 we introduce the fundamental propagation equations which involve the propagation of a global R-matrix, $\mathfrak{R}$, across sub-regions within a 2-D plane. The optimal route through this plane is identified in §3. The propagation equations are dominated by matrix multiplications involving submatrices of $\mathfrak{R}$. However, the multiplications are not straightforward in the sense that $\mathfrak{R}$ dynamically changes the designations of its rows and columns and increases in size as the propagation proceeds. The effect of this dynamic behaviour on the use of PBLAS is discussed in §4. Two sets of timings are presented in §5. The first involves propagating relatively small matrices over a large distance (600 a.u.) while the second involves propagating larger matrices over a smaller distance (200 a.u.). Finally, in §6, we make some concluding remarks.

## 2 Outline of 2-D R-Matrix Propagation Theory

This section highlights the salient features of the 2-D R-matrix propagation theory. A more thorough treatment of the theory can be found in [4], [12] and [13].

The 2-D R-matrix propagator method proceeds by subdividing the $(r_1, r_2)$ plane within the R-matrix inner-region into a set of connected sub-regions as shown in Figure 1(a). As illustrated in Figure 1(b) each sub-region has four edges $i \in \{1, 2, 3, 4\}$. Within each sub-region the motion of the scattering electron and the target electron are described in terms of a set of one-electron basis functions. These basis functions are in turn used to construct surface amplitudes $\omega_{inl_1l_2k}$ associated with each edge $i$. A set of local R-matrices $(R_{ji})$ relating the

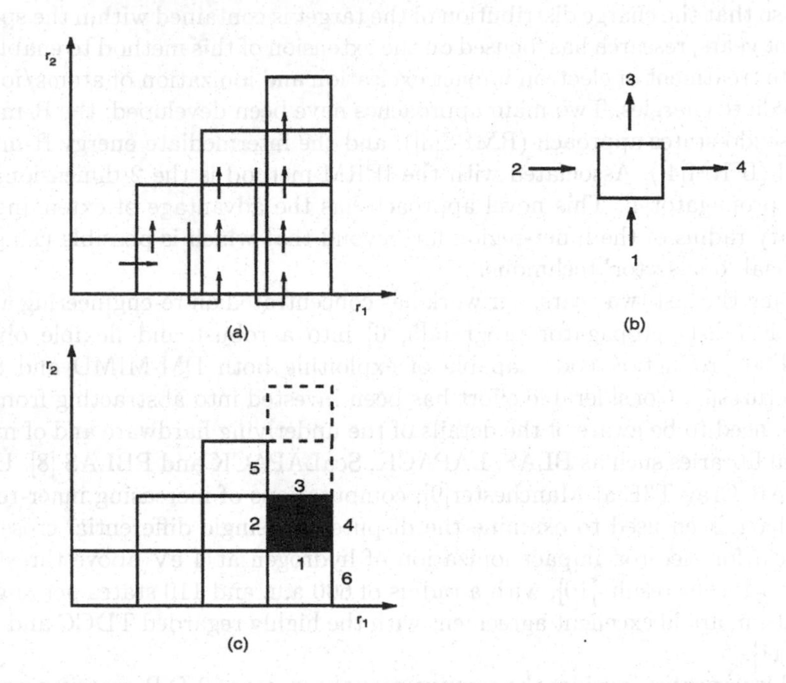


**Fig. 1.** (a)Subdivision of the $(r_1, r_2)$ plane into a set of connected sub-regions. (b) The input(2,1) and output(3,4) edges of a general sub-region. (c) Propagation across the shaded sub-region involves a global input $\Re$-matrix on edges 5($X$),2($I$),1($I$) and 6($X$) and a global output $\Re$-matrix on edges 5($X$),3($O$),4($O$) and 6($X$): $I$ are input edges, $O$ are output edges and $X$ are common edges

channels $n'l_1'l_2'$ associated with edge $j$ to the channels $nl_1l_2$ associated with edge $i$ can be constructed from the surface amplitudes as follows:

$$(R_{ji})_{n'l_1'l_2'nl_1l_2} = \frac{1}{2a_i} \sum_k \frac{\omega_{jn'l_1'l_2'k}\omega_{inl_1l_2k}}{E_k - E}, j, i \in \{1, 2, 3, 4\} \qquad (1)$$

Here $a_i$ is the radius of the $i^{th}$ edge, $E$ is the total energy of the two-electron system and $E_k$ are the eigenenergies obtained by diagonalising the two-electron Hamiltonian in the sub-region.

The fundamental task is to propagate the global $\Re$-matrix, $\Re$, across the subregions as illustrated in Fig. 1(a). Because of symmetry only the lower half of the $(r_1, r_2)$ plane needs to be considered. To see how propagation across one subregion is achieved we consider the shaded sub-region in Fig. 1(c). It is assumed that the global input $\Re$-matrix, $\Re^I$, associated with the input boundary defined by edges 5,2,1 and 6, is known. We wish to evaluate the global output $\Re$-matrix, $\Re^O$, associated with edges 5, 3, 4 and 6 on the output boundary. The edges 5 and

6 are common to both boundaries and are denoted by $X$. We write the global input $\Re$-matrix $\Re^I$ and the global output $\Re$-matrix $\Re^O$ in the form

$$\Re^I = \begin{pmatrix} \Re^I_{II} & \Re^I_{IX} \\ \Re^I_{XI} & \Re^I_{XX} \end{pmatrix}, \Re^O = \begin{pmatrix} \Re^O_{OO} & \Re^O_{OX} \\ \Re^O_{XO} & \Re^O_{XX} \end{pmatrix}, \tag{2}$$

and the local R-matrices in the sub-region as

$$r_{II} = \begin{pmatrix} R_{11} & R_{12} \\ R_{21} & R_{22} \end{pmatrix}, r_{IO} = \begin{pmatrix} R_{13} & R_{14} \\ R_{23} & R_{24} \end{pmatrix}, \tag{3}$$

$$r_{OI} = \begin{pmatrix} R_{31} & R_{32} \\ R_{41} & R_{42} \end{pmatrix}, r_{OO} = \begin{pmatrix} R_{33} & R_{34} \\ R_{43} & R_{44} \end{pmatrix}. \tag{4}$$

Using equations which relate the local and global R-matrices to the wave-function and its derivative on edges $I$,$O$ and $X$, it is straightforward to derive the following fundamental propagation equations which apply to an arbitrary off-diagonal sub-region. These equations are simplified when propagating across axis and diagonal sub-regions.

$$\Re^O_{OO} = r_{OO} - r_{OI}(r_{II} + \Re^I_{II})^{-1} r_{IO}, \tag{5}$$
$$\Re^O_{OX} = r_{OI}(r_{II} + \Re^I_{II})^{-1} \Re^I_{IX}, \tag{6}$$
$$\Re^O_{XO} = \Re^I_{XI}(r_{II} + \Re^I_{II})^{-1} r_{IO}, \tag{7}$$
$$\Re^O_{XX} = \Re^I_{XX} - \Re^I_{XI}(r_{II} + \Re^I_{II})^{-1} \Re^I_{IX}. \tag{8}$$

The solution of equations (5)-(8) in an efficient manner on distributed memory HPC machines is the principal focus of this paper.

## 3 Routes Through the Inner-Region

The route through the inner-region is not unique. For example, it is possible to proceed horizontally across rows rather than vertically in columns as illustrated in Fig. 1(a). However, the amount of computation required for each route is different. It is important, therefore, to ensure that the route taken minimizes the amount of computation. In this section we provide an informal proof that the route indicated in Fig. 1(a) is optimal.

Consider a sub-region $(p, q)$ shown in Fig. 2, where $p \geq 1$ and $q > 1$. To generate the south input edge to $(p, q)$ one must have at least propagated across sub-region $(p-1, q)$ generating at least one common east output edge. Applying this argument recursively, and terminating at $(1, q)$ which does not have a south input edge, there are at least $p-1$ common east edges. To generate the west input edge to $(p, q)$ one must have at least propagated across sub-region $(p, q-1)$ generating at least one common north output edge. Applying this argument recursively, and terminating at $(p, p)$ which does not have a west input edge or a north output edge(due to symmetry in diagonal sub-regions), there are at least $q-p-1$ common north edges when $p \neq q$ and 0 otherwise. This implies that

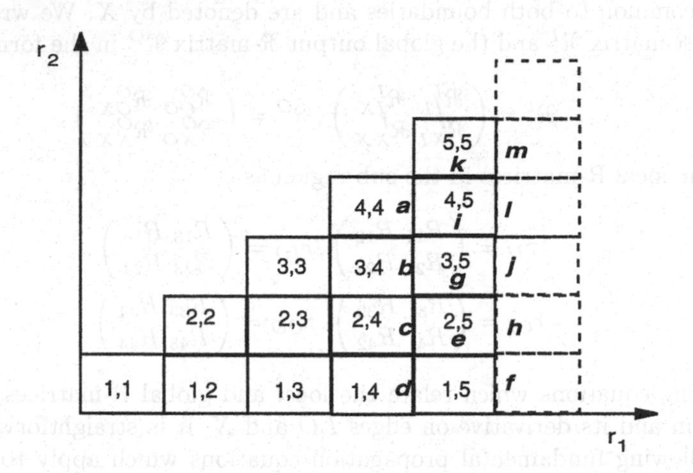


**Fig. 2.**

when propagating $\Re$ across a sub-region $(p, q)$, $\Re^I$ must have at least $q$ edges. [1] It is clear that the route illustrated in Fig. 1(a) is optimal in that each sub-region computation involves the minimum possible number of edges.

It should be noted that propagation is inherently serial. For example, in Fig. 2 propagation across $(1,2)$ can be followed by either $(2,2)$ or $(1,3)$. However, it is not possible to propagate across these two sub-regions concurrently. This is because the north edge of $(1,2)$, which is an input edge to $(2,2)$, will be changed during propagation across $(1,3)$.

## 4 The Dynamic Nature of $\Re$ during Propagation

Equations (5)-(8) are dominated by matrix multiplications. In addition, they appear to be of a form that can easily exploit the PSGEMM subroutine in the PBLAS library

$$C = \alpha AB + \beta C. \tag{9}$$

However, the dynamic nature of $\Re$ in equations (5)-(8) complicates the use of PSGEMM. These difficulties are explored in this section.

First, from §3, it can be seen that the size of $\Re$ increases as the propagation progresses from one column to the next. Therefore, the optimum number of processors over which to spread the computation varies from column to column.

[1] $(q-2)$ common edges and 2 input edges when $p \neq q$ and $(q-1)$ common edges and 1 input edge when $p = q$.

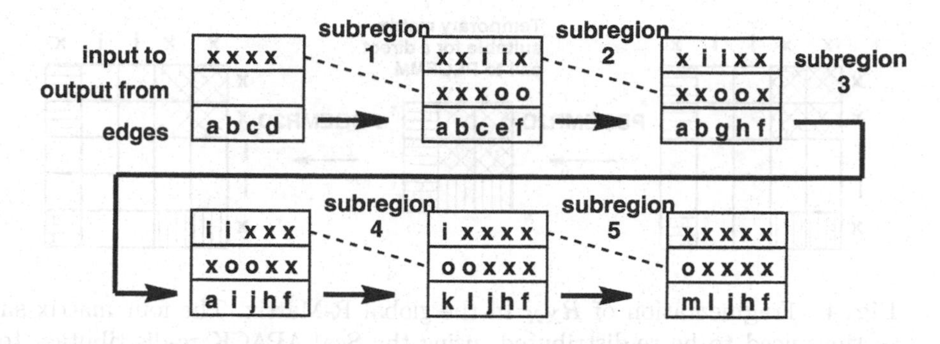


**Fig. 3.** Propagation across a strip. The edges associated with each sub-region at the input and output boundaries in column 5 of Fig. 2 are illustrated

Second, the designations of the rows and columns of $\Re$ change from one sub-region to the next. For example, when propagating across $(2,5)$ and $(3,5)$ the edges of $\Re^I$ are ordered as $XXIIX$ and $XIIXX$ respectively. These features are illustrated in Fig. 3 which describes the propagation through column 5 in Fig. 2.

Unfortunately, the approach outlined in Fig. 3 exhibits bad fragmentation of the the input R-Matrix $\Re^I$ rendering direct use of Eq.(9) impossible. Consider the solution to Eq. (11) in sub-region $(1,5)$. Fig. 4 shows that the four matrix sub-sections need to be re-distributed, using the ScaLAPACK re-distribution tool PSGEMR2D, into a single temporary matrix before $\Re^I_{XX}$ is in a suitable form for a direct call to PSGEMM. After the call to PSGEMM the four matrix sub-sections, which now represent $\Re^O_{XX}$, must be re-distributed back to their respective positions in the original matrix. This double re-distribution must also be performed for the computation of $\Re^O_{XO}$ and $\Re^O_{OX}$.

To avoid this potentially expensive communication overhead a new matrix-mapping method was devised. The algorithm is illustrated in Fig. 5. Here the output edges are always stored in the first two columns. The input edges are then located in the first and last columns. Before the call to PSGEMM, $\Re^I$, is re-distributed to a second matrix where the four submatrices are not fragmented e.g. $IXXXI \rightarrow IIXXX$. After the call to PSGEMM, $\Re^O$, is automatically in the correct location. Hence a second re-distribution is not necessary.

Unfortunately, when operating on sub-matrices PSGEMM can sometimes suffer from complicated *alignment restrictions*[14]. Alignment restrictions are constraints in the type of distributions and the indexing into the matrix that the user may utilize when calling a particular routine. For example, some routines will not accept submatrices whose starting index is not a multiple of the physical blocking factor. Absence of these restrictions can never be guaranteed as the size of the R-matrices are determined by the physics of the computation and are not directly under user control.

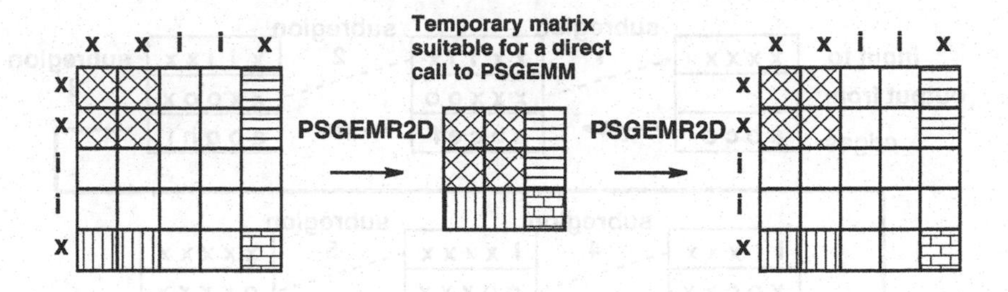


**Fig. 4.** Fragmentation of $R_{XX}$ in the global R-Matrix. The four matrix subsections need to be re-distributed, using the ScaLAPACK re-distribution tool PSGEMR2D, into a single temporary matrix before $\Re^I_{XX}$ is in a suitable form for a direct call to PSGEMM. After the call to PSGEMM the four matrix subsections, which now represent $\Re^O_{XX}$, must be re-distributed back to their respective positions in the original matrix

In the following section we investigate two approaches to solving the alignment restrictions problem. In the first approach the standard PBLAS library is used but array section references are not passed directly into the PBLAS routines. The sub-matrices $\Re_{XX}$,$\Re_{IX}$, $\Re_{XI}$ and $\Re_{OO}$ of $\Re^I$ are re-distributed, in similar fashion to that described by Fig. 4 , to independent matrices before the call to PSGEMM, thereby avoiding the boundary alignment restrictions. After the call, however, the independent matrices must be re-distributed back. This, in effect, removes much of the advantage of the new matrix-mapping approach discussed above. The second solution makes use of PBLAS V2, a prototype library using logical algorithmic blocking techniques to produce an alignment restric-

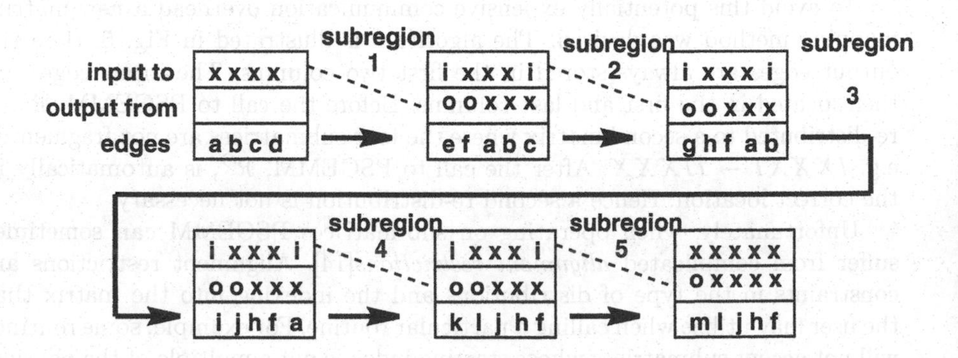


**Fig. 5.** Propagation across a strip using the new matrix-mapping method

**Table 1.** Total wall-clock times for the 600 a.u. case using 1,2,4,8,16 and 32 processors. The figures in brackets indicate the speed-up relative to 2PEs

| | BLAS | Standard PBLAS | | | | |
|---|---|---|---|---|---|---|
| | 1 | 2 | 4 | 8 | 16 | 32 |
| Propagation Time | 1664 | 787 | 440 (1.79) | 282 (2.79) | 201 (3.92) | 165 (4.77) |
| Re-distribution Time | 511 | 881 | 502 (1.75) | 380 (2.32) | 238 (3.70) | 206 (4.28) |
| Local R-matrix Time | 333 | 273 | 149 (1.83) | 155 (1.76) | 124 (2.20) | 98 (2.79) |
| Total Time | 2508 | 1941 | 1091 (1.78) | 817 (2.38) | 563 (3.45) | 469 (4.14) |
| | BLAS | PBLAS V2 | | | | |
| | 1 | 2 | 4 | 8 | 16 | 32 |
| Propagation Time | 1673 | 871 | 505 (1.72) | 328 (2.66) | 259 (3.36) | 194 (4.49) |
| Re-distribution Time | 274 | 541 | 315 (1.72) | 213 (2.54) | 139 (3.89) | 97 (5.58) |
| Local R-matrix Time | 336 | 287 | 151 (1.90) | 168 (1.71) | 155 (1.85) | 100 (2.87) |
| Total Time | 2283 | 1699 | 971 (1.75) | 709 (2.40) | 553 (3.07) | 391 (4.35) |

tion free PBLAS[15]. Here the advantage of the new matrix-mapping approach is exploited.

## 5 Illustrative Timings

In this section we present timings from two types of electron scattering simulations each with different computational characteristics. The first is from electron impact ionization and the second is from electron impact excitation. The computations were performed on a Cray T3E at CSAR in Manchester using untuned BLAS/PBLAS libraries on 1,2,4,8,16 and 32 processors.

Electron impact ionization is characterized by propagating moderately sized matrices over large distances. In this particular calculation, $\mathfrak{R}$, is propagated 600 a.u. across 946 sub-regions (43 columns or strips ). In strip 1, $\mathfrak{R}$, is a small $(57 \times 57)$ matrix. However, by the time it reaches strip 43 it has grown to $(2451 \times 2451)$. Each local R-matrix, defined by Eq. (1), is of order $(57 \times 57)$ and is constructed from surface amplitudes, $\omega$, of order $(1085 \times 57)$. Unlike the global R-matrix, $\mathfrak{R}$, the amount of computation involved in constructing a local R-matrix is constant and independent of the strip involved[2].

Table 1 summarizes, for both standard PBLAS and PBLAS V2, the total wall-clock time for: the propagation (Eqs. (5)-(8)); the associated re-distribution using PSGEMR2D; and the construction of the local R-matrices (Eqs. (1),(3) and (4)). Fig. 6 shows the average propagation plus re-distribution wall-clock time for sub-regions in each of the 43 strips using the standard PBLAS library. The corresponding re-distribution time alone is shown in Fig. 7.

[2] The amount of computation does depend on the sub-region type. However, the computation is dominated by off-diagonal sub-regions with 4 edges and hence 16 local R-matrices.

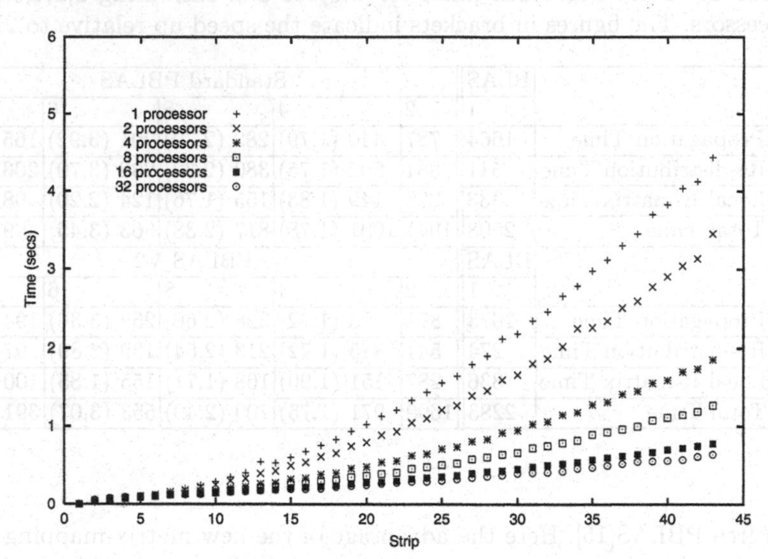


**Fig. 6.** 600 a.u. case: average propagation plus re-distribution wall-clock time for sub-regions in each of the 43 strips using standard PBLAS

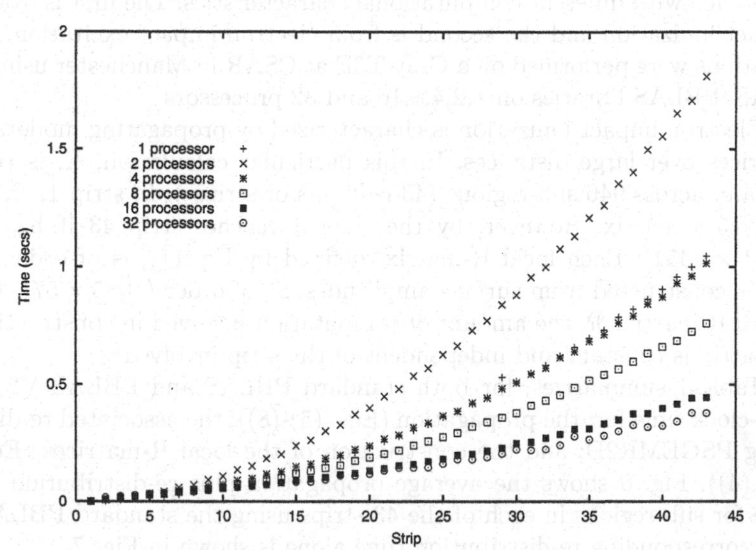


**Fig. 7.** 600 a.u. case: average re-distribution time for sub-regions in each of the 43 strips using PSGEMR2D

A striking feature of this simulation is that the time for re-distribution, using the standard PBLAS approach, is comparable to the time for propagation. The time taken to compute the local R-matrices is not insignificant, but relatively small by comparison.

The single processor times were obtained using BLAS. Here the re-distribution involved local sub-matrix copying rather than the block-cyclic re-distribution across processors required by PBLAS. Comparison of the times for 1 and 2 processors in Fig. 7 clearly shows the overhead incurred by using PS-GEMR2D.

Fig. 6 shows that the average time required to propagate across a sub-region increases as we move from column to column. This reflects the dynamic nature of $\Re$ as it grows in size during the propagation. In addition, the computation exhibits decreasing scalability as the number of processors increases. This is principally due to the increased communication and decreased computation per processor as more processors are added. The local R-matrix computation shows particularly bad scalability because of the small size of matrix involved.

The timings in Table 1 show a degradation in the performance of PSGEMM when using PBLAS V2. However, this is more than offset by the removal of the second re-distribution call to PSGEMR2D resulting in an improved overall performance throughout.

The second simulation involves electron impact excitation which is characterized by propagating larger matrices over significant but smaller distances. The example is taken from the $^{1,3}S$ symmetry in electron scattering from hydrogen where all target states up to n=7 inclusive are included. Here, $\Re$, is propagated 200 a.u. across 136 sub-regions (16 columns or strips). In strip 1, $\Re$, is a $(164 \times 164)$ matrix. When it reaches strip 16 it has rapidly grown to $(2624 \times 2624)$. Each local R-matrix is of order $(164 \times 164)$ and is constructed from surface amplitudes of order $(3404 \times 164)$.

**Table 2.** Total wall-clock times for the 200 a.u. case. The figures in brackets indicate the speed-up relative to 2PEs

| | Standard PBLAS | | | | |
|---|---|---|---|---|---|
| | 2 | 4 | 8 | 16 | 32 |
| Propagation Time | 313 | 179 (1.75) | 107 (2.93) | 68 (4.60) | 48 (6.52) |
| Re-distribution Time | 108 | 61 (1.77) | 32 (3.38) | 17 (6.35) | 12 (9.00) |
| Local R-matrix Time | 253 | 132 (1.92) | 91 (2.78) | 49 (5.16) | 53 (4.77) |
| Total Time | 674 | 372 (1.81) | 230 (2.93) | 134 (5.03) | 113 (5.96) |
| | PBLAS V2 | | | | |
| | 2 | 4 | 8 | 16 | 32 |
| Propagation Time | 333 | 194 (1.72) | 119 (2.80) | 82 (4.06) | 61 (5.46) |
| Re-distribution Time | 68 | 39 (1.74) | 20 (3.40) | 11 (6.18) | 8 (8.50) |
| Local R-matrix Time | 285 | 136 (2.10) | 111 (2.57) | 58 (4.91) | 67 (4.25) |
| Total Time | 686 | 369 (1.86) | 250 (2.74) | 151 (4.54) | 136 (5.04) |

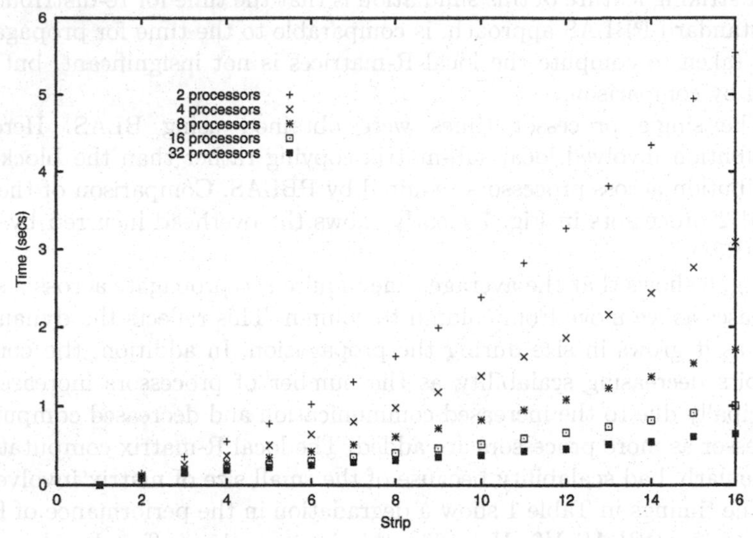


**Fig. 8.** 200 a.u. case: average propagation plus re-distribution times for sub-regions in each of the 16 strips using standard PBLAS

In this example the amount of memory required to perform the simulation cannot be accommodated by a single Cray T3E node(256 Mb). The simulation must, therefore, be distributed over at least two processors.

Table 2 summarizes, for both standard PBLAS and PBLAS V2, the total wall-clock time for: the propagation (Eqs. (5)-(8)); the associated re-distribution using PSGEMR2D; and the construction of the local R-matrices (Eqs. (1),(3) and (4)). Fig. 8 shows the average propagation plus re-distribution wall-clock time for sub-regions in each of the 16 strips using the standard PBLAS library. The corresponding re-distribution wall-clock time alone is shown in Fig. 9.

This example shows a different distribution of activity. The computation of the local R-matrices is now comparable to the propagation. The re-distribution time, while not insignificant, is much smaller.

Again the graphs show an increase in propagation and re-distribution time as we move from strip to strip. And, as before, the scalability decreases as the number of processors increases.

The timings for PBLAS V2 in Table 2 again show a degradation in the performance of PSGEMM. However, this time it is not offset by the removal of the second re-distribution call to PSGEMR2D, resulting in a small decline in overall performance throughout[3].

[3] Apart from the 4 processors case where there is a small improvement.

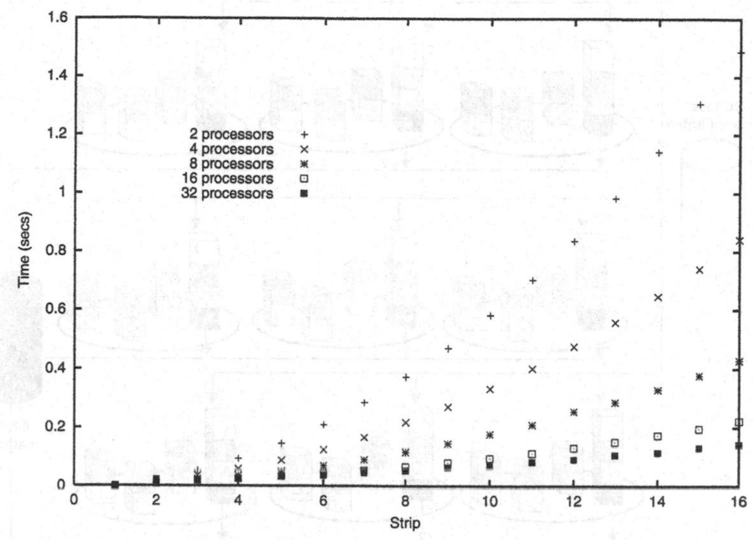


**Fig. 9.** 200 a.u. case: average re-distribution times for sub-regions in each of the 16 strips using PSGEMR2D

## 6 Concluding Remarks

In this paper we have described the use of the PBLAS Library within the context of an important problem in atomic physics. The computation involves the construction of local R-matrices within the sub-regions of a 2-D plane and the propagation of a global R-matrix, $\Re$, across the sub-regions. While the local R-matrices are of constant size within each sub-region the global R-matrix, $\Re$, changes the designations of its rows and columns and grows in size as the propagation proceeds.

We have identified the optimum route through the 2-D plane and established techniques to enable the use of PBLAS with and without alignment restrictions.

Timings have been presented for two complementary simulations with very different distributions of activity. Both simulations exhibit satisfactory scalability using up to about 4 procesors and declining scalability thereafter. This is not unexpected given the relatively small final size of $\Re$.

Both simulations involved the propagation of a single global R-matrix defined at a single total energy $E$. In a complete simulation many independent global R-matrices of different sizes need to be propagated over hundreds of independent energies. For example, in the $^{1,3}F$ symmetry in electron scattering from hydrogen (where all target states up to n=7 are included) $\Re$ grows rapidly from $(637 \times 637)$ to $(10,192 \times 10,192)$ in 16 strips. In these circumstances the use of a high performance computer is justified.

The timings presented in §5 suggest that rather that distributing a single propagation over many processors a more efficient and flexible strategy is to

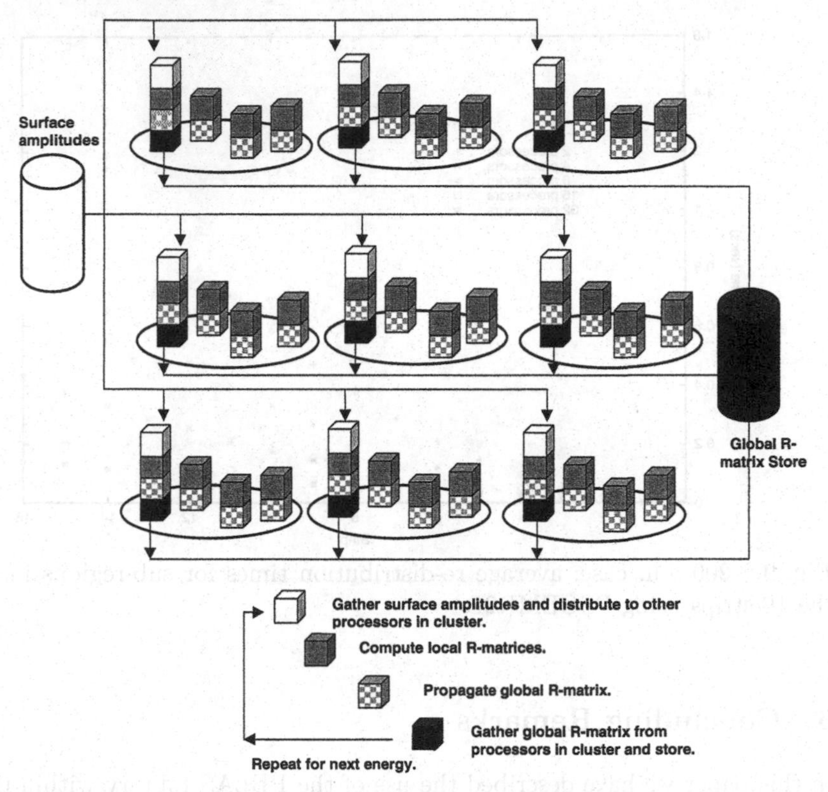


**Fig. 10.** Propagation of $\Re$ over a collection of energies using 9 clusters with 4 processors in each

choose a small number of processors that can accommodate the memory requirements and compute as many energies in parallel as the system configuration will allow. This is illustrated in Fig. 10 where 9 clusters, with 4 processors in each, are used to propagate a global R-matrix over a collection of energies.

## Acknowledgements

The authors are grateful to the UK EPSRC for their support through grants GR/M01784/01 and GR/R89073/01. They also wish to thank the CSAR staff at Manchester for their assistance in building the various BLAS/PBLAS libraries needed to carry out this work. TS acknowledges the receipt of an ESF/Queen's University postgraduate studentship.

## References

[1] Becker, K. H., McCurdy, C. W., Orlando, T. M., Rescigno, T. N.: Workshop on :Electron-Driven Processes:Scientific Challenges and Technological Opportunities. http://attila.stevens-tech.edu/physics/People/Faculty/Becker/EDP/EDP-FinalReport.pdf
[2] Burke, P. G., Berrington, K. A.(eds): Atomic and Molecular Processes and R-matrix Approach. IOP Publishing, Bristol (1993)
[3] Bartschat, K., Hudson, E. T., Scott, M. P., Burke, P. G., Burke, V. M.: J.Phys.B:At.Mol.Opt.Phys. **29** (1996) 115-13
[4] Burke, P. G., Noble, C. J., Scott, M. P.: Proc. Roy. Soc. **A 410**(1987) 287-310
[5] Heggarty, J. W., Burke, P. G., Scott, M. P., Scott, N. S.: Compt. Phys. Commun. **114** (1998) 195-209
[6] Heggarty, J. W.: ParalleL R-matrix Computation. Ph.D. Thesis, Queen's University Belfast (1999)
[7] Stitt, T.: Ph.D. Thesis, Queen's University Belfast, in preparation (2003)
[8] http://www.netlib.org/scalapack/scalapack_home.html
[9] http://www.csar.cfs.ac.uk/
[10] Scott, M. P., Burke, P. G., Scott, N. S. and Stitt, T. In: Madison, D. H. and Schultz, M.(eds): Correlations, Polarization, and Ionization in Atomic Systems. AIP Conference Proceedings, Vol.604. (2002) 82-89
[11] Baertschy, M., Rescigno, T. N., McCurdy, W. C., Coglan, J., Pindzloa,M. S.: Phys. Rev. A. **63** (2001) R50701
[12] Dunseath, K. M., Le Dourneuf, M., Terao-Dunseath, M., Launay, J. M.: Phys. Rev. A. **54** (1996) 561-572
[13] Le Dourneuf, M., Launay, J. M., Burke, P. G.: J. Phys. B. **23** (1990) L559
[14] Blackford, L. S., Choi, J., Cleary, A., D'Azevedo, E., Demmel, J., Dhillon, I., Dongarra, J., Hammarling, S., Henry, G., Petitet, A., Stanley, K., Walker, D., Whaley, R. C.: ScaLAPACK Users' Guide.
http://www.netlib.org/scalapack/slug/scalapack_slug.html
[15] http://netlib.bell-labs.com/netlib/scalapack/prototype/

# A Parallel Implementation of the Atkinson Algorithm for Solving a Fredholm Equation

Filomena Dias d'Almeida[1] and Paulo Beleza Vasconcelos[2]

[1] Faculdade Engenharia Univ. Porto
Rua Dr. Roberto Frias, 4200-465 PORTO, Portugal
falmeida@fe.up.pt

[2] Faculdade Economia Univ. Porto
Rua Dr. Roberto Frias 4200-464 PORTO, Portugal
pjv@fep.up.pt

**Abstract.** In this paper we study the parallelization of the Atkinson iterative refinement method. This algorithm will be used in conjunction with a projection method to produce an approximate solution of a radiation transfer problem in Astrophysics. The transfer problem is modelled via a Fredholm integral equation of the second kind with weakly singular kernel.

## 1 Introduction

The Atkinson algorithm is an iterative refinement method that can be used to solve systems of linear equations. In each iteration it only requires matrix-vector operations and the solution of a moderate size linear system obtained by projection techniques. It can be regarded as a two-grid method and it is well suited for the solution of large and sparse linear systems.

In this paper we will deal with the parallelization of this algorithm on a distributed memory architecture using MPI (Message Passing Interface) as message passing protocol.

The remainder of the paper is organized as follows. Section 2 describes the problem to be solved, give references to previous work and point out the motivation for the parallelization of the Atkinson method. In section 3 we recall how to solve it by projection techniques and describe how to construct the initial approximation to be refined by the Atkinson's method. The description of the Atkinson's method is done in section 4, where we present the sequential algorithm and one possible parallelization of it. Numerical results and conclusions will be presented.

## 2 Motivation

The study and development of such a method was motivated by the need to solve a Fredholm integral equation of the second kind with weakly singular kernel in an Astrophysics problem. This equation represents a linear integral formulation

J.M.L.M. Palma et al. (Eds.): VECPAR 2002, LNCS 2565, pp. 368–376, 2003.

of a transfer problem that belongs to a nonlinear and strongly coupled system of equations dealing with radiative transfer in stellar atmospheres [1]. The system includes transfer, structural and energy equations, and it becomes linear by considering given (or computed) temperature and pressure [2].

The kernel of the integral operator on a Banach space is expressed in terms of the first exponential-integral function [3].

The transfer problem is described by

$$T\varphi = z\varphi + f \tag{1}$$

where $T$ is a Fredholm integral operator of the second kind, $T : X \rightarrow X$, $X = L^1([0, \tau_0])$, defined by

$$(T\varphi)(\tau) = \frac{\varpi}{2}\int_0^{\tau_0} E_1\left(|\tau - \tau'|\right)\varphi\left(\tau'\right)d\tau', \tag{2}$$

where $z$ is in the resolvent set of $T$, $\tau_0$ is the optical depth of a stellar atmosphere, $\varpi \in ]0, 1[$ is the albedo and

$$E_1(\tau) = \int_1^{\infty} \frac{\exp(-\tau\mu)}{\mu} d\mu, \tau > 0 \tag{3}$$

is the first exponential-integral function. In this paper we consider the source term $f \in L^1([0, \tau_0])$ to be

$$f(\tau) = \begin{cases} -1 \text{ if } 0 \leq \tau \leq \frac{\tau_0}{2} \\ \;\;0 \text{ if } \frac{\tau_0}{2} < \tau \leq \tau_0 \end{cases} . \tag{4}$$

A finite number of linearly independent functions in the Banach space $X$ are taken, spanning a finite dimensional subspace $X_n$ where the integral operator is projected. The solution of the approximate equation corresponds to the solution of a large system of linear equations.

The solution by (block band) LU factorization becomes prohibitive for very large dimensional problems, pointing out the need for the use of iterative refinement schemes [1].

## 3 Projection Method for the Initial Approximation

To solve ( 1) a finite rank approximation $T_n$ of $T$ is considered

$$T_n x = \sum_{j=1}^{n} \langle x, l_{n,j}\rangle\, e_{n,j} \tag{5}$$

with $x$ and $e_{n,j} \in X$ and $l_{n,j} \in X^*$ (the topological adjoint space). For that we take a grid $0 := \tau_{n,0} < \cdots < \tau_{n,n} = \tau_0$ and define $h_{n,j} := \tau_{n,j} - \tau_{n,j-1}$ and $e_{n,j}(\tau)$ equal to zero except for $\tau \in ]\tau_{n,j-1}, \tau_{n,j}[$ where it is equal to one.

The solution of the approximate equation

$$T_n\varphi_n = z\varphi_n + f \tag{6}$$

leads to a linear system of $n$ equations

$$(A - zI)x = b, \tag{7}$$

where

$$A(i,j) = \frac{\varpi}{2h_{n,i}} \int_{\tau_{n,j-1}}^{\tau_{n,j}} \int_0^{\tau_0} E_1(|\tau - \tau'|)\, e_{n,j}(\tau')\, d\tau' d\tau, \tag{8}$$

$$b(i) = \frac{\varpi}{2h_{n,i}} \int_{\tau_{n,j-1}}^{\tau_{n,j}} \int_0^{\tau_0} E_1(|\tau - \tau'|)\, f(\tau')\, d\tau' d\tau \tag{9}$$

and $x(j) := \langle \varphi_n, l_{n,i} \rangle$. To recover the solution of ( 6) we need to compute

$$\varphi_n = \frac{1}{z}\left(\sum_{j=1}^{n} x(j) e_{n,j} - f\right). \tag{10}$$

For more details see [4].

The error bound for the solution obtained by this projection technique is given in [5].

To obtain a high precision solution for this projection method this system is required to have a large dimension. Alternatively we can compute a moderate size projection solution and refine it by Atkinson's iterative refinement method.

# 4 Atkinson Method for the Iterative Refinement

The Atkinson's algorithm is a Newton-type method to solve $Tx - zx - f = 0$, where the Jacobian is approximated by the resolvent operator of $T_n$ :

$$x^{(k+1)} = x^{(k)} - (T_n - zI)^{-1}(Tx^{(k)} - zx^{(k)} - f). \tag{11}$$

In practise $T$ is not used. We consider another grid $0 := \tau_{m,0} < \cdots < \tau_{m,m} = \tau_0$ and define $h_{m,j} := \tau_{m,j} - \tau_{m,j-1}$ and $m$ functions $e_{m,j}(\tau)$ equal to zero except for $\tau \in ]\tau_{m,j-1}, \tau_{m,j}[$ where it is equal to one. These functions span $X_m$. A finer projection discretization of $T$, $T_m$ $(m >> n)$, is used instead of $T$. Its restriction to the subspace $X_m$ is represented by $A_m$. The matrices $C$ and $D$ are, respectively, the restriction of $T_n$ to $X_m$ and the restriction of $T_m$ to $X_n$ [1]. Matrices $A_n$ and $A_m$ are sparse band matrices.

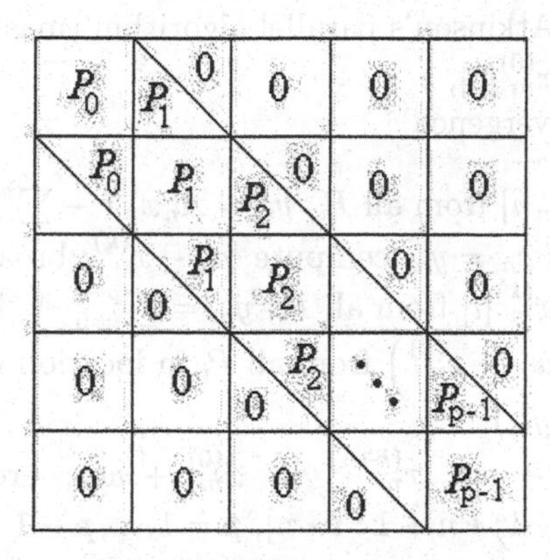


**Fig. 1.** Data distribution between processors: cyclic along diagonals

**Algorithm 1:** Atkinson's sequential algorithm
given $A_n$, $A_m$, $C$, $D$, $x_n^{(0)}$, $x_m^{(0)}$, $z$
repeat until convergence

1. $y_n = A_n x_n^{(k)} - C x_m^{(k)}$
2. solve $(A_n - zI)\, w_n = y_n$
3. $w_m = \frac{1}{z}\left(D(w_n - x_n^{(k)}) + A_m x_m^{(k)}\right)$
4. $x_n^{(k+1)} = x_n^{(0)} + w_n$
5. $x_m^{(k+1)} = x_m^{(0)} + w_m$
6. $k = k + 1$

For the parallelization of the previous algorithm let us assume $p$ processors, the master processor being $P_0$ and the slaves $P_i$, $i = 1, ..., p - 1$. Let us consider the distribution of the blocks of matrix $A_m$ cyclic by block diagonals as in Figure 1. We will assume that the matrix $A_n$ fits in the master processor's memory.

Matrices $C$ and $D$ are divided into blocks of dimension $n$ wich are cyclic distributed by the processors. Note that $C$ has dimension $m \times n$ and $D$, $n \times m$, and so $C$ is a row block-matrix and $D$ is a column block-matrix. This static distribution gives a good data balance strategy.

The notation $M[i]$ stands for the block part of matrix $M$ that is stored on the local memory of processor $P_i$, $M$ is either $C$, $D$ or $A_m$.

**Algorithm 2:** Atkinson's parallel algorithm (master processor $P_0$)
given $A_n$, $x_n^{(0)}$, $x_m^{(0)}[0]$, $z$
repeat until convergence

1. receive $C[i] * x_m[i]$ from all $P_i$; $y_n = A_n x_n^{(k)} - \sum_{i=1}^{p-1} C[i] * x_m[i]$
2. solve $(A_n - zI)\, w_n = y_n$; compute $w_n - x_n^{(k)}$; broadcast $w_n - x_n^{(k)}$
3. receive $A_m[i] * x_m^{(k)}[i]$ from all $P_i$; $y_m = \sum_{i=0}^{p-1} A_m[i] * x_m^{(k)}[i]$
4. receive $D[i] * \left(w_n - x_n^{(k)}\right)$ from all $P_i$ in location $w_m[(i-1)*n+1 : i*n]$
5. $w_m = \frac{1}{z}\left(w_m + y_m\right)$
6. $x_n^{(k+1)} = x_n^{(0)} + w_n$; $x_m^{(k+1)} = x_m^{(0)} + w_m$; broadcast $x_n^{(k+1)}$ and send to $P_i$ $x_m^{(k+1)}[(i-1)*n+1 : i*n]$, $p = 1, ..., p-1$
7. $k = k+1$

**Algorithm 3:** Atkinson's parallel algorithm (slave processor $P_i$)
given $A_m[i]$, $C[i]$, $D[i]$, $z$, $x_n^{(0)}$, $x_m^{(0)}[i]$
repeat until convergence

1. compute $C[i] * x_m^{(k)}[i]$; send $C[i] * x_m^{(k)}[i]$ to $P_0$
2. compute $A_m[i] * x_m^{(k)}[i]$; send $A_m[i] * x_m^{(k)}[i]$ to $P_0$
3. receive $\left(w_n - x_n^{(k)}\right)$ from $P_0$
4. compute $D[i] * \left(w_n - x_n^{(k)}\right)$; send $D[i] * \left(w_n - x_n^{(k)}\right)$ to $P_0$
5. receive $x_n^{(k+1)}$ and $x_m^{(k+1)}[i]$ from $P_0$
6. $k = k+1$

The algorithm is presented on a master/slave paradigm only for a more clear exposition. The SPMD (Single Program Multiple Data) parallel version of this algorithm is easy to derive. All the results that will be presented in the next section were obtained using a SPMD version of the above algorithm.

## 5 Numerical Results

The machine used for the computations was the "Beowulf Cluster" from the Faculdade de Engenharia da Universidade do Porto. The configuration of this machine consists on a front-end based on a 550 MHz dual Pentium III processor with 512MB RAM. Each one of the 22 nodes consists of a 450 MHz Pentium III processor with 512 KB L2 cache and 128MB RAM. The nodes are connected via a Fast Ethernet switch (100 Mbps). The operating system was Linux Slackware 7.0.

For message passing software, it was used MPI (Message Passing Interface) from the ANL-Argonne National Laboratory distribution (mpich 1.2). MPI is a standard specification for message-passing libraries and mpich is a portable implementation of the full MPI specification.

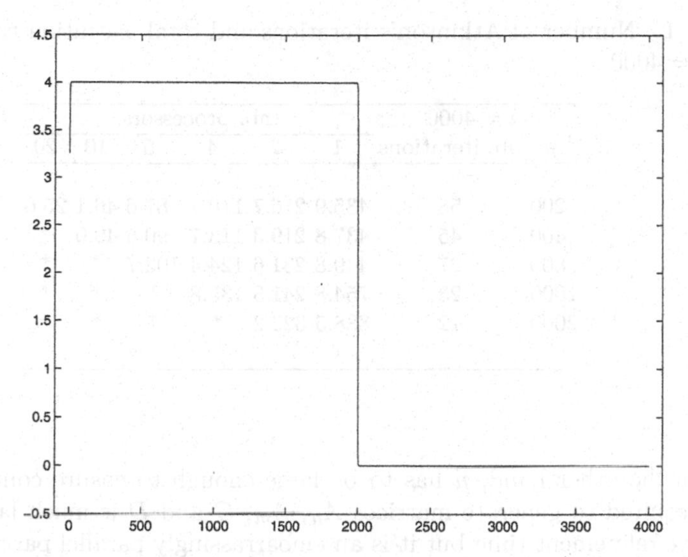

**Fig. 2.** Solution obtained by Atkinson's refinement formulae, $\tau_0 = 4000$, $m = 4000$ and $n = 200$

The solution of the problem ( 6) involves the generation of matrices $A_n$ and $A_m$ as defined in ( 8), the restriction matrices $C$ and $D$ as referred in section 4, the right-hand side defined in ( 9), and the computation of the solution of ( 7) by the Atkinson's iterative refinement method. All the computations involved in the algorithms were based on sparse data format CSR (compressed sparse row). Block band LU factorization was used to solve step 2. of the algorithm. The LU factorization (of a moderate size matrix $n$) is done only once and each $w_n$ is obtained by triangular solves inside the iterative refinement procedure.

All the tests considered the case $\varpi = 0.75$ and the stopping criterion for the refinement process was that of the relative residual norm on the discretization points less than $10^{-12}$.

In Figure 2 we show the vector solution for $\tau_0 = 4000$, $m = 4000$ and $n = 200$. The computation of the Atkinson's refinement method took 52 iterations for convergence. The same number of iterations was obtained using sparse Krylov iterative methods to solve the moderate size linear systems instead of block band LU factorization.

In Table 1 we summarize for $\tau_0 = 4000$, $m = 4000$ and for several values of $n$ the number of iterations required by the refinement iterative process and the total time (wall time) in seconds for several number of processors. By total time we mean the generation time plus the iterative refinement time.

As expected the number of iterations needed for the convergence of the Atkinson's iterative refinement method increases with the ratio $m/n$. It is important to maintain $n$ small to deal with small systems in step 2 of Algorithms 1 and 2.

**Table 1.** Number of Atkinson's iterations and total execution times (seconds) for $\tau_0 = 4000$

| $m = 4000$ | | nb. processors | | | | | |
|---|---|---|---|---|---|---|---|
| $n$ | nb. iterations | 1 | 2 | 4 | 5 | 10 | 20 |
| 200 | 58 | 435.9 | 216.7 | 110.7 | 87.6 | 46.1 | 25.6 |
| 400 | 45 | 437.8 | 219.3 | 113.7 | 90.5 | 49.0 | * |
| 800 | 27 | 449.8 | 231.6 | 124.4 | 102.7 | * | * |
| 1000 | 23 | 754.8 | 241.3 | 135.8 | * | * | * |
| 2000 | 12 | 888.5 | 322.2 | * | * | * | * |

but, on the other hand, $n$ has to be large enough to ensure convergence. The time required to generate matrices $A_n$, $A_m$, $C$ and $D$ is much larger than the iterative refinement time but it is an embarrassingly parallel part of the code.

Another important remark is that the number of refinement steps is kept constant when the number of processors grows.

Table 1 also shows large gains obtained by the parallelization of the code. The speedup is very close to the ideal one and for large values of $n$ (case $n = 1000$ and $n = 2000$) we get superlinear speedup. This is probably due to memory managements required in the serial version, such as cache miss and swap.

In Figure 3 we plot the speedup obtained using up to 20 processors for the case $n = 200$ and we can observe that the increase in the number of processors still keeps a speedup close to the ideal one.

In Table 2 we report the time in seconds for larger problems and the number of nonzeros concerning matrices $A_n$, $A_m$, $C$ and $D$. We can conclude that the parallel code shows scalability since a constant memory use per node allows efficiency to be maintained (comparing for instance the times for $\tau_0 = 1000$ and $\tau_0 = 10000$ vs. the number of processors we can observe isoefficiency).

**Table 2.** Number of nonzeros ($nnz$) and total execution times (seconds) for several $\tau_0$ values on several number of processors

| | | nb. processors | | | | | |
|---|---|---|---|---|---|---|---|
| $\tau_0$ | $nnz$ | 1 | 2 | 4 | 5 | 10 | 20 |
| 1000 | 242.832 | 43.3 | 12.8 | 6.8 | 5.8 | 3.6 | 2.2 |
| 4000 | 1070.532 | 431.9 | 216.7 | 110.7 | 87.6 | 46.1 | 25.6 |
| 10000 | 2725.932 | 4526.3 | 1385.9 | 695.6 | 559.8 | 286.9 | 151.7 |
| 20000 | 5484.932 | 18120.9 | 5589.8 | 2816.1 | 2248.6 | 1146.6 | 598.4 |

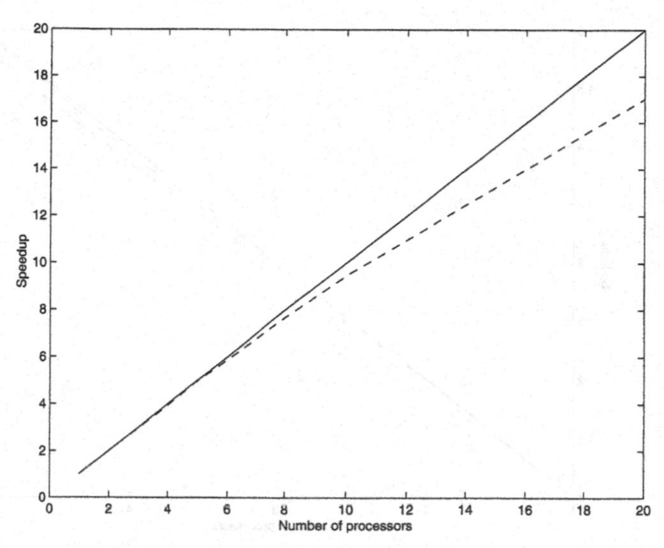


**Fig. 3.** Speedup obtained for the solution of the problem size $\tau_0 = 4000$ with albedo= 0.750 using up to 20 processors and iterative refinement parameters $m = 4000$, $n = 200$. The solid line is the linear speedup and the dashed line is the computed speedup

In Figure 4 we plot the speedups obtained using up to 5 processors for different values of $n$ for $\tau_0 = 4000$ and $m = 4000$ and we can observe almost linear speedups for $n$ equal to 200, 400 and 800. As already mentioned, the parallelization of the code allowed to increase the ratio $m/n$ at the cost of using more iterations on the refinement process, as we can also see from the speedups in Figure 4.

Empirical study of the optimal ratio $m/n$ is an ongoing work. Larger values of $n$ allows a convergence on less number of iterations on the iterative refinement procedure but implies the solution of larger linear systems.

In conclusion, we showed the parallel properties of this method, its scalability and efficiency for this problem.

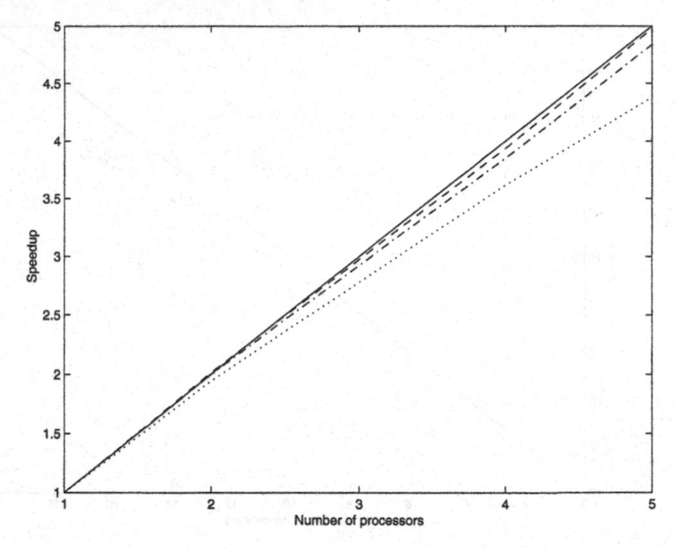


**Fig. 4.** Speedups obtained for the solution of the problem size $\tau_0 = 4000$ with albedo= 0.750 using up to 5 processors. The solid line is the linear speedup and the computed speedups for the iterative refinement parameters $n = 200$, $n = 400$ and $n = 800$ are, respectively, represented by the dashed line, the dashdot line and the doted line

## References


[1] Ahues, M., D'Almeida, F. D., Largillier, A., Titaud, O., Vasconcelos, P.: An $L1$ Refined Projection Approximate Solution of the Radiative Transfer Equation in Stellar Atmospheres. Journal of Computational and Apllied Mathematics **140** (2002) 13–26

[2] Busbridge, I. W.: The Mathematics of Radiative Transfer. Cambridge University Press (1960)

[3] Abramowitz, M., Stegun, I. A.: Handbook of Mathematical Functions. Dover, New York (1965)

[4] Ahues, M., D'Almeida, F. D., Largillier, A., Titaud, O., Vasconcelos, P.: Iterative Refinement Schemes for an Ill-Conditioned Transfer Equation in Astrophysics. Algorithms for Approximation IV **33** (2002) 70–77

[5] Ahues, M., Largillier, A., Titaud, O.: The roles of a weak singularity and the grid uniformity in the relative error bounds. Numer. Funct. Anal. and Optimiz., **22** (2001) 789–814

# SLEPc: Scalable Library for Eigenvalue Problem Computations

Vicente Hernández, Jose E. Román, and Vicente Vidal

Departamento de Sistemas Informáticos y Computación
Universidad Politécnica de Valencia, Camino de Vera, s/n, E-46022 Valencia, Spain
{vhernand,jroman,vvidal}@dsic.upv.es
Tel. +34-96-3877356, Fax +34-96-3877359

**Abstract.** The eigenvalue problem is one of the most important problems in numerical linear algebra. Several public domain software libraries are available for solving it. In this work, a new PETSc-based package is presented, which is intended to be an easy-to-use yet efficient object-oriented parallel framework for the solution of standard and generalised eigenproblems, either in real or complex arithmetic. The main objective is to allow the solution of real world problems in a straightforward way, especially in the case of large software projects.

**Topics.** Numerical methods, parallel and distributed computing.

## 1 Introduction

Together with linear systems of equations, eigenvalue problems are a very important class of linear algebra problems. The need for the numerical solution of these problems arises in many situations in science and engineering. There is a strong demand for solving problems associated with stability and vibrational analysis in practical applications, which are usually formulated as large sparse eigenproblems.

Computing eigenvalues is essentially more difficult than solving linear systems of equations. This has resulted in a very active research activity in the area of computational methods for eigenvalue problems in the last years, with many remarkable achievements. However, these state-of-the-art methods and algorithms are not easily transferred to the scientific community, and, apart from a few exceptions, scientists keep on using traditional well-established techniques.

The reasons for this situation are manifold. First, new methods are increasingly complex and difficult to implement and therefore robust implementations must be provided by computational specialists, for example as software libraries. The development of such libraries requires to invest a lot of effort but sometimes they do not reach normal users due to a lack of awareness (some initiatives [10] try to address this issue).

In the case of eigenproblems, using libraries is not straightforward. It is usually recommended that the user understands how the underlying algorithm works and typically the problem is successfully solved only after several cycles of testing

J.M.L.M. Palma et al. (Eds.): VECPAR 2002, LNCS 2565, pp. 377–391, 2003.

and parameter tuning. Methods are often specific for a certain class of eigenproblems (e.g. complex symmetric) and this leads to an explosion of available algorithms from which the user has to choose. Not all these algorithms are available in the form of software libraries, even less frequently with parallel capabilities.

A further obstacle appears when these methods have to be applied in a large software project developed by inter-disciplinary teams. In this scenery, libraries must be able to interoperate with already existing software and with other libraries, possibly written in a different programming language. In order to cope with the complexity associated with such large software projects, libraries must be designed carefully in order to overcome hurdles such as different storage formats. In the case of parallel software, care must be taken also to achieve portability to a wide range of platforms with good performance and still retain flexibility and usability. In this work, we present a new library for eigenvalue problems that aims to achieve these goals by making use of the good properties of PETSc. This new library offers a growing number of solution methods as well as interfaces to well-established eigenvalue packages.

The text is organised as follows. Section 2 gives a short overview of the mathematical problem and the methods used to solve it. A list of currently available software is also included. Section 3 discusses some software design issues and reviews the PETSc package. In section 4 the new library, SLEPc, is described, and some examples of usage are given. Section 5 presents some experimental results. Finally, the last section gives some conclusions.

## 2 The Eigenvalue Problem

The eigenvalue problem is a central topic in numerical linear algebra. In the standard formulation, the problem consists in the determination of $\lambda \in \mathcal{C}$ for which the equation

$$Ax = \lambda x \tag{1}$$

has nontrivial solution, where $A \in \mathcal{C}^{n \times n}$ and $x \in \mathcal{C}^n$. The scalar $\lambda$ and the vector $x$ are called eigenvalue and eigenvector, respectively. Quite often, the problem appears in generalised form, $Ax = \lambda Bx$. Other related linear algebra problems such as the quadratic eigenvalue problem or the singular value decomposition can be formulated as standard or generalised eigenproblems.

Many methods have been proposed to compute eigenvalues and eigenvectors [5]. Some of them, such as the QR iteration, are not appropriate for large sparse matrices because they are based on modifying the matrix by certain transformations which destroy sparsity. On the other hand, most applications require only a few selected eigenvalues and not the entire spectrum. In this work, we consider only methods that compute a few eigenpairs of large sparse problems.

Methods for sparse eigenproblems [2] usually obtain the solution from the information generated by the application of the matrix to various vectors. Matrices are only involved in matrix-vector products. This not only preserves sparsity but also allows the solution of problems in which matrices are not available explicitly.

The most basic method of this kind is the Power Iteration, in which an initial vector is repeatedly premultiplied by matrix $A$ and conveniently normalised. Under certain conditions, this iteration converges to the dominant eigenvector, the one associated to the largest eigenvalue in module. After this eigenvector has converged, deflation techniques can be applied in order to retrieve the next one. A variation of this method, called Inverse Iteration, can be used to obtain the eigenpair closest to a certain scalar $\sigma$ by applying $(A - \sigma I)^{-1}$ instead of $A$.

The Subspace Iteration is a generalisation of the Power Method, in which the matrix is applied to a set of $m$ vectors simultaneously, and orthogonality is enforced explicitly in order to avoid the convergence of all of them towards the dominant eigenvector. This method is often combined with a projection technique. The idea is to compute approximations to the eigenpairs of matrix $A$, extracting them from a given low-dimensional subspace on which the problem is projected. In the case of the Subspace Iteration, this subspace is the one spanned by the set of iteration vectors.

The projection scheme is common to many other methods. In particular, so-called Krylov methods use a projection onto a Krylov subspace,

$$\mathcal{K}_m(A, v) \equiv \text{span}\left\{v, Av, A^2v, \ldots, A^{m-1}v\right\} \quad . \tag{2}$$

The most basic algorithms of this kind are Arnoldi, Lanczos and non-symmetric Lanczos. These methods are the basis of a large family of algorithms.

The Arnoldi algorithm can be used for non-symmetric problems. Its main drawback with respect to Lanczos is the increment of cost and storage as the size of the subspace increases, since all the basis vectors must be kept. Several restart techniques have been proposed to avoid this.

There are several issues which are common to the methods mentioned above. One of them is the orthogonalisation strategy when constructing the basis. Different schemes are available and usually try to minimise problems with round-off errors. Another important aspect is the management of convergence. Locking already converged eigenvalues can considerably reduce the cost of an algorithm.

Convergence problems can arise in the presence of clustered eigenvalues. Selecting a sufficiently large number of basis vectors can usually avoid the problem. However, convergence can still be very slow and acceleration techniques must be used. Usually, these techniques consists in computing eigenpairs of a transformed operator and then recovering the solution of the original problem.

The most commonly used spectral transformation is called shift-and-invert and operates with the matrix $(A - \sigma I)^{-1}$. The value of the shift, $\sigma$, is chosen so that the eigenvalues of interest are well separated in the transformed spectrum thus leading to fast convergence. When using this approach, a linear system of equations must be solved whenever a matrix-vector product is required in the algorithm.

There are several parallel software libraries which approach the problem by some variant of the methods mentioned above. The most complete of these libraries is ARPACK [8, 11], which implements the Arnoldi/Lanczos process with Implicit Restart for standard and generalised problems, in both real and com-

**Table 1.** Available parallel software packages for sparse eigenproblems

| Package | Algorithm | Interface[a] |
|---|---|---|
| PLANSO | Lanczos with partial re-orthogonalisation | MV |
| BLZPACK | Block Lanczos with partial and selective re-orthogonalisation | RC |
| ARPACK | Implicitly Restarted Arnoldi/Lanczos | RC |
| TRLAN | Dynamic thick-restart Lanczos | MV |

[a] MV=matrix-vector product subroutine, RC=reverse communication.

plex arithmetic. Other available libraries are BLZPACK [9], PLANSO [13] and TRLAN [12]. All of them either require the user to provide a matrix-vector subroutine or use a reverse communication scheme. These two interfacing strategies are format-independent and allow the solution of problems with implicit matrices. Table 1 shows the algorithms implemented in these packages.

## 3 Object Oriented Software

In the commercial arena, object oriented methodologies have demonstrated to be beneficial, enabling transportability and re-usability. Object oriented codes are faster to write, easier to maintain, and easier to reuse.

In the case of scientific computing, these aspects are also important, but one of the most attractive claims is that software designed in this way is better able to interoperate with other codes. This is a major concern in large software projects developed by inter-disciplinary teams, such as multiphysics simulation programs, in which the complexity can be quite high especially in the case of parallel codes. In these projects, new software usually must be integrated into an existing base of software. The cost of rewriting existing codes can be very high, not only for the programming effort but also for validation. On the other hand, new software is often written in a different language than existing software, so language interoperability is also important.

A possible solution is to write wrappers that change a code's interface. Careful software design can bring solutions to all the issues mentioned above, and this can be achieved by applying object oriented techniques to create packages that improve and extend the traditional concept of subroutine library. Many of these packages are what might be called frameworks, which unlike traditional subroutine libraries, allow the user flexibility either in what specific operations the library performs, or in how the data is structured.

Benefits of object oriented design are not priceless. Most of the time there is a consequent loss of performance, due to the design in itself or to new features of object oriented languages (tailored syntax, polymorphism, etc.). One of the key challenges is to achieve high performance (including parallel performance) and retain flexibility.

SLEPc aims at providing a software library for eigenvalue computations which addresses all these issues. This is achieved by having its foundations on PETSc.

### 3.1 Review of PETSc

PETSc [4] is one of the most widely accepted parallel numerical frameworks, due to a large extent to the level of support to the end users. Feedback from users has in turn allowed its developers to constantly incorporate many improvements. PETSc focuses on components required for the solution of partial differential equations and it provides support for problems arising from discretisation by means of regular meshes as well as unstructured meshes. Its approach is to encapsulate mathematical algorithms using object-oriented programming techniques in order to be able to manage the complexity of efficient numerical message-passing codes. All the PETSc software is freely available and used around the world in many application areas (a list is available in [3]).

PETSc is built around a variety of data structures and algorithmic objects (see Figure 1). The application programmer works directly with these objects rather than concentrating on the underlying data structures. The three basic abstract data objects are index sets (`IS`), vectors (`Vec`) and matrices (`Mat`). Built on top of this foundation are various classes of solver objects, including linear, nonlinear and time-stepping solvers. These solver objects encapsulate virtually all information regarding the solution procedure for a particular class of problems, including the local state and various options such as convergence tolerances, etc. Options can be specified by means of calls to subroutines in the source code and also as command-line arguments.

PETSc is written in C, which lacks direct support for object-oriented programming. However, it is still possible to take advantage of the three basic principles of object-oriented programming to manage the complexity of such a large package. PETSc uses *data encapsulation* in both vector and matrix data objects. Application code access data through function calls. Also, all the operations are supported through *polymorphism*. The user calls a generic interface routine which then selects the underlying routine which handles the particular data structure. This is implemented by structures of function pointers. Finally, PETSc also uses *inheritance* in its design. All the objects are derived from an abstract base object. From this fundamental object, an abstract base object is defined for each PETSc object (`Mat`, `Vec` and so on) which in turn has a variety of instantiations that, for example, implement different matrix storage formats.

The key is to cleanly separate the interface to the user from the implementation. This separation is particularly important in PETSc because a single interface (e.g., `SLESSolve` for solving linear systems of equations) may invoke one of many different implementations (e.g., parallel or sequential, different Krylov methods and preconditioners and different matrix storage formats). Since PETSc provides a uniform interface to all of its linear solvers —the Conjugate Gradient, GMRES and others— and a large family of preconditioners —block Jacobi, overlapping additive Schwarz and others—, one can compare several combinations of method and preconditioner by simply specifying them at execution time.

PETSc focuses on portability, extensibility and flexibility and includes other good properties such as support for debugging and profiling and also a Fortran programming interface.

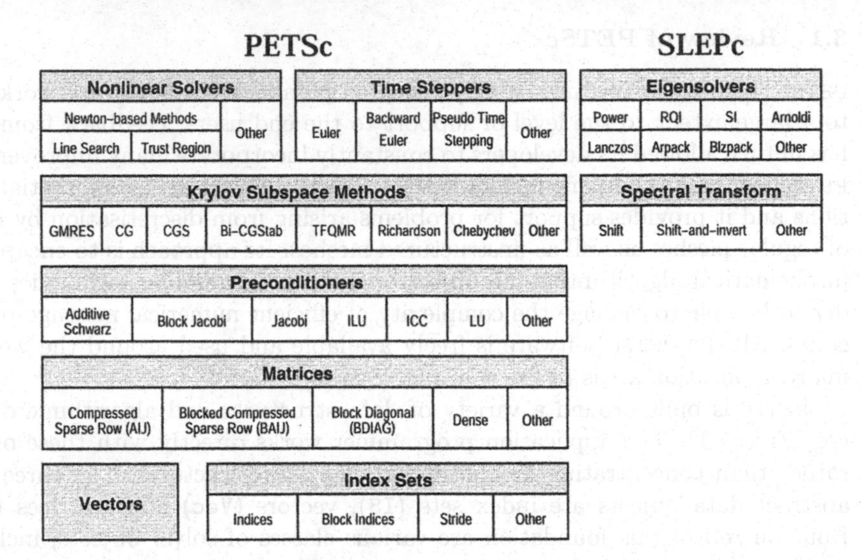


**Fig. 1.** Structure of the PETSc (on the left) and SLEPc (on the right) packages

## 4 The SLEPc Library

SLEPc, the Scalable Library for Eigenvalue Problem Computations, is a new library which extends the functionality of PETSc to solve large sparse eigenvalue problems. Some of the features of the library are the following:

- It can be used for standard and generalised problems.
- It can solve problems in real and complex arithmetic.
- It includes solvers for either Hermitian and non-Hermitian problems.
- The problem can be specified with any instance of matrix object provided by PETSc, including matrix-free problems.
- It provides a Fortran interface.

The library has been designed in such a way to exploit all the functionality of PETSc. Also, other design issues have been taken into account:

- To allow the user to choose among a list of algorithms.
- To keep acceleration techniques isolated from solution algorithms so that the user can specify any combination of them.
- To allow the selection of different variants of the methods, for example, to change the orthogonalisation technique or switch between symmetric or non-symmetric versions.
- To set reasonable default values for the parameters, some of which can be different depending on the method.
- To allow method-specific parameters in addition to general parameters.
- Interoperability with already available packages for eigenvalue problems.

The installation of SLEPc is straightforward. Assuming that PETSc is already installed in the system and that the environment variables $PETSC_DIR and $PETSC_ARCH are set appropriately, the user only has to set the $SLEPC_DIR variable and then type make in this directory. Optionally, the user can edit a configuration file to specify the location of external libraries (ARPACK, BLZPACK, PLANSO or TRLAN) installed in the system.

The new functionality provided by the SLEPc library is organised around two objects, EPS and ST, as shown in the diagram of Figure 1.

### 4.1 EPS: Eigenvalue Problem Solver

The Eigenvalue Problem Solver (EPS) is the main object provided by SLEPc. It is used to specify an eigenvalue problem, either in standard or generalised form, and provides uniform and efficient access to all of the package's eigensolvers. Conceptually, the level of abstraction occupied by EPS is similar to other solvers in PETSc such as SLES for solving linear systems of equations.

Apart from the usual object management functions (EPSCreate, EPSDestroy, EPSView, EPSSetFromOptions), the EPS object provides functions for setting several parameters such as the number of eigenvalues to compute, the dimension of the subspace, the requested tolerance and the maximum number of iterations allowed. The user can also specify other things such as the orthogonalisation technique or the portion of the spectrum of interest.

The solution of the problem is obtained in several steps. First, the matrices associated to the eigenproblem are specified via EPSSetOperators. Then, a call to EPSSolve is done which invokes the subroutine for the selected eigensolver. EPSGetConverged can be used afterwards to determine how many of the requested eigenpairs have converged to working precision. Finally, EPSGetSolution is used to retrieve the eigenvalues and eigenvectors.

The solution method can be specified procedurally or via the command line. The currently available methods are the following:

- Power Iteration with deflation. When combined with shift-and-invert (see below), it is equivalent to the Inverse Iteration.
- Rayleigh Quotient Iteration.
- Subspace Iteration with non-Hermitian projection and locking.
- Arnoldi method with explicit restart and deflation.

In addition to these methods, there are also wrappers for ARPACK, BLZPACK, PLANSO, TRLAN and also a wrapper for some dense methods in LAPACK.

### 4.2 ST: Spectral Transformation

The other main SLEPc object is the Spectral Transformation (ST), which encapsulates the functionality required for acceleration techniques based on the transformation of the spectrum. The user does not usually need to create an ST object explicitly. Instead, every EPS object internally sets up an associated ST.

**Table 2.** Operators used in each spectral transformation mode

| ST | Standard problem | Generalized problem |
|---|---|---|
| `none` | $A$ | $B^{-1}A$ |
| `shift` | $A+\sigma I$ | $B^{-1}A+\sigma I$ |
| `sinvert` | $(A-\sigma I)^{-1}$ | $(A-\sigma B)^{-1}B$ |

One of the design cornerstones of SLEPc is to separate spectral transformations from solution methods so that any combination of them can be specified by the user. To achieve this, all the eigensolvers contained in `EPS` must be implemented in a way that they are independent of which transformation has been selected by the user. That is, the solver algorithm has to work with a generic operator, whose actual form depends on the transformation used. After convergence, eigenvalues are transformed back appropriately, if necessary. Table 2 lists the operator used in each case, either for standard or generalised eigenproblems.

By default, no transformation is done (`none`). The other two options represent shift of origin (`shift`) and shift-and-invert (`sinvert`) transformations. In both cases, the value of the shift can be specified at run time.

The expressions shown in Table 2 are not built explicitly. Instead, the appropriate operations are carried out when applying the operator to a certain vector. The inverses imply the solution of a linear system of equations which is managed by setting up an associated `SLES` object. The user can control the behaviour of this object by adjusting the appropriate options, as will be illustrated in the examples below.

### 4.3 Supported Matrix Objects and Extensibility

Methods implemented in SLEPc merely require vector operations and matrix-vector products. In PETSc, mathematical objects such as vectors and matrices are defined in a uniform fashion without making unnecessary assumptions about internal representation. In particular, the parallelism within SLEPc methods are handled completely by PETSc's vector and matrix modules. This implies that SLEPc can be used with any of the matrix and vector representations provided by PETSc, including dense and the different parallel sparse formats.

In many applications, the matrices that define the eigenvalue problem are not available explicitly. Instead, the user knows a way of applying these matrices to a vector. An intermediate case is when the matrices have some block structure and the different blocks are stored separately. There are numerous situations in which this occurs, such as the discretisation of equations with a mixed finite-element scheme. An example is the eigenproblem arising in the stability analysis associated with Stokes problems,

$$\begin{bmatrix} A & C \\ C^H & 0 \end{bmatrix} \begin{bmatrix} x \\ p \end{bmatrix} = \lambda \begin{bmatrix} B & 0 \\ 0 & 0 \end{bmatrix} \begin{bmatrix} x \\ p \end{bmatrix} , \tag{3}$$

where $x$ and $p$ denote the velocity and pressure fields. Similar formulations also appear in many other situations. Many of these problems can be solved by reformulating them as a reduced-order standard or generalised system, in which the matrices are equal to certain operations of the blocks.

All these cases can be easily handled in SLEPc by means of shell matrices. These are matrices which do not require explicit storage of the component values. Instead, the user must provide subroutines for all the necessary matrix operations, typically only the application of the linear operator to a vector. Shell matrices are a simple mechanism of extensibility, in the sense that the package is extended with new user-defined matrix objects. Once the new matrix has been defined, it can be used by SLEPc as a regular matrix.

SLEPc further supports extensibility by allowing application programmers to code their own subroutines for unimplemented features such as a new eigensolver. It is possible to register these new methods to the system and use them as the rest of standard subroutines. In the case of spectral transformations, ST, there is also a `shell` transformation provided for implementing user-defined transformations in a similar way as shell matrices.

## 4.4 Examples of Usage

In this section some examples are presented to illustrate the usage of the package, in terms of programming as well as from the command line.

The following C source code implements the solution of a simple standard eigenvalue problem. Code for setting up the matrix $A$ is not shown and error-checking code is omitted.

```
 1 #include "slepceps.h"
 2
 3 Vec         *x;            /* basis vectors */
 4 Mat         A;             /* operator matrix */
 5 EPS         eps;           /* eigenproblem solver context */
 6 PetscReal   *error;
 7 Scalar      *kr,*ki;
 8 int         its, nconv;
 9
10 EPSCreate( PETSC_COMM_WORLD, &eps );
11 EPSSetOperators( eps, A, PETSC_NULL );
12 EPSSetFromOptions( eps );
13 EPSSolve( eps, &its );
14 EPSGetConverged( eps, &nconv );
15 EPSGetSolution( eps, &kr, &ki, &x );
16 PetscMalloc( nconv*sizeof(PetscReal), &error );
17 EPSComputeError( eps, error );
18 EPSDestroy( eps );
```

The header file included in the first line loads all the object definitions as well as function prototypes.

All the operations of the program are done over a single EPS object, the eigenvalue problem solver. This object is created in line 10. In line 11, the operator of the eigenproblem is specified to be the object A of type Mat, which could be a user-defined shell matrix.

At this point, the value of the different options could be set by means of a function call such as EPSSetTolerances. After this, a call to EPSSetFromOptions should be made as in line 12. The effect of this call is that options specified at runtime in the command line are passed to the EPS object appropriately. In this way, the user can easily experiment with different combinations of options without having to recompile.

Line 13 launches the solution algorithm. The subroutine which is actually invoked depends on which solver has been selected by the user. At the end, the number of iterations carried out by the solver is returned in its and all the data associated to the solution of the eigenproblem is kept internally. Line 14 queries how many eigenpairs have converged to working precision. The solution of the eigenproblem is retrieved in line 15. The $j$-th eigenvalue is kr[j]+$i\cdot$ki[j] and the $j$-th eigenvector is stored in the Vec object x[j]. In line 17 the relative residual error $\|Ax_j - \lambda_j Bx_j\|/|\lambda_j|$ associated to each eigenpair is computed. Finally, the EPS context is destroyed.

The command line for executing the program could be the following (MPI directives are omitted),

```
$ program -eps_nev 10 -eps_ncv 24
```

where the number of eigenvalues and the dimension of the subspace have been specified. In this other example, the solution method is given explicitly and the matrix is shifted with $\sigma = 0.5$

```
$ program -eps_type subspace -st_type shift -st_shift 0.5
          -eps_mgs_orthog -eps_monitor
```

The last two keys select the modified Gram-Schmidt orthogonalisation algorithm and instruct to activate the convergence monitor, respectively.

A slight change in the above code is sufficient to solve a generalised problem:

```
12 EPSSetOperators( eps, A, B );
```

In this case, the user can additionally specify options relative to the solution of the linear systems. For example,

```
$ program -st_type none -st_ksp_type gmres -st_pc_type ilu
          -st_ksp_tol 1e-7
```

In the above example, the prefix st_ is used to indicate an option for the linear system of equations associated to the ST object of type none. Similarly, for using the shift-and-invert technique with $\sigma = 10$:

```
$ program -st_type sinvert -st_shift 10 -st_pc_type jacobi
          -st_ksp_type cg -st_sinvert_shift_mat
```

The last option can be used to force the construction of $(A - \sigma I)$ or $(A - \sigma B)$, which are kept implicit otherwise. The user must consider if this is appropriate for the particular application. On the other hand, it is necessary in order to be able to use certain types of preconditioners for the linear systems.

## 5 Experimental Results

This section presents some experimental results aimed at assessing the performance of the SLEPc library. All the experiments have been run on a cluster of PC's made up of 12 nodes with Pentium III processors at 866 MHz connected by a Fast Ethernet network and Linux 2.4 operating system.

The experiments have been carried out with matrices taken from the NEP collection [1], which is available at the Matrix Market[1]. SLEPc has also been used to successfully solve other problems arising from real applications such as the one described in [6].

### 5.1 Convergence Tests

With SLEPc it is very easy to compare the behaviour of different solution methods when applied to a particular problem, as well as with the available spectral transformations. This can be done simply by selecting them at the command line. Table 3 compares the convergence behaviour of several combinations of solution method and spectral transformation for three different test cases. The figures shown in the table correspond to the number of operator-vector products carried out by the solution method when computing the first eigenvalue.

The results show that in all the cases the performance of the **arpack** solver is superior to the rest of solution methods. This result is not surprising and will hold in general for other eigenproblems as well, since the method implemented in ARPACK is much more advanced compared to the rest.

With respect to the spectral transformation, it should be noted that the first column cannot be compared with the other two since the converged eigenpairs are different in both cases. The shift-and-invert transformation reduces the number of operator-vector products, although in this case the cost of these products is considerably higher.

### 5.2 Parallel Performance

For the analysis of parallel performance, the wall time corresponding to the solution of the eigenvalue problem has been measured for a different number of processors. The experiment has been carried out for test case QC of order 2534, which is a complex symmetric eigenproblem arising from quantum chemistry. The times are shown in table 4, together with speedup and efficiency.

The Subspace Iteration solver was used to compute one eigenvalue with 9 basis vectors. The experiment also used the shift-and-invert technique with $\sigma = 2$

[1] `http://gams.nist.gov/MatrixMarket`.

**Table 3.** Comparison of different solution strategies for several test cases taken from the Matrix Market, where $n$ is the dimension of the matrix and $\sigma$ is the shift

| Test Case | eps_type | ST | | |
|---|---|---|---|---|
| | | none | shift | sinvert |
| RDB | power | 146 | 402 | 327 |
| $n = 200$ | subspace | 293 | 653 | 293 |
| $\sigma = 40$ | arnoldi | 61 | 91 | 31 |
| | arpack | 30 | 39 | 24 |
| DW | power | 16354 | 15262 | 77 |
| $n = 2048$ | subspace | 2309 | 2885 | 221 |
| $\sigma = -0.5$ | arnoldi | 4096 | 3891 | 31 |
| | arpack | 174 | 246 | 12 |
| QC | power | 90 | 8066 | 2962 |
| $n = 324$ | subspace | 221 | 11669 | 4721 |
| $\sigma = -1$ | arnoldi | 70 | 988 | 281 |
| | arpack | 27 | 693 | 174 |

**Table 4.** Execution time (in seconds), speedup and efficiency (in %)

| $p$ | $T_p$ | $S_p$ | $E_p$ | $T_p$ | $S_p$ | $E_p$ |
|---|---|---|---|---|---|---|
| 1 | 101.29 | 1.00 | 100 | 80.01 | 1.00 | 100 |
| 2 | 53.09 | 1.91 | 95 | 41.19 | 1.94 | 97 |
| 4 | 28.89 | 3.51 | 88 | 31.69 | 2.52 | 63 |
| 8 | 19.14 | 5.29 | 66 | 20.79 | 3.85 | 48 |
| 11 | 14.13 | 7.17 | 65 | 12.89 | 6.21 | 56 |

and the associated linear systems were solved with GMRES and Jacobi (on the left side of the table) or Block Jacobi preconditioning (on the right).

In the case of Jacobi preconditioning, the efficiency is maintained above 65 %. When using Block Jacobi preconditioning, the solution is faster in 1 processor but speedup is not so regular as in the other case, due to the fact that the effectiveness of the preconditioner depends on the number of processors used.

### 5.3 Overhead of the Library

Another test has been carried out in order to estimate the overhead of the SLEPc library, that is, which percentage of the execution time can be attributed to the utilisation of object-oriented programming compared to a traditional programming approach.

The experiment consists in executing two different programs, one of them is a Fortran program which solves a certain problem by calling ARPACK, and the other one is a C program which uses SLEPc's built-in ARPACK wrapper to solve the same problem. Since the underlying computations are exactly the same in both cases, the difference of the turnaround times gives us an idea of the overhead

**Table 5.** Turnaround times (in seconds)

| $n$ | ARPACK | SLEPC |
|---|---|---|
| 10000 | 22.91 | 22.83 |
| 22500 | 142.60 | 142.87 |

introduced by SLEPc, not only caused by object management and storage formats but also by other issues such as dynamic linkage of libraries.

The test case used for this experiment is one of the examples provided in the ARPACK distribution file: the eigenpairs of the 2-dimensional Laplacian on the unit square. Table 5 shows turnaround times for computing the four dominant eigenpairs of two matrices of different dimensions. The executions were repeated several times to avoid random effects. The figures demonstrate that in this case the overhead is negligible, even in the smaller case the SLEPc program seems to be faster.

## 6 Conclusions

This paper introduces SLEPc, a new library for the solution of eigenvalue problems, which is based on PETSc and has been designed to be able to cope with problems arising in real applications by using well-known techniques such as shift-and-invert transformations and state-of-the-art methods. It offers a growing number of solution methods as well as interfaces to well-established eigenvalue packages.

One key factor for the success of a new library is that its learning curve is not too much steep. For those users acquainted with PETSc, the use of SLEPc is straightforward. With little programming effort it is possible to easily test different solution strategies for a given eigenproblem. Once the problem has been specified, it is extremely easy to experiment with different solution methods and to carry out studies by varying parameters such as the value of the shift.

SLEPc inherits all the good properties of PETSc, including portability, scalability, efficiency and flexibility. Due to the seamless integration with PETSc, the user has at his disposal a wide range of linear equation solvers. In particular, the user holds control over the linear solvers associated with eigenproblems as if they were stand-alone objects.

The use of shell matrices allows easy formulation of complicated block-structured eigenproblems, and this is one of the claims for superiority of SLEPc compared to existing alternatives.

*Future Work.* Future versions of SLEPc will include advanced methods (e.g. Jacobi-Davidson, Rational Krylov Sequence, Truncated RQ) as well as more spectral transformations (e.g. Cayley transform). Further testing of the library needs to be done, in particular using parallel direct linear solvers (provided in the latest versions of PETSc) for the solution of shift-and-invert systems.

*Related Work.* Trilinos [7] is an effort to bring modern object-oriented software design to high-performance parallel solver libraries for the solution of large-scale complex multi-physics applications. It will include classes for the solution of eigenvalue problems.

*Software Availability.* The SLEPc distribution file is available for download at the following address: `http://www.grycap.upv.es/slepc`.

## Acknowledgements

SLEPc was developed in part during a three-month stay at Argonne National Laboratory (in collaboration with the PETSc team) and a short visit to Lawrence Berkeley National Laboratory.

## References

[1] Z. Bai, D. Day, J. Demmel, and J. Dongarra. A test matrix collection for non-Hermitian eigenvalue problems. Technical report CS-97-355, University of Tennessee, Knoxville, March 1997. LAPACK Working Note 123.
[2] Z. Bai, J. Demmel, J. Dongarra, A. Ruhe, and H. van der Vorst, editors. *Templates for the solution of Algebraic Eigenvalue Problems: A Practical Guide.* SIAM, Philadelphia, 2000.
[3] Satish Balay, William D. Gropp, Lois Curfman McInnes, and Barry F. Smith. PETSc home page. See `http://www.mcs.anl.gov/petsc`.
[4] Satish Balay, William D. Gropp, Lois Curfman McInnes, and Barry F. Smith. PETSc Users Manual. Technical Report ANL-95/11 - Revision 2.1.1, Argonne National Laboratory, 2001.
[5] Gene H. Golub and Henk A. Van der Vorst. Eigenvalue computation in the 20th century. *Journal of Computational and Applied Mathematics*, 123(1-2):35–65, November 2000.
[6] Vicente Hernández, José E. Román, Antonio M. Vidal, and Vicent Vidal. Calculation of Lambda Modes of a Nuclear Reactor: a Parallel Implementation Using the Implicitly Restarted Arnoldi Method. *Lecture Notes in Computer Science*, 1573:43–57, 1999.
[7] M. Heroux et al. Trilinos Project home page. See `http://www.cs.sandia.gov/~mheroux/Trilinos`.
[8] R.B. Lehoucq, D.C. Sorensen, and C. Yang. *ARPACK Users' Guide, Solution of Large-Scale Eigenvalue Problems by Implicitly Restarted Arnoldi Methods.* SIAM, Philadelphia, PA, 1998.
[9] O.A. Marques. BLZPACK: Description and user's guide. Technical Report TR/PA/95/30, CERFACS, Toulouse, France, 1995.
[10] O.A. Marques and L.A. Drummond. Advanced Computational Testing and Simulation (ACTS) Toolkit home page. See `http://acts.nersc.gov`.
[11] K.J. Maschhoff and D.C. Sorensen. PARPACK: An Efficient Portable Large Scale Eigenvalue Package for Distributed Memory Parallel Architectures. *Lecture Notes in Computer Science*, 1184:478–486, 1996.

[12] K. Wu and H. D. Simon. Thick-restart Lanczos method for symmetric eigenvalue problems. *Lecture Notes in Computer Science*, 1457:43–55, 1998.
[13] Kesheng Wu and Horst Simon. A parallel Lanczos method for symmetric generalized eigenvalue problems. Technical Report LBNL-41284, Lawrence Berkeley National Laboratory, 1997.

# Parallelization of Spectral Element Methods

Stéphane Airiau[1], Mejdi Azaïez[2], Faker Ben Belgacem[3], and Ronan Guivarch[1]

[1] LIMA-IRIT, UMR CNRS 5505
2 rue Charles Camichel, 31071 Toulouse Cedex 7, France
ronan.guivarch@enseeiht.fr
http://www.enseeiht.fr/lima
[2] IMFT, UMR CNRS 5502
118, route de Narbonne, 31062 Toulouse Cedex 4, France
[3] MIPS, UMR CNRS 5640
118, route de Narbonne, 31062 Toulouse Cedex 4, France

**Abstract.** Spectral element methods allow for effective implementation of numerical techniques for partial differential equations on parallel architectures. We present two implementations of the parallel algorithm where the communications are performed using MPI. In the first implementation, each processor deals with one element. It leads to a natural parallelization. In the second implementation certain number of spectral elements are allocated to each processor. In this article, we describe how communications are implemented and present results and performance of the code on two architectures: a PC-Cluster and an IBM-SP3.

## 1 Introduction

Spectral element methods [2] allow for effective implementation of numerical techniques for partial differential equations on parallel architectures.

The method is based on the assumption that the given computational domain, say $\Omega$, is partitioned into non overlap subdomains $\Omega_i$, $i \in \{1, \ldots, N\}$. Next, the original problem can be reformulated upon each subdomain $\Omega_i$ yielding a family of subproblems of reduced size which are coupled each others through the values of the unknown solution at subdomain interfaces. Our approach is to interpret the domain decomposition as iterative procedures for an interface equation which is associated with the given differential problem, here the Laplacian one. This interface problem is handled by the Steklov-Poincaré operator (see [1] and [3]).

Two steps are considered for our algorithm: in the first one a family of local and independent problems is solved and the second one is to complete the solution by solving the interface problem. Sections 2 and 3 present the method and the algorithm to solve the two-dimensional Poisson problem.

We describe in section 4 two implementations of the parallel algorithm where the communications are performed using MPI. In the first implementation, each processor deals with one element. It leads to a maximal parallelism. The main disadvantage of this implementation is the ratio between the computations and

J.M.L.M. Palma et al. (Eds.): VECPAR 2002, LNCS 2565, pp. 392–403, 2003.

the communications. In a second implementation certain number of spectral elements are allocated to each processor. This implementation permits flexibility in the ratio computation/communication ; various granularities and elements repartitions can be explored.

Section 5 presents results and performance of the code on two architectures: an IBM-SP3[1] and a PC Cluster[2].

## 2 Presentation of Spectral Element Methods

This study is limited to $\Omega$, a square domain of the plan. The boundary of $\Omega$ is designed by $\partial\Omega$. This domain is splitted in $n$ square elements. Each element is discretized by $N_p * N_p$ points.

We want to solve the Poisson problem on $\Omega$:

$$\begin{cases} -\Delta u = f \text{ for } x \in \Omega \\ u = 0 \text{ for } x \in \partial\Omega \end{cases} \tag{1}$$

Decomposition of $\Omega$ into $n$ non overlap elements $\Omega_i$ leads to interior egdes between elements and we call $\gamma$ this interface.

We split the initial problem in two subproblems. The first one is the solution of local Poisson problem on each element; the second one concerns the interface to join the solution between elements. This second problem exibits a new unknown function $\lambda : \gamma \mapsto \mathbb{R}$. This function is the solution of the Poisson problem on the interface.

This process leads to decompose the unknown function $u$ in two functions: for each element $i$, $u_i$ the restriction of $u$ on $i$ is splitted in $\overline{u_i}$, which only depends on the element $i$, and in $\widetilde{u_i}(\lambda)$, which depends on $\lambda$, and so on the neighboring elements. So we have two sets of subproblems (2) and (3) to solve.

The $n$ subproblems on each element:

$$\begin{cases} -\Delta\overline{u_i} = f_i \text{ for } x \in \Omega_i \\ \overline{u_i} = 0 \text{ for } x \in \partial\Omega_i \end{cases} \tag{2}$$

Each subproblem $i$ of (2) doesn't depend on others subproblems, so all subproblems can be solved in parallel.

The second subproblems are harmonic problems:

$$\begin{cases} -\Delta\widetilde{u_i}(\lambda) = 0 \text{ for } x \in \Omega_i \\ \widetilde{u_i}(\lambda) = \lambda \text{ for } x \in \gamma \\ \widetilde{u_i}(\lambda) = 0 \text{ for } x \in \partial\Omega_i/\gamma \end{cases} \tag{3}$$

To solve these subproblems, we have to know the function $\lambda : \gamma \to \mathbb{R}$, continuous on $\gamma$ which verifies:

$$\begin{cases} -\Delta u_i(\lambda) = f \text{ for } x \in \Omega_i \\ u_i(\lambda) = \lambda \text{ for } x \in \gamma \\ u_i(\lambda) = 0 \text{ for } x \in \partial\Omega_i/\gamma \end{cases} \tag{4}$$

[1] We would like to thank IDRIS(Institut du Développement et de Ressources en Informatique Scientifique) for computation hours allocated to our project

[2] and INPT for the credits which allowed the cluster acquisition

and on each edge between $\Omega_i$ and $\Omega_j$,

$$\frac{\partial u_i(\lambda)}{\partial n} = \frac{\partial u_j(\lambda)}{\partial n} \tag{5}$$

After a function decomposition, we obtain:

$$\frac{\partial \widetilde{u_i}(\lambda)}{\partial n_i} - \frac{\partial \widetilde{u_j}(\lambda)}{\partial n_j} = \frac{\partial \overline{u_j}}{\partial n_j} - \frac{\partial \overline{u_i}}{\partial n_i} \tag{6}$$

A discretization with spectral elements, leads to a linear system

$$S\lambda = b \tag{7}$$

where $S$ is the Schur complement matrix. Once $\lambda$ is computed, we have to solve a Poisson problem on each element. For the same reason than previously, these problems can be solved in parallel.

# 3 Algorithm for the Solution of Problem on the Interface

The Schur complement matrix is a symetric matrix and we use Conjugate Gradient method. The parallel algorithm to solve problem (7) is presented in Fig. 1.

$\beta = 0$ ; $\Lambda = 0$ ; $\rho_0 = (r_0, r_0)$ ; — Initializations

**While** the residual is greater than epsilon **do**

Solve

$$\begin{cases} -\Delta \tilde{u}_i(\Lambda) &= 0 \quad \text{for } x \in \Omega_i \\ \tilde{u}_i(\Lambda) &= \Lambda \quad \text{for } x \in \gamma \\ \tilde{u}_i(\Lambda) &= 0 \quad \text{for } x \in \partial\Omega_i/\gamma \end{cases}$$ — Local solutions

$v = S\Lambda$ ; — Steklov-Poincaré operator (we need contributions of each element)

$ps_1 = (\Lambda, v)$ ; $ps_2 = (r_0, r_0)$ ; — Global computation

$\alpha = \frac{ps_2}{ps_1}$ ; $\lambda = \lambda + \alpha\Lambda$ ; $r_1 = r_0 - \alpha v$ ; — Local computation

$\rho_1 = (r_1, r_1)$ — Global computation

$\beta = \frac{\rho_1}{\rho_0}$ ; $r_0 = r_1$ ; $\Lambda = r_0 + \beta\Lambda$ — Local computation

**End do**

**Fig. 1.** Parallel algorithm

# 4 Parallelization and Implementations

## 4.1 First Implementation: One Spectral Element by Processor

In the spectral element method, we solve a local problem on each element. In the partitioning phase, a natural way to distribute the work the solver is to give an

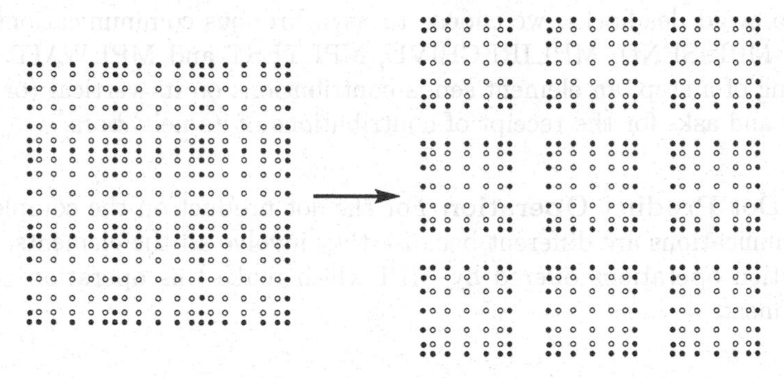

**Fig. 2.** *9-element domain on 9 processors*

element to each processor. As a consequence, all local solutions will be performed in a single processor. An inherent disadvantage of this approach is that we have to duplicate values of the interior edges.

Because we place an element on each processor, we use indifferently the terms element and processor in this subsection.

Looking at the algorithm 1, two different tasks need communications between processors: the computation of the image of a vector by Steklov-Poincaré operator and the computation of three dot products.

In the first task, the computation for a point requires values hold on the neighbor points. For interior points, neighbor points belong to the same element. For points on the edges, we need the contribution of several elements. This operation, called gathering, can be separated in two cases: for points of vertices, four elements are concerned, while for the others points, only two elements interfere.

For the second task, the computation of the dot product needs all the elements; this operation can be carried using a global operation.

**The Gathering Operation** For each edge of the interface, we have to make the sum of the contribution of each element. A difficulty arises for interior vertices; for these points four elements are concerned. We must be careful with the order of communications between elements: a deadlock could easily occur or we could not take account of all contributions.

In order to minimize the number of communications, we do not choose to separate vertices from the others points of the edges. The solution is to decompose the gathering in two steps: the first one treats vertical egdes and the second horizontal ones.

For example, in the first step, each processor sends the contribution of its right edge and receives the contribution for its left edge, then send data of its left edge and receives data for its right edge.

Using this process, a copy of the intermediate sum in proper vectors between the two steps, and, at the end, a summation of the contribution of all elements, the vertices are treated like others points.

To avoid deadlocks, we choose to asynchronous communications offered by MPI: MPI_ISEND, MPI_IRECEIVE, MPI_TEST and MPI_WAIT. At the beginning of a step, an element sends contributions on its vertical (or horizontal) edges and asks for the receipt of contributions of its neighbors.

**The Dot Product Operation** For the dot product on the complete domain, communications are different because they involve all the elements. We use the reduction operations offered by MPI which make this operation very easy to implement.

### 4.2 Second Implementation: Several Spectral Elements by Processor

It seemed natural to associate an element to each processor. This choice, although it permits a maximal parallelism, has some disadvantages: when computing with a great number of elements, we may not have all the processors we wanted. Furthermore, if the size of an element is too small, communication will take more time than computation.

In this second implementation, we choose to associate several elements to a processor (see Fig. 3). Increasing the granularity permits to improve the computation communication ratio. The code is more flexible: we can perform some tests with a great number of elements and for a fixed number of elements we can choose different repartitions (see Table 1).

Concerning the gathering operation, we do not have to communicate data of edges inside the domain treated by a processor. Communications only concern the edges of the boundary. Instead treating communications at the level of an element, we can gather in an interface, all edges to be sent by a processor to one of its neighbors (see Fig. 4). We then reduce the number of messages (messages are bigger) and thus reduce communication time.

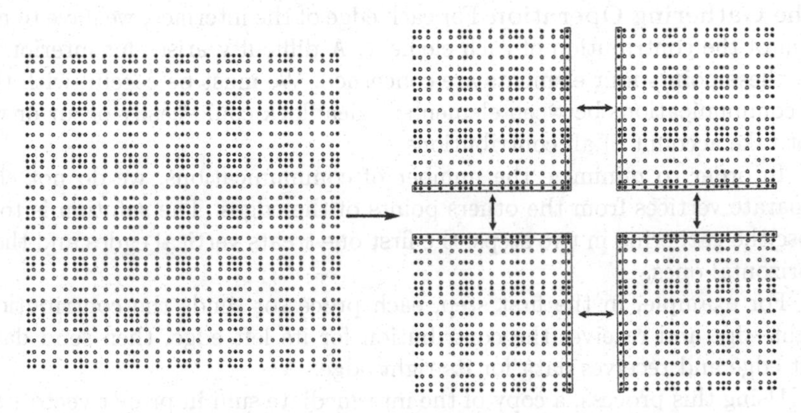

**Fig. 3.** *36 elements on 4 processeurs*

**Table 1.** Repartitions of $64 \times 64$ elements

| Number of processors | Number of elements on a processor in X-direction | Number of elements on a processor in Y-direction |
|---|---|---|
| 1 | 64 | 64 |
| 2 | 64 | 32 |
| 4 | 32 | 32 |
| 4 | 64 | 16 |
| 8 | 32 | 16 |
| 8 | 64 | 8 |
| 16 | 16 | 16 |
| 16 | 32 | 8 |
| 16 | 64 | 4 |
| 32 | 16 | 8 |
| 32 | 32 | 4 |
| 32 | 62 | 2 |

Nothing changes for the dot product operation; we still use reduction operations.

There is another advantage of this repartition regarding asynchronous communication. A processor can start its communications and overlay them with the computation of the sum on the interior edges. To sum contributions on the interface, it waits the end of communications and hopes that computations have recovered communications. This advantage will be shown in next section with numerical experiments.

## 5 Numerical Experiments

Numerical experiments were carried on two architectures: an IBM-SP3 and a cluster of 8 biprocessors PC. We only show performance of the second implementation which permits several elements on each processor.

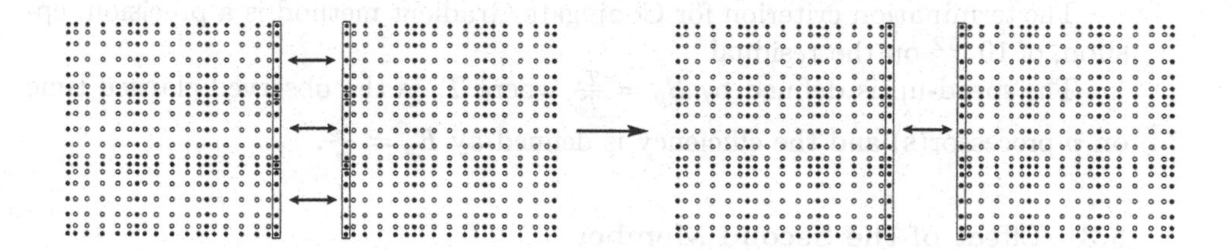

**Fig. 4.** *Communications* between two processors: a message by edge against a message by interface

**Table 2.** Effect of the second member

| Second member | Number of iterations | Efficiency on IBM-SP3 | Efficiency on Cluster |
|---|---|---|---|
| $f : (x, y) \mapsto 4\pi^2 \sin(\pi x) sin(\pi y)$ | 15 | 1.14 | 0.96 |
| $f : (x, y) \mapsto 4 - 2x^2 - 2y^2$ | 516 | 1.16 | 0.96 |
| $f : (x, y) \mapsto \frac{2}{(4 + x + y)^2}$ | 921 | 1.07 | 0.96 |
| $f : (x, y) \mapsto \frac{1}{2}$ | 835 | 1.05 | 0.96 |

### 5.1 Specifications of the Two Architectures

**IBM-SP3 Brodie of IDRIS** Each Node of the SP3 has 16 Power3 processors at 375 MHz with 2 Gbytes of shared memory. Inter-node communications are handled with the IP protocol.

**PC-Cluster Tarasque of INPT–ENSEEIHT** Each node of the cluster has 2 Pentium III processors at 800 MHz with 265 Mbyte of shared memory. The network is a Commute Ethernet. The operating system is Linux (SuSE 7.2 distribution) and we use a free implementation of MPI (LAM 6.5.1). The software has been compiled with GNU compilers.

### 5.2 Description of the Numerical Tests

The problem to solve is a Poisson problem. The difficulty depends on the nature of the second member. We choose four kinds of functions:

- sinus product: $f : (x, y) \mapsto 4\pi^2 \sin(\pi x) sin(\pi y)$
- polynomial: $f : (x, y) \mapsto 4 - 2x^2 - 2y^2$
- rationnal: $f : (x, y) \mapsto \frac{2}{(4 + x + y)^2}$
- constant: $f : (x, y) \mapsto \frac{1}{2}$

The termination criterion for Conjugate Gradient method is a precision, epsilon, of $10^{-12}$ on the residual.

The speed-up is defined by $S_p = \frac{T_1}{T_p}$ where $T_p$ is the observed elapsed time on $p$ processor(s) and the efficiency is defined by $E_p = \frac{S_p}{p}$.

### 5.3 Effect of the Second Member

We report the efficiency for the four considered second members on the two architectures. The parameters are set to:

**Table 3.** Number of iterations with different discretizations

| Number of discretization points | Number of iterations |
|---|---|
| 4 | 239 |
| 8 | 372 |
| 12 | 473 |
| 16 | 555 |
| 20 | 631 |

- $64 \times 64$ elements,
- $15 \times 15$-point discretization on an element,
- 16 processors
- $16 \times 16$-element repartition on each processor,

Table 2 shows that the efficiency does not depend on the nature of the second member. Whatever the number of iterations is (and so the computation amount), the efficiency remains the same for both architectures.

### 5.4 Effect of the Number of Discretization Points

We present in Figures 5 and 6, the efficiency of our code with different numbers of discretization points on each spectral element.

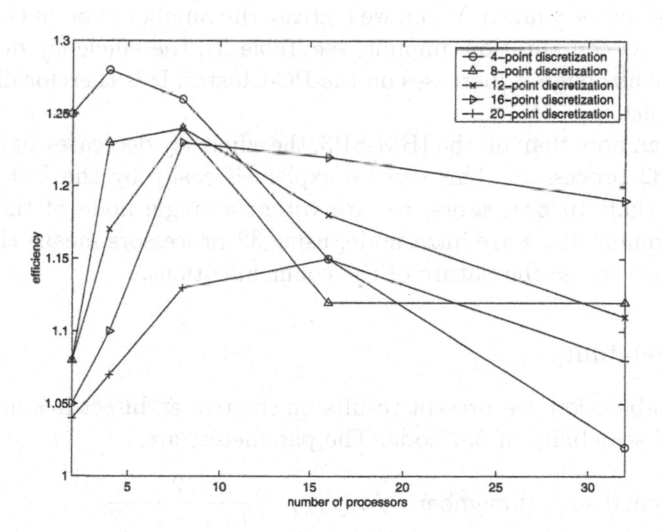


**Fig. 5.** *Efficiency* on the IBM-SP3 Brodie with different *numbers of discretization points* on each element

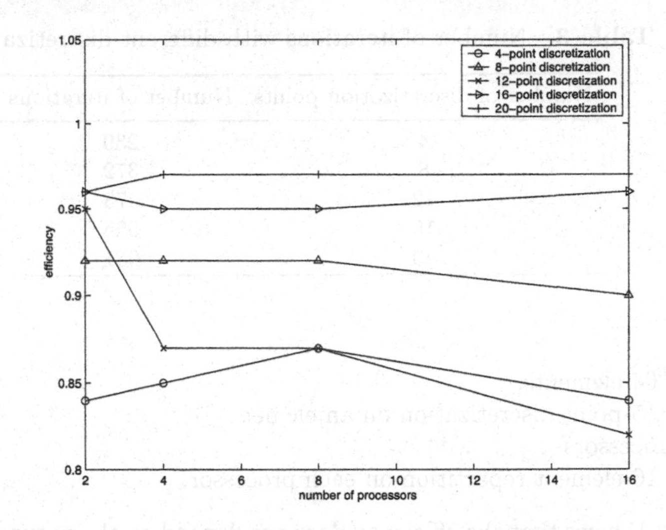


**Fig. 6.** *Efficiency* on the PC-Cluster Tarasque with different *numbers of discretization points* on each element

The number of elements is $64 \times 64$, the second member is constant ($f : (x, y) \mapsto \frac{1}{2}$) and results are obtained with 2, 4, 8, 16 and 32 processors on IBM-SP3 and 2, 4, 8 and 16 processors on PC-Cluster.

These results show that the number of discretization points does not influence the efficiency very much. When we increase the number of points on each element (and so the computation amount, see Table 3), the efficiency decreases on the IBM-SP3 although it increases on the PC-Cluster. It is therefor difficult to draw conclusions.

We can note than on the IBM-SP3, the efficiency decreases significantly when we use 32 processors. This can be explained easily by the fact that when we use less than 16 processors, we are within a single node of the machine and the communications are intra-node; using 32 processors means that we use two nodes and change the nature of the communications.

### 5.5 Scalability

In this subsection, we present results on the two architectures in order to show the good scalability of our code. The parameters are:

- rationnal second member : $f : (x, y) \mapsto \dfrac{2}{(4 + x + y)^2}$
- $64 \times 64$ elements,
- $15 \times 15$-point discretization on an element,

We use the different repartitions of Table 1.

**Table 4.** Efficiency and speed-up on the IBM-SP3 Brodie

| Number of processors | Repartition | Time | Speed-up | Efficiency |
|---|---|---|---|---|
| 1 | 64 × 64 | 1040.11 | - | - |
| 2 | 64 × 32 | 512.55 | 2.03 | 1.02 |
| 4 | 32 × 32 | 243.80 | 4.26 | 1.07 |
| 4 | 64 × 16 | 249.33 | 4.17 | 1.04 |
| 8 | 32 × 16 | 115.42 | 9.01 | 1.13 |
| 8 | 64 × 8 | 115.97 | 8.97 | 1.12 |
| 16 | 16 × 16 | 60.73 | 17.13 | 1.07 |
| 16 | 32 × 8 | 60.68 | 17.14 | 1.07 |
| 16 | 64 × 4 | 62.36 | 16.68 | 1.04 |
| 32 | 16 × 8 | 35.45 | 29.34 | 0.92 |
| 32 | 32 × 4 | 33.36 | 31.18 | 0.97 |
| 32 | 64 × 2 | 33.69 | 30.87 | 0.96 |

**Table 5.** Efficiency and speed-up on the PC-Cluster Tarasque

| Number of processors | Repartition | Time | Speed-up | Efficiency |
|---|---|---|---|---|
| 1 | 64 × 64 | 6251.44 | - | - |
| 2 | 64 × 32 | 3264.49 | 1.91 | 0.96 |
| 4 | 32 × 32 | 1628.61 | 3.84 | 0.96 |
| 4 | 64 × 16 | 1628.84 | 3.84 | 0.96 |
| 8 | 32 × 16 | 814.18 | 7.68 | 0.96 |
| 8 | 64 × 8 | 813.43 | 7.68 | 0.96 |
| 16 | 16 × 16 | 408.58 | 15.30 | 0.96 |
| 16 | 32 × 8 | 411.00 | 15.21 | 0.95 |
| 16 | 64 × 4 | 408.35 | 15.31 | 0.96 |

Tables 4 and 5 presents observed elapsed time, speed-up and efficiency for different number of processors.

The results exhibit the good scalability of the code ; efficiency remains the same when we increase the number of processors. As we said previously, computations overlay communications.

A second remark concerns the different repartitions for a fixed number of processors. The results are quite the same whatever the repartition is (expecially on the cluster) ; repartition only concern the size of the interface between processors. When we change repartition, some interfaces of a processors are increased and some others reduced; the number of values to communicate remain the same because our elements are square.

The Figures 7 and 8 synthetize these results with the linear curve of the speed-up; for number of processors with several repartitions, we only plot the speed-up of the first repartition.

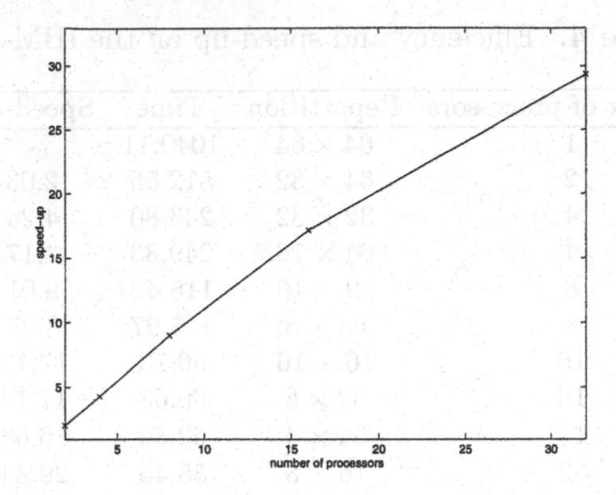


**Fig. 7.** *Speed-up* on the IBM-SP3 Brodie with different *numbers of processors*

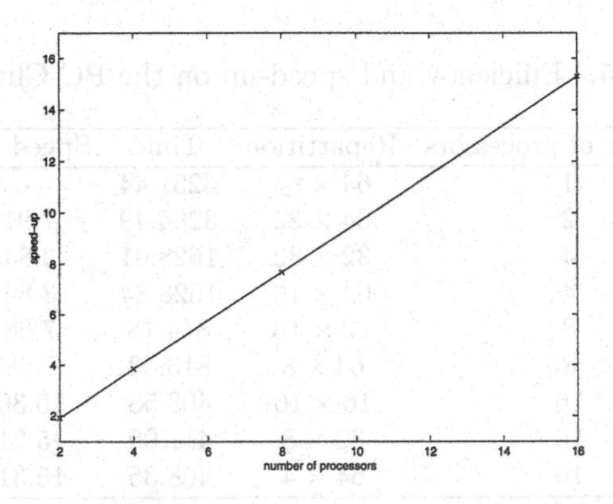


**Fig. 8.** *Speed-up* on the PC-Cluster Tarasque with different *numbers of processors*

## 6 Conclusion

In this work, we have proposed a parallel implementation of spectral element methods to solve a 2D Poisson problem. The performance of this implementation have been studied on two architectures, an IBM-SP3 and a PC-Cluster following several criteria: different second members, different numbers of discretization points, different repartitions of the elements on the processors.

It appears that our approach exhibits good parallel performance on both platforms and our code shows a good scalability.

Future works can study parallel implementation of much complicated problems: convection diffusion problem, Navier-Stokes problem in two-dimensional or three-dimensional domains.

## References

[1] V. I.Agoshkov, *Poincaré-Steklov's operators and domain decomposition methods in finite dimensional spaces*, First International Symposium on Domain Decomposition Method for Partial Differential Equations, R. Glwinski, G. H. Golub, G. A. Meurant and J. Périaux eds., SIAM, Philadelphia, pp. 73-112 (1988).
[2] C. Bernardi and Y. Maday, *Approximations spectrales de problèmes aux limites elliptiques*, Paris, Springer Verlag (1992).
[3] A.Quarteroni and A.Valli, *Domain Decomposition Methods for Partial Differential Equations*, Oxford University Press, Ofxord (1999).

# Chapter 6

# Cluster Computing

# Mapping Unstructured Applications into Nested Parallelism

## Best Student Paper Award: First Prize

Arturo González-Escribano[1], Arjan J.C. van Gemund[2], and Valentín Cardeñoso-Payo[1]

[1] Dept. de Informática, Universidad de Valladolid
E.T.I.T. Campus Miguel Delibes, 47011 - Valladolid, Spain
Phone: +34 983 423270
arturo@infor.uva.es

[2] Faculty of Information Technology and Systems (ITS)
P.O.Box 5031, NL-2600 GA Delft, The Netherlands
Phone: +31 15 2786168
a.j.c.vangemund@its.tudelft.nl

**Abstract.** Nested parallel programming models, where the task graph associated to a computation is series-parallel are easy to program and show good analysis properties. These can be exploited for efficient scheduling, accurate cost estimation or automatic mapping to different architectures. Restricting synchronization structures to nested series-parallelism may bring performance losses due to a less parallel solution, as compared to more generic ones based in unstructured models (e.g. message passing).
A new algorithmic technique is presented which allows automatic transformation of the task graph of any unstructured application to a series-parallel form (nested-parallelism). The tool is applied to random and irregular application task graphs to investigate the potential performance degradation when conveying them into series-parallel form. Results show that a wide range of irregular applications can be expressed using a structured coordination model with a small loss of parallelism.

**Topic:** Parallel and distributed computing.

## 1 Introduction

A common practice in high-performance computing is to program applications in terms of the low-level concurrent programming model provided by the target machine, trying to exploit its maximum possible performance. Portable APIs, such as message passing interfaces (e.g. MPI, PVM) propose an abstraction of the machine architecture, still obtaining good performance. However, all these unrestricted coordination models can be extremely error-prone and inefficient, as the synchronization dependencies that a program can generate are complex

J.M.L.M. Palma et al. (Eds.): VECPAR 2002, LNCS 2565, pp. 407–420, 2003.

and difficult to analyze by humans or compilers. Moreover, important implementation decisions such as scheduling or data-layout become extremely difficult to optimize.

Facing all these shortcomings, more abstract parallel programming models have been proposed and studied which restrict the possible synchronization and communication structures available to the programmer (see e.g. [1]). Restricted synchronization structures lead to models easier to understand and program, and enable the design of powerful tools and techniques to help in mapping decisions.

Nested-parallelism provides a good compromise between expressiveness, complexity and ease of programming [1]. They restrict the coordination structures and dependencies to those that can be represented by series-parallel (SP) task graphs (see definitions in [2]). Nested-parallelism is also called SP programming. The inherent properties of SP structures provide clear semantics and analyzability properties (see discussion about SP algebras and automatas in [3]), a simple compositional cost model [4, 5, 6] and efficient scheduling [7, 8]. All these properties enable the development of automatic compilation techniques which increase portability and performance. Examples of parallel programming models based in nested-parallelism include BSP [9], nested BSP (e.g. NestStep [10], PUB [11]), BMF [6], skeleton based (e.g. SCL [12], Frame [13]), SPC [4] and Cilk [14]). Despite the obvious advantages of SP programming models, a critical question arises as to what extent the ability to express parallelism is sacrificed when restricting parallelism to SP form, and how existing applications can be mapped into nested-parallelism efficiently.

In previous work [15] we focused in the potential performance degradation in highly regular applications, providing analytic and empirical evidence that this performance loss is typically outweighed by the benefits of SP models. We identified the most important topological DAG properties that are responsible for the performance loss when using an SP model, and quantitatively captured this loss in terms of an analytical cost model. We also measured the actual loss of parallelism at the actual machine execution level, thus including the possible performance degradation due to, e.g., additional barrier synchronization overhead. The study included well-known regular programming paradigms and applications selected amongst the worst possible cases.

In this paper we extend our results to unstructured graphs and applications. In particular, the following contributions are highlighted:

- A new algorithmic technique to efficiently transform generic graphs to SP form, with improved complexity bounds and other nice characteristics is presented.
- Using this new technique we experimentally check a random sample of the possible graph space, relating performance degradation to simple graph parameters.
- We also show that these results are supported by experiments with task graphs from real irregular applications. We present results obtained when applying the transformation algorithm to task graphs with up to thousands of nodes, generated by an unstructured domain decomposition and sparse-

matrix factorization software [16, 17]. Real workloads are considered in these graphs.

The paper is organized as follows: Section 2 presents a disscussion about the problems faced when mapping applications to series-parallel form. In section 3 the new algorithm is introduced and analyzed. Section 4 presents the results of experiments carried out with the algorithm, with random and real graph topologies. Conclusions are outlined in section 5.

## 2 Mapping Applications to Series-Parallel Form

Consider a parallel algorithm whose task graph is given by the non-SP (NSP) DAG on the left-hand side of Fig. 1. When using an NSP programming model (such as a message-passing model) the program level DAG will be identical to the algorithm level DAG. Expressing the original parallel computation in terms of an SP model, however, essentially involves *adding* synchronization arcs in order to obey all existing precedence relations, thus possibly *increasing* the critical path of the DAG. At the same time, porting or reprogramming existent unstructured applications to this recursive parallelism model is not straightforward. There exist different combinations of new added dependencies that transform the original DAG in SP form, and each of them can lead to a different increase in the critical path. In Fig. 1 the right-hand side DAG represents one SP programming solution where every dashed line represents a local synchronization barrier. While for a normally balanced task workload distribution the SP solution has an equal execution time compared to the original NSP version, in the pathological case where the black nodes have a delay value of $\tau$ while the white ones have zero workload, the critical path of the SP solution would be increased by a factor of 3. Despite the negligible probability of the above case, there clearly exists a class of DAGs for which the performance loss of the typical SP programming solution may be noticeable.

The introduction of nested parallelism is done at a highly abstract level, bringing analyzability benefits all along the implementation and mapping phases. However, this implies that the structure of an application should be mapped to series-parallel form when knowledge is not available about the workloads.

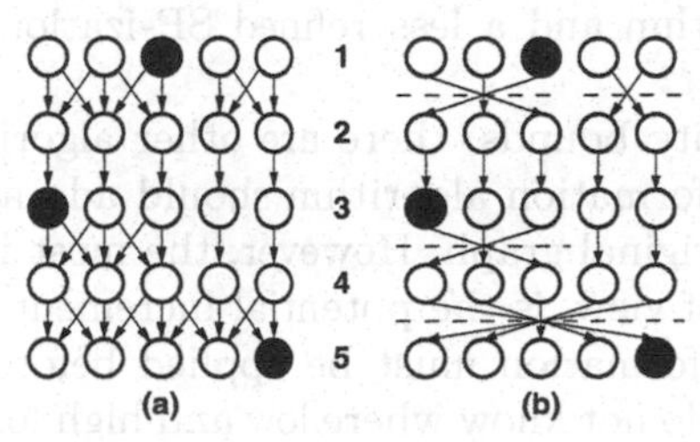


**Fig. 1.** Original NSP DAG and an SP DAG solution

For highly structured applications, simple layering techniques (synchronizing with full barriers between each layer of nodes) can be used to flatten synchronization structures to nested parallelism, obtaining similar results as if flattening to BSP (Bulk Synchronous Parallelism) structure. In contrast to the relative simplicity of imposing nested parallelism to structured applications, finding the optimal mapping (i.e., with minimal increase of critical path) is not always trivial. Furthermore, for unstructured applications it is virtually impossible to obtain a nested parallel solution without the aid of a sophisticated algorithm.

A simple and typical model of the structure of a parallel application is a task graph. Nodes represent tasks, and are annotated with a cost, load or execution time value. In previous work we presented a study on how to use simple task graphs with different colored nodes to represent computations and communications mapped to specific architectures at a fine grained implementation detail [18]. Nodes could represent computations or communications and, then, edges will only represent dependencies between nodes and will have no associated load.

Manual techniques for specific well-known structures based on simple heuristics can help in nested parallel or bulk synchronous application design [19, 20]. Automatic techniques to map synchronization structures to SP form are scarcely found and most of the times restricted. An automatic flattening technique based on nested data structures can be found in [21]. Non work-preserving techniques for UTC (Unit Time Cost) graphs, can be devised duplicating a linear number of tasks [22, 23]. However, determining the best duplication factor is dependent on input graph topology and it is still an on-going research. In fact, the optimal factor cannot be computed for a generic message-passing model as LogP [24].

In [25, 26] analytic results are presented to show that generic message passing based applications (represented by *LogP*) can be flattened to bulk-synchronous parallelism within logarithmic performance degradation in the worst case. Nevertheless, a study of worst and mean case have not yet been presented for nested parallelism models. The algorithm technique presented in this paper can fill the gap, providing, at least, an empirical evaluation procedure of potential performance degradation when mapping unstructured applications to nested parallelism.

A work-preserving algorithmic technique that transforms any non-SP graph to an SP form is called an *SP-ization*. In this work we will compare a new automatic SP-ization algorithm with two other ones [15]: Simple layering (bulk-synchronous) algorithm and a less refined SP-ization algorithm we had introduced previously.

Besides complexity bounds, there are other algorithm characteristics to be considered. A transformation algorithm should add a minimum number of dependencies to the original graph. However, the most important factor from the performance point of view, is the potential increment of the critical path value. Typically, the transformation must be applied before workload information is available. While we do not know where low and high loaded nodes are, the fairest assumption is to consider all tasks i.i.d. (independent identically distributed). In this situation, a technique that do not increase the critical path of UTC (Unit Time Cost) graphs will have a good chance of success.

## 3 Transformation Algorithm

A simplified description of the main algorithm phases along with some hints to its correctness proof, its complexity measures and comparison with previous algorithms is presented in this section. For a full formal description and analysis see [27].

### 3.1 Graph Preliminaries

We present some basic notation used throughout the following sections. Let $G = (V_G, E_G, \tau)$ be a DAG (direct acyclic graph) in which $V_G$ is a finite set of nodes and $E_G$ a finite set of edges (or ordered node pairs) and $\tau(v \in V_G)$ the workload distribution function that assigns to every node its load value. We work with *multidigraphs*, which allow several instances of the same edge $(u, w)$ in $E_G$ at the same time. Let $indeg(v \in V_G)$ be the number of edges $(u, v)$ for which $v$ is the *target*, and $outdeg(v \in V_G)$ be the number of edges $(v, w)$ for which $v$ is the *source.* $R(G)$ is the set of *roots* of $G$ (nodes $v : indeg(v) = 0$), and $L(G)$ is the set of *leaves* of $G$ (nodes $v : outdeg(v) = 0$). An StDAG is a DAG with only one root and one leaf ($|R(G)| = 1, |L(G)| = 1$). Any DAG can be transformed to an StDAG by a simple algorithm that adds a new unique root connected to all original roots and a new unique leaf connected to all original leaves.

A *path* between two nodes ($p(u, w)$) is a collection of nodes $u, v_1, v_2, ..., v_n, w$ such that $(u, v_1), (v_1, v_2), ..., (v_{n-1}, v_n), (v_n, w) \in E_G$. The *length* of a path is the number of edges determined by the path sequence. The *critical path value* ($cpv(G)$) is the maximum, over all possible paths in $G$, of the accumulated load along the path.

The *depth level* of a node ($d(v)$) is the maximum length of a path from a root to that node. The *maximum depth level* of a graph ($\hat{d}(G)$) is the maximum length of a path from a root to a leaf. We define *d-edges* as the subset of edges $(u, v) : d(v) - d(u) > 1$. We say that a node $v$ is *previous* to $w$ or $w$ *depends on* $v$ ($v \prec w$) if there is a path $p(v, w)$. We say that two nodes $v, w$ are *connected* ($v \prec\succ w$) if $v$ depends on $w$ or $w$ depends on $v$.

A *series reduction* is an operation that substitutes a node $v : indeg(v) = outdeg(v) = 1$ and its two incident edges $(u, v), (v, w) \in E_G$ for an edge $(u, w)$. A *parallel reduction* is an operation that substitutes several instances of an edge $(u, w)$ for a unique instance of the same edge $(u, w)$.

The *minimal series-parallel reduction* of a graph, is another graph obtained after iteratively applying all posible series and parallel reductions. A *trivial graph* $G_t = (V_{Gt}, E_{Gt})$ is a graph with only two nodes and only one edge connecting them ($V_{Gt} = \{u_1, u_2\}, E_{Gt} = \{(u_1, u_2)\}$). An StDAG is SP iff its minimal reduction is a trivial graph $G_t$ [28, 2].

### 3.2 Algorithm Description

**Initialization Phase:**

1. Transform the input DAG into an StDAG along the lines previously discussed.
2. Layering of the graph. Compute a partition of $V_G$, grouping nodes with the same depth level.

$$L_i \subset V_G; L_i = \{v \in V_G : d(v) = i\}$$

3. Initialize an ancillary tree $T = (V_T, E_T)$ to $L_0$. This tree will represent the minimal series-parallel reduction of the step by step processed subgraphs.

**Graph Transformation:** For all layers (sorted) $i$ from 0 to $\hat{d}(G) - 1$:

1. **Split Layer in Classes of *Relatives*:** Let us consider the subgraph $S \subset G$ formed by $L_i \cup L_{i+1}$ and all edges from $G$ incident to two nodes in this subset. We construct the partition of this nodes into connected subgraphs. We define *relatives* classes as the subsets of nodes that belong to the same connected component of $S$ and the same layer. $\mathcal{P}_U = \{U_1, U_2, ..., U_n\}$ will be the *up* classes (of nodes in $L_i$) and $\mathcal{P}_D = \{D_1, D_2, ..., D_n\}$ will be the *down* classes (of nodes in $L_{i+1}$). Each class $U \in \mathcal{P}_U$ induces a class $D \in \mathcal{P}_D$ that belongs to the same connected component ($U \to D$).
2. **Tree Exploration to Detect *Handles* for Classes of Relatives:** We look in the tree for *handles*. For each class, the *U-handle* is the nearest common ancestor of all nodes in $U$. We define $K_T(U)$ as the set of source nodes to $D$ (it includes $U$ and source nodes of d-edges targeting $D$). Sources of edges showing transitive dependency to $D$ through the U-handle are to be discarded from $K_T(U)$.
   The *handle* node of class $U$, $h(U)$ is defined as:

$$H(U) = \{v \in V_T : \forall w \in K_T(D), v \preceq_T w\}$$

$$h(U) = h \in H(U) : \forall h' \in H(U) : d(h) \geq d(h')$$

   We also define the forest of a class, as the set of complete sub-trees below $h(U)$ that include nodes in $K_T(U)$:

$$\mathit{SubF}(U) = \{u \in V_T : v \preceq_T u, (h(U), v) \in E_T, v \preceq_T w : w \in K_T(U)\}$$

3. **Merge Classes with Overlapping Forests:** Classes with overlapping forests are merged in a unique $U$ and $D$ class. They will be synchronized with the same barrier.
4. **Capture *Orphan Nodes*:** We define *orphan nodes* as the leaves of the tree $T$ that are not in any $U$ class (they are nodes in layers previous to $i$ with only d-edges to layers further than $i + 1$). These nodes are included in the $U$ class of the forest they belong to.

5. **Class Barrier Synchronization:** For each final $U \rightarrow D$ classes:
   - Create a new synchronization node $b_U$ in the graph and the tree.
   - In $G$, eliminate all edges targeting a node in $D$. Add edges from every node in $U$ to $b_U$ and from $b_U$ to every node in $D$ (barrier synchronization).
   - Substitute every d-edge $(v, w)$ with source $v \in SubF(U)$ and targeting a node $w \in L_k : k > i + 1$ (a further layer) for an edge $(b_U, w)$. This operation eliminate d-edges from the new synchronized SP subgraph, avoiding the loss of dependencies in the original graph.
   - Substitute the forest $SubF(U)$ in $T$ for an edge $(h(U), b_U)$ representing the minimal series-parallel reduction of the new synchronized SP subgraph.

*Correctness:* Hints to a full correctness proof (see [27]) can be given as follows: since any tree can be easily transformed to a trivial SP StDAG, any graph which minimal series-parallel reduction is a tree, will be SP. As can be easily shown by induction on the depth of the StDAG, the minimal series-parallel reduction graph at each step is always a tree and, thus, has the SP property.

*Space Complexity:* Let $n$ be the number of nodes and $m$ the number of edges in the original graph. The number of nodes in the graph increases with one more node for each $U$ class. Every node appears just once in an $U$ class over the full algorithm run. Thus, the total number of nodes is upper bounded by $2n$. The number of edges is upper bounded because the processed subgraph (after each iteration) is SP, and the number of edges in an SP graph is bounded by $m \leq 2(n - 2)$. Other ancillary structures (as the tree) store graph nodes and/or edges. Thus, space complexity is $O(m + n)$.

*Time Complexity:* StDAG construction can be done in $O(n)$ and getting layers information in $O(m)$ with a simple graph search. Classes of relatives for two consecutive layers can be computed testing a constant number of times each edge. Thus, all the classes along the algorithm run are computed in $O(m)$. Exploration of the tree for handles can be self-destructive: Nodes are eliminated during the search. While searching for the handle of a class, all the forest can be eliminated and orphan nodes and other classes to be merged detected. Check and eliminate a transitive edge can be done in $O(1)$ if appropriate data structures are used for the tree [29], but assuming tree modifications are done in $O(\log n)$. $O(n)$ nodes and edges are inserted and eliminated in the tree. Thus, all tree manipulation has a time complexity $O(n \log n)$. The synchronization phase adds $O(n)$ nodes, eliminate $O(m)$ edges and add a bounded number of edges ($O(n)$ because it is an SP graph). The movement of d-edges can be traced in $O(n \log n)$ with a tree-like groups joining structure to avoid real edge manipulation (see [27] for details). Thus, time complexity is $O(m + n \log n)$.

### 3.3 An Example

We show an example application of the algorithm in Fig. 2. The first graph represents one of the domain decomposition and sparse-matrix factorizations used in following sections. The second graph is the result of the algorithm. Darker nodes represent more loaded nodes.

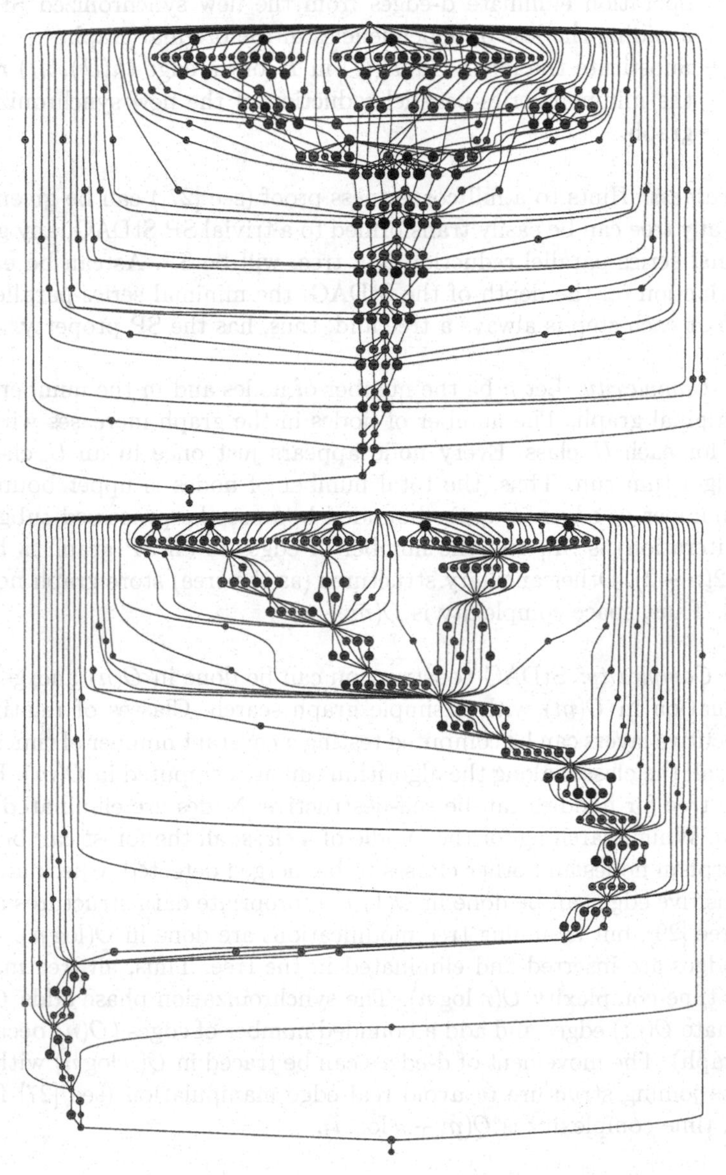

**Fig. 2.** Example of the algorithm result for a real application graph

### 3.4 Critical Path Value Property for UTC Graphs

For an UTC (Unit Time Cost) input graph $G$, the result $G'$ is not UTC (nodes added by the algorithm carry no load), but despite the added dependencies, the critical path value has not increased. For UTC graphs, the critical path value is equal to the maximum number of nodes that can be traversed from a root to a leaf ($cpv(G) = 1 + \hat{d}(G)$).

The algorithm keeps the layers structure of the original graph, adding zero loaded nodes between layers. In the resulting graph, all even layers are populated by zero loaded nodes and odd layers by nodes in the original layers. The longest path from the root to the leaf alternatively crosses nodes with unit and zero time cost. The number of unit time cost nodes in the longest path is at most $1 + \hat{d}(G)$, and, thus, the critical path value in $G'$ is the same as in $G$.

### 3.5 Algorithm Comparison

We compare the algorithm just described (SP-ization2) with the simple layering technique (Layering) and our first algorithm presented in [30] (SP-ization1). Results are summarized in Table 1.

For most highly regular applications the result of all three techniques is the same. While the best complexity bounds are for the simple layering technique, it does not offer good results for irregular structures (see following sections for experimental measures). For these irregular cases both SP-ization algorithms offer very similar results (if not the same). Nevertheless, the SP-ization1 algorithm could generate different results depending on the input graph nodes labeling. Our new SP-ization algorithm presents two interesting improvements: the time complexity of the new algorithm is tightly bounded, and the output is always the same, ensuring no critical path value increase for UTC graphs. We conclude that for highly regular applications, a layering technique (or bulk synchronous parallelism) is similar to a nested parallelism solution, but a layering technique is faster. For more irregular problems, nested parallelism is more appropriate and our new algorithm can avoid a cpv increase in UTC graphs at only a logarithmic time complexity increase on the number of graph nodes.

**Table 1.** Algorithms comparison

| Algorithm | Space | Time | UTC-cpv | Regular app. | Irregular app. |
|---|---|---|---|---|---|
| Layering | $O(m+n)$ | $O(m+n)$ | Yes | Good | Bad |
| SP-ization1 | $O(m+n)$ | $O(m \times n)$ | No | Good | Good |
| SP-ization2 | $O(m+n)$ | $O(m+n\log n)$ | Yes | Good | Good |

# 4 Experimental Results

We study the increase of the critical path value when sample graphs are transformed by the layering and SP-ization algorithm techniques. We relate results to simple graph parameters: $P$ the maximum degree of parallelism, $D$ the maximum depth level, and $S$ the synchronization density (a parameter related to the chromatic index and defined by $S = |E_G|/|V_G|$). Specifically, we are interested in extremely low values of the $S$ parameter ($S < 2$), not found in highly regular applications. In this paper we introduce a new parameter $Rs = |E_G|/|V_G|^2$ that measures the *relative synchronization density*, the $S$ value relative to the graph size.

As an indicator of the performance loss, we define $\gamma$, as the relative increment of the critical path value after an SP-ization. Let $G$ be a graph, and $G'$ the SP transformation of $G$:

$$\gamma = \frac{cpv(G')}{cpv(G)}$$

## 4.1 Random Sampling of Graph Space

Random task graphs with small (32) to big (1024) number of nodes were generated, using well-known topology generation techniques [31], appropriate for irregular graphs. For each size and edge density value (around 10 different values), we have generated sets of 100 random topologies. Thus, around 1000 topologies for each graph size were tested.

Sample workloads are also randomly generated with the i.i.d. premise. We use 4 gaussian distributions to represent computations with balanced and unbalanced workloads (mean=1.0, deviation=0.1, 0.2, 0.5, 1.0). For each topology we draw the workload of every task 25 times for each of the 4 distributions.

We measure the parameter $S$ after transforming the generated DAGs to connected StDAGs. Although $S$ is similar to the original edge density, it is slightly modified due to added edges when connecting the graph in an StDAG form. For low values of edge density, many edges are added to connect the highly sparse generated graphs. As we show in Fig. 3*(a)*, in these irregular topologies, the $P$ and $D$ parameters are highly correlated with $S$. If $S$ is low, many nodes or subgraphs will be unconnected, and thus, parallel in the StDAG (high $P$ and low $D$). As $S$ increases, more nodes and subgraphs are serialized (low $P$ and high $D$).

Fig. 3*(b)* illustrates how the algorithm improves critical path value results as compared with the layering technique. Specifically, for $S$ values below 2, the highly unstructured graph is much better transformed by the new algorithm. For small $S$ values the distribution of the loads across the same topology has a great impact on $\gamma$.

The plot in Fig. 3*(c)* shows $\gamma$ values obtained for medium size random graphs transformed with the new algorithm for different workload models (from unbalanced computations $G(1,1)$ to highly balanced computations $G(1,0.1)$. The workload balance is a basic factor for the impact of SP-ization. In the plot, it

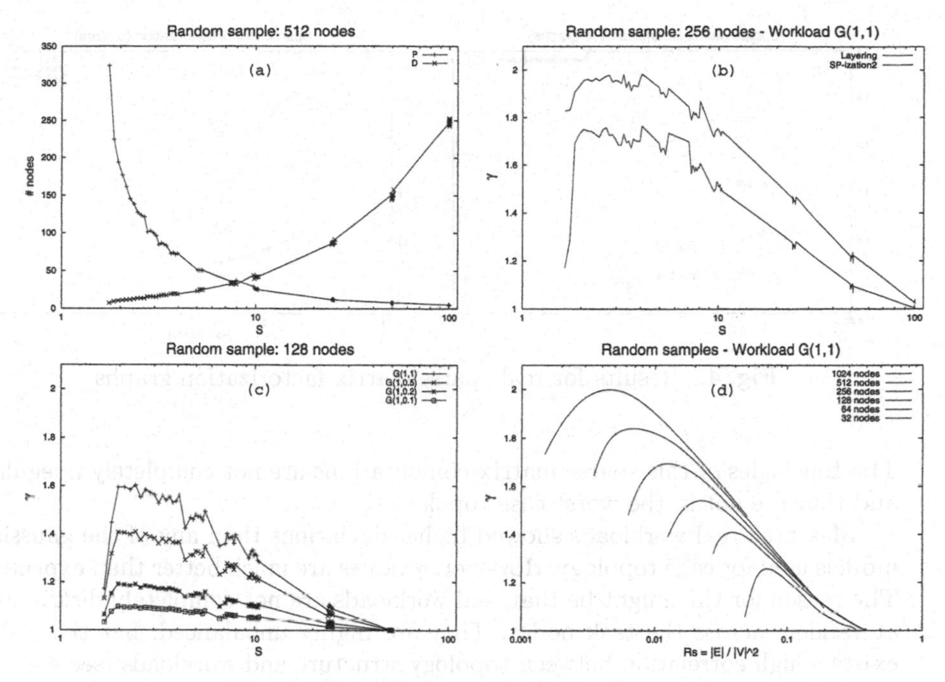


**Fig. 3.** Results for random graphs sample

can be appreciated how the new SP-ization algorithm offers good solutions for graphs with an $S$ value lesser than 2.

The plots have different slopes for different graph sizes. We propose a new parameter $Rs = |E_G|/|V_G|^2$ to predict the behavior of $\gamma$ more independently of the graph size. In Fig. 3*(d)* we present smoothed curves for mean $\gamma$ relative to $Rs$, for all graph sizes and normal workload distribution. Curves drop again to the left due to our new algorithm treatment for $S$ values below 2.

## 4.2 Example of Real Applications

As an example of the algorithm behavior with graphs obtained from real computations, we apply it to some example graphs generated by monitoring the execution of an unstructured domain decomposition and sparse-matrix factorization software for finite elements problems [16, 17]. The real workloads obtained during execution are also available. The number of nodes in these graphs are: 59, 113, 212, 528, 773 and 2015. (In Fig. 2 we showed the 113 nodes graph before and after the transformation).

In Fig. 4*(a)* we show $\gamma$ results for the graphs considered with real workloads measured during execution. Information about the number of nodes $n$ and the relative deviation ($dev = \sigma/\mu$) is added to each point. Fig. 4*(b)* shows $\gamma$ results when real workloads are substituted by modeled workloads with gaussian distributions. The points are below the expected mean values for random samples.

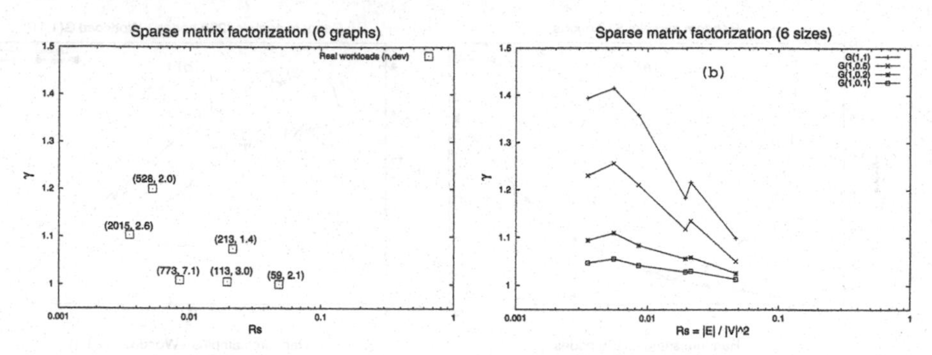


**Fig. 4.** Results for real sparse matrix factorization graphs

The topologies of this sparse matrix computations are not completely irregular, and they are not in the worst case topologies.

Measured real workloads showed higher deviations than any of the gaussian models used for each topology. However, $\gamma$ values are much better than expected. The reason for this might be that real workloads are not completely distributed at random across the task nodes. They are highly unbalanced, but there still exists a high correlation between topology structure and workloads (see e.g. the position of darker highly loaded nodes in Fig. 2).

Workload parameters $\mu, \sigma$ are not enough to get an accurate estimation of the impact of SP-ization on a given unstructured topology. In general, the correlation between highly loaded nodes and layers will produce an improvement in performance when mapping to nested parallelism structures.

## 5 Conclusion

We have presented an efficient algorithmic technique to automatically transform any existing application task graph to SP form. This technique can be useful for compiler technology that gets profit of the SP analyzability properties to automatically optimize an application for a given target machine.

We conclude that unstructured applications can be efficiently mapped to nested-parallelism or SP programming models with a small loss of parallelism. Performance degradation is related to simple graph size parameters. However, information about the workload distribution and its correlation with the topology will be necessary to accurately predict the impact of such a mapping.

## References

[1] Skillicorn, D., Talia, D.: Models and languages for parallel computation. ACM Computing Surveys **30** (1998) 123–169

[2] Valdés, J., Tarjan, R., Lawler, E.: The recognition of series parallel digraphs. SIAM Journal of Computing **11** (1982) 298–313

[3] Lodaya, K., Weil, P.: Series-parallel posets: Algebra, automata, and languages. In: Proc. STACS'98. Volume 1373 of LNCS., Paris, France, Springer (1998) 555–565
[4] Gemund, A.v.: The importance of synchronization structure in parallel program optimization. In: Proc. *11th* ACM ICS, Vienna (1997) 164–171
[5] Sahner, R., Trivedi, K.: Performance and reliability analysis using directed acyclic graphs. IEEE Trans. on Software Eng. **13** (1987) 1105–1114
[6] Skillicorn, D.: A cost calculus for parallel functional programming. Journal of Parallel and Distributed Computing **28** (1995) 65–83
[7] Blumofe, R., Leiserson, C.: Scheduling multithreaded computations by work stealing. In: Proc. Annual Symposium on FoCS. (1994) 356–368
[8] Finta, L., Liu, Z., Milis, I., Bampis, E.: Scheduling UET–UCT series–parallel graphs on two processors. Theoretical Computer Science **162** (1996) 323–340
[9] Valiant, L.: A bridging model for parallel computation. Comm.ACM **33** (1990) 103–111
[10] Kessler, C.: NestStep: nested parallelism and virtual shared memory for the BSP model. In: Int. Conf. on Parallel and Distributed Processing Techniques and Applications (PDPTA'99), Las Vegas (USA) (1999)
[11] Bonorden, O., Juurlink, B., von Otte, I., Rieping, I.: The Paderborn University BSP (PUB) library - design, implementation, and performance. In: Proc. IPPS/SPDP'99, San Juan, Puerto Rico, Computer Society, IEEE (1999)
[12] Darlington, J., Guo, Y., To, H., Yang, J.: Functional skeletons for parallel coordination. In: Europar'95. LNCS (1995) 55–69
[13] Cole, M.: Frame: an imperative coordination language for parallel programming. Technical Report EDI-INF-RR-0026, Division of Informatics, University of Edinburgh (2000)
[14] Blumofe, R., Joerg, C., Kuszmaul, B., Leiserson, C., Randall, K., Zhou, Y.: Cilk: An efficient multithreaded runtime system. In: Proc. of 5th PPoPP, ACM (1995) 207–216
[15] González-Escribano, A., Gemund, A.v., Cardeñoso-Payo, V., Alonso-López, J., Martín-García, D., Pedrosa-Calvo, A.: Measuring the performance impact of SP-restricted programming in shared-memory machines. In J. M. L. M. Palma, J. Dongarra, V. H., ed.: VECPAR 2000. Number 1981 in LNCS, Porto (Portugal), Springer (2000) 128–728
[16] Lin, H.: A general approach for parallelizing the FEM software package DIANA. In: Proc. High Performance Computing Conference'94, National Supercomputing Research Center. National University of Singapur (1994) 229–236
[17] Lin, H., Gemund, A.v., Meijdam, J., Nauta, P.: Tgex: a tool for portable parallel and distributed execution of unstructured problems. Volume 1067 of LNCS., Berlin, Springer (1996) 467–474
[18] González-Escribano, A., Gemund, A.v., Cardeñoso, V.: Predicting the impact of implementation level aspects on parallel application performance. In: CPC'2001 Ninth Int. Workshop on Compilers for Parallel Computing, Edinburgh, Scotland UK (2001) 367–374
[19] Gerbessiotis, A., Valiant, L.: Direct bulk-synchronous parallel algorithms. Technical Report TR-10-92, Center for Research in Computing Technology, Harvard University, Cambridge, Massachussets (1992)
[20] Malony, A., Mertsiotakis, V., Quick, A.: Automatic scalability analysis of parallel programs based on modeling techniques. In Haring, G., Kotsis, G., eds.: Comp. Perf. Eval.: Modelling Techniques and Tools *(LNCS 794)*, Berlin, Springer-Verlag (1994) 139–158

[21] Blelloch, G., Chatterjee, S., Hardwick, J., Sipelstein, J., Zagha, M.: Implementation of a portable nested data-parallel language. Journal of Parallel and Distributed Computing **21** (1994) 4–14

[22] Boeres, C., Rebello, V., Skillicorn, D.: Static scheduling using task replication for LogP and BSP models. In: EuroPar'98. Volume 1480 of LNCS., Springer (1998) 337–346

[23] Munier, A., Hanen, C.: Using duplication for scheduling unitary tasks on $m$ processors with unit communication delays. Technical Report LITP 95/47, Laboratoire Informatique Théorique et Programmation, Institut Blaise Pascal, Université Pierre et Marie Curie, 4, place jussieu, 75252 Paris cedex 05 (1995)

[24] Eisenbiegler, J., Lówe, W., Wehrenpfennig, A.: On the optimization by redundancy using an extended LogP model. In: Proc. Advances in Parallel and Distributed Computing Conference (APDC'97), IEEE (1997)

[25] Bilardi, G., Herley, K., Pietracaprina, A.: BSP vs. LogP. In: Proc. 8th ACM symposium on Parallel algorithms and architectures (SPAA'96), Padua, Italy, ACM (1996) 25–32

[26] Ramachandran, V., Grayson, B., Dahlin, M.: Emulations between QSM, BSP and LogP: A framework for general-purpose parallel algorithm design. In: Proc. ACM-SIAM SODA'99. (1999) 957–958

[27] González-Escribano, A., Gemund, A.v., Cardeñoso, V.: A new algorithm for mapping DAGs to series-parallel form. Technical Report IT-DI-2002-2, Dpto. Informática, Univ. Valladolid (2002)

[28] Bein, W., Kamburowski, J., Stallman, F.: Optimal reductions of two-terminal directed acyclic graphs. SIAM Journal of Computing **6** (1992) 1112–1129

[29] Bodlaender, H.: Dynamic algorithms for graphs with treewidth 2. In: Proc. Workshop on Graph-Theoretic Concepts in Computer Science. (1994)

[30] González-Escribano, A., Cardeñoso, V., Gemund, A.v.: On the loss of parallelism by imposing synchronization structure. In: Proc. 1st Euro-PDS Int'l Conf. on Parallel and Distributed Systems, Barcelona (1997) 251–256

[31] Tobita, T., Kasahara, H.: A standard task graph set for fair evaluation of multiprocessor scheduling algorithms. In: ICS'99 Workshop. (1999) 71–77

# An Efficient Parallel and Distributed Algorithm for Counting Frequent Sets

Salvatore Orlando[1], Paolo Palmerini[1,2], Raffaele Perego[2], and Fabrizio Silvestri[2,3]

[1] Dipartimento di Informatica, Università Ca' Foscari, Venezia, Italy
[2] Istituto CNUCE, Consiglio Nazionale delle Ricerche (CNR), Pisa, Italy
[3] Dipartimento di Informatica, Università di Pisa, Italy

**Abstract.** Due to the huge increase in the number and dimension of available databases, efficient solutions for counting frequent sets are nowadays very important within the Data Mining community. Several sequential and parallel algorithms were proposed, which in many cases exhibit excellent scalability. In this paper we present ParDCI, a distributed and multithreaded algorithm for counting the occurrences of frequent sets within transactional databases. ParDCI is a parallel version of DCI (Direct Count & Intersect), a multi-strategy algorithm which is able to adapt its behavior not only to the features of the specific computing platform (e.g. available memory), but also to the features of the dataset being processed (e.g. sparse or dense datasets). ParDCI enhances previous proposals by exploiting the highly optimized counting and intersection techniques of DCI, and by relying on a multi-level parallelization approach which explicitly targets clusters of SMPs, an emerging computing platform. We focused our work on the efficient exploitation of the underlying architecture. Intra-Node multithreading effectively exploits the memory hierarchies of each SMP node, while Inter-Node parallelism exploits smart partitioning techniques aimed at reducing communication overheads. In depth experimental evaluations demonstrate that ParDCI reaches nearly optimal performances under a variety of conditions.

## 1 Introduction

Association Rule Mining (ARM) [6, 7] is one of the most popular topic in the Data Mining field. The process of generating association rules has historically been adopted for *Market-Basket Analysis*, where transactions are records representing point-of-sale data, while items represent products on sale. The importance for marketing decisions of association rules like "the 80% of customers who buy products $X$ with high probability also buy $Y$" is intuitive, and explains the strong interest in ARM.

Given a database of transactions $\mathcal{D}$, an association rule has the form $X \Rightarrow Y$, where $X$ and $Y$ are sets of items (*itemsets*), such that $(X \cap Y) = \varnothing$. A rule $X \Rightarrow Y$ holds in $\mathcal{D}$ with a minimum confidence $c$ and a minimum support $s$, if at least the $c\%$ of all the transactions containing $X$ also contain $Y$, and $X \cup Y$ is

J.M.L.M. Palma et al. (Eds.): VECPAR 2002, LNCS 2565, pp. 421–435, 2003.

present in at least the $s\%$ of all the transactions of the database. It's important to distinguish among support and confidence. While confidence measures the probability of having $Y$ when $X$ is given, the support of a rule measures the number of transactions in $\mathcal{D}$ that contain both $X$ and $Y$.

The ARM process can be subdivided into two main steps. The former is concerned with the Frequent Set Counting (*FSC*) problem. During this step, the set $\mathcal{F}$ of all the *frequent* itemsets is built, where an itemset is frequent if its support is greater than a fixed support threshold $s$, i.e. the itemset occurs in at least *minsup* transactions ($minsup = s/100 \cdot n$, where $n$, is the number of transaction in $\mathcal{D}$). In the latter step the association rules satisfying both minimum support and minimum confidence conditions are identified. While generating association rules is straightforward, the first step may be very expensive both in time and space, depending on the support threshold and the characteristic of the dataset processed. The computational complexity of the FSC problem derives from the size of its search space $\mathcal{P}(M)$, i.e. the power set of $M$, where $M$ is the set of items contained in the various transactions of $\mathcal{D}$. Although $\mathcal{P}(M)$ is exponential in $m = |M|$, effective pruning techniques exist for reducing it. The capability of effectively pruning the search space derives from the intuitive observation that none of the superset of an infrequent itemset can be frequent. The search for frequent itemsets can be thus restricted to those itemsets in $\mathcal{P}(M)$ whose subsets are all frequent. This observation suggested a level-wise, or breadth-first, visit of the lattice corresponding to $\mathcal{P}(M)$, whose partial order is specified by the subset relation ($\subseteq$) [16].

*Apriori* [3] was the first effective algorithm for solving FSC. It iteratively searches frequent itemsets: at each iteration $k$, the set $F_k$ of all the frequent itemsets of $k$ items ($k$-itemsets), is identified. In order to generate $F_k$, a *candidate* set $C_k$ of potentially frequent itemsets is first built. By construction, $C_k$ is a superset of $F_k$, and thus in order to discover frequent $k$-itemsets, the supports of all candidate sets are computed by scanning the entire transaction database $\mathcal{D}$. All the candidates with minimum support are then included in $F_k$, and the next iteration is started. The algorithm terminates when $F_k$ becomes empty, i.e. when no frequent set of $k$ or more items is present in the database. *Apriori* strongly reduces the number of candidate sets generated on the basis of a simple but very effective observation: a $k$-itemset can be frequent only if all its subsets of $k - 1$ items are frequent. $C_k$ is thus built at each iteration as the set of all $k$-itemsets whose subsets of $k-1$ items are all included in $F_{k-1}$. Conversely, $k$-itemsets that at least contain an infrequent $(k-1)$-itemset are not included in $C_k$.

Several variations to the original *Apriori* algorithm, as well as many parallel implementations, have been proposed in the last years. We can recognize two main methods for determining itemset supports: *counting*-based [1, 3, 5, 8, 13] and *intersection*-based [14, 16]. The former one, also adopted by *Apriori*, exploits a *horizontal* dataset, where the transactions are stored sequentially. The method is based on *counting* how many times each candidate $k$-itemset occurs in every transaction. The intersection–based method, on the other hand, exploits a *vertical* dataset, where a *tidlist*, i.e. a list of the identifiers of all the transactions which contain a given item, is associated with the identifier of the item itself. In this case the support of any $k$-itemset can be determined by computing the cardinality of the tidlist resulting from the $k$-way intersection of the $k$ tidlists associated with the corresponding $k$ items. If we are able to buffer the tidlists of previously computed frequent $(k-1)$-itemsets, we can speedup the computation since the support of a generic candidate $k$-itemset $c$ can be simply computed by intersecting the tidlists of two $(k-1)$-itemsets whose union produces $c$. The *counting*-based approach is, in most cases, quite efficient from the point of view of memory occupation, since it only requires enough main memory to store $C_k$ along with the data structures exploited to make the access to candidate itemsets faster (e.g. hash-trees or prefix-trees). On the other hand, the *intersection*-based method is more computational effective [14]. Unfortunately, it may pay the reduced computational complexity with an increase in memory requirements, in particular to buffer the tidlists of previously computed frequent $(k-1)$-itemsets.

In this paper we discuss ParDCI, a parallel and distributed implementation of DCI (Direct Count & Intersect), which is an effective FSC algorithm previously proposed by the same authors [12]. DCI resulted faster than previously proposed, state-of-the-art, sequential algorithms. The very good results obtained for sparse and dense datasets, as well as for real and synthetically generated ones, justify our interest in studying parallelization and scalability of DCI. ParDCI works similarly to DCI. It is based on a level-wise visit of $\mathcal{P}(M)$, and adopts a *hybrid* approach to determine itemset supports: it exploits an effective *counting*-based method during the first iterations, and a very fast *intersection*-based method during the last ones. When the counting method is employed, ParDCI relies on optimized data structures for storing and accessing candidate itemsets with high locality. The database is partitioned among the processing nodes, and a simple but effective database pruning technique [11, 12] is exploited which allows to trim the transaction database partitions as execution progresses. When the pruned dataset is small enough to fit into the main memory, ParDCI changes its behavior, and adopts an intersection-based approach to determine frequent sets. The representation of the dataset is thus transformed from horizontal into vertical, and the new dataset is stored in-core on each node. Candidates are then partitioned in a way that grants load balancing, and the support of each candidate itemset is locally determined on-the-fly by intersecting the corresponding tidlists. Tidlists are actually represented as vectors of *bits*, which can be accessed with high locality and intersected very efficiently. Differently from other proposals, our intersection approach only requires a *limited* and *configurable* amount of memory. To speedup the intersecting task, ParDCI reuses most of the intersections previously done, by caching them in a fixed-size buffer for future use. Moreover, ParDCI adopts several heuristic strategies to adapt its behavior to the features of datasets processed. For example, when a dataset is dense, the sections of tidlists which turn out to be identical are aggregated and clustered in order to reduce the number of intersections actually performed. Conversely, when a dataset is sparse, the runs of zero bits in the intersected tidlists are promptly identified and skipped. More details on these strategies can be found in [12]. ParDCI implementation is explicitly targeted towards the efficient use of clusters of SMP nodes, an emerging computing platform. Inter-Node parallelism exploits the MPI communication library, while Intra-Node parallelism uses multithreading and out-of-core techniques in order to effectively exploit the memory hierarchies of each SMP node. To validate our proposal we conducted several experiments on a cluster of three dual-processor PCs running linux. Different synthetic datasets and various support thresholds were used in order to test ParDCI under different conditions. The results were very encouraging since the performances obtained were nearly optimal.

This paper is organized as follows. Section 2 introduces FSC parallelization techniques and discusses same related work. In Section 3 we describe ParDCI in depth, while Section 4 presents and discusses the results of the experiments conducted. Finally in Section 5 we draw future works and some conclusions.

## 2 Related Work

Several parallel algorithms for solving the FSC problem have been proposed in the last years [2, 8]. Zaki authored a survey on ARM algorithms and relative parallelization schemas [15]. Most proposals regard parallelizations of the well-known *Apriori* algorithm. Agrawal *et al.* in [2] proposes a broad taxonomy of the parallelization strategies that can be adopted for *Apriori* on distributed-memory architectures. These strategies, summarized in the following, constitute a wide spectrum of trade–offs between computation, communication, memory usage, synchronization, and exploitation of problem–domain knowledge.

The *Count Distribution* strategy follows a data-parallel paradigm according to which the transaction database is statically partitioned among the processing nodes, while the candidate set $C_k$ is replicated. At each iteration every node counts the occurrences of candidate itemsets within the local database partition. At the end of the counting phase, the replicated counters are aggregated, and every node builds the same set of frequent itemsets $F_k$. On the basis of the global knowledge of $F_k$, candidate set $C_{k+1}$ for the next iteration is then built. Inter-Node communication is minimized at the price of carrying out redundant computations in parallel.

The *Data Distribution* strategies attempts to utilize the aggregate main memory of the whole parallel system. Not only the transaction database, but also the candidate set $C_k$ are partitioned in order to permit both kinds of partitions to fit into the main memory of each node. Processing nodes are arranged in a logical ring topology to exchange database partitions, since every node has to count the occurrences of its own candidate itemsets within the transactions of the whole database. Once all database partitions have been processed by each node, every node identifies the locally frequent itemsets and broadcasts them to all the other nodes in order to allow them to build the same set $C_{k+1}$. This approach clearly maximizes the use of node aggregate memory, but requires a very high communication bandwidth to transfer the whole dataset through the ring at each iteration.

The last strategies, *Candidate Distribution*, exploits problem–domain knowledge in order to partition both the database and the candidate set in a way that allows each processor to proceed independently. The rationale of the approach is to identify, as execution progresses, disjoint partitions of candidates supported by (possibly overlapping) subsets of different transactions. Candidates are subdivided on the basis of their prefixes. This trick is possible because candidates, frequent itemsets, and transactions, are stored in lexicographical order. Depending from the resulting candidate partitioning schema, the approach may suffer from poor load balancing. The parallelization technique is however very interesting. Once the partitioning schema for both $C_k$ and $F_k$ is decided, the approach does not involve further communications/synchronizations among the nodes.

The results of the experiments described in [2] demonstrate that algorithms based on Count Distribution exhibits optimal scale-up and excellent speedup, thus outperforming the other strategies. Data Distribution resulted the worst approach, while the algorithm based on Candidate Distribution obtained good performances but paid a high overhead caused by the need of redistributing the dataset.

## 3 ParDCI

During its initial *counting*-based phase, ParDCI exploits a *horizontal* database with variable length records. During this phase, ParDCI trims the transaction database as execution progresses. In particular, a pruned dataset $\mathcal{D}_{k+1}$ is written to the disk at each iteration $k$, and employed at the next iteration [11]. Dataset pruning is based on several criteria. The main criterium states that transactions that do not contain any frequent $k$-itemset will not surely contain larger frequent itemsets and can thus be removed from $\mathcal{D}_{k+1}$. Pruning entails a reduction in I/O activity as the algorithm progresses, since the size of $\mathcal{D}_k$ is always smaller than the size of $\mathcal{D}_{k-1}$. However, the main benefits come from the reduced computation required for subset counting at each iteration $k$, due to the reduced number and size of transactions. As soon as the pruned dataset becomes small enough to fit into the main memory, ParDCI adaptively changes its behavior, builds a *vertical-layout* in-core database, and adopts the intersection-based approach to determine larger frequent sets. Note, however, that ParDCI continues to have a level-wise behavior, so that the search space for finding frequent sets is still traversed breadth-first [10].

ParDCI uses an *Apriori*-like technique to generate $C_k$ starting from $F_{k-1}$. Each candidate $k$-itemset is generated as the union of a pair of $F_{k-1}$ itemsets sharing a common $(k-2)$-prefix. Since itemsets in $F_{k-1}$ are lexicographically ordered, the various pairs occur in close positions within $F_{k-1}$, and ParDCI can generate candidates by exploiting high spatial and temporal locality. Only during the counting-based phase, $C_k$ is pruned by checking whether all the other $(k-1)$-subsets of a candidate $k$-itemset are frequent, i.e. are included in $F_{k-1}$. Conversely, during the intersection-based phase, since our intersection method is able to quickly determine the support of a candidate itemsets, we found more

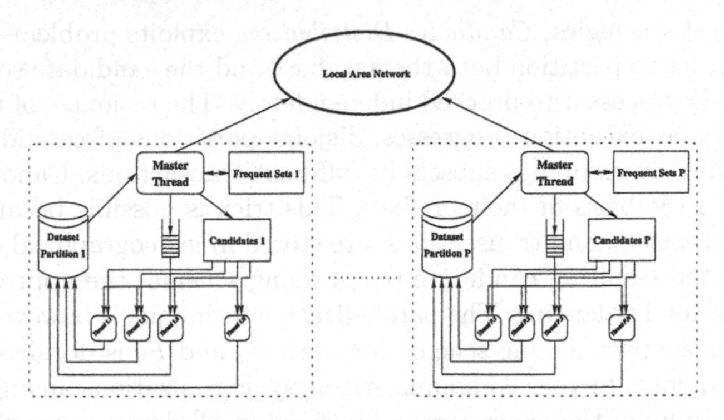


**Fig. 1.** ParDCI: threads and processes interaction schema

profitable to avoid this check. As soon a candidate $k$-itemset is generated, ParDCI determines on-the-fly its support by intersecting the corresponding tidlists. As a consequence, while during its counting-based phase ParDCI has to maintain $C_k$ in main memory to search candidates and increment associated counters, this is no longer needed during its intersection-based phase. This is an important improvement over other FSC algorithms, which suffer from the possible huge memory requirements due to the explosion of the size of $C_k$ [9].

In designing ParDCI we exploited effective out-of-core techniques, so that the algorithm is able to adapt its behavior to the characteristics of the dataset and of the underlying architecture. The efficient exploitation of memory hierarchies received particular attention: datasets are read/written in blocks, to take advantage of I/O prefetching and system pipelining [4]; at each iteration the frequent set is written to a file which is then `mmap`-ped into memory in order to access it for candidate generation during the next iteration.

In the following we describe the different parallelization techniques used in the *counting-* and *intersection*-based phases of ParDCI. In both phases we have to further distinguish between the *Intra-Node* and the *Inter-Node* level of parallelism exploitation. At the Inter-Node level we used the message–passing paradigm through the MPI communication library, while in the Intra-Node level we exploited multi-threading through the *Posix Thread* library. A *Count Distribution* approach is adopted to parallelize the *counting*-based phase, while the *intersection*-based phase exploits a very effective and original implementation of the *Candidate Distribution* approach [2].

### 3.1 The Counting-Based Phase

A *Count Distribution* approach was adopted for the *counting*-based phase of ParDCI. Since the counting-based approach is used only for a few iterations (in all the experiments conducted ParDCI starts using intersections at the third or fourth iteration), in the following we only sketch the main features of the counting method adopted (interested readers can refer to [11]). In the first iteration, as all FSC algorithms, ParDCI directly counts the occurrences of items within all the transactions. For $k \geq 2$, instead of using complex data structures like hash-trees or prefix-trees, ParDCI uses a novel *Direct Count technique* that can be thought as a generalization of the technique used for $k = 1$. The technique uses a *prefix table*, $\text{PREFIX}_k[\,]$, of size $\binom{m_k}{2}$, where $m_k$ is the number of different items contained in the pruned dataset $\mathcal{D}_k$. In particular, each entry of $\text{PREFIX}_k[\,]$ is associated with a distinct *ordered prefix* of two items. For $k = 2$, $\text{PREFIX}_k[\,]$ directly contains the counters associated with the various candidate 2-itemsets, while, for $k > 2$, each entry of $\text{PREFIX}_k[\,]$ points to the contiguous section of ordered candidates in $C_k$ sharing the associated prefix. To permit the various entries of $\text{PREFIX}_k[\,]$ to be directly accessed, we devised an order preserving, minimal perfect hash function. This prefix table is thus used to count the support of candidates in $C_k$ as follows. For each transaction $t = \{t_1, \ldots, t_{|t|}\}$, we select all the possible 2-prefixes of all $k$-subsets included in $t$. We then exploit $\text{PREFIX}_k[\,]$ to find the sections of $C_k$ which must be visited in order to check set-inclusion of candidates in transaction $t$.

At the Inter-Node level, candidate set $C_k$ is replicated, while the transaction database is statically split in a number of partitions equal to the number of SMP nodes available. Every SMP node at each iteration performs a scan of the whole local dataset partition. When the occurrence of a candidate itemset $c \in C_k$ is discovered in a transaction, the counter associated with $c$ is incremented. At the end of the counting step, the counters computed by every node are aggregated (via a MPI_Allreduce operation). Each node then produces the same set $F_k$ and generates the same candidate set $C_{k+1}$ employed at the next iteration. These operations are however inexpensive, and their duplication does not degrade performances.

As depicted in Figure 1, at the Intra-Node level each node uses a pool of threads. They have the task of checking in parallel each of the candidate itemset against chunks of transactions of the local dataset partition. The task of subdividing the local dataset in disjoint chunks is assigned to a particular thread, the *Master Thread*. It loops reading blocks of transactions and forwarding them to the *Worker Threads* executing the counting task. To overlap computation with I/O, minimize synchronization, and avoid data copying overheads, we used an optimized producer/consumer schema for the cooperation among the Master and Worker threads. A prod/cons buffer, which is logically divided into $Npos$ sections, is shared between the Master (producer) and the Workers (consumers). We also have two queues of pointers to the various buffer positions: a *Writable Queue*, which contains pointers to free buffer positions, and a *Readable Queue*, which contains pointers to buffer positions that have been filled by the Master with transactions read from the database. The operations that modify the two queues (to be performed in critical sections) are very fast, and regard the attachment/detachment of pointers to the various buffer positions. The Master thread detaches a reference to a free section of the buffer from the *Writable Queue*, and uses that section to read a block of transactions from disk. When reading is completed, the Master inserts the reference to the buffer section into the (initially empty) *Readable Queue*. Symmetrically, each Worker thread self-schedules its work by extracting a reference to a chunk of transactions from the *Readable Queue*, and by counting the occurrences of candidate itemsets within such transactions. While the transactions are processed, the Worker also performs transaction pruning, and uses the same buffer section to store pruned transactions to be written to $\mathcal{D}_{k+1}$. At the end of the counting step relative to the current chunk of transactions, the worker writes the transactions to disk and reinserts the reference to the buffer section into the *Writable Queue*. When all transactions (belonging to the partition of $\mathcal{D}_k$) have been processed, each Master thread performs a local reduction operation over the various copies of counters (reduction at the Intra-Node level), before performing via MPI the global counter reduction operation with all the other Master threads running on the other nodes (reduction at the Inter-Node level). Finally, to complete the iteration of the algorithm, each Master thread generates $F_k$, writes this set to the local disk, and generates $C_{k+1}$.

### 3.2 The Intersection-Based Phase

When the size of the pruned dataset $\mathcal{D}_k$ becomes small enough to fit into the main memory of all nodes, ParDCI changes its behavior, and starts the *intersection*-based phase, by first transforming the dataset from horizontal into vertical. Tidlists in the vertical dataset are actually represented as bit-vectors, where the bit $i$ is set within tidlist $j$ if transaction $i$ contains item $j$. This representation enhances locality exploitation, and allows intersections to be efficiently performed by using simple Boolean *and* instructions.

In order to speedup tidlist intersection, ParDCI stores and reuses most of the intersections previously done by caching them in a fixed-size buffer. In particular, it uses a small "cache" buffer to store all the intermediate intersections that have been computed to determine the support of $c \in C_k$, where $c$ is the last evaluated candidate. The cache buffer used is a simple bi-dimensional bit-array $Cache[\,][\,]$, where the bit vector $Cache[j][\,]$, $2 \leq j \leq (k-1)$ is used to store the results of the intersections relative to the first $j$ items of $c$. Since candidate itemsets are generated in lexicographic order, with high probability two candidates consecutively generated, say $c$ and $c'$, share a common prefix. Suppose that $c$ and $c'$ share a prefix of length $h \geq 2$. When we consider $c'$ to determine its support, we can save work by reusing the intermediate result stored in $Cache[h][]$. Even if, in the worst case, the tidlists corresponding to all the $k$ items included in a candidate $k$-itemset have to be intersected ($k$-way intersection), our caching method is able to strongly reduce the number of intersections actually performed [12].

During the intersection-based phase, an effective Candidate Distribution approach is adopted at both the Inter and Intra-Node levels. This parallelization schema makes the parallel nodes completely independent: communications and synchronization are no longer needed for all the following iterations of ParDCI. Let us first consider the Inter-node level, and suppose that the intersection-based phase is started at iteration $\overline{k}+1$. Therefore, at iteration $\overline{k}$ the various nodes build the bit-vectors representing their own portions of the vertical in-core dataset. The construction of the vertical dataset is carried out on-the-fly, while transactions are read from the horizontal dataset for subset counting. The partial vectors are then broadcast to obtain a complete replication of the whole vertical dataset on each node. The frequent set $F_{\overline{k}}$ (i.e., the set computed in the last counting-based iteration) is then statically partitioned among the SMP nodes by exploiting problem-domain knowledge, thus entailing a Candidate Distribution schema for all the following iterations. Partitioning is done in a way that allows each processing node $P^i$ to generate a unique $C_k^i$ $(k > \overline{k})$ independently of all the other nodes, where $C_k^i \cap C_k^j = \emptyset, i \neq j$, and $\bigcup_i C_k^i = C_k$. To this end $F_{\overline{k}}$ is partitioned as follows. First, it is split into $l$ *sections*, on the basis of the prefixes of the lexicographically ordered frequent itemsets included. All the frequent $\overline{k}$-itemsets that share the same $\overline{k}-1$ prefix (i.e. those itemsets whose first $\overline{k}-1$ items are identical) are assigned to the same section. Since ParDCI builds each candidate $(\overline{k}+1)$-itemsets as the union of two $\overline{k}$-itemsets sharing the first $\overline{k}$ items, we are sure that each candidate $\overline{k}$-itemset can be independently generated starting from one of the $l$ disjoint sections of $F_{\overline{k}}$. The various partitions of $F_{\overline{k}}$ are then created by assigning the $l$ sections to the various processing nodes, by adopting a simple greedy strategy that considers the number of itemsets contained in each section in order to build well-balanced partitions. From our tests, this policy suffices for balancing the workload at the Inter-Node level. Once completed the partitioning of $F_{\overline{k}}$, nodes independently generate the associated candidates and determine their support by intersecting the corresponding tidlists of the replicated vertical dataset. Finally they produce disjoint partitions of $F_{\overline{k}+1}$. Nodes continue to work according to the schema above also for the following iterations. It is worth noting that, although at iteration $\overline{k}$ the whole vertical dataset is replicated on all the nodes, as the execution progresses, the implemented pruning techniques trim the vertical dataset in a different way on each node.

At the Intra-Node level, the same Candidate Distribution parallelization schema is employed, but at a finer granularity and by exploiting a dynamic scheduling strategy to balance the load among the threads. In particular, at each iteration $k$ of the intersection-based phase, the Master thread initially splits the local partition of $F_{k-1}$ into $x$ sections, $x >> t$, where $t$ is the total number of active threads. This subdivision entails a partitioning of the candidates generated on the basis of these sections (Candidate Distribution). The information to identify every section $S_i$ are inserted in a shared queue. Once this initialization is completed, also the Master thread becomes a Worker. Thereinafter, each Worker thread loops and self-schedules its work by performing the following steps:

1. access in mutual exclusion the queue and extract information to get $S_i$, i.e. a section of the local partition of $F_{k-1}$. If the queue is empty, write $F_k$ to disk and start a new iteration.
2. generate a new candidate $k$-itemset $c$ from $S_i$. If it is not possible to generate further candidates, go to step 1.

3. compute on-the-fly the support of $c$ by intersecting the vectors associated with the $k$ items of $c$. In order to reuse effectively previous work, each thread exploits a private cache for storing the partial results of intersections. If $c$ turns out to be frequent, put $c$ into $F_k$. Go to step 2.

The various sections $S_i$ are not created on the basis of prefixes. So, since in order to generate candidates we have to pick pairs of frequent itemsets, once selected a section $S_i$ only the first element of each pair must belong to $S_i$, while the second element must be searched in the following elements of the local partition of $F_k$.

## 4 Experimental Evaluation

ParDCI is the parallel version of DCI, a very fast sequential FSC algorithm previously proposed by the same authors [12]. In order to show both the efficiency of the counting method exploited during early iterations, and the effectiveness of the intersection-based approach used when the pruned vertical dataset fits into the main memory, we report the result of some tests where we compared DCI performances with those of two other FSC algorithms: FP-growth, currently considered one of the fastest algorithm for FSC[1], and the Christian Borgelt's implementation of *Apriori*[2]. For these tests we used publicly available datasets, and a Windows-NT workstation equipped with a Pentium II 350 MHz processor, 256 MB of RAM memory and a SCSI-2 disk. The datasets used are the `connect-4` dense dataset, which contains many frequent itemsets also for high support thresholds, and the T25I20D100K synthetic dataset[3].

Figure 2 reports the total execution times of *Apriori*, FP-growth, and DCI on these two datasets as a function of the support threshold $s$. As it can be seen, DCI significantly outperforms both FP-growth and *Apriori*. The efficiency of our approach is also highlighted by the plots reported in Figure 3, which show per iteration execution times for DCI and *Apriori* on dataset T25I20D100K with support thresholds equal to 0.3 and 1. As it can be seen DCI significantly outperforms *Apriori* in every iteration but the second one due to the additional time spent by DCI to build the vertical dataset used for the following intersection-based iterations. Complete performance tests and comparisons of DCI are discussed in depth in [12]. In that work we analyzed benchmark datasets characterized by different features, thus permitting us to state that the design of DCI is not focused on specific datasets, and that our optimizations are not over-fitted only to the features of these datasets.

For what regards parallelism exploitation, we report an experimental evaluation of ParDCI on a Linux cluster of three two-way SMPs, for a total of six

[1] We acknowledge Prof. Jiawei Han for having kindly provided us the last optimized version of FP-growth.

[2] `http://fuzzy.cs.uni-magdeburg.de/~borgelt`

[3] `http://www.almaden.ibm.com/cs/quest.`

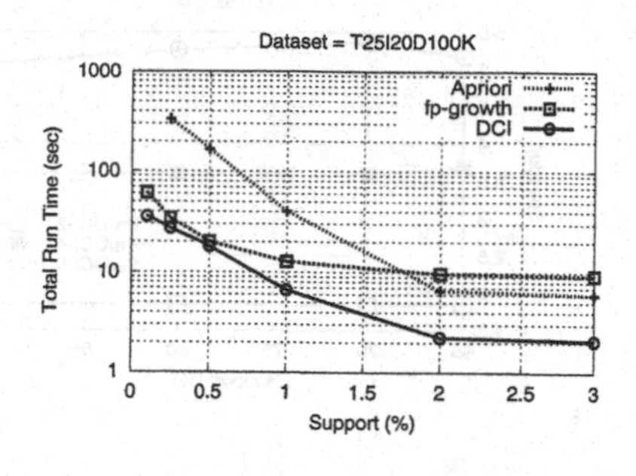


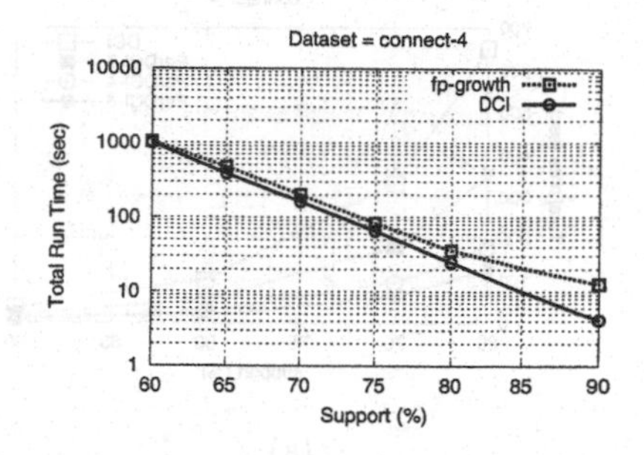


**Fig. 2.** Total execution times for DCI, *Apriori*, and FP-growth on datasets T25I20D100K and connect-4 as a function of the support threshold

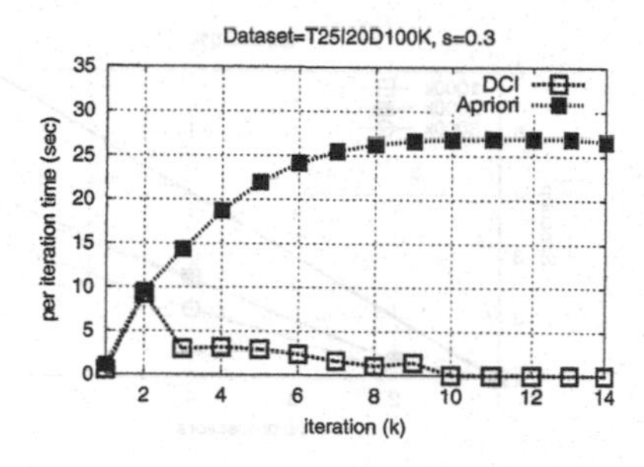


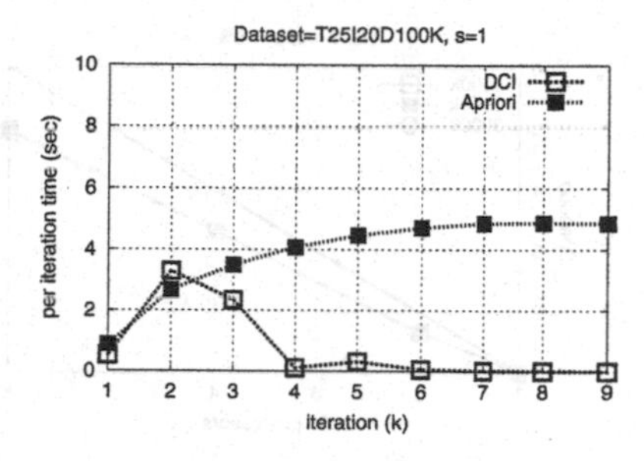


**Fig. 3.** Per iteration execution times for DCI and *Apriori* on dataset T25I20D100K with support thresholds 0.3 and 1

processors. Each SMP is equipped with two Pentium II 233MHz, 256 MB of main memory, and a SCSI disk.

First we compared the performance of DCI and ParDCI on the dense dataset `connect-4`. Figure 4 plots total execution times and speedups (nearly optimal ones) as functions of the support thresholds $s$ (%). ParDCI-2 corresponds to the pure multithread version running on a single 2-way SMP, while ParDCI-4 and ParDCI-6 also exploit inter-node parallelism, and run, respectively, on two and three 2-way SMPs. Note that, in order to avoid exponential explosion in the number of frequent itemsets, we used relatively high supports for the tests with the dense `connect-4` dataset.

Figure 5 plots the speedups obtained on three synthetic datasets for two fixed support thresholds ($s = 1.5\%$ and $s = 5\%$), as a function of the number of processors used. The datasets used in these tests were all characterized by an average transaction length of 50 items, a total number of distinct items of 1000, and a size of the maximal potentially frequent itemset of 32. We only varied the total number of transactions from 500K to 3000K, so we can identify them on

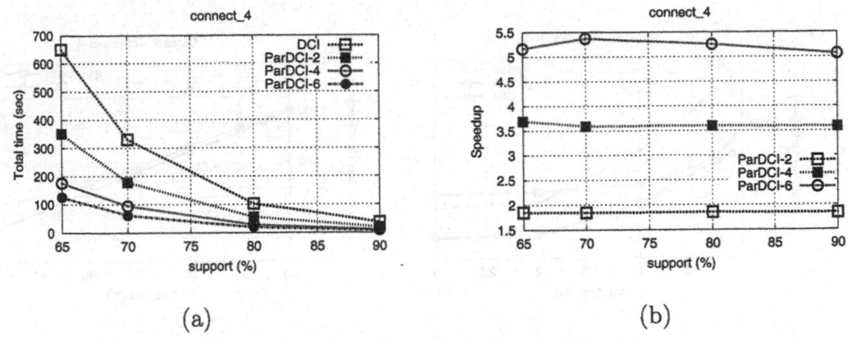


**Fig. 4.** Dense dataset `connect-4`: completion times of DCI and ParDCI (a) and speedups of ParDCI (b), varying the minimum support threshold

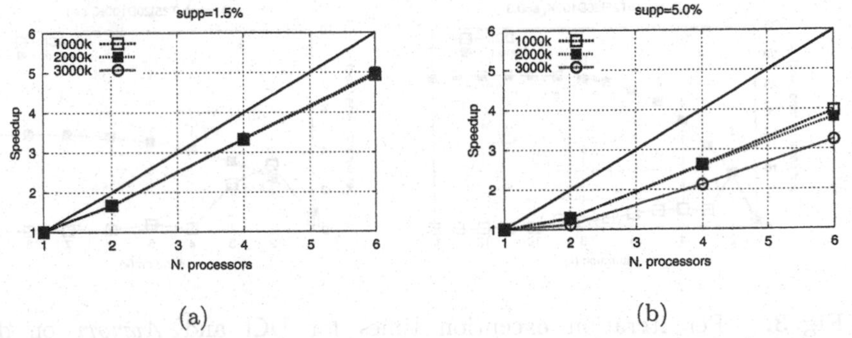


**Fig. 5.** Speedup for the three datasets 1000K 2000K and 3000K with *support* = 1.5% (a) and *support* = 5%(b)

the basis of their number of transactions. It is important to remark that since our cluster is composed of 2-way SMPs, we mapped tasks on processors always using the minimum number of nodes (e.g., when we use 4 processors, we actually use 2 SMP nodes). This implies that experiments performed on either 1 or 2 processors actually have the same memory and disk resources available, whereas the execution on 4 processors benefits from double amount of such resources. According to these experiments, ParDCI shows a quasi linear speedup. As it can be seen by considering the results obtained with one or two processors, the slope of the speedup curve turns out to be relatively worse than its theoretical limit, due to resource sharing and thread implementation overheads at the Inter-Node level. Nevertheless, when several nodes are employed, the slope of the curve improves. For all the three datasets, when we fix $s = 5\%$, we obtain a very small number of frequent itemsets. As a consequence, the CPU-time decreases, and becomes relatively smaller than I/O and interprocess communication times.

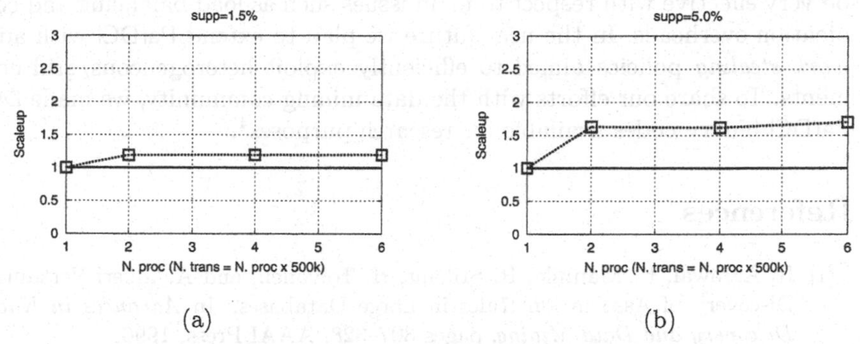


(a) (b)

**Fig. 6.** Scaling behavior obtained varying the dataset size together with the number of processors for *support* = 1.5% (a) and *support* = 5%(b)

Figure 6 plots the scaleup, i.e. the relative execution times measured by varying, at the same time, the number of processors and the dataset size. We can observe that the scaling behavior remains constant, although slightly above one. This is again due to thread management overheads and resource sharing (mainly disk sharing).

The strategies adopted for partitioning the dataset and the candidates well balanced the workload among the processing nodes. In all the tests conducted on our homogeneous cluster of SMPs used as a dedicated resource, the differences in the completion time between the fastest and the slowest processing were always lower than the 1% of the total execution time. Clearly, since we used a static partitioning strategy at the inter-node level, in the case of a heterogeneous pool of computational resources, or a non-dedicated environment, a different partitioning strategy should be necessary.

## 5 Conclusions and Future Works

Originally used as a *Market Baskets* analysis tool, ARM is today used in various fields such as Web Mining, where it is adopted to discover the correlations among the various pages visited by users, Web Searching, where association rules can be used to build a statistical thesaurus or to design intelligent caching policies. Due to the impressive growth rate of data repositories, only efficient parallel algorithms can grant the needed scalability of ARM solutions.

ParDCI originates from DCI, a very fast sequential FSC algorithm previously proposed [12]. Independently of the dataset peculiarities, DCI outperforms not only *Apriori*, but also FP-growth [9], which is currently considered one of the fastest algorithm for FSC. ParDCI, the multithreaded and distributed version of DCI, due to a number of optimizations and to the resulting effective exploitation of the underlying architecture, exhibits excellent Scaleup and Speedup under a variety of conditions. Our implementation of the Count and Candidate Distribution parallelization approaches at both Inter and Intra-Node levels resulted to be very effective with respect to main issues such as load balancing and communication overheads. In the near future we plan to extend ParDCI with adaptive *work stealing* policies aimed to efficiently exploit heterogeneous/grid environments. To share our efforts with the data mining community, we made DCI and ParDCI binary codes available for research purposes[4].

## References

[1] R. Agrawal, H. Mannila, R. Srikant, H. Toivonen, and A. Inkeri Verkamo. Fast Discovery of Association Rules in Large Databases. In *Advances in Knowledge Discovery and Data Mining*, pages 307–328. AAAI Press, 1996.

[2] R. Agrawal and J. C. Shafer. Parallel Mining of Association Rules. *IEEE Transaction On Knowledge And Data Engineering*, 8:962–969, 1996.

[3] R. Agrawal and R. Srikant. Fast Algorithms for Mining Association Rules in Large Databases. In *Proc. of the 20th VLDB Conf.*, pages 487–499, 1994.

[4] R. Baraglia, D. Laforenza, S. Orlando, P. Palmerini, and R. Perego. Implementation Issues in the Design of I/O Intensive Data Mining Applications on Clusters of Workstations. In *Proc. of the 3rd Work. on High Performance Data Mining, (IPDPS-2000), Cancun, Mexico*, pages 350–357. LNCS 1800 Spinger-Verlag, 2000.

[5] R. J. Bayardo. Efficiently Mining Long Patterns from Databases. In *Proc. of the ACM SIGMOD Intl. Conf. on Management of Data*, pages 85–93, 1998.

[6] U. M. Fayyad, G. Piatetsky-Shapiro, P. Smith, and R. Uthurusamy, editors. *Advances in Knowledge Discovery and Data Mining.* AAAI Press, 1998.

[7] V. Ganti, J. Gehrke, and R. Ramakrishnan. Mining Very Large Databases. *IEEE Computer*, 32(8):38–45, 1999.

[8] E. H. Han, G. Karypis, and Kumar V. Scalable Parallel Data Mining for Association Rules. *IEEE Transactions on Knowledge and Data Engineering*, 12(3):337–352, May/June 2000.

[9] J. Han, J. Pei, and Y. Yin. Mining Frequent Patterns without Candidate Generation. In *Proc. of the ACM SIGMOD Int. Conf. on Management of Data*, pages 1–12, Dallas, Texas, USA, 2000.

[10] Hipp, J. and Güntzer, U. and Nakhaeizadeh, G. Algorithms for Association Rule Mining – A General Survey and Comparison. *SIGKDD Explorations*, 2(1):58–64, June 2000.

[11] S. Orlando, P. Palmerini, and R. Perego. Enhancing the Apriori Algorithm for Frequent Set Counting. In *Proc. of the $3^{rd}$ Int. Conf. on Data Warehousing and Knowledge Discovery, LNCS 2114*, pages 71–82, Germany, 2001.

[12] S. Orlando, P. Palmerini, R. Perego, and F. Silvestri. Adaptive and Resource-Aware Mining of Frequent Sets. In *Proc. of the 2002 IEEE Int. Conference on Data Mining (ICDM'02)*, Maebashi City, Japan, Dec. 2002.

[13] J. S. Park, M.-S. Chen, and P. S. Yu. An Effective Hash Based Algorithm for Mining Association Rules. In *Proc. of the 1995 ACM SIGMOD Int. Conf. on Management of Data*, pages 175–186, 1995.

[14] A. Savasere, E. Omiecinski, and S. B. Navathe. An Efficient Algorithm for Mining Association Rules in Large Databases. In *Proc. of the 21th VLDB Conf.*, pages 432–444, Zurich, Switzerland, 1995.

[4] Interested readers can download the binary codes at address `http://www.miles.cnuce.cnr.it/~palmeri/datam/DCI`

[15] M. J. Zaki. Parallel and Distributed Association Mining: A Survey. *IEEE Concurrency*, 7(4):14–25, 1999.
[16] M. J. Zaki. Scalable Algorithms for Association Mining. *IEEE Transactions on Knowledge and Data Engineering*, 12:372–390, May/June 2000.

# A Framework for Integrating Network Information into Distributed Iterative Solution of Sparse Linear Systems*

Devdatta Kulkarni[1] and Masha Sosonkina[2]

[1] Department of Computer Science, University of Minnesota
Minneapolis, MN 55414 USA
dkulk@cs.umn.edu

[2] Department of Computer Science, University of Minnesota
Duluth, MN 55812 USA
masha@d.umn.edu

**Abstract.** Recently, we have proposed a design of an easy-to-use network information discovery tool that can interface with a distributed application non-intrusively and without incurring much overhead. The application is notified of the network changes in a timely manner and may react to the changes by invoking the adaptation mechanisms encapsulated in notification handlers. Here we describe possible adaptations of a commonly used scientific computing kernel, distributed sparse large-scale linear system solution code.

## 1 Introduction

Distributing computation and communication resources is already a well established way of solving large-scale scientific computing tasks. Communication library standards, such as Message Passing Interface (MPI) [14], make applications portable across various distributed computing platforms. However, for high-performance applications in distributed environments, the efficiency and robustness are difficult to attain due to the varying distributed resource – such as communication throughput and latency – availability at any given time. This problem is especially acute in computational grids, in which the resource pool itself may vary during the application execution. One way to handle this situation is to incorporate into application execution run-time adaptive mechanisms, which may complement the static (compile-time) adjustments. At present, however, there is a lack of easy-to-use tools for the application to learn the network information and to request particular network resources dynamically. It is desirable to have a mechanism that provides the network information transparently to the application programmer or user, so that the burden of handling the low level network information is shifted to a network developer. With a knowledge

* This work was supported in part by NSF under grants NSF/ACI-0000443 and NSF/INT-0003274, and in part by the Minnesota Supercomputing Institute

J.M.L.M. Palma et al. (Eds.): VECPAR 2002, LNCS 2565, pp. 436–450, 2003.

of network performance, the application may adapt itself to perform the communication more efficiently. The adaptation features are, of course, application-specific. For a scientific application, it may be beneficial to perform more local computations (iterations) waiting for the peer processors. This paper presents a general way to incorporate the network information and adaptation procedures into an application, that can be used by a wide range of scientific applications. Section 3 provides an overview of our design and implementation of the Network Information Collection and Application Notification (NICAN) tool [16]. It will be used to notify the chosen application of the changes in network status. In Section 4, we outline the distributed sparse linear system solution code used in the experiments. Specifically, the application under consideration is Parallel Algebraic Multilevel Solver (pARMS) [6], developed at the University of Minnesota and shown to be effective for solving large sparse linear systems. Section 5 describes an adaptation scenario for pARMS invoked in response to certain network conditions. We summarize the work in Section 7.

## 2 Related Work

Providing network information to distributed applications has been widely recognized as an important task. Remos [7] makes available the runtime network information for the applications through an interface. A network independent API is provided which enables applications to get network topology information and per flow information. Netlogger [4] gives a detail logging of all the events occurring during the execution of a distributed application. Authors of Congestion Manager (CM) [1] show an effective way to adapt network applications based on a kernel module which enables similar flows, between same source and destination, to share the network congestion information. It allows applications to adapt by providing them with an API for querying the network status. While CM tries to capture the adaptability concept within the kernel bounds, it puts the responsibility of finding out the relevant network information upon the application. HARNESS [2] is a distributed virtual machine system having the capability of supporting plug-ins for tasks like communication, process control and resource monitoring [2]. CANS infrastructure [3] achieves the application adaptation through monitoring the system and application events and re-planning the data paths through the use of semantics preserving modules of code. Network Weather Service (NWS) [18], monitors network behavior and shows that this information can be successfully used for scheduling in distributed computational grid environments.

Our approach is different from these systems mainly in the scope of seamless interaction with application at runtime. We concentrate on specifically providing the adaptive capabilities to scientific distributed applications. Our results are rather application specific but the framework is general enough to be used with scientific applications having alternating computation communication characteristics. In [8], authors show dynamic adaptation of Jacobi-Davidson eigensolver based on the memory thrashing competing applications and CPU based loads.

Their approach is similar to our approach but they do not yet address the network interface related issues. Decoupling of network information collection from the application execution and providing a mechanism to encapsulate application adaptations are the main features of NICAN. The developed tool is light weight to complement high computation and memory demands of large-scale scientific application.

## 3 Providing Dynamic Network Information to Applications

A major design goal is to augment the application execution with the knowledge of the network while requiring minimum modifications of the application and without involving the user/programmer into the network development effort. Indeed, the network information collector should have a negligible overhead and not compete with application for resource usage. This design requirement is especially vital since we target high-performance distributed applications which often demand full capacity of computer resources.

The NICAN accepts the request from the application and delivers the obtained network characteristics to the application. This enables supplying an application with the network information only *if this information becomes critical*, i.e., when the values for the network characteristics to be observed fall outside of some feasible bounds. The feasibility is determined by an application and may be conveyed to the network information collector as parameter. This *selective notification* approach is rather advantageous both when there is little change in the dynamic network characteristics and when the performance is very changeable. In the former case, there is no overhead associated with processing unnecessary information. In the latter, the knowledge of the network may be more accurate since it is obtained more frequently. Multiple probes of the network are recorded to estimate the network performance over a longer period of time. They may also be useful for the prediction of network performance in such common cases as when an iterative process lies at the core of application.

In NICAN, the process of collecting the network information is separated from its other functions, such as notification, and is encapsulated into a module that can be chosen depending on the types of the network, network software configuration, and the information to be collected. For example, assume that the current throughput is requested by an application during its execution. Then, if the network has the Simple Network Management Protocol (SNMP) [9] installed, NICAN will choose to utilize the SNMP information for throughput calculation. Otherwise, some benchmarking procedure – more general than probing SNMP but also more costly – could be applied to determine the throughput. To determine the latency between two hosts, the system utilities such as `ping` and `traceroute` can be used. NICAN collects latency independently of throughput. Whenever a network parameter value becomes available, NICAN processes it immediately without waiting for the availability of the other parameter values. Delaying the processing would cause the excessive overhead for NICAN and would lead NICAN to notify an application with possibly obsolete or wrong data. Figure 1 shows a modular design of NICAN/application interface. The NICAN implementation consists of two parts, the NICAN front end and the NICAN back end. NICAN front end provides for the application adaptation and the NICAN back end performs the network data collection. The application starts the NICAN back end by invocation of the NICAN back end thread. NICAN back end thread consists of separate threads for collecting different network parameters. Note that for simplicity NICAN does not attempt to perform a combined parameter analysis: each network parameter is monitored and analyzed separately from others. The modular design, shown in Figure 1, enables an easy augmentation of the collection process with new options, which ensures its applicability to a variety of network interconnections.

Figure 1 shows that the application starts the NICAN back end thread and passes the monitoring request to the NICAN informs the application about the changes in the network conditions in a timely fashion such that there is no instrumenting of an application with, say, call-queries directed to the network interface. In fact, the initialization of the NICAN tool may be the *only* non-application specific modification required in the application code to interface with NICAN. Upon the notification from the NICAN the application may need to engage its adaptive mechanisms. To minimize changes inside the application code, we propose to encapsulate application adaptation in a notification handler invoked when NICAN informs the application about the changes in the network conditions. This handler can contain an adaptation code with a possible access to some application variables. NICAN front end also provides for preparing the execution environment for better application execution by dynamically selecting relatively less loaded nodes from among the pool of nodes. A more detailed description of NICAN design and implementation can be found in [5].

For a distributed application that uses Message Passing Interface (MPI), the communication overhead also includes the overhead for MPI. Since most of the high performance computing applications use MPI to ensure portability across distributed environments, measuring and monitoring the MPI overheard may be useful for performance tuning. NICAN provides a way to interact with MPI-based distributed applications.

## 4 Distributed Sparse Linear System Solution

Among the techniques for solving large sparse linear systems is a recently developed Parallel Algebraic Recursive Multilevel Solver (pARMS) [6]. This is a distributed-memory *iterative method* (see, e.g., [10] for the description of modern iterative methods) that adopts the general framework of distributed sparse matrices and relies on solving the resulting distributed Schur complement systems [11]. pARMS focuses on novel linear system transformation techniques, called *preconditioning* [10], which aim to make the system easier to solve by iterative methods. In particular, pARMS combines a set of domain decomposition techniques [13], frequently used in parallel computing environments, with

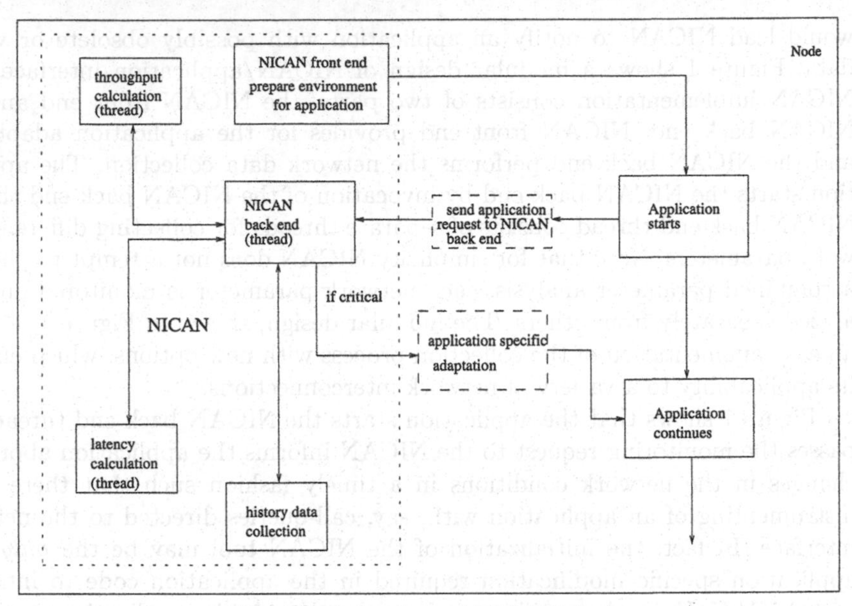


**Fig. 1.** Overview of NICAN design

multi-level preconditioning [17], which leads to a scalable convergence process with increase in problem size.

An iterative solution method can be easily implemented in parallel, yielding a high degree of parallelism. Consider, for example, a parallel implementation of FGMRES [6], a variation of a popular solution method, restarted Generalized Minimum RESidual algorithm (GMRES) [10]. If the classical Gram-Schmidt procedure is used in its orthogonalization phase, an iteration of the parallel algorithm has only two synchronization points, in which all-to-all processor communications are incurred.

One way to partition the linear system $Ax = b$ is to assign certain equations and corresponding unknowns to each processor. For a graph representation of sparse matrix, graph partitioner may be used to select particular subsets of equation-unknown pairs (sub-problems) to minimize the amount of communication and to produce sub-problems of almost equal size. It is common to distinguish three types of unknowns: (1) Interior unknowns that are coupled only with local equations; (2) Inter-domain interface unknowns that are coupled with both non-local (external) and local equations; and (3) External interface unknowns that belong to other sub-problems and are coupled with local equations. Thus each local vector of unknowns $x_i$ is reordered such that its sub-vector $u_i$ of internal components is followed by the sub-vector $y_i$ of local interface components. The right-hand side $b_i$ is conformly split into the sub-vectors $f_i$ and $g_i$ , i.e.,

$$x_i = \begin{pmatrix} u_i \\ y_i \end{pmatrix} \;\; ; \;\; b_i = \begin{pmatrix} f_i \\ g_i \end{pmatrix} .$$

When block-partitioned according to this splitting, the local matrix $A_i$ residing in processor $i$ has the form

$$A_i = \left( \begin{array}{c|c} B_i & F_i \\ \hline E_i & C_i \end{array} \right),$$

so the local equations can be written as follows:

$$\begin{pmatrix} B_i & F_i \\ E_i & C_i \end{pmatrix} \begin{pmatrix} u_i \\ y_i \end{pmatrix} + \begin{pmatrix} 0 \\ \sum_{j \in N_i} E_{ij} y_j \end{pmatrix} = \begin{pmatrix} f_i \\ g_i \end{pmatrix}.$$

Here, $N_i$ is the set of indices for sub-problems that are neighbors to the sub-problem $i$. The term $E_{ij}y_j$ reflects the contribution to the local equation from the neighboring sub-problem $j$. The result of the multiplication with external interface components affects only the local interface unknowns, which is indicated by zero in the top part of the second term of the left-hand side.

## 5 Integration of Network Information into pARMS

The iterative nature of the pARMS execution process is a typical example of most distributed iterative linear system solutions. In each iteration, local computations are alternated with the data exchange phase among all neighboring processors following the pattern of sparse matrix-vector multiplication. This pattern is preserved if a domain decomposition type preconditioner (see e.g., [13]) is used. For such a preconditioner, it is possible to change the amount of local computations in each processor depending on local sub-problem or computing platform characteristics. For varying sub-problem complexity, this issue has been considered in [12] and extended to encompass unequal processor loads in [15]. It has been shown that performing more local iterations in the less loaded processors and thus computing a more accurate local solution would eventually be beneficial for the overall performance. In other words, the accuracy would eventually propagate to other processors, resulting in a reduction of the number of iterations to converge. Here we describe how, with the information provided by NICAN, these adaptations can be carried out based on the changing network conditions.

pARMS uses MPI for communication between participating processors. The rendezvous of all the peers might not be at the same time since each processor might have a different computational load or incur delays in sending or receiving data on its network interface resulting in *low network interface throughput.* Thus, the knowledge of how busy the processor is and how much network interface throughput is available for the processor communications can help in devising adaptation strategies for pARMS.

### 5.1 Adaptation of pARMS Based on Network Condition

If the local computations are balanced, then each processor completes its local computations and then waits, at a communication rendezvous point, for others to complete. Consider what happens after the first exchange of data. The processor which has more network interface throughput available for communication would complete its data transfer earlier relative to the other processors. Therefore this processor will start (and finish) its local computations early and incur an idle time waiting for the other processors. For a distributed iterative linear system solution performed on two processors, Figures 2 and 3 depict, respectively, the ideal scenario of balanced computations and communications and a scenario in which Processor 1 has a low network interface throughput. Instead of idling, Processor 2 can perform more local computations to obtain a more accurate solution and arrive at the rendezvous point later.

pARMS source code is instrumented with an initialization call for NICAN after the initialization of MPI. Within this function call the parameters to be monitored are passed. We have used network interface throughput as the parameter, and the notification criterion is set to reporting global maximum *achieved* throughput and local *achieved* throughput only when their values differ *substantially.* Note that a large, relative to the link nominal bandwidth, value for the achieved throughput would indicate reaching the capacity of the network interface and would lead to communication delays.

After NICAN initialization, pARMS continues its normal operation. NICAN, on the other hand, starts monitoring the achieved throughput for each proces-

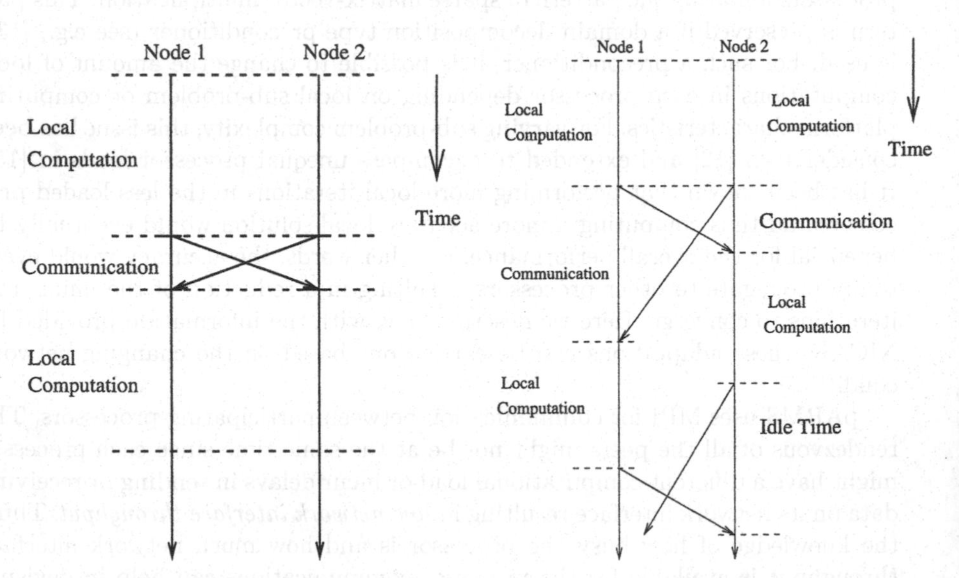


**Fig. 2.** Ideal scenario: Balanced computations and communications

**Fig. 3.** A more realistic scenario: Balanced computations and unbalanced communications

sor using the SNMP protocol. Then the maximum achieved throughput is computed by NICAN among all the participating processors without interfering with pARMS communication or computation tasks. Upon meeting the notification condition, the values obtained for the global maximum and local throughput are signaled to the local application process and passed to the respective notification handler.

The user may incorporate a desired adaptation strategy into the notification handler, thus avoiding direct changes to the application source code. If certain application variables need to be adjusted in response to the network performance, they can be shared with the handler, which contains the adaptation code. For example, when the pARMS adaptation consists of adjusting the number of local iterations $n^i$ in the preconditioner application on processor $i$, the variable $n^i$ is shared with the handler local to processor $i$.

Algorithm 51 outlines a procedure of incorporating adaptive features in pARMS using NICAN. For both sets of experiments, we construct a rather simple notification handler which, upon invocation, changes the value of a shared variable $n^{inner}$ depending on which experiment is performed.

ALGORITHM **51** *Incorporating adaptive features in pARMS using NICAN*

in pARMS:

1. *Declare* $n^{inner} = n^i$ *as shared between pARMS and NICAN.*
2. *Start pARMS outer iteration*
3. *Do pARMS* $n^{inner}$ *inner iterations;*
4. *Exchange interface variables among neighbors;*
5. *End pARMS outer iteration.*

in NICAN:

1. *Start handler when application is notified*
2. *If (adaptation condition for experiment set One)*
3. $n^{inner} = n^{inner}$ *+ constant;*
4. *If (adaptation condition for experiment set Two)*
5. $n^{inner} = n^{inner}$ *+ variable;*
6. *End handler.*

The time ${T_c}^i$ each node $i$ spends in a communication phase may be determined knowing the current achieved throughput $\tau^i$ and the amount of data $D^i$ to be communicated. The throughput for each node is calculated using SNMP. The amount of data $D^i$ to be communicated is readily available from one of the pARMS data structures in each node. (The value of $D^i$ is made accessible via the adaptation handler as well.) Specifically, ${T_c}^i$ is computed as follows:

$${T_c}^i = \frac{D^i}{B^i - \tau^i},$$

where $B^i$ is the link nominal bandwidth, the difference of which with $\tau^i$ gives the throughput available for the communication of data $D^i$. This formula will give us the time required for communicating $D^i$ data values for the node. Among the neighbors the node having the largest value for this time will have the maximum usage of the network interface reflecting the competing network process running on that node. The maximum communication time $T_{\max}$ over all the nodes is calculated. The adjustment of the local iteration number is similar to the strategy proposed in [12]. In the (next) $j$th outer iteration of the iterative process

$$n_j^i = n_{j-1}^i + \Delta_j^i,$$

where $\Delta_j^i$, the number of iterations that node $i$ can fit into the time to be wasted in idling otherwise at the $j$th outer iteration, is determined as follows:

$$\Delta_j^i = \frac{(T_{\max} - {T_c}^i)}{{T_c}^i + {T^i}_{comp}}, \tag{1}$$

where ${T^i}_{comp}$ is the computation time per iteration in node i. The computation time ${T^i}_{comp}$ varies more as convergence approaches. This is because the preconditioning time required at different phases of computation depends upon the characteristics of the matrix to be preconditioned (number of non-zero elements, amount of fill-in).

## 6 Experiments

We present two sets of experiments to demonstrate a seamless incorporation of adaptations into pARMS using NICAN on high performance computing platform and to show that the adaptations based on the communication waiting time enhance overall performance of the application.

### 6.1 Adaptations for a Regularly Structured Problem on IBM SP

The first set of experiments was conducted on the IBM SP at the Minnesota Supercomputing Institute using four WinterHawk+ nodes (375 MHz Power3 node) with 4 GB of main memory. Although each WinterHawk+ node has four processors, only one processor per node was used in the experiments. All the nodes run the AIX operating system version 4. They use a switch having peak bi-directional bandwidth of 120 MBps between each node for communication. The problem, defined in [15], is as follows: The problem is modeled by a system of convection-diffusion partial differential equations (PDE) on rectangular regions with Dirichlet boundary conditions, discretized with a five-point centered finite-difference scheme. If the number of points in the $x$ and $y$ directions (respectively) are $m_x$ and $m_y$, excluding the boundary points, then the mesh is mapped to a virtual $p_x \times p_y$ grid of nodes, such that a sub-rectangle of $m_x/p_x$ points in the $x$ direction and $m_y/p_y$ points in the $y$ direction is mapped to a node. Each of the sub-problems associated with these sub-rectangles is generated in parallel. This problem is solved by FGMRES(100) using a domain decomposition pARMS preconditioning. The combining phase uses Additive Schwarz procedure [10]. To

**Table 1.** Adaptation based on the observation of previous waiting times: On IBM SP for PDE problem

| $n_0^i$ | Adapt. yes/no | Outer Iter. | Solution, s |
|---|---|---|---|
| 5 | no | 2,000 (0.0) | 1,065.0 (30.33) |
| | yes | 713.3 (63.105) | 607.89 (100.0) |

**Table 2.** Waiting Time for nodes: On IBM SP for PDE problem

| $n_0^i$ | Adapt. yes/no | Tot. Wait., s | | | |
|---|---|---|---|---|---|
| | | R0 | R1 | R2 | R3 |
| 5 | no | 111.5 (23.33) | 50.5 (21.92) | 39.0 (18.38) | 75.0 (7.77) |
| | yes | 56.3 (42.5) | 38.0 (13.0) | 24.66 (11.23) | 44.0 (36.29) |

make the problem challenging for this computational environment, we consider the convection term of 2,400 for this system of PDEs.

To trigger pARMS adaptation, the amount of the time $T_c{}^i$ spent in communications has been considered in each node. Specifically, we use the criterion that the $T_c{}^i$ normalized over the previous $w$ outer iterations, called the window size, be nonzero. When this condition is met the number of inner iterations are increased by a constant value called dynamic addition value. The rationale behind using this criteria is that if a node shows waiting time in this iteration then it is more likely for it to show waiting time in the subsequent iteration. Also experimental observations for the waiting times for non adaptive runs showed that the waiting times are incurred in chunks with phases showing waiting times and phases showing no waiting times. Thus a rather simplistic approach of observing the previous waiting time can be used to find out the possibility of next waiting time.

Note that this criterion is local to a particular node and thus reflects only the interaction among neighboring nodes, which is acceptable due to the regular partitioning of this problem among the nodes, as shown in [15]. Otherwise NICAN may be used to exchange the global communication time data as seen in the next set of experiment. Table 1 shows results of executing pARMS with and without adaptation (column `Adapt`) on the IBM SP. In this experiment the window size $w$ used was three. The total number of outer pARMS iterations is shown in the column `Outer Iter` followed by the total solution time `Solution`. Table 2 shows the total waiting time for each node. $R0$, $R1$, $R2$, and $R3$ indicate the ranks of the nodes. In each table, the average values over several runs are mentioned with the standard deviations shown in the brackets. Figure 4 plots the total waiting times of the nodes.

The effect of different dynamic addition values is seen in Figure 5, left. It is seen that increasing the dynamic addition value (from 1 to 3) decreases the solution time. Figure 5, right shows that increasing the window size ($w = 10$) also decreases the solution time. From these two figures it can be concluded that

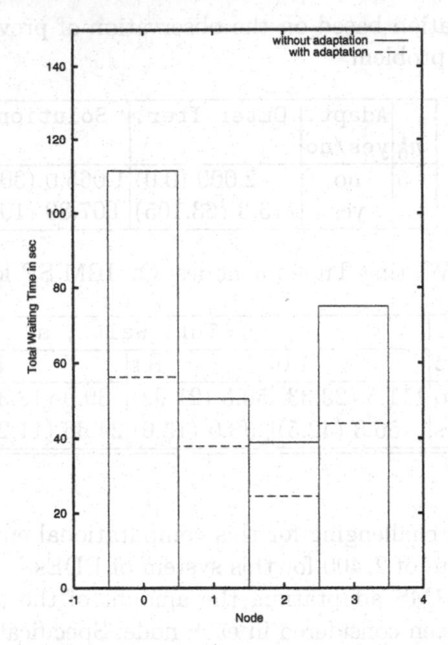


**Fig. 4.** Total Waiting time for nodes

either increasing the window size or increasing the dynamic addition value for a particular window size reduces the total solution time on this platform. Increasing the window size increases the amount of history information and thus helps in predicting more accurately the possibility of the next waiting time for a node. The window size determines when the adaptations are to be employed. The amount of dynamic addition determines the number of adaptations. For a given window size increasing this value balances the nodes in their inner iteration phase thus reducing the waiting time in the subsequent iterations and causing the reduction in the solution time.

A better convergence of the system of PDEs with adaptations is explained by more computations at the inner-level Krylov solver per each outer iteration. With the pARMS preconditioning the linear system is solved level by level and at the last level the system is solved using ILUT-FGMRES [10]. By using this strategy of observing the previous waiting time for a node and increasing the number of iterations for this last level solution step helps in engaging the node for more time so that it arrives at the rendezvous point later. The nodes showing waiting time are considered "faster" nodes as compared to other nodes. More local computations on nodes with "faster" communications result in the generation of more accurate local solutions per outer iteration. Upon exchange in the subsequent communication phases, this accuracy propagates to other nodes resulting in a better convergence overall. The total execution time is decreased also due to reducing of the waiting times in each outer iteration for the nodes. Consider the total waiting times in Table 2. It is seen that, with dynamic adaptation, the

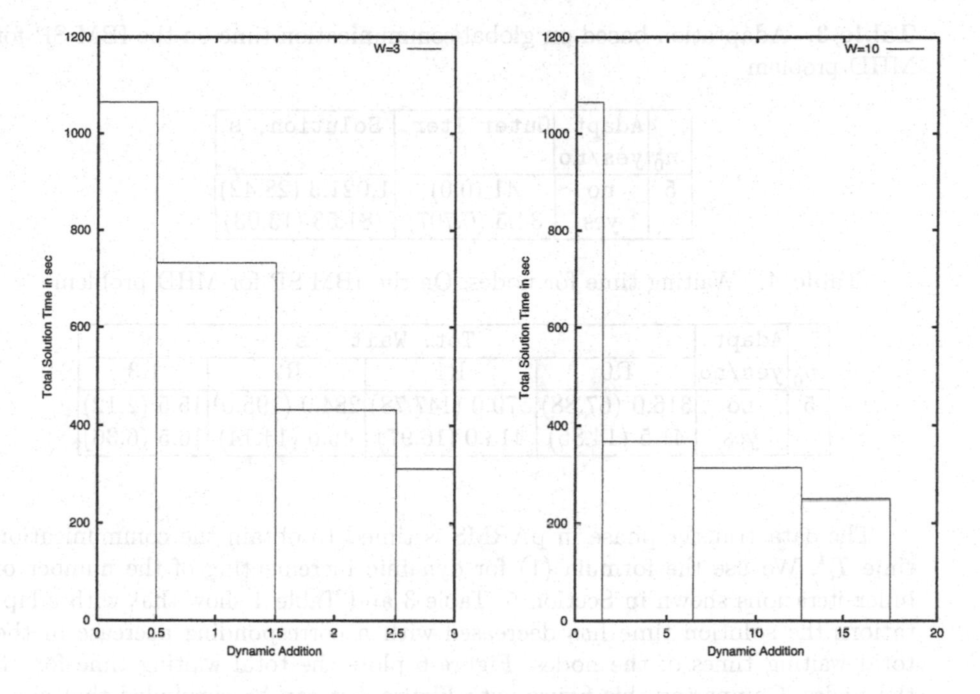


**Fig. 5.** Total time reduction for different dynamic additions: window sizes 3 (left) and 10 (right)

total waiting time has decreased in all the cases. Load balance is achieved in terms of total waiting time incurred by each node. Without adaptation the total waiting time for all the nodes show a large range. With adaptation, however, this range is narrowed.

### 6.2 Adaptations for Irregularly Structured Problem on IBM SP

For the second set of experiments the problem used is taken from the field of Magnetohydrodynamic (MHD) flows. The problem as described in [6] is as follows: "The flow equations are represented as coupled Maxwell's and the Navier-Stokes equations. We solve linear systems which arise from the Maxwell equations only. In order to do this, a pre-set periodic induction field is used in Maxwell's equation. The physical region is the three-dimensional unit cube $[-1,1]^3$ and the discretization uses a Galerkin-Least-Squares discretization. The magnetic diffusivity coefficient is $\eta = 1$. The linear system has $n = 485,597$ unknowns and $24,233,141$ nonzero entries. The gradient of the function corresponding to Lagrange multipliers should be zero at steady-state. Though the actual right-hand side was supplied, we preferred to use an artificially generated one in order to check the accuracy of the process. A random initial guess was taken" [6]. For the details on the values of the input parameters see [6].

**Table 3.** Adaptation based on global communication time on the IBM SP for MHD problem

| $n_0^i$ | Adapt. yes/no | Outer Iter. | Solution, s |
|---|---|---|---|
| 5 | no | 41 (0.0) | 1,021.3 (28.42) |
| | yes | 34.5 (0.707) | 781.53 (13.03) |

**Table 4.** Waiting time for nodes: On the IBM SP for MHD problem

| $n_0^i$ | Adapt. yes/no | Tot. Wait., s | | | |
|---|---|---|---|---|---|
| | | R0 | R1 | R2 | R3 |
| 5 | no | 316.0 (67.88) | 570.0 (147.78) | 284.0 (195.0) | 15.5 (2.12) |
| | yes | 41.5 (14.85) | 413.0 (16.97) | 46.0 (14.14) | 16.5 (6.36) |

The data transfer phase in pARMS is timed to obtain the communication time ${T_c}^i$. We use the formula (1) for dynamic incrementing of the number of inner iterations shown in Section 5. Table 3 and Table 4 show that with adaptations the solution time has decreased with a corresponding decrease in the total waiting times of the nodes. Figure 6 plots the total waiting time for all the nodes. Comparing this figure with Figure 4 it can be concluded that since this problem is irregular, the overall balance in the waiting time of the nodes is not achieved. The differing time spent by each node in the preconditioning computation phase makes it difficult for the waiting time to balance for all the nodes though decrease in the waiting time for individual node is observed. Even though the problem is harder and non uniform as compared to the problem in experiments of subsection 6.1, the adaptation strategies are seen to be as useful as in the previous case.

## 7 Conclusion

We have proposed a framework for a distributed scientific application to learn and make use of dynamic network information through notification handlers mechanisms. The framework is implemented in a network information collection tool that is lightweight and requires little modifications to the user application code. The proposed framework is rather flexible since a variety of adaptation strategies can be used in a notification handler. Our network information collection tool is capable of expanding beyond existing data collecting strategies due to its modular design. The experiments show that there is an improvement in the solution time by about 43% with adaptations based on measuring previous waiting time of a node. The adaptations based on peer communication time show an improvement of about 23% as compared to non-adaptive case. Applications having pARMS like behavior can benefit from such adaptations.

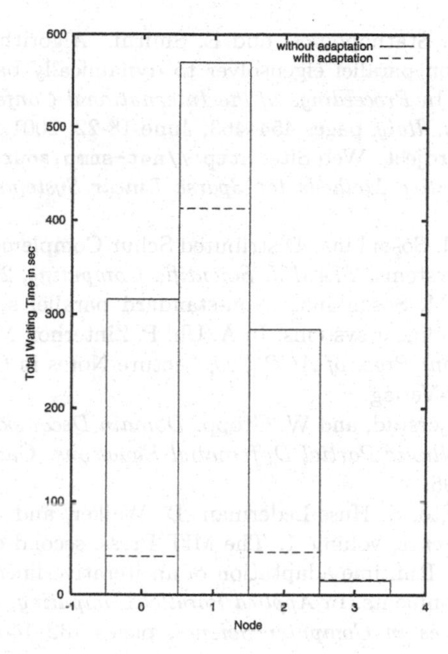


**Fig. 6.** Total Waiting time for: MHD problem

# References


[1] D. Andersen, D. Bansal, D. Curtis, S. Seshan, and H. Balakrishnan. System support for bandwidth management and content adaptation in Internet applications. In Proceedings of 4th Symposium on Operating Systems Design and Implementation San Diego, CA, October 2000. USENIX Association.:213–226, 2000.

[2] J. Dongarra, G. Fagg, A. Geist, and J. A. Kohl. HARNESS: Heterogeneous adaptable reconfigurable NEtworked systems. pages 358–359, citeseer.nj.nec.com/327665.html, 1998.

[3] X. Fu, W. Shi, A. Akkerman, and V. Karamcheti. CANS: Composable, adaptive network services infrastructure. In *Proceedings of the USENIX Symposium on Internet Technologies and Systems*, 2001.

[4] D. Gunter, B. Tierney, B. Crowley, M. Holding, and J. Lee. Netlogger: A toolkit for distributed system performance analysis. In *Proceedings of the IEEE Mascots 2000 Conference*, 2000.

[5] D. Kulkarni and M. Sosonkina. Using dynamic network information to improve the runtime performance of a distributed sparse linear system solution. Technical Report UMSI-2002-10, Minnesota Supercomputer Institute, University of Minnesota, Minneapolis, MN, 2002. accepted in VECPAR 2002.

[6] Z. Li, Y. Saad, and M. Sosonkina. pARMS: A parallel version of the algebraic recursive multilevel solver. Technical Report UMSI-2001-100, Minnesota Supercomputer Institute, University of Minnesota, Minneapolis, MN, 2001.

[7] B. Lowekamp, N. Miller, D. Sutherland, T. Gross, P. Steenkiste, and J. Subhlok. A resource query interface for network-aware applications. *Cluster Computing*, 2:139–151, 1999.

[8] R. T. Mills, A. Stathopoulos, and E. Smirni. Algorithmic modifications to the Jacobi-Davidson parallel eigensolver to dynamically balance external CPU and memory load. In *Proceedings of the International Conference on Supercomputing 2001, Sorrento, Italy*, pages 454–463, June 18-22, 2001.
[9] NET SNMP project. Web Site, `http://net-snmp.sourceforge.net/`.
[10] Y. Saad. *Iterative Methods for Sparse Linear Systems.* PWS publishing, New York, 1996.
[11] Y. Saad and M. Sosonkina. Distributed Schur Complement techniques for general sparse linear systems. *SIAM J. Scientific Computing*, 21(4):1337–1356, 1999.
[12] Y. Saad and M. Sosonkina. Non-standard parallel solution strategies for distributed sparse linear systems. In A. Uhl P. Zinterhof, M. Vajtersic, editor, *Parallel Computation: Proc. of ACPC'99*, Lecture Notes in Computer Science, Berlin, 1999. Springer-Verlag.
[13] B. Smith, P. Bjørstad, and W. Gropp. *Domain Decomposition: Parallel Multilevel Methods for Elliptic Partial Differential Equations.* Cambridge University Press, New York, 1996.
[14] M. Snir, S. Otto, S. Huss-Lederman, D. Walker, and J. Dongarra. *MPI - The complete Reference*, volume 1. The MIT Press, second edition, 1998.
[15] M. Sosonkina. Runtime adaptation of an iterative linear system solution to distributed environments. In *Applied Parallel Computing, PARA'2000*, volume 1947 of *Lecture Notes in Computer Science*, pages 132–140, Berlin, 2001. Springer-Verlag.
[16] M. Sosonkina and G. Chen. Design of a tool for providing network information to distributed applications. In *Parallel Computing Technologies PACT2001*, volume 2127 of *Lecture Notes in Computer Science*, pages 350–358. Springer-Verlag, 2001.
[17] C. Wagner. Introduction to algebraic multigrid - course notes of an algebraic multigrid. University of Heidelberg 1998/99.
[18] R. Wolski. Dynamically forecasting network performance using the network weather service. *Cluster Computing*, 1(1):119–132, 1998.

# Efficient Hardware Implementation of Modular Multiplication and Exponentiation for Public-Key Cryptography

Nadia Nedjah and Luiza de Macedo Mourelle

Department of Systems Engineering and Computation
Faculty of Engineering, State University of Rio de Janeiro
(nadia,ldmm)@eng.uerj.br

**Abstract.** Modular multiplication and modular exponentiation are fundamental operations in most public-key cryptosystems such as RSA and DSS. In this paper, we propose a novel implementation of these operations using systolic arrays based architectures. For this purpose, we use the Montgomery algorithm to compute the modular product and the left-to-right binary exponentiation method to yield the modular power. Our implementation improves time requirement as well as the time×area factor when compared that of Blum's and Paar's.

## 1 Introduction

RSA [1] is a strong public-key cryptosystem, which is generally used for authentication protocols. It consists of a set of three items: a *modulus* $M$ of around 1024 bits and two integers $D$ and $E$ called *private* and *public* keys that satisfy the property $T^{DE} \equiv T \bmod M$. Plain text $T$ obeying $0 \leq T < M$. Messages are encrypted using the public key as $C = T^E \bmod M$ and uniquely decrypted as $T = C^D \bmod M$. So the same operation is used to perform both processes: encryption and decryption. The modulus $M$ is chosen to be the product of two large prime numbers, say $P$ and $Q$. The public key $E$ is generally small and contains only few bits set (i.e. bits = 1), so that the encryption step is relatively fast. The private key $D$ has as many bits as the modulus $M$ and is chosen so that $DE = 1 \bmod (P-1)(Q-1)$. The system is secure as it is computationally hard to discover $P$ and $Q$. It has been proved that it is impossible to break an RSA cryptosystem with a modulus of 1024-bit or more.

The modular exponentiation is a common operation for scrambling and is used by several public-key cryptosystems, such as the RSA encryption scheme [1]. It consists of a repetition of modular multiplications: $C = T^E \bmod M$, where $T$ is the plain text such that $0 \leq T < M$ and $C$ is the cipher text or vice-versa, $E$ is either the public or the private key depending on whether $T$ is the plain or the cipher text. The decryption and encryption operations are performed using the same procedure, i.e. using the modular exponentiation.

The performance of such cryptosystems is primarily determined by the implementation efficiency of the modular multiplication and exponentiation. As the operands (the plain text of a message or the cipher or possibly a partially ciphered) text are usually large (i.e. 1024 bits or more), and in order to improve time requirements of the encryption/decryption operations, it is essential to attempt to minimise the number of modular multiplications performed and to reduce the time requirement of a single modular multiplication. Hardware implementations of RSA cryptosystems are widely studied as in [2, 3, 4, 5, 6, 7, 8].

Most proposed implementations of RSA system are based on Montgomery modular algorithm. There have been various proposals for systolic array architectures for modular multiplication [3, 5, 6, 7] and for modular exponentiation [5, 7]. But as far as we know, only one physical implementation has been reported including numerical figures of space and time requirements for different sizes of the exponent. It is that of Blum and Paar [6]. Tiountchik and Trichina [7] provided only one numerical figure, which is for a 132-bit modulus cryptosystem. It needs 4 Kgates FPGA chip.

In contrast with Tiountchik's and Trichina's and Walter's systolic version for modular multiplication which uses two full-adders per processing element, in our our implementation, a processing element uses at most one full adder. Some processing elements use only a half adder or a simple and-gate as it will be shown in details in Section 3. So, in general, our processing element has a simpler architecture and thus saves hardware space as well as response time.

All hardware implementations of modular exponentiation reported so far as well as the present one use the binary method as it constitutes the unique method suitable for hardware implementation. Other exponentiation methods exist (see Section 4).

Unlike Tiountchik's and Trichina's systolic version for modular exponentiation, our implementation uses more hardware space but avoids extra inputs and analysis to decide what to do: multiplication or squaring. In contrast with Blum's and Paar's implementation of exponentiation is that theirs is iterative and thus consumes more time but less space area.

In the rest of this paper, we start off by describing the algorithms used to implement the modular operation. Then we present the systolic architecture of the modular multiplier. Then we describe the systolic architecture for the exponentiator, which based on the binary exponentiation method. Finally, we compare the space and time requirements of our implementation, produced by the Xilinx project manager with those produced for Blum's and Paar's implementation.

## 2 Montgomery Algorithm

Algorithms that formalise the operation of modular multiplication generally consist of two steps: one generates the product $P = A \times B$ and the other reduces this product modulo $M$, i.e. $R = P \bmod M$.

The straightforward way to implement a multiplication is based on an iterative adder-accumulator for the generated partial products. However, this solution is quite slow as the final result is only available after $n$ clock cycles, $n$ is the size of the operands [7].

A faster version of the iterative multiplier should add several partial products at once. This could be achieved by *unfolding* the iterative multiplier and yielding a combinatorial circuit that consists of several partial product generators together with several adders that operate in parallel [8] and [9].

One of the widely used algorithms for efficient modular multiplication is the Montgomery's algorithm [10, 11] and [12]. This algorithm computes the product of two integers modulo a third one without performing division by $M$ and it is described in Figure 1. It yields the reduced product using a series of additions

Let $A$, $B$ and $M$ be the multiplicand and multiplier and the modulus respectively and let $n$ be the number of digit in their binary representation, i.e. the *radix* is 2. So, we denote $A$, $B$ and $M$ as follows:

$$A = \sum_{i=0}^{n-1} a_i \times 2^i,\ B = \sum_{i=0}^{n-1} b_i \times 2^i \text{ and } M = \sum_{i=0}^{n-1} m_i \times 2^i$$

The pre-conditions of the Montgomery algorithm are as follows:

- The modulus $M$ needs to be relatively prime to the *radix*, i.e. there exists no common divisor for $M$ and the radix; (As the radix is 2, then $M$ must be an odd number.)
- The multiplicand and the multiplier need to be smaller than $M$.

As we use the binary representation of the operands, then the modulus $M$ needs to be odd to satisfy the first pre-condition.

The Montgomery algorithm uses the least significant digit of the accumulating *modular partial product* to determine the multiple of $M$ to subtract. The usual multiplication order is reversed by choosing multiplier digits from least to most significant and shifting down. If $R$ is the current modular partial product, then $q$ is chosen so that $R+q\times M$ is a multiple of the radix $r$, and this is right-shifted by one position, i.e. divided by $r$ for use in the next iteration. Consequently, after $n$ iterations, the result obtained is $R = A\times B\times r^{-n}$ mod $M$. A version of Montgomery algorithm is given in Figure 1 where $r_0$ is the less significant bit of the partial modular product or residue $R$.

In order to yield the exact result, we need an extra Montgomery modular multiplication by the constant $r^n$ mod $M$. As we use binary representation of numbers, we need to compute $(2^n \bmod M)\times R$ mod $M$. We look at in the next section.

```
algorithm Montgomery(A, B, M)
      int R ← 0;
      1: for i = 0 to n-1
      2:        R ← R + a_i×B;
      3:        if R mod 2 = 0 then
      4:                R ← R div 2
      5:        else
      6:                R ← (R + M) div 2;
return R;
```

**Fig. 1.** Montgomery modular algorithm

## 3 Systolic Montgomery Modular Multiplication

A modified version of Montgomery algorithm is that of Figure 2. The least significant bit of $R + a_i{\times}B$ is the least significant bit of the sum of the least significant bits of $R$ and $B$ if $a_i$ is 1 and the least significant bit of $R$ otherwise. Furthermore, new values of $R$ are either the old ones summed up with $a_i{\times}B$ or with $a_i{\times}B + q_i{\times}M$ depending on whether $q_i$ is 0 or 1.

Consider the expression $R + a_i{\times}B + q_i{\times}M$ of line 2 in the algorithm of Figure 2. It can be computed as indicated in the last column of the Table 1 depending on the value of the bits $a_i$ and $q_i$.

A bit-wise version of the algorithm of Figure 2, which is at the basis of our systolic implementation, is described in Figure 3.

**Table 1.** Computation of $R + a_i{\times}B + q_i{\times}M$

| $a_i$ | $q_i$ | $R + a_i{\times}B + q_i{\times}M$ |
|---|---|---|
| 1 | 1 | $R + MB$ |
| 1 | 0 | $R + B$ |
| 0 | 1 | $R + M$ |
| 0 | 0 | R |

```
algorithm ModifiedMontgomery(A, B, M) {
        int R ← 0;
        1: for i = 0 to n-1
        2:      qi  ←  (r0 + ai×b0) mod 2;
        3:      R   ←  (R + ai×B + qi×M) div 2;
return R;
```

**Fig. 2.** Modified Montgomery modular algorithm

```
algorithm SystolicMontgomery(A, B, M, MB)
        int  R ← 0; bit  carry ← 0, x;
        0: for i = 0  to n
        1:      qi ← r0^(i) ⊕ ai.b0;
        2:      for j = 0 to n
        3:           switch ai, qi
        4:                1,1: x ← mbi;
        5:                1,0: x ← bi;
        6:                0,1: x ← mi;
        7:                0,0: x ← 0;
        8:           rj^(i+1)  ← rj+1^(i) ⊕ xi ⊕ carry;
        9:           carry     ← rj+1^(i).xi + rj+1^(i).carry + xi.carry;
return R;
```

**Fig. 3.** Systolic Montgomery modular algorithm

All algorithms, i.e. those of Figure 1, Figure 2 and Figure 3 are equivalent. They yield the same result. In the algorithm above *MB* represents the result of $M + B$, which has at most has $n$+1 bits.

Assuming the algorithm of Figure 3 as basis, the main processing element (PE) of the systolic architecture of the Montgomery modular multiplier computes a bit $r_j$ of residue $R$. This represents the computation of line 8. The left border PEs of the systolic arrays perform the same computation but beside that, they have to compute bit $q_i$ as well. This is related to the computation of line 1. The duplication of the PEs in a systolic form implements the iteration of line 0. The systolic architecture of the modular Montgomery multiplier is shown in Figure 4.

The architecture of the basic PE, i.e. $cell_{i,j}$ $1 \leq i \leq n-1$ and $1 \leq i \leq n-1$, is shown in Figure 5. It implements the instructions of lines 2-9 in systolic Montgomery algorithm of Figure 3.

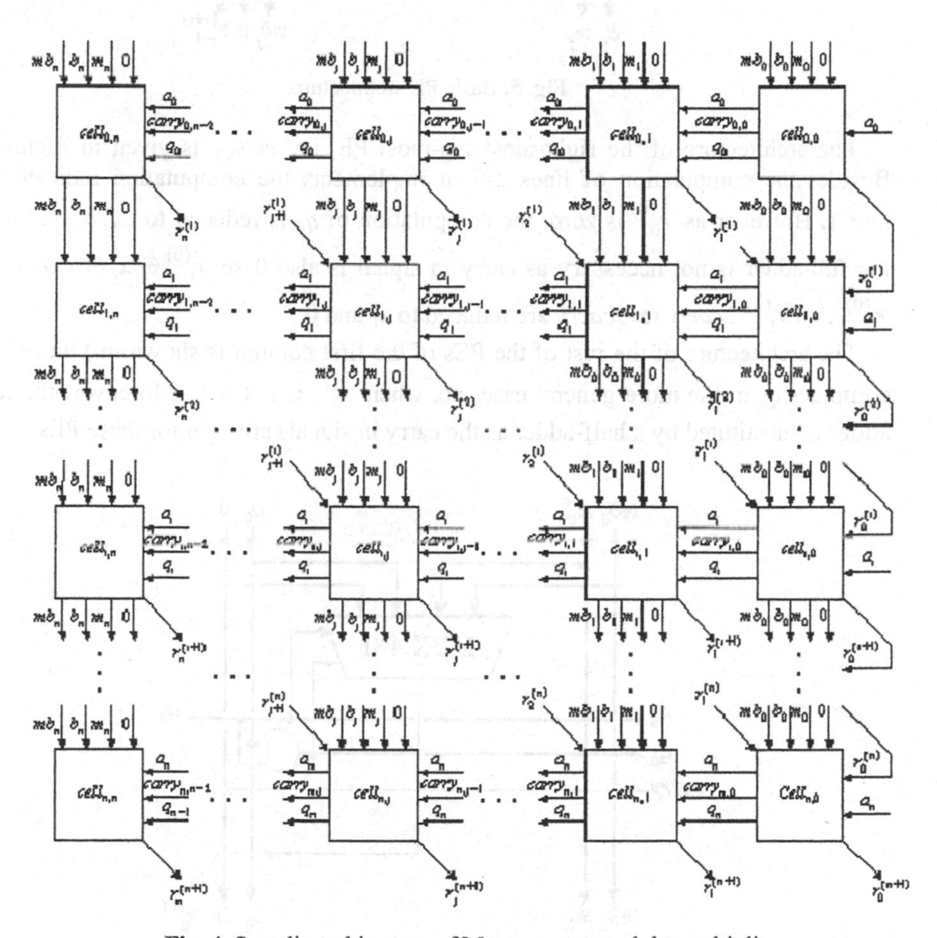

**Fig. 4.** Systolic architecture of Montgomery modular multiplier

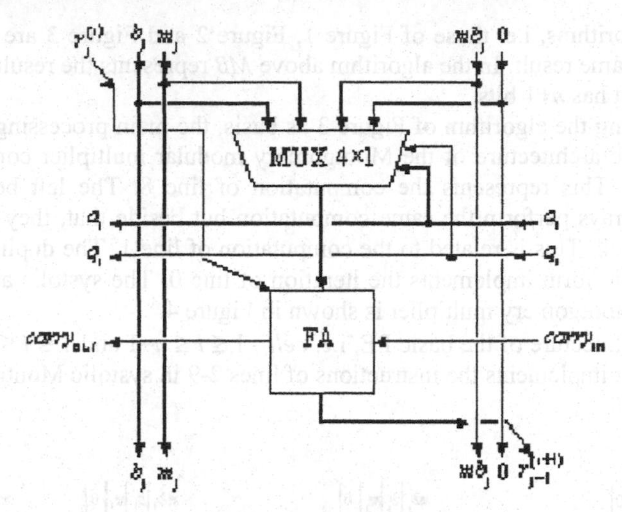


**Fig. 5.** Basic PE architecture

The architecture of the right-most top-most PE, i.e. $cell_{0,0}$, is given in Figure 6. Besides the computation of lines 2-9, it implements the computation indicated in line 1. However as $r_0^{(0)}$ is zero, the computation of $q_0$ is reduced to $a_0 . b_0$. Besides, the full-adder is not necessary as carry in signal is also 0 so $r_1^{(0)} \oplus x_i \oplus carry$ and $r_1^{(0)} .x_i + r_1^{(0)} .carry + x_i .carry$ are reduced to $x_i$ and 0.

The architecture of the rest of the PEs of the first column is shown in Figure 7. It computes $q_0$ in the more general case, i.e. when $r_0^{(i)}$ is not null. Moreover, the full-adder is substituted by a half-adder as the carry in signals are zero for these PEs.

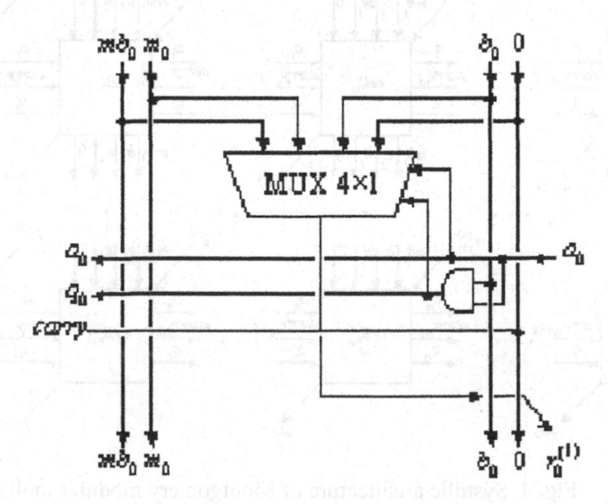


**Fig. 6.** Right-most top-most PE – $cell_{0,0}$

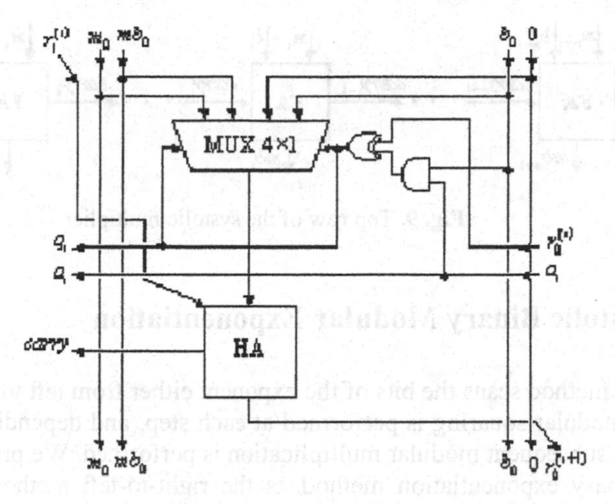


**Fig. 7.** Right border PEs – $\text{cell}_{i,0}$

The architecture of the architecture of the left border PEs, i.e. $cell_{0,j}$, is given in Figure 8. As $r_n^{(i)} = 0$, the full-adder is unnecessary and so it is substituted by a half-adder.

The sum $M$+$B$ is computed only once at the beginning of the multiplication process. This is done by a row of full adder as depicted in Figure 9.

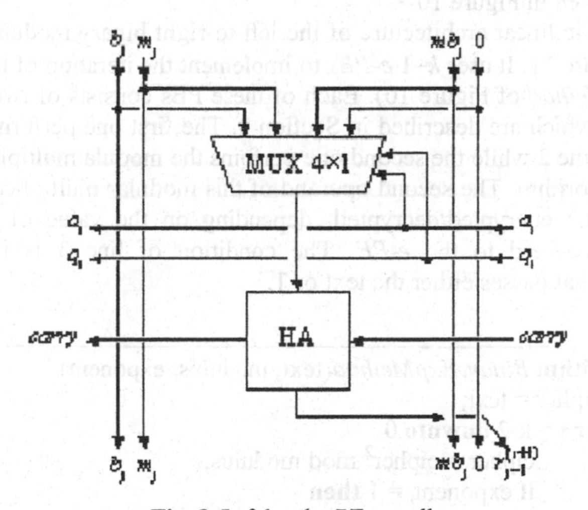


**Fig. 8.** Left border PEs – $\text{cell}_{0,j}$

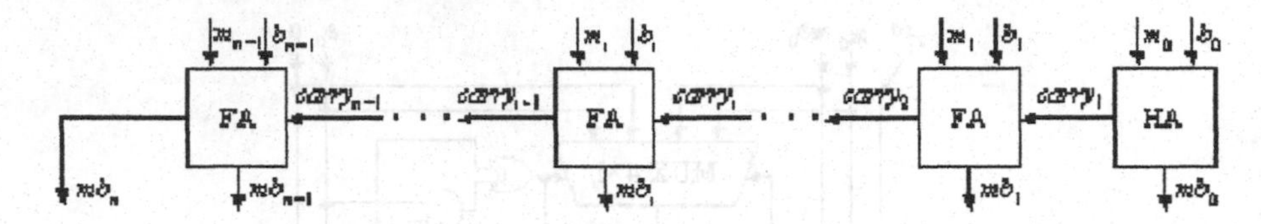

**Fig. 9.** Top row of the systolic multiplier

## 4 Systolic Binary Modular Exponentiation

The binary method scans the bits of the exponent either from left to right or from right to left. A modular squaring is performed at each step, and depending on the scanned bit value, a subsequent modular multiplication is performed. We prefer to use the left-to-right binary exponentiation method as the right-to-left method needs and extra variable (i.e. an extra register) to bookkeeping the powers of the text to be encrypted/decrypted (for details see [13]). There other more sophisticated methods to compute modular exponentiation. A comprehensive list of such methods can be found in Gordon's survey [14]. These methods, however, are more suitable for software implementations rather that hardware ones as they need a lot of memory space to store results of necessary pre-computations as well as an extensive pre-processing of the exponent before starting the actual exponentiation process.

Let $k$ be the number of bits in the exponent. We assume that the most significant bit of the exponent is 1. The algorithm of the left-to-right binary exponentiation method is given in Figure 10.

The systolic linear architecture of the left-to-right binary modular exponentiator is given in Figure 11. It uses $k$–1 *e-PEs* to implement the iteration of line 1 (in algorithm *BinaryExpMethod* of Figure 10). Each of these PEs consists of two systolic modular multipliers, which are described in Section 3. The first one performs the the modular squaring of line 2 while the second one perfoms the modula multiplication of line 4 of the same algorithm. The second operand of this modular multiplication is either 1 or the text to be encrypted/decrypted, depending on the value of the bit (from the exponent) provided to the *e-PE*. The condition of line 3 is implemented by a multiplexer that passes either the text or 1.

---

**algorithm** *BinaryExpMethod*(text, modulus, exponent)
 0: cipher = text;
 1: **for** i = k-2 **downto** 0
 2: cipher = cipher$^2$ mod modulus;
 3: **if** exponent$_i$ = 1 **then**
 4: cipher = cipher×text mod modulus;
**return** cipher;

---

**Fig. 10.** Left-to-right binary exponentiation algorithm

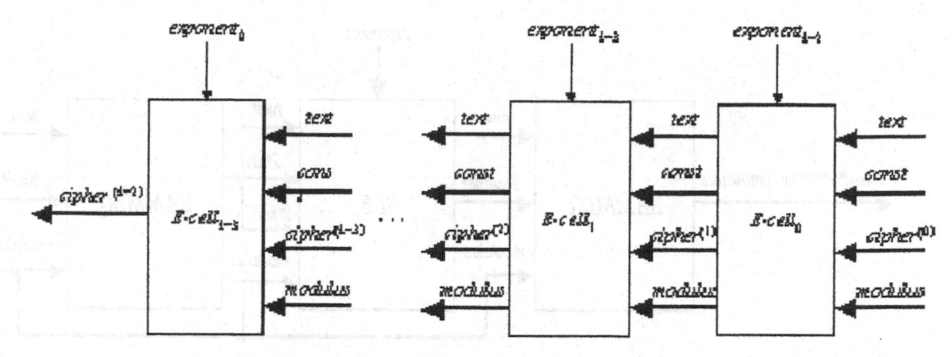


**Fig. 11.** The systolic linear architecture of the binary modular exponentiator

The architecture of an *e-PEs* is shown in Figure 12 wherein *SAMMM* stands for *s*ystolic *a*rray *M*ontgomery *m*odular *m*ultiplier.

The Montgomery algorithm (of Figure 1, Figure 2 and Figure 3) yields the result $2^{-n} \times A \times B$ mod $M$. To compute the right result, we need to further Montgomery multiply the result by the constant $2^n$ mod $M$. However, as we are interested rather in the exponentiation result than a simple product, we only need to pre-Montgomery multiply the operands by $2^{2n}$ mod $M$ and post-Montgomery multiply the obtained result by 1 to get rid of the factor $2^n$ that is carried by every partial result. So the final architecture of the Montgomery exponentiator is that of Figure 12 augmented with two extra *SAMMM* PEs. The first PE performs the Montgomery multiplication $2^{2n} \times text$ mod *modulus* and the second performs the Montgomery multiplication of 1 and the ciphertext yield by the binary method. This architecture is shown in Figure 13, wherein *SLE* stands for *s*ystolic *l*inear *e*xponentiator, i.e. that of Figure 11, $two^{2n}$, $2^n text$, $2^n cipher$, $two^n$, and *one* represent $2^n$ mod *modulus*, $text \times two^{2n}$ mod $M$, $2^n \times text^e$ mod *modulus*, $1\underbrace{0\ldots0}_{n-1}$ and $\underbrace{0\ldots0}_{n-1}1$ respectively.

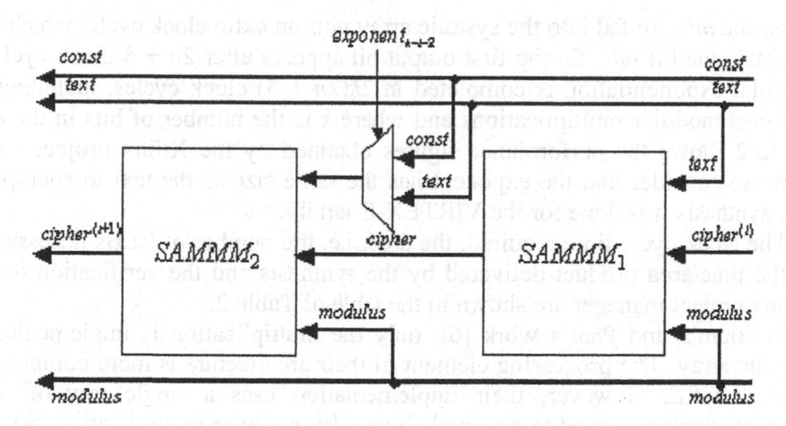


**Fig. 12.** Architecture of the exponentiator PE

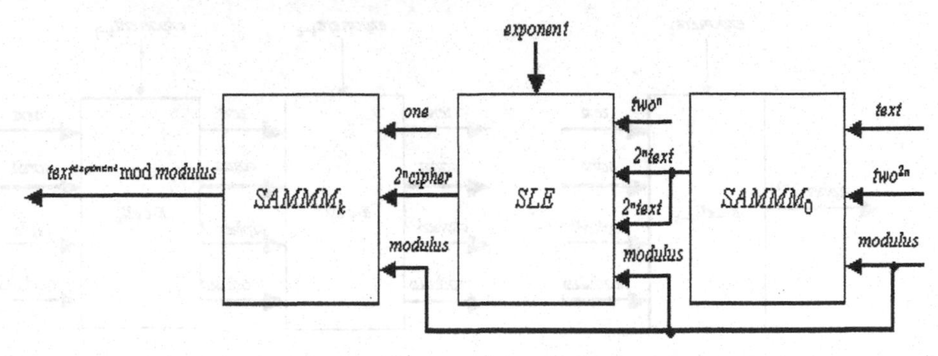


**Fig. 13.** Final architecture of the exponentiator

# 5 Time and Area Requirements

The entire design was done using the Xilinx Project Manager (version Build 6.00.09) [15] through the steps of the Xilinx design cycle shown in Figure 14. The design was elaborated using VHDL [16]. The *synthesis* step generates an optimised netlist that is the mapping of the gate-level design into the Xilinx format: *XNF*. Then, the *simulation* step consists of verifying the functionality of the elaborated design. The *implementation* step consists of partitioning the design into logic blocks, then finding a near optimal placement of each block and finally selecting the interconnect routing for a specific device family. This step generates a logic PE array file from which a bit stream can be obtained. The implementation step provides also the number of configurable logic blocks (CLBs). The *verification* step allows us to verify once again the functionality of the design and determine the response time of the design including all the delays of the physical net and padding. The *programming* step consists of loading the generated bit stream into the physical device.

The output bit $r_j^{(n+1)}$ of the modular multiplication is yield after $2n+2+j$ after bits $b_j$, $m_j$ and $mb_j$ are fed into the systolic array plus an extra clock cycle, which is needed to obtain the bit $mb_j$. So the first output bit appears after $2n + 3$ clock cycles. A full modular exponentiation is completed in $2k(2n + 3)$ clock cycles, including the pre- and post-modular multiplications and where $k$ is the number of bits in the exponent. Table 2 shows the performance figures obtained by the Xilinx project synthesiser. Here we consider that the exponent has the same size as the text to encrypt/decrypt. The synthesis was done for the VIRTEX-E family.

The clock cycle time required, the area, i.e. the number of CLBs necessary as well as the time/area product delivered by the synthesis and the verification tools of the Xilinx project manager are shown in the table of Table 2.

In Blum's and Paar's work [6], only the multiplication is implemented using a systolic array. The processing element of their architecture is more complex than our multiplier PE. However, their implementation uses a single row of processing elements that are reused to accomplish an $n$-bit modular multiplication. So Blum and Paar reduce the required hardware area at the expense of response time as they have to include a synchronised in-control. The implementation of the exponentiation operation is not a systolic implementation, which should again reduce area at the expense of response time. In the contrary to that of Blum's and Paar's, our implementation attempts to minimise time requirements at the expense of hardware area as we think that one can afford hardware area if one can gain in encryption/decryption time. The clock cycle time required, the area, i.e. the number of CLBs necessary as well as the time/area product delivered by Blum and Paar [7] project manager are shown in the table of Table 3.

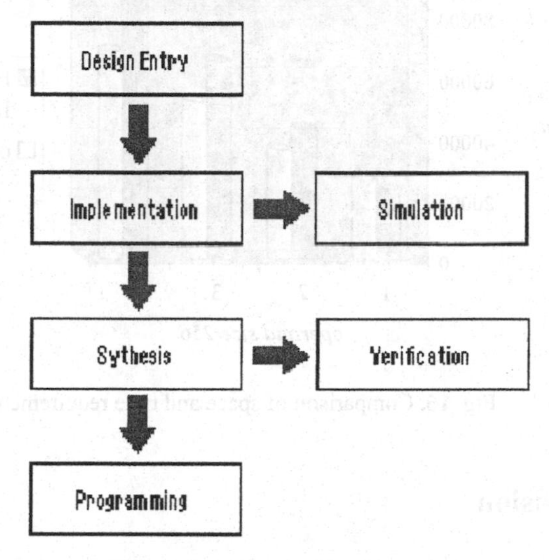


**Fig. 14.** Design cycle

**Table 2.** Performance figures of our implementation

| operand size | *Area* (CLBs) | clock cycle time (ns) | area×time |
|---|---|---|---|
| 128 | 3179 | 3.3 | 10490 |
| 256 | 4004 | 4.2 | 26426 |
| 512 | 5122 | 5.1 | 36366 |
| 768 | 6278 | 5.9 | 52107 |
| 1024 | 7739 | 6.6 | 68877 |

**Table 3.** Performance figures of Blum's and Paar's implementation

| operand size | *Area* (CLBs) | clock cycle time (ns) | area×time |
|---|---|---|---|
| 256 | 1180 | 19.7 | 23246 |
| 512 | 2217 | 19.5 | 43231 |
| 768 | 3275 | 20.0 | 65500 |
| 1024 | 4292 | 22.1 | 8039 |

The chart of Figure 15 compares the area/time product of Blum's and Paar's implementation vs. our implementation. It shows that our implementation improves the product as well as time requirement while theirs improves area at the expense of both time requirement and the product.

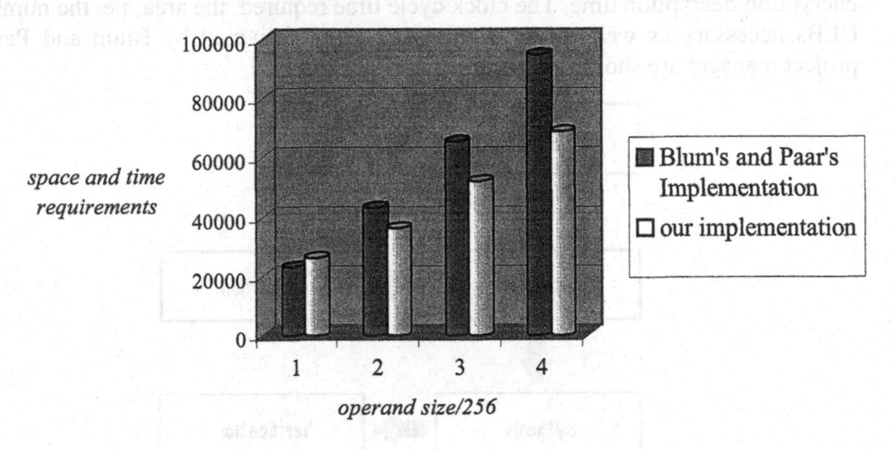


**Fig. 15.** Comparison of space and time requirements

## 6 Conclusion

In this paper, we described a systolic architecture of the modular multiplier. Then we engineered a novel systolic architecture for the exponentiator, which based on the binary exponentiation method. We compared the space and time requirements of our implementation, produced by the Xilinx project manager against those produced for Blum's and Paar's implementation. The results show clearly that despite of requiring much more hardware area, our implementation improves substantially the time requirement and the area/time product when the operand size is bigger than 256 bits. This is almost always the case in RSA encryption/decryption systems.

The authors' related work can be found in [17] and [18]. In [17] we compare a parallel implementation of the Montgomery modular multiplier vs. a sequential one and in [18] an alternative parallel binary exponentiator is described and evaluated in terms of area and performance.

## References

[1] R. Rivest, A. Shamir and L. Adleman, *A method for obtaining digital signature and public-key cryptosystems*, Communications of the ACM, **21**:120-126, 1978.

[2] E. F. Brickell, *A survey of hardware implementation of RSA*, In G. Brassard, ed., Advances in Crypltology, Proceedings of CRYPTO'98, Lecture Notes in Computer Science **435**:368-370, Springer-Verlag, 1989.

[3] C. D. Walter, *Systolic modular multiplication*, IEEE Transactions on Computers, **42**(3):376-378, 1993.

[4] C. D. Walter, *An improved linear systolic array for fast modular exponentaition*, IEE Computers and Digital Techniques, **147**(5):323-328, 2000.

[5] A. Tiountchik, *Systolic modular exponentiation via Montgomery algorithm*, Electronic Letters, **34**(9):874-875, 1998.

[6] K. Iwamura, T. Matsumoto and H. Imai, *Montgomery modular multiplictaion and systolic arrays suitable for modular exponentiation*, Electronics and Communications in Japan, 77(3):40-51, 1994.

[7] T. Blum and C. Paar, *Montgmery modular exponentiation on econfigurable hardware,* 14th IEEE Symposium on Computer Arithmetic, April 14-16, 1999, Adelaide, Australia.

[8] J. Rabacy, Digital integrated circuits: A design perspective, Prentice-Hall, 1995.

[9] N. Nedjah and L. M. Mourelle, *Yet another implementation of modular multiplication*, Proceedings of 13$^{th}$. Symposium of Computer Architecture and High Performance Computing, IFIP, Brasilia, Brazil, pp. 70-75, September 2001.

[10] N. Nedjah and L. M. Mourelle, *Simulation model for hardware implementation of modular multiplication*, Proceedings of WSES/IEEE International. Conference on Simulation, Knights Island, Malta, September 2001.

[11] P.L. Montgomery, *Modular multiplication without trial division*, Mathematics of Computation 44, pp. 519-521, 1985.

[12] L. M. Mourelle and N. Nedjah, *Reduced hardware Architecture for the Montgomery modular multiplication*, Proceedings of the third WSEAS Transactions on Systems, **1**(1):63-67, January 2002.

[13] Ç.K. Koç, *High-speed RSA implementation*, Technical report, RSA Laboratories, RSA Data Security, Inc. Redwood City, CA, USA, November 1994.

[14] D. M. Gordon, *A survey of Fast exponentiation methods*, Journal of Algorithms 27, pp. 129-146, 1998

[15] Xilinx, Inc. *Foundation Series Software*, Http://www.xilinx.com.

[16] Z. Navabi, *VHDL - Analysis and modeling of digital systems*, McGraw Hill, Second Edition, 1998.

[17] N. Nedjah and L. M. Mourelle, *Two Hardware Implementations for the Montgomery Modular Multiplication: Parallel vs. Sequential*, Proceedings of 15$^{th}$. Symposium on Integrated Circuits and Systems Design, IEEE Computer Society, Porto Alegre, Brazil, pp. 3-8, September 2002.

[18] N. Nedjah and L. M. Mourelle, *Reconfigurable Hardware Implementation of Montgomery Modular Multiplication and Parallel Binary Exponentiation*, Proceedings of Euromicro Symposium on Digital Systems Design, IEEE Computer Society, Dortmund, Germany, pp. 226-233, September 2002.

# Real-Time Visualization of Wake-Vortex Simulations Using Computational Steering and Beowulf Clusters*

Anirudh Modi[1], Lyle N. Long[2], and Paul E. Plassmann[3]

[1] Ph.D. Candidate, Department of Computer Science and Engineering
Pennsylvania State University, University Park, PA 16802
anirudh@anirudh.net
http://www.anirudh.net/phd/

[2] Professor, Department of Aerospace Engineering
Pennsylvania State University, University Park, PA 16802
lnl@psu.edu
http://www.personal.psu.edu/lnl/

[3] Associate Professor, Department of Computer Science and Engineering
Pennsylvania State University, University Park, PA 16802
plassman@cse.psu.edu
http://www.cse.psu.edu/~plassman/

**Abstract.** In this paper, we present the design and implementation of POSSE, a new, lightweight computational steering system based on a client/server programming model. We demonstrate the effectiveness of this software system by illustrating its use for a visualization client designed for a particularly demanding real-time application—wake-vortex simulations for multiple aircraft running on a parallel Beowulf cluster. We describe how POSSE is implemented as an object-oriented, class-based software library and illustrate its ease of use from the perspective of both the server and client codes. We discuss how POSSE handles the issue of data coherency of distributed data structures, data transfer between different hardware representations, and a number of other implementation issues. Finally, we consider how this approach could be used to augment AVOSS (an air traffic control system currently being developed by the FAA) to significantly increase airport utilization while reducing the risks of accidents.

## 1 Introduction

Parallel simulations are playing an increasingly important role in all areas of science and engineering. As the areas of applications for these simulations expand and their complexity increases, the demand for their flexibility and utility grows. Interactive computational steering is one way to increase the utility of these high-performance simulations, as they facilitate the process of scientific

* This work was supported by NSF grants EIA–9977526 and ACI–9908057, DOE grant DG-FG02-99ER25373, and the Alfred P. Sloan Foundation.

J.M.L.M. Palma et al. (Eds.): VECPAR 2002, LNCS 2565, pp. 464–478, 2003.

discovery by allowing the scientists to interact with their data. On yet another front, the rapidly increasing power of computers and hardware rendering systems has motivated the creation of visually rich and perceptually realistic virtual environment (VE) applications. The combination of the two provides one of the most realistic and powerful simulation tools available to the scientific community.

As an example of such an important application here is the problem of maximizing airport efficiency. National Aeronautics and Space Administration (NASA) scientists predict that by the year 2022, three times as many people will travel by air as they do today [9]. To keep the number of new airports and runways to a minimum, there is an urgent need to increase their efficiency while reducing the aircraft accident rate. Today, the biggest limiting factor for airport efficiency is the wait between aircraft take-offs and landings which are necessary because of the wake-vortices generated by the moving aircraft. Moreover, according to the predictions by the United States Federal Aviation Administration (FAA), if by the year 2015, the wake-vortex hazard avoidance systems do not improve in any significant way, there is the potential for a significant increase in the number of aviation accidents [16]. The ultimate goal of the work presented in this paper is to create a wake-vortex hazard avoidance system by realistically simulating an airport with real-time visualization of the predicted wake-vortices. If implemented, such a system has the potential to greatly increase the utilization of airports while reducing the risks of possible accidents. In this work, we utilize an easy-to-use, yet powerful computational steering library to deal with the complexities of real-time wake-vortex visualization.

To enable such a complex simulation, we will require a computational steering system. A significant amount of work has been done on computational steering over the past few years. Reitinger [18] provides a brief review of this work in his thesis. Some of the well known steering systems are *Falcon* from Georgia Tech [7], *SCIRun* from Scientific Computing and Imaging research group at University of Utah [15], ALICE Memory Snooper from Argonne National Laboratory [2], *VASE* (Visualization and Application Steering) from University of Illinois [10], *CUMULVS* from Oak Ridge National Laboratory [5], *CSE* (Computational Steering Environment) from the Center for Mathematics and Computer Science in Amsterdam [24], and *Virtue* from University of Illinois at Urbana-Champaign [19]. While they are all powerful, the major drawback of these systems is that they are often too complex, are not object-oriented and have a steep learning curve. To be productive with these systems by using them in existing scientific codes is not an easy task, and may take a significant amount of time, especially for the large number of computational scientists with no formal education in computer science or software systems.

To address these problems, we have developed a new lightweight computational steering system based on a client/server programming model. In this paper, we first discuss computational steering in section 2, then the details of wake-vortex simulations in section 3, and finally some experimental results in section 4.

## 2 Computational Steering

While running a complex parallel program on a high-performance computing system, one often experiences several major difficulties in observing computed results. Usually, the simulation severely limits the interaction with the program during the execution and makes the visualization and monitoring slow and cumbersome (if at all possible), especially if it needs to be carried out on a different system (say a specialized graphics workstation for visualization).

For our simulations, it is very important for the predictions by the wake-vortex code to be known in real-time by the Air-Traffic Control (ATC) in order for it to take appropriate action. This activity is referred to as "monitoring," which is defined as the observation of a program's behavior at specified intervals of time during its execution. On the other hand, the weather conditions at the airport may keep changing and both the number and the trajectories of the aircraft can change as they take-off and land. Thus, there is a need to modify the simulation based on these factors by manipulating some key characteristics of its algorithm. This activity is referred to as "steering," which is defined as the modification of a program's behavior during its execution.

Software tools which support these activities are called computational steering environments. These environments typically operate in three phases: instrumentation, monitoring, and steering. Instrumentation is the phase where the application code is modified to add monitoring functionality. The monitoring phase requires the program to run with some initial input data, the output of which is observed by retrieving important data about the program's state change. Analysis of this data gives more knowledge about the program's activity. During the steering phase, the user modifies the program's behavior (by modifying the input) based on the knowledge gained during the previous phase by applying steering commands, which are injected on-line, so that the application does not need to be stopped and restarted.

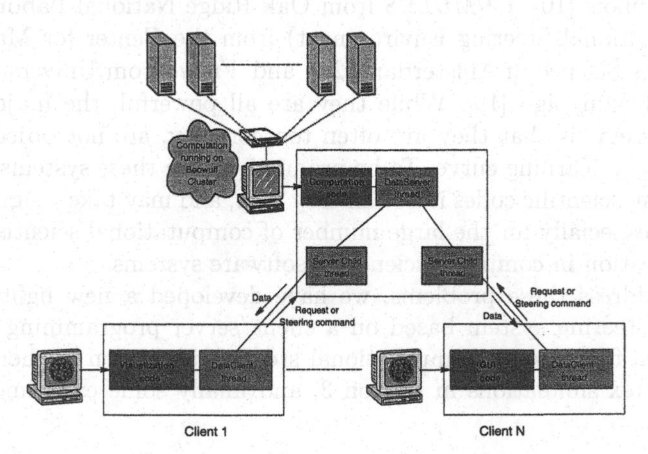


**Fig. 1.** A schematic view of POSSE

Our steering software, the *Portable Object-oriented Scientific Steering Environment* (POSSE) [14], is very general in nature and is based on a simple client/server model. It uses an approach similar to Falcon [7] (an on-line monitoring and steering toolkit developed at Georgia Tech) and ALICE Memory Snooper [2] (an application programming interface designed to help in writing computational steering, monitoring and debugging tools developed at Argonne National Lab). Falcon was one of the first systems to use the idea of threads and shared memory to serve registered data efficiently. POSSE consists of a steering server on the target machine that performs steering, and a steering client that provides the user interface and control facilities remotely. The steering server is created as a separate execution thread of the application to which local monitors forward only those "registered" data that are of interest to steering activities. A steering client receives the application run-time information from the application, displays the information to the user, accepts steering commands from the user, and enacts changes that affect the application's execution. Communication between a steering client and server are done via UNIX sockets and threading is done using POSIX threads. POSSE has been completely written in C++, using several of C++'s advanced object-oriented features, making it fast and powerful, while hiding most of the complexities from the user. Fig. 1 shows

```
#include "dataserver.h"

int dummyInt = 0, n1, n2;
double **dyn2D;

REGISTER_DATA_BLOCK()   // Register global data
{
  REGISTER_VARIABLE("testvar", "rw", dummyInt);
  REGISTER_DYNAMIC_2D_ARRAY("dyn2D", "ro", dyn2D, n1, n2);
}

int main(int argc, char *argv[])
{
  DataServer *server = new DataServer;

  if (server->Start(4096) != POSSE_SUCCESS) // Start Server thread
      {
       delete server;
       exit(-1);
      }
  n1 = 30; n2 = 40;
  ALLOC2D(&dyn2D, n1, n2);

  for (int iter = 0; iter < MAX_ITER; iter++) {
        server->Wait("dyn2D"); // Lock DataServer access for dyn2D

             Compute(dyn2D);    // Update dyn2D with new values

        server->Post("dyn2D"); // Unlock DataServer access for dyn2D
    }
  FREE2D(&dyn2D, n1, n2);
  delete server;
}
```

**Fig. 2.** A simple, complete POSSE server application written in C++

```
#include "dataclient.h"

int main(int argc, char *argv[])
{
  DataClient *client = new DataClient;
  double **dyn2D;

  if (client->Connect("cocoa.ihpca.psu.edu", 4096) != POSSE_SUCCESS) // Connect to DataServer
      {
       delete client;
       exit(-1);
      }
  client->SendVariable("testvar", 100); // Send new value for "testvar"
  int n1 = client->getArrayDim("dyn2D", 1);
  int n2 = client->getArrayDim("dyn2D", 2);
  ALLOC2D(&dyn2D, n1, n2);
  client->RecvArray2D("dyn2D", dyn2D);

  Use(dyn2D);  // Utilize dyn2D

  FREE2D(&dyn2D, n1, n2);
  delete client;
}
```

**Fig. 3.** A simple, complete POSSE client application written in C++

a schematic view of how POSSE can be used. An on-going scientific simulation is running on a remote Beowulf computing cluster. Any number of number of remote clients can query/steer registered data from the simulation from the DataServer thread. Two clients are shown, a visualization client and a GUI client that provides a simple user interface to all registered simulation data.

POSSE is designed to be extremely lightweight, portable (runs on all Win32 and POSIX-compliant Unix platforms) and efficient. It deals with byte-ordering and byte-alignment problems internally and also provides an easy way to handle user-defined classes and data structures. It is also multi-threaded, supporting several clients simultaneously. It can also be easily incorporated into parallel simulations based on the Message Passing Interface (MPI) [4] library. The biggest enhancement of POSSE over existing steering systems is that it is equally powerful, yet extremely easy to use, making augmentation of any existing C/C++ simulation code possible in a matter of hours. It makes extensive use of C++ classes, templates and polymorphism to keep the user Application Programming Interface (API) elegant and simple to use. Fig. 2 and Fig. 3 illustrate a simple, yet complete, POSSE client/server program in C++. As seen in the figures, registered data on the steering server (which are marked *read-write*) are protected using binary semaphores when they are being updated in the computational code. User-defined data structures are handled by a simple user-supplied pack and unpack subroutine that call POSSE data-packing functions to tackle the byte-ordering and byte-alignment issues. The programmer does not need to know anything about the internals of threads, sockets or networking in order to use POSSE effectively. POSSE also allows a simulation running on any parallel or serial computer to be monitored and steered remotely from any machine on the network using a cross-platform Graphical User Interface (GUI) utility. Among other applications, we have successfully used POSSE to enhance our existing parallel Computational Fluid Dynamics (CFD) code to perform visualizations of large-scale flow simulations [12].

## 3 Wake-Vortex Simulation

One of the main problems facing the ATC today is the "wake-vortex" hazard. Just as a moving boat or a ship leaves behind a wake in the water, an aircraft leaves behind a wake in the air. These wake-vortex pairs are invisible to the naked eye and stretch for several miles behind the aircraft and may last for several minutes. The aircraft wake is generated from the wings of the aircraft and consists of two counter-rotating swirling rolls of air which are termed "wake-vortices". In Fig. 4, we show a photograph depicting the smoke flow visualization of wake-vortices generated by a Boeing 727. It is to be noted that these are not contrails (i.e., condensation trail left behind by the jet exhausts). The strength of these vortices depends on several factors, including weight, size and velocity of the aircraft. The strength increases with the weight of the aircraft. The life of the vortex depends on the prevailing weather conditions. Typically, vortices last longer in calm air and shorter in the presence of atmospheric turbulence. The study of these vortices is very important for aircraft safety [11]. The rapid swirling of air in a vortex can have a potentially fatal effect on the stability of a following aircraft. Currently, the only way to deal with this problem is the use of extremely conservative empirical spacing between consecutive take-offs and landings from the same runway, which has been laid down by the International Civil Aviation Organization (ICAO) and FAA. In instrument flying conditions, aircraft may follow no closer than three nautical miles, and a small aircraft must

**Fig. 4.** B-727 vortex study photo (Courtesy: NASA Dryden Flight Research Center)

follow at least six nautical miles behind a heavy jet such as a Boeing 747. But, despite these spacings being extremely conservative, they are not always able to prevent accidents owing to the several unknowns involved, primarily the exact location and strength of the vortices. The US Air Flight 427 (Boeing 737) disaster which occurred on September 8, 1997 near Pittsburgh is attributed to this phenomenon, wherein the aircraft encountered the wake-vortices of a preceding Boeing 727 [1]. The more recent Airbus crash on November 12, 2001 in New York is also believed to be, at least partially, a result of wake-vortex encounter from a preceding Boeing 747.

To tackle this problem of reduced airport capacity which is a direct fallout of these overly conservative spacing regulations, and to address the concerns of the aircraft in circumstances when these regulations fail to meet the safety requirements, NASA researchers have designed a system to predict aircraft wake-vortices on final approach, so that the aircraft can be spaced more safely and efficiently. This technology is termed AVOSS or Aircraft VOrtex Spacing System (AVOSS) [8]. AVOSS, in spite of performing a rigorous simulation of the wake-vortices, does not implement any system for their visualization. It only provides the ATC with the aircraft spacing time for each aircraft which is all the current ATC systems can handle. Thus, at present, it is unable to provide alternate trajectories for the take-off and landing of aircraft.

This work attempts to fill in the gaps left by AVOSS by creating a wake-vortex hazard avoidance system by realistically simulating an airport with real-time 3D visualization of the predicted wake-vortices generated by the moving aircraft. Aircraft will be able to adjust their flight trajectory based on the information obtained from the visualization system to avoid the wake-vortices and operate more safely and efficiently.

### 3.1 Wake-Vortex Theory

For the wake-vortex simulations described in this paper, we use potential theory to predict the strength of the wake-vortex elements [17]. The circulation generated by the lift is assumed to be contained in two vortices of opposite signs trailing from the tips of the wing. The wake is assumed to consist of a pair of vortices which are parallel and the longitudinal axis of the tracked airplane is assumed to be parallel to the vortex pair. The centers of the vortices are on a horizontal line separated by a distance of $b_s = \frac{\pi}{4} b_g$, a result of assuming an elliptic distribution, where $b_s$ is the separation of the vortices in the wake-vortex pair, and $b_g$ is the span of the airplane wing generating the wake vortex [22]. The magnitude of the circulation of each vortex is approximately

$$|\Gamma| = \frac{4}{\pi} \frac{L_g}{\rho V_g b_g},$$

where $L_g$ and $V_g$ are the lift and the velocity of the aircraft, respectively. References [23, 20, 17] deal with more details on the numerical simulation of these aircraft vortices.

```
V ← ∅
t ← 0
Foreach aircraft A on a different processor
   While (A in specified range from airport) do
      read updated aircraft position from airport data server
      read updated weather condition from airport data server
      V ← V + {newly created vortex element from wing using potential theory}
      Foreach vortex element (v_i ∈ V)
         v_i.inducedvel ← 0
         Foreach vortex element ((v_j ∈ V) ≠ v_i & |j − i| ≤ k)
            v_i.inducedvel ← v_i.inducedvel + InducedVelocity(v_i, v_j)
         Endforeach
         v_i.position ← v_i.position + Δt × v_i.inducedvel
         v_i.position ← v_i.position + Δt× (prevailing wind velocity)
         v_i.strength ← v_i.strength − DecayFunction(Δt, Weather Conditions)
         If (v_i.strength < threshold) then
            V ← V − v_i
         Endif
      Endforeach
      t ← t + Δt
   Endwhile
Endforeach
```

**Fig. 5.** Algorithm for Wake-Vortex prediction

After the strength of these vortices are computed, the effect due to the prevailing weather data is applied to the prediction. The vortex filaments propagate with the freestream wind conditions and the induced velocity due to the other vortex elements. The decay of the vortex strength is based on a simplified version of the model suggested by Greene [6]:

$$\Gamma_{t+\Delta t} = \Gamma_t \left(1 - \frac{\Delta t V_t}{8 b_g}\right),$$

where $V_t$ is the vortex velocity at time $t$ and is given by

$$V_t = \frac{\Gamma_t}{2\pi b_g}.$$

Here $\Gamma_t$ represents the strength of the vortex element at time $t$ and $\Gamma_{t+\Delta t}$ represents the strength of the vortex at time $t + \Delta t$ (next time-step).

### 3.2 Simulation Complexity

The wake-vortex prediction for an entire fleet of aircraft taking-off and landing at a busy airport is an extremely computationally intensive problem. As such, a parallel solution for the same is required to maintain a real-time response of the simulation. For example, a typical metropolitan airport in the US is extremely busy with several take-offs and landings occurring every few minutes. Dallas/Fort Worth, the country's third busiest airport, has seven runways that handle nearly 2,300 take-offs and landings every day. For the wake-vortex code to track the vortices shed by an aircraft for 5 miles after take-off, assuming that a vortex core is stored every 5 meters, $5 \times 1,600/5 \times 2 = 3,200$ vortex filaments have to be tracked. For 2,300 take-offs and landings every day, it implies that $3,200 \times 2,300/24 = 306,667$ vortex filaments have to be tracked every hour. Since the vortices may take as long as 15 minutes to decay significantly, vortices due to typically half the take-offs and landings every hour need to be tracked at any given time. This amounts to roughly 153,333 vortex filaments. While this may not seem to be a very large number on its own, the problem gets complicated by the presence of an $O(N^2)$ calculation for the induced velocity of every vortex element on every other vortex element, where $N$ represents the number of vortex elements. Even if the induced velocity effect due to vortices from the other aircraft are ignored, this still amounts to as much as $3,200 \times 3,200 = 10.24$ million computations for each airplane at every timestep. For 2,300 planes/day, this comes out to $10.24 \times 2,300/24/2 = 490.7$ million calculations per timestep for the induced velocity, a very large number indeed for a conventional uniprocessor system. And with each timestep being, say 0.2 seconds, this amounts to 2.45 billion calculations per second. Although this number can be reduced by as much as a factor of 100 by making simplifying assumptions for the induced velocity calculations (wherein, we say that any vortex element is only affected by a fixed number of neighboring elements, say $k$, rather than all the other elements), this still amounts to a large computation considering that each induced velocity calculation consists of 200 − 300 floating point operations. This takes our net computational requirement to approximately 5 − 8 Gigaflops, necessitating the

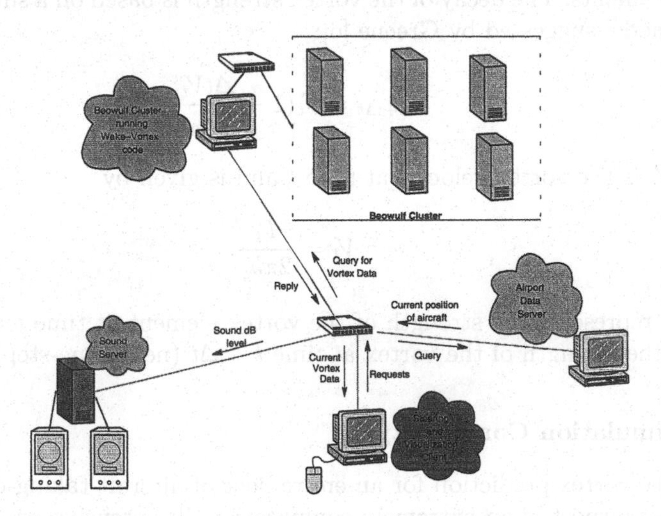


**Fig. 6.** Wake-Vortex Simulation System

**Fig. 7.** Screenshot of the Wake-Vortex simulation for a single aircraft from a visualization client

need of a parallel computer. Hence, our wake-vortex prediction code, based on the potential flow theory described above, is written in C++ with MPI for parallelization.

Pseudocode for the simulation is given in Fig. 5. Each vortex element has two main properties associated with it, strength and position. The initial strength ($\Gamma$) is calculated based on the potential flow theory and the initial position is based on the position of the aircraft. The strength then decays as a function of time and the prevailing weather conditions, and the position changes due to the velocity induced by neighboring vortex elements and the prevailing wind velocity. Fig. 6 depicts a diagram of the complete client/server simulation system. The simulation system consists of the *Wake-vortex Server*, *Airport Data Server* and the *Sound Server*. The Wake-vortex Server is the actual simulation code enhanced using POSSE. The Airport Data Server is another POSSE server that serves the positions of the aircraft in the vicinity of the airport as well as the prevailing weather conditions. The Sound Server is an optional component in the system for simulating the noise-level at the airport. The wake-vortex code has been parallelized to track vortex elements from each aircraft on a different processor in such a way that we get an almost real-time solution to this problem with tolerable lag no more than the time-step $\Delta t$ in our simulation. The first processor acts as the master doing a round-robin scheduling of any new aircraft to be tracked among the available processors (including itself). The master, therefore does the additional work of distributing and collecting vortex data from the slave nodes. It is also ensured that the master is always running on a Symmetric Multi-Processor (SMP) node with at least two processors so that the POSSE server thread runs on an idle processor and does not slow down

**Fig. 8.** Screenshot of the Wake-Vortex simulation for several aircraft flying above the San Francisco (SFO) airport

the master node because of the constant monitoring of the vortex data by the visualization client.

For the real-time simulation, the parallel wake-vortex code has been augmented with POSSE, so that it can remotely run on our in-house 40 processor PIII-800 Mhz *Beowulf Cluster*[1] [21], the COst-effective COmputing Array 2 (COCOA-2) [13]. For a steering client, a visualization tool has been written in C++ using the OpenGL API for graphics and CAVELib [25] API for stereo-graphics and user interaction. The monitoring code runs as a separate thread in the visualization client retrieving new vortex data whenever the simulation on the remote cluster updates. A screenshot of the program depicting the Wake-Vortex simulation for a single aircraft is shown in Fig. 7. Another screenshot (Fig. 8) shows several aircraft flying above the San Francisco International airport (SFO). The colors represent the relative strength of the vortices with red being maximum and blue being minimum. The Reconfigurable Automatic Virtual Environment (RAVE) from FakeSpace Systems [3] driven by an HP Visualize J-class workstation is then used as the display device.

## 4 Experimental Results

POSSE has been extensively tested using various platforms for stability and performance. Tests demonstrating both the single and multiple client performance for POSSE are discussed here.

[1] A cluster of commodity personal computers running the `LINUX` operating system.

### 4.1 Single Client Performance

Fig. 9 shows a plot of the effective network bandwidth achieved by varying the size of a dynamic 1-D array requested by a steering client. These tests were carried between machines connected via a Fast Ethernet connection having a peak-theoretical network bandwidth of 100 Mbps. The communication time used to calculate the effective bandwidth includes the overheads for byte-ordering, data packing and other delays introduced by the querying of the registered data. The average latency for query of any registered data by the client has been found to be 38 ms. As can be seen, there is a noticeable decrease in the bandwidth (about 10 Mbps) when communicating between machines with different byte-ordering (i.e., Little Endian vs. Big Endian) as opposed to machines with the same byte-ordering. This reflects the overhead involved in duplicating the requested data and converting it into the byte-order of the client machine for communication. In the same byte-order case, as the size of the requested data increases to about 5 MB, the effective bandwidth approaches 80 Mbps, which is 80% of the peak-theoretical bandwidth.

### 4.2 Multiple Client Performance

Fig. 10 shows a plot of the effective bandwidth achieved by varying the number of clients simultaneously requesting data. In this test, both the clients and the server were machines with the same byte-ordering. The server had a registered 4-D array with 200,000 *double* elements (1.6 MB of data). All the clients were then run simultaneously from two remote machines on the same network and

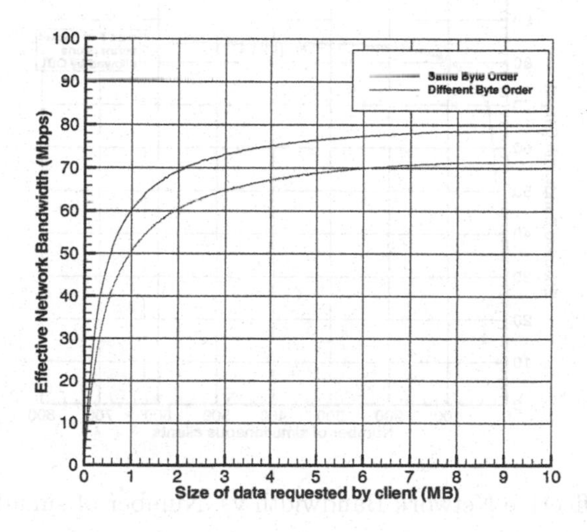


**Fig. 9.** Effective Network Bandwidth vs. Size of data requested by client

were programmed to request the 1.6 MB 4-D array from the server. The effective bandwidth in this case is obtained by dividing the total amount of data served by the server with the total wall-clock time required to serve all the requests. It can be seen that the network performance of POSSE is very good (84 Mbps) even when dealing with over 500 client requests simultaneously.

For the wake-vortex simulation system, the amount of data communicated to the client after every update is 420 bytes for every aircraft and 56 bytes for every vortex element. For 10 aircrafts each having 2,000 elements tracked, this amounts to 1.12 MB of data. From Fig. 9, we can see that this corresponds to a data-rate of approximately 62 Mbps, or 145 ms of communication time. Thus, we can get updated data at a rate of almost 7 fps from the server. The wake-vortex simulation runs with a $\Delta t$ of 0.2 seconds which can be maintained for up to 2,000 vortex elements per aircraft on COCOA-2. The parallel code has very good scalability for up to 15 processors (tracking 15 aircrafts) after which it linearly deteriorates due to the overhead borne by the master for distributing and collecting data from the slave nodes. At this point, the simulation has only been qualitatively checked and seems to be consistent with the theory. The simplification of using $k$ neighbors for induced-velocity computation works very well with an error of less than 1% when compared to the original $O(N^2)$ case. Since the weather conditions play a substantial role in the determination of the vortex strength, a more sophisticated weather model like the one used in AVOSS will definitely improve the accuracy of the simulations.

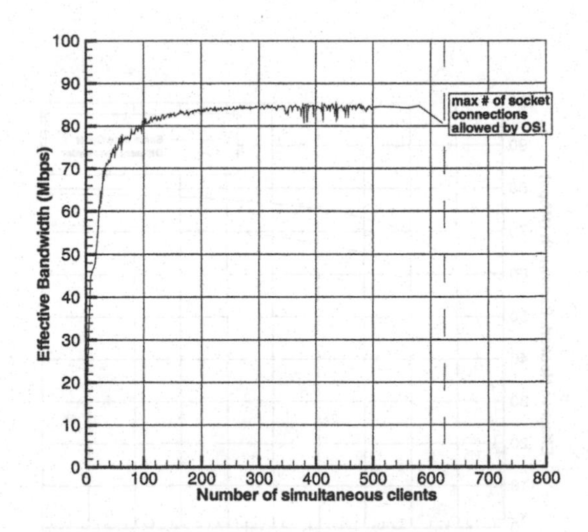


**Fig. 10.** Effective Network Bandwidth vs. Number of simultaneous clients

## 5 Conclusions

The coupling of computational steering to our parallel simulation makes the real-time visualization of the wake-vortex simulations possible. It opens a new way for the ATC to effectively deal with the wake-vortex hazard problem and to improve the capacity and safety of large airports. Our steering system, POSSE, has proven to be a very powerful, yet easy to use software with a high rate of acceptance and approval in our research group. If scientists are given an easy to use software system with a mild learning curve, they will use it. At a more basic level, this ability to interact and visualize a complex solution as it unfolds and the real-time nature of the computational steering system opens a whole new dimension to the scientists for interacting with their simulations.

## References


[1] Airdisaster.com. Investigation: USAir Flight 427. `http://www.airdisaster.com/investigations/us427/usair427.shtml`, 1997.

[2] I. Ba, C. Malon, and B. Smith. Design of the ALICE Memory Snooper. `http://www.mcs.anl.gov/ams`, 1999.

[3] CAVE. Fake Space Systems. `http://www.fakespacesystems.com`, 2001.

[4] Message Passing Interface Forum. MPI: A Message-Passing Interface Standard. (UT-CS-94-230), 1994.

[5] G. A. Geist, II, James Arthur Kohl, and Philip M. Papadopoulos. CUMULVS: Providing Fault Tolerance, Visualization, and Steering of Parallel Applications. *The International Journal of Supercomputer Applications and High Performance Computing*, 11(3):224–235, Fall 1997.

[6] G. C. Greene. An approximate model of vortex decay in the atmosphere. *Journal of Aircraft*, 23(7):566–573, July 1986.

[7] W. Gu, G. Eisenhauer, E. Kraemer, K. Schwan, J. Stasko, J. Vetter, and N. Mallavarupu. Falcon: On-line Monitoring and Steering of Large-Scale Parallel Programs. *Proceedings of the Fifth Symposium on the Frontiers of Massively Parallel Computation*, pages 433–429, February 1995.

[8] D. A. Hinton. Aircraft Vortex Spacing System (AVOSS) Conceptual Design. *NASA TM-110184*, August 1995.

[9] Hypotenuse Research Triangle Institute. Wake Vortex Detection System: Engineered for Efficiency and Safety. `http://www.rti.org/hypo_etc/winter00/vortex.cfm`, 2001.

[10] D. Jablonowski, J. Bruner, B. Bliss, and R. Haber. VASE : The Visualization and Application Steering Environment. *Proceedings of Supercomputing '93*, pages 560–569, 1993.

[11] Barnes W. McCormick. Aircraft Wakes: A Survey of the Problem. *Keystone Presentation at FAA Symposium on Aviation Turbulence*, March 1971.

[12] A. Modi, N. Sezer, L. N. Long, and P. E. Plassmann. Scalable Computational Steering System for Visualization of Large Scale CFD Simulations. *AIAA 2002-2750*, June 2002.

[13] Anirudh Modi. COst effective COmputing Array-2. `http://cocoa2.ihpca.psu.edu`, 2001.

[14] Anirudh Modi. POSSE: Portabale Object-oriented Scientific Steering Environment. `http://posse.sourceforge.net`, 2001.

[15] S. G. Parker, M. Miller, C. D. Hansen, and C. R. Johnson. An Integrated Problem Solving Environment: The SCIRun Computational Steering System. *IEEE Proceedings of the Thirty-First Hawaii International Conference on System Sciences*, 7:147–156, 1998.

[16] T. S. Perry. In Search of the Future of Air Traffic Control. *IEEE Spectrum*, 34(8):18–35, August 1997.

[17] Fred H. Proctor and George F. Switzer. Numerical Simulation of Aircraft Trailing Vortices. *Ninth Conference on Aviation, Range and Aerospace Meteorology*, September 2000.

[18] Bernhard Reitinger. On-line Program and Data Visualization of Parallel Systems in a Monitoring and Steering Environment. Dipl.-Ing. Thesis, Johannes Kepler University, Linz, Austria, Department for Graphics and Parallel Processing, `http://eos.gup.uni-linz.ac.at/thesis/thesis.pdf`, January 2001.

[19] E. Shaffer, D. A. Reed, S. Whitmore, and B. Schaeffer. Virtue: Performance Visualization of Parallel and Distributed Applications. *IEEE Computer*, 32(12):44–51, December 1999.

[20] S. Shen, F. Ding, J. Han, Y. Lin, S. P. Arya, and F. H. Proctor. Numerical Modeling Studies of Wake Vortices: Real Case Simulations. *AIAA 99-0755*, January 1999.

[21] T. Sterling, D. Savarese, D. J. Becker, J. E. Dorband, U. A. Ranawake, and C. V. Packer. BEOWULF: A Parallel Workstation for Scientific Computation. *Proceedings of the 24th International Conference on Parallel Processing*, pages 11–14, 1995.

[22] Eric C. Stewart. A Comparison of Airborne Wake Vortex Detection Measurements With Values Predicted From Potential Theory. *NASA TP-3125*, November 1991.

[23] G. F. Switzer. Validation Tests of TASS for Application to 3-D Vortex Simulations. *NASA CR-4756*, October 1996.

[24] Robert van Liere and Jarke J. van Wijk. CSE : A Modular Architecture for Computational Steering. In M. Göbel, J. David, P. Slavik, and J. J. van Wijk, editors, *Virtual Environments and Scientific Visualization '96*, pages 257–266. Springer-Verlag Wien, 1996.

[25] VRCO. CAVELib Users Manual. `http://www.vrco.com/CAVE_USER/caveuser_program.html`, 2001.

# PALM: A Dynamic Parallel Coupler

Andrea Piacentini and The PALM Group

CERFACS
Centre Européen de Recherche et Formation Avancée en Calcul Scientifique
F-31057 Toulouse, France
{andrea,palm}@cerfacs.fr
http://www.cerfacs.fr/~palm

**Abstract.** The PALM project aims to implement a dynamic MPMD coupler for parallel codes. It has originally been designed for oceanographic data assimilation algorithms, but its application domain extends to every kind of parallel dynamic coupling.
The main features of PALM are the dynamic launching of the coupled components, the full independence of the components from the application algorithm, the parallel data exchanges with redistribution and the separation of the physics from the algebraic manipulations performed by the coupler.
A fully operational SPMD prototype of the PALM coupler has been implemented and it is now available for research purposes.

## 1 Introduction

PALM aims to provide a general structure for the modular implementation of a data assimilation algorithm. In PALM, an algorithm is described by a set of sequences of elementary "units" such as the forecast model, the observation operator, the adjoint model, the operators approximating the error covariance matrices, etc. This approach allows for an effective separation of the physical part of the problem from the algebraic part.

PALM performs the scheduling and synchronization of the units, the communication of the fields between the units and the algebra. These tasks have to be achieved without a significant loss of performance when compared to a standard implementation. This approach proves to be useful for any kind of dynamic coupling not only for data assimilation algorithms.

In section 2 the reader will find a summary of the studies which led to the PALM coupler definition. Section 3 deals with the main PALM concepts, detailing how an algorithm is represented in PALM, the most relevant features of the data exchange model and the extent to which PALM handles the algebra. In section 4 we give some examples of PALM applications. The main features of the PALM coupler and the future perspectives are summarized in section 5.

J.M.L.M. Palma et al. (Eds.): VECPAR 2002, LNCS 2565, pp. 479–492, 2003.

## 2 Genesis

The main impetus for the PALM project has come from the French oceanography project MERCATOR whose aim is to implement a high resolution ocean forecasting system [8]. The data assimilation algorithm is the most costly component of the system. Therefore the assimilation methods have to be thoroughly investigated and their implementation should be very efficient. These requirements impose a double constraint on the implementation: modularity and high performance. On this basis, CERFACS has been developing a high performance coupler to combine the different data assimilation components into a high performance application.

The PALM project began in 1997 with a feasibility study aimed at proving that the coupling approach is well suited for data assimilation and at determining the most appropriate technical solution for the coupler implementation. The theoretical part of the study has led to the definition of a flow-charting based graphical formalism to represent data assimilation algorithms. This formalism is the object of a separate scientific publication [1]. The study proved that the main algorithms can all be obtained as different execution sequences of the same basic operators. Moreover, it appeared that all the algorithms require heavy algebraic computations such as minimization, linear algebra and eigenvector/eigenvalue computation and therefore that a particular effort had to be made on the efficiency of the algebraic libraries.

These considerations have been taken into account in the definition of the implementation strategy. The MPMD coupling approach has been retained because it allows for the full independence of the coupled components. These components can thus be developed in different laboratories and reused for different applications. For the same reason, the coupling interfaces have to be completely independent of the specific application for which the components are used. From the technical point of view, different possible implementations have been considered. Unix based solution like shared memory segments have been rejected because they are not adapted to distributed memory machines or cluster configurations. CORBA was not sufficiently ported on supercomputers. For this reason we chose a message passing implementation based on MPI. We rejected PVM since it has no standard implementation on many of the supercomputer and the public domain version does not take advantage of the hardware communication mechanisms. MPI, on the other hand, is standardized and all the constructors have an optimized MPI implementation. Moreover, public domain implementations ensure portability on basic configurations like workstation clusters. The results of the feasibility studies are summarized in a technical report [2].

The MPI2 standard provides all the features that are required by MPMD coupling [4], in particular the ability od dynamically spawning tasks. However, at the time of the initial implementation of PALM, it was not available on all the targeted platforms. For this reason, the PALM implementation was organized in two phases: during the initial phase MPI1 was used. Since MPI1 only supports the SPMD programming model in which all the tasks are implemented as subroutines of a single executable, we developed a SPMD prototype emulating a MPMD coupler. This prototype is intended to act as a functional prototype more than a technical one. It is used for the prototype forecasting system in the MERCATOR project and for some other research activities. The second phase concerns the implementation of the final version of the MPMD coupler.

The SPMD prototype is now complete and it is available for research purposes [6, 7]. This paper is devoted to its description.

## 3 The PALM Concepts

The programming model underlying the PALM applications is closely related to the approach used for coupling models but it has been enhanced to ensure modularity and parallelism. The whole application is composed of basic pieces of code that are assembled in a certain sequence. Some of these codes can run concurrently. We call the basic components *units* and the sequences *branches.*

Units can be thought of as independent processes that consume, transform and produce data. The exchanged data are organized into elementary chunks of information that we call *objects.* Every object is identified by its *name*, its *time stamp* and an optional integer *tag.* To ensure modularity, one of the main assumptions of the PALM philosophy is that units are independent from each other. For this reason, the nomenclature of the objects is local to each unit and it is declared in an *identity card* of the unit. This is the principle of the *end-point* communication protocol. The correspondence between the two sides of the communication is set via the PrePALM interface.

The PALM approach allows for a clean separation of the physics of the application (which are inside the units) from the description of the algorithm. The structure of the algorithm can be defined via the branches and communications. The algebraic operators which appear in the algorithm (linear combinations, linear system solvers, eigenvector/eigenvalue solvers, minimizers, ...) can be treated as units. They represent a class of special units which do not belong to branches and which are made available to the user in a toolbox.

### 3.1 The Algorithm

An algorithm can be split into elementary operations and the corresponding computation can be performed by the coupling of independent pieces of codes that represent the elementary operations. The *unit* is the piece of code that implements an elementary operation. It is up to the user to choose what "elementary" means, i.e. to choose the granularity of the coupling. For example, a unit can be a sum or a complex high resolution ocean model.

The data exchange between units is well represented by the data flow model: output of a unit is piped into the input of other units.

What makes the difference between two algorithms built from the same units is the sequence of their execution. This consideration leads us to introduce the concept of *branch*, i.e. the execution sequence of the units. A branch in PALM is

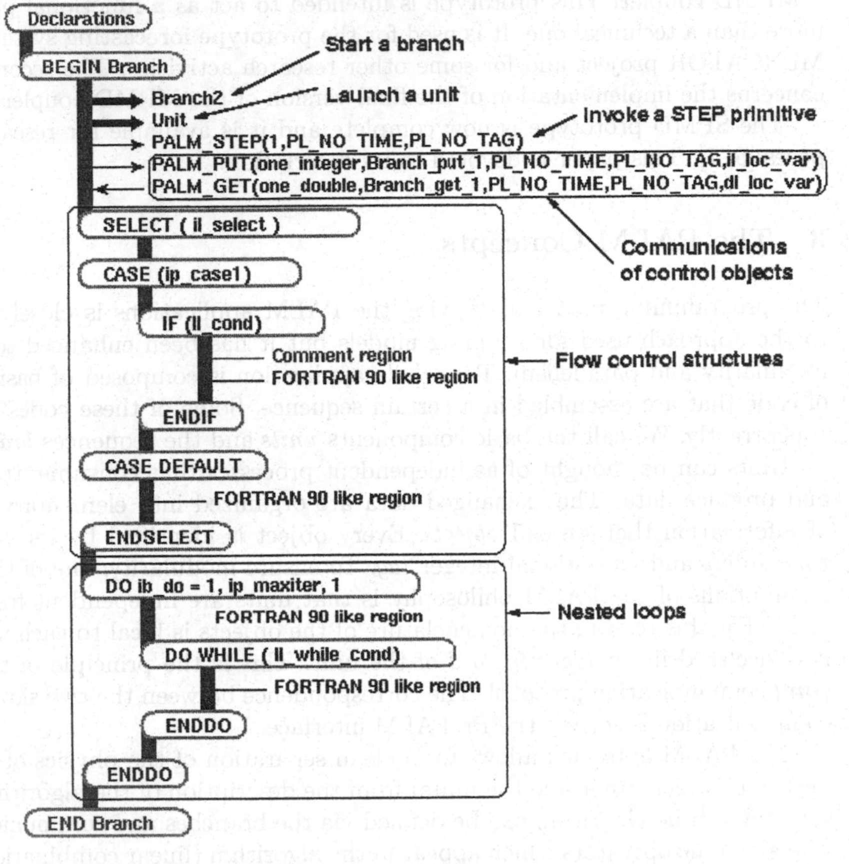


**Fig. 1.** The actions invoked by a branch code

something more than a simple sequence because it can include control structures like loops, "if" constructs or "select" switches.

The branch code commands are entered by the PrePALM graphical interface. An example is shown in Fig. 1.

Performance can be increased if we take advantage of the potential parallelism of the algorithm. For the 'highly parallel' algorithm $r = (a + b) - (c + d)$, the computation can be split into $r_1 = (a + b)$, $r_2 = (c + d)$ and finally $r = r_1 - r_2$. Since the computation of $r_1$ and $r_2$ are independent, we can define two concurrent branches as shown in Fig. 2.

Since units can be distributed codes, there are two levels of parallelism in PALM: parallelism at the algorithmic level (concurrent branches) and parallelism at the units level (distributed units). PALM helps in parallelizing the application at the algorithmic level, meanwhile the parallelization of the units is entirely done by the unit implementer. The modularity of PALM gives the user great freedom for the implementation of the units.

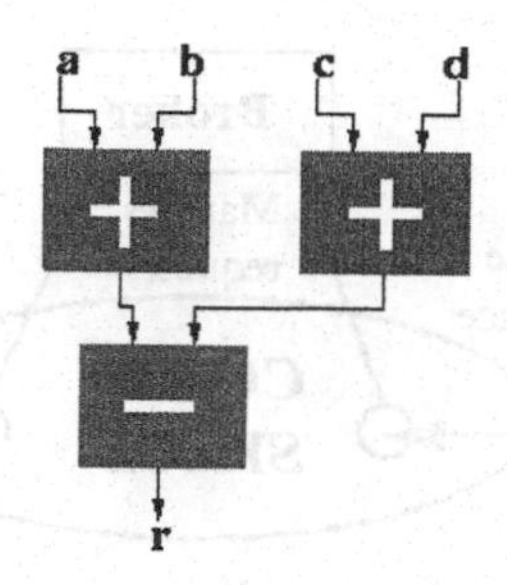


**Fig. 2.** Flow chart representation of $r = (a + b) - (c + d)$

The SPMD emulation of MPMD dynamic coupling is based on a pool of idle processes, driven by a master process, called the *driver*. At least one branch starts at the beginning of the application: the driver allocates one idle process to each branch with the START_ON attribute. When a branch terminates, its process joins back the idle pool. The same mechanism applies when a running branch asks the driver to start a new branch. If no idle processes are available, the start request is queued until a process is released.

When a branch asks the driver to launch an $n$-process parallel unit, the branch code suspends its execution and the process is allocated to the unit. The driver allocates $n - 1$ idle processes to the unit. At the unit termination, $n - 1$ processes join back the idle pool and the branch code resumes on its previous process. If not enough idle processes are available, the launch request is queued until enough processes are released. This mechanism is summarized in Fig. 3.

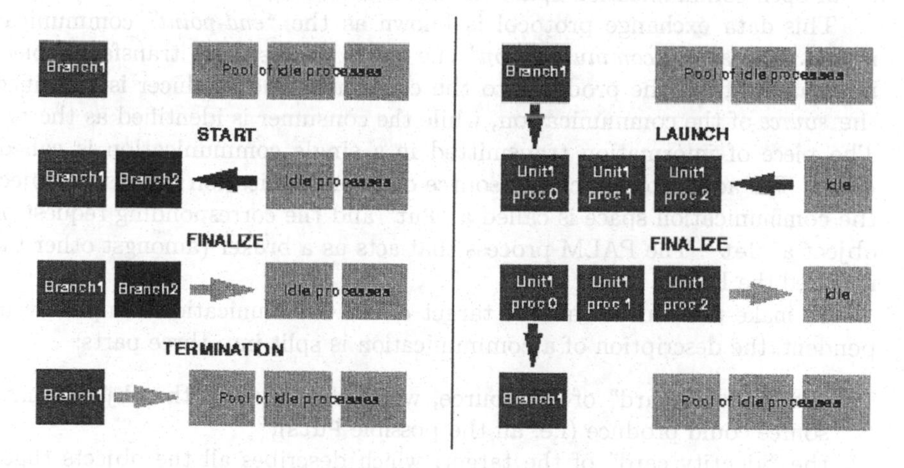


**Fig. 3.** Process management for branches (on the left) and for units (on the right)

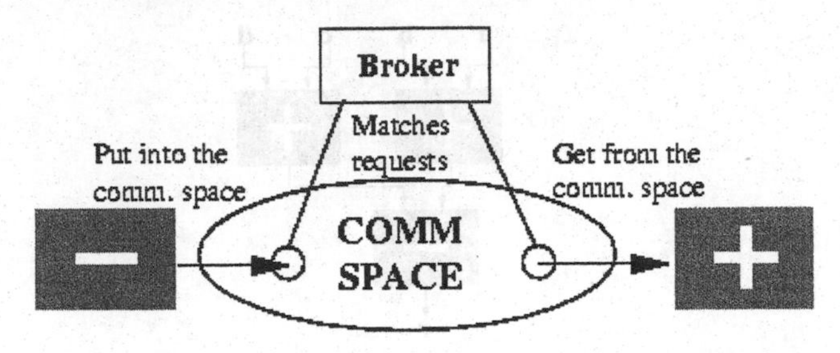


**Fig. 4.** A communication with broker and communication space

Finally, process can be allocated to PALM tasks mainly for memory storage. On distributed memory machines, if the processor memory usage reaches a user defined threshold, new idle processes are allocated to the PALM driver. When memory is freed a garbage collection mechanism is triggered and memory occupation is concentrated on the minimum number of processes. Freed processes join back the idle pool.

### 3.2 The Data Exchanges

The data exchange model in PALM has been designed to grant full modularity with the minimum loss of performance. Modularity is ensured only if the unit code is independent of the communication pattern in the application, i.e. only if inside the units there is no reference to the origin of the input and to the destination of the output. For this reason we have designed PALM to be a broker in an open communication space as shown in Fig. 4.

This data exchange protocol is known as the "*end-point*" communication model. We call a "*communication*" the whole process that transfers a piece of information from the producer to the consumer. The producer is identified as the *source* of the communication, while the consumer is identified as the *target*. The piece of information transmitted in a single communication is called an *object*. The action by which the source of a communication injects an object in the communication space is called a "`Put`" and the corresponding request of an object a "`Get`". The PALM process that acts as a broker (amongst other tasks) is called the PALM *driver*.

To make the source and the target of the communication completely independent, the description of a communication is split into three parts:

1. the "identity card" of the source, which describes all the objects that the source could produce (i.e. all the possible `Puts`);
2. the "identity card" of the target, which describes all the objects that the target could require (i.e. all the possible `Gets`);
3. the information on the matching between productions and requests. This information is known by the PALM driver and depends on the specific application.

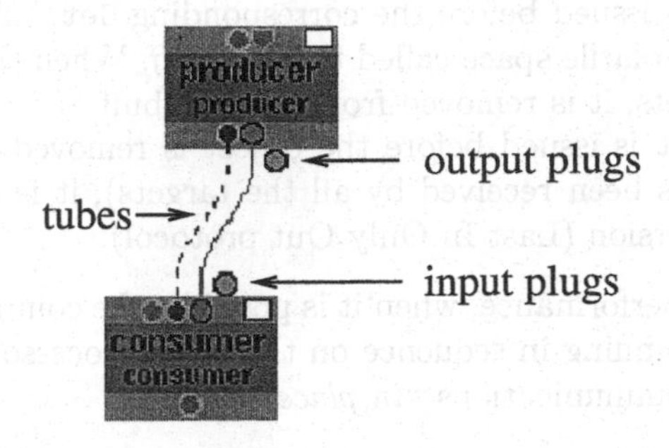


**Fig. 5.** The graphical representation of communications

A communication takes place only if the source issues a `Put`, the target issues a `Get` and a possible pattern between a `Put` and a `Get` has been defined. In the PALM language, we call such a pattern a *tube*. The user enters the description of all possible communication tubes with the graphical interface PrePALM. Fig. 5 shows that a tube is defined by a simple line joining two unit plugs. The plugs correspond to the potential `Put`s and `Get`s as they have been described in the identity cards. Notice that not all the plugs must be used. If multiple `Get`s refer to the same unique `Put`, the driver starts the right number of communications and serves all the targets.

The source and target of a communication can be parallel units and there is no requirement that the distribution of the object to be exchanged be the same on the source and sides. PALM takes care entirely of the change of the distribution of the object during the communication. This action is called *remapping*.

There exists in PALM a permanent storage space, which is called the *buffer*. The user can indicate to the driver that a produced object has to be stored in the buffer. It is much like a standard communication having as target a special unit, always running, which keep copies of the permanently stored objects. Similarly, the user can indicate to the driver that a required object has to be recovered from the buffer. When the object is not needed anymore, the driver sends an order to the buffer to delete the object and free its place. With the graphical interface PrePALM, the user describes to the driver the relevant points of a computation (the *steps*) and the *actions* to be taken on buffered objects.

PALM tries to minimize the number of MPI messages to exchange and the memory usage. This is achieved by the following strategy.

- A `Get` is blocking: the unit issuing a request waits until the object is received.
- A `Put` is never blocking: since the `Get` is blocking, this choice is mandatory to ensure the independence of the object production and reception order.
- If a `Get` is issued before the corresponding `Put` (i.e. the one associated to the `Get` in a communication), it is immediately and directly routed from the source to the target unit when the object is produced.

- If the `Put` is issued before the corresponding `Get`, the object is temporarily stored in a volatile space called the *mailbuff*. When the object is received by all the targets, it is removed from the mailbuff.
- If a new `Put` is issued before the object is removed from the mailbuff (i.e. before it has been received by all the targets), it is overridden by the new produced version (Last In Only Out protocol).

For the sake of performance, when it is possible, the communications established between units running in sequence on the same processors use a local mailbuff. We call these communications "*in place*".

Communications are activated and controlled by invoking PALM primitives from inside the units. For the sake of simplicity the primitives set reduces to:

**PALM_Get, PALM_Put:** trigger PALM end-point communications.
**PALM_Query_get, PALM_Query_put:** query if `PALM_Get` or `PALM_Put` calls correspond to actual communications i.e. *tubes* exist.
**PALM_Time_convert:** converts GMT dates to and from integer PALM time stamps according to the selected calendar.
**PALM_Error_explain:** prints out a mnemonic explanation of returned error codes.
**PALM_Abort:** clean shutdown of a PALM application.
**PALM_Verblevel_set:** sets the verbosity level for a specific category of messages.
**PALM_Verblevel_overall_set, PALM_Verblevel_overall_off:** set the overall verbosity level.

During the exchange, the object can undergo some treatments:

**Remapping:** if the object is distributed on the source and/or on the target side of the communication, it is automatically redistributed, if necessary, during the exchange phase. The most efficient communication pattern is computed by PALM.
**Track:** according to the selected verbosity level, it is possible to track the communications. It means that some information on the ongoing communications are printed in the standard output files. Since the amount of information printed for all the communications would be overwhelming, it is possible to set an attribute to indicate whether the communication has to be tracked or not.
**Debug:** in order to diagnose the results of part of a simulation or to check the correctness of some data, the user can apply specific procedures to the objects exchanged via a communication. These user-defined procedures are implemented in the `palm_debug.f90` file which is user provided and which has to be linked with the application. Debug is activated or not according to a communication attribute set via the PrePALM interface. The debug procedures can produce diagnostic output in the standard PALM output files and can return an error code which is propagated back to the user code as an output argument of `PALM_Get` or `PALM_Put`.

If the communication recovers the object from the buffer, the time argument of `PALM_Get` needs not to correspond to the time of an object actually stored in the buffer. It is possible to indicate as an attribute of the communication if and how a time interpolation is to be performed. In order to be compliant with the different techniques used in data assimilation applications, PALM provides four kinds of time interpolation:

**PL_GET_EXACT** No interpolation is done. If the required time is not (and will not be) stored in the buffer, `PALM_Get` fails with non zero error code.

**PL_GET_NNBOR** Next Neighbour interpolation. If the required time is not (and will not be) stored in the buffer, the closest time will be selected instead. When `PALM_Get` returns, the "time" argument contains the time stamp actually selected. It can be converted back to a date with a call to `PALM_Time_convert`.

**PL_GET_LINEAR** Linear interpolation. If the required time is not (and will not be) stored in the buffer, PALM linearly interpolates the two closest neighbours.

**PL_GET_CUSTOM** User defined interpolation. If the required time is not (and will not be) stored in the buffer, PALM interpolates the two closest neighbours using a rule provided by the user in the `palm_time_int.f90` file.

### 3.3 The Algebra

One of the purposes of the PALM software is to give the user the possibility of separating the physical and the algebraic parts of the problem. There are two main ways in which PALM handles the algebra. The simplest one concerns the assembling of objects by addition or subtraction of contributions. For example, consider an object $J$ which is the sum of several contributions $J_1 + J_2 + J_3 + \ldots$ In such a case, we can consider that the object $J$ is the assembling of the contributions $J_i$. The assembling is automatically performed in the *buffer*, the permanent storage space of PALM. The object $J$ cannot be recovered until all the contributions have been added up.

More complicated operations are performed by specialized predefined units. These units are provided with PALM and comprised in what we call the *algebra toolbox*. These units provide the user with an interface to the main computing libraries (BLAS, LAPACK, ScaLAPACK and more) which fits into the PALM formalism and data exchange model.

The main difference between the specialized algebra units and the user defined units is that the algebra units do not belong to any branch. They are automatically executed when the PALM driver decides that it is the best moment to do it. This strategy is based on the availability of the operands, on the resource allocation and on the communication strategy that minimizes the number of object transfers.

## 4 Applications

Current applications are mainly oriented toward data assimilation. The PALM prototype is used in routine for the pre-operational ocean state forecast of the MERCATOR project, for oceanic data assimilation research at CERFACS and for atmospheric chemistry data assimilation research at EDF and at Météo-France, in the framework of the European project ASSET. The use of PALM for other data assimilation and climate simulation applications is under study at the Canadian Meteorological Service and at the U.K. Met. Office.

As an example of the type of the applications which can be handled by PALM, we show in Fig. 6 the PrePALM canvas representation of a 4D-VAR data assimilation scheme [1] applied to a Shallow Water model [9]. The data assimilation method is formulated as a minimization problem. The PALM algebraic toolkit contains the minimization routine as well as the other algebra routines needed for computing scalar products and vector sums or subtractions. The cost function and the gradient computations are organized on two branches: this allows for an overlapping of the model integration and of the model/observation misfit computation. A user unit has been coded for each operator: the nonlinear model, the adjoint model, the observation operators, the data loader, the error covariance matrices and so on. This example is taken from a comparison of different data assimilation methods applied to the same case. Different methods like 3D-FGAT, 3D-PSAS, 4D-VAR, incremental 4D-VAR [1] have been implemented using the same basic operators. PALM allows us to pass from one algorithm to another simply by changing the coupling of the basic units. A selection of these implementations is illustrated in the PALM prototype user guide [6].

This example allows us to show how PALM takes advantage of the two levels of parallelism of an algorithm. In the 4D-VAR scheme, the forward model integration can be overlapped with the observation branch execution. In particular, as soon as a model state has been produced for a time step corresponding to an observation slot, the state is passed to the observation branch and treated while the model advances to the next observation time. The adjoint model, on the other hand, is integrated backward in time and can start only when the last observation time slot has been treated. It implies that the resources allocated to the observation branch are idle during the adjoint model integration.

If we look at the pie-chart performance graphs in Fig. 7, we can observe that processor 2 is idle for almost three quarters of the total integration time. On the time-line we find the confirmation that processor 2 is idle while processor one is allocated to the adjoint model **M_lin_T**. It gives a total integration time of 417 seconds on a Compaq alpha-server for a minimization loop with 45 iterations, followed by a three-week forecast.

It is natural to optimize the algorithm by parallelizing the adjoint model. Running the adjoint model on two processors allows the use of processor 2 for the adjoint model integration when the observation branch frees it. We see on the pie charts of Fig. 8 that with the parallel version of the adjoint model the resources are well used and that the total integration time is reduced to less than 310 seconds.

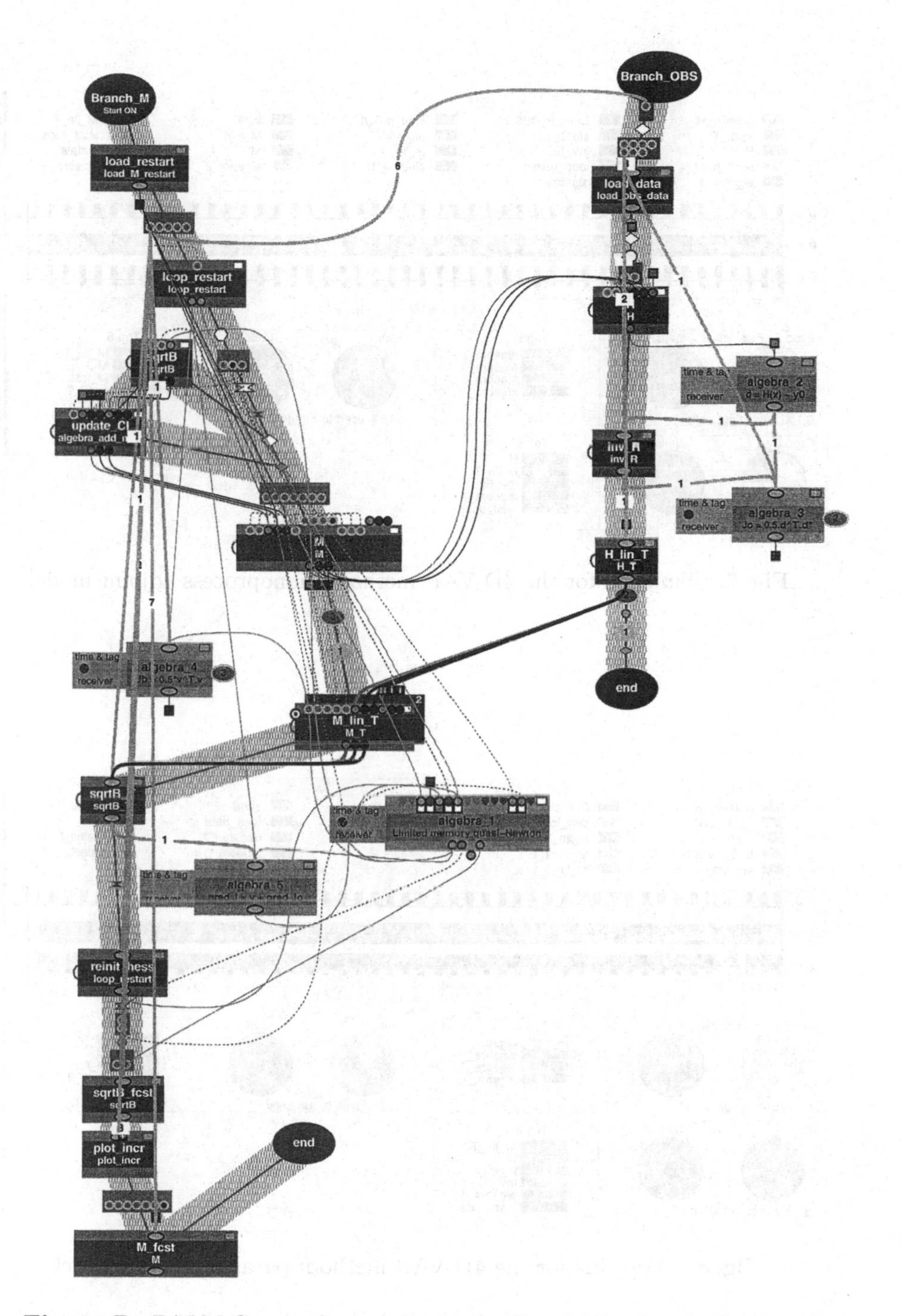


Fig. 6. PrePALM Canvas for a 4 dimensional variational assimilation scheme

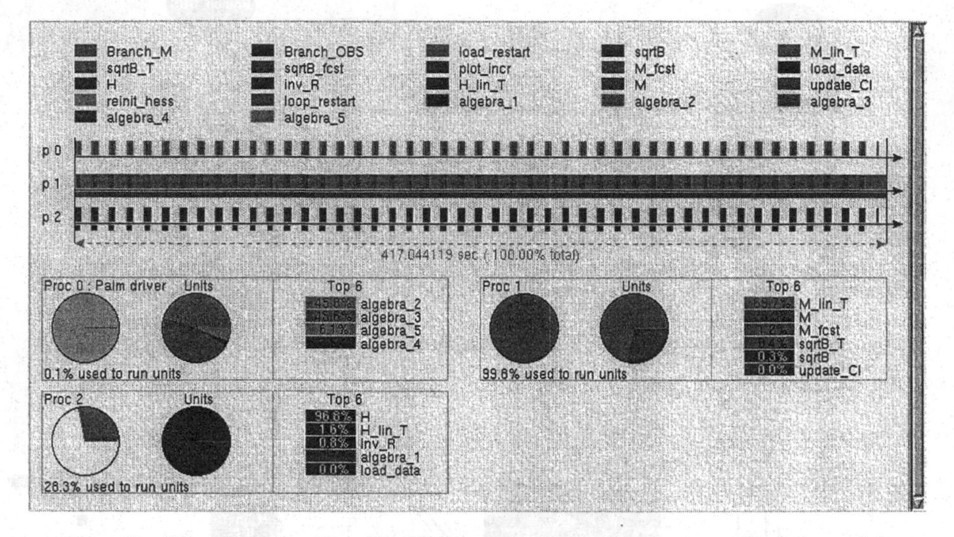


**Fig. 7.** Timeline for the 4D-VAR method: monoprocess adjoint model

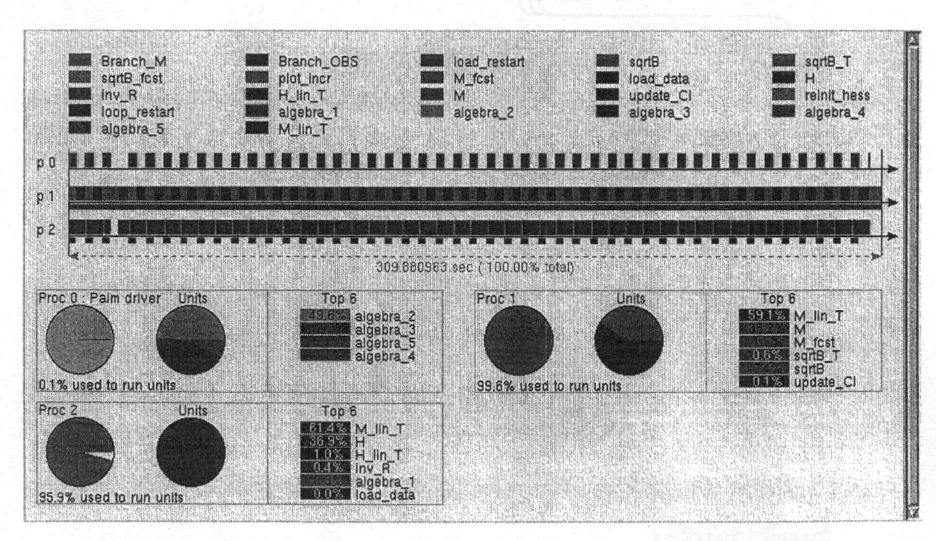


**Fig. 8.** Timeline for the 4D-VAR method: parallel adjoint model

## 5 Conclusions

The SPMD version of the PALM coupler has been implemented and ported to the main platforms and is now distributed for research purposes under the name of PALM Research. It is an emulation of the final MPMD version of PALM, which is now being implemented.

The main features of the PALM coupler are as follows.

- Dynamic coupling: coupled components can start and terminate at any moment of the application. Loop and conditional execution are supported.
- Two levels of parallelism: coupling of parallel codes running on concurrent execution threads.
- Automatic handling of data redistribution during data exchanges.
- Time dependency of the exchanged fields: the fields are identified by a time stamp and fields for any time instance can be obtained by interpolation.
- Use of high performance devices for data exchanges via MPI.
- Independence of the coupled components from the algorithm: the end-point communication scheme makes the unit implementation independent of the algorithm. All the complementary information is entered via the PrePALM graphical interface.
- Separation of the physical and algebraic components of the problem. The algebra can be handled by a toolkit provided by PALM.
- Basic performance analysis.
- Portability on the main supercomputers.

More information about the portings, the code availability and the documentation can be found on the the PALM web site (`http://www.cerfacs.fr/~palm`) or can be obtained by contacting the PALM group (`mailto:palm@cerfacs.fr`).

The experiences that we have carried on in data assimilation research applications with the PALM Research prototype have demonstrated the flexibility and modularity of the PALM approach. The pre-operational applications have proven its efficiency and have provided us with important hints on the way of optimizing the communication scheme. The experience obtained with the prototype is now integrated in the development of the final version PALM_MP. In particular, the dynamic aspects of the task management and the run-time redefinition of the communication patterns and features will greatly improve the flexibility. The optimization of the remapping patterns computation and the use of one-sided MPI2 communications for internal control will improve the efficiency.

## References

[1] Lagarde, Th., Piacentini, A., Thual, O.: A new representation of data assimilation methods: the PALM flow charting approach. Q. J. R.Meteorol.Soc. **127** (2001) 189–207

[2] The PALM group: Etude de faisabilité du projet PALM. Technical Report TR/CMGC/98/50, CERFACS, Toulouse, France (available on the web site)

[3] The PALM group: A coding norm for PALM. Technical Report TR/CMGC/02/10, CERFACS, Toulouse, France (available on the web site)
[4] The PALM group: Implementation of the PALM MP. Working Note WN/CMGC/00/71 CERFACS, Toulouse, France
[5] D. Déclat and the PALM group: The Coding Conventions for the PALM MP. Technical Report TR/CMGC/02/9, CERFACS, Toulouse, France
[6] The PALM Team: PALM and PrePALM Research QuickStart Guide. Technical Report TR/CMGC/02/38, CERFACS, Toulouse, France (available on the web site)
[7] The PALM Team: PALM and PrePALM Research Installation and Configuration Guide. Technical Report TR/CMGC/02/39, CERFACS, Toulouse, France
[8] André, J. C., Blanchet, I., Piacentini, A.: Besoins algorithmiques et informatiques en océanographie opérationnelle. Le programme Mercator. Calculateurs Parallèles, Réseaux et Systèmes répartis **11** No. 3 (2000) 275–294
[9] Vidard, P., Blayo, E., Le Dimet, F. X., Piacentini, A.: 4D Variational Data Analysis with Imperfect Model. Journal of Flow, Turbulence and Combustion (2001), in press

# A Null Message Count of a Conservative Parallel Simulation

Ha Yoon Song[1], Sung Hyun Cho[2], and Sang Yong Han[3]

[1] College of Information and Computer Engineering, Hongik University, Seoul, Korea
song@cs.hongik.ac.kr

[2] School of Software and Game, Hongik University, Chungnam, Korea
scho@wow.hongik.ac.kr

[3] School of Computer Science, Seoul National University, Seoul, Korea
syhan@pplab.snu.ac.kr

**Abstract.** A performance analysis of the conservative parallel simulation protocol is presented in this study. We analyze several performance factors of a simulation model with an arbitrary number of logical processes. The analysis probabilistically identifies the critical path in a conservative model, and it also estimates the number of null messages, which is the major overhead factor of a conservative simulation. Apart from the factors of the hardware platform on which a simulation system is running, null message count is completely based on logical factors such as properties of simulation topology and simulation protocol. Null message count can be estimated in terms of lookahead of logical processes, simulation time, and the connectivity of simulation topology. Several experimental results have been presented to verify the null message count model. Capturing the null message count will lead to a new idea of how to revise the simulation model in order to improve the performance of a conservative parallel simulation.

## 1 Introduction

The growing capacity of simulation models has highlighted the use of parallel simulation technology as a viable solution for high-performance simulation. PDES (Parallel and Distributed Event Simulation) is spotted as a reasonable solution of complex DES (Discrete Event Simulation). Currently, several simulation systems and protocols are available for PDES, primarily based on one of two basic parallel simulation protocols - the optimistic [1] and conservative [2] [3]. The benefit from parallel execution of a model is directly related to the overhead incurred by the protocol to implement the necessary synchronization.

The overhead factors are usually divided into two categories: protocol-independent and protocol-dependent factors. In the protocol-independent case, idle time exists in an LP (Logical Process) because the computational load may not be distributed evenly among processes. Additional source of overhead includes factors such as context switching, scheduling between LPs on a CPU, event queue manipulation, etc. For the protocol-dependent factors, conservative protocols have a unique overhead due to excessive blocking of an LP until

J.M.L.M. Palma et al. (Eds.): VECPAR 2002, LNCS 2565, pp. 493–506, 2003.

an available message is identified as being safe to process, while the optimistic system has rollback and state-saving overheads.

In this paper, we will study protocol-dependent performance factors of a conservative parallel simulation. Our goal is to estimate the null message count of a conservative parallel simulation. To date, there has been a limited amount of research in analytical approach. Felderman and Kleinrock used the queuing theory to analyze the performance of optimistic and conservative simulation systems [4]. In the conservative version, they applied the queueing theory to analyze a two-processor system running a conservative distributed simulation protocol. The models enables quantitative approaches in speedup, the count of null messages, and degradation due to the cost of breaking deadlocks. However, these researchers did not generalize their case to that of more than two processors. Another stochastic methodology came from Nicol [5], who analyzed the performance and cost of synchronous conservative protocols. His model used the time window as a basis for synchronization; no processor can exceed the window's time boundary. Nicol analyzed the overhead of synchronization, event list manipulation, lookahead calculation, and the processor idle time. Baley and Pagels [6] categorized overheads involved in a conservative simulation into null message overhead, context switching overhead, and blocking overhead. They tried to establish the quantitative profile of overheads by experiments. Song [7] proposed an analytical model of the blocking overhead for a conservative simulation with probability theory based on the concept of critical path. However, little effort has been paid for the analysis of the null message count. To our best knowledge, this work is the first one proposing an analytical model for the null message count of a conservative simulation.

Distributed systems have suffered from the slower communication speed than the processing speed: larger memory access time in case of shared memory processors or high transmission costs of messages in case of fully distributed systems. For the parallel simulation, the cost of message passing still remains in considerable amount, which affects the performance of parallel simulation. Focusing on these phenomena, we study the inherent null message overhead of a conservative parallel simulation. The null message transmission cost is dependent on the memory access time or network transmission time, which depends on the physical factors of hardware implementation. However, some factors of the null message transmission overhead are clearly dependent on the conservative parallel simulation protocol and the simulation model. The count of null messages of a protocol and a model determines the null message cost. The count of null messages can be distinguished from the physical factors of hardware and can be identified in the logical world. From the simulation model properties and the definition of a null message algorithm, we will derive the estimation of the null message count for a conservative parallel simulation.

In this paper, our target protocol is a variant of the Chandy-Misra null message protocol. Unlike preceding approaches, our study analyzes an asynchronous conservative parallel protocol for an arbitrary number of LPs. Our analysis is done in the following sequence: We first distinguish the logical factors from the physical factors in a simulation system in section 2. In section 3, we introduce the concept of critical path of LPs, which may vary dynamically. For a given simulation time, we divide the logical time axis (or virtual time axis) into a set of disjointed sequence of intervals, and we finally determine the count of null messages. Several experimental results will be presented to verify this count in section 4 before the conclusion section.

## 2 Conservative Simulation Protocol

The discrete even simulation algorithm is intended to manipulate the logical timestamps of events in order to prevent or recover from the causality errors. While the events are processed to progress logical time, the physical time is required to progress logical times depending on the hardware factors. The simulation finishes usually when the logical time exceeds the maximum simulation time, and the corresponding physical time determines the total execution time of the simulation. For parallel execution of a simulation, LPs of a conservative system process events concurrently. With Chandy-Misra-Bryant protocol, one of the conservative parallel algorithms, we will begin with the description of our null message protocol. A PDES model is composed of a set of logical processes that are connected in a (possibly dynamic) topology. For the LP named $LP_i$, we cite the following general definition [8]:

- $SS_i$ (Source Set): the set of LPs which send messages to $LP_i$.
- $DS_i$ (Destination Set): the set of LPs to which $LP_i$ sends messages.
- $EOT_i$ (Earliest Output Time): the earliest time at which $LP_i$ might send a message to any LP in $DS_i$.
- $EIT_i$ (Earliest Input Time, *safe time*): the earliest time at which $LP_i$ might receive a message, calculated as $min(EOT_s), LP_s \in SS_i$. Any events with timestamps earlier than $EIT_i$ are safe to process.
- $Lookahead_i$: the difference between $EIT_i$ and $EOT_i$. Lookahead is the degree to which an object can predict its future messages. It is also termed as a simpler form, $la_i$.

The EIT of an LP advances in discrete steps as shown in figure 1, with receiving the EOT information from predecessors (parent LPs) in its source set. When the EIT of an LP is updated, the LP executes events as they become safe, and the LP is subsequently blocked until the next EIT update. Whenever an EIT update occurs, the LP updates its EOT as soon as possible, and a new EOT must be sent to each LP in its destination set. The sending of EOT is done through the use of null messages and it is done transparently to the user. Upon receiving a null message, the recipient LP might be able to update its EIT and thus determine safe events.

The above scheme is called as the aggressive null message protocol. In this protocol, it is possible that an incoming null message does not change the receiver's EIT, since the EIT is the minimum of all EOTs from parent LPs. Therefore, we will use the term *EIT update* to denote a situation where a null message

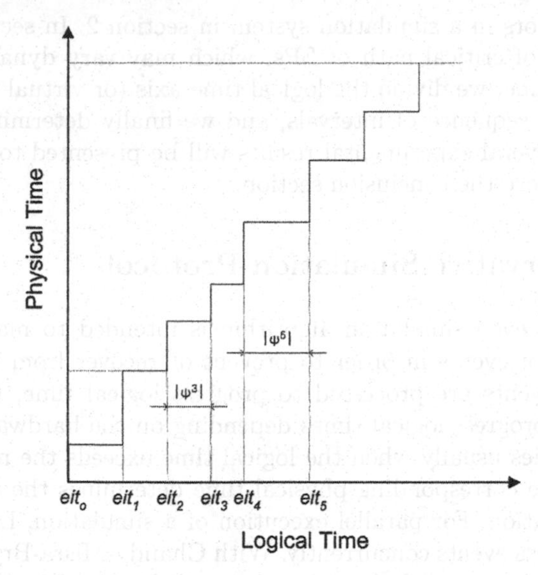


**Fig. 1.** The progression of EIT update

actually causes the EIT of the recipient LP to advance. The null message causing the EIT update is called a *critical null message* for the EIT update and the sender of the critical null message is called a *critical parent.* After updating the EIT, the LP checks if any safe events exist in the incoming message queue and processes the safe events. It then checks to see if there are new null message(s) from parent LPs and tries to update its EIT.

The behavioral abstraction of LPs is as follows. Like any other simulation model, each LP has its event processing time distribution and event generation time distribution. An event processed by an LP may require some physical time to be processed (which is dependent on hardware factors), and will generate other (child) event(s) with a logical timestamp. We assume that the processing time and the timestamp of a child event have independent distributions defined by an LP. For simplicity, we assume that the simulation topology is static and each LP processes homogeneous events dedicated to the LP. Each LP has its own non-zero lookahead, which is static, LP-specific, and can be precomputed. The child events are sent to other LPs as messages. We assume that the message transmission time has its own distribution across the whole simulation topology and is dependent on hardware systems. Message delivery is assumed to be FIFO. The number of child events generated by processing an event (birth rate) is also an LP-specific factor, and the ratio of child events being sent to a specific successor to all child events is called the *branching factor* of an LP. Every outgoing event has a unique timestamp, and there is no delayed message; i.e., the distribution of processing time of an event is the same as that of an event generation time in the logical time domain.

**Table 1.** Notations for logical factors

| | |
|---|---|
| $LP_i$ | A logical process named 'i' |
| $la_i$ | Lookahead of $LP_i$ |
| $SS_i$ | Source set of $LP_i$ |
| $DS_i$ | Destination set of $LP_i$ |
| $eit_i$ | A logical time value of Earliest Incoming Time of an $LP_i$ |
| $eot_i$ | A logical time value of Earliest Outgoing Time of an $LP_i$ |
| $PCP_i$ | Precedence Critical Path of $LP_i$ |
| $\psi_i$ | An EIT update Interval of $LP_i$ |
| $CNM_i$ | Count of Null Messages of $LP_i$ |
| $NI_i$ | Number of EIT update intervals of $LP_i$ |

Table 1 shows the notations used in this paper. Since the null message count is a totally logical one, it contains only logical factors, which are the model related factors.

## 3 Probabilistic Null Message Count

### 3.1 EIT Update

The EIT of an LP is the minimum of the incoming EOTs from its parents (predecessor LPs). The EIT update starts a new interval till the next EIT update, and the EIT of the LP remains unchanged during the interval.

Let the EIT of an (arbitrary) $LP_i$ at the EIT update be $eit_i$. Let the updated EOT value of an $LP_i$ at the EIT update be $eot_i$. When the EIT is updated, there must be an $LP_{critical} \in SS_i$ which sends a null message to $LP_i$ so that $eot_{critical} = eit_i$. The $LP_{critical}$ is referred to as the critical parent of $LP_i$ for this EIT update, i.e., the $EIT_i$ can be determined when $LP_i$ receives the value $eot_{critical}$ from the critical parent $LP_{critical}$. In order to develop a model of the probabilistic length of the EIT update interval, we can estimate that $eot_i = eit_i + la_i$, which is not true in general, can be a useful approximation. For a given $LP_i$ and its critical parent $LP_{critical}$, the value of the EIT of $LP_i$ can be expressed as follows:

$$eit_i = eot_{critical} = eit_{critical} + la_{critical} .$$

The above expression is recurrent since $eit_{critical}$ can be expressed in the same manner, unless $LP_{critical}$ is the source of the simulation topology. Sources do not have any predecessors and autonomously generate events. This expression holds for every $LP_{critical} \in SS_i$. We next define the concept of the precedence closure (PC) and the precedence critical path (PCP) of an LP. The $PC_i$ of $LP_i$ is the set of all LPs that lie on any path from a source of a simulation to $LP_i$. In other words, $PC_i$ contains every possible path from the sources to $LP_i$. Every LP in $PCP_i$ must be the critical parent of its successor or has no successor ($LP_i$ itself). Thus, $PCP_i$ is a set of LPs in the critical path and in $PC_i$, that is, the set of every critical path of LPs leading to the $LP_i$ from the source LP. If $LP_i$ is in the cycle and the cycle is the critical path, there must be at least one source that autonomously generates event(s) without processing one. Expanding the above expression using $PCP_i$, we get

$$eot_{critical} = eot_{source} + \sum_{LP_k \in PCP_{critical} - \{LP_{source}\}} la_k \,.$$

Note that $eot_{source}$ is specific to a simulation model. It is related to the event generation rate at the simulation source.

### 3.2 Precedence Critical Path of LPs

Except for the simple result in the previous subsection, the critical parent changes from time to time in general. Whenever an LP has a new critical parent, the LP starts a new EIT update. Therefore, we must differentiate each EIT update. Let $eit_i^r$ be the value of EIT at the *r-th* EIT update for $LP_i$. Also, let $PCP_i^r$ be the corresponding precedence critical path for $eit_i^r$. We define $\psi_i^r = [eit_i^{r-1}, eit_i^r]$ as the *r-th* EIT update interval. Note that $eit_i^0 = 0$ for any $LP_i$. Figure 1 shows the progression of EIT updates and selected EIT update intervals.

For a parent being a critical parent of an $LP_i$, $critical(LP_i)$, a parent in a PCP with smaller sum of lookaheads has higher probability to be a critical one. Thus, one of the possible models is:

$\Pr(critical(LP_i) \equiv LP_s \in SS_i) = K_i / \sum_{LP_k \in PCP_s} la_k$

which is the same probability that a path leading to $LP_s$ is the $PCP_s$. We can find the value of constant $K_i$ from the properties of probability as follows:

$$\sum_{LP_s \in SS_i} \Pr(LP_s \equiv \mathrm{critical}(LP_i)) = \sum_{LP_s \in SS_i} \sum_{PCP_s \in PC_i} \frac{K_i}{\sum_{LP_k \in PCP_s} la_k} = 1 \,.$$

$$K_i = \frac{1}{\sum_{LP_s \in SS_i} \sum_{PCP_s \in PC_i} \frac{1}{\sum_{LP_k \in PCP_s} la_k}} \,.$$

Note that the values of lookaheads can be replaced with the simulation time progressed by processing an event if we have exact knowledge of the event processing time, or this probability model can be further developed with the distribution of the event processing time.

### 3.3 Average Size of EIT Update Interval

Let $|\psi_i^n|$ be the size of the interval $\psi_i^n$ for $LP_i$. The EOT from a critical parent determines $|\psi_i^n|$. Under the condition of given two consequent critical parents, $EOT_{critical}$ increase is bounded by $eot_{source}$ and the sum of lookaheads. Of course the two consecutive sources cannot be always unique with multiple sources on the simulation topology. On the contrary, the average difference of the size of the interval can always be calculated from the properties of sources since each source emits events autonomously and independently. With two different sources, the size of EIT update interval is derived as follows:

$$|\psi_i^n| = eot_{critical}^n - eot_{critical}^{n-1} = eot_{source}^n - eot_{source}^{n-1} + \sum_{LP_k \in PCP_{critical}^n - \{LP_{source}^n\}} la_k - \sum_{LP_k \in PCP_{critical}^{n-1} - \{LP_{source}^{n-1}\}} la_k .$$

With the probability of a specific PCP, we can estimate the average length of the EIT update interval for all possible pairs of PCPs.

$$\overline{|\psi_i^n|} = \sum_{PCP_i^n \in PC_i} \sum_{PCP_i^{n-1} \in PC_i} (eot_{source}^n - eot_{source}^{n-1} + \sum_{LP_k \in PCP_i^n} la_k - \sum_{LP_k \in PCP_i^{n-1}} la_k) \times K_i^2 / (\sum_{LP_k \in PCP_i} la_k)^2 . \quad (1)$$

Note that $\overline{|\psi_i^n|}$ does not need to be differentiated by each interval, and is dependent on known factors such as lookaheads of LPs, EOT difference of source(s), and possible PCPs.

When an LP is at simulation time t, the number of EIT update intervals left behind by an $LP_i$ can be drawn from the definition of expectation with estimated $\overline{|\psi_i|}$ from equation (1) above.

$$t = \sum_{k=0}^{n} |\psi_i^k| = NI_i \times \overline{|\psi_i|}, \; NI_i = \frac{t}{\overline{|\psi_i|}} . \quad (2)$$

Note that the values of lookaheads here can also be replaced with the simulation time progressed by processing an event if we have exact knowledge of the event processing time, similarly to the probability model.

### 3.4 Relationship of Two LPs in Terms of EIT Update Interval

Not all of the LPs have the same lookahead or the same processing speed. It implies that every LP has the different size of EIT update intervals even if two LPs have the same PCP. In other words, there is the speed gap between two LPs in terms of simulation time. Figure 2 further illustrates this phenomenon. The critical parent $LP_{critical}$ starts a new interval $\psi_{critical}^r$. Since it has a new EIT to start a new interval, it can calculate its new EOT and send the EOT, say, $eot_{critical}^r$ to the descendents as soon as possible. After the physical time of message transmission, $LP_i$ receives $eot_{critical}^r$, and since the EOT is from the critical parent $LP_{critical}$, it then calculates its $eot_i$ and starts a new interval $\psi_i^n$. The events processed on $LP_{critical}$ can generate and send out events.

From the fact that $eot_{critical}^r = eit_i^n$, we can establish the (probabilistic) relationship between r and n. At a given $eot_{critical}^r$ and average of $|\psi_i|$, from equation (2), we can get $r = \frac{eot_{critical}^r}{|\psi_{critical}|}$ and $n = \frac{eit_i^n}{|\psi_i|} = \frac{eot_{critical}^r}{|\psi_i|}$. Thus, $n = \frac{r \times |\psi_{critical}|}{|\psi_i|}$. This expression shows the logical relationship in the speed, in terms of the number of EIT updates, between the parents and child LPs.

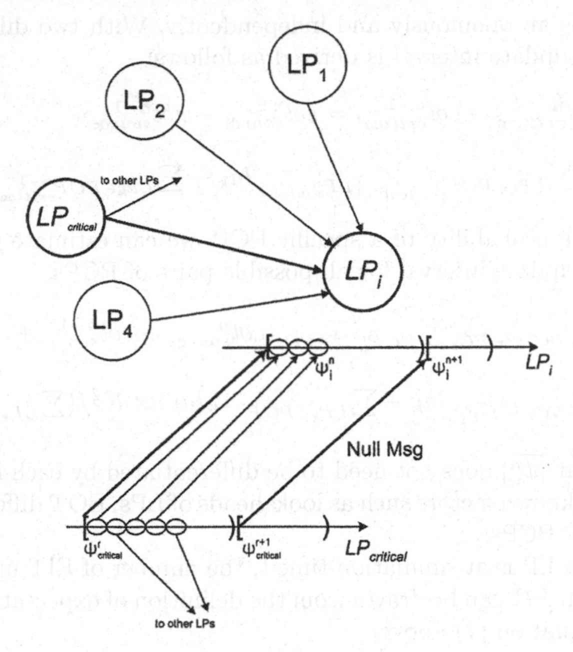


**Fig. 2.** The EIT update of an LP

### 3.5 A Probabilistic Null Message Count

Now, it is simple to determine the null message count of a conservative simulation. According to the definition of our EIT update interval, a null message will be sent to every successor for every EIT update. An LP will send null messages for every EIT update interval. With a given simulation time t, $NI_i = \frac{t}{|\psi_i|}$ denotes the number of EIT update intervals of $LP_i$. Therefore, the count of null messages for an $LP_i$, which is denoted as $CNM_i$, is given as follows:

$$CNM_i = NI_i * |DS_i|$$

where $NI_i$ can be drawn from equation (2). Thus the sum of a null message count for each LP is the total null message count of a simulation. It is clear that the count of null messages of an LP is affected by i) the size of its destination set (also known as connectivity of simulation topology), ii) the total simulation time, iii) the critical path to the LP and iv) the lookaheads of LPs in the PCP.

## 4 Verification

From the aforementioned analytical model, it is clear that the number of null messages for an LP can be drawn by lookaheads of LPs in PCP, the total simulation time, and the connectivity of a simulation. In this section, we will verify our null message count model with the results of our real experiments. The CQNF (Closed Queueing Network with FIFO queues) models with grid topologies will be used. The models have almost all features of real simulation models and virtually every property of these models can be controllable.

Figure 3 shows the CQNF model that is composed of homogeneous LPs in the grid topology, which is similar to the PHOLD model addressed in [9]. We will call this model as CQNF-GRID. Every LP in the CQNF-GRID topology acts as a source at an initial state, sending out 8 events to itself. Whenever it receives an event, it then calculates lookahead and sends the event to its equally likely selected neighbors after specified (inter-departure) time calculated from the exponential distribution. That is, the arrival of events to an LP is Poisson arrival. It can be easily found that there are numerous PCPs for an LP, mostly in the form of two-dimensional random work. We can easily calculate and reduce the possible PCPs from the homogeneity of LP properties and random work. The size of destination set of an LP can be controlled by adding or deleting links to or from other LPs. This connectivity factor can be controlled by using the communicable distance. Suppose that the LPs are mapped on a two-dimensional coordinate system with integer points. For example, an LP can have four neighbors if the communicable distance between two LPs is one, it can have eight neighbors if the distance is 1.5, or it can have twelve neighbors if the distance is two, and so on.

Since its topology shows one of the most general cases, CQNF-GRID is selected to compare between the analytic and experimental results. The experiments are performed on IBM SP2 Nighthawk supercomputer systems with 375MHz POWER3 CPU. It is a fully distributed computing system whose OS is AIX4.3. The simulation system has a strict null message algorithm with the EOT preferential scheduling addressed in [10]. All of the experiments are done using four CPUs. Even though our theory shows that the null message count is in a logical domain and cannot be varied with the number of CPUs used, we utilized four CPUs in our experiments.

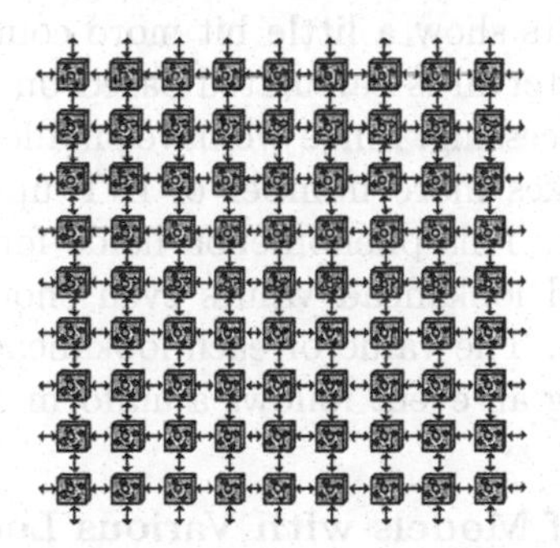

**Fig. 3.** CQNF model with the grid of queues

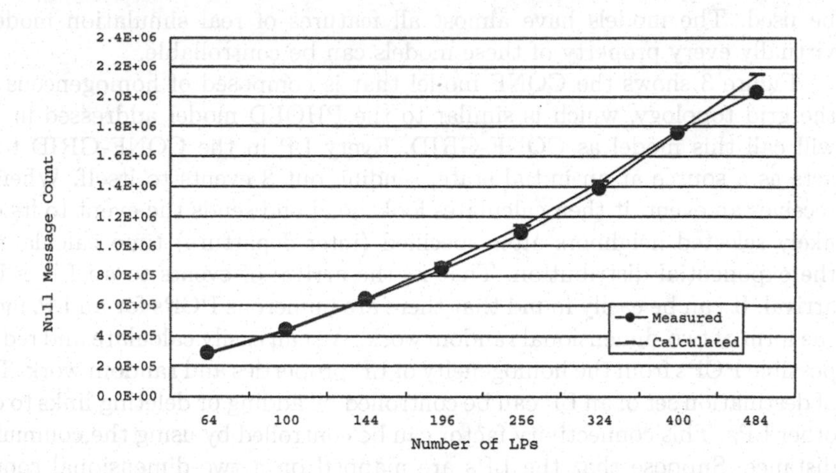


**Fig. 4.** Null message count versus number of LPs

### 4.1 Verification of a Basic Model

With the analytical results, we verified the null message count with CQNF-GRID topology. Figure 4 shows the calculated and measured counts of null messages while the number of LPs varies in the simulation topology. The property of a two-dimensional grid shows that the number of neighbors increases quadratically as the communicable distance increases. We changed the total number of LPs by changing the edge size of a rectangular grid. In this experiment, the communicable distance of every LP is one and thus every LP has four neighbors.

Intuitively, we can predict that the null message count will be proportional to the number of LPs. Figure 4 shows that the number of null messages increases almost linearly as we expected.

The analytic results show a little bit more count of null messages since the size of EIT update interval is calculated based on lookaheads rather than time progress by event processing. Since we have smaller size of EIT update interval than real one, it makes more number of EIT update interval with the same given simulation time. This phenomenon holds for all other results in general. We intentionally used lookahead values even though we have actual value of event processing time. The value of each lookahead is set to 45 while the time progress by processing an event follows a uniform distribution over [45, 47).

### 4.2 Verification of Models with Various Lookahead

Figure 5 shows the case of restricted lookahead. Every LP in this case has restricted portion of perfect lookahead. If we have restricted knowledge to the models, we have only restricted lookahead, which will increase the number of null messages intuitively. In these experiments, the number of null messages increases when we have poor lookahead among all LPs. The restricted lookahead

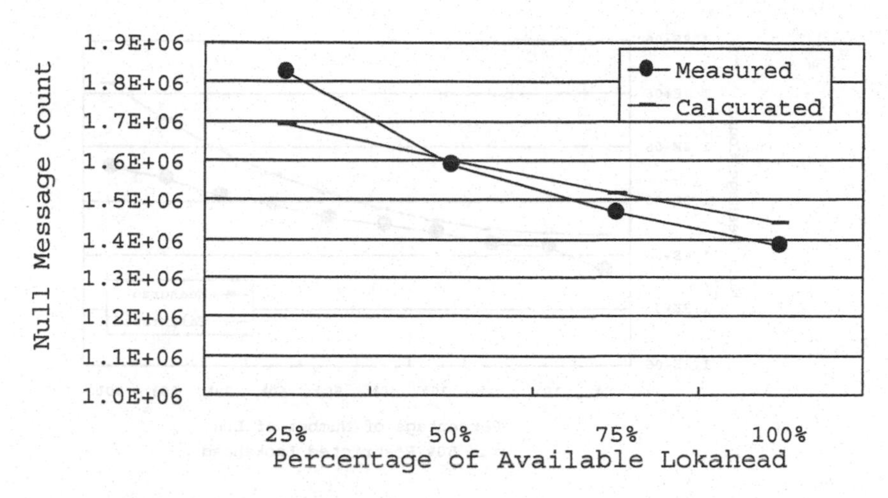


**Fig. 5.** Null message count versus restricted lookahead of LPs

will increase the probability of a predecessor being a critical LP and shorten the length of one EIT update interval. This short size of an EIT update interval increases the number of EIT update intervals for a simulation with constant simulation time and thus increases the null message count. With extreme restriction of lookahead down to 25%, the estimated null message count shows smaller value than the measured one. This phenomenon explains the importance of lookahead for the conservative parallel simulation. We experienced 30% increment of the null message count with 25% of lookahead case comparing to 100% lookahead case while estimation showed 17% increment of the null message count.

Figure 6 shows the case of partial knowledge to the simulation model. This is the common phenomenon when representing the real world. In the model of real world, we can find perfect lookahead for some LPs while we can find only partial lookahead for the most of other LPs. Once we can have LPs that have partial lookaheads only, we will expect larger number of null messages that will degrade the performance of a simulation. LPs with smaller lookahead have a tendency to be a critical parent and make its successor having shorter EIT update interval, and the sum of restricted lookahead on the PCP will highly affect the length of an EIT update interval. Then the total number of EIT update interval will be larger than that of perfect lookahead models. We set the restriction of lookahead to 50% since results of measured and estimated null message count in figure 5 with 50% of partial lookahead shows the best match between theory and phenomena. As we expected, the count of null messages increases as the number of LPs with partial lookahead knowledge increases.

### 4.3 Verification of Models with Various Connectivity

We focused CQNF-GRID model here in order to verify the effect of connectivity between LPs. As mentioned already, LPs in CQNF-GRID can change its connec-

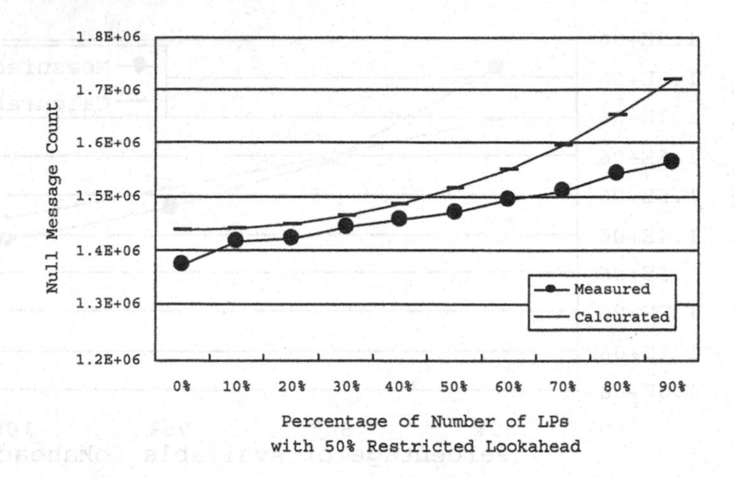


**Fig. 6.** Null message count versus number of LPs with 50% restricted lookahead

tivity by changing the communicable distance between two LPs. With the linear increase of communicable distance, the number of communicable neighbors increases quadratically. The number of LPs in the destination set will increase and therefore the number of PCPs will increase so that we can check the various cases of different PCP sets. Equation (1) shows that the average length of an EIT update interval is determined by two consequent PCPs. Thus the size of an EIT update interval is proportional to the inverse of square of the PCP cardinality and the count of null messages is proportional to the square of the PCP cardinality. The grid nature of CQNF-GRID accurately reflects this fact. Figure 7 shows the count of null messages with various numbers of neighbors. Actually we vary the communicable distance ranging from 1 to 5 and thus calculate the number of neighbors. As we expected, the increment of neighbors leads to the increment of PCP cardinality. Thus the count of null messages shows quadratic increase. As a result, even the slight increase of connectivity will increase the null message count drastically for a conservative parallel simulation.

## 5 Conclusion

We developed a probabilistic model of a null message count for the simulations with a conservative parallel null message algorithm. The null message count has been regarded as a dominant overhead factor for the simulations based on the null message algorithm in conservative parallel simulation systems. Our motivation is from the fact that the null message count is totally dependent on the logical factors of a simulation model and the null message algorithm.

Starting from the concept of a critical parent, we developed the concept of a closure of critical paths. With the approximation of EOT using lookahead of each LP, a probability model for a critical path was developed. The time-to-time

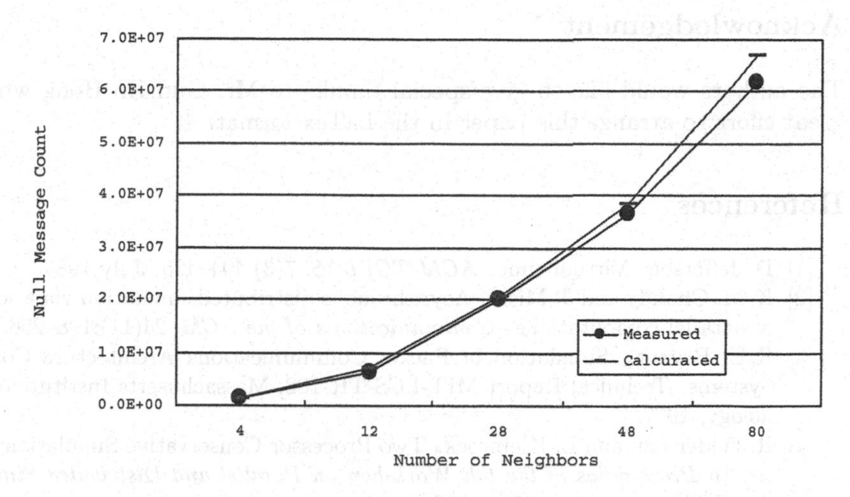


**Fig. 7.** Null message count versus number of neighbors

variation of a critical path introduced the concept of the EIT update interval thus leading to the stochastic model for a length of an EIT update interval. From the logical factors such as the number of LPs in the simulation topology, given simulation time, the cardinality of the destination set of each LP and the length of the EIT update interval, we was able to estimate the count of null messages.

There could be difficulties in applying this model in case of a huge number of PCPs in a simulation topology. However, with the grid topology which has a concept of two dimensional random walk and is one of the most general topologies, we verified that our null message count model is effective enough to predict the null message count of a simulation model with reasonable gap between measured counts and estimated counts. We used exponential distribution for the logical time increment by processing an event. It is beyond our study to generalize the probability distributions of time increment in processing events, with which we will make this null message count model more robust. In case of CQNF-GRID topology, the event time progression shows a little but definite difference between measured and estimated results.

Our study may provide a guideline to simulation model developers to enhance the performance of a parallel simulation. With the null message count, model developers can estimate the number of null messages in their own model and might revise their model to reduce the number of null messages so that they can achieve better performance for their simulation. As shown with experimental results and our null message count model, simulation model designers must take care of simulation model factors such as poor lookahead and dense simulation topology. The more the number of LPs is, the higher the parallelism will be, however high connectivity should be avoided since the performance of simulation with the null message algorithm may degrade drastically.

## Acknowledgement

The authors would like to give special thanks to Mr. Gun Ho Hong who paid great effort to arrange this paper in the LaTex format.

## References

[1] D. Jefferson. Virtual time. *ACM TOPLAS*, 7(3):404–425, July 1985.

[2] K. M. Chandy and J. Misra. Asynchronous distributed simulation via a sequence of parallel computations. *Communications of the ACM*, 24(11):198–206, 1981.

[3] R. E. Bryant. Simulation of Packet Communications Architecture Computer Systems. Technical Report MIT-LCS-TR-188, Massachusetts Institute of Technology, 1977.

[4] R. Felderman and L. Kleinrock. Two Processor Conservative Simulation Analysis. In *Proceedings of the 6th Workshop on Parallel and Distributed Simulation (PADS92)*, pages 169–177, 1992.

[5] David M. Nicol. The cost of conservative synchronization in parallel discrete event simulations. *Journal of the ACM, 1967-*, 40, 1993.

[6] M. L. Bailey and Pagels M. A. Empirical measurements of overheads in conservative asynchronous simulations. *ACM Transactions on Modeling and Computer Simulation*, 4(4):350–367, 1994.

[7] Ha Yoon Song. A probabilistic performance model for conservative simulation protocol. In *Proceedings of the 15th Workshop on Parallel and Distributed Simulation (PADS2001)*, Lake Arrowhead, CA, USA, 2001.

[8] Rajive Bagrodia, Richard Meyer, Mineo Takai, Yu an Chen, Xiang Zeng, Jay Martin, and Ha Yoon Song. Parsec: A parallel simulation environment for complex systems. *Computer*, 31(10):77–85, October 1998.

[9] R. M. Fujimoto. Performance of time warp under synthetic workloads. In *Proceedings of the SCS Multiconference on Distributed Simulation*, San Diago, CA, USA, 1990.

[10] H. Y. Song, R. A. Meyer, and R. L. Bagrodia. An empirical study of conservative scheduling. In *Proceedings of the 14th Workshop on Parallel and Distributed Simulation (PADS2000)*, Bologna, Italy, 2000.

# Chapter 7

# Imaging

# Static Scheduling with Interruption Costs for Computer Vision Applications

Francisco A. Candelas, Fernando Torres, Pablo Gil, and Santiago T. Puente

Physics, Systems Engineering and Signal Theory Department
University of Alicante, Spain
{fcandela,ftorres,pgil,spuente}@disc.ua.es

**Abstract.** It is very difficult to find pre-emptive scheduling algorithms that consider all the main characteristics of computer vision systems. Moreover, there is no generic algorithm that considers interruption costs for such systems. Taking the interruption of tasks into account scheduling results can be improved. But it is also very important to take the costs that arise from interruptions into account because they not only increase the total execution time, but also because the scheduler can evaluate whether it is adequate to interrupt certain tasks or not. Thus, the result can be more realistic. Therefore, we present an extension to the static algorithm SASEPA for computer vision which considers interruption costs.

## 1 Introduction

Of the many specific characteristics of computer vision applications regarding scheduling, we would like to emphasize the following:

1. Multi-processor systems with different kinds of processors are generally employed. Thus, techniques for spatial allocation and temporal scheduling which consider it are needed. It is possible to differentiate between generic *CPUs* and *IAPBs* (Image Acquiring and Processing Boards). This classification involves three basic kinds of tasks: *cpu*, *iapb* and *cpu/iapb*. The last one is a communication task between processors of different kinds [1][2].
2. There are precedence and exclusion relations among tasks. The first ones are determined by the executing order and the data flow between the operations. The second ones establish what tasks can be interrupted by others and which cannot be, and they are considered in pre-emptive scheduling techniques [3][4].
3. The elementary tasks are generally sporadic, and their creation times depends on precedence relations.

Taking task interruptions into account, allow more flexibility in distributing tasks in scheduling [3]. This can improve the results since the execution time of the processors is better used and the parallelism of the execution is greater [4][5].

J.M.L.M. Palma et al. (Eds.): VECPAR 2002, LNCS 2565, pp. 509–522, 2003.

Though many scheduling strategies are been developed and implemented, few are directly designed for computer vision applications. Moreover the majority are designed for specific architectures and they do not take all of the important characteristics of such applications into account [1][2]. As an example, we can consider techniques such as the PREC 1 [6], the Empty-Slots Method [7] or the Critical Path [1][2]. The first one takes the precedence relations, sporadic tasks and interruptions into account, but it does not consider different kinds of processors. The second one considers the different characteristics, but it works with sporadic tasks and is designed for RT-LANs. The third takes all of the characteristics mentioned above into account, but it does not make spatial allocation for a multi-processor system. Also, there are scheduling strategies for computer vision as the one proposed in [8] and [9], but this does not consider interruptions or different kinds of processors.

References [10] and [2] describe the static algorithm SASEPA (Simultaneous Allocation and Scheduling with Exclusion and Precedence Relations Algorithm), which carries out a spatial allocation and a temporal scheduling over a multi-processor system considering all of the above-mentioned characteristics for computer vision systems. This algorithm also does a pre-emptive scheduling and considers the task interruptions to make the resulting scheduling better. But it does not take the temporal costs derived from interruptions into account. Because it is static, it is suitable for the research and design steps of an computer vision application [3].

In this paper, a SASEPA extension that considers interruption costs is proposed. It represents a new approach in the scheduling algorithms for computer vision applications which can be also applied to other systems. After describing the basic aspects of SASEPA below, Section 2 explains how interruptions and their costs are modeled. Next, Section 3 describes how interruption costs are considered in the SASEPA extension. The practical evaluation of the proposed extension is described in Section 4. Finally, our conclusions are presented in Section 5.

### 1.1 Basic Aspects of SASEPA

Each high-level operation of a computer vision application can be divided into a set of elementary tasks. The scheduling algorithm takes a DAG (Directed Acyclic Graph) which contains the attributes of the tasks and the relations between them as an input [6]:

$$\begin{gathered} \mathrm{G} = (\mathrm{T}, \mathrm{A}). \\ \mathrm{T} = \{\tau_1, \tau_2, \ldots, \tau_N\}, \quad \tau_i = (\mathrm{c_i}, \mathrm{k_i}, \mathrm{IntCost_i}), \quad \mathrm{i} = 1, 2, \ldots, \mathrm{N}. \\ \mathrm{A} = \{(\tau_\mathrm{i}, \tau_\mathrm{j}) / \tau_\mathrm{i} \text{ precedes } \tau_\mathrm{j}\}, \quad \mathrm{i}, \mathrm{j} = 1, 2, \ldots, \mathrm{N}. \end{gathered} \tag{1}$$

Each task has its computation time $c_i$ and its kind $k_i$ (which can be *cpu*, *iapb* or *cpu/iapb*) associated. It also has a set o function $IntCost_i$ that gives interruption costs. The creation time of each task is not explicitly specified and it is determined by means of the precedence relations. Nor are the deadlines considered.

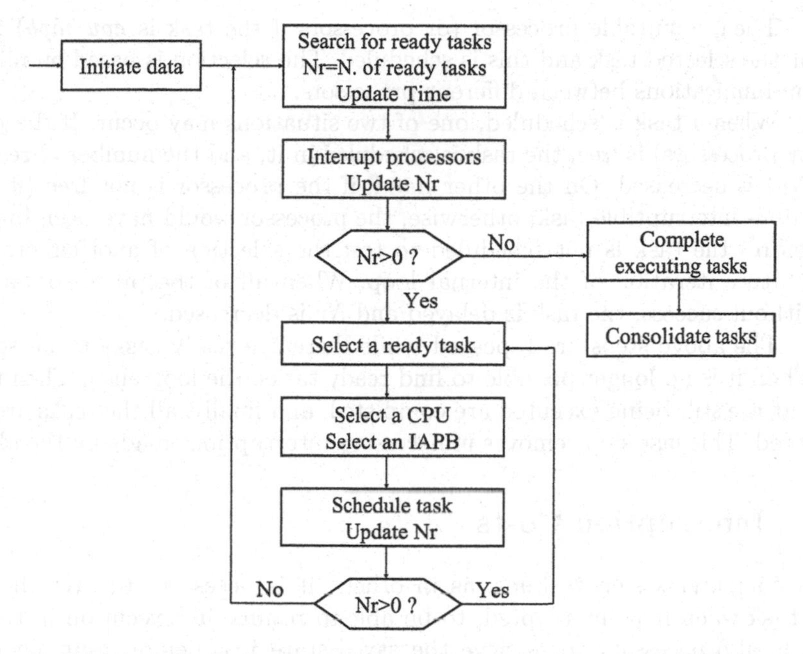


**Fig. 1.** Block diagram of the main procedure of SASEPA

The result of the algorithm is the spatial allocation and the temporal scheduling of tasks in the available processors (CPU and IAPB) taking the kinds of tasks and the relations between them into account. The algorithm also minimizes the total execution time, making the necessary interruptions of tasks to do so.

Fig. 1 shows the main steps of the SASEPA. After initiating the algorithm, ready tasks are searched for among the unfinished tasks. These are tasks that have all their preceding tasks finished and the minimum creation time. The creation time of a task depends on the finishing time of its preceding tasks and the accumulated delay due to interruptions and deferments of the task due to a lack of free processors.

Next, all the processors executing tasks that can be interrupted are interrupted, and corresponding tasks also become ready tasks. The interruption costs are not considered.

If there are ready tasks ($Nr > 0$), one is selected to be scheduled. The criterion for selecting a task is based on finding the *critical task* first, which is the task with a maximum finishing time. Then a ready task that is a predecessor of the critical task is selected. In order to resolve ties among several tasks, a weight is associated to each one, and the task with the highest weight is selected. The weight of a task expresses the current computation time required to execute it and all of its successors with the maximum parallelism. Thus, the selected task is the one that delays the total execution most.

Then, a suitable processor (or processors if the task is *cpu/iapb*) is chosen for the selected task and this is scheduled. The selection is based on minimizing communications between different processors.

When a task is scheduled, one of two situations may occur. If the processor (or processors) is free, the task is scheduled in it, and the number of ready tasks (*Nr*) is decreased. On the other hand, if the processor is not free (it executes a non- interruptible task; otherwise, the processor would have been interrupted before) the task is not scheduled to test the selection of another processor in a future iteration of the internal loop. When all of the processors are tested without success, the task is delayed and *Nr* is decreased.

The above steps are repeated while there are ready tasks to be scheduled. When it is no longer possible to find ready tasks, the loop ends. Then the tasks that are still being executed are completed, and finally, all the tasks are consolidated. This last step removes unnecessary interruptions made by the algorithm.

## 2 Interruption Costs

In computer vision systems, as in others, it is necessary to save the state of a task when it is interrupted, to be able to resume its execution in the future. It is also necessary to retrieve the saved state just before resuming the task. These operations, which are called context switching, involve time costs which may become important if many interrupts are generated. Because of this, it is advisable to bear these costs in mind. Moreover, if the scheduler takes the interruption costs into account it can evaluate whether it is suitable to make an interruption or not.

Fig. 2 shows how the interruptions affect any task $\tau_i$. Due to the interruptions, the execution of a task may be broken down into several intervals of time. If a generic interval $j$ is considered, with $j \in \{1, 2, \ldots, \mathrm{M}\}$, a reading cost $r_{i,j-1}$ is required to retrieve the original state of the task at the beginning of the interval. The cost $r_{i,j-1}$ depends on the previous interruption. Furthermore, a writing cost $w_{i,j}$ is required to save its the state at the end of the interval. For the first interval of a task $r_{i,0}$ is 0, and for the last interval $w_{i,M}$ is 0.

Because of the interruption costs, the effective time spent in computing a task ($ec_{i,j}$) is shorter than the duration of the interval ($d_{i,j}$), and the finishing time of the task is postponed. Thus, if $c'_i$ is defined as the remaining time of computation of the task $\tau_i$, this value is increased in each interruption of an interval $j$ according to $w_{i,j} + r_{i,j}$:

$$\begin{aligned} \mathrm{c}'_{\mathrm{i},0} &= \mathrm{c}_{\mathrm{i}}. \\ \mathrm{c}'_{\mathrm{i},\mathrm{j}} &= \mathrm{c}'_{\mathrm{i},\mathrm{j}-1} - \mathrm{d}_{\mathrm{i},\mathrm{j}} + \mathrm{w}_{\mathrm{i},\mathrm{j}} + \mathrm{r}_{\mathrm{i},\mathrm{j}} \quad \mathrm{j} = 1, 2, \ldots, \mathrm{M}. \end{aligned} \tag{2}$$

It is noteworthy that the writing of the state is performed within the corresponding interval before the instant of interruption. That is, given a desired instant of interruption $t_int$, the writing of state $w_{i,j}$ is considered just before this instant, beginning at $t_w$. In this way the scheduler can get the desired length of time $d_{i,j}$ for the interval. This approach simplifies the interruption management.

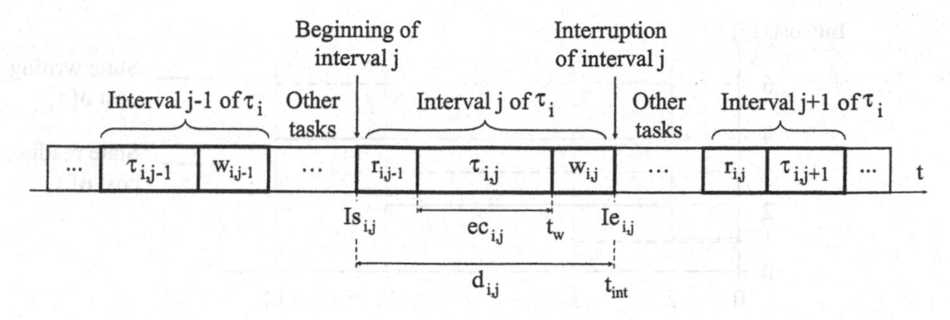


**Fig. 2.** A task broken down into several execution intervals, due to the interruptions

However, it is necessary that the scheduler algorithm is static to be able to carry it out.

To simplify the modeling and the management of interruptions by the scheduler, the costs of writing and reading the state can be considered constant for each task $\tau_i$:

$$\begin{aligned} \mathrm{w}_{\mathrm{i},1} = \mathrm{w}_{\mathrm{i},2} = \ldots = \mathrm{w}_{\mathrm{i}}. \\ \mathrm{r}_{\mathrm{i},1} = \mathrm{r}_{\mathrm{i},2} = \ldots = \mathrm{r}_{\mathrm{i}}. \end{aligned} \tag{3}$$

## 2.1 Interruption Cost Function

However, costs for writing and restoring the state of a task are not constant, but depend on the instant of time $t_w$ in which the task interruption begins. This instant is measured relative to the effective computation time of the task. For example, let us consider an operation for computer vision that searches for some characteristics of an image and processes them all at the end. The more advanced the operation is, the more information about characteristics detected will have to be saved temporarily in case of interruption.

Thus, a more realistic but more complex model is considering a function for each task that returns the writing and reading costs for it. The parameter of these functions is the instant in which the interruption begins in relation to the effective computation time of the task:

$$(\mathrm{w_i}, \mathrm{r_i}) = \mathrm{IntCost_i}(\mathrm{t_w}), \qquad \mathrm{t_w} \in [0, \mathrm{c_i}). \tag{4}$$

This cost function can be defined by the different intervals of time that involve different writing and reading costs. As an example, let us consider the function that Fig. 3 shows, which can be expressed in this way:

$$(\mathrm{w_i}, \mathrm{r_i}) = \mathrm{IntCost_i}(\mathrm{t_w}) = \begin{cases} (1,1)\ 0 \le t_w < 2 \\ (2,1)\ 2 \le t_w < 5 \\ (2,3)\ 5 \le t_w < 11 \\ (4,6)\ 11 \le t_w < 14 \end{cases}. \tag{5}$$

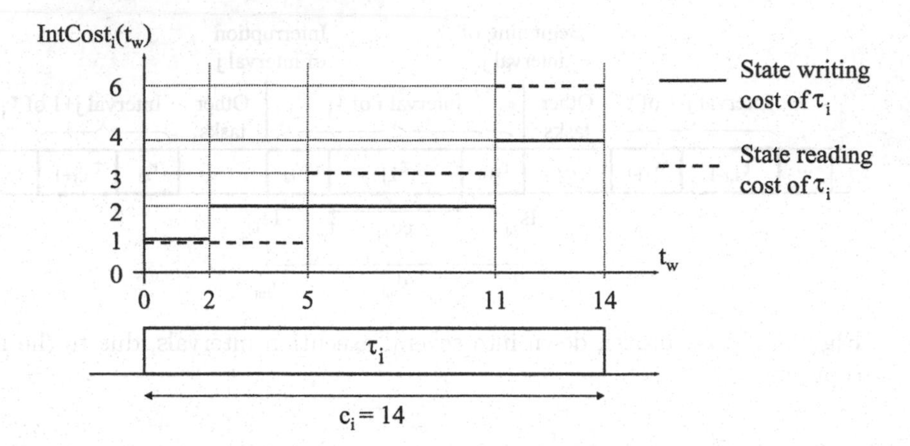


**Fig. 3.** Example of a cost function for a task $\tau_i$

## 2.2 Considerations about the Interruption Cost Function

When the cost function is used to determine interruption costs in a static scheduling, two problematic situations may arise. To illustrate the first situation, let us suppose that the scheduler needs to interrupt the first interval of a task $\tau_i$ in instant $t_{int} = 10$. The task has the following cost function associated:

$$\mathrm{IntCost_i}\left(\mathrm{t_w}\right) = \begin{cases} (9,3)\ 0 \le t_w < 6 & (2,2)\ 8 \le t_w < 9 \\ (4,3)\ 6 \le t_w < 7 & (3,4)\ 9 \le t_w < 10 \\ (3,4)\ 7 \le t_w < 8 & (4,3)\ 10 \le t_w < 12 \end{cases} . \tag{6}$$

For the sake of simplicity, the interval starts at instant 0 of time. The scheduler disposes of the following options to carry out that interruption: to initiate the state writing at $t_w = 8$ which involves a writing cost $w_{i,1} = 2$; to initiate state writing at $t_w = 7$ with $w_{i,1} = 3$; or to consider $t_w = 6$ with $w_{i,1} = 4$. These cases are illustrated in Fig 4.

In the previous example, the best option is *A* since it maximizes the effective computation time of the task for the interrupted interval.

Now let us consider a new cost function for $\tau_i$:

$$(\mathrm{w_i}, \mathrm{r_i}) = \mathrm{IntCost_i}(\mathrm{t_w}) = \begin{cases} (4,3)\ 0 \le t_w < 5 \\ (5,4)\ 5 \le t_w < 8 \\ (3,2)\ 8 \le t_w < 12 \end{cases} . \tag{7}$$

In this case, if the scheduler wants to interrupt the first interval at $t_w = 10$ there are no possible options to finish the interval at that precise instant. The best option is to begin the writing at $t_w = 4$ and finish the interval at $t_{int} = 8$ as shown in Fig. 5.

The criterion that the scheduler must apply to solve the former situations when it needs to interrupt a task is not just to determinate the instant $t_w$ which

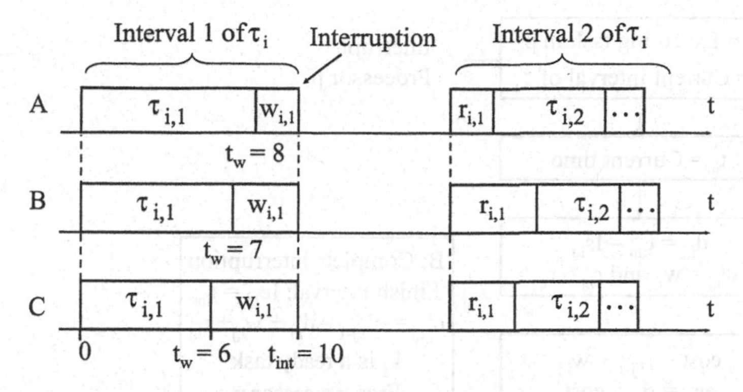


**Fig. 4.** Task with three interruption options for instant $t_{int} = 10$

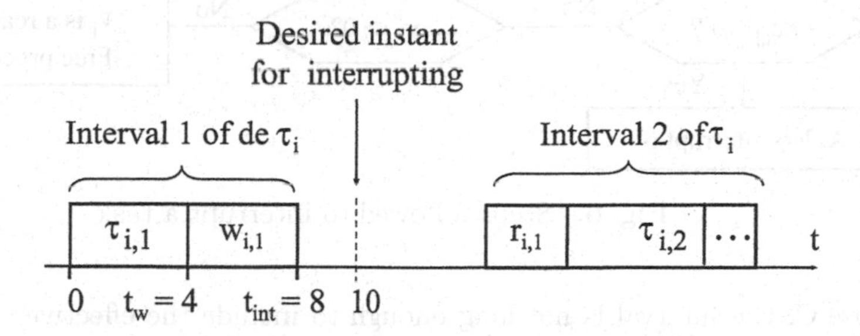


**Fig. 5.** Interval which can not be interrupted at the desired instant

involves a interval that finishes before the desired instant of interruption, but also to maximize the effective computation time for the task.

## 3 SASEPA with Interruption Costs

We have developed an extension of the SASEPA algorithm explained in Section 1.1. This extension considers the aspects related to the task interruptions that were described in Section 2. As described in that section, the interruption costs must be considered when a task is being scheduled or interrupted. Thus, these two operations will be the next procedures to be described.

### 3.1 Interruption of a Task

Fig. 6 shows the steps required to interrupt a processor, considering that interruption costs are constant, as (3) expresses. Three different situations can be distinguished depending on the duration of the interval that has been interrupted in relation to the costs of the interruptions. In case A, the interval is just long enough to include the costs of the interruption, but not the effective computation time. In such a case, the execution of the task is allowed to continue.

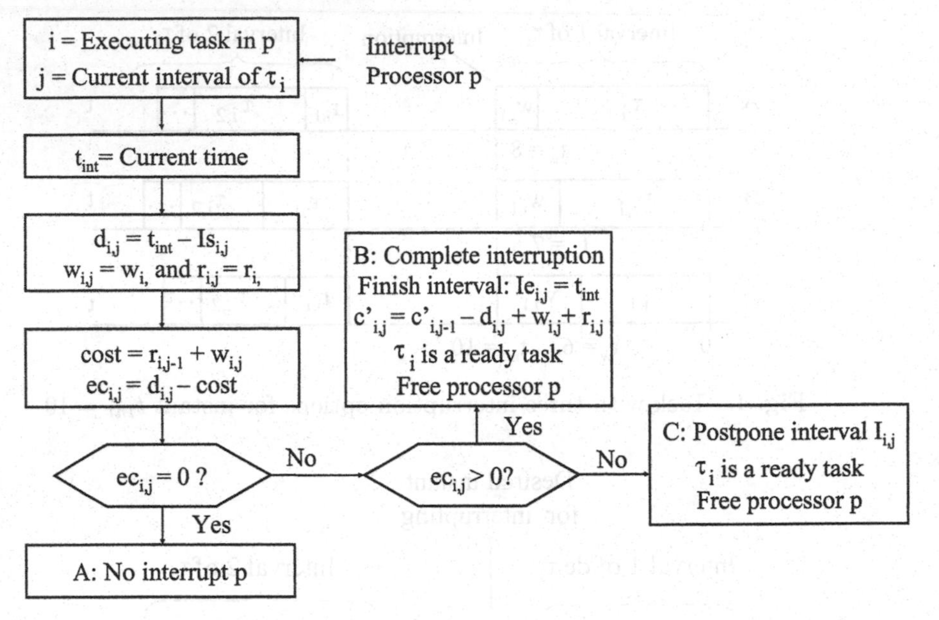


**Fig. 6.** Steps followed to interrupt a task

In case C, the interval is not long enough to include the effective computation of the task, and so it is temporarily postponed.

The same steps are followed in quite a similar way to consider the cost functions where they are specified. The only difference is that now it is necessary to calculate the instant $t_w$ by means of the cost function of a task before calculating $t_{int}$. If $t_w$ is found, then $r_{i,j}$, $w_{i,j}$, $t_{int}$ and $d_{i,j}$ are determined from it. On the other hand, if $t_w$ is not found, the interval is postponed. Furthermore, it is necessary to time the effective duration of the computation of each task to be able to apply the cost functions.

### 3.2 Scheduling a Task

The procedure shown in Fig. 7 is followed to schedule a pre-emptable task, considering constant interruption costs. If the assigned processor is free then one of two basic situations can occur: either the previous interval can be continued or it is necessary to start a new interval, depending on how and where the previous interval of the task was finished.

If the previous interval was interrupted in the same processor just before the instant which is being scheduled, it is possible to continue that interval (case A). In this case, the interruption costs that were considered before must be subtracted from $c_{i,j}$. If the previous interval was postponed in the same processor it is also possible to continue it, to achieve a longer interval (case C). In this case, it is not necessary to subtract the interruption costs, since they were not considered before. In other cases, a new interval must be considered.

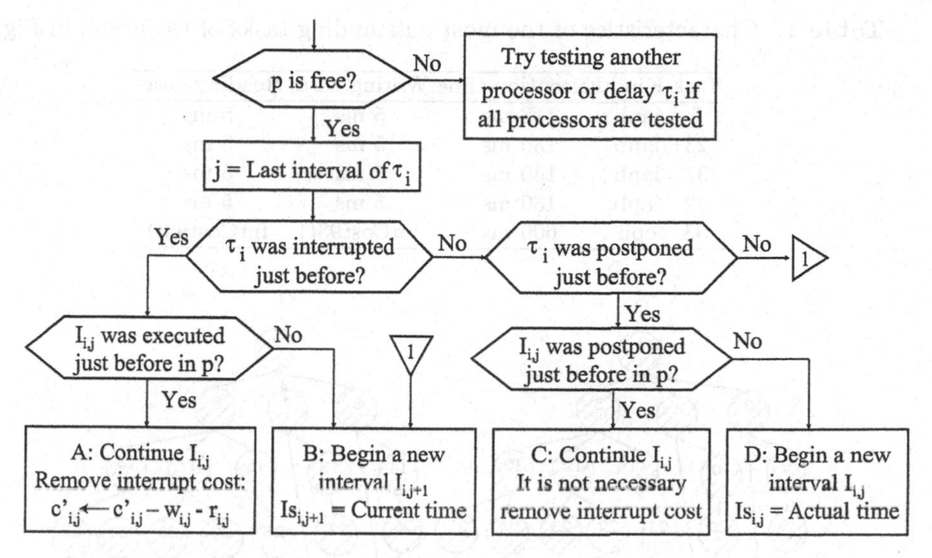


**Fig. 7.** Step followed to schedule (begin or continue) a task

The same steps are considered to take the cost functions into account during scheduling, but adding a new feature: it must be possible to continue an interval which has been interrupted before the desired instant.

## 4 Evaluation

To evaluate the proposed SASEPA extension that considers interruption costs, a real computer vision application has been considered; a correspondence algorithm for the characteristics of two images captured with a pair of stereoscopic cameras.

The first step was to define the tasks and to estimate their characteristics, including state writing and reading costs for each task. We should point out that the developed extension can manage both constant costs and function costs models for each task, and that the two models can be used in the very execution of the algorithm. The cost function has only been defined for the more complex tasks.

Afterwards, the application was specified as a high-level scheme using the tools described in [2] and [11]. These tools also generate the task graph that has been used as the input for the static scheduling algorithms tested. Fig 3. shows this DAG. Table 1 shows the main characteristics of the most outstanding tasks of the task graph which will be discussed later on.

The tasks have been scheduled using four different scheduling algorithms considering a target architecture with a CPU and two IAPBs. The four algorithms were the PREC 1 [6], the Critic Path [1], the SASEPA [10] and the SASEPA extension with interruption costs. In order to apply the two first algorithms for the

**Table 1.** Characteristics of the most outstanding tasks of the graph in Fig. 8

| Task | Kind | Execution time | Writing costs | Reading costs |
|---|---|---|---|---|
| 5 | iapb | 118 ms | 5 ms | 5 ms |
| 23 | iapb | 130 ms | 5 ms | 5 ms |
| 31 | iapb | 130 ms | 5 ms | 5 ms |
| 32 | iapb | 130 ms | 5 ms | 5 ms |
| 93 | cpu | 600 ms | IntCost93() | IntCost93() |

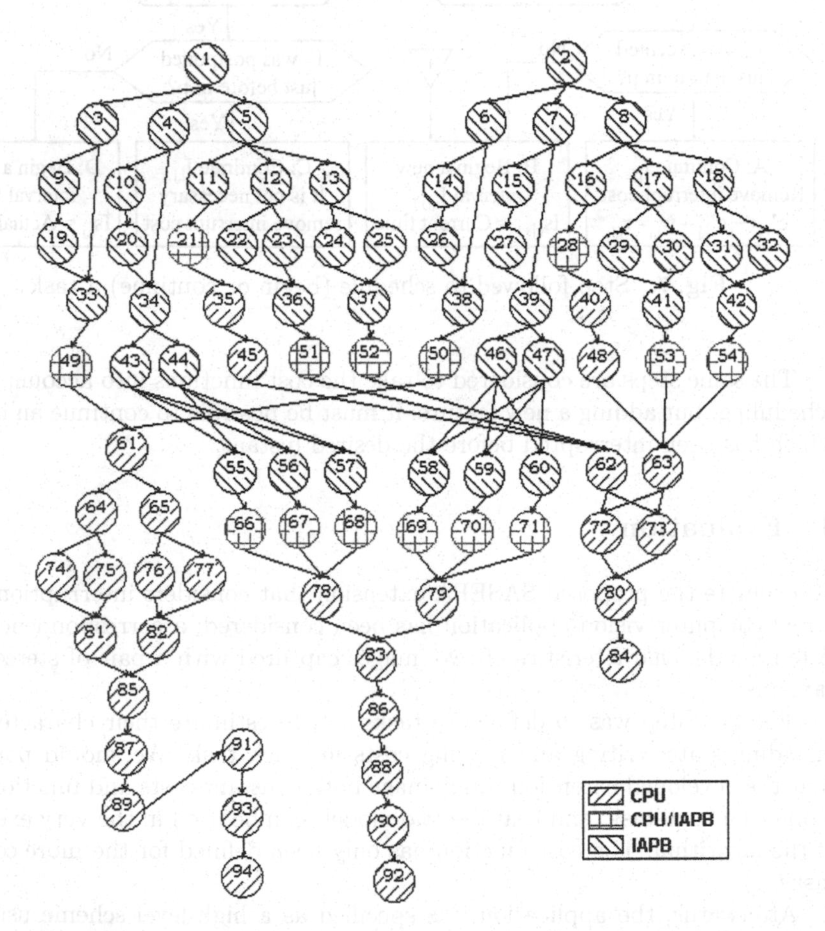


**Fig. 8.** Tasks graph used to evaluate the scheduling algorithms

target architecture it was necessary to improve them with several new features (kinds of tasks, sporadic tasks, spatial allocation...).

The main results for the four algorithm are shown in Table 2. We can see that the SASEPA executes all of the tasks in less time and with a higher processor occupation than the PREC 1 and the Critic Path. Moreover, the SASEPA makes

**Table 2.** Result of scheduler algorithms for the graph in Fig. 8

| Scheduler | Execution time | Processor occupation | Number of interruptions | Interrupted tasks |
|---|---|---|---|---|
| PREC1 - M | 2494 ms | 56% | 0 | - |
| Critic Path - M | 1774 ms | 79% | 8 | 43, 46, 83 (3), 93 (3) |
| SASEPA | 1766 ms | 79% | 5 | 5, 31 (2), 32 (2) |
| SASEPA with int. costs | 1838 ms | 77% | 5 | 23, 93 (4) |

fewer interruptions. Regarding the SASEPA extension, it takes more time to execute all of the tasks and decreases the processor occupation. This it is logical because it takes the interruption costs into account. Furthermore, the SASEPA extension interrupts different tasks than the previous algorithm. This shows how this algorithm considers interruption costs to decide what tasks it can interrupt. This important aspect is explained in more detail below.

The resulting scheduling of the SASEPA is shown in Fig. 9, and Table 3 details the intervals into which the interrupted tasks are broke down. We can verify that intervals $I_{5,1}$, $I_{31,1}$ y $I_{32,2}$ are not long enough to be able to execute the state reading and writing in accordance with the values shown in Table 1. Some intervals are even just one or two milliseconds long, in contrast with the total duration of over a hundred milliseconds of the task. As such, they are invalid intervals for an implementation in practice.

In contrast, Fig. 10 shows how the resulting scheduling, when the SASEPA extension is considered, is different from the result shown in Fig. 9, previously commented. This is because this algorithm has decided to interrupt other tasks, which have been broken down into the intervals detailed in Table 4. In this case, the intervals are sufficiently long to execute the context switching, in addition to a portion of the task. Thus, it is possible to implement the resulting scheduling in practice. Table 4 also shows how the durations of the tasks are increased by including the interruption costs.

It must be remembered that the scheduling algorithm employed is static, and the scheduling of the tasks is done in an off-line manner before they are executed.

**Table 3.** Intervals of tasks interrupted by SASEPA

| Task | Intervals (ms) | Length of intervals (ms) | Total length (ms) |
|---|---|---|---|
| 5 | (182,184), (186,302) | 2, 116 | 118 |
| 31 | (96, 1083), (1084,1092), (1100,1101) | 121, 8, 1 | 130 |
| 32 | (96, 1084), (1085,1092), (1096,1097) | 123, 7, 1 | 130 |

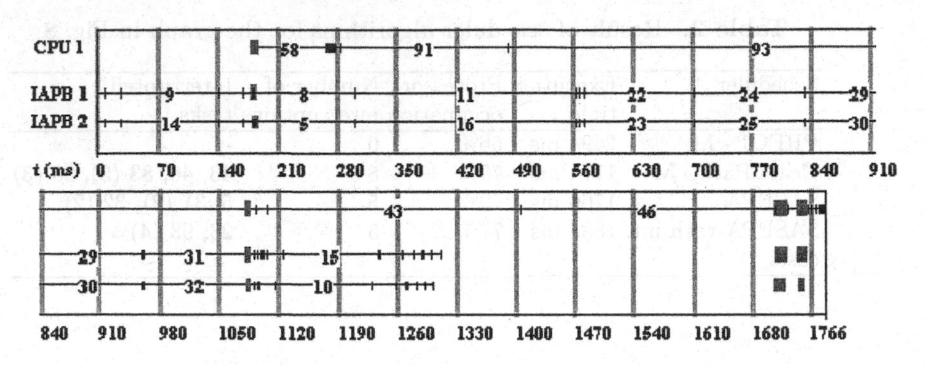


**Fig. 9.** Resulting scheduling of SASEPA

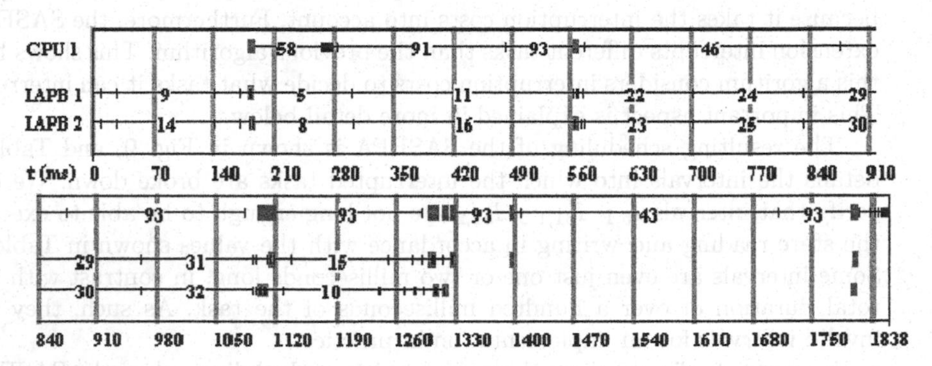


**Fig. 10.** Resulting scheduling of SASEPA with interruption costs extension

Even the system in which the scheduler is executed can be different from the target system. As such, the cost of scheduling does not influence the final execution. Furthermore, the algorithm has been originally designed for computer vision systems in which the static scheduling is done in the first stages of the design, and in this case, the costs in time and space are not much more important than other aspects like the friendliness that the interface of the design tools should offer. As such, we have not considered a detailed study of the costs or a comparison with other algorithms necessary. In any case, other static scheduling algorithms with the same features as the one we propose here do not exist and, as such, a direct comparison would not be useful.

However, we have verified, in practice, that the SASEPA static scheduler with the model presented here has a lower time cost. So much so, that a dynamic version of the scheduler is now being studied. For example, with an implementation of the algorithm for MS Windows 95 being used in a Pentium III of 450MHz PC, a complete scheduling of the computer vision application mentioned in Section 4 can be done in 170ms.

**Table 4.** Intervals of tasks interrupted by SASEPA with interruption costs

| Task | Intervals (ms) | Length of intervals (ms) | Total length (ms) |
|---|---|---|---|
| 23 | (575,702), (1096,1109) | 127, 13 | 130+10 |
| 93 | (483,562), (875,1083), (1117,1290), (1326,1396), (1707,1797) | 479, 208, 173, 70, 90 | 600+420 |

## 5 Conclusions

In this paper we have presented a new model which allows us to consider the costs involved in reading and writing the state, derived from interruptions made by a static scheduling algorithm. In this particular case, a SASEPA extension for computer vision applications has been considered. However, the model can be applied to other pre-emptable static scheduling algorithms.

Interruption costs can be modeled as constant values or as a cost function for each task. Although the second approach is more realistic, it is usually difficult to estimate such functions for all tasks in practice, and it is easier to considerer constant costs. However, the extension developed allows us to use the two models simultaneously, choosing the most suitable one to express the costs over tasks, according to the characteristics of the tasks.

The proposed model is interesting for other applications that use static and pre-emptive scheduling, since it offers a more realistic result, as it does not create any interruptions that are impossible to carry out later on in practice. However, to be able to apply the model directly, the costs of task interruptions must be known or estimated some how. This way, the model is useful in applications whose task are well defined before their scheduling, such as computer vision applications like the industrial inspection of products, in which the characteristics of the tasks to be done and the images to be processed are known in advance. In other words, the task algorithms and how their execution depends on the images that they process are known. This way, the costs of storing and retrieving the state of the tasks can be estimated, even in relation to the part of the task that has already been carried out at a given moment.

When interruption costs are considered in a pre-emptable static scheduling algorithm, not only is a more realistic and generally longer execution time obtained, but also the tasks can be scheduled in a more intelligent way. In other words, the scheduler can avoid interruptions that can not be implemented in practice or that are not the most appropriate.

We should point out that it is not necessary to know the exact costs of interruption to enjoy the advantages of the model proposed here. By estimating the values of the costs of interruption concerning the execution of the task and taking them into account in the scheduling, a certain "intelligence" can be afforded to the pre-emptive scheduler so that it can decide whether it is more convenient to interrupt a certain task at a given moment or not.

## References


[1] Torres, F., Candelas, F. A., Puente, S. T., Jim'enez, L. M. et al.: Simulation and Scheduling of Real-Time Computer Vision Algorithms. Lecture Notes In Computer Science, Vol. 1542. Springer-Verlag, Germany (1999) 98-114

[2] Candelas, F. A.: Extensi'on de T'ecnicas de Planificaci'on Espacio-Temporal a Sistemas de Visi'on por Computador. P. D. Thesis. University of Alicante, Spain (2001)

[3] Nissanke, N.: Realtime Systems. Prentice Hall Europe, Hertfordshire (1997)

[4] Zhao. W, Ramamritham, K., Stankovic, J. A.: Preemptive scheduling under time and resource constrains. IEEE Transactions on Computers, Vol. 36 (1987) 949-960

[5] Xu, J., Parnas, D. L.: On satisfying timing constrains in hard-real-time systems. IEEE Transactions on Software Engineering, Vol. 19 (1993) 74-80

[6] Krishna, C. M., Shin, K. G. : Real-Time Systems. McGraw-Hill (1997)

[7] Santos, J., Ferro, E., Orozco, J., Cayssials, R.: A Heuristic Approach to the Multitask-Multiprocessor Assignment Problem using Empty-Slots Method and Rate Monotonic Scheduling. Real-time Systems, Vol. 13. Kluwer Academic Publishers, Boston (1997) 167-199

[8] Lee, C., Wang, Y.-F., Yang, T.: Static Global Scheduling for Optimal Computer Vision and Image Processing Operations on Distributed-Memory Multiprocessors. Technical Report TRC94-23, University of California, Santa Barbara, California (1994)

[9] Lee, C., Wang, Y.-F., Yang, T.: Global Optimization for Mapping Parallel Image Processing Task on Distributed Memory Machines. Journal of Parallel & Distributed Computing, Vol. 45. Academic Press, Orlando, Fla. (1997) 29-45

[10] Fern'andez, C., Torres, F., Puente, S. T.: SASEPA: Simultaneous Allocation and Scheduling with Exclusion and Precedence Relations Algorithm. Lecture Notes In Computer Science, Vol. 2328. Springer, Germany (2001) 65-70

[11] Torres, F., Candelas, F. A., Puente, S. T. Ort'iz, F. G.: Graph Models Applied to Specification, Simulation, Allocation and Scheduling of Real-Time Computer Vision Applications. International Journal of Imaging Systems & Technology, Vol. 11. John Wiley & Sons, Inc., USA (2000) 287-291

# A Parallel Rendering Algorithm Based on Hierarchical Radiosity

Miguel Caballer, David Guerrero, Vicente Hernández, and Jose E. Román

Departamento de Sistemas Informáticos y Computación
Universidad Politécnica de Valencia
Camino de Vera, s/n, E-46022 Valencia, Spain
{micafer,dguerrer,vhernand,jroman}@dsic.upv.es
Tel. +34-96-3877356, Fax +34-96-3877359

**Abstract.** The radiosity method is a well-known global illumination algorithm for rendering scenes which produces highly realistic images. However, it is a computationally expensive method. In order to reduce the response time required for rendering a scene, a parallel version of a hierarchical radiosity algorithm has been developed, which takes profit of both shared and distributed memory architectures. The parallel code is used in an electronic commerce application and, therefore, the main focus is on achieving the fastest response rather than being able to render large scenes.

## 1 Introduction

One of the applications of computer graphics which is attracting more attention is the photorealistic visualisation of complex environments. Advances in graphics technology coupled with the steady increment of computers' power have enabled many new applications in architecture, engineering design, entertainment and other fields.

Global illumination methods like ray tracing and radiosity algorithms are currently available in image synthesis systems. These methods provide the necessary rendering quality, but suffer from an exceedingly high computational cost and enormous memory requirements. Parallel computing can help overcome these time and space restrictions, by using either massively parallel systems or simply networks of PC's.

Among all of these methods, radiosity techniques are becoming more and more popular due to the good quality of the images they produce. Since the introduction of the classical radiosity algorithm in the mid 1980s [8], many variants of the method have been proposed, such as progressive radiosity or hierarchical radiosity (see for example [4]).

In this work, a parallel algorithm has been implemented to reduce the time needed to render an image by the radiosity method. This algorithm is based on the hierarchical version of the method. It has been designed in a way that allows to exploit both distributed and shared memory computers.

J.M.L.M. Palma et al. (Eds.): VECPAR 2002, LNCS 2565, pp. 523–536, 2003.

The rest of the paper is divided in the following parts. Section 2 briefly describes the application for which the parallel radiosity algorithm is intended. The next section gives an introduction to radiosity methods and a description of the sequential algorithm on which the parallel version has been based. Section 4 focuses on the parallelisation of the code. Experimental results obtained with this parallel implementation are presented in section 5. Finally, some concluding remarks are given.

## 2 The Application

The parallel radiosity algorithm presented in this paper is one of the components of the software developed within the VRE-Commerce project, a EU-funded research project which aims at developing all the necessary technology for the integration of low-cost fast synthetic image generation in electronic commerce applications.

The project aims at promoting the use of electronic commerce in small and medium sized enterprises in sectors in which the use of that type of tools is not usual, such as the furniture and ceramics industries. In order to accomplish this objective it is necessary to incorporate innovative added-value features by integrating advanced technologies.

The system developed in this project incorporates remote rendering capabilities which can be used by retailers of furniture or ceramics to support the selling process of their products. Also, the final client can later connect from home and explore more possibilities to make a final decision.

### 2.1 Creation of Scenes in Electronic Commerce

The main new functionality added in the VRE-Commerce system is the possibility of creating virtual ambients by means of a room planner. The room planner is an intuitive and easy-to-use tool which enables the customer to easily design a virtual ambient and then request the generation of a realistic picture of the scene.

The room planner is an applet which is embedded in a traditional electronic commerce store. The user can design the layout of the room and then place some elements in it. These elements are either products selected from the store's catalogue or auxiliary elements such as doors, lights and so on. In the case of ceramics, the customer can also design the pattern of tiles covering the floor and walls.

Once the user is satisfied with the design of the room, it is possible to request the generation of a high quality synthetic image from a certain viewpoint. This is carried out by the radiosity code, which is described in detail in the rest of the paper. Figure 1 shows an example of a generated image and its corresponding room layout.

The fact of using the radiosity method within an electronic commerce application has two important implications which are not present in other situations.

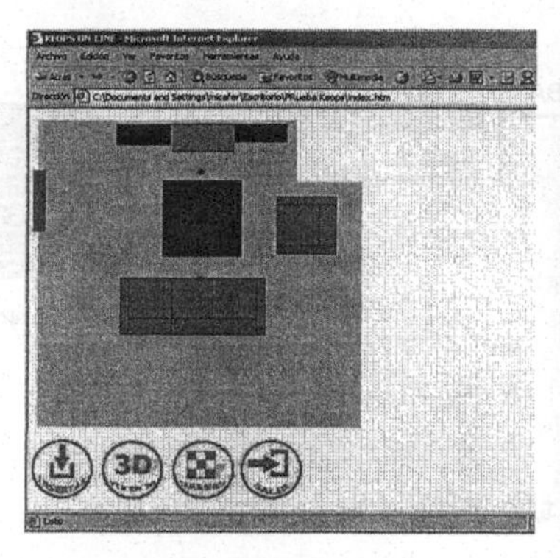

**Fig. 1.** Layout of the room designed by the user (on the left) and image generated by the system (on the right)

First of all, the response time is critical in this kind of application. Turnaround time must be kept below two or three minutes because otherwise the user will soon give up using the system. Due to the computational complexity of the radiosity calculations, parallel computing must be used to achieve this short response time, together with some simplifications of the algorithm.

On the other hand, the client is interacting with the system by means of an Internet browser in a regular computer and therefore the computational power required for the parallel rendering must be located somewhere else in the net. Access to this computing power must be ubiquitous and transparent to the user, very much in the spirit of modern grid computing systems.

The VRE-Commerce system is similar to other already available systems [10], but enhancing some aspects such as ease of use and integration of all the software components.

### 2.2 System Architecture

The architecture of the VRE-Commerce system is depicted in figure 2. The client accesses the system via Internet with a standard browser. From his point of view, the system looks much like a traditional virtual store, with many tools for searching and browsing the catalogue of products.

The two main parts of the system are the following:

- The web server and the electronic commerce system installed in it. This virtual store has an up-to-date catalogue of products and might be also connected to the company's back-office. The virtual store also contains the room planner tool.
- The computational server in charge of generating the realistic picture that will be shown to the user, produced by the parallel radiosity kernel. It can be a cluster of PC's, for example.

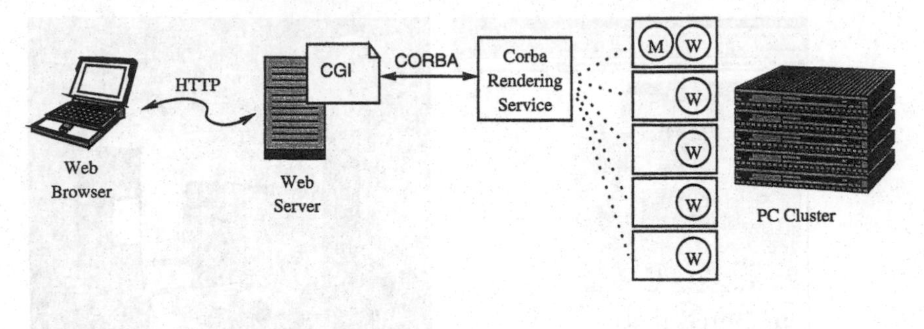


**Fig. 2.** Architecture of the VRE-Commerce system

The request for the generation of a render is made in the room planner by means of a "submit" button. This button activates a program which is executed in the web server via the Common Gateway Interface (CGI). This is a very small program that is in charge of transferring the request to the computational server. Once the image is generated, the CGI program will display it in the client's browser. In the meantime, it also displays some information about the progress of the computation.

The CGI program does not contact the computational server directly. Instead, the request is routed via an intermediate agent: a CORBA service. This additional element is included in order to add flexibility to the system. The rendering service will be able to receive several request for renders, either from the same web server or from different web servers. Also, it will be able to route these requests to one or several computational servers, allowing for sophisticated scheduling policies.

In the simplest case, there will be only one computational server, as shown in the figure. The CORBA service will launch the requests, one at a time, for example via a queueing system. The parallel job comprises a number of processes organised in a manager-worker scheme, as explained in section 4.

## 3 Image Generation

The objective of realistic image synthesis is to generate pictures with a maximum degree of realism. The calculation of realistic images requires a precise treatment of lighting effects like indirect illumination, penumbras and soft shadows from light sources. This can be achieved by simulating the underlying physical phenomena of light emission, propagation and reflection. Two different classes of algorithms exist which try to simulate the global illumination; that is, they model the inter-reflection of light between all objects in a 3D scene.

The first class are ray tracing methods, which simulate the motion of light photons either tracing the photon paths backwards from the position of an observer or forwards from the light sources. These methods are viewpoint dependent, which implies that the whole calculation must be repeated when the position of the observer changes.

The other class of methods are radiosity techniques. In these methods, the diffuse illumination of the scene is described by means of a system of linear equations, which is independent of the position of an observer of the scene.

### 3.1 Radiosity

The basic radiosity algorithm consists in solving the radiosity equation for every patch in the scene. A patch is a portion of a polygon with known physical and geometrical properties and which is assumed to have a constant radiosity. The number of patches in a scene is a trade off between the quality of the obtained solution and the cost to compute that solution.

A solution to a scene is given by the values of radiosity for all the patches in the scene. With these values, the visualisation of the scene is a trivial process, since the radiosity of a patch represents the quantity of light that patch is contributing to the scene, which is proportional to the intensity with which this patch must be rendered. Once the radiosity solution has been computed, it is possible to render the scene from the desired viewpoint using an algorithm which only must take care of drawing correctly the hidden/visible portions of polygons. This can be easily accomplished, for example, by the hardware of the majority of the currently available graphics cards.

The radiosity equation represents the interrelations among the different patches of the scene via their radiosity quantities given by

$$B_i = E_i + \rho_i \int B_j F_{j,i} \quad , \tag{1}$$

where

- $B_i$ represents the radiosity of patch i in the scene,
- $E_i$ represents the radiosity emitted by patch i in the scene (zero unless patch i is a light source),
- $F_{j,i}$ is the form factor from patch j to patch i and it represents the percentage of radiosity from patch j that arrives to patch i; and
- $\rho_i$ is the reflectivity of patch i.

The integral symbol is used to indicate that the radiosity interactions with all other patches (j) in the scene have to be taken into account. The radiosity equation (1) can be rewritten as a discrete equation by changing the integral by a summation for all the patches in the scene,

$$B_i A_i = E_i A_i + \rho_i \sum_{1 \leq j \leq n} A_j B_j F_{j,i} \quad , \tag{2}$$

where $A_i$ represents the area of patch i.

This equation can be rewritten by using the relation between the form factors of two patches i and j

$$F_{i,j} = F_{j,i}\frac{A_j}{A_i} \quad , \tag{3}$$

giving the equation

$$B_i = E_i + \rho_i \sum_{1 \le j \le n} B_j F_{i,j} \quad . \tag{4}$$

The instantiation of this equation for every patch in the scene gives a linear system of equations that can be expressed as

$$\begin{bmatrix} 1-\rho_1 F_{1,1} & -\rho_1 F_{1,2} & \cdots & -\rho_1 F_{1,n} \\ -\rho_2 F_{2,1} & 1-\rho_2 F_{2,2} & \cdots & -\rho_2 F_{2,n} \\ \cdots & \cdots & \cdots & \cdots \\ -\rho_n F_{n,1} & -\rho_n F_{n,2} & \cdots & 1-\rho_n F_{n,n} \end{bmatrix} \begin{bmatrix} B_1 \\ B_2 \\ \cdots \\ B_n \end{bmatrix} = \begin{bmatrix} E_1 \\ E_2 \\ \cdots \\ E_n \end{bmatrix} \quad . \tag{5}$$

Therefore, the problem of rendering a scene using the radiosity method requires to solve this system of linear equations to obtain the values of $B_i$ (radiosity of patch i). The coefficients of the system are given by parameters $\rho_i$, $E_i$ and $F_{i-j}$. The first two are entries to the problem, but the latter, the form factors between each pair of patches of the scene, must be computed by the algorithm using the geometrical data of the patches.

There are several variants of the basic radiosity algorithm. The most commonly used ones are the techniques known as progressive radiosity [2] and hierarchical radiosity [9].

The progressive radiosity algorithm is based on altering the order in which operations are carried out in the Gauss-Seidel process so that contributions of patches with largest radiosity values are spread to the rest of patches as soon as possible. In this way, after a few steps the illumination is already close to the final solution.

The algorithm for hierarchical radiosity is based on the idea of adaptively subdividing the scene in order to obtain better quality without increasing too much the computational cost. Better quality is usually accomplished by increasing the number of patches and using smaller ones. Patches are successively subdivided into sub-patches until the desirable precision is achieved. This splitting is done only in case of large radiosity gradients. Patches are organised in a hierarchical tree, in an attempt to minimise the number of interactions, that is, dependencies between patches. The solution process becomes an adaptive process which will modify interactions in each iteration. The results obtained by the hierarchical radiosity algorithm are as good as those obtained by the classical algorithm for finely meshed scenes, but with a much lower computational cost.

### 3.2 Sequential Implementation

Among the different possible variants of the radiosity algorithm, the hierarchical approach was chosen for the project. A scheme of the algorithm is depicted in Figure 3.

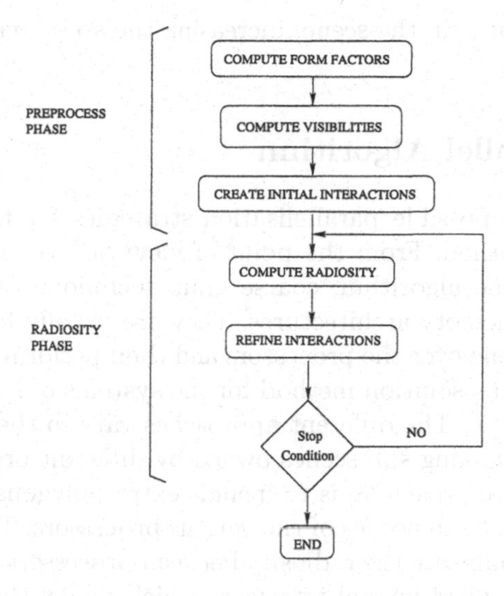


**Fig. 3.** Diagram of the sequential algorithm

The input to the sequential algorithm is the geometrical description of the scene to be rendered. The algorithm has two main stages: the preprocess phase, in which the initial interactions are completed, and the radiosity phase, in which the patches and interactions are processed and refined iteratively. One interaction is refined if the radiosity emitted form source to destination patch is greater than a predefined threshold value. When an interaction is refined the larger patch is partitioned in four smaller patches. At the end of each iteration the total amount of radiosity must be computed in order to evaluate the stopping criterion.

In the proposed algorithm, the form factors $F_{j,i}$ of equation (1) are implemented by means of two separate concepts called form factors and visibilities. Now the form factor between two patches is computed using only their geometry and position, but not taking into account the other patches which may be in between them thus decreasing the form factor. This possible influence is computed as the visibility between the patches, which represents the percentage of the patch seen from the other one. Visibility of patch A from patch B is the same as that of patch B from patch A. The form factor between two patches, as explained in the previous subsection, is mathematically equivalent to the product of this new form factor and the visibility.

In order to achieve the lowest turnaround time, the first action is to optimise the sequential algorithm and also make simplifications which help reduce the number of computations. This simplifications are always done having in mind the characteristics of the particular application. One example of such simplification is to avoid the computation of the radiosity three times, one per each component of the colour representation, red, green and blue. The proposed algorithm will solve the problem only for one spectrum which represents the intensity of the light for every patch in the scene, increasing the speed without a significant loss of quality.

# 4 The Parallel Algorithm

There are many possible parallelisation strategies for the hierarchical radiosity rendering method. From the point of view of level at which parallelisation is exploited in the algorithm, coarse grain techniques are usually best suited for distributed memory architectures. They are usually based on distributing in some way the scene over the processors and then perform the different iterations of the Gauss-Seidel solution method for the systems of linear equations in a distributed fashion [5]. The different approaches vary in the way they perform the flux of radiosity among sub-scenes owned by different processors.

One possible approach [3] is to include extra polygons representing walls between the different sub-scenes of the various processors. These polygons are used in order to communicate the radiosity between processors. Another approach [12] makes use of so-called virtual interfaces which model the radiosity interchange between sub-scenes.

Finer grain approaches include the computation of the visibilities in parallel by distributing the scene data structures into the available processors.

The decision of replicating or not the scene in all the processors has also an important impact in the parallelisation. If the scene is not replicated, only a subset of the scene is stored in each processor thus requiring less amount of memory and allowing larger scenes. The drawback of this solution is that more communications will be required when accessing data of the scene which are stored in other processors. The replication of the scene data makes this communication unnecessary, but on the other hand requires a larger amount of memory in each processor.

## 4.1 Parallel Implementation

The proposed parallel algorithm is a combination of two different parallel approaches, oriented to distributed and shared memory processors respectively. This solution is tailored to the use of clusters of multiprocessors (e.g. biprocessor PC's).

**Distributed Memory.** The approach chosen for the project is a coarse grain strategy, similar to those which distribute the scene among the different processors and perform the different iterations in parallel. As in those cases, some kind of synchronisation has to be done in order to compute the radiosity flux to or from patches in other processors. This operation consists in packaging all the radiosity values which are needed by other processors and exchanging them after each iteration.

With respect to the replication of the geometrical data of the scene in all the processors, a hybrid solution has been implemented. The initial scene is replicated in all the processors providing them with all the necessary information about the scene. However the refinements produced during the iterations of the hierarchical algorithm are stored only in the processors which own the interaction being refined. In this way, memory requirements do not grow in a processor unless it is necessary.

The Message Passing Interface (MPI) standard [14] has been chosen as the underlying layer to make communications among the different nodes in the parallel program.

**Load Balancing.** In the case of hierarchical radiosity, achieving a good work balancing is a critical issue for obtaining good parallel performance. Since the solution process is an adaptive scheme which modifies the number of patches in every iteration, the quantity of work to perform is not always proportional to the number of initially distributed patches.

The simplest load balancing scheme is to try to distribute statically the scene across the processors. With this approach no communication is necessary, since each processor has its own patches, and processes them. However, the workload may be severely unbalanced, because it is difficult to estimate at the beginning how much work will be needed by each initial polygon.

Therefore a dynamic load balancing scheme is necessary in order to distribute the workload at run time. Many different load balancing schemes have been proposed in the literature (see for example [6]). The following four alternatives are quite usual:

- Manager-Worker. A manager process is in charge of distributing the load in a balanced way among the different worker processes. The worker processes ask the manager a unit of workload to process once they have finished the previous one.
- Synchronous Work Stealing. Synchronous blocking all-to-all communication is used to periodically update the global workload state information. Every idle processor, if there are any, "steals" some units of workload from the processor with the heaviest workload.
- Synchronous Diffusive Load Balancing. Like in the previous one, there is a global state that is maintained with periodical communications, by which all the processors involved try to compensate their workloads by sending some units from the processors with the heaviest workload to the processors with less workload.
- Asynchronous Diffusive Load Balancing. In this case, the processors do not maintain periodical communications. Instead, when approaching the idle state, a processor sends out a request for work to all the other processors, and waits for any processor's donation of work.

The strategy selected in the project has been the *manager-worker* scheme, due to its simplicity and easy implementation. In our particular application, this load balancing approach gives quite good results, as will be demonstrated in the experimental results.

The manager process is in charge of distributing the load in a balanced way between the different worker processes. This manager is a simple process that waits a message from any of the workers. It knows the total number of initial polygons and their sizes, so it does not need to load the geometrical data of the scene. When a request from a worker is received, the remaining polygons are sorted by its size and the largest one is delivered. The worker processes ask the manager one polygon at a time. Since the geometrical data are already distributed in all the nodes, sending the index of the polygon in the scene is enough. Once the number has been received, the worker processes it, then requests a new polygon and so on. In this way a balanced distribution of the processing between the different nodes is obtained, with a minimum shipment of data. This distribution makes possible to use the algorithm efficiently in a network with processors of different speed, obtaining nearly 100% utilisation in all the nodes. This approach obtains worse results if the number of processors increases. In that case, the Asynchronous Diffusive Load Balancing scheme mentioned above can be a better solution.

The manager-worker strategy can be difficult to deploy if there are few large polygons. In order to improve the distribution in this case, initially the largest polygons of the scene are divided, to avoid that the work of a single polygon is so large, that the distribution of the load is unbalanced.

The implemented approach is only effective in the first iteration (the most time-consuming one) of the radiosity algorithm, since when this iteration finishes the geometrical data loaded in each processor is different due to the refinements produced by the hierarchical algorithm. If we wanted to use the same approach in the other iterations it would be necessary to send all the new geometrical data of each polygon, and communication time and memory usage would increase too much.

**Shared Memory.** Assuming that each node is a shared memory multiprocessor, in order to take advantage of parallelism in each node a method based on the execution of the program with several threads is proposed. The number of threads is set to the number of processors available in each node. The different threads have to distribute the local work and synchronise themselves when necessary.

In order to balance the load between the processors in a shared memory computer a different approach has been used, based on a global queue for all processors (a similar approach is used in [13]). This queue is used to store the subpolygons to be processed. Initially the original polygons of the scene are inserted in this queue. Then, due to the refinements produced by the hierarchical algorithm, each polygon is divided into smaller subpolygons. Each subpolygon generated in that way is appended to that queue. The process ends when the queue is empty. This scheme yields a balanced distribution. Mutual exclusion must be guaranteed in order to avoid problems when two or more threads try to add or remove elements from this queue at the same time.

**Table 1.** Scenes used for performance evaluation

| Scene | Polygons | Patches | Lights | Elements |
|---|---|---|---|---|
| Ex. 1 | 694 | 62980 | 2 | 15 |
| Ex. 2 | 1550 | 92610 | 1 | 21 |
| Ex. 3 | 1558 | 109750 | 2 | 22 |

## 5 Experimental Results

A parallel platform consisting in a network of several PC's has been used to test the implemented algorithm. The available platform is a set of 12 biprocessor nodes connected by a Gigabit Ethernet network, each of them having two Pentium III processors at 866 MHz with 512 MB of RAM. The installed operating system is Red Hat Linux version 7.1.

Since the platform has potentially two levels of parallelism (shared memory parallelism between processors in each node, and distributed memory parallelism among different nodes), both of them have been exploited.

Three scenes have been used for analysing the parallel performance of the code. These test cases are considered to be representative of the normal usage of the VRE-Commerce system. Table 1 shows some parameters of the three scenes in order to give an indication of their sizes.

Tables 2 and 3 show the execution times and speed-up obtained when running the parallel rendering kernel in the target platform, using one processor per node and the two available processors per node, respectively.

The time measurements shown in the table demonstrate a good performance of the code. As an example, the execution time for test case 3 is reduced in a factor of more than 22 when using 24 processors.

Figure 4 shows a graph of the efficiencies obtained for test case 3, both using 1 processor per node and using 2 processors per node. It can be noted that efficiencies are higher when using 1 processor per node. This is because the distributed memory algorithm assigns polygons to processors in a balanced way,

**Table 2.** Execution time (in seconds) and speed-up when using one processor per node

| | Ex. 1 | | Ex. 2 | | Ex. 3 | |
|---|---|---|---|---|---|---|
| $p$ | $T_p$ | $S_p$ | $T_p$ | $S_p$ | $T_p$ | $S_p$ |
| 1 | 104.9 | 1.00 | 138.6 | 1.00 | 249.9 | 1.00 |
| 2 | 52.6 | 1.99 | 69.9 | 1.98 | 125.6 | 1.99 |
| 4 | 26.3 | 3.99 | 35.2 | 3.93 | 62.8 | 3.98 |
| 6 | 17.7 | 5.93 | 24.0 | 5.87 | 41.8 | 5.98 |
| 8 | 13.3 | 7.89 | 18.4 | 7.53 | 31.5 | 7.93 |
| 10 | 10,8 | 9.71 | 15.1 | 9.17 | 25.4 | 9,84 |
| 12 | 9,0 | 11.66 | 12.8 | 10.83 | 21.5 | 11,62 |

**Table 3.** Execution time (in seconds) and speed-up when using two processors per node

| | Ex. 1 | | Ex. 2 | | Ex. 3 | |
|---|---|---|---|---|---|---|
| $p$ | $T_p$ | $S_p$ | $T_p$ | $S_p$ | $T_p$ | $S_p$ |
| 2 | 53.5 | 1.96 | 70.1 | 1.98 | 126.3 | 1.96 |
| 4 | 26.6 | 3.90 | 35.3 | 3.93 | 63.1 | 3.91 |
| 8 | 13.3 | 7.83 | 17.8 | 7.79 | 31.5 | 7.81 |
| 12 | 8.9 | 11.66 | 12.4 | 11.18 | 21.2 | 11.57 |
| 16 | 6.7 | 15.43 | 9.4 | 14.74 | 15.8 | 15.52 |
| 20 | 5.4 | 19.43 | 7.7 | 18.00 | 12.7 | 19.22 |
| 24 | 5.0 | 20.53 | 7.0 | 19.80 | 10.7 | 22.72 |

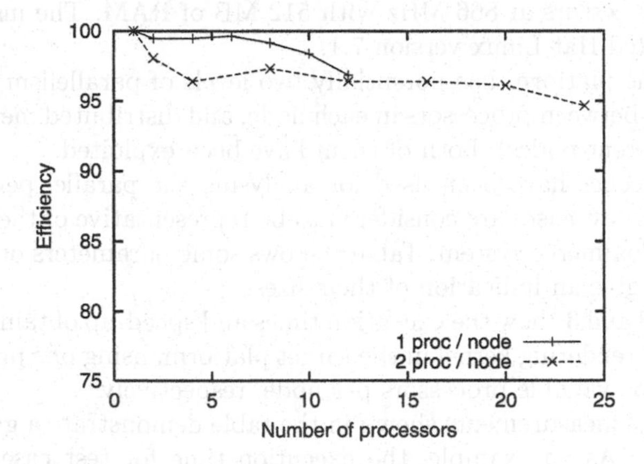


**Fig. 4.** Efficiencies obtained for test case 3

with a minimum shipment of data. In this case, the contention time produced by the memory competition of the processors in the shared memory algorithm, is larger than the communication time. However, when the number of nodes increases, the efficiency in the pure distributed memory case falls faster than in the hybrid one.

## 6 Conclusions

A parallel algorithm has been implemented for the hierarchical radiosity method on distributed and shared memory architectures. It tries to take advantage of each type of parallelism.

In distributed memory machines the main objective is to distribute the workload among all the nodes in a balanced way. But on hierarchical radiosity methods, this is a difficult problem because the computing time required by a polygon to compute its radiosity is not known a priori.

To solve this problem a *manager-worker* approach has been used in order to distribute the polygons and patches to the processing nodes dynamically. This solution is only used in the first iteration (the most time-consuming one) of the radiosity algorithm. In the rest of iterations, the same distribution made in the initial iteration is used. With this approach good load balance and acceptable scalability has been achieved.

The algorithm has been designed both optimising communications in a distributed memory platform and minimising synchronisations in a shared memory platform, thus giving good results for both types of platforms and also in hybrid configurations.

Some improvements to the algorithm would probably help get a better load balancing. A possible line of work for the future is to use the dynamic load balancing in all the iterations of the algorithm, trying to minimise the communication time, and comparing the manager-worker approach with an asynchronous diffusive load balancing technique.

## Acknowledgements

This work has been supported by the 5th Framework Programme of the European Commission under contract IST-1999-20785.

## References

[1] M. Amor, E. J. Padrón, J. Touriño, and R. Doallo. Scheduling of a Hierarchical Radiosity Algorithm on a Distributed-Memory Multiprocessor. *4th International Meeting on Vector and Parallel Processing*, 581–591, 2000.

[2] M. Cohen, S. E. Chen, J. R. Wallace and D. P. Greenberg. Progressive Refinement Approach for Fast Radiosity Image Generation. *Computer Graphics (SIGGRAPH'88 Proceedings)*, 22(4):31–40, 1988.

[3] Chen-Chin Feng and Shi-Nine Yang. A parallel hierarchical radiosity algorithm for complex scenes *IEEE Parallel Rendering Symposium*, 71–78 , November 1997.

[4] James D. Foley, Andries van Dam, Steven K. Feiner and John F. Hughes. Computer Graphics: Principles and Practice. 2nd ed. *Addison-Wesley*, 1996.

[5] T. Funkhouser. Coarse-Grained Parallelism for Hierarchical Radiosity Using Group Iterative Methods. *Computer Graphics (SIGGRAPH 96)*, August 1996.

[6] Chao-Yang Gau and Mark A. Stadtherr. Parallel branch-and-bound for chemical engineering applications: Load balancing and scheduling issues (invited talk). *Lecture Notes in Computer Science*, 1981:273–300, 2001.

[7] S. Gibson, R. J. Hubbold. A perceptually-driven parallel algorithm for efficient radiosity simulation. *IEEE Trans. on Visualisation and Computer Graphics*, 6(3):220–235, 2000.

[8] Cindy M. Goral, Kenneth K. Torrance, Donald P. Greenberg, and Bennett Battaile. Modelling the interaction of light between diffuse surfaces. *Computer Graphics*, 18(3):213–222, July 1984.

[9] P. Hanrahan, D. Salzman and L. Aupperle. A Rapid Hierarchical Radiosity Algorithm. *Computer Graphics*, 25(4):197–206, 1991.

[10] Reinhard Lüling and Olaf Schmidt. Hipec: High performance computing visualization system supporting networked electronic commerce applications. *Lecture Notes in Computer Science*, 1470:1149–1152, 1998.

[11] D. Meneveaux K. Bouatouch. Synchronization and load balancing for parallel hierarchical radiosity of complex scenes. *Computer Graphics Forum*, 18(4):201–212, 1999.

[12] Luc Renambot, Bruno Arnaldi, Thierry Priol and Xavier Pueyo. Towards efficient parallel radiosity for DSM-based parallel computers using virtual interfaces. *IEEE Parallel Rendering Symposium*, 79–86 , November 1997.

[13] Francois Sillion and J.-M. Hasenfratz. Efficient Parallel Refinement for Hierarchical Radiosity on a DSM Computer. *Proceedings of the Third Eurographics Workshop on Parallel Graphics and Visualisation*, September 2000.

[14] Marc Snir, Steve Otto, Steven Huss-Lederman, David Walker and Jack Dongarra. MPI: The Complete Reference *The MIT Press*, 1996.

# A High–Performance Progressive Radiosity Method Based on Scene Partitioning

Antonio P. Guerra, Margarita Amor, Emilio J. Padrón, and Ramón Doallo

Dept. Electronic and Systems
Univ. da Coruña, 15071 A Coruña, Spain
{margamor,doallo}@udc.es
emilioj@mail2.udc.es

**Abstract.** The radiosity method is one of the most popular global illumination models in which highly realistic synthetic images can be achieved. In this paper we present a parallel algorithm for computing the progressive radiosity method on distributed memory systems. Our implementation allows to work with large scenes using a partition of the input scene among the processors; besides good results in terms of speedup have been obtained.

## 1 Introduction

The rendering of realistic synthetic images [1] is one aim of the research in computer graphics. In order to achieve a precise treatment of lighting effects in synthetic images, it is important to use a global illumination algorithm. These algorithms compute the color at a point in the environment in terms of light directly emitted by light sources and light that reaches that point after its reflection and transmission through its own and other surfaces. Two different global illumination methods are mainly used to simulate the light transfer through the environment: radiosity method [1] and ray tracing [2].

Radiosity methods are based on the realistic behavior of light in a closed environment in a view-independent way. Advanced algorithms for the radiosity method have been proposed such as progressive radiosity [3], hierarchical radiosity [4] and wavelet radiosity algorithms [5]. However, these algorithms have a significant computational cost and high memory requirements, being the complete resolution of the radiosity equation for large images highly time consuming. For this reason, despite high quality images can be obtained, high performance computing techniques such as parallel computing ought to be considered in order to decrease execution times, allowing to these methods to be more widely used in image rendering.

Several parallel implementations for solving the radiosity methods have been previously proposed. We can classify these implementations according to the type of access to the scene allowed to each processor. In a first group, parallel solutions where all the processors have access to the whole scene are considered. Parallel algorithms based on the hierarchical radiosity method on distributed

J.M.L.M. Palma et al. (Eds.): VECPAR 2002, LNCS 2565, pp. 537–548, 2003.

memory systems were proposed in [6, 7]. In [8] a parallel algorithm based on the progressive method was implemented on a distributed memory system. In the second group the scene is distributed among the different local memories of the processors. Parallel algorithms based on the progressive method were proposed in [9, 10] on distributed memory systems and in [11, 12] on shared memory systems. In [13] a fine–grain parallelism was applied to a hierarchical radiosity method using a master–slave paradigm; and in [14, 15] parallel algorithms based on the wavelet radiosity method were proposed.

The main drawback affecting the first approach (allowing access to the whole scene) is that they have to replicate the radiosity and geometric data in the local memory of each processor, which limits the scene's size. On the other hand, the second approach (working with a distributed scene) allows to work with larger scenes. Nevertheless, the existing implementations of this approach on distributed memory systems present some problems. As some examples, the implementation of [9] is not suited for all kinds of scenes and [10] uses a visibility algorithm that is not very accurate.

In this paper, we propose a parallel implementation of the progressive radiosity algorithm on a distributed memory multicomputer. Although our approach uses the message–passing paradigm, the communication overhead is minimized because nonblocking communications are used to overlap communication and computation. On the other hand, the scene is partitioned into sub-environments which are distributed among the processors, which allows to work with large scenes. Each processor performs the whole computation of the radiosity for the set of data into its assigned sub-environment where the visibility values are calculated using ray casting.

The paper is organized as follows: in Section 2 we describe the progressive radiosity algorithm; in Section 3 we present the partitioning algorithm for subdividing the scene into sub-environments; the parallel implementation is presented in Section 4; experimental results on a SGI Origin 2000 multicomputer are shown in Section 5; finally, in Section 6 we present the main conclusions.

## 2 The Progressive Radiosity Algorithm

The radiosity method, initially developed for its application on engineering problems regarding radiative heat transfer, is nowadays one of the most popular global illumination models in computer graphics. The radiosity method is based on applying to image synthesis the concepts of thermodynamics that rule the balance of energy in a closed environment. This method solves a global illumination problem expressed by the rendering equation [16], simplified by considering only ideal diffuse surfaces.

Furthermore, the radiosity method is a finite element method that approximates the radiosity function by subdividing the domain of this function into small elements or patches across which the function can be approximated by a sum of relatively simple functions, called *basis functions*. Then, the resultant equation system, known as the discrete radiosity equation, is:

$$B_i = E_i + \rho_i \sum_{j=1}^{N} B_j F_{ij} \tag{1}$$

where $B_i$ is the radiosity of patch $P_i$, $E_i$ is the emittance and $\rho_i$ the diffuse reflectivity. The summation represents the contributions of the other patches of the scene, where $F_{ij}$ is called the form factor between patches $P_i$ and $P_j$. It is an adimensional constant that only depends on the geometry of the scene and represents the proportion of the radiosity leaving patch $P_i$ that is received by patch $P_j$. There is one radiosity equation for each patch and one form factor between all pairs of $N$ patches, therefore the radiosity equation system is liable to be very large with complexity $O(N^2)$ and relatively dense.

Progressive radiosity, in which we have focused, is a slightly different form of Southwell's [1] algorithm; its goal is to display results after each step of the iterative process, in order to provide immediate feedback. The requirement is thus not only to provide an algorithm that converges quickly, but one that makes as much progress at the beginning of the algorithm as possible. The basic idea behind progressive radiosity is to *shoot* in every step the energy of the patch with the greatest unshot radiosity. This way we forget about the "strict" concept of iteration and are able to obtain a quite good illuminated scene even after the first steps. Thus, the complexity would be $O(k \times N)$, being $N$ the number of patches and $k$ the number of steps performed, with $k << N$.

In the sequential progressive algorithm we can distinguish the following steps:

- **Model environment**. The first step is to obtain a mesh of patches from the environment using a meshing algorithm. The finer the mesh is the better the obtained illumination will be. Our implementation uses both triangular and rectangular patches, allowing to choose parameters as mesh density or T-vertices elimination.
- **Iterative process**. In the context of progressive radiosity it is not easy to mark the boundaries between iterations since some patches may be revisited many times before another one is operated on. Therefore we now talk about iterations in the sense of steps. Thus, at the beginning of each iteration (step) the patch with the greatest unshot radiosity, $P_i$, is chosen and its radiosity is shot through the environment. As a result of that shot, the other patches, $P_j$, may receive some new radiosity and the patch $P_i$ has no unshot radiosity. Each shooting step can be viewed as multiplying the value of radiosity to be "shot" by a column of the form factor matrix. Therefore at this moment we need to compute the required form factors, and the visibility value between the chosen patch and the rest of the patches.
  Regarding to the form factor computation, we have used the analytic method called *differential area to convex polygon* [1], also known as *point to patch.* This is a commonly used method when handling with polygonal models.
  For visibility determination a space partitioning of the 3D environment is carried out. We have implemented an algorithm based on *SEADS* (Spatial Enumerated Auxiliary Data Structure), an uniform spatial division scheme

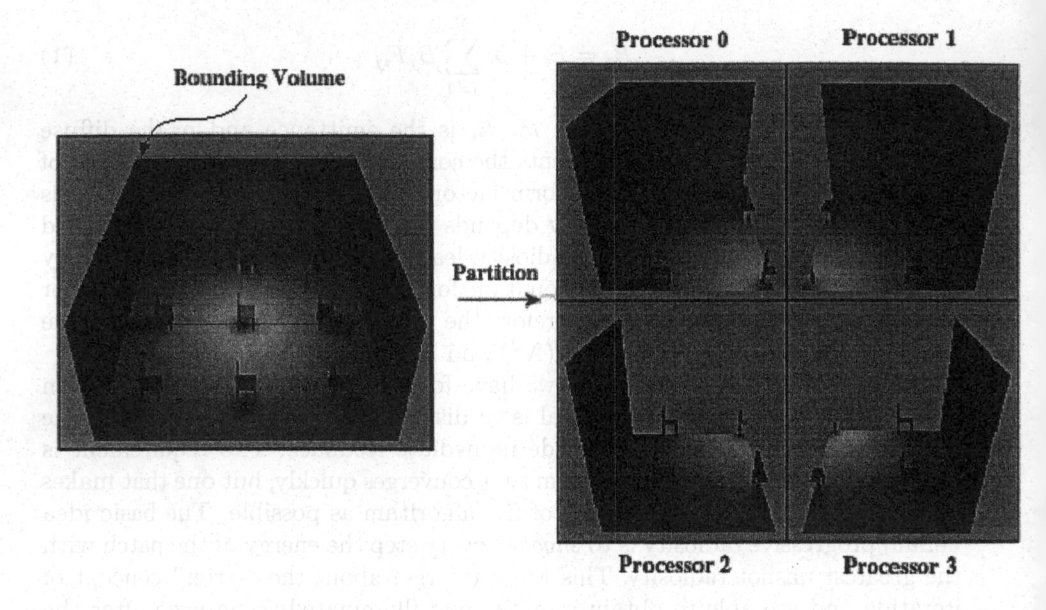


**Fig. 1.** Scene partitioning into 4 sub-environments

developed by A. Fujimoto *et al.* [17] with the aim of increasing the speed with which ray-polygons intersections are computed in ray tracing. In this scheme, space is simply divided into a fixed number of equally sized cubic regions called voxels. Thus, ray-scene's objects intersections can be computed by traversing voxel space and checking just the polygons within the traversed voxels. This method has a significant increased speed compared with, for example, methods using BSP trees or octrees. However, due to the uniform partition, a drawback arises when objects are not uniformly distributed around the scene since that means regions of high object density are partitioned at the same level as regions of the scene of lower object density.

## 3 Partitioning of the Scene

In this section the partitioning algorithm to subdivide the whole scene into sub-environments is presented. The partitioning algorithm we use computes an uniform parallel partitioning of the input scene via a geometric technique [18]. This way, the partitioning algorithm subdivides the bounding volume of the scene into a regular grid of equal-sized sub-environments. These partitioning schemes are extremely fast, easy to parallelize, and present high data locality. As we will see the main drawback of uniform partitioning is that poor load balancing could be achieved.

Our partitioning algorithm consists of two steps:

- **Sub-environment partitioning**. The original scene is replicated in each processor; this fact does not represent a drawback due to the low storage requirements for the input scene. First, each processor calculates the bounding volume of the scene. The bounding volume construction is based on the minimum parallelepiped search, which contains all objects of the scene (see Figure 1) and defines the environment of the scene. Next, each processor must divide the bounding volume into $p$ uniform disjoint sub-environments, where $p$ is the number of processors. Then, it calculates the number of partitions ($p_x$, $p_y$ and $p_z$) each coordinate axis is going to be subdivided. The bounding volume will be divided on each dimension by so many planes as the number of partitions minus one. These numbers are calculated as the prime factors ($d_i$) of $p$,

$$p = \prod_{i=1}^{n} d_i \tag{2}$$

In order to avoid large sub-environments, the factors $d_i$ are assigned to $\{p_x, p_y, p_z\}$ according to the following steps: First, the $n$ factors are sorted in decreasing order and $p_x = p_y = p_z = 1$ are set; next, the list of factors is traversed and each factor $d_i$ is assigned to the minimun of $\{p_x, p_y, p_z\}$ so that:

$$p_j = p_j \cdot d_i \tag{3}$$

As an example let us consider a 36 processor configuration, then the list of prime factors is $\{3, 3, 2, 2\}$ so we would obtain the following partitions: $p_x = 3, p_y = 3, p_z = 2 \cdot 2 = 4$.

- **Sub-environment distribution**. Each input polygon is assigned to a sub-environment according to its geometrical position; thus each polygon interacts to a greater extent with those polygons that are closer. Polygons belonging to several sub-environments must be clipped into the same number of sub-environments. Then, these new polygons are assigned to the different sub-environments which now completely contain them.
Once the list of objects belonging to each sub-environment is completed, each sub-environment must be assigned to a processor (see Figure 1). During the evaluation of the algorithm on the multicomputer, processors need to exchange data, using a protocol we present in the next section. Then, for a greater degree of simplification in that scheduling, we map adjacent partitions in the physical domain into neighboring processors. Once scene distribution is done, each processor has a disjoint set of polygons and only computes the meshing of these polygons.

## 4 Parallel Implementation

The parallel version of the sequential progressive radiosity algorithm has been implemented using a message-passing paradigm (specifically we have used the message-passing library MPI) and following a coarse-grain approach, that is, each processor performs the whole computation of the radiosity for the set of patches of its sub-environment.

The communication among processors takes place at the end of each iteration of local radiosity computation. Each communication includes information about the shot patch and its visibility in the sub-environment. For the latter we have used a visibility mask technique [11]. In the next subsections we can see these steps in detail.

### 4.1 Local Radiosity Computation

In each iteration, processors choose the patch with the greatest unshot energy and its radiosity is shot through the rest of patches in the sub-environment. Once the local radiosity is updated in the sub-environment, the visibility masks are computed and packed along with the shot patch's information to be sent. It is very important to note that the visibility of the patch in its local environment must be sent, since it may exist another patch which hides the energy of the selected patch to several environments.

The initial visibility mask [11] is a structure that stores all the visibility information encountered during the processing of a selected patch in its local environment. In our new proposal, we define an interior frontier as the face of the parallelepiped which divides two local environments. A frontier has to be decomposed into uniform $n \times n$ patches where each patch corresponds to a boolean

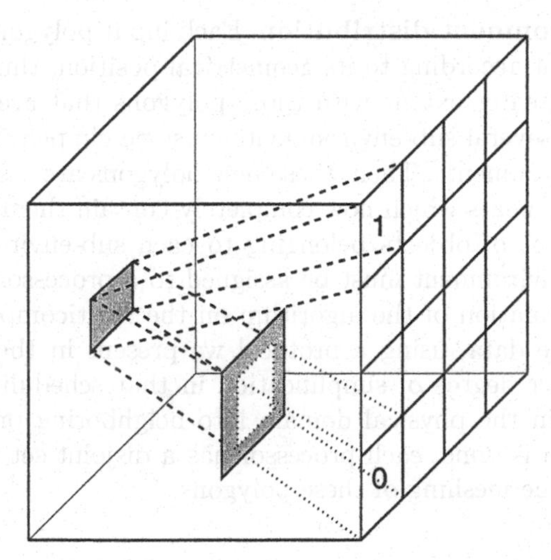


**Fig. 2.** Computation of the starting visibility mask

value. The value is 1 if the frontier patch is visible from the selected patch, otherwise the value is 0 (see Figure 2). This visibility information between the selected patch and each frontier patch is calculated using the SEADS algorithm. It is important to note that the visibility structure we propose has low storage requirements.

### 4.2 Communication Stage

Once the energy of a selected patch is distributed in the local environment in which it is contained, its energy is propagated to the whole scene. In this stage, the processor sends the visibility information and a copy of the shooting patch to the other processors.

First, the processor sends a copy of the patch to the rest of processors using a global communication. Then, each processor receives the patch being appended to a pending queue while its total visibility information is not achieved. Next, the processor sends to the processors with common frontier the initial visibility mask. Then, it is tested if the visibility information of the received patch is complete. In this case, the patch moves from the pending queue to the queue of unshot patches and its visibility mask is updated. The visibility information is calculated by casting rays from the received patch through its visibility mask to the interior frontier of the new environment. Finally, the visibility mask is sent to the neighbor processors. In order to avoid sending the visibility mask of a patch twice to the same processor, we attach a flag with each mask, containing a bit of data for each processor. These bits are set if a patch is sent to a processor. After the communication stage, the processor can continue to perform the selection of the most unshot patch while there are any patch in the queue of unshot patches. During this communication stage, nonblocking communications are used to overlap communication and computation.

As an example let us consider the routing of the visibility mask of the patch_1 of Figure 3.a. Processor 1 selects the patch_1 as the current shooting patch. After the patch_1's radiosity is shot through its environment, the patch_1 is sent to the other processors by a global communication (see Figure 3.b). In Figure 3.c the initial visibility mask of patch_1 ($m_{11}$, where the first number is the patch which the mask correspond to and the second one the processor that has computed it) is computed by the processor 1 and sent to its neighbors, processors 2 and 3. Then, these processors update their queue of unshot patches and calculate the new visibility masks of patch_1 ($m_{12}$ and $m_{13}$), using the received visibility mask ($m_{11}$). After, each processor sends the computed visibility mask to the processor 4 (see Figure 3.d) which can move the patch_ 1 to the queue of unshot patches when it receives the information from both the processors 2 and 3 (at that moment it has already available the patch_1's visibility value).

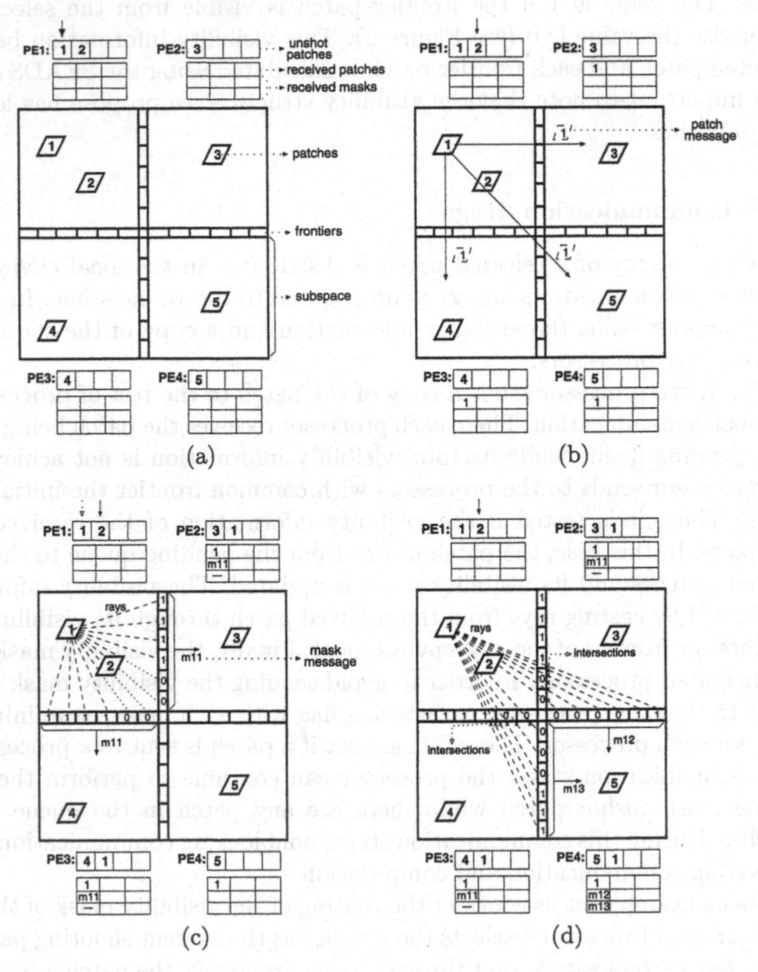


**Fig. 3.** The routing of the visibility mask on four processors

# 5 Experimental Results

The progressive parallel algorithm has been implemented on a SGI Origin 2000, using the message–passing paradigm. We have used the MPI programming environment.

We have rendered a test scene to demonstrate the performance of our parallel algorithm: *Office* scene (see Figure 4.a) with 977 input polygons, and about 13000 patches after the model environment step. The resulting illuminated scene is depicted in Figure 4.b.

In order to evaluate the degree of error due to the visibility mask employed in the calculation of form factors, we have used the residual error of the radiosity. In Table 1 the residual error and the execution time for 8 processors with

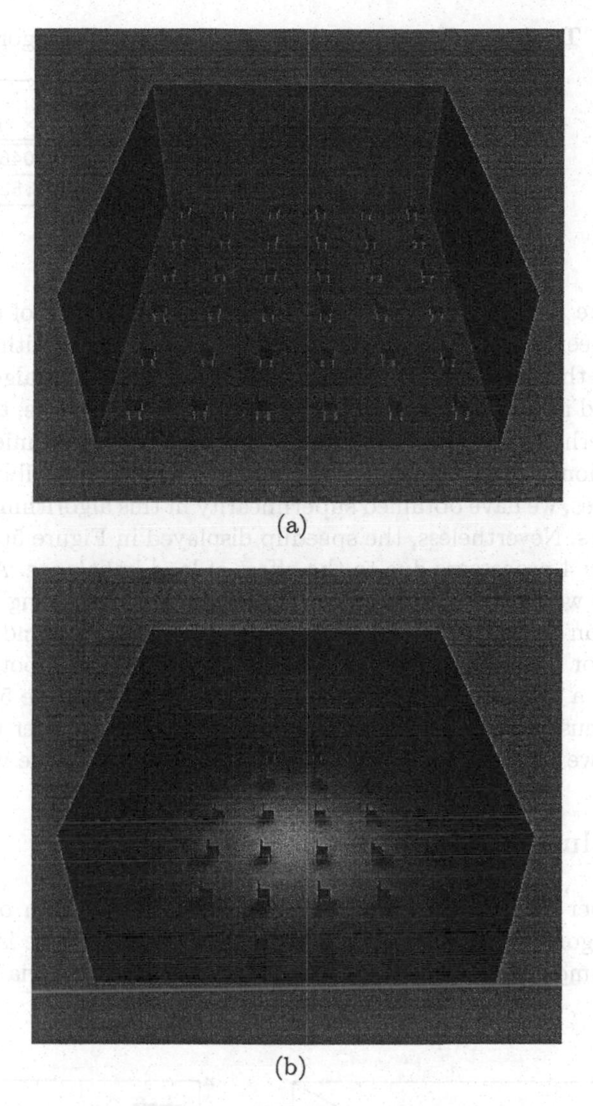

Fig. 4. Original and resulting illuminated scene

different number of patches in the frontier is indicated. As we can see on this table, errors are small, which shows the accuracy of the parallel algorithm and that the approximation of the visibility information with the visibility mask has little impact on the quality of the rendering. On the other hand, the increasing in the number of patches in the frontiers does not produce a reduction in the error because the visibility relationships are locally dense but globally sparse. Nevertheless, this increasing produces an increase in the execution time so we have used the version with $6 \times 6$ patches in the frontier between local environments.

**Table 1.** The residual errors of the parallel algorithm

| Number of patches in the frontier | 6 × 6 | 8 × 8 | 10 × 10 |
|---|---|---|---|
| Error | 0.0004627 | 0.0004617 | 0.0004617 |
| Time | 19.37 | 22.14 | 22.53 |

In Figure 5(a) we show the speedup for several number of processors. These data have been measured with regard to the sequential algorithm before the partitioning of the data. The execution time of the sequential algorithm is 150.831 seconds, and it is 9.6 seconds on 24 processors. As we can see, these results show the low overhead associated with our method and the techniques employed in parallelization, providing an excellent performance on distributed memory systems. Besides, we have obtained superlinearity in this algorithm for a low number of processors. Nevertheless, the speedup displayed in Figure 5(a) shows a severe knee around 4 processors due to the effect of load imbalance. As an example, in Figure 5(b) we have checked the load balancing by measuring in each processor the execution time, both for two different configurations: 4 and 8 processors. For a 8-processor configuration, processors 5, 6, 7 and 8 are a bottleneck. Configuration with a low number of processors, according to Figure 5(b), balances the load and, thus, minimizes idle times. As future work, in order to get better load balancing, we intend to apply a non uniform partition of the whole scene.

## 6 Conclusions

In this paper we have described a parallel implementation of the progressive radiosity algorithm on a multicomputer. The original scene is partitioned into sub-environments employing an uniform parallel algorithm via a geometric tech-

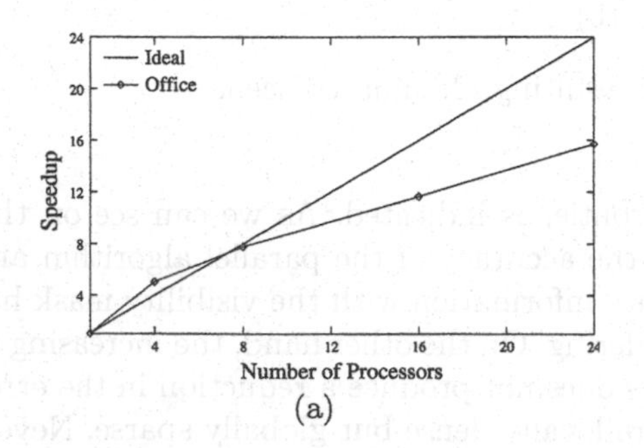


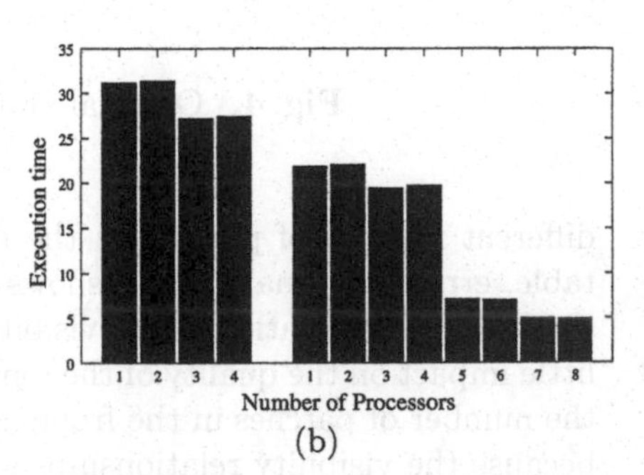


**Fig. 5.** (a) Speedups on the SGI Origin 2000 (b) Execution time (in seconds) for each processor with 4 and 8 processors configuration

nique. Afterwards, the sub-environments are assigned to each processor and these perform the whole computation of the radiosity for the set of assigned patches.

The parallel algorithm maintains a high quality in the illuminated images. Furthermore, a good speedup was achieved through this approach, as we have experimentally shown on the SGI Origin 2000, exploiting data locality and reducing and optimizing the communications. The algorithm presents, as it is currently designed, some workload problems that should be addressed in future implementations; we think applying a non uniform partition of the whole scene should be our next work step.

## Acknowledgements

We gratefully thank Centro de Supercomputación Complutense (Madrid, Spain) for providing access to the SGI Origin 2000. This work was supported in part by the Ministry of Science and Technology (MCYT) of Spain under the project TIC 2001-3694-C02-02. We acknowledge University of A Coruña (Spain) for the research funds that also contributed to support this work.

## References

[1] Cohen, M. F., Wallace, J. R.: Radiosity and Realistic Image Synthesis. Academic Press Professional (1993)

[2] Glassner, A. S.: Principles of Digital Image Synthesis. Morgan Kaufmann Publishers (1995)

[3] Cohen, M., Chen, S. E., Wallace, J. R., Greenberg, D. P.: A Progressive Refinement Approach to Fast Radiosity Image Generation. Computer Graphics (Proc. of the SIGGRAPH'88) **22** (1988) 31–40

[4] Hanrahan, P., Salzman, D., Aupperle, D.: A Rapid Hierarchical Radiosity Algorithm. Computer Graphics (Proc. of the SIGGRAPH'91) **25** (1991) 197–206

[5] Schröder, P.: Wavelet Algorithms for Illumination Computations. Master's thesis, Univ. of Princeton (1994)

[6] Padrón, E., Amor, M., Touriño, J., Doallo, R.: Hierarchical Radiosity on Multicomputers: a Load–Balanced Approach. In: Proc. of the Tenth SIAM Conference on Parallel Processing for Scientific Computing. (2001)

[7] Garmann, R.: Spatial Partitioning for Parallel Hierarchical Radiosity on Distributed Memory Architectures. In: Proc. of the Third Eurographics Workshop on Parallel Graphics and Visualisation. (2000)

[8] Cerruela, G.: Radiosidad en Multiprocesadores. Master's thesis, Univ. of Málaga (1999)

[9] Renambot, L., Arnaldi, B., Priol, T., Pueyo, X.: Towards Efficient Parallel Radiosity for DSM-based Parallel Computers using Virtual Interfaces. In: IEEE Parallel Rendering Symposium. (1997) 79–86

[10] Schmidt, O., Reeker, L.: New Dynamic Load Balancing Strategy for Efficient Data-Parallel Radiosity Calculations. In: Proceedings of Parallel and Distributed Processing Techniques and Applications (PDPTA '99). Volume 1. (1999) 532–538

[11] Arnaldi, B., Priol, T., Renambot, L., Pueyo, X.: Visibility Masks for Solving Complex Radiosity Computations on Multiprocessors. In: Proc. of the First Eurographics Workshop on Parallel Graphics and Visualisation. (1996) 219–232
[12] Gibson, S., Hubbold, R. J.: A Perceptually-Driven Parallel Algorithm for Efficient Radiosity Simulation. IEEE Transactions on Visualisation and Computer Graphics **6** (2000) 220–235
[13] Funkhouser, T. A.: Coarse-grained Parallelism for Hierarchical Radiosity using Group Iterative Methods. In: Proc. of the SIGGRAPH'96. (1996) 343–352
[14] Cavin, X., Alonso, L., Paul, J. C.: Parallel Wavelet Radiosity. In: Proc. of the 2nd Eurographics Workshop on Parallel Graphics and Visualisation. (1998) 61–75
[15] Meneveaux, D., Bouatouch, K.: Synchronization and Load Balancing for Parallel Hierarchical Radiosity of Complex Scenes on a Heterogeneous Computer Network. Computer Graphics Forum **18** (1999) 201–212
[16] Kajiya, J. T.: The Rendering Equation. Computer Graphics **20** (1986) 143–150
[17] Fujimoto, A., Tanaka, T., Iwata, K.: Arts: Accelerated Ray Tracing. IEEE Computer Graphics and Aplications **6** (1986) 16–26
[18] Schloegel, K., Karypis, G., Kumar, V.: Graph Partitioning for High Performance Scientific Simulations. In: CRPC Parallel Computing Handbook. Morgan Kaufmann (2000)

# Wavelet Transform for Large Scale Image Processing on Modern Microprocessors*

Daniel Chaver, Christian Tenllado, Luis Piñuel,
Manuel Prieto, and Francisco Tirado

Departamento de Arquitectura de Computadores y Automatica
Facultad de Ciencias Fisicas, Universidad Complutense, 28040 Madrid, Spain
{dani02,tenllado,lpinuel,mpmatias,ptirado}@dacya.ucm.es

**Abstract.** In this paper we discuss several issues relevant to the vectorization of a 2-D Discrete Wavelet Transform on current microprocessors. Our research is based on previous studies about the efficient exploitation of the memory hierarchy, due to its tremendous impact on performance. We have extended this work with a more detailed analysis based on hardware performance counters and a study of vectorization, in particular, we have used the Intel Pentium SSE instruction set. Most of our optimizations are performed at source code level to allow automatic vectorization, though some compiler intrinsic functions have been introduced to enhance performance. Taking into account the abstraction at which the optimizations are performed, the results obtained on an Intel Pentium III microprocessor are quite satisfactory, even though further improvement can be obtained by a more extensive use of compiler intrinsics.

## 1 Introduction

Over the last few years, we have witnessed an important development in applications based on the discrete wavelet transform. The most outstanding success of this technology has been achieved in image and video coding. In fact, standards such as MPEG-4 or JPEG-2000 are based on the discrete wavelet transform (DWT). Nevertheless, it is without doubt a valuable tool for a wide variety of applications in many different fields [1][2]. This growing importance makes a performance analysis of this kind of transformation of great interest.

Our study focuses on general-purpose microprocessors. In these particular systems, the main aspects to be addressed are the efficient exploitation of the memory hierarchy, especially when handling large images, and how to structure the computations to take advantage of the SIMD extensions available on modern microprocessors.

---

* This work has been supported by the Spanish research grant TIC 99-0474.

With regard to the memory hierarchy, the main problem of this transform is caused by the discrepancies between the memory access patterns of two principal components of the 2-D wavelet transform: the vertical and the horizontal filtering [2]. This difference causes one of these components to exhibit poor data locality in the straightforward implementations of the algorithm. As a consequence, the performance of this application is highly limited by the memory accesses.

The platform on which we have chosen to study the benefits of the SIMD extensions is an Intel Pentium-III based PC. However, we should remark that most of the optimizations that we have performed to take advantage of this kind of parallelism do not depend on the particular characteristics of the Intel Pentium's SSE instruction set [3]. In fact, due to portability reasons, we have avoided the assembly language programming level. All the optimizations have been performed at the source code level. Basically, we have introduced some directives which inform the compiler about pointer disambiguation and data alignment, and some code modifications, such as changing the scope of the variables in order to allow automatic vectorization. Furthermore, we have also compared this approach with a hand-tuned vectorization based on language intrinsics, in order to measure the quality of the compiler.

This paper is organized as follows. The investigated wavelet transform and some related work are described in sections 2 and 3 respectively. The experimental environment is covered in section 4. In Section 5 we discuss the memory hierarchy optimizations, then in section 6 our automatic vectorization technique is explained and some results are presented. Finally, the paper ends with some conclusions.

## 2 2-D Discrete Wavelet Transform

The discrete wavelet transform (DWT) can be efficiently performed using a pyramidal algorithm based on convolutions with Quadrature Mirror Filters (QMF). The wavelet representation of a discrete signal *S* can be computed by convolving *S* with the lowpass filter *H(z)* and highpass filter *G(z)* and downsampling the output by 2. This process decomposes the original image into two sub-bands, usually denoted as the coarse scale approximation (lower band) and the detail signal (higher band) [2].

This transform can be easily extended to multiple dimensions by using separable filters, i.e. by applying separate 1-D transforms along each dimension. In particular, we have studied the most common approach, commonly known as the *square* decomposition. This scheme alternates between operations on rows and columns, i.e. one stage of the 1-D DWT is applied first to the rows of the image and then to the columns. This process is applied recursively to the quadrant containing the coarse scale approximation in both directions. In this way, the data on which computations are performed is reduced to a quarter in each step (see figure 1) [2].

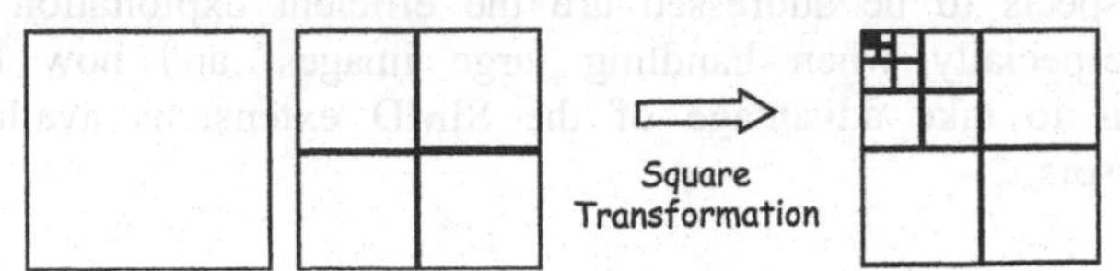


**Fig. 1.** The *Square* variant of the 2-D DWT

From a performance point of view, the main bottleneck of this transformation is caused by the vertical filtering (the processing of image columns) or the horizontal one (the processing of image rows), depending on whether we assume a row-major or a column-major layout for the images. In particular, all the measurements taken in our research have been obtained performing the whole wavelet decomposition using a (9,7) tap biorthogonal filter [2]. Nevertheless, qualitatively our results are filter-independent.

# 3 Related Work

A significant amount of work on the efficient implementation of the 2-D DWT has already been done for all sorts of computer systems. However, most previous research has concentrated on special purpose hardware for mass-market consumer products [4][5][6]. Focusing on general purpose microprocessors, S. Chatterjee *et al.* [7] and P. Meerwald *et al.* [8] proposed several optimizations aimed at improving cache performance. Basically, [8] investigates the benefits of traditional loop-tiling techniques, while [7] investigates the use of specific array layouts as an additional means of improving data locality.

The thesis of [7] is that row-major or column-major layouts (canonical layouts) are not advisable in many applications, since they favor the processing of data in one direction over the other. As an alternative, they studied the benefits of two non-linear layouts, known in the literature as 4-D (see fig. 2) and Morton [7]. In these layouts the original $m \times n$ image is conceptually viewed as an $\lceil m/tr \rceil \times \lceil n/tc \rceil$ array of $tr \times tc$ tiles. Within each block, a canonical (row-major or column-major) layout is employed. For a benchmark suite composed of different dense matrix kernels and two different wavelet transforms, both layouts have low implementation costs (2-5% of the total running time) and high performance benefits. In particular, focusing on the wavelet transform, the running time improvements achieved on a DEC workstation (equipped with a 500 MHz Alpha 21164 microprocessor and 2 MB of L3 cache) reached up to 60% compared to a more traditional version of the code [7] (the Morton layout performance was slightly better).

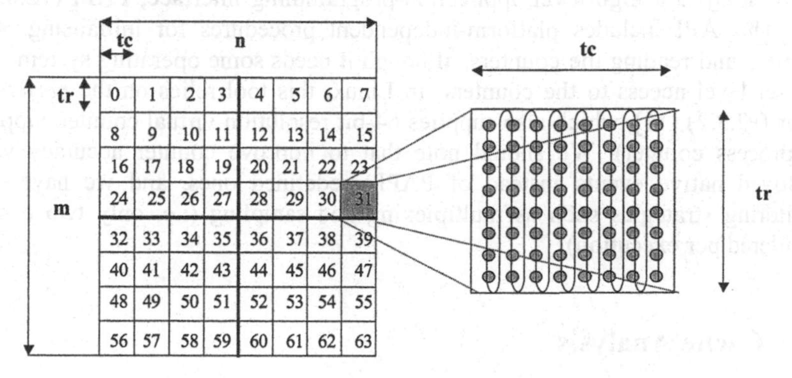


**Fig. 2.** 4-D layout

The approach investigated in [8] is less aggressive. Nevertheless, they addressed the memory exploitation problem in the context of a whole application, the JPEG2000 image coding, which is more tedious to optimize than a wavelet kernel. In particular, they considered the reference implementations of the standard (known as jasper and jj2000). By default, both implementations use a five-level wavelet decomposition with (7,9) biorthogonal filters as the intra-component transform of the coding [8]. The solution investigated by these authors consists in applying a loop-tiling strategy to the vertical filtering (the reference implementations used a row-major layout), which they dubbed "aggregation". In this scheme, instead of processing every image column all the way down in one step, which produces very low data locality (on a row-major layout, rows are aligned along cache lines), the algorithm is improved by processing several columns concurrently so that the spatial locality can be more effectively exploited.

We have extended these previous studies by assessing the influence of the SIMD extensions and including a more elaborate analysis based on the Intel PIII's performance counters. In this first version of our study we have followed the kernel approach chosen by S. Chatterjee *et al.* although, as a future research area, we intend to introduce the proposed optimizations on a entire well-known application such as the JPEG-2000.

## 4 Experimental Environment

The performance analysis presented in this paper has been carried out on a Pentium-III 866 MHz (0,18microns) based PC running under Linux, the main features of which are summarized in [20]. The programs have been compiled using the Intel C/C++ Compiler for Linux (v5.0.1) [9] and the compiler switches (-O3 –tpp6 -xK) described also in [20].

Our measurements have been made using the performance-monitoring counters available on the P6 processor family [3]. This micro-architecture provides two 40-bit performance counters, allowing two types of events to be monitored simultaneously. In order to avoid assembly language programming and due to portability reasons, we have employed a high-level application-programming interface, PAPI (v2.0.1 beta) [10]. This API includes platform-independent procedures for initialising, starting, stopping, and reading the counters, although it needs some operating system support for user level access to the counters. In Linux, this tool relies on the *perfctr* kernel driver (v2.3.2) [11], which also supplies 64-bit resolution virtual counter support (i.e. per process counter). We should note that to improve counter accuracy we have employed native events instead of PAPI predefined ones, and we have avoided monitoring strategies such as multiplexing and sampling (i.e. only two events are considered per execution).

## 5 Cache Analysis

As mentioned before, the wavelet transform poses a major hurdle for the memory hierarchy, due to the discrepancies between the memory access patterns of the two main components of the 2-D wavelet transform: the vertical and horizontal filterings [2]. Consequently, the improvement in the memory hierarchy use represents the most important challenge of this algorithm from a performance perspective. In this section, we have exhaustively analyzed the cache behavior of the three different approaches introduced previously, namely the 4-D and Morton layouts (non-linear layouts) and the row-major layout combined with the aggregation technique (see section 3). We have divided this section into 3 different parts. First, we describe the different ways to apply the vertical and horizontal filters and their relation to the image layout. Section 5.2 discusses some implementation issues and the experimental results are presented in section 5.3.

## 5.1 Tile Layout and Block Processing Type

The following algorithms show four reasonable ways of applying a 1-D filter to a tile of $tr \times tc$ elements, which we have denoted as vertical, horizontal, N and Z element processing:

* *Vertical:* foreach column{ foreach row{ foreach coef{ filter }}}
* *Horizontal:* foreach row{ foreach column{ foreach coef{ filter }}}
* *N:* foreach column{ foreach coef{ foreach row{ filter }}}
* *Z:* foreach row{ foreach coef{ foreach column{ filter }}}

Depending on the tile memory layout, some processing types are preferable over others due to data locality. It is relatively obvious that for the row-major layout the best access pattern is produced when elements are processed horizontally, for either vertical or horizontal filtering. On the other hand, for the column-major layout, it is better to process the elements vertically. The Z and N approaches represent a hybrid approach that was considered in the previous versions of our codes since they are easily vectorizable, as will be explained later in section 6.

In order to make a fair comparison of the different alternatives analyzed in this section, we employed the best processing type for each kind of block layout. Given that for the 4-D and Morton layouts we have opted to use the same approach as that followed in [7] (within each block a column-major layout is employed), we have chosen vertical processing in these cases. For the row-major layout combined with the aggregation technique we have employed horizontal processing. Nevertheless, we should remark that due to the symmetry of the problem, these particular choices have no effect on the overall performance.

Figure 3 graphically describes the processing of the image tiles in the 4-D and Morton approaches (for simplicity, boundary data have been ignored). The vertical filtering does not need any special treatment since the image columns are stored contiguously in memory. In this case, the main problem is due to the horizontal filtering, since processing the tile row by row does not take advantage of the spatial locality. In order to remedy this situation, the tile is swept column by column in a similar way to the aggregation technique proposed in [8].

Figure 4 illustrates both filtering types when a row-major layout is employed for the whole image. The horizontal filtering does not cause any problem whereas the vertical one has to be improved by means of aggregation [8].

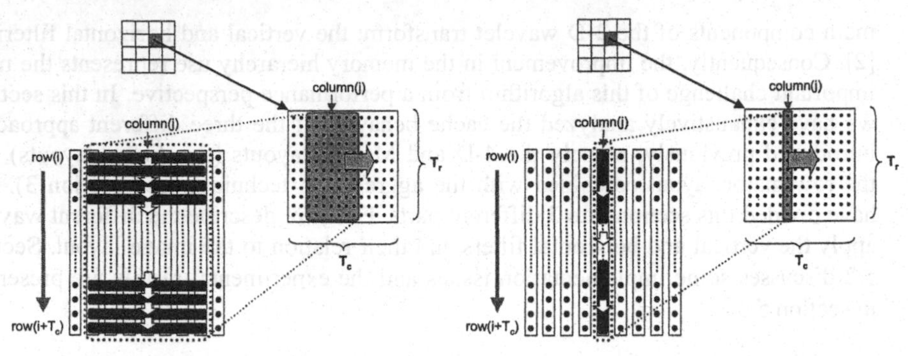


**Fig. 3.** Horizontal (left-hand chart) and vertical (right-hand) filtering employed in the 4-D and Morton approaches

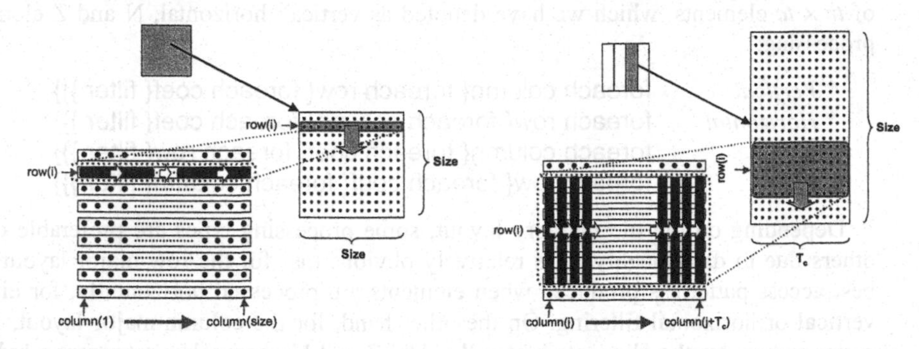


**Fig. 4.** Horizontal filtering (left-hand chart) and vertical filtering (right-hand chart) for the row-major layout

## 5.2 Implementation Issues

* *Filter loop unroll:* We have observed that the compiler does not automatically unroll the filter loop in any components of the transform. Due to its reduced number of iterations (filter length), this modification can be easily performed by hand. In addition, this modification allows the compiler to perform further optimizations and, as section 6.1 explains, also permits automatic vectorization to be employed in the vertical and horizontal processing.

* *Data alignment:* Strictly speaking, data alignment [12][13] is not required in our codes since the SSE instruction set includes instructions that allow unaligned data to be copied into and out of the vector registers. However, such operations are much slower than aligned accesses, which may cause a significant overhead. To avoid this drawback we have employed 16-byte aligned data in all our codes, although for the scalar versions this optimization has no significant effect.

## 5.3 Experimental Results

The results reported in this section have been obtained using the experimental framework explained in section 4. Before analyzing them, we should briefly explain the metrics involved. The execution time measurements have been obtained using the PAPI virtual time routines [10], which are context-switch independent. The memory hierarchy behavior has been monitored through the "DCU LINES IN" and "L2 LINES IN" events, which represent the number of lines that have been allocated in the L1 data cache, and the number of L2 allocated lines respectively. These events are strongly related with the number of L1 and L2 cache misses.

Figures 5 and 6 represent memory hierarchy behavior and the execution time for an image size of $8192^2$ pixels. When employing horizontal filtering all the approaches obtain, for the optimal block size, comparable results in execution time and in both L1 and L2 allocated lines. However, with vertical filtering, Morton and 4-D produce a significantly lower number of misses than the row-major layout in both levels of the memory hierarchy, as well as a slower running time.

This relatively bad row-major layout behavior is due to the poor spatial locality of its memory access pattern. Furthermore, for the L2 cache, elements belonging to the same wavelet coefficient computation are using the same block set, resulting in a high number of conflicts (we use a 9-coefficient filter, and the Pentium-III has only 8 blocks per set). In [8], this problem is overcome using array padding (dubbed "row extensions technique" by the authors) to force the image width not to be a power of two. However, as the authors themselves suggest, this simple technique has the disadvantage that the original input image has to be modified. In an application such as the JPEG-2000, inserting dummy data changes the final coded bitstream [8]. Using 4-D or Morton, array padding is not necessary, since in these cases conflict misses are not a function of the image size but of the tile size.

Regarding the L1 data cache behavior we should mention that, although less influential on performance, the improvements of both the 4-D and Morton layouts on the row-major layout are also significant. We should also remark that the similarity of the curves for the execution time and L2 allocated lines suggests a strong relationship between performance and L2 behavior. As a result, the 4-D and Morton approaches produce a speedup gain of about 2.25 over the row-major layout. Comparing the two non-linear approaches, we observe that the results are almost the same (1.3% of difference), especially for the optimum tile sizes.

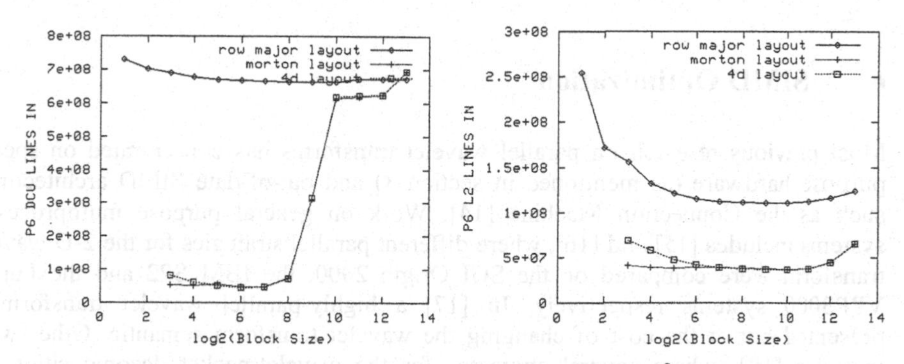


**Fig. 5.** L1 data cache and L2 cache behavior for an $8192^2$ pixels image

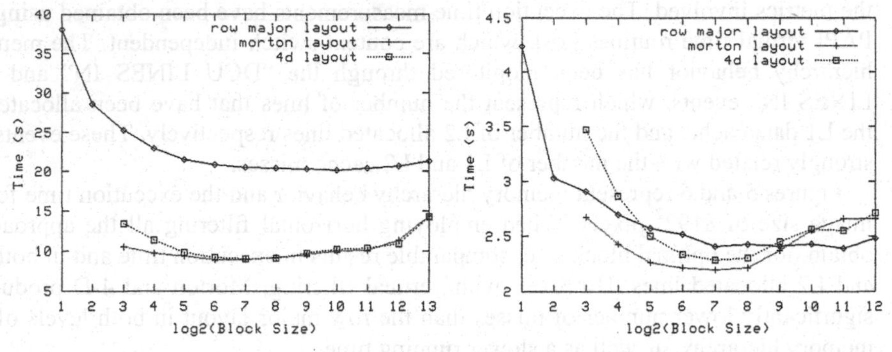


**Fig. 6.** Execution time for an $8192^2$ pixels image (left chart) and for a $4096^2$ pixels image (right chart)

Figure 6 (right chart) shows the execution time for an $4096^2$ pixel image. Performance is strongly related to the L2 behavior, although differences in the L1 behavior now translate as small variations in execution time (with a $8192^2$ image they were almost inappreciable in terms of time). The benefits of non-linear layouts are less significant in this case, since the stride of the row-major layout in the vertical filtering is lower, resulting in a smaller number of L2 cache conflicts (only half of the elements of wavelet coefficient computation are competing for the same block set). Comparing the behavior of the Morton and 4-D layouts, we remark once again that they are very similar (4% of difference).

From the previous results we can conclude that, in general, Morton and 4-D are preferable to the row-major layout, since their memory access patterns exhibit more locality. The memory hierarchy is therefore more efficiently exploited and thus the execution time is significantly reduced. The running time benefits of these approaches are higher for image sizes of $8192^2$ pixels and above, mainly due to the L2 behavior explained earlier. Taking into account that the 4-D layout is simpler to implement (it does not need a lookup table to handle blocks [7]) and achieves a similar performance to the Morton, we have chosen this method to study the vectorization.

## 6 SIMD Optimization

Most previous research on parallel wavelet transforms has concentrated on special purpose hardware (as mentioned in section 3) and out-of-date SIMD architectures, such as the Connection Machine [14]. Work on general purpose multiprocessor systems includes [15] and [16], where different parallel strategies for the 2-D wavelet transform were compared on the SGI Origin 2000, the IBM SP2 and the Fujitsu VPP3000 systems respectively. In [17] a highly-parallel wavelet transform is presented but at the cost of changing the wavelet transform semantic. Other work includes [18], where several strategies for the wavelet-packet decomposition are studied.

We have focused our research on the potential benefits of Single Instruction Multiple Data (SIMD) extensions. Among related work, we can mention [19], where an assembly language vectorization of real and complex FIR filters is introduced based on Intel SSE. Our main interest is to assess whether it is possible to take advantage of such extensions to exploit the data parallelism available in the wavelet transform, though in a filter-independent way and avoiding low level programming.

Most of the results reported in this work have been obtained by using automatic vectorization. As expected, the compiler was not able to vectorize any loop by itself, so both code modifications and guided compilation were necessary. However, it should be noted that the analysis of the vectorization inhibitors provided by the Intel compiler has been a considerable aid. We have also optimized the code using the intrinsic functions that the Intel compiler offers. This technique involves additional improvements at the expense of a greater coding effort, although it is more portable than coding at the assembly level since most compilers provide similar functions.

This section is divided into two parts. In the first, we have studied how to vectorize the horizontal filtering. In the second, we have extended the vectorization method to the whole transform. For the sake of simplicity, we have only considered the 4-D column-major layout, although analogous results can be obtained for the Morton approach (see [20]).

## 6.1 Horizontal Filtering Vectorization

Depending on the memory layout, either column-major or row-major, we can vectorize either the horizontal or vertical filtering using the methodology presented below. In particular, we have applied this technique to the horizontal filtering since we are focusing on the 4-D column-major layout.

### 6.1.1 Methodology

Loops must fulfill some requirements in order to be automatically vectorized. Primarily, only loops with simple array index manipulation (i.e. unit increment) and which iterate over contiguous memory locations are considered (thus avoiding non-contiguous accesses to vector elements). Obviously, only inner loops can be vectorized. In addition, global variables must be avoided since they inhibit vectorization. Finally, if pointers are employed inside the loop, pointer disambiguation is mandatory (this must be done by hand using compiler directives).

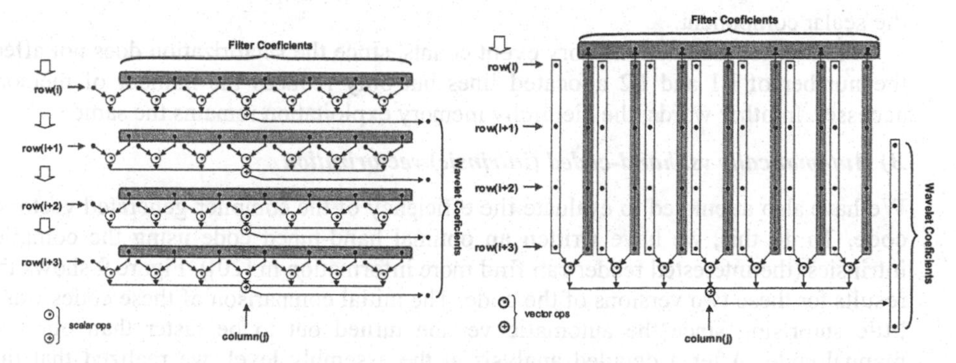


**Fig. 7.** Horizontal filtering using vertical sweep for the scalar version (left chart) and Horizontal filtering using vertical sweep for the vector version (right chart)

Considering these restrictions, it is obvious that not all the processing types and filtering components can be vectorized automatically. In particular, for a Row-major Layout only the Vertical Filtering using *Horizontal* or *Z Processing* can be automatically vectorized, while for a Column-major Layout only the Horizontal Filtering using *Vertical* or *N Processing* do. Nevertheless, we should remark that when using either assembly language or function intrinsics these limitations can be overcome at the expense of more coding effort.

### 6.1.2 Vectorization

This technique consists in calculating the wavelet coefficients following the element layout in the memory. We shall suppose that to evaluate a certain wavelet coefficient we must center the filter on element *j* of row *i* in one of the 4-D layout tiles. The optimum processing consists in moving the filter downwards, from row to row, to calculate all the wavelet coefficients of column *j*. From figure 7 (left chart) it seen that, in this particular case (a 7-tap filter), each coefficient requires 7 floating point multiplications and 6 floating point additions. Consequently, to calculate 4 coefficients, 28 floating point multiplications and 24 floating point additions are necessary. However, if vectorization is enabled (figure 7, right chart) the calculations of every 4 coefficients can be performed concurrently. Since the elements of each column are stored contiguously, the compiler is able to load the matrix elements into the SSE registers in groups of four (thus using less instructions).

### 6.1.3 Experimental Results

***A) Vectorized vs. Non-vectorized horizontal Filtering***

Figure 8 shows the execution time for $8192^2$ and $4096^2$ pixel images using both vectorized and non-vectorized versions of the code. We observed that for every configuration and image size under study, the vectorized horizontal filtering beats the scalar running time. In particular, for the optimum block size it achieves a speedup of about 2 (for both image sizes), which translates to a speedup of about 1.4 for the whole transform. Obviously, the vertical filtering behavior has not changed since this part of the program has not been modified. Considering the entire transform, we should note that the optimum block size for each version of the code is different, because the contribution of the vectorized horizontal component is lower than that of the scalar component.

We have not included memory event counts, since the vectorization does not affect the number of L1 and L2 allocated lines but only reduces the number of memory accesses. In other words, the hierarchy memory exploitation remains the same.

***B) Automatically vs. hand-coded (intrinsic) vectorization***

We have also attempted to evaluate the efficiency of the compiler-generated vectorial code. To do this, we have written an optimal hand-tuned code using the compiler intrinsics (the interested reader can find more information in [20]). Figure 9 shows the results for these two versions of the code. The initial comparison of these codes was a little surprising since the automatic version turned out to be faster than our best manual code. After a detailed analysis at the assembly level, we realized that this difference is caused by the prefetching introduced by the compiler when automatic vectorization is enabled. We have verified this conclusion by removing the prefetch instructions from the assembler code, the results of which are also shown in figure 9. As can be seen, with regard to vectorization the automatic code is worse than the manual (about 21% worse).

We should remark that the compiler does not perform automatic prefetching in the hand-tuned code. In addition, introducing manual prefetching is a tough task and the resulting code is highly platform-dependent, which makes the automatic vectorization preferable since it produces a higher speedup with a minor programming effort.

Thus, returning to our original comparison (automatic vectorizable vs. scalar), the speedup of the automatic version (about 2) over the scalar code is due not only to vectorial operations but also to pre-fetching, each with the same contribution to the overall gain in speedup.

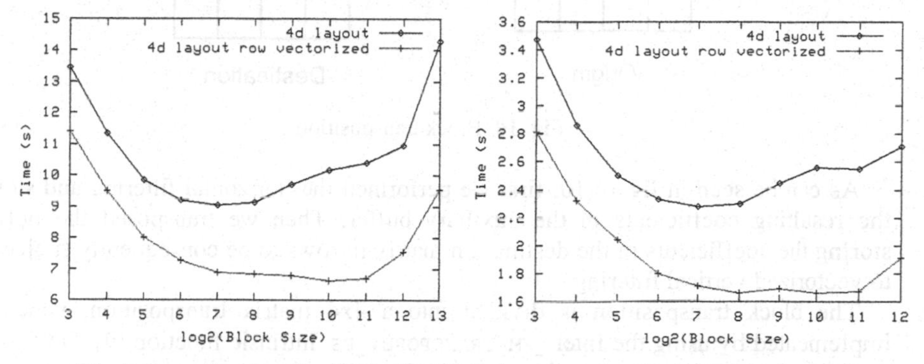


**Fig. 8.** Execution time for 8192x8192 (left chart) and for 4096x4096 (right chart)

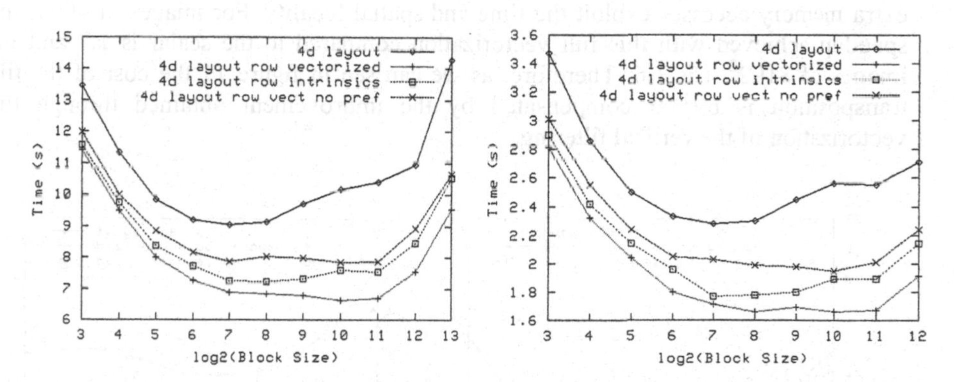


**Fig. 9.** Execution time for 8192x8192 (left chart) and for 4096x4096 (right chart)

## 6.2 Full Vectorization

We have obtained excellent results from horizontal filtering vectorization due to the matrix elements being stored contiguously in columns. In order to apply the same technique to the vertical filtering, we needed the elements to be stored contiguously in rows. One possible solution was to apply the horizontal filtering followed by a transposition of the resulting wavelet coefficients. This then allowed us to use the same vectorization technique vertically. To carry out the transposition efficiently we could not work with the whole matrix at the same time. Therefore, this transform was performed tile by tile taking advantage of the 4-D layout, which required the use of an auxiliary buffer of tile size. Note that this kind of transposition is not feasible when using the row-major layout since the image is not divided into tiles.

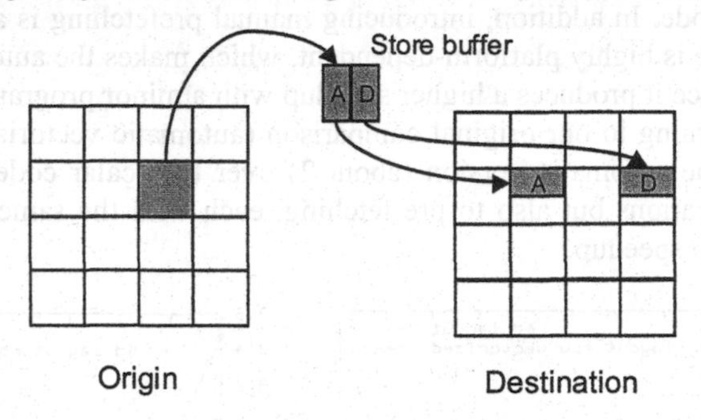


**Fig. 10.** Block transposition

As can be seen in figure 10, first we performed the horizontal filtering and stored the resulting coefficients in the auxiliary buffer. Then we transposed the buffer, storing the coefficients in the destination matrix in rows to be consequently subjected to vectorized vertical filtering.

The block transposition is divided into a 4x4 matrix transposition, which is implemented by using the Intel _MM_TRANSPOSE4_PS intrinsic function [9]. Obviously the transpose computation implies a cost in time since it is necessary to use extra load and store instructions. However, the smaller the tile size the more efficiently all these extra memory accesses exploit the time and spatial locality. For images of $4096^2$ the speedup achieved with this full vectorization compared to the scalar is 1.7 and for images of $8192^2$ it is 1.6. Therefore, as we can see in figure 11 the cost of the tile transposition is by far compensated by the improvement obtained through the vectorization of the vertical filtering.

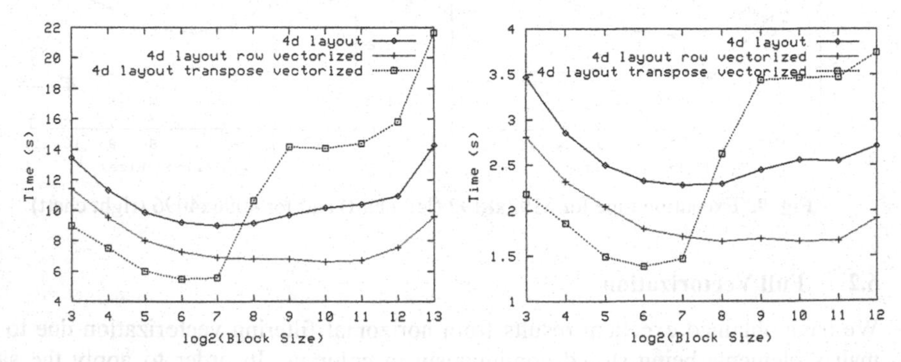


**Fig. 11.** Execution time for the whole wavelet transform for 8192x8192 (left chart) and for 4096x4096 (right chart)

## 7 Conclusions

In this paper we have introduced a novel approach to optimize the computation of the 2D DWT for large scale image processing based on non-linear data layouts, and automatic prefetching and vectorization. The main conclusions can be summarized as follows:

1. As shown by previous research [7], an increase in speedup is obtained through non-linear accesses, such as the 4-D or Morton layouts, which exploit spatial locality more effectively. The difference in speedup between the 4-D and Morton layouts is insignificant. Due to the simplicity of the 4-D compared to Morton, we recommend the former.
2. We have introduced a novel approach to structure the computation of the wavelet coefficients that allows automatic vectorization and pre-fetching, which is independent of the filter size and the computing platforms (assuming that similar SIMD extensions are available).
3. Our hand-tuned code achieves a better exploitation of SIMD parallelism at the expense of more coding effort. However, the compiler cannot perform prefetching in this code. In addition, introducing prefetching by hand is a tough task and it is highly platform-dependent. As a consequence, the automatic version is strongly preferable since the lower SIMD exploitation is by far compensated by prefetching.
4. In order to apply the vectorization to both filterings (horizontal and vertical) a block transposition is required. However, the performance gain achieved through vectorization by far compensated the transposition overhead.

## References


[1] Z. Zhang and R. S. Blum. A Categorization of Multiscale-Decomposition-Based Image Fusion Schemes with a Performance Study for a Digital Camera Application. Proceeding of the IEEE, Vol. 87(8):1315-1325, August 1999

[2] E. J. Stollnitz, T. D. DeRose and D. H. Salesin. Wavelets for Computer Graphics: Theory and Applications. Computer Graphics and Geometric Modeling, Morgan Kaufmann Publishers, Inc. San Francisco, 1996

[3] Intel Corp. Pentium-III processor. http://developer.intel.com/design/PentiumIII

[4] C. Chakrabarti and C. Mumford. Efficient realizations of encoders and decoders based on the 2-D discrete wavelet transforms. IEEE Trans. VLSI Syst., pp. 289-298, September 1999

[5] T. Denk and K. Parhi. LSI Architectures for Lattice Structure Based Orthonormal Discrete Wavelet Transforms. IEEE Trans. Circuits and Systems, vol. 44, pp. 129-132, February 1997

[6] C. Chrysafis and A. Ortega. Line Based Reduced Memory Wavelet Image Compression. IEEE Trans. on Image Processing, Vol 9, No 3, pp. 378-389, March 2000

[7] S. Chatterjee, V. V. Jain, et al. Nonlinear Array Layouts for Hierarchical Memory Systems. Proceedings of 1999 ACM International Conference on Supercomputing, pp. 444-453, Rhodes, Greece, June 1999

[8] P. Meerwald, R. Norcen, et al. Cache issues with JPEG2000 wavelet lifting. In C.-C. Jay Kuo, editor, Visual Communications and Image Processing 2002 (VCIP'02), volume 4671 of SPIE Proceedings, San Jose, CA, USA, January 2002

[9] Intel Corp. C/C++ Compiler. http://www.intel.com/software/products/compilers

[10] K. London, J. Dongarra, et al. End-user Tools for Application Performance Analysis, Using Hardware Counters. Presented at International Conference on Parallel and Distributed Computing Systems. August 2001

[11] Perfctr Linux driver. Info. available at http://www.csd.uu.se/~mikpe/linux/perfctr

[12] Intel Corp. Data Alignment and Programming Issues for the Streaming SIMD Extensions with the Intel C/C++ Compiler. Intel Application Note AP-833. Available at http://developer.intel.com

[13] Intel Corp. Intel Architecture Optimization. Reference Manual. Available at http://developer.intel.com

[14] M. Holmström. Parallelizing the fast wavelet transform. Parallel Computing, 1(21):1837-1848, April 1995

[15] D. Chaver, M. Prieto, L. Piñuel, F. Tirado. Parallel Wavelet Transform for Large Scale Image Processing. Proceedings of the International Parallel and Distributed Processing Symposium (IPDPS'2002). Florida, USA, April 2002

[16] O.M. Nielsen and M. Hegland. Parallel Performance of Fast Wavelet Transform. International Journal of High Speed Computing, 11 (1): 55-73, June 2000

[17] L. Yang and M. Misra. Coarse-Grained Parallel Algorithms for Multi-Dimensional Wavelet Transforms. The journal of Supercomputing 11:1-22 , 1997

[18] M. Feil and A. Uhl. Multicomputer algorithms for wavelet packet image decomposition. Proceedings of the International Parallel and Distributed Processing Symposium (IPDPS'2000), pages 793-798, Cancun, Mexico, 2000

[19] Intel Corp. Real and Complex FIR Filter Using Streaming SIMD Extensions. Intel Application Note AP-809. Available at http://developer.intel.com

[20] D. Chaver, C. Tenllado, L. Piñuel, M. Prieto and F. Tirado. Vectorizing the Wavelet Transform on the Intel Pentium III Microprocessor. Technical Report 02-001. Dept. of Computer Architecture. Complutense University, 2002

# Chapter 8

# Software Tools and Environments

# An Expandable Parallel File System Using NFS Servers

Félix García, Alejandro Calderón, Jesus Carretero,
Jose M. Pérez, and Javier Fernández

Computer Architecture Group, Computer Science Department
Universidad Carlos III de Madrid, 28911 Leganés, Madrid, Spain
fgarcia@arcos.inf.uc3m.es

**Abstract.** This paper describes a new parallel file system, called Expand (Expandable Parallel File System)[1], that is based on NFS servers. Expand allows the transparent use of multiple NFS servers as a single file system. The different NFS servers are combined to create a distributed partition where files are declustered. Expand requires no changes to the NFS server and uses RPC operations to provide parallel access to the same file. Expand is also independent of the clients, because all operations are implemented using RPC and NFS protocol. Using this system, we can join heterogeneous servers (Linux, Solaris, Windows 2000, etc.) to provide a parallel and distributed partition. The paper describes the design of Expand and the evaluation of a prototype of Expand. This evaluation has been made in Linux clusters and compares Expand, NFS and PVFS.

## 1 Introduction

Traditional network and distributed file systems support a global and persistent name space that allow multiple clients to share the same storage devices, but does not provide parallel access to data, becoming the file servers in a major bottleneck in the system. The use of parallelism in file systems alleviates the growing disparity in computational and I/O capability of the parallel and distributed architectures.

Parallelism in file systems is obtained using several independent server nodes supporting one o more secondary storage devices. Data are *declustered* among these nodes and devices to allow parallel access to different files, and parallel access to the same file. This approach increases the performance and scalability of the system. Parallelism has been used in some parallel file system and I/O libraries described in the bibliography (Vesta [3], HFS [13], PIOUS [16], Scotch [10], ParFiSys [2, 9], Galley [19], PVFS [1], Armada [20], and ViPIOS [8]). However, current parallel file systems and parallel I/O libraries lack generality

[1] This work has been partially support by the Spanish Ministry of Science and Technology under the TIC2000-0469 and TIC2000-0471contracts, and by the Community of Madrid under the 07T/0013/2001 contract.

J.M.L.M. Palma et al. (Eds.): VECPAR 2002, LNCS 2565, pp. 565–578, 2003.

and flexibility for general purpose distributed environments. Furthermore, all parallel file systems do not use standard servers, thus it is very difficult to use these systems in heterogeneous environments as, for example, clusters of workstations.

This paper shows a new approach to the building of parallel file systems for heterogeneous distributed and parallel systems. The result of this approach is a new parallel file system, called Expand (*Expandable Parallel File System*). Expand allows the transparent use of multiple NFS servers as a single file system. Different NFS servers are combined to create a distributed partition where files are declustered. Expand requires no changes to the NFS server and it uses RPC operations to provide parallel access to the same file. Using this system, we can join different servers (Linux, Solaris, Windows 2000, etc.) to provide parallel access to files in a heterogeneous cluster.

The rest of the paper is organized as follows. Section 2 shows systems related with Expand. Section 3 presents the motivations and goals of this work. The Expand design is shown in Section 4, where we describe the data distribution, the structure of the files, the naming and metadata management and the parallel access to files. Performance evaluation is presented in Section 5. The evaluation compares Expand with the performance obtained in NFS and PVFS. Finally, Section 6 summarizes our conclusions and the future work.

## 2 Related Work

The use of parallelism in the file systems is based on the fact that a distributed and parallel system consists of several nodes with storage devices. The performance and bandwidth can be increased if data accesses are exploited in parallel [25]. Parallelism in file systems is obtained using several independent server nodes supporting one o more secondary storage devices. Data are *declustered* among these nodes and devices to allow parallel access to different files, and parallel access to the same file. Initially, this idea was used in RAID (*Redundant Array of Inexpensive Disks*) [21]. However, when a RAID is used in a traditional file server, the I/O bandwidth is limited by the server memory bandwidth. But, if several servers are used in parallel, the performance can be increased in two ways:

1. Allowing parallel access to different files by using several disks and servers.
2. Declustering data using distributed partitions [2], allowing parallel access to the data of the same file.

The use of parallelism in file systems is different, however, to the use of replicated file systems. In a replicated file system, each disk into each server stores a full copy of a file. Using parallel I/O, each disk into each server stores only a part of the file. This approach allows parallel access to the same file.

Three different parallel I/O software architectures can be distinguished [7]: application libraries, parallel file systems and intelligent I/O systems.

*Application libraries* basically consist of a set of highly specialized I/O functions. These functions provides a powerful development environment for experts with specific knowledge of the problem to model using this solution. Representative examples are MPI-IO [18], an I/O extension of the standarized message passing interface MPI [17], and ADIO [23], a standard API yielding an abstract device interface for portable I/O.

*Parallel file systems* operate independently from the applications, thus allowing more flexibility and generality. Examples of parallel file systems are: Bridge [6], CFS [22], nCUBE [5], SFS [14], Vesta [3], HFS [13], PIOUS [16], Scotch [10], PPFS [11], ParFiSys [2], Galley [19], and PVFS [1].

Finally, an *intelligent I/O system* hides the physical disk access to the application developer by providing a transparent logical I/O environment. The user describes what he wants and the system tries to optimize the I/O requests applying optimization techniques. This approach is used in Armada [20], ViPIOS [8], and PPFS [15].

The main problem with application libraries and intelligent I/O systems are that they often lack generality and flexibility by creating only tailor-made software for specific problems. By the other hand, parallel file systems are specially conceived for multiprocessors and multicomputers, and does not integrate appropriately in general purpose distributed environments as cluster of workstations.

## 3 Motivation and Goals

The main motivation of this work is to build a parallel file system for heterogeneous general purpose distributed environments. To satisfy this goal, authors are designing and implementing a parallel file system using NFS servers, whose first prototype is described in this paper.

Network File System (NFS) [26] supports the NFS protocol, a set of remote procedure call (RPC) that provides the means for clients to perform operations on a remote file server. This protocol is operating system independent. Developed originally for being used in networks of UNIX systems, it is widely available today in many systems, as LINUX or Windows 2000, two operating systems very frequently used in clusters.

Figure 1 shows the architecture of Expand. This architecture shows how multiple NFS servers can be used as a single file system. File data are declustered by Expand among all NFS servers, using blocks of different sizes as stripping unit. Processes in clients use an Expand library to access an Expand distributed partition.

Using the former approach offers the following advantages:

1. No changes to the NFS server are required to run Expand. All aspects of Expand operations are implemented on the clients. In this way, we can use several servers with different operating system to build a distributed partition. Furthermore, declustered partitions can coexist with NFS traditional partitions without problems.

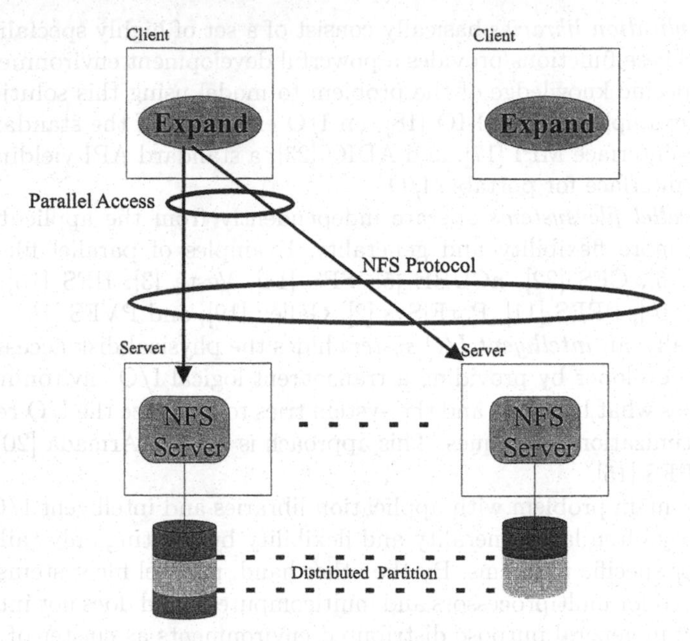


**Fig. 1.** Expand Architecture

2. Expand is independent of the operating system used in the client. All operations are implemented using RPC and NFS protocol.
3. The parallel file system construction is greatly simplified, because all operations are implemented on the clients. This approach is completely different to that used in many current parallel file system, that implement both client and server sides.
4. It allows parallel access to both data of different files and data of the same file, reducing the bottleneck that represents the traditional distributed file systems.
5. It allows the using of servers with different architectures and operating systems, because the NFS protocol hides those differences. Because of this feature, Expand is very suitable for heterogenous systems, as cluster of workstations.
6. Simplifies the configuration, because NFS is very familiar to users. Server only need to export the appropriate directories and clients only need a little configuration file that explain how is the distributed partition.

There are other systems that use NFS servers as basis of their work, similarly to Expand. Bigfoot-NFS [12], for example, also combines multiples NFS servers. However, this system uses files as the unit of interleaving (i.e., all data of a file reside in one server). Although files in a directory might be interleaved across several machines, it does no allow parallel access to the same file. Other similar system is Slice file system [4]. Slice is a storage system for high-speed networks that uses a packet filter $\mu$proxy to virtualize the NFS, presenting to NFS clients a unified shared file volume. This system uses the $\mu$proxy to distribute file service requests across a server ensemble and it offers compatibility with file system clients. However, the $\mu$proxy can be a bottleneck that can affect the global scalability of the system.

# 4 Expand Design

The user vision of the file is usually a byte stream, whereas the physical vision is a set of scattered blocks. The main goal of a file system is to translate user's logical I/O requests into system's physical orders providing high performance.

Expand provides high performance I/O exploiting parallel accesses to files stripped among several NFS servers. Expand is designed as a client-server system with multiple NFS servers, with each Expand file striped across some of the NFS servers. All operations in Expand clients are based on RPCs and NFS protocol. The first prototype of Expand is a user-level implementation, through a library that must be linked with the applications.

Next sections describe data distribution, file structure, naming, metadata management, and parallel access to files in Expand.

## 4.1 Data Distribution

To provide large storage capacity and to enhance flexibility, Expand combines several NFS servers to provide a generic stripped partition which allows to create several types of file systems on any kind of parallel I/O system. A partition in Expand is defined as follows:

$$DistributedPartition = \bigcup_{i=1}^{N} \{server_i, exportpath_i\}$$

Each server exports one o more directories that are combined to build a distributed partition. All files in the system are declustered across all NFS servers to facilitate parallel access, with each server storing conceptually a subfile of the parallel file.

A Partition in Expand can be expanded adding more servers to an existing distributed partition:

$$NewDistributedPartition = OldPartition \bigcup \{ \bigcup_{i=1}^{K} \{server_i, exportpath_i\}\}$$

This feature increases the scalability of the system and it also allows to increase the size of partitions. When new servers are added to an existent partition, the partition must be rebuilt to accommodate all files. An easy algorithm can be used to rebuild a partition:

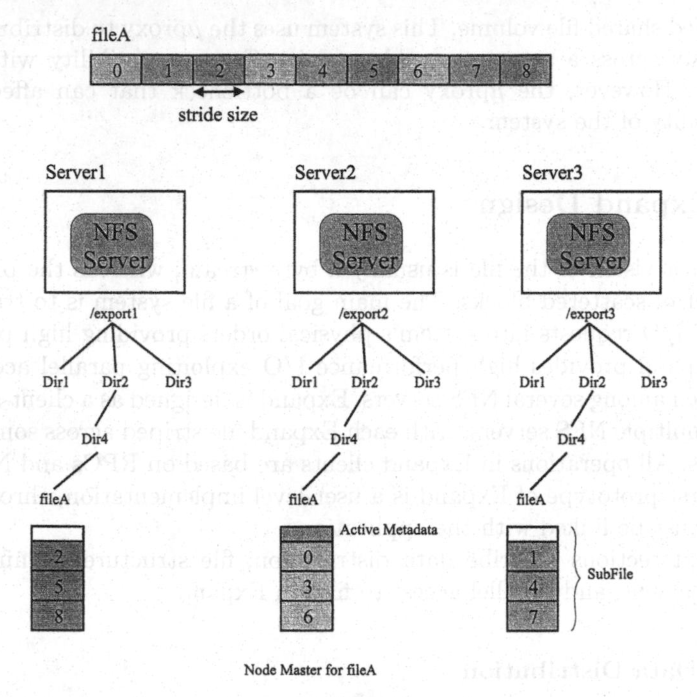


**Fig. 2.** Expand file with cyclic layout

```
rebuild_partition{old_partition, new_partition}
{
    for each file in old_partition
    {
        copy the file into the new_partition
        unlink the file in old_partition
    }
}
```

In this algorithm, when a file is copied into the new partition, the new file is declusterd across all NFS server of the new partition.

### 4.2 The Structure of Expand Files

A file in Expand consists in several *subfiles*, one for each NFS partition. Thus, a Expand file can be defined as:

$$File = \bigcup_{i=1}^{N} \{file_i\}$$

where $file_i$ is a file with the same name in the NFS partition $i$. All subfiles are fully transparent to the Expand users. Expand hides this subfiles, offering to the clients a traditional view of the files.

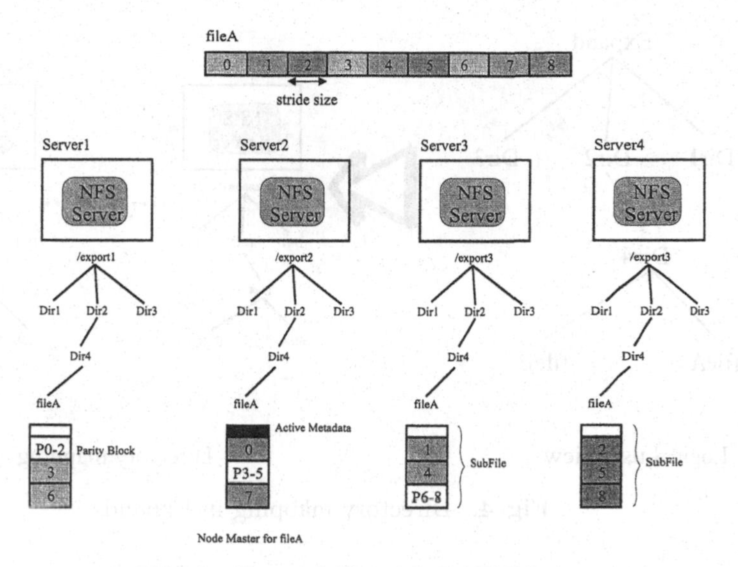


**Fig. 3.** Expand file with RAID5 layout

On a distributed partition, the user can create several types of files:

- Stripped files with cyclic layout. In these files, blocks are distributed across the partition following a round-robin pattern. This structure is shown in Figure 2.
- Fault tolerant files. These files can use RAID4 or RAID5 schemes to offer fault tolerance for files. Figure 3 shows how would be a file with a RAID5 configuration.

Each subfile of a Expand file (see Figure 2 and Figure 3) has a small header at the beginning of the subfile. This header stores the file's metadata. This metadata includes the following information:

- *Stride size.* Each file can use a different stride size. This parameter can be configured in open operation.
- *Kind of file*: cyclic, RAID4, or RAID5. This parameter also can be specified in the open operation.
- *Base node.* This parameter identifies the NFS server where the first block of the file resides.
- *Round-robin pattern.* All files in Expand, including RAID4 and RAID5 files, are striped using a round-robin pattern. Metadata stores the order used to decluster the file across the servers.

All subfiles has a header for metadata, although only one node, called *master node* (described below) stores the current metadata. The master node can be different from the base node.

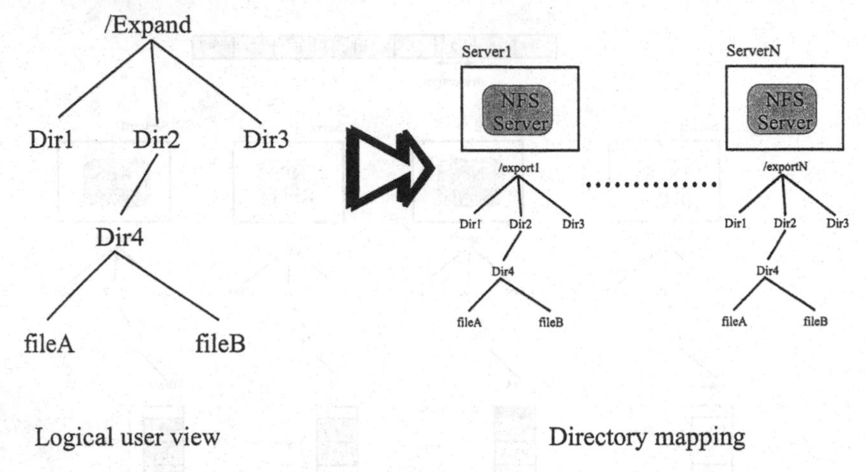


**Fig. 4.** Directory mapping in Expand

### 4.3 Naming and Metadata Management

To simplify the naming process and reduce potential bottlenecks, Expand does not use any metadata manager, as the used in PVFS [1]. Figure 4 shows how directory mapping is made in Expand. The Expand tree directory is replicated in all NFS servers. In this way, we can use the lookup operation of NFS without any change to access to all subfiles of a file. This feature also allows access to fault tolerance files when a server node fails.

The metadata of a file resides in the header of a subfile stored in a NFS server. This NFS server is the *master node* of the file, similar to the mechanism used in the Vesta Parallel File System [3]. To obtain the master node of a file, the file name is hashed into the number of node:

$$hash(namefile) \Rightarrow NFSserver_i$$

Fault tolerance files have the metadata replicated in several subfiles.

Initially the base node for a file agrees with the master node. The use of this simple scheme allows to distribute the master nodes and the blocks between all NFS servers, balancing the use of all NFS servers and, hence, the I/O load.

Because the determination of the master node is based on the file name, when a user renames a file, the master node for this file must be changed. The algorithm used in Expand to rename a file is the following:

```
rename(oldname, newname)
{
    oldmaster = hash(oldname)
    newmaster = hash(newname)
    move the metadata from oldmaster to newmaster
}
```

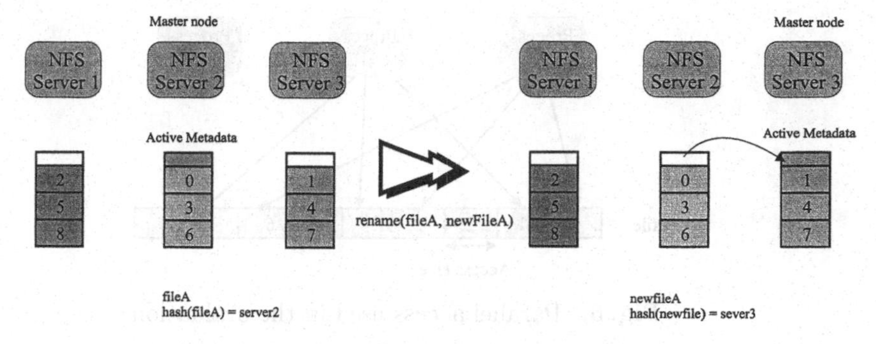


**Fig. 5.** Renaming a file in Expand

This process is shown in Figure 5. Moving the metadata is the only operation needed to maintain the coherence of the base node system for all the Expand files.

### 4.4 Parallel Access

NFS clients use *filehandle* to access the files. A NFS filehandle is an opaque reference to a file or directory that is independent of the filenane. All NFS operations use a filehandle to identify the file or directory which the operation applies to. Only the server can interpret the data contained within the filehandle. Expand uses a *virtual filehandle*, that is defined as follows:

$$virtualFilehandle = \bigcup_{i=1}^{N} filehandle_i$$

where $filehandle_i$ is the filehandle used for the NFS server $i$ to reference the subfile $i$ belonging to the Expand file. The virtual filehandle is the reference used in Expand to reference all operations. When Expand needs to access to a subfile, it uses the appropriated filehandle. Because filehandles are opaque to clients, Expand can use different NFS implementations for the same distributed partition.

To enhance I/O, user requests are split by the Expand library into parallel subrequests sent to the involved NFS servers. When a request involves $k$ NFS servers, Expand issues $k$ requests in parallel to the NFS servers, using threads to parallelize the operations. The same criteria is used in all Expand operations. A parallel operation to $k$ servers is divided in $k$ individual operations that use RPC and the NFS protocol to access the corresponding subfile.

### 4.5 User Interface

Expand offers an interface based of POSIX system call. This interface, however, is not appropriate for parallel applications using strided patterns with small access size [19]. For parallel applications, we are integrating Expand inside ROMIO [24] to support MPI-IO interface.

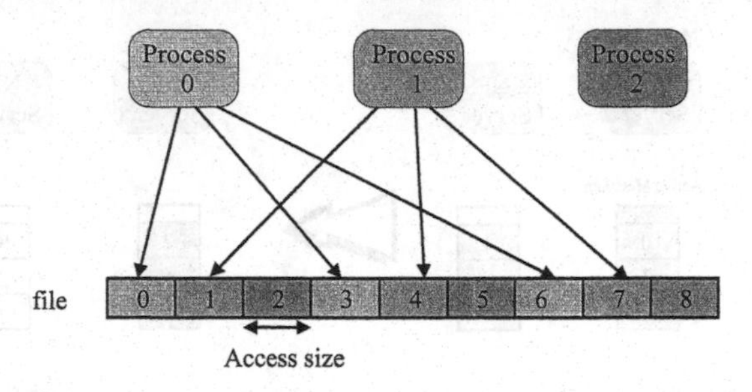


**Fig. 6.** Parallel access used in the evaluation

## 5 Evaluation

To evaluate Expand we have used a parallel program that writes a file of 100 MB in an interleaved fashion (see Figure 6), using 4, 8 and 16 processes. The whole file is written by all process in an interleaved fashion. The file is written using data access sizes from 128 bytes to 128 KB. The same program is used for reading a file. The result of the evaluation is the aggregated bandwidth obtained when executing the tests.

The platform used for the evaluation has been a cluster with 8 Pentium III biprocessors, each one with 1GB of main memory, connected through a Fast Ethernet, and executing Linux operating system (kernel 2.4.5). All experiments have been executed in Expand, NFS, and PVFS [1]. For Expand and PVFS, a distributed partition with 4 servers have been used. The block size used in all files, both in Expand as PVFS tests, has been 8 KB. All clients have been executed in 4 machines; thus, for 16 processes, 4 are executed in each machine. NFS implementation uses an I/O cache, while Expand and PVFS do not have a client cache.

Figure 7 shows the aggregated bandwidth obtained for write operations , and Figure 8 shows the same result for the read ones.

As can be seen in the Figures, NFS offers the worst performance, because all clients use one server. The performance offered by the server does not scale with the number of clients. Figures also show that the use of a parallel file system as Expand or PVFS offers a better performance that the provided by a traditional distributed file system as NFS. A special detail of the evaluation is the poor performance obtained by PVFS in the cluster used in the evaluation for short access size (lower than 1 KB).

Figures also show that the best performance is obtained for Expand, specially when the number of clients is increased, and even for short access sizes.

## 6 Conclusions and Future Work

In this paper we have presented the design of a new parallel file system, named Expand, for clusters of workstations. Expand is built using NFS servers as basis.

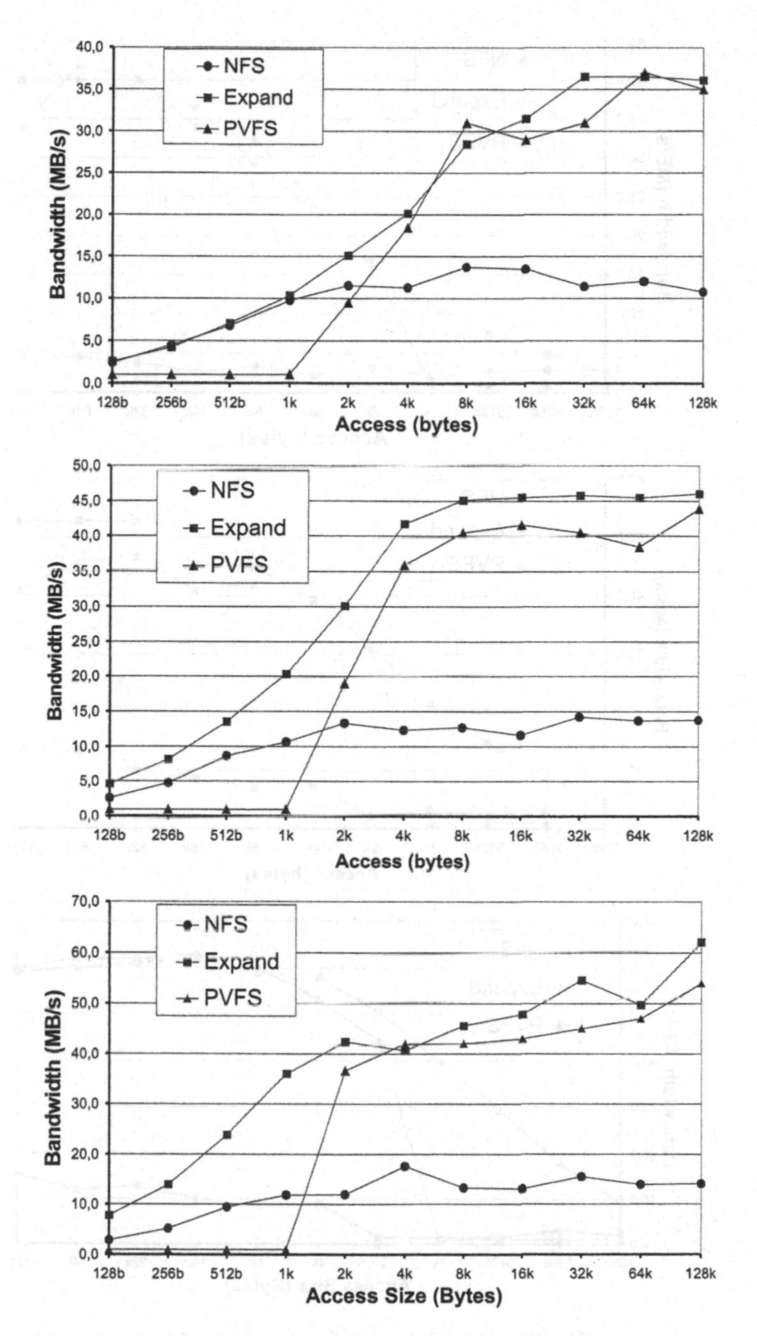


**Fig. 7.** Parallel write for 4, 8, and 16 clients

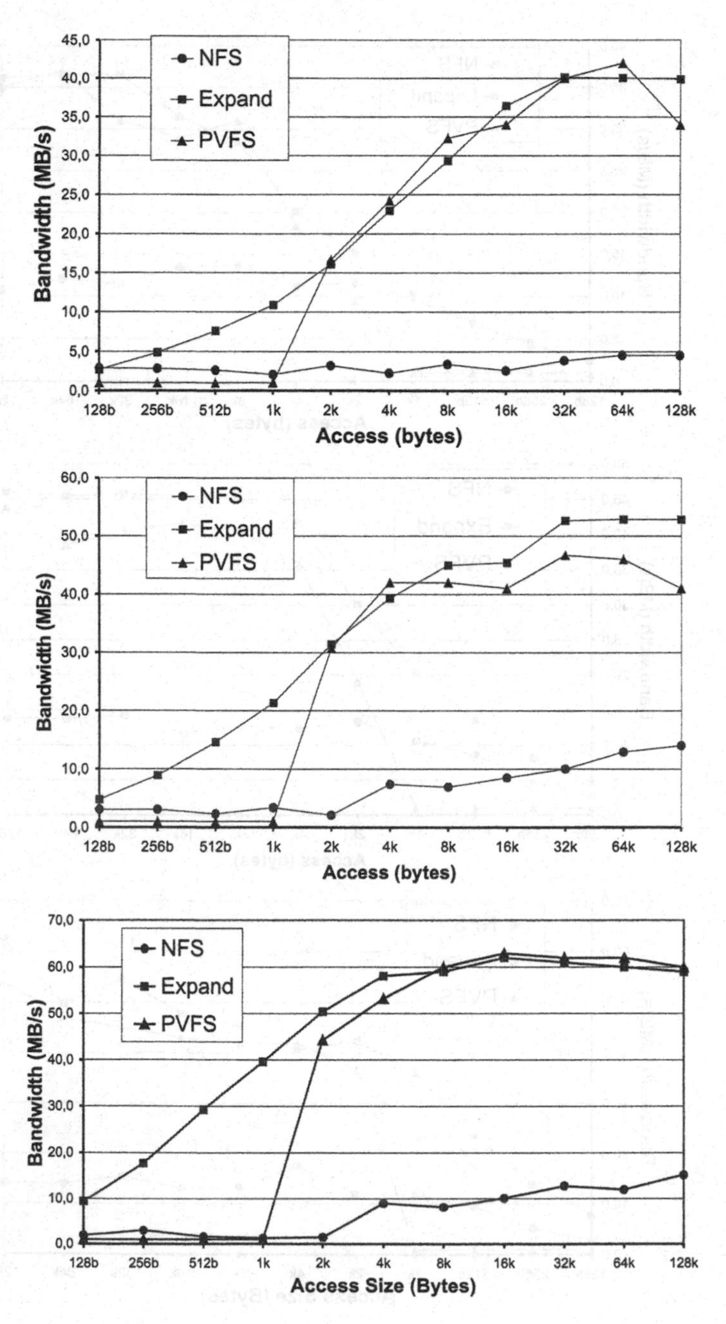


**Fig. 8.** Parallel read for 4, 8, and 16 clients

This solution is very flexible because no changes to the NFS servers are required to run Expand. Furthermore, Expand is also independent of the operating system used in the clients, due to the use of RPC and NFS protocol. Expand combines several NFS servers to provide a distributed partition where files are declustered. Expand improves the scalability of the system because the distributed partitions can be expanded adding more NFS servers to the partition.

The current prototype of Expand is a user-level implementation, through a library. This prototype only implements files with cyclic layout. The evaluation has compared the performance of Expand with the obtained in NFS and PVFS. The evaluation shows that Expand offers the best results, specially when the number of clients is increased, and even for small access size.

Further work is going on to optimize Expand adding a cache to clients and a cache coherence protocol similar to the one described in [9]. Other future work will be to include the fault tolerance files implementation. We also want to include Expand into ROMIO [24] implementation of MPI-IO.

## References

[1] Carns, P. H., Ligon III, W. B., Ross, R. B., Takhur, R. PVFS: A Parallel File System for Linux Clusters. Technical Report ANL/MCS-P804-0400, 2000.

[2] Carretero, J., Perez, F., de Miguel, P., Garcia, F., Alonso, L. Performance Increase Mechanisms for Parallel and Distributed File Systems. *Parallel Computing: Special Issue on Parallel I/O Systems. Elsevier*, (23):525–542, April 1997.

[3] Corbett, P., Johnson, S., Feitelson, D. Overview of the Vesta Parallel File System. *ACM Computer Architecture News*, 21(5):7–15, December 1993.

[4] Chase, F. S., Anderson, D. C., Vahdat, A. M. Interposed Request Routing for Scalable Network Storage. In *Fourth Symposium on Operating System Design and Implementation (OSDI2000)*, 2000.

[5] DeBenedictis, E., del Rosario, J. M. Ncube Parallel I/O Software. In *Eleventh Annual IEEE International Phoenix Conference on Computers and Communications (IPCCC)*, pages 117–124, April 1992.

[6] Dibble, P., Scott, M., Ellis, C. BRIDGE: A High Performance File System for Parallel Processors. In *Proceedings of the IEEE Eighth ICDCS*, pages 154–161. IEEE, June 1988.

[7] Wanek, H., Schikuta, E. Parallel I/O. In *Cluster Computing White Paper*, December 2000.

[8] Fuerle, T., Jorns, P., Schikuta, E., Wanek, H. Meta-ViPIOS: Harness Distributed I/O resources with ViPIOS. *Journal of Research Computing and Systems, Special Issue on Parallel Computing*, 1999.

[9] Garcia, F., Carretero, J., Perez, F., de Miguel, P., Alonso, L. High Performance Cache Management for Parallel File Systems. *Lecture Notes in Computer Science*, Vol. 1573, 1999.

[10] Gibson, G. The Scotch Paralell Storage Systems. Technical Report CMU-CS-95-107, Scholl of Computer Science, Carnegie Mellon University, Pittsburbh, Pennsylvania, 1995.

[11] Huber, J., Elford, C. L., et al. PPFS: A High Performance Portable Parallel File System. In *Proceedings of the 9th ACM International Conference on Supercomputing*, pages 385–394. IEEE, July 1995.

[12] Kim, G. H., Minninch, R. G. Bigfoot-NFS: A Parallel File-Striping NFS Server. Technical report, Sun Microsystems Computer Corp., 1994.
[13] Krieger, O. *HFS: A Flexible File System for Shared-Memory Multiprocessors.* PhD thesis, Department of Electrical and Computer Engineering, University of Toronto, 1994.
[14] LoVerso, S., Isman, M., Nanopoulos, A., Nesheim, W., Milne, E., Wheeler, R. *sfs*: A Parallel File System for the CM-5. In *Proceedings of the 1993 Summer Usenix Conference*, pages 291–305, 1993.
[15] Madhyastha, T. *Automatic Classification of Input/Output Access Patterns.* PhD thesis, niversidad de Illinois, Urbana-Champaign, 1997.
[16] Moyer, S. A., Sunderam, V. S. PIOUS: A Scalable Parallel I/O System for Distributed Computing Environments. In *Proceedings of the Scalable High-Performance Computing Conferece*, pages 71–78, 1994.
[17] MPI: A Message-Passing Interface Standard, 1995. http://www.mpi-forum.org.
[18] MPI-2: Extensions to the Message-Passing Interface, 1997. http://www.mpi-forum.org.
[19] Nieuwejaar, N., Kotz, D. The Galley Parallel File System. In *Proceedings of the 10th ACM International Conference on Supercomputing*, May 1996.
[20] Olfield, R., Kotz, D. The Armada Parallel File System, 1998. http://www.cs.dartmouth.edu/ dfk/armada/design.html.
[21] Patterson, D., Gibson, G., Katz, R. A Case for Redundant Arrays of Inexpensive Disks (RAID). In *Proceedings of ACM SIGMOD*, pages 109–116. ACM, June 1988.
[22] Pierce, P. A Concurrent File System for a Highly Parallel Mass Storage Subsystem. In John L. Gustafson, editor, *Proceedings of the Fourth Conference on Hypercubes Concurrent Computers and Applications*, pages 155–161. HCCA, March 1989.
[23] Gropp, W., Takhur, R., Lusk, E. An Abstract-Device Interface for Implementing Portable Parallel-I/O Interfaces. In *Proceedings of the 6th Symposium on the Frontiers of Massively Parallel Computation*, pages 180–187, October 1996.
[24] Gropp, W., Takhur, R., Lusk, E. On Implementing MPI-IO Portably and with High Performance. In *of the Sixth Workshop on I/O in Parallel and Distributed Systems*, pages 23–32, 1999.
[25] del Rosario, J. M., Bordawekar, R., Choundary, A. Improved Parallel I/O via a Two-phase Run-time Access Strategy. *ACM Computer Architecture News*, 21(5):31–39, December 1993.
[26] Sandberg, R., Goldberg, D., Kleiman, S., Walsh, D., Lyon, B. Design and Implementation of the SUN Network Filesystem. In *Proc. of the 1985 USENIX Conference*. USENIX, 1985.

# Scalable Multithreading in a Low Latency Myrinet Cluster*

Albano Alves[1], António Pina[2], José Exposto, and José Rufino

[1] Instituto Politécnico de Bragança
albano@ipb.pt
[2] Universidade do Minho
pina@di.uminho.pt

**Abstract.** In this paper we present some implementation details of a programming model – pCoR – that combines primitives to launch remote processes and threads with communication over Myrinet. Basically, we present the efforts we have made to achieve high performance communication among threads of parallel/distributed applications. The expected advantages of multiple threads launched across a low latency cluster of SMP workstations are emphasized with a graphical application that manages huge maps consisting of several JPEG images.

## 1 Introduction

Cluster computing is a new concept that is emerging with the new advances in communication technologies; several affordable heterogeneous computers may be interconnected through high performance links like Myrinet.

Using these new computing platforms several complex problems, which in the past have required expensive mainframes, may now be solved using low cost equipment. Particularly, we are interested in providing cluster solutions for informational problems that require a combination of massive storage and moderate computing power.

### 1.1 Resource Oriented Computing – CoR

The CoR computing model has been primarily motivated by the need of creating a parallel computer environment to support the design and evaluation of applications conforming to the MC$^2$ (Cellular Computation Model) [17].

A full specification of CoR and an initial prototype – pCoR – were presented in [14] and [18]. CoR paradigm extends the process abstraction to achieve structured fine-grained computing using a combination of message passing, shared memory and POSIX threads. Specification, coordination and execution of applications lie on the definition of a variety of physical and logical resources, such as domains, tasks, data, ports, synchronizers, barriers, topologies, etc.

* Research supported by FCT/MCT, Portugal, contract POSI/CHS/41739/2001, under the name "SIRe – Scalable Information Retrieval Environment".

J.M.L.M. Palma et al. (Eds.): VECPAR 2002, LNCS 2565, pp. 579–592, 2003.

First attempts to introduce high performance communication into CoR, exploiting Myrinet, were presented in [1]. Preliminary results and validation were obtained with the development of a distributed hash system [19] and a global file system to exploit a local Myrinet cluster particularly for information retrieval.

### 1.2 Multithreading and Message Passing

Multithreading and message passing are two fundamental low-level approaches to express parallelism in programs. The first approach proved to be convenient in SMP workstations and the latter is widely used to program applications that distribute computations across networked machines.

Considering that most clusters are built from multiprocessor machines, there is a strong motivation to use a hybrid approach, combining multithreading, shared memory and message passing. This is not an easy task because message-passing primitives of most communication libraries are not thread safe. For instance, the device driver to interface Myrinet and the set of primitives provided by Myricom are not thread safe. However, we do believe that programmers could benefit from hybrid approaches because some applications can be easily structured as a set of concurrent/parallel tasks. That was the major motivation that led us to the investigation of a scalable communication strategy to support massive multithreaded applications in a cluster environment.

## 2 Background

Last decade many projects aimed to exploit the full computing power of networks of SMP workstations. In what follows we briefly present some key ideas that influenced nowadays cluster computing.

### 2.1 Distributed Multithreaded Programming

To run a distributed multithreaded program it is necessary to have a runtime system and a set of primitives to interface it[1]. Those primitives and their functionality highly influence the way programmers structure distributed applications.

MPI [21] programmers structure their applications according to the SPMD model and they are familiar with processor-to-processor message passing. PVM [10] permits some high level abstractions by introducing the notion of task. Communication takes place between tasks. The runtime system maps tasks to hosts.

Other platforms like TPVM [8], LPVM [23], a modified version of P4 [7], Chant [12] and Athapascan-0 [6] allow the creation of multiple threads. Communication occurs between threads using thread identifiers and send/receive primitives. Athapascan adds the concept of ports and requests: ports are independent

[1] Some distributed programming environments also include specific compilers.

from threads and so any thread can receive a message sent to a particular port; requests are used to test termination of asynchronous communication.

Panda [4], PM2 [16] and Nexus [9] also include thread support but they manage communication in a different manner; messages are delivered executing handlers previously registered by the user. This way programs are not forced to explicitly receive messages (via blocking or nonblocking primitives). These runtime systems are also able to automatically launch threads to execute handlers.

Remote service requests are another paradigm for remote execution and data exchange that some platforms do support. RPCs are asynchronous and match perfectly the communication paradigm of Panda, PM2 and Nexus, which obviously support this facility. Chant and Athapascan also implement RPCs.

Nexus provides an extra abstraction - the context - used to group a set of threads, which is an important concept for structuring applications. A context is mapped to a single node.

For thread support two different approaches may be used: developing a thread library or selecting an existent one. Panda and PM2 developed specific thread libraries in order to integrate communication and multithreading in a more efficient way. Chant manipulates the scheduler of existing thread packages (pthreads, cthreads, etc) to take message polling into account when scheduling ready threads. Cilk [5], which provides an abstraction to threads in explicit continuation-passing style, includes a work-stealing scheduler.

## 2.2 Efficient Message Handling

Using recent communication hardware, it is possible to send a message from one host to another in a few microseconds while throughput between hosts can achieve hundreds of Mbytes[2].

However, operating systems usually take advantage of internal buffers and complex scheduling techniques to deliver data to user level programs. For that reason low-level communication libraries have been developed to directly interface the hardware. GM [15], BIP [11] and LFC [2] are communication libraries that take full advantage from Myrinet technology, by means of zero-copy communication.

On the other hand, distributed applications manipulate complex entities and use several threads/processes of control. Messages incoming to a specific host must be redirected to the right end-point and so context-switching overheads may decrease performance. Active messages [22] are a well-known mechanism to eliminate extra overheads on message handling. Upcalls and popup threads are two techniques to execute message handlers [3] used in Panda.

The choice between polling or interrupts for message reception [13] may also have significant impact on program performance. LFC uses both mechanisms, switching from one to another according to the system status.

[2] Myrinet latency is less then $10\mu s$ and one-way throughput is near 250MB/s.

### 2.3 pCoR Approach

pCoR aims to efficiently combine existent POSIX threads implementations (kernel Linux Threads, for example) and low-level communication facilities provided by hardware vendors (GM, for example). The goal is to provide a platform suitable for the development and execution of complex applications but we do not intend to directly support threads or to develop specific communication drivers.

Using Linux Threads we can take full advantage of multiprocessor systems and ensure compatibility with existent sequential routines. By implementing traditional send/receive primitives over a well-supported low-level communication facility as GM we guarantee performance and extendibility.

## 3 Thread-to-Thread Communication

pCoR runtime system distinguishes between inter and intra-node communication. Intra-node communication may occur between threads sharing the same address space (intra-process communication) or between threads from different processes (inter-process communication).

To manage communication, pCoR runtime system must be aware of thread location in order to select the most efficient mechanism for data sending. As a consequence the communication subsystem must be perfectly integrated on pCoR runtime system. It would be particularly difficult to use an existent thread-to-thread communication facility in an efficient manner because it would be necessary to integrate it with pCoR naming service.

At present we support two ports: UDP (for Ethernet devices) and GM (for Myrinet hardware).

### 3.1 Communication Channels

The development of a communication library to overcome pCoR communication needs must address two main issues:

1. identification – global process and thread identifiers, provided by pCoR resource manager, must be combined to produce unique identifiers to assign to communication end-points;
2. port virtualisation – low-level communication libraries to interface network adapters provide port abstractions to identify receivers and senders, but those abstractions are limited in number (GM library, for instance, only supports up to 8 ports).

In pCoR, identification is handled by a distributed service running on every process belonging to the same application. Basically, this is a directory service responsible to map pCoR identifiers into low-level identifiers used to route data at network interface level. To route information between components of the directory service, pCoR uses alternative communication facilities over TCP/IP. The impact of that solution is minimized through the use of local caches.

Port virtualisation will be explained in section 4.

### 3.2 Low-Level Interface

Communication between pCoR remote entities is implemented through a few primitives that use GM facilities to send and receive data. Although CoR specifies high-level abstractions to interconnect computing resources, it is possible to use these primitives to transmit and receive data in pCoR applications.

Senders must specify the destination using a pair `<pCoR process id, pCoR thread id>`, a tag and the address of the data to be sent. Data can be copied from its original address or it can be directly delivered to the GM library if it resides on a DMAble memory block. The reciprocal is valid for receivers.

Because both send and receive primitives are asynchronous, a test communication primitive with two modes – blocking or non-blocking – is provided.

```
int hndl   = sendCopy(int trgt_pid, int trgt_thid, int tag, void *data,
                 int size)
             sendDMA(...)
int hndl   = recvCopy(int src_pid, int src_pid, int tag, void *data,
                 int size, int *apid, int *athid, int *atag, int *asize)
             recvDMA(..., void **data, ...)
int status = testHandle(int handle, int mode)
```

## 4 Message Dispatching

Port virtualisation introduces the need to create a dispatching mechanism to handle messages from/to an arbitrary number of entities. Our approach uses a dispatcher thread per port to make possible several threads to share the same communication facility.

### 4.1 Dispatcher Thread

Send and receive primitives, executed by concurrent/parallel threads, interact with the dispatcher thread through queues. The send primitive enqueues messages for future dispatch whereas the receive primitive dequeues messages if any is available. Synchronous operation is supported through thread blocking mechanisms. Figure 1 shows the main aspects of message dispatching.

The dispatcher thread detects message arrival, via GM, using polling or blocking primitives. Every new message arriving to a port is enqueued in the receive queue and blocked threads (waiting for specific messages) are awakened. Whenever pending messages are detected in the send queue, the dispatcher thread initiates their transmission via GM.

Since we provide two approaches[3] to interface GM – polling and blocking primitives – the dispatcher operates in one of two modes: non-blocking or blocking.

[3] Currently available as compile options.

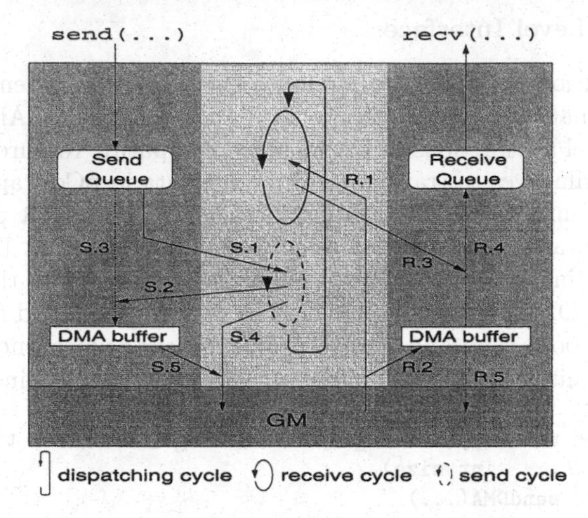


**Fig. 1.** Message dispatching mechanism

The non-blocking dispatcher uses a sole thread to execute an infinite loop sending and receiving messages. After polling the GM port for events[4] the dispatcher tries to find messages to transmit.

The blocking dispatcher must overcome a basic problem: if pCoR blocks itself waiting for message arrival, it will be impossible to send out any messages until a network event occurs because GM primitives are not thread safe. Experience proved that if a thread is blocked (after calling a GM blocking primitive) it is possible for another thread to send out messages if we use an additional GM port. Thus the blocking dispatcher uses two threads and two ports – one to receive and another to send messages. A thread waits for messages (from other nodes) issuing a GM blocking primitive while the other blocks itself waiting for messages to be sent to other nodes.

## 4.2 Segmentation and Retransmission

To transmit messages over GM, it is necessary to copy data into DMAble memory blocks[5]. pCoR supports the transmission of arbitrary size messages, i.e., the communication layer must allocate DMAble arbitrary size buffers. Because processes cannot register all their memory as DMAble, we use buffers up to 64kbytes requested on library start-up. This means that long messages must be segmented.

Segmentation involves reassembling message fragments at destination and it implies that sequence numbering to identify fragments belonging to the same message is needed. Sequence numbers are managed by the interface developed to

[4] GM events signal network activity (message arrival, acknowledgment, etc).

[5] Program data stored in DMAble memory is transmitted as a zero copy message.

manage the queues used by the dispatcher. Every fragment is handled as a simple message by the dispatcher; dequeue and enqueue operations are responsible for fragmentation and reassembling.

Message sequencing is used to overcome another problem: fragment/message retransmission. Although GM guarantees the correct delivery of messages, the lack of resources at destination may not permit reception at a specific time. In those cases it is necessary to retry transmission after a certain period of time.

### 4.3 Multiple Routes and Message Forwarding

Cluster nodes may have installed multiple network interfaces from different vendors[6]. It is also possible that not all nodes from a cluster share a common communication technology. Even clusters on different locations may be interconnected using Internet protocols.

For those scenarios, it is desirable to allow computing entities to select at runtime the appropriate communication protocol and to provide forwarding capabilities to overcome cluster partitions (Madeleine [20] addresses these topics). It is also important to provide mechanisms to choose the better location for computing threads according to host-to-host link capabilities. For instance, for a cluster fully connected with Fast Ethernet but having part of the nodes connected with Myrinet, it would be desirable to have the runtime system responsible to start on Myrinet nodes those threads with higher communication demands.

pCoR uses a straightforward mechanism to provide multiple routes on heterogeneous clusters. At start-up each node registers its communication ports and builds a simple routing table containing information about protocols and gateways available to reach each remote node. As pCoR allows to dynamically add nodes to an application, the runtime system rebuilds the routing table at each node every time a start-up event is received.

Message forwarding is accomplished by the dispatcher thread. pCoR message headers include the final destination (process id) of the message along with the information pointed out in figure 2.

## 5 Data Structures

Message dispatching requires appropriate data structures to guarantee low-latency reception and sending. The pCoR communication layer architecture uses two main queues per port to store messages. Those queues must minimize required memory size and must permit fast access to store/retrieve data.

### 5.1 Message Reception

The `recv` primitive used in pCoR, executed concurrently by an arbitrary number of threads, searches for messages according to certain attributes: originator

[6] It's common to connect cluster nodes to both Ethernet and Myrinet switches.

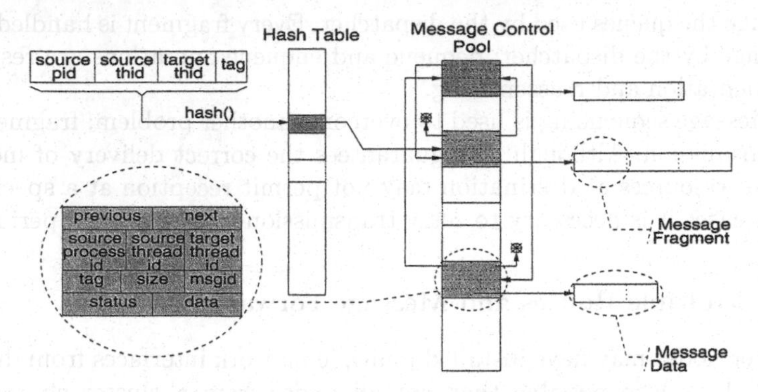


**Fig. 2.** Data structures for message reception

process, originator thread and message tag. As we use an only receive queue per process the destination thread identifier is also automatically included to search for a specific message.

A tuple `<source process, src. thread, target thread, tag>` is used to calculate a hash index to access a vector of pointers to message control blocks. The message control blocks are stored in a fixed size array which means that a limited number of messages can be pending for reception. Collisions resulting from the application of the hash function and messages addressed to the same thread from the same origin and with the same tag are managed as a linked list of message control blocks as shown in figure 2.

Message control blocks contain message attributes, a pointer to the message data, sequencing information and fragmentation status. For fragment tracking 32 bits are used – 1 bit for each fragment – supporting messages up to 2095872 bytes[7].

### 5.2 Message Sending

The `send` primitive enqueues messages for future dispatch whereas the dispatcher thread dequeues those messages for sending over GM. Because message dispatching uses FIFO order, at first sight we might think that a simple queue would be adequate to hold pending messages. However, since segmentation and retransmission are provided, the dispatcher needs some mechanism to access a specific message. Actually, segmentation requires the ability to dequeue single message fragments whereas delivery acknowledgment events from GM layer, handled by the dispatcher, require the ability to set message status for a specific message.

For short, data structures for message sending will be analogous to those used for message reception, but it is necessary to have a dequeue operation performing according to FIFO.

[7] Maximum message size results from $(64k - (fragmentheadersize)) * 32$.

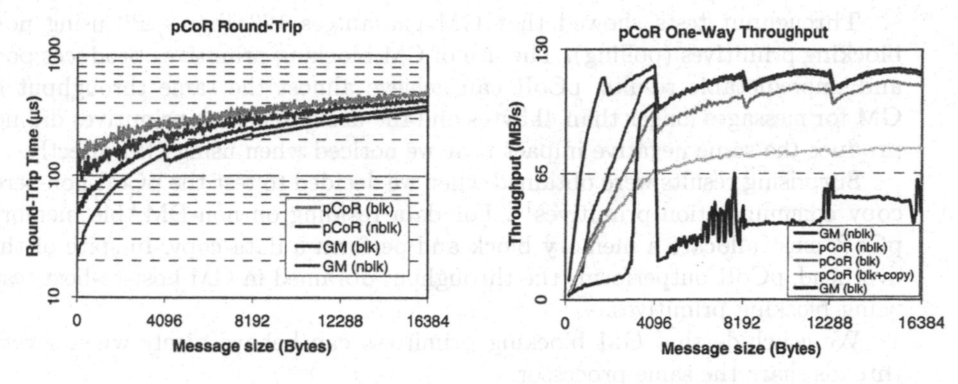


**Fig. 3.** Round-Trip and Throughput in pCoR

## 6 pCoR Raw Performance

Although pCoR provides high-level abstractions, like remote thread creation and definition of complex organizers, it is important to evaluate the raw performance obtained when transferring data between threads.

The results we present were obtained using a simple benchmark program that engages on communication two pCoR tasks (threads) executing on different machines. We used two dual PIII 733MHz workstations, connected by Myrinet (LANai9 64bits/66MHz interfaces), running Linux RedHat 7.2 (kernel 2.4.7-10smp). The tests were performed with no additional load at each workstation.

Figure 3 presents round-trip times and one-way throughput for messages from 1byte to 16kbytes. Values for the GM low-level interface performance (host-to-host communication) are also presented to better understand the overhead of thread-to-thread communication. The experiments took into account the two mechanisms GM provides to receive events - polling and blocking receives[8].

It is important to note the impact of message dispatching[9]. For each message exchange, the pCoR runtime system must wake up two blocked threads; the dispatcher must signal the message arrival to a specific thread. Using some simple POSIX threads tests, we evaluated the overhead of waking up a thread blocked on a condition variable (using linuxthreads-0.9, libc-2.2.4). We concluded that this overhead exceeds 35$\mu$s. This explains round-trip times obtained in pCoR; a message exchange in pCoR incurs in a 70$\mu$s penalty due to thread wake up.

It is also important to note the result of using blocking primitives to interface the GM library. Although the use of blocking primitives has the advantage of freeing the processor for useful computation, message reception incurs in a 15$\mu$s penalty (30$\mu$s for a message exchange) due to interrupt dispatching.

[8] In the charts legends *blk* and *nblk* stands for blocking and non-blocking.

[9] Legend items order correspond to the placement of chart curves; the top curve corresponds to the first legend item and vice-versa.

Throughput tests showed that GM guarantees 120Mbytes/s[10] using non-blocking primitives (polling). The use of GM blocking primitives produces poor and unpredictable results. pCoR can achieve almost the same throughput as GM for messages longer than 4kbytes and the use of blocking primitives did not produce the same negative impact that we noticed when using GM directly.

Surprising results were obtained when we decided to test the pCoR non-zero-copy communication primitives[11]. For data residing on non-DMAble memory, pCoR must allocate a memory block and perform a data copy. In spite of this overhead, pCoR outperforms the throughput obtained in GM host-to-host tests using blocking primitives.

We conclude that GM blocking primitives can behave nicely when several threads share the same processor.

## 7 Case Study

To emphasize the importance of thread-to-thread communication we present an application intended to manage (display) huge maps. Those maps are composed of several 640x480 pixel JPEG images.

In our case study we used a 9600x9600 pixel map consisting of a 15x20 matrix of JPEG images. The main objective is the visualization of arbitrarily large map regions. Regions larger than the window size require the images to be scaled down.

The architecture we propose to manage this kind of maps takes into account the following requisites: high computing power to scale down images, large hard disk capacity to store images and high I/O bandwidth to load JPEG images from disk.

### 7.1 Multithreading

Assuming we have an SMP machine with enough resources to handle those huge maps a multithreaded solution can be developed to accelerate the decompression of JPEG images and the reduction of image bitmaps.

Figure 4 shows two C++ classes used to model a simple multithreaded solution. An object `imgViewer` is used to display a map region, according to a specified window size, after creating the corresponding bitmap. The bitmap is created using an object `imgLoader` which creates a thread to execute the method `startFragLoad`. The `imgViewer` calls the method `startFragLoad` from class `imgLoader` for each JPEG image required to create the final bitmap.

To display a 9600x9600 pixel map region, for instance, 300 threads will be created to load the corresponding JPEG images and to scale them down. Using a 600x600 window to display the final bitmap, each thread will scale down 16

[10] Our Myrinet configuration would reach 1.28Gbits/s, due to switch constraints, but the workstations PCI bus cannot guarantee such performance.

[11] In the legend of the throughput graph *blk+copy* stands for blocking with buffer copy.

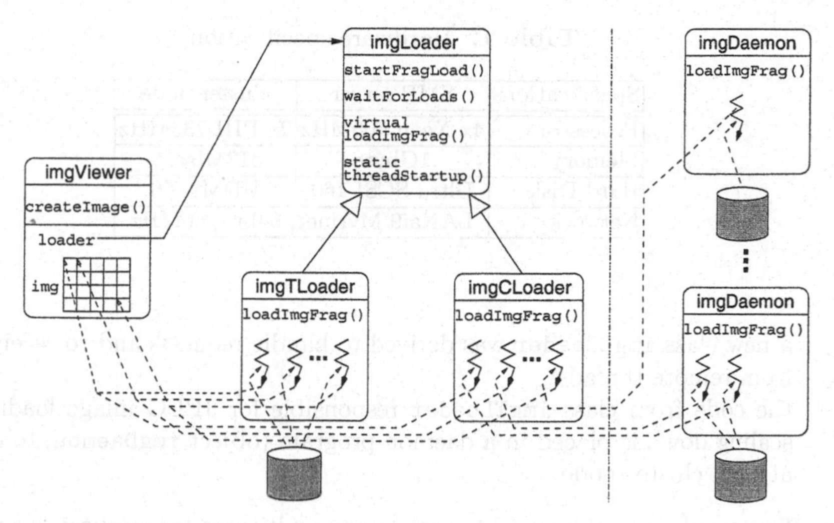


**Fig. 4.** Object model for multithread loading of huge maps

times an original 640x480 JPEG image in order to produce a 40x30 bitmap fragment. The object `imgViewer` is responsible for bitmap fragment reassembling.

## 7.2 Scalable Multithreading

Assuming we have a cluster with enough disk space at each computing node it is possible to spread all the JPEG images across all the nodes. Thus we will overcome disk capacity limitations and each node will be able to produce local results, without requesting images from a centralized server, taking advantage from cluster nodes computing power and local I/O bandwidth. Of course we will need some mechanism to discover which node holds a specific image, but it can be done using a simple hash function.

Figure 4 depicts `imgDaemon` object instances corresponding to daemons running on each cluster node to load and transform images according to requests received from a remote `imgLoader`. The `imgLoader` used in our cluster environment requests bitmap fragments from remote cluster nodes instead of loading it directly from disk.

The `imgLoader` class is in fact a virtual class used to derive two classes:

1. `imgTLoader` – multithreaded loader to use in a single SMP machine;
2. `imgCLoader` – multithreaded broker to use in a cluster.

Note that the development of the multithreaded solution to use in a cluster environment, assuming we had already developed a solution to use in a single SMP machine, was a trivial task:

- a virtual class `imgLoader` was introduced to permit the use of the same `imgViewer` object;

**Table 1.** Hardware specifications

| Specifications | SMP server | cluster node |
|---|---|---|
| Processor | 4x Xeon 700MHz | 2x PIII 733MHz |
| Memory | 1Gbyte | 512Mbytes |
| Hard Disk | Ultra SCSI 160 | UDMA 66 |
| Network | LANai9 Myrinet, 64bits/66MHz | |

– a new class `imgCLoader` was derived to handle requests and to receive data from remote threads;
– the code from class `imgTLoader` responsible for JPEG image loading and scaling down is placed in a daemon program (object `imgDaemon`) to execute at each cluster node.

This approach can be used to scale many multithreaded applications primarily developed to use in a single SMP machine.

### 7.3 Performance Evaluation

Performance evaluation was undertaken using a four node Myrinet cluster and an SMP server connected to cluster nodes, all running Linux. Table 1 summarises hardware specifications for our test bed.

Figure 5 presents computation times required to display 7 different map regions using a 600x600 window. The left side of the figure shows 7 map regions consisting of 2 to 300 JPEG images. Those regions, marked from 1 to 7, correspond respectively to 1:1, 1:2, 1:4, 1:5, 1:8, 1:10 and 1:16 scaling factors. The right side of the figure presents the results obtained using:

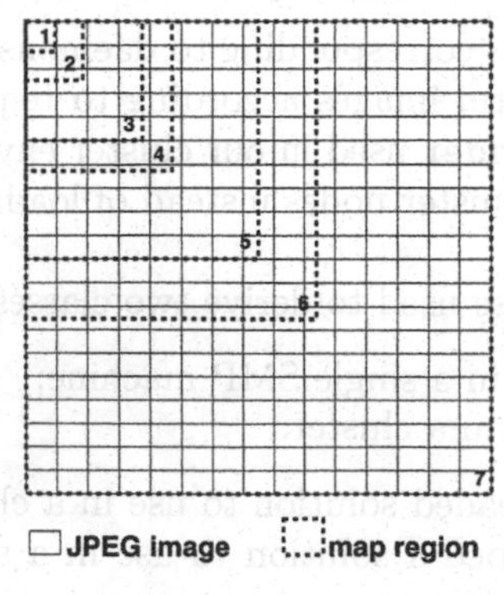


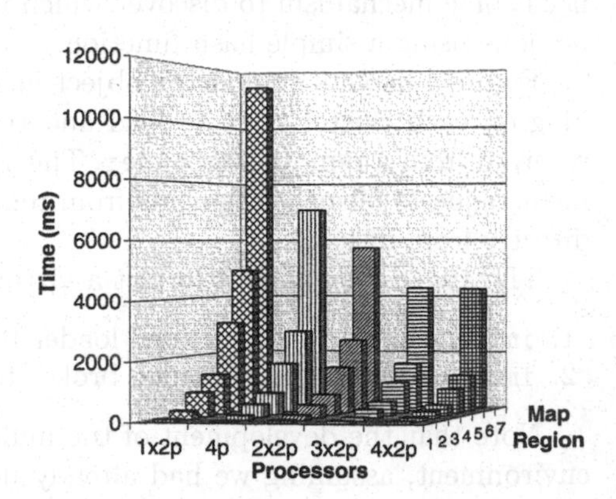


**Fig. 5.** Performance measurements for differrent scenarios

- a single 4-processor SMP machine (an `imgTLoader` object instance is used by the application) identified as `4p`;
- 1 to 4 cluster nodes (an `imgCLoader` object instance is used by the application) identified as `1x2p`, `2x2p`, `3x2p` and `4x2p`.

It is important to point out that the results obtained using the cluster solution based on 2 nodes (4 processors) supersede the results from the multithreaded solution based on a 4-processor SMP server. The main cause is the higher bandwidth available to load JPEG images from disk.

It is also important to emphasize the results obtained using the cluster solution based on 4 nodes (8 processors). As expected better performance was achieved for the majority of region maps tested, but it was not possible to outperform the result achieved with 3 cluster nodes for 9600x9600 region maps. That happens because the object `imgCLoader`, executing 300 threads to receive results from cluster nodes, is not fast enough to process incoming messages because of thread contention accessing communication library.

## 8 Conclusions

Using the current pCoR implementation it is possible to achieve communication between threads residing on any node of a cluster.

Thread scheduling is still a high CPU consuming task, particularly when using Linux Threads. Port virtualisation is consequently somewhat inefficient. Nevertheless, we do believe that it is convenient to program multithreading solutions to run in a cluster environment using Linux kernel threads because they can take full advantage of multiprocessor systems and I/O can easily overlap computation.

For applications demanding a high level of parallelism it is possible to develop traditional multithreaded solutions (to use in a single SMP machine). Considering that in most cases data sharing among threads is not a high requisite, because data can be easily spread among computational entities, it is possible to implement thread synchronization using messages. For those applications pCoR provides support for scalable multithreading.

## References

[1] A. Alves, A. Pina, V. Oliveira, and C. Moreira. CoR's Faster Route over Myrinet. In *MUG '00 - First Myrinet User Group Conference*, pages 173–179, 2000.

[2] R. Bhoedjang. *Communication Architectures for Parallel-Programming Systems*. PhD thesis, Advanced School for Computing and Imaging, Vrije Universiteit, 2000.

[3] R. Bhoedjang and K. Langendoen. Friendly and Efficient Message Handling. In *29th Hawaii International Conference on System Science*, pages 121–130, 1996.

[4] R. Bhoedjang, T. Rühl, R. Hofman, K. Langendoen, and H. Bal. Panda: A Portable Platform to Support Parallel Programming Languages. In *USENIX Symposium on Experiences with Distributed and Multiprocessor Systems (SEDMS IV)*, 1993.

[5] R. Blumofe, C. Joerg, B. Kuszmaul, C. Leiserson, K. Randall, and Y. Zhou. Cilk: An Efficient Multithreaded Runtime System. *Journal of Parallel and Distributed Computing*, 37(1):55–69, 1996.

[6] J. Briat, I. Ginzburg, and M. Pasin. *Athapascan-0 User Manual*, 1998.

[7] A. Chowdappa, A. Skjellum, and N. Doss. Thread-safe message passing with p4 and MPI. Technical report, Computer Science Department and NSF Engineering Research Center, Mississippi State University, 1994.

[8] J. Ferrari and V. Sunderam. TPVM: Distributed Concurrent Computing with Lightweight Processes. In *4th IEEE Int. Symposium on High Performance Dist. Computing - HPDC '95*, 1995.

[9] I. Foster, C. Kesselman, and S. Tuecke. The Nexus Approach to Integrating Multithreading and Communication. *Journal of Parallel and Distributed Computing*, 37(1):70–82, 1996.

[10] A. Geist, A. Beguelin, J. Dongarra, W. Jiang, R. Manchek, and V. Sunderam. *PVM: Parallel Virtual Machine. A User's Guide and Tutorial for Networked Parallel Computing.* Scientific and Engineering Computation. MIT Pres, 1994.

[11] P. Geoffray, L. Prylli, and B. Tourancheau. BIP-SMP: High Performance Message Passing over a Cluster of Commodity SMPs. In *SC99: High Performance Networking and Computing Conference*, 1999.

[12] M. Haines, D. Cronk, and P. Mehrotra. On the Design of Chant: A Talking Threads Package. In *Supercomputing '94*, 1994.

[13] K. Langendoen, J. Romein, R. Bhoedjang, and H. Bal. Integrating Polling, Interrupts, and Thread Management. In *6th Symp. on the Frontiers of Massively Parallel Computing*, 1996.

[14] C. Moreira. CoRes - Computação Orientada ao Recurso - uma Especificação. Master's thesis, Universidade do Minho, 2001.

[15] Myricom. *The GM Message Passing System*, 2000.

[16] R. Namyst and J. Méhaut. PM$^2$: Parallel Multithreaded Machine. A computing environment for distributed architectures. In *ParCo'95*, 1995.

[17] A. Pina. *$MC^2$ - Modelo de Computação Celular. Origem e Evolução.* PhD thesis, Departamento de Informática, Universidade do Minho, Braga, Portugal, 1997.

[18] A. Pina, V. Oliveira, C. Moreira, and A. Alves. pCoR - a Prototype for Resource Oriented Computing. In *Seventh International Conference on Applications of High-Performance Computers in Engineering*, 2002.

[19] A. Pina, J. Rufino, A. Alves, and J. Exposto. Distributed Hash-Tables. PADDA Workshop, Munich, 2001.

[20] B. Planquelle, J. Méhaut, and N. Revol. Multi-protocol communications and high speed networks. In *Euro-Par '99*, 1999.

[21] M. Snir, S. Otto, S. Huss-Lederman, D. Walker, and J. Dongarra. *MPI - The Complete Reference.* Scientific and Engineering Computation. MIT Pres, 1998.

[22] T. von Eicken, D. Culler, S. Goldstein, and K. Schauser. Active Messages: A Mechanism for Integrated Communication and Computation. In *19th International Symposium on Computer Architecture*, pages 256–266, Gold Coast, Australia, 1992.

[23] H. Zhou and A. Geist. LPVM: A Step Towards Multithread PVM. *Concurrency: Practice and Experience*, 10(5):407–416, 1998.

# Minimizing Paging Tradeoffs Applying Coscheduling Techniques in a Linux Cluster*

Francesc Giné[1], Francesc Solsona[1], Porfidio Hernández[2], and Emilio Luque[2]

[1] Departamento de Informática e Ingeniería Industrial
Universitat de Lleida, 25001 Lleida, Spain
{sisco,francesc}@eup.udl.es
phone: +34 973 702729, fax: +34 973 702702
[2] Departamento de Informática
Universitat Autònoma de Barcelona, 08193 Bellaterra, Barcelona, Spain
{p.hernandez,e.luque}@cc.uab.es
phone:+34 93 581 1356, fax:+34 93 581 2478

**Abstract.** Our research is focused on keeping both local and parallel jobs together in a non-dedicated cluster or NOW (Network of Workstations) and efficiently scheduling them by means of coscheduling mechanisms.
The performance of a good coscheduling policy can decrease drastically if memory requirements are not kept in mind. The overflow of the physical memory into the virtual memory usually provokes a severe performance penalty. A real implementation of a coscheduling technique for reducing the number of page faults across a non-dedicated Linux cluster is presented in this article. Our technique is based on knowledge of events obtained during execution, such as communication activity, page faults and memory size of every task. Its performance is analyzed and compared with other coscheduling algorithms.

**Topics:** Cluster and Grid Computing, Distributed Computing

## 1 Introduction

The studies in [3] indicate that the workstations in a NOW are normally underloaded. Basically, there are two methods of making use of these CPU idle cycles to run parallel jobs, task migration [4, 3] and time-slicing scheduling [5, 7]. Systems based on task migration have been created to use the available resources when owners are away from their workstations. However, there are more available resources that such systems do not harvest. This is due to the fact that even when the user is actively working on the machine, the resource usage is very low. Thus, we propose running parallel jobs together with local tasks by means of time-slicing scheduling while limiting the slowdown of the owner's workload to

* This work was supported by the MCyT under contract TIC 2001-2592 and partially supported by the Generalitat de Catalunya -Grup de Recerca Consolidat 2001SGR-00218.

J.M.L.M. Palma et al. (Eds.): VECPAR 2002, LNCS 2565, pp. 593–607, 2003.

an acceptable level. In a time-slicing environment, two issues must be addressed: how to coordinate the simultaneous execution of the processes of a parallel job and how to manage the interaction between parallel and local user jobs.

If the processes of a parallel application communicate and synchronize frequently, it may be beneficial to schedule them simultaneously on different processors. The benefits are obtained when the overhead of frequent context switches is saved and the need for buffering during communication is reduced. Such simultaneous execution, combined with time slicing, is provided by *gang scheduling* [6], which is quite popular in MPP environments. However, gang scheduling implies that the participating processes are known in advance. The alternative is to identify them during execution [7, 11, 15]. Thus, only a subset of the processes are scheduled together, leading to *coscheduling* rather than gang scheduling. Coscheduling decisions can be made taking account of the implicit runtime information of the jobs, basically CPU cycles and communication events [6, 7, 8, 9, 11, 15].

The performance of a good coscheduling policy can decrease drastically if memory requirements are not kept in mind [1, 2, 10, 13]. Most of the reference researchers [1, 10] have proposed different techniques to minimize the impact of the real job memory requirements on the performance of a gang scheduling policy. However, to our knowledge, there is a lack of research into minimizing the impact of the memory constraints by applying coscheduling techniques. We are interested in proposing implicit coscheduling techniques with memory considerations. That is, to coschedule distributed applications taking the dynamic allocation of memory resources in each cluster node into account.

In a non-dedicated system, the dynamic behavior of local applications or a distributed job mapping policy without memory considerations cannot guarantee that parallel jobs have enough resident memory as would be desirable throughout their execution. In these conditions, the local scheduler must coexist with the operating system's demand-paging virtual memory mechanism. In distributed environments (cluster or NOW), the traditional benefits that paging provides on uniprocessors may decrease depending on various factors, such as the interaction between the CPU scheduling discipline, the synchronization patterns within the application programs, the page reference patterns of these applications and so on. Thus, our main aim is to reduce the number of page faults in a non-dedicated system giving more execution priority to the distributed tasks with the lowest fault page rates, letting them finish as soon as possible. Therefore, on completion of these, the released memory will be available for the remaining (local or distributed) applications. Consequently, major opportunities arise for advancing execution of all the remaining tasks.

The execution of the distributed tasks must not disturb local task interactivity, so excessive local-task response time should be avoided. This means that a possible starvation problem of this kind of task must be taken into account.

In this paper, a new coscheduling environment over a non-dedicated cluster system is proposed. The main aim of this new scheme is to minimize the impact of demand paged virtual memory, with capabilities of prevention local-task starvation.

The rest of the paper is organized as follows. In section 2, a coscheduling algorithm under memory constraints is explained. The implementation of the proposed algorithm is presented in section 3. Its performance is evaluated and compared in section 4. Finally, the conclusions and future work are detailed.

## 2 CSM: Coscheduling Algorithm under Memory Constraints

In this section, a local coscheduling algorithm under memory constraints, denoted as CSM, is proposed and discussed. The following subsection gives an overview of the notation and main assumptions taken into account.

### 2.1 Assumptions and Notation

Given that our aim is the implementation of CSM in the Linux operating system, due to its free property condition, some of the following assumptions are made based on the properties of Linux.

Our framework is a non-dedicated cluster (or NOW), where all nodes are under the control of our coscheduling scheme. Every local coscheduler takes autonomous decisions based on local information provided by the underlying time-sharing operating system.

The Memory Management Unit (MMU) of each node is based on pages, with an operating system which provides demand-paging virtual memory. This means that if the page referenced by a task is not in its *resident set* (allocated pages in the main memory), a *page fault* will occur. This fault will suspend the task until the missing page is loaded in the resident set. Next, such a task will be reawakened and moved to the Ready Queue (RQ). Meanwhile, another task could be dispatched into the CPU for execution.

When the kernel is dangerously low on memory, swapping is performed. We assume a swapping mechanism based on the Linux page replacement algorithm (v 2.2.15), in which the process with most pages mapped in memory is chosen for swapping. The pages to be swapped out from the located process are selected according to the *clock replacement algorithm,* an approximation to *the Least Recently Used (LRU) algorithm,* in which the page chosen for replacement is the one that has not been referenced for the longest time.

A priority scheduling policy based on the position of the RQ (the *top* task has the highest priority whereas the *bottom* task has the lowest) is assumed. The basic notation used throughout this paper is summarized as follows:

- $RQ[k]$: pointer to the task on the position $k$ of the Ready Queue (RQ). Special cases are k = 0 (*top*) and $k = \infty$ (*bottom*) of the RQ.
- $task.c\_mes$: number of current receiving-sending messages for the *task*. This is defined as:

$$task.c\_mes = task.rec + task.send, \quad (1)$$

where *task.rec* is the number of receiving messages and *task.send* is the number of sending messages for the *task*.
- *task.mes*: number of past receiving-sending messages for the *task*. This is defined as:

$$task.mes = P * task.mes + (1 - P) * task.c\_mes, \quad (2)$$

where $P$ is the percentage assigned to the past messages (*task.mes*) and $(1 - P)$ is the percentage assigned to the current messages (*task.c_mes*). This field will be used to distinguish between local tasks ($task.mes = 0$) and distributed tasks ($task.mes \neq 0$). Based on experimental results [15], a $P = 0.3$ has been chosen.
- *task.de*: number of times that the *task* has been overtaken in the RQ by another one due to a coscheduling cause, since the last time such a task reached the RQ. In order to avoid the starvation of the local and distributed tasks, the maximum number of overtakes suffered by a task, denoted as *MNO*, will be restricted.
- *task.flt_rate*: accumulative number of pages read from the swap and file space due to a page fault during a time interval (experimentally, a period of 10 sec. has been chosen).
- $N_k.M$: main memory size of node k.
- $N_k.mem$: it is the sum of the memory requirements of all the resident processes into node $k$.

## 2.2 CSM Algorithm

The CSM algorithm must decide which task is going to run next, according to three different goals:

1. The minimization of the number of page faults throughout the cluster. Paging is typically considered [2, 10] to be too expensive due to its overhead and adverse effect on communication and synchronization.
2. The coscheduling of the communication-synchronization processes. No processes will wait for a non-scheduled process for synchronization/communication.
3. The performance of the local jobs. The CSM algorithm should avoid the starvation of the local processes.

According to these goals, CSM is implemented inside a generic routine (called *insert_RQ*), which inserts every ready-to-run task into the Ready Queue. The flow graph of the CSM algorithm is shown in figure 1. Its behavior is explained below.

When the RQ is empty, the *task* is inserted at the top. If it is not, the algorithm works to find the position in the RQ where the task should be inserted in

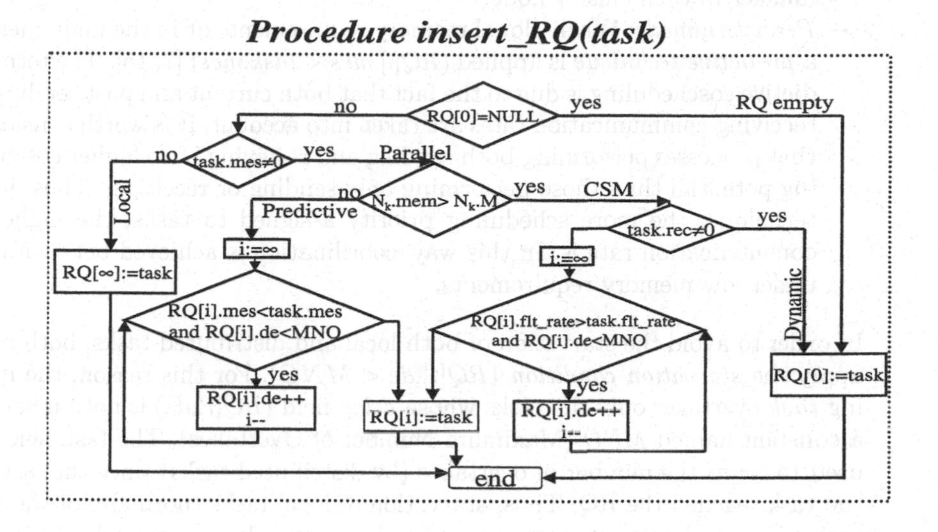


**Fig. 1.** Flow graph of the CSM algorithm

accordance with its communication and page fault rates as well as the starvation condition.

Under CSM, local tasks (generally tasks without remote communication, $task.mes = 0$) are inserted at the bottom of the RQ ($RQ[\infty] := task$). On the other hand, the insertion of distributed tasks (those with $task.mes \neq 0$) can be done according to two different modes:

– *CSM mode*: When global memory requirements in such a node ($N_k.mem$) exceed the main memory size ($N_k.M$), the scheduling will depend on the *page faults* of the distributed tasks ($RQ[i].flt\_rate > task.flt\_rate$) together with the *starvation condition* ($RQ[i].de < MNO$). So, CSM gives more execution priority to the tasks with fewer page faults. In accordance with the page replacement algorithm assumed in the previous section, every time a page fault happens, the least recently used pages belonging to the process with more pages mapped into memory and, as a consequence, with the highest page fault rate will be replaced by the missing pages. This means that with time, more memory resources will be allocated to distributed tasks with a lower page fault rate. In order to maintain the coscheduling between nodes under CSM mode, a *dynamic technique* [8] is applied. In dynamic coscheduling, the inserted task is led to the top of the RQ ($RQ[0]$) if it has any incoming message in the reception queue ($task.rec \neq 0$). Thus, the CSM algorithm ensures that fine-grained distributed applications are coscheduled under high memory requirements. So, global coordination (through the whole cluster) of the

tasks making up the distributed applications is performed in an independent manner in each cluster node.
- *Predictive mode*: When global memory requirements fit in the main memory, a *predictive technique* is applied ($RQ[i].mes < task.mes$) [7, 15]. The term predictive coscheduling is due to the fact that both current and past sending and receiving communication rates are taken into account. It is worth considering that processes performing both sending and receiving have higher coscheduling potential than those performing only sending or receiving. Thus, in this technique, the more scheduling priority assigned to tasks, the higher the communication rate is. In this way, coordination is achieved between nodes under low memory requirements.

In order to avoid the starvation of both local and distributed tasks, both modes apply the *starvation condition* ($RQ[i].de < MNO$). For this reason, the inserting *task* overtakes only the tasks whose *delay* field ($RQ[i].de$) is not higher than a constant named *MNO* (Maximum Number of Overtakes). The task field *de* is used to count the number of overtakes (by distributed tasks) since the last time the task reached the RQ. Thus, starvation of local tasks (normally of the interactive kind and without communication) is avoided. It's worthwhile to highlight the need also to evaluate the *de* field for distributed tasks in order to avoid the starvation of a parallel job with a high page fault rate. For instance, if this was not taken into account, a continuous stream of parallel jobs with small memory requirements could lead to that a parallel job with a high page fault rate always being pushed to the end of the RQ.

The performance of CSM was evaluated by simulation in [16]. The good results obtained encouraged us to implement it in a real operating system. We chose the Linux operating system due to its free property characteristics.

## 3 Implementing the CSM Algorithm over the Linux Kernel

CSM has been implemented in the Linux Kernel v.2.2.15. Implementing coscheduling in the kernel space provides transparency to the overall system and allows the application of coscheduling regardless of the message passing environment (PVM, MPI, etc ...). The drawback is in portability. A patch must be introduced into each cluster node in the Linux source. Implementation involved minimum modifications to the scheduler and discarding pages mechanism used by the original Linux. In this section, the main characteristics and changes introduced into both subsystems will be explained.

### 3.1 The Linux Scheduler

In Linux [12], the Ready Queue (RQ) is implemented by a double linked list of *task_struct* structures, the Linux PCB (Process Control Block). The fields used in our implementation are:

- *policy*: scheduling policy. There are three scheduling policies in Linux. The "normal" time shared tasks policy and two "real-time" scheduling policies (with more scheduling priority than normal policies).
- *priority*: "*static*" scheduling priority between normal tasks. it ranges from 1 (low priority) to 40 (high priority). Default value = 20.
- *counter*: "*dynamic*" normal tasks priority. The initial value is set to *priority* one. When the task is executing, each tick[1], the *counter* is decremented towards 0 by one unit, then the CPU is yielded. Thus, the maximum *time slice* for a normal task with a default static priority is 210 ms.
- *maj_flt*: pages read from the swap and file space due to a page fault.
- *nswap*: pages out to the swap.
- *mm*: every process virtual memory is represented by an *mm_struct*. In this structure the *rss* field (resident set size of the process) has been used.

Furthermore, the following fields, explained in section 2.1, have been added to the PCB structure: *rec* (current receiving messages); *send* (current sending messages); *c_mes* (current receiving-sending messages); *mes* (past sending and receiving messages ); and flt_rate (page fault rate).

Tasks making up distributed applications are executed under the CSM algorithm as local (owner or interactive) ones, so no explicit information for differentiating between both kinds of tasks is assumed in advance. This way, all these tasks will have a "normal" scheduling policy.

The Linux scheduler inserts every ready-to-run task into the RQ by means of the internal function *add_to_runqueue*. After that, the Linux scheduler picks up the next process with the highest *counter* to be executed from the RQ. When all the processes in the RQ have exhausted their quantum, that is, all of them have a zero *counter* field, the scheduler assigns a fresh quantum to all the existing processes (not only to the ready-to-run tasks), whose duration is the sum of the *priority* value plus half the *counter* value. In this way, suspended or stopped processes have their dynamic priorities periodically increased. The rationale for increasing the counter value of suspended or stopped processes is to give preference to I/O-bound processes, usually local tasks. Thus, the Linux scheduler avoids the starvation of interactive tasks.

Given that the CSM algorithm is intended to modify the *priority* of the tasks when they are inserted into the RQ, the *add_to_runqueue function* was chosen to implement our algorithm. The main modifications introduced into this function are shown in algorithm 1. The original *add_to_runqueue* would have only line 15, where the task is inserted into the bottom of the RQ. Our implementation has added the 14 lines above. Firstly, in order to check if the main memory is overloaded, the *tempswap* variable counts the pages sent into the swap from all the processes mapped into memory (lines 1-3). If the *tempswap* is not zero, then the *CSM mode* (lines 4-11) is activated. In this mode, and according to the *dynamic technique*, the inserted task is promoted to the top when its *c_mes* field is different from zero. Note that we have implemented a new function called

[1] 1 tick $\simeq$ 10 ms.

**Algorithm 1** CSM algorithm. Inside function add_to_runqueue(*task*)

```
1  tempswap=0;
2  q=&init_task;
3  for(p=q-¿next_task; p≠q; p-¿next_task) tempswap=tempswap+p-¿nswap;
4  if(tempswap¿0)  //CSM mode
5     task-¿c_mes=number_packets();
6     if(task-¿c_mes≠0)  //Dynamic technique
7        task-¿counter = 40;
8     else
9        task-¿priority= assign_priority(task-¿flt_rate,CSM);
10    endif
11 else  //Predictive mode
12       task-¿mes =P ∗ task -¿mes + (1 − P) ∗ task− > c_mes;
13       task-¿priority= assign_priority(task-¿mes,PRED);
14 endif
15 RQ[∞] := task;
```

*number_packets* (line 5) to collect the sending/receiving messages from the socket packet-buffering queues in the Linux system layer. Otherwise, if the memory is not overloaded, then the *Predictive mode* is activated (lines 11-13). A new kernel function called *assign_priority* has been implemented to assign the priority proportionally to the value of the argument - ($flt\_rate$) (line 9) or *mes* (line 13).

In order to maintain the performance of local applications, *CSM* should guarantee a minimum of memory resources for the local user. With this aim, some changes have been introduced into the Linux memory management.

### 3.2 Page Replacement Mechanism

The Linux kernel (v.2.2) provides demand-paging virtual memory[2]. Linux uses the *Buddy* algorithm to allocate and deallocate pages. In doing so, it uses the *free_area* vector, which includes information about free physical pages. Periodically (once per second) the swap kernel daemon (swapd) checks if the number of free pages on the free_area vector falls below a system constant. If so, swapd will try to increase the number of free pages applying the page replacement mechanism explained in section 2.1.

Thus, in some situations and due to the memory requirements of distributed tasks, the swapping algorithm cannot guarantee a minimal resident set size (RSS) for local tasks. For this reason, we have modified the swapping algorithm. The modified algorithm for selecting the process from which pages can be stolen when memory is required, is shown in algorithm 2.

In this algorithm, firstly, the memory requested for active local tasks is computed and stored in the *memloc* variable (lines 3-6). In the same loop, if there

[2] Page size = 4KB.

**Algorithm 2** Selection of the best candidate for swapping

```
1 pbest=NULL; memloc=0; max_cnt=0; distri=0;
2 q=&init_task;
3 for(p=q-¿next_task;p!=q;p-¿next_task)
4   if(p-¿mes==0) memloc=memloc+p-¿mm-¿rss;
5   else distri=1;
6 endfor
7 for(p=q-¿next_task;p!=q;p-¿next_task)
8      if(0¡memloc¡Mem_Threshold) && (distri==1) && (p-¿mes≠0) && (p-¿mm-
¿rss¿max_cnt )
9        max_cnt=p-¿mm-¿rss;
10       pbest=p;
11   else if((memloc¿Mem_Threshold) —— (distri==0)) && (p-¿mm-¿rss¿max_cnt )
12       max_cnt=p-¿mm-¿rss;
13       pbest=p;
14   endif
15 endfor
```

is any distributed task in such a node, the variable *distri* is activated (line 5). After that, if the requested memory (*memloc*) is under the *Mem_Threshold,* in order to preserve memory resources for local tasks, the distributed task that owns the largest number of page frames is selected (lines 8-11). Otherwise, all the active processes (distributed and local) are candidates to be selected to swap out pages according to the same rules explained above (lines 12-14). The relationship between the *Mem_Threshold* value and local task performance will be analyzed experimentally in the next section.

## 4 Experimentation

The experimental environment used in this study was composed of eight 350MHz Pentium II with 128MB of memory. They were all connected through a 100Mbps bandwidth Ethernet network and a minimal latency of 0.1 ms. The performance of the coscheduling implementation was evaluated by running three PVM distributed applications *(IS, EP* and *MG*), from the NAS parallel benchmarks suite [14] with class A. In all the benchmarks, the communication between remote nodes was done through *RouteDirect* PVM mode.

The local workload characterization was carried out by means of running an application (called *local*) which performs floating point operations indefinitely (or a variable number of times when the local task overhead is measured) over an array with a size set by the user. Although *local* is not very representative of a realistic local workload, it represents the most unfavorable case for parallel task performance.

Five different workloads, with different memory and communication requirements, were chosen to verify the performance of the CSM algorithm. Table 1

**Table 1.** Workloads. Overload($x$) means that $x$ nodes had the main memory overloaded whereas Underload($y$) means that $y$ nodes had the main memory underloaded during the trial. Local($z$-$y$) means that one local task of $yMB$ of memory requirements had been executed in $z$ different nodes

| Workload (Wrk) | | 4 | 8 |
|---|---|---|---|
| 1 | IS+MG | Overload(4) | Underload(8) |
| 2 | MG+MG + local(2-50) | Overload(4) | Overload(2)+Underload(6) |
| 3 | IS+MG+EP | Overload(4) | Underload(8) |
| 4 | IS+local(2-80) | Overload(2)+Underload(2) | Overload(2)+Underload(6) |
| 5 | MG+local(2-60) | Overload(2)+Underload(2) | Overload(2)+Underload(6) |

shows the memory state of every node of the cluster when every workload was running in a cluster of 4 and 8 PCs, respectively.

Three different environments were evaluated and compared, the plain Linux scheduler (denoted as *LINUX*), a Predictive coscheduling (denoted as *PRED*) and the *CSM* algorithm. The PRED policy corresponds to the predictive mode of algorithm 1, whereas the CSM policy is the whole algorithm 1. In [15], the reader can find a real implementation of the predictive coscheduling without memory constraints over Linux o.s. together with a detailed experimental analysis of its performance in NOWs.

The performance of the CSM policy with respect to other coscheduling techniques is validated by means of three different metrics:

- Mean Page Faults (MPF): This is defined as the page fault experimented by a task averaged over all the jobs of the cluster.
- Mean Waiting Time (MWT): The average additional time spent by a task waiting on communication events in the multi-programming environment. This parameter shows how good the coordination is between the decisions taken by the CSM algorithm in different nodes.
- Slowdown: This is the ratio of the response time of a job in a non dedicated system in relation to the time taken in a system dedicated solely to this job. It was averaged over all the jobs in the cluster.

### 4.1 Case Study: A Detailed Analysis of the CSM Behavior under Workload 1

Starting from *Workload 1 (Wrk1)* when it is executed in 4 nodes, the performance of our algorithm is analyzed in detail. Figure 2 (left) and (right) shows the resident set size and page faults of every job through time when *Wrk 1* is executed in LINUX and CSM environments for one specific node of the cluster. Given that the distributed benchmarks used are SPMD programs (Same Program Multiple Data), the behavior (memory requirements and communication patterns) in the other nodes is expected to be similar [13].

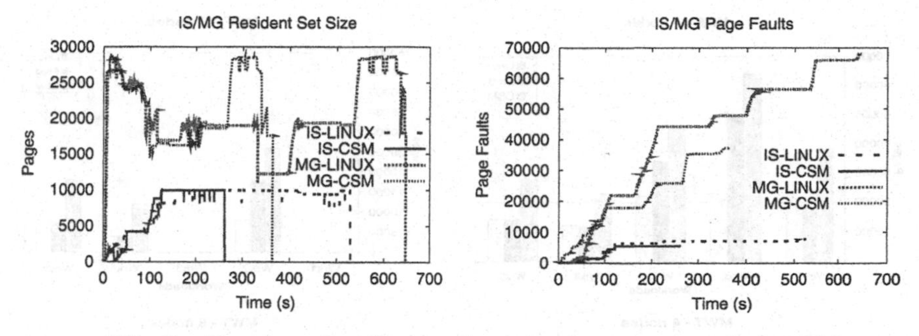


**Fig. 2.** Resident Set Size vs Page Faults in one node for Wrk1

In figure 2 (left), it can be seen that without any memory control policy (LINUX), both benchmarks (IS and MG) finish after $540s$ and $650s$, respectively. However, when CSM is applied and provided that the main memory is overloaded, CSM gives priority to IS execution -it has fewer page faults than MG- and, as a consequence, its resident set size is increased with time until it has all its address space residents in the main memory, at the expense of memory space reduction for task MG. Thus, IS finishes its execution sooner than in the LINUX case (after $260s$). When IS finishes execution, it frees its assigned memory and so the memory available for task MG is considerably increased, leading to a speedy execution of such task (MG finishes after 390s). So, MG also finishes considerably sooner in CSM than in the LINUX case. Figure 2 (right) verifies how CSM drastically diminishes the number of page faults for both benchmarks.

The global metrics Mean Page Faults (MPF), Mean Waiting Time (MWT) and Slowdown have been computed from the results obtained in every single node. Fig. 3 (left) shows the values obtained for all the Workloads in a cluster of 4 nodes. Focusing on the *Wrk1* results, we can see in the MPF parameter that the number of page faults applying CSM was reduced to approximately half. As a consequence of the low page fault rate achieved by CSM -together with the coordination mechanisms implemented in such an algorithm (Dynamic mode)-, it delivers the best coordination between remote nodes. Finally, the improvements obtained in the MPF and MWT metrics are reflected in the global slowdown one, which was reduced by half.

In next section, the analysis will be extended to all the defined workloads with the aim of checking the performance of CSM under different environments.

### 4.2 Distributed Task Performance

Fig. 3 shows the MPF (at the top), MWT (in the middle) and Slowdown (at the bottom) metrics obtained for all the workloads when they were executed in a cluster made up of 4 (fig. 3 left) and 8 nodes (fig. 3 right), respectively. All these trials were done with a *Mem_Threshold* equal to $M/3$, where $M$ is the main memory size.

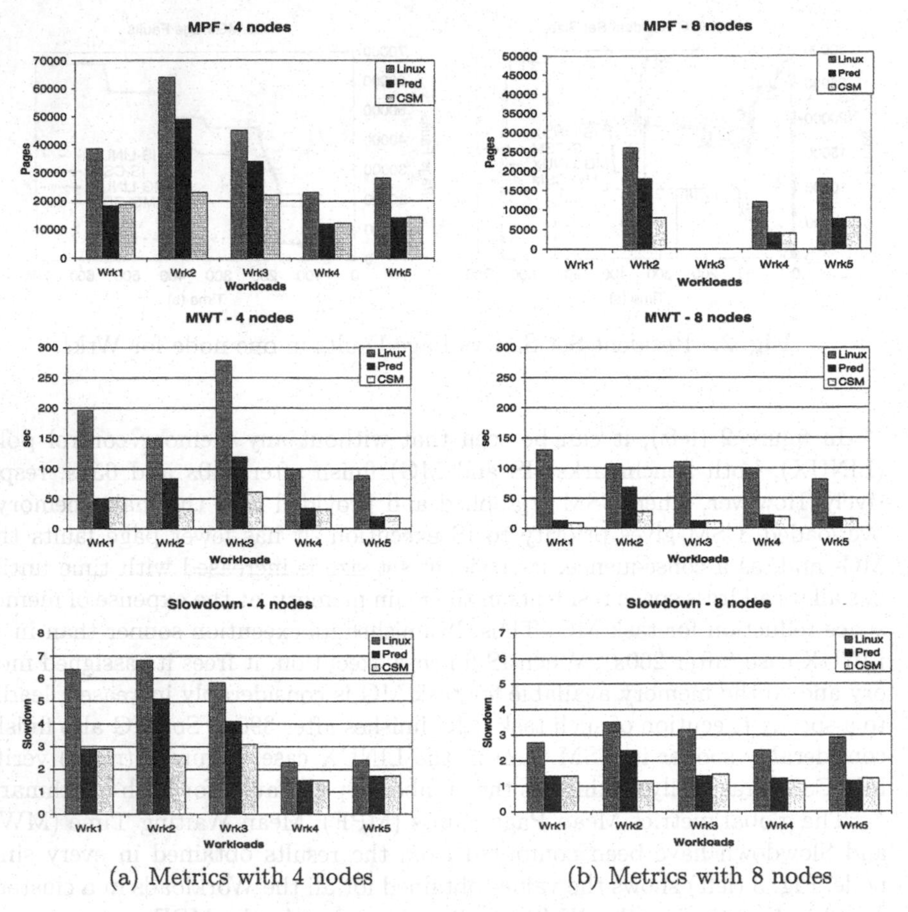

(a) Metrics with 4 nodes (b) Metrics with 8 nodes

**Fig. 3.** MPF, MWT and Slowdown metrics with 4 and 8 nodes

The behavior of the MPF metric reflects how CSM and PRED cause a similar number of page faults in all cases, excluding the *Wrk3* and *Wrk2* cases. As for *Wrk1*, this similarity is due to the benchmark characteristics; that is both environments give priority to the same benchmark (IS) because it is the benchmark with the highest communication rate (in the PRED case) and the lowest page fault rate (in the CSM case). Nevertheless, the similarity in the *Wrk4* and *Wrk5* cases is due to the fact that both techniques (CSM and PRED) run equally when there is only one distributed job executing concurrently through the cluster. The analysis of the *Wrk3 and Wrk2* cases reveals that the execution priority order in accordance with these page fault rate (CSM case) is the best under high memory requirements. Specifically, *Wrk2* behavior shows that CSM leads to a FIFO scheduling policy when all distributed applications have similar memory requirements. As a conclusion, CSM fully achieves its purpose of diminishing the number of page faults when the memory is overloaded.

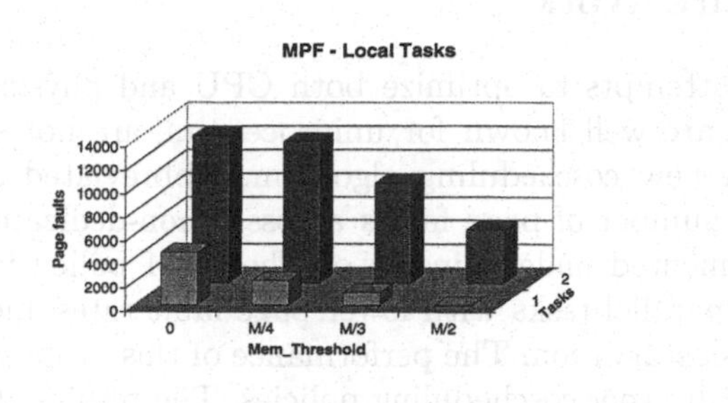


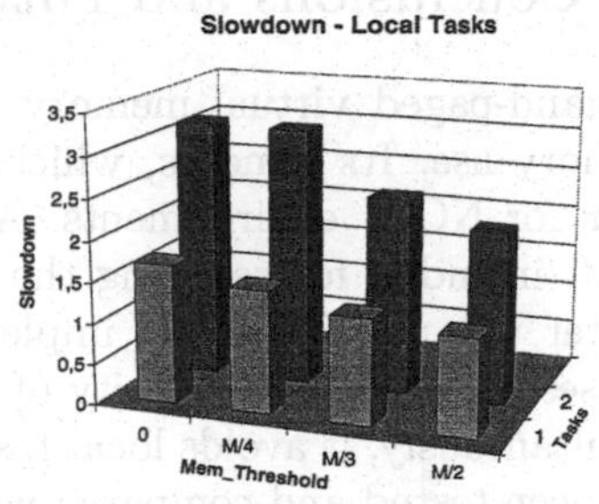


**Fig. 4.** (left) *MPF*, (right) *Slowdown* for local tasks

The analysis of the MWT metric with 4 nodes shows the same trends explained with the MPF metric. This behavior is due to the high overhead of page faults caused by synchronization delays. This results in high waiting time obtained in the LINUX environment with all the workloads and with *Wrk3* and *Wrk2* in the case of the PRED environment. For the same reason, the fall in the page fault number in the execution under 8 nodes leads to better coordination between the nodes.

Given that the access time of a page swapped to disk is two orders of magnitude slower than the access time of the same page in the main memory (it has been measured by means of the LMBENCH benchmarks suite [17]), the slowdown rises proportionally to the number of page faults. Thus, the analysis of the Slowdown metric reveals how CSM reaches the best performance in all the cases as a consequence of the good *MPF and MWT* metrics obtained.

### 4.3 Local Task Overhead

This section evaluates how the local task performance is influenced by the *Mem_Threshold.* Taking this aim into account, the *MPF* and *Slowdown (*fig. 4 left and right) metrics for local tasks were measured with different values of the *Mem_Threshold.* In these trials, we executed different instances of the local task (50MB of memory requirement) together with the MG benchmark.

As was expected, both figures reflect the necessity of keeping memory resources for local tasks in order to maintain their performance. This way, an agreement should be acquired between the performance of local and distributed tasks. We can see that the local task performance in the case of high memory requirements (2 instances of *local*) improves considerably with a *Mem_Threshold* greater than M/4. According to these results and the performance reached for distributed tasks with a Mem_Threshold equal to M/3 (see section 4.2), this *Mem_Threshold* has been selected by default.

## 5 Conclusions and Future Work

Demand-paged virtual memory attempts to optimize both CPU and physical memory use. Its benefits, which are well known for uniprocessors, are not so clear for NOW environments. A new coscheduling algorithm, abbreviated as CSM, intended for reducing the number of page faults across a non-dedicated cluster was presented and implemented under Linux s.o. The CSM policy increases the execution priority of parallel tasks with lower page fault rates and, simultaneously, it avoids local-task starvation. The performance of this proposal has been tested and compared with other coscheduling policies. The results obtained demonstrated its good behavior by reducing the number of fault pages and the Slowdown of distributed tasks and maintaining the performance of local tasks with respect to other policies.

The performance of CSM was evaluated by means of SPMD programs, so future work will be directed towards extending our analysis to other parallel programming paradigms. A new improvement to be introduced would be to adjust the length of the quantum to the real necessities of the distributed – memory and communication requirements– and local tasks –response time–.

The absence of interchanging control information between nodes to take scheduling decisions by CSM provides a big chance for scalability. Accordingly, one of our challenges will be to prove the scalability behavior of our approach across a much larger NOW than the one tested in this study.

## References


[1] S. Setia, M. S. Squillante and V. K. Naik. "The Impact of Job Memory Requirements on Gang-Scheduling Performance". *In Performance Evaluation Review*, 1999.

[2] D. Burger, R. Hyder, B. Miller and D. Wood. "Paging Tradeoffs in Distributed Shared-Memory Multiprocessors". *Journal of Supercomputing,* vol. 10, 1996.

[3] T. Anderson, D. Culler, D. Patterson and the Now team." A case for NOW (Networks of Workstations)". *IEEE Micro*, 1995.

[4] M. Litzkow, M. Livny and M. Mutka." Condor - A Hunter of Idle Workstations". 8th Int'l *Conference of Distributed Computing Systems*, 1988.

[5] R. H. Arpaci, A. C. Dusseau, A. M. Vahdat, L. T. Liu, T. E. Anderson and D. A. Patterson. "The Interaction of Parallel and Sequential Workloads on a Network of Workstations". *ACM SIGMETRICS'95*, 1995.

[6] J. K. Ousterhout. "Scheduling Techniques for Concurrent Systems." In *3rd. Intl. Conf. Distributed Computing Systems*, pp.22-30, 1982.

[7] P. G. Sobalvarro and W. E. Weihl. "Demand-based Coscheduling of Parallel Jobs on Multiprogrammed Multiprocessors". *IPPS'95 Workshop on Job Scheduling Strategies for Parallel Processing*, 1995.

[8] P. G. Sobalvarro, S. Pakin, W. E. Weihl and A. A. Chien. "Dynamic Coscheduling on Workstation Clusters". *IPPS'98 Workshop on Job Scheduling Strategies for Parallel Processing,* 1998.

[9] F. Solsona, F. Giné, P. Hernández and E. Luque. "Implementing Explicit and Implicit Coscheduling in a PVM Environment". *6th International Euro-Par Conference' 2000*, Lecture Notes in Computer Science, vol. 1900, 2000.

[10] A. Batat and D. G. Feitelson. "Gang Scheduling with Memory Considerations". *Intl. Parallel and Distributed Processing Symposium*, pp. 109-114, 2000.
[11] A. C. Arpaci-Dusseau, D. E. Culler and A. M. Mainwaring. "Scheduling with Implicit Information in Distributed Systems". *ACM SIGMETRICS'98*, 1998.
[12] D. Bovet and M. Cesati. "Understanding the Linux Kernel". *O'Reilly,* 2001.
[13] K. Y. Wang and D. C. Marinescu. "Correlation of the Paging Activity of Individual Node Programs in the SPMD Execution Model". In *28th Hawaii Intl. Conf. System Sciences*, vol. I, 1995.
[14] Parkbench Committee. Parkbench 2.0. http://www.netlib.org/parkbench, 1996.
[15] F. Solsona, F. Giné, P. Hernández and E. Luque. "Predictive Coscheduling Implementation in a non-dedicated Linux Cluster". *7th International Euro-Par Conference'2001*, Lecture Notes in Computer Science, Vol. 2150, 2001.
[16] F. Giné, F. Solsona, P. Hernández and E. Luque. "Coscheduling Under Memory Constraints in a NOW Environment". *7th Workshop on Job Scheduling Strategies for Parallel Processing*, Lecture Notes in Computer Science, Vol. 2221, 2001.
[17] LMbench: http://www.bitmover.com/lmbench/lmbench.html.

# Introducing the Vector C

Patricio Bulić and Veselko Guštin

Faculty of Computer and Information Science, University of Ljubljana
Tržaška cesta 25, 1000 Ljubljana, Slovenia
patricio.bulic@fri.uni-lj.si
http://lra-1.fri.uni-lj.si/index.html

**Abstract.** This paper presents the vector C (VC) language, which is designed for the multimedia extensions included in all modern microprocessors. The paper discusses the language syntax, the implementation of its compiler and its use in developing multimedia applications. The goal was to provide programmers with the most natural way of using multimedia processing facilities in the C language. The VC language has been used to develop some of the most frequently used multimedia kernels. The experiments on these scientific and multimedia applications have yielded good performance improvements.

## 1 Introduction

Fortran 90 introduced a lot of new features and facilities, many more than C. A primary objective of FORTRAN 90 is that it should be efficiently executable on modern high-performance supercomputers, i.e. those with vector and single-instruction multiple data (SIMD) architecture, however the increasing need for multimedia applications has prompted the addition of a multimedia extension to most existing general-purpose microprocessors [8, 9, 10]. These extensions introduce short SIMD instructions to the microprocessor's "scalar" instruction set. This instruction set is supported by special hardware that enables the execution of one instruction on multiple data sets. Such vectored instruction sets are primarily used in multimedia applications, and it seems that their use will grow rapidly over the next few years. The Fortran and C languages do not support this kind of data parallelism. The compiler designers find it difficult to translate the C language directly into SIMD code. An additional problem is that the C language does not contain constructs or explicit parallel constructs; however, it does contain features that could express data parallelism, namely, its array features.

In Fortran 90 we have operation on multiple data items:

```
INTEGER A(10), B(10), C(10)
```

the following array assignment statement:

```
A=B+C
```

is equivalent to a set of 10 elemental assignment statements:

J.M.L.M. Palma et al. (Eds.): VECPAR 2002, LNCS 2565, pp. 608–621, 2003.

```
A(1) = B(1) + C(1)
A(2) = B(2) + C(2)
....
A(10) = B(10) + C(10)
```

These ten statements can be executed in any order, even in parallel.

We would like to do the same in C, however, we do not have any syntactic possibility to do this, and there remains the problem of how to introduce the SIMD instruction set. The short SIMD processing is exploited on larger words (from 64 to 128 bits) than usual. To store these words there are large register files inside the CPU, as well as all the necessary logic, which enables the processing in parallel of 8 bytes, 4 words, 4 floating-point values, etc.

As a consequence of the above we decided to extend the syntax of C and to redefine the existing semantics in such a way that we could use vector processing facilities in C. We redefined the semantics of the existing operators, we added a class of new ones, and we also added some new syntax, or more precisely, some new constructs which redefine the declaration of the arrays and the access to their elements. There were some attempts to specify vector C in [2], [5], and [7]. We would like to go further and implement the SIMD instruction set. We named this extended C as VC.

This paper is organized as follows. In Section 2 we make comparisons with related studies. In Section 3 we analyse the algorithms that are most frequently used in multimedia processing. In Section 4 we describe the VC programming language. In Section 5 we describe the implementation of the VC compiler, in Section 6 we give real example from multimedia applications and the performance results.

## 2 Comparisons with Other Studies

The language presented in [5] was designed and implemented on the CDC Cyber 205 at Purdue University. It extends C by allowing arrays, in effect, to be treated as first-class objects (vectors) by using a special subscripting syntax to select array slices. This language targets general vector machines with many vector processing facilities that multimedia-enhanced processors do not have. The syntax introduced in [5] allows periodic scatter/gather operations and compress/expand operations. Two new data types, the vector descriptor (which acts as a pointer to the array but is extended in such a way that it can handle non-stride-1 vectors) and the bit vector, as well as vector function call and multidimensional parallelism are also introduced. Vector conditional expressions in that language are handled with the bit vector. For example, the elements from the vector are changed if corresponding bits in the bit vector are one, otherwise they remains unchanged. Multimedia hardware does not support this kind of operation which depends on the bit vector, thus in the VC language we had to redefine the act of conditional assignment. VC also differs from [5] in providing fewer special facilities for vector manipulation and in preserving the interchangeability of arrays and pointers.

The **C[]** language [2] also targets general vector machines and it introduces a large number of new vector operators that have no analogue in ordinary C and are not supported by the existing multimedia hardware extensions. But we found the syntax notation introduced in the C[] language the most suitable for VC expressions of multimedia operations over packed data within a register.

The **C*** [7] language is a commercial data-parallel language from Thinking Machine Corporation, which was compiled onto their SIMD CM-2 machine. The main difference between our work and C* is that C* targets large-scale SIMD machines while VC targets the multimedia extension. C* also adds to C additional overloaded meanings of existing operators and new library functions. These overloaded operators provide patterns of communication. The authors of C* have added two new parallel operators (min, max). Both could easily be expressed in VC through semantically extended C operators. C* also differs from VC in adding a new type of statement to C, the selection statement, which is used to activate multiple processors.

And finally, VC tries to incorporate as much as possible of multimedia processing facilities and in addition to provide as few as possible new operators and type extensions to ANSI C.

## 3 Algorithms and Multi-media Processing

A short overview of the algorithms used in multimedia (MM) processing is presented.

1. monadic (unary) operations, intensity shift: b[i] = a[i] + k; *)
2. intensity multiply: a[i] = a[i] * k; **)
3. negate value: a[i] =' a[i];
4. threshold: a[i] = (a[i] >= k)?a[i] : 0;
5. highlight: a[i] = (a[i] >= k1)?k2 : a[i];
6. dyadic (binary) operations, addition of vectors, vector average:
   c[i] = a[i] + b[i]; *)
   c[i] = (a[i] + b[i])/2; ***)
7. subtraction of vectors: c[i] = a[i] - b[i]; *)
8. multiply: c[i] = a[i] * b[i]; and divide c[i] = a[i] / b[i]; two vectors **)
9. local operators on arrays, or matrices: these operators have to be constructed using the above set of operators, and are varying from example to example.

*) in the SIMD instruction set there are two additional instructions for add, or subtract, unsigned add (subtract), and unsigned add (subtract) with saturation,
**) multiplication, or division is permitted if and only if (iff) k = 1,2,4,8,16,32 (i.e. shift left), and
***) there is a special SIMD instruction for calculating averages.

## 4 Syntax Summary of VC

VC language is an extended ANSI C language with multimedia (short vector or SIMD within a register) processing facilities. It keeps all the ANSI C syntax plus the syntax rules for vector processing. It extends the ANSI C syntax only in the access possibilities for the array elements and in the new vector operators. The syntax notation is mostly based on the notation that was first introduced in [2].

## 4.1 Arrays

Let us present some basic definitions for an array (vector) and a vector strip in VC.

**Definition 1.** *In the VC language an array (or vector) is a data structure that consists of sequentially allocated elements of the same type with a strictly positive unit step.*

## 4.2 Vector Strips

Because of hardware limitations, especially the multimedia execution hardware and the multimedia register set within a microprocessor, not all the lengths of the array components are permitted. So we will define some notations, which we will use throughout this paper and which represent different vector strips.

**Definition 2.** *A vector strip is a subset of an array where all of the components have the same type. These components can be as long as 8 bits (or a byte), 16 bits (or a word), 32 bits (or a double-word), 64 bits (or a quadword) and 128 bits (or a superword). The size of the vector strip is also constant, it is limited to the length of the multimedia register in a microprocessor, and for most modern microprocessors this length is 64 or 128 bits.*

**Definition 3.** *We can define the following possible vector strips:*

1. *A VB vector strip is an array slice composed of 8(16) byte components.*
2. *A VW vector strip is an array slice composed of 4(8) word components.*
3. *A VD vector strip is an array slice composed of 2(4) double-word components.*
4. *A VQ vector strip is an array slice composed of 1(2) quadword component(s).*
5. *A VS vector strip is an array slice composed of 1 superword component.*
6. *A VSF vector strip is an array slice composed of 4 single-precision floating-point components.*
7. *A VDF vector strip is an array slice composed of 2 double-precision floating-point components.*

## 4.3 Access to the Array Elements

To access the elements of an array or a vector we can use one of the following expressions:

1. `expression[expr1]` - with this expression we can access the `expr1`-th element of an array object `expression`. Here, the `expr1` is an integral expression and `expression` has a type "array of type".

2. `expression[expr1:expr2, expr3:expr4]` - with this expression we can access the bits `expr4` through `expr3` of the elements `expr2` to `expr1` of an array object `expression`. Here, the `expr1`, `expr2`, `expr3`, `expr4` are integral expressions and `expression` has a type "array of type". The `expr1` denotes the last accessed element, `expr2` denotes the first accessed element, `expr3` denotes the last accessed bit and `expr4` denotes the first accessed bit.

   *Example 1.* We define two arrays of 100 integers:

   ```
   int A[100], B[100];
   ```

   We can now put the upper 16 bits of the first 50 elements of the array A into the lower 16 bits of the last 50 elements of the array B with the expression statement:

   ```
   B[99:50, 15:0] = A[49:0, 31:16];
   ```

   If a programmer specifies something unusual like access to the `array[7:3, 11:4]`, where array is of the byte type, the VC compiler should divide this operation into several memory accesses (actually, the current laboratory version of the VC compiler will only report an error). We have enabled such irregular access as we believe that the language should be designed for longevity and 'look to the future'. If these multimedia operations are to remain important in the future, some sort of bit scatter/gather hardware will become available on many platforms.
3. `expression[,expr1:expr2]` - with this expression we can access the bits `expr1` through `expr2` of all the elements of an array object `expression`. Here, the `expr1` and `expr2` are integral expressions and `expression` has a type "array of type". The `expr1` denotes the last accessed bit and `expr2` denotes the first accessed bit.

   *Example 2.* We define two arrays of 100 integers:

   ```
   int A[100], B[100];
   ```

   We can now move the low-order bits of all the elements in the array A into the high-order bits of all the elements of the array B with the expression statement:

   ```
   B[, 31:16] = A[, 15:0];
   ```

   The operation for one element is shown in Figure 1.
4. `expression[]` - with this expression we can access the whole array object `expression`. Here, the `expression` has a type "array of type".

   *Example 3.* We define two arrays of 100 integers:

```
int A[100], B[100];
```

We can now copy the array A into the array B with the expression statement:

```
B[] = A[];
```

The operator [] was first introduced in the C[] language as described in [2]. It is called the *block operator* because blocks (forbids) the conversion of the operand to a pointer. We found it suitable to denote the whole array object and thus avoid any possible confusion of arrays with pointers.

To support these new access types we have to redefine the syntax of an array access expression. The original production from ANSI C for postfix expression is extended as follows:

```
postfix_expression : primary_expression
| postfix_expression '[' expression ']'
| postfix_expression '(' ')'
| postfix_expression '(' argument_expression_list ')'
| postfix_expression '.' IDENTIFIER
| postfix_expression PTR_OP IDENTIFIER
| postfix_expression INC_OP
| postfix_expression DEC_OP
| postfix_expression '[' vector_access_expression ']'
```

and these new productions are added:

```
vector_size_expression : expression ':' conditional_expression

vector_access_expression : vector_size_expression
| ',' vector_size_expression
| vector_access_expression ',' vector_size_expression
```

And finally, we have to rewrite the production for the conditional expression in order to avoid ambiguity. So the original ANSI C production for the conditional expression is replaced with:

```
conditional_expression : logical_or_expression
| logical_or_expression '?' vector_size_expression
```

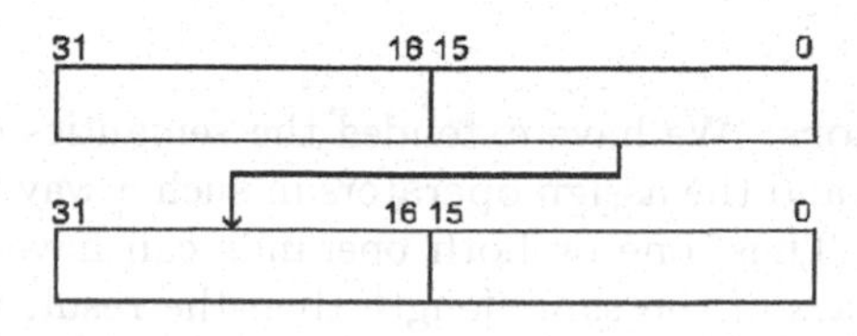


**Fig. 1.** Move the lower part of A[i] into the upper part of B[i]

## 4.4 Operators

**Unary Operators.** We extended the semantics of the existing ANSI C unary operators `&, *, +, -, ~, !` in the sense that they may now have both scalar- and vector-type operands.

*Example 4.* First, we declare two arrays and then we assign negative values of the elements from the first array to the elements from the second array:

```
int A[] = {1, 2, 3, 4};
int B[4];
...
B[] = -A[4];
```

After that, array B will have the following values: $\{-1, -2, -3, -4\}$.

We have also, in a similar way to [2], added new reduction unary operators `[+], [-], [*], [&], [|], [^ ]`. These operators are overloaded existing binary operators `+, -, *, &, |, ^` and are only applicable to the vector operands. These unary operators perform the given arithmetic/logic operation between the components of the given vector. The result is always a scalar value. Again, we believe that `[op]` notation, which was introduced in [2], in a more "natural" way indicates that the operation is to be performed over all vector components.

*Example 5.* First, we declare array A and then we make the sum of all its components:

```
int sum;
int A[] = {1, 2, 3, -4};
...
sum = [+]A[];
```

After that, the `sum` will be 2.

We have also added one new vector operator `|/`, which calculates the square root of each component in the vector (please note, that this works only with floating-point vectors, although the VC compiler does not perform any type checking, and if we apply this operator to integer vectors the result my be undetermined).

**Binary Operators.** We have extended the semantics of the existing ANSI C binary operators and the assign operators in such a way that they can now have vector operands. Thus, one or both operands can have an array type. If both operands are arrays of the same length then the result is an array of the same length (note that the length is measured in the number of components and not in the number of bits!). If one array operand has N elements and another array operand has M elements and $N < M$ then the operation is only performed over N elements. If arrays have different types then the VC compiler reports an error. If one of the operands is of the scalar type then it is internally converted by the VC compiler into a vector strip of the corresponding type and length. This type of element in the vector strip and its length strongly depends on the processor for which we compile our program. For example, if the array operand consists of word components then for the Intel Pentium processor the scalar operand is converted into the VW vector strip (vector of four 16-bit values).

*Example 6.* We can now make the sum of two arrays :

```
short A[4] = {1, 2, 3, 4};
short B[4] = {4, 3, 2, 1};
short C[4] = {0, 0, 0, 0};
short d = 7;
...
C[] = A[] + B[];
A[] = d + A[];
```

Note, that the + operator has both operands with an array type. Now, the components from array C will have the following values: `(5, 5, 5, 5)` and the components from array A will have the following values: `(8, 9, 10, 11)`. Variable `d` is expanded internally by the VC compiler into the vector `[0007h, 0007h, 0007h, 0007h]`.

*Example 7.* The Intel SIMD instruction `PMADDWD` first carries out the component-wise product of the integers, and then, second, makes the sum of the products into an integer. This can also be written in the VC language as:

```
short A[4], B[4];
int R[2];
...
R[1:1,31:0] = [+] ( A[2:1,15:0] * B[2:1,15:0] );
R[2:2,31:0] = [+] ( A[4:3,15:0] * B[4:3,15:0] );
```

We have overloaded the existing binary operators with 3 new operators:

? this operator overloads the binary operators in such a way that the given binary operator performs the operation with saturation,

@ this operator overloads the binary add operator in such a way that the given binary operator first performs addition over adjacent vector elements and then averages (shift right one bit) the result. Let `A[]` and `B[]` be two vectors of the same dimension N and the same type. Then expression `A[] @+ B[]` has the same semantics as the expression `(A[0] + B[0])/2, ..., (A[N-1] + B[N-1])/2`,

~ this operator overloads the multiply operator in such a way that the result is the high part of the product,

_ this operator overloads the multiply operator in such a way that the result is the low part of the product.

Thus, we may have the following operations:

?+ for add with saturation
(in the grammar denoted as VEC_ADD_SAT),
?− for subtract with saturation
(in the grammar denoted as VEC_SUB_SAT),
@+ for average add
(in the grammar denoted as VEC_ADD_AVG),
~* multiply, the result is the high part of the product
(in the grammar denoted as VEC_MUL_HI),
_* multiply, the result is the low part of the product
(in the grammar denoted as VEC_MUL_LO).

Besides the existing binary operators we have added one new, binary operator, which we found to be important in multimedia applications. This operator is applicable only on vector operands (if any operand has a scalar type then it is expanded into an appropriate vector strip) and is as follows:

|−| absolute difference
(in the grammar denoted as VEC_SUB_ABS).

*Example 8.* The Intel SIMD instruction PSADBW computes the absolute differences of the packed unsigned byte vector strips (VB). Differences are then summed to produce an unsigned word integer result. This can also be written in the VC language as:

```
unsigned char A[8], B[8]; /* components are 8 bits long */
unsigned short c;
...
c = [+] (A[] |-| B[]) ;
```

**Conditional Expression.** The conditional operator from ANSI C '?:' which is used in the conditional expression:

```
conditional_expression :
  logical_or_expression '?' expression ':' assignment_expression
```

can now have array-type operands. If the first operand is a scalar or an array and the second and third are arrays then the result operand has the same array type as both operands. If the array operands have different array lengths or different types of components then the behavior of the conditional expression is undetermined. If the second or third operand is scalar then it is converted into a vector (the same conversion as for binary operators). If all the operands are arrays of the same length the operation is performed component-wise.

*Example 9.* The Intel SIMD instruction `PMAXUB` returns the greater vector components between two byte vectors (VB). This can also be written in the VC language as:

```
int A[100], B[100];
int C[100];
...
C[] = (A[] > B[]) ? A[] : B[] ;
```

Tables 1 and 2 summarize the multimedia instruction set supported by the Intel, Motorola and SUN processor families and the associated VC expression statements.

**Table 1.** Relations between integer multimedia instructions and VC expressions

| VC expression | Intel MMX | SUN UltraSpark VIS | Motorola Altivec PowerPC |
|---|---|---|---|
| R[] = A[] + B[]; | PADDB \| W \| D | vis fadd[16, 32] | vec add[8, 16, 32] |
| R[] = A[] ?+ B[]; signed | PADDSB \| W | | vec adds[8, 16, 32] |
| R[] = A[] ?+ B[]; unsigned | PADDUSB\| W | | |
| R[] = A[] - B[]; | PSUBB \| W \| D | vis fpsub[16, 32] | vec sub[8, 16, 32] |
| R[] = A[] ?- B[]; signed | PSUBSB \| W \| D | | |
| R[] = A[] ?- B[]; unsigned | PSUBUSB\| W \| D | | |
| R[] = A[] ~* B[]; signed | PMULHW | | |
| R[] = A[] ~* B[]; unsigned | PMULHUW | | |
| R[] = A[] * B[]; signed | PMULLW | | |
| R[] = A[] * B[]; unsigned | PMULLUW | | |
| A[] = A[] << count; | PSLLW\| PSLLD | | vec sl[8, 16, 32] |
| A[] = A[] >> count; | PSRAW\| D | | vec sra[8, 16, 32] |
| A[] = A[] >> count; | PSRLW\| PSRLD | | vec sr[8, 16, 32] |
| R[] = A[] op B[]; | POR \| AND \| XOR \| ANDN | | vec or\| and\| xor[64] |
| R[] = (A[] = = B[])? 0xFF : 0; | PCMPEQB\| W\| D | vis fcmpeq[16] | vec cmpeq[8, 16, 32] |
| R[] = (A[] > B[])? 0xFF : 0; | PCMPGTB\| W\| D | vis fcmpgt[16] | vec cmpgt[8, 16, 32] |
| R[] = A[] > B[] ? A[] : B[]; | PMAXUB \| W | | vec max[8, 16, 32] |
| R[] = A[] < B[] ? A[] : B[]; | PMINUB \| W | | vec min[8, 16, 32] |
| R[] = A[] @+ B[]; | PAVGB \| W | | vec avg[8, 16, 32] |
| R[] = A[] \|-\| 0; | | | vec abs[8, 16, 32] |
| R = [+] (A[] \|-\| B[]); unsigned | PSADBW | | |

op ={ |, &, ^, ~&}

**Table 2.** Relations between floating-point multimedia instructions and VC expressions

| VC expression | Intel SSE | SUN VIS | Motorola Altivec |
|---|---|---|---|
| R[] = A[] + B[]; | ADDPS | | vec add[4x32] |
| R[] = A[] - B[]; | SUBPS | | vec sub[4 x 32] |
| R[] = A[] * B[]; | MULPS | | |
| R[] = A[] / B[]; | DIVPS | | |
| R[] = 1 / A[]; | RCPPS | | vec rc[4 x 32] |
| R[] = \|/^A[]; | SQRTPS | | |
| R[] = 1 / ( \|/A[] ); | RSQRT | | vect rsqrte [4 x 32] |
| R[] = A[] log_op B[]; | POR, PAND, PXOR, PANDN | | vec or \| and \| xor \| andn [4 x 32] |
| R[] = (A[] rel_op B[])? 0xFF : 0; | CMPPS, rel_op1 | | vec cmp [rel_op2] [4 x 32] |
| R[] = A[] > B[] ? A[] : B[]; | MAXPS | | vec max[8, 16, 32] |
| R[] = A[] < B[] ? A[] : B[]; | MINPS | | vec min[8, 16, 32] |

log_op ={ |, & , ^ , ~&}    rel_op1 = {= =, <, <=, !=, !<, !<=, !? }    rel_op2 = {= =, <, <=, !=, >, >=}

## 5 Translation of a Program Written in VC into C

The laboratory version of the VC compiler is implemented for Intel Pentium III and Intel Pentium IV processors. It is implemented as a translator to ordinary C code that is then compiled by an ordinary C compiler (in our example with Intel C++ Compiler for Linux [11]).

The VC compiler parses input VC code, performs syntax and semantics analysis, builds its internal representation, and finally translates the internal representation into ANSI C with macros written in a particular assembly language instead of the VC vector statements.

Here we will only show the macro library for the integer operations for the Intel Pentium class of microprocessor and that is used by the VC compiler to translate VC code into ordinary C code. The macro library itself is written in assembly language, and can be listed in some separate sets:

1. Arithmetic and logic instructions on vectors:
   (a) vectors $\mathbf{A} + \mathbf{B}$, $\mathbf{A}? + \mathbf{B}$ to **RESULT**
   `ADDB, ADDW, ADDD, ADDUB, ADDUW, ADDUD ADDUSB, ADDUSW, ADDUSD (RESULT, A, B);`
   (b) vectors $\mathbf{A} - \mathbf{B}$, $\mathbf{A}? - \mathbf{B}$ to **RESULT**
   `SUBB, SUBW, SUBD, SUBUB, SUBUW, SUBUD, SUBUSB, SUBUSW, SUBUSD (RESULT, A, B);`
   (c) vectors $\mathbf{A\&B}, \overline{\mathbf{A}}\mathbf{\&B}, \mathbf{A|B}, \mathbf{A\nabla B}$ to **RESULT**
   `ANDQ/O, ANDNQ/O, ORQ/O, EXORQ/O (RESULT, A, B);`
   (d) vectors $\mathbf{A}_{\_} * \mathbf{B}$, $\mathbf{A}^{\sim} * \mathbf{B}$ to **RESULT**
   `MULLW, MULLUW, MULHW, MULHUW (RESULT, A, B);`
   (e) `AVGB, AVGW (RESULT, A, B);` // $([+](\mathbf{A} - \mathbf{B}))/2$ to **RESULT**
   (f) `SRL, SLLW/D/Q (A, NOSHIFTS);` //shift R/L vector **A** for NOSHIFTS
   (g) `SRAW/D (A, NOSHIFTS);` //shift R vector **A** for NOSHIFTS
2. Miscellaneous instruction set:
   (a) `SUMMULTWD(RESULT, A, B);` // $[+](\mathbf{A} * \mathbf{B})$ to scalar RESULT
   (b) `SUMABSDIFFBW(RESULT, A, B);` // $[+](\mathbf{A} - \mathbf{B})$ to scalar RESULT
3. Compound (control) instructions:
   (a) `IFMGTB/W/FP(MASK, A, B, R);` // $\mathbf{R} = (\mathbf{MASK} > \mathbf{A})?\mathbf{A} : \mathbf{B}$
   (b) `IFMEQB/W/FP(MASK, A, B, R);` // $\mathbf{R} = (\mathbf{MASK} == \mathbf{A})?\mathbf{A} : \mathbf{B}$
   (c) `IFGTB/W/FP(A, B, D, E, R);` // $\mathbf{R} = (\mathbf{B} > \mathbf{A})?\mathbf{D} : \mathbf{E}$
   (d) `IFEQB/W/FP(A, B, D, E, R);` // $\mathbf{R} = (\mathbf{B} == \mathbf{A})?\mathbf{D} : \mathbf{E}$

The complete macro library can be downloaded from `http://lra-1.fri.uni-lj.si/vect/MacroVect.c`.

The library MacroVect.c is under development, and during its use we will be able to evaluate its pros and cons. From this perspective we have developed a general set of mostly usable macros.

*Example 10.* The VC statement:

```
R[] = (A[] > MASK)? A[] : B[];
```

is evaluated during the compilation process into macro `IFGTB(MASK, A, B, R);` which is defined as:

```
#define IFMGTB(MASK, A, B, R); __asm{ mov eax, B \
   movq mm3, [eax] \
   mov ebx, A~\
   movq mm2, [ebx] \
   mov ecx, MASK \
   movq mm1, [ecx] \
   pcmpgtb mm1, mm2 \
   movq mm4, mm1 \
   pand mm1, mm3 \
   pandn mm4, mm2 \
   por mm1, mm4 \
   mov edx, R \
   movq [edx], mm1 };
```

In Section 6 we can see the use of this macro in a real application.

## 6 Implementation of the Vectorizable Operations

We used our VC for processing a b/w signal from a video frame-grabber. For a comparison we first used the program written in ANSI C and then the program written in the VC language.

*Example 11.* We mix the two images, one a live picture from the video camera, and the second, the background map, which is in the memory. The principle of mixing is very simple, and is as follows:

```
if (Threshold > Live_picture) then Show =  Background;
else Show = Live_picture;
```

As the color white is 'FF' we put the Threshold a little lower, i.e. about 'F0'. The above statement we rewrite in the VC language:

```
char Show[NOITEMS];
char Live_picture[NOITEMS];
char Background_map[NOITEMS];
char Live_picture[NOITEMS];
...

Show[]=Thr > Live_picture[] ? Background_map[] : Live_picture[];
```

Where the constant `Thr` is `0xF0`.

These statements in VC are translated into the C code in such a way that the used scalar program can run in vectored mode :

**Table 3.** Execution results

| Example in / Compiled with | Number of total instructions | Time [%] byte | Time [%] short | Time [%] int |
|---|---|---|---|---|
| C / Intel C++ | 22 | 1 | 1.6 | 1.7 |
| VC / VCC, Intel C++ | 22 + 9 SIMD | 0.65 | 1.37 | 1.65 |

```
// create new symbols for loop indexes:
int i00001, i00002;
// expand Thr into vector:
int ThrMASK[8] = {Thr, Thr, Thr, Thr, Thr, Thr, Thr, Thr} ;

for( i00001 = 0; i00001 < div(NOITMES, 8).quot; i00001+=8 ) {
   A~= Live_picture + i00001;    // prepare addresses for macro
   B = Background_map + i00001;
   R = Show + i00001;
   D = ThrMASK;
   IFMGTB(D, A, B, R);           // macro insertion
}

for( i00002=div(NOITMES,8).quot;i00002 < NOITEMS;i00002++ ) {
  if (Thr > Live_picture[i00002]) {
    Show[i00002] =  Background[i00002];
  }
  else Show[i00002] = Live_picture[i00002];
}
```

We used our VC for processing a b/w signal from a video frame-grabber. The processing kernel is the mixing function presented in the previous example. For a comparison we first used the program written in ANSI C and then the program written in the VC language. In Table 3 we can see the results of both tests, the number of instructions, and the improvement of execution times for processing the array of 442368 bytes, shorts, and integers. We made the test on the Pentium III microprocessor.

We see that the execution time improves by about 30% on larger arrays (int), and by about 100% on smaller arrays (byte), even though the total number of instructions is greater.

## 7 Conclusion

We have developed a VC programming language which is able to use hardware-level multimedia execution capabilities. The VC language is an upward extension of ANSI C and it saves all the ANSI C syntax. In this way it is suitable for use by programmers who want to extract SIMD parallelism in a high-level programming language and also by programmers who do not know anything about multimedia processing facilities and who are using the C language. The presented extension to C also preserves the interchangeability of arrays and pointers and adds as few as possible new operators. All added operators have an analogue in ordinary C. The declarations of arrays are left unchanged and also no new types have been added.

We obtained good performance for several application domains. Experiments on scientific and multimedia applications have significant performance improvements.

Our VC, its compiler and macro library are still in their infancy. Although successful, we believe their effectiveness can be further improved.

## References


[1] Patricio Bulić, Veselko Guštin. Macro extension for SIMD processing. in *Proc. 7th European Conference on Parallel Processing EURO PAR 2001, Manchester, UK, 28-31 August, 2001, Lecture Notes in Computer Science 2150*, pp. 448-451, 2001.

[2] Sergey Gaissaryan, Alexey Lastovetsky. An ANSI C for Vector and Superscalar Computers and Its Retargetable Compiler, *Journal of C Language Translation*, 5(3), pp. 183-198, 1994.

[3] Veselko Guštin, Patricio Bulić. Extracting SIMD Parallelism from "for" Loops. in *Proceedings of the 2001 ICPP Workshop on HPSECA, ICPP Conference, Valencia, Spain, 3-7 September, 2001*, pp. 23-28. 2001.

[4] Veselko Guštin, Patricio Bulić. Introducing the vector C. *Proc. 5th International Meeting VECPAR 2002, Porto, Portugal, 26-28 June, 2002.* pp. 253-266. 2002.

[5] Kuo-Cheng Li. A note on the vector C language. *ACM SIGPLAN Notices*, Vol. 21, No. 1, pp. 49-57, 1986.

[6] Millind Mitall, Alex Peleg, Uri Weiser. MMX Technology Architecture Overview, *Intel Technology Journal*, 1997.

[7] John R. Rose, Guy L. Steele. C* : An extended C Language for Data Parallel Programming. *Proceedings of the Second International Conference on Supercomputing ICS87, May, 1987*, pp. 2-16, 1987.

[8] -. Intel Architecture Software Developer's Manual Volume 1: Basic Architecture, `http://download.intel.nl/design/pentiumii/manuals/24319002.pdf`.

[9] -. Intel Architecture Software Developer's Manual Volume 2: Instruction Set Reference, `http://download.intel.nl/design/pentiumii/manuals/24319102.pdf`.

[10] -. Intel Architecture Software Developer's Manual Volume 3: System Programming, `http://download.intel.nl/design/pentiumii/manuals/24319202.pdf`.

[11] -. Intel C++ Compiler for Linux 6.0. `http://www.intel.com/software/products/compilers/c60l/`.

# Mobile Agent Programming for Clusters with Parallel Skeletons*

Rocco Aversa, Beniamino Di Martino, Nicola Mazzocca, and Salvatore Venticinque

Dipartimento di Ingegneria dell' Informazione - Second University of Naples
Real Casa dell'Annunziata - via Roma, 29, 81031 Aversa (CE) - Italy
{beniamino.dimartino,n.mazzocca}@unina.it
{rocco.aversa,salvatore.venticinque}@unina2.it

**Abstract.** Parallel programming effort can be reduced by using high-level constructs such as algorithmic skeletons. Within the Magda toolset, supporting programming and execution of mobile agent based distributed applications, we provide a *skeleton-based* parallel programming environment, based on specialization of *Algorithmic Skeleton* Java interfaces and classes. Their implementation include mobile agent features for execution on heterogeneous systems, such as clusters of WSs and PCs, and support reliability and dynamic workload balancing. The user can thus develop a parallel, mobile agent based application by simply specialising a given set of classes and methods and using a set of added functionalities.

## 1 Introduction

The Mobile Agents model [11] has the potential to provide a flexible framework to face the challenges of High Performance Computing, especially when targeted towards heterogeneous distributed architectures such as clusters. Several characteristics of potential benefit for cluster computing can be provided by the adoption of the mobile agent technology, as shown in the literature [7, 8, 9]: they range from network load reduction, heterogeneity, dynamic adaptivity, fault-tolerance to portability to paradigm-oriented development.

We are developing a framework for supporting programming and execution of mobile agent based distributed applications, the *MAGDA* (Mobile AGents Distributed Applications) toolset. It enhances the mobile agent technology with a set of features for supporting parallel programming on a dynamic heterogeneous distributed environment.

Nevertheless, as a matter of fact, developing distributed applications using the mobile agent model remains a non trivial task because the mobility feature

* This work has been supported by the Italian Ministry for University and Research (MURST) (P.R.I.N. Project *ISIDE* - "Dependable reactive computing systems for industrial applications") and by the CNR - Consiglio Nazionale delle Ricerche, Italy (Agenzia 2000 Project *METODOLOGIE E STRUMENTI PER LABORATORI VIRTUALI DISTRIBUITI*).

J.M.L.M. Palma et al. (Eds.): VECPAR 2002, LNCS 2565, pp. 622–634, 2003.

introduces additional difficulties in designing coordination, synchronization and communications among the different tasks. In practice, many applications share the same communication and synchronization structure, independently from the application-specific computations. So, especially when the starting point is an available sequential code, using the concept of the algorithmic skeletons [4], that is separating the communication/synchronization structure from the application-dependent functions, can ease the programming task, and improve the mapping for performance on parallel systems. Within the framework of Magda, we have defined and implemented a set of Java packages, which enable to program distributed applications by adopting a skeletons-like approach, exploiting the peculiar features of both Object Oriented and Mobile Agents programming models. By means of the provided skeletons interfaces the programmer is able to implement his own application by specialising an assigned structure and utilizing the set of functionalities that the mobile agents framework offers. In addiction such approach allows to reuse a great deal of the sequential code, when available. A predefined algorithmic skeleton allows to follow the sequential programming model by filling some methods, classes and interfaces and to hide the difficulties involved by an explicit parallel programming paradigm. The difficulties and the features of the Mobile Agent programming paradigm can be managed at a lower level, hidden to the user.

Two algorithmic skeletons have been implemented and tested. In *Processor Farm* skeleton the master process creates a number of slaves and assigns some work to everyone of them. The slaves compute their work and return the results to the master. The second algorithmic skeleton we have considered belongs to *Divide and Conquer*-like skeleton class, but not to the highest abstraction level. It is an example of *Tree computation* algorithmic skeleton. It solves the initial problem by dividing it in several subproblems assigned to different agent workers logically connected according to a tree topology.

In order to illustrate the expressiveness and effectiveness of the above mentioned approach we utilize two case-study applications: the simple and well known Quick Sort algorithm, and a real combinatorial optimization application implementing the Branch & Bound (B&B) technique. These are typical examples of irregularly structured problems that expose a highly dynamic execution behaviour.

The remainder of the paper proceeds as follows: section 2 is devoted to the description of the Magda environment; section 3 and 4 describe the two considered skeletons, their implementation within the Magda framework and the description of the set of classes and methods provided to the user to fit the application to the skeletons. The two mentioned case studies are illustrated in these sections. Section 5 presents and discuss experimental results for the combinatorial optimization application. Finally (section 6) we present the related work and give some concluding remarks.

## 2 MAGDA

The *MAGDA* (Mobile AGents Distributed Applications) toolset [2] is a framework for supporting programming and execution of mobile agent based distributed applications. It supplements mobile agent technology with a set of features for supporting parallel programming on a dynamic heterogeneous distributed environment.

The (prototype) framework currently supports the following features:

- *collective communication* among mobile agents, with provision of a set of collective communication primitives, implemented through multicast interplace messaging.
- *dynamic workload balancing* among mobile agents, through centralized or distributed mechanisms;
- *dynamic workload estimation* for each running agent;
- *dynamic system parameters estimation* for each node of the cluster;
- *automated mechanisms for agents' migration and cloning*, driven by policies based on architecture nodes' and network monitoring, in terms of utilization, idle time, memory usage and different run-time events (like shutdown);
- *Agents' authentication*, which can be used in security critical applications;
- *remote agents' creation*, which provides the possibility to activate an agent from different terminals, including wireless handheld devices;
- a *skeleton-based* parallel programming environment, based on specialization of *Skeleton* Java interfaces and classes;
- *integration of MA (Mobile Agents) programming paradigm and OpenMP*, for programming hierarchical distributed-shared memory multiprocessor architectures, in particular heterogeneous clusters of SMPs (and uniprocessor) nodes.

We based the development of our framework on the basic agent mechanisms provided by the Aglet Workbench, developed by IBM Japan research group [10]. An *aglet* (agile applet) is a lightweight Java object that can move to any remote host that supports the Java Virtual Machine. An Aglet server program (*Tahiti*) provides the agents with an execution environment that allows for an aglet to be created and disposed, to be halted and dispatched to another host belonging to the computing environment, to be cloned, and, of course, to communicate with all the other aglets.

## 3 The Tree Computation Skeleton

### 3.1 Skeletons' Description

The skeleton frequently referred to as *tree computation* consists of a set of processes connected by communication channels according to a tree structure. Each process receives a problem to be solved, tests for a condition and then either splits the problem into k subproblems that are sent to k child processes, or does some processing work. When a process terminates its job, it remains waiting for a reply from each child, then combines these replies and sends back the new result to its parent. In other words, according to this skeleton, the data flow from the root into the leaves and the solutions flow back up towards the root. In practice a number of questions (how to choose the type, the degree and the depth of the tree, the processes have to be created dynamically or the entire tree must be instantiated statically, etc.) have to be answered before a working program can be produced out of this skeleton. However, they do not depend on the particular nature of the computation to be parallelized, but rather they are part of the skeleton and can be solved once and for all in the context of the skeleton itself.

### 3.2 Implementation in Magda

Our implementation mainly follows the general structure depicted in the previous section, except that processes are replaced with agents and we have chosen to implement a binomial algorithm to build our tree, so its shape and the results recombination procedure is consequentially determined. Initially, the first agent of the application starts cloning itself and generating a first son, splitting and assigning half of the initial workload to it. At each following step, every agent clones itself once, thus generating a son; at each step all agents simultaneously clone themselves, until the number of agents matches a target number. The tree shape and the communication pattern among agents are consequentially determined, as is shown in Figure 1. The number of levels of the tree is equal to

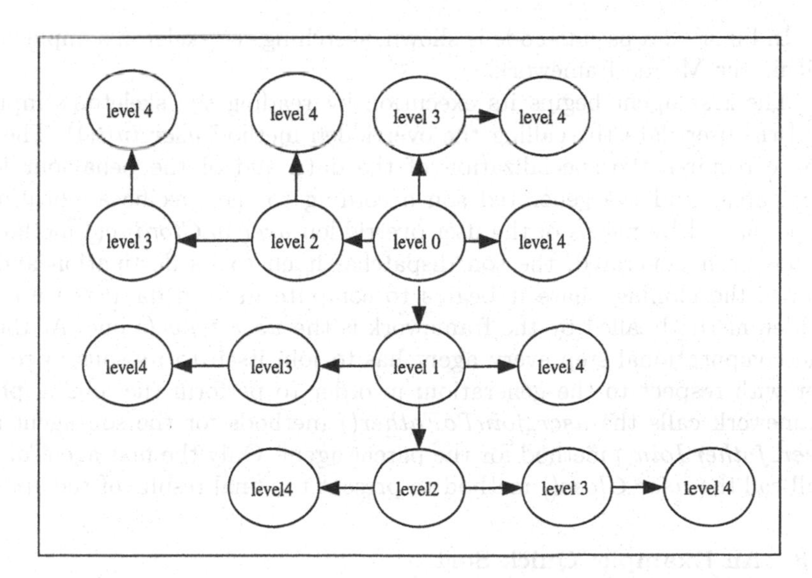


**Fig. 1.** Cloning/communication graph of the Tree Computation skeleton (binomial tree case)

the number of cloning steps. The number of nodes belonging to the $i^{th}$ level is equal to $2^{nc-i}$, where nc is the number of cloning steps and $i$ is the level index. A the end of this cloning step each agent can start solving its subproblem. Once the work is performed each agent collects and joins the results from its sons (from the youngest to the oldest one) and in its turn pass them to its parent.

An agent returns its results only if it is not a parent or it has collected the results from all its own sons (from the youngest to the oldest one).

We have defined a *BinomialTree* class which performs the basic actions of the described skeleton behavior and activates the specific methods that the user should override in order to specialize the user agent's class behaviour. All the names of such methods begin with the word *user_*.

The following methods belong to the *BinomialTree* class and should be overridden:

- The *user_init()* method, invoked by the first agent in order to initialize the application data.
- The *user_onCloning()* method, invoked in order to generate a new agent;
- The *user_Solve()* method, invoked in order to solve the agent's task;
- The *handler_message()* method, invoked in order to handle a message from the application;
- The *user_joinToFather()* method, invoked by an agent in order to return its results to its parent;
- The *user_fatherJoining()* method, invoked by an agent in order to join its own results to its son's results;
- The *user_Close()* method, invoked, by the first agent, on the occurrence of its disposing, at the end of the application, in order to present the final results.

In Figure 2 a pseudo-code is shown, sketching the skeleton's implementation within the Magda framework.

The first agent begins its execution by reading the skeleton's input data, and the user data (by calling the overridden method *user_init*()). The cloning phase requires the specialization of the data and of the behaviour for both the parent and the generated son according to the specific application. This is performed by means of the user overridden *user_onCloning*() method. After it has been generated, the son dispatches itself to its destination and at the end of the cloning phase it begins to compute its own data: the user's overridden method called by the framework is the *user_Solve*() one. At the end of the computational step every agent has to join itself to its sons in reverse order with respect to the generation; in order to perform the joining phase the framework calls the *user_joinToFather*() methods for the son agent and the *user_fatherJoin*() method for the parent agent. Only the first agent of the tree will call the *user_Close*() method to present the final results of the application.

### 3.3 An Example: Quick Sort

We now show, through a simple example, how the above described class and methods are to be utilized in order to develop a distributed mobile tree com-

```
OnCreation()

  user_init();

run()
 if(!dispatched)
    dispatch(myDestination);
 while(clones<nodes)

    clones=2*clones;
    user_onCloning();
     clone();


 user_compute(myObject);

 while(!mySons.isEmpty())

    next=mySons.next();
    reply= next.send("join to father");
    user_fatherJoin(reply);

 wait("jointofather");
    myFather.reply(user_joinToFather());

  if(!master)  dispose();
  else user_Close();
    dispose();
```

**Fig. 2.** BinomialTree class pseudo-code

putation. The chosen computation is the simple and well known Quick Sort algorithm, applied to an array of integers.

In the user-defined *QuickSort* class the user declares and initializes all the data that each agent needs in order to solve its own sub-problem: in the QuickSort example the data are an array of integers to be sorted, and the sub-arrays' bounds, by means of which each agent is able to know which part of the original array is assigned to. The first agent fills up its array by calling the user-overridden *user_init*() method and sets its lower bound to 0, and the upper one equal to the length of the array.

The *user_onCloning*() method calls the user-defined *split*() method, in order to split the array into two sub-arrays, representing the parent's sub-problem and its son's one. The method returns two new QuickSort objects for the two agents (the parent and the son). When the cloning phase ends up the user-overridden *user_Solve*() method is called, performing a recursive sorting of the array (by calling the *sort*() method of the *SeqQuickSort* object).

In this example the *handler_message*() method is not used because the sorting computation doesn't require communication among agents. The *user_joinToFather*() method is called by the son when the parent collects its results: in the QuickSort example this method returns the sorted sub-array. The *user_fatherJoining*() method is called by the parent when the son's results arrive; in the example this method performs the fusion between the array owned by the current agent and the ones returned by the sons.

The *user_Close*() method finally prints the sorted array.

## 4 The Task Queue (Processor Farm) Skeleton

### 4.1 Skeletons' Description

The skeleton usually referred to as *processor farm* consists of a *coordinator* process and a set of *worker* processes that act as slaves of the coordinator. In a processor farm, the coordinator decomposes the work to be done in subproblems and assigns a different subproblem to each worker. Upon receipt of a subproblem, each worker solves it and returns a result to the coordinator. Again some details have to be defined before the skeleton can become a working program, and slightly different organizations can be selected for the processor farm (for instance workers may or may not be allowed to communicate with each other). However, even in this case, these issues do pertain to the skeleton definition and can be entirely dealt with in the skeleton context. We have chosen to implement the *Task Queue*Skeleton, that is the most general *Farm*-like skeleton: every slave may produce new work to be performed by itself or by other slaves.

The task queue skeleton is illustrated in figure 3.

### 4.2 Implementation in Magda

For this skeleton the Magda framework makes available the *TaskQueue* Java class, and the *Task_Interface* interface.

The TaskQueue class describes the behaviour of an agent worker. When the user extends this class it needs to declare and initialise all the data that each worker needs in order to solve the generic task. The user has to implement its worker by extending the class and overriding the following methods:

- The *user_init()* method, invoked by the first agent worker in order to initialize the application data;
- the *user_split_task()* method, invoked by the first worker in order to fill the bag with more sub-problems, which will be distributed among all the workers;
- the *user_handle_message()* method, invoked in order to handle a message from the application;
- the *user_Close()* method, invoked by the first agent worker at disposal, in order to present the final results.

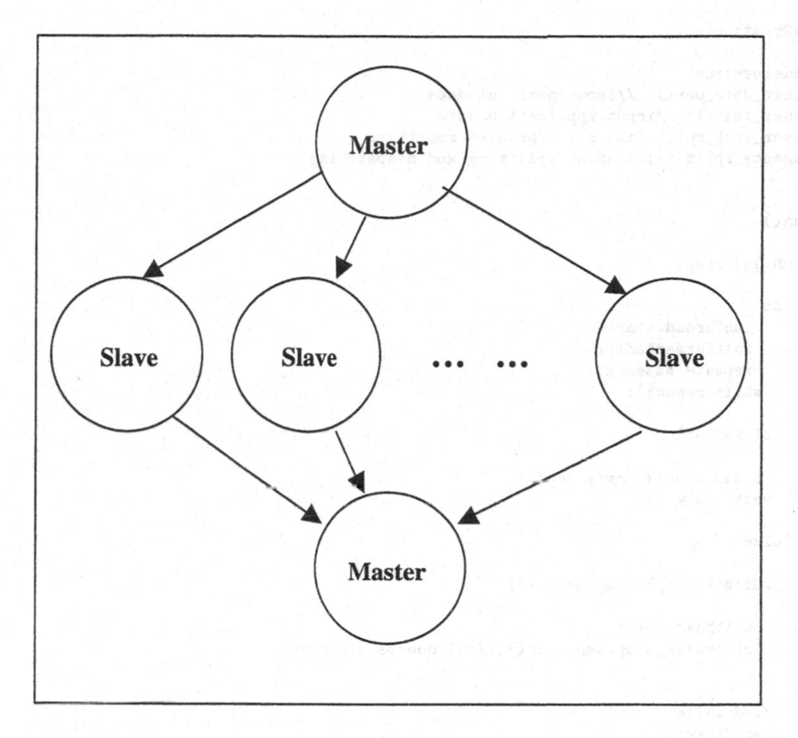


**Fig. 3.** The communication graph of the Task Queue skeleton

In Figure 4 a pseudo-code is shown, sketching the skeleton's implementation within the Magda framework.

The first agent begins its execution by reading the skeleton's input data, and the user data (by calling the overridden method *user_init*()). The splitting of the initial problem is performed by the user, by overriding the *user_split_task*() method. The effect of this phase is filling of the agent's queue with a certain number of subproblems. Then the master clones itself by distributing the queue among the different clones and dispatches the workers to the target hosts. All the global data is cloned with the agents. Arrived at destination each agent activates a thread for the extraction and the computation of the tasks stored into the queue. When the queue is empty a request for new tasks is sent. If no more tasks are available the worker signals the empty queue condition to the master. The master, after its computation ends, receives the empty queue signal from all the workers and asks for the termination condition by calling the user-overridden *user_stopCondition*() method. The return of the true value allows the master to dispose the workers and itself. When a message is received by an asynchronous activaction the message handler is called. The handler reads the message tag and if it is not able to handle it, because it is not a service message, the user-overridden *user_handle_message*() is called.

```
OnCreation()

 master=true;
 init_data_par(); //input parallel data
 user_init(); //input application data
 user_task_split (int n); //problem splitting
 agents_split (); //agent splitting and dispatching


run()

  while(!stop)

   do
      runThread.start();
      waitThreadEnd();
      repeat= balance();
     while(repeat);

   if(!master)

     sendMessage("empty_bag");
    waitMessage();

   else

    while((!stop)&(bag.empty())

      waitMessage();
      stop=user_stopCondition();//all queues are empty


   //end while
   user_Close();
    sendToAll("Dispose");

    dispose();
//end run
handleMessage(Message msg)

  if(msg.sameKind("service msg")
...
   else
user_handleMessage(msg);

Class runThread
run()

  while(!bag.empty())

   newTask=(Task_interface) bag.pop();
   newTask.solve();
```

**Fig. 4.** TaskQueue class pseudo-code

In order to specify how the specific task extracted from the queue has to be consumed, the user has to define a *Workload* class that implements the *Task_Interface* interface.

The Task_Interface is implemented as follows:

```
public interface task_interface

    public void compute();
```

The *compute*() method implemented by the user is called by the agent worker on the user's object in order to consume the task.

Other methods, which need not to be overrridden and can be used by the programmer, are defined in the *TaskQueue* class.

As the skeleton is designed in order to hide the real distribution of the application, the programmer is not able to locate each single worker. It means that only collective communication are allowed, in order tho share global values and update them, or in order to synchonize all the workers. The *procFarmBroad*(*Message*) method allows the programmer to send a message to all the other workers.

The *bag_push(Object obj)* method allows the programmer to insert some new tasks into the queue according to the Task Queue skeleton.

### 4.3 An Example: A Combinatorial Optimization Application

We now show, through a quite complex real application, how the above described skeleton's classes and methods can be utilized. The chosen application is a combinatorial discrete optimization, performed with the well known Branch and Bound method.

A *discrete optimization problem* consists in searching the optimal value (maximum or minimum) of a function $f : \boldsymbol{x} \in \mathcal{Z}^n \rightarrow \mathcal{R}$, and the solution $\boldsymbol{x} = \{x_1, \ldots, x_n\}$ in which the function's value is optimal. $f(\boldsymbol{x})$ is said *cost function*, and its domain is generally defined by means of a set of $m$ constraints on the points of the definition space. Constraints are generally expressed by a set of inequalities:

$$\sum_{i=1}^{n} a_{i,j} x_i \leq b_j \qquad \forall j \in \{1, \ldots, m\} \tag{1}$$

and they define the set of feasible values for the $x_i$ variables (the *solutions space* of the problem).

Branch & Bound [13] is a class of methods solving such problems according to a *divide & conquer* strategy. The initial solution space is recursively divided in subspaces, until attaining to the individual solutions; such a recursive division can be represented by a (abstract) tree: the nodes of this tree represent the solution subspaces obtained by dividing the parent subspace, the leaf nodes represent the solutions of the problem, and the tree traversal represents the recursive operation of dividing and conquering the problem.

The method is enumerative, but it aims to a non-exhaustive scanning of the solutions space. This goal is achieved by estimating the best feasible solution for each subproblem, without expanding the tree node, or trying to prove that there are no feasible solutions for a subproblem, whose value is better than the *current* best value. (It is assumed that a best feasible solution *estimation function* has been devised, to be computed for each subproblem.) This latter situation corresponds to the so called *pruning* of a search subtree. The Task queue skeleton is thus very well suited to a parallel implementation of the Branch and Bound method.

According the skeleton's description given above we have developed two java classes: the *BB_Application*, which extends the *TaskQueue* main class, and the *BB_Task* class, which implements the *Task_Interface* interface.

In the example the class data are the two arrays of weights and the matrix of bounds. The first array is used in order to compute the best value; the second one and the matrix are used to check if the bounds are satisfied. The local best value and the corresponding solution are member variables too.

The *user_split_tasks*() reads from a file the elements of the arrays of data described before. In the B&B implementation the *user_split_tasks*() method fills the queue with a number of tasks greater or equal to the number of workers. Each task is a node belonging to the same level of the tree of solutions.

In B&B implementation *handler_message*() method updates the optimum value that is communicated by the workers through a collective communication. The *user_Close*() method prints the final optimum value and the correspondent solution.

The *BB_Task* class implements the Task_interface, which is used by the worker in order to manage the single task of the bag. An object of this class declares only those data that are local to the single task. In B&B implementation the data are:

- *int sc[]* : an array of integers, which represents a subset of solutions, (a node of the tree),
- *int lsc* : an integer, which represents the level of the tree where the above node is placed.

The *compute*() method, is invoked by the worker in order to solve the task. The *compute*() method reproduces the sequential code for the visit of the tree. It starts from the node correspondent to *sc*[] array, of the the *BB_Task* object extracted form the bag. When a node of the tree should be processed later a new *BB_Task* object is built, and it is pushed in the worker's queue. When the leaf of the node is reached the task is computed and another one will be extracted from the queue by the worker.

## 5 Experimental Results

In this section we discuss experimental results of execution of the mobile agent based branch and bound optimization application described in the previous section. The target architecture is a cluster of 8 Pentium Celeron processors, 800 MHz clock frequency and 128M RAM, connected via an 100Mb/s ethernet switch. Experimental figures are provided, with varying the problem size and

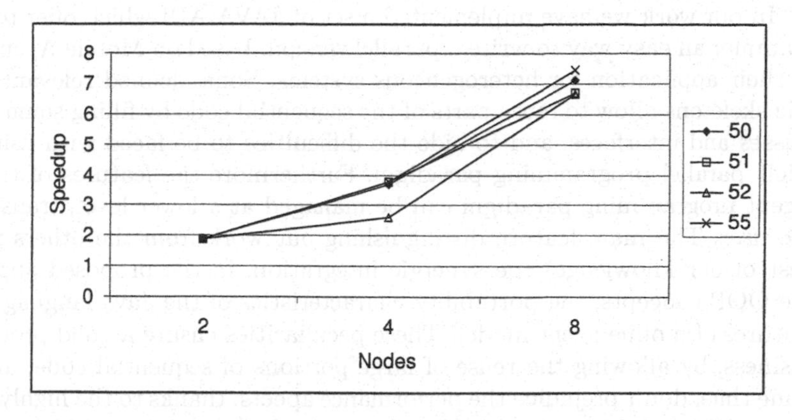


**Fig. 5.** Normalized speed up values for four different problem sizes

the number of computing nodes. Because of the highly irregular behavior of this kind of algorithms, due to the branching effect, the simple speedup measure is meaningless for our purposes. The distributed branch biasing effect can be taken into account by considering the total number of solved subproblems, which is largely variable with different executions, even with the same input data. We have thus defined $PRi$, as being the ratio between the number of solved problems during the parallel execution of the application and the number of problems solved by the sequential version of the application. We then define a *normalized speedup* as being $Sn = Speedup * PRi$; this latter is considered in order to filter the branching effect in experimental results.

In Figure 5 the normalized speed up values are shown, for four different problem sizes, and from 2 to 8 processors.

The normalized speedup scales well with the number of processors for all the considered problem sizes.

## 6 Related Work and Conclusions

Many authors presented taxonomies of distributed algorithmic skeletons [1, 4, 3]; in particular, we refer to Campbell's work that examined the classification of algorithmic skeletons and proposed a general one. According to this classification many kinds of algorithms can be included in the Farm-like and Divide and Conquer-like skeleton. Other works [4] use the algorithmic skeleton approach mainly to separate the communication/synchronization structure from the application-specific computation in order to optimize the mapping of the skeletons on different underlying distributed architectures. Similarly many programming environments [5, 12, 6], based on functional or OOP languages, have been developed to provide, through the skeletons, implicit parallel programming mechanisms. A mobile agent based approach for Paradigm-Oriented Distributed Computing is described in the paper [9].

In our work we have implemented a set of JAVA API which offer to the programmer an easy way to write a parallel version, based on Mobile Agents model, of their application for heterogeneous systems. Some defined relevant algorithmic skeletons allow to reuse parts of the sequential code by filling some methods, classes and interfaces, and to hide the difficulties to be faced with using an explicit parallel programming paradigm. Furthermore the features of the Mobile Agent programming paradigm can be managed at a lower level, transparent to the user. The main feature distinguishing our work from the others is (to the best of our knowledge) the synergic integration, in the proposed approach, of the OOP concepts, the portability characteristics of the Java language and the features of mobile agent model. These peculiarities ensure a good programming easiness, by allowing the reuse of large portions of sequential code, and at the same time, don't prejudice the performance apects, thanks to the highly dynamic adaptability of the implemented skeletons to the underlying architecture.

## References


[1] G. R. Andrews, "Paradigms for Process Interaction in Distributed Programs", *ACM Computing Surveys*, Vol. 23, N. 1, March 1991.

[2] R. Aversa, B. Di Martino, N. Mazzocca, S. Venticinque, "MAGDA: a software environment for Mobile AGent based Distributed Applications", submitted to: Int. Conf. on High Performance Distributed Computing (HPDC-11), 2002.

[3] D. Campbell, "Towards the Classification of Algorithmic Skeletons", Tech. Rep. YCS-276, Dept. of Comp. Science, Univ. of York, 1996.

[4] Murray Cole. Algorithmic Skeletons: Structured Management of Parallel Computation. Pitman, 1989.

[5] M. Danelutto, R. Di Meglio, S. Orlando, S. Pelagatti, M. Vanneschi, "The $P^3L$ Language: an Introduction", Technical Report HPL-PSC-91-29, Hewlett-Packard Laboratories, Pisa Science Centre, Dec. 1991.

[6] J. Darlington, A. J.Field, P. G. Harrison, P. H. J. Kelly, D. W. N. Sharp, Q. Wu and R. L.Whie, "Parallel Programming Using Skeleton Functions", in PARLE'93, LNCS 694, pp. 146-160, Springer-Verlag, 1993.

[7] T. Drashansky, E. Houstis, N. Ramakrishnan, J. Rice, "Networked Agents for Scientific Computing", Communications of the ACM, vol. 42, n. 3, March 1999.

[8] Gray R., Kotz D., Nog S., Rus D., Cybenko G., "Mobile agents: the next generation in distributed computing" Proc. of Int. Symposium on Parallel Algorithms/Architecture Synthesis, 1997.

[9] H. Kuang, L. F. Bic, M. Dillencourt, "Paradigm-oriented distributed computing using mobile agents", Proc. of. 20th Int. Conf. on Distributed Computing Systems, 2000.

[10] D. Lange and M. Oshima, *Programming and Deploying Java Mobile Agents with Aglets*, Addison-Wesley, Reading (MA), 1998.

[11] V. A. Pham, A. Karmouch, "Mobile software agents: an overview", IEEE Communications Magazine, Vol. 36(7), July 1998, pp. 26 -37.

[12] F. A. Rabhi. Exploiting Parallelism in Functional Languages: a Paradigm - Oriented Approach. In Abstract Machine Models for Highly Parallel Computers, pages 118-139, April 1993.

[13] H. W. J. Trienekens, "Parallel Branch&Bound Algorithms", *Ph.D. Thesis at Erasmus Universiteit-Rotterdam*, Nov. 1990.

# Translating Haskell$_{\#}$ Programs into Petri Nets

Francisco Heron de Carvalho Junior[1], Rafael Dueire Lins[2], and Ricardo Massa Ferreira Lima[3]

[1] Centro de Informática, Universidade Federal de Pernambuco
Av. Professor Luis Freire s/n, Recife, Brazil
fhcj@cin.ufpe.br

[2] Depart. de Eletrônica e Sistemas, Universidade Federal de Pernambuco
Av. Acadêmico Hélio Ramos s/n, Recife, Brazil
rdl@ee.ufpe.br

[3] Depart. de Sistemas Computacionais, Escola Politécnica de Pernambuco
Rua Benfica, 455, Madalena, Recife, Brazil
rmfl@poli.upe.br

**Abstract.** Haskell$_{\#}$ is a concurrent programming environment aimed at parallel distributed architectures. Haskell$_{\#}$ programs may be automatically translated to Petri nets, an important formalism for analysis of properties of concurrent and non-determinisc systems. This paper motivates and formalizes the translation of Haskell$_{\#}$ programs into Petri nets, providing some examples of their usage.

## 1 Introduction

Haskell$_{\#}$[14, 16] is a concurrent extension to Haskell[21], a pure non-strict functional language, based on the coordination paradigm[8], aimed at distributed architectures. Haskell$_{\#}$ main features are:

- Targeted at supporting *explicit*, *static* and *coarse-grain* parallelism, for maximizing performance of parallel programs over high-latency distributed architectures, such as *clusters*[1];
- Easy and modular implementation, by using well disseminated tools for underlying management of parallelism and compilation of sequential functional code. For these purposes, MPI (*Message Passing Interface*) standard[7] and GHC (*Glasgow Haskell Compiler*)[22] were chosen, respectively, due to their *efficiency* and *portability*;
- *Clean* separation of concurrent programs in two *orthogonal* components: a *sequential* one, pure *Haskell* modules implementing computation, and a *parallel* one, configuring the interaction between the functional processes. The parallel component is programmed with a declarative language called HCL (*Haskell$_{\#}$ Configuration Language*). *Process hierarchy* increases *modularity* and *abstraction*, motivating a disciplined approach to the development of *large-scale concurrent programs* and making easier their formal analysis;

J.M.L.M. Palma et al. (Eds.): VECPAR 2002, LNCS 2565, pp. 635–649, 2003.

– Well defined process semantics based on Hoare's CSP[10], with parallel constructors similar to OCCAM[11]. Functional processes are connected through *unidirectional*, *point-to-point*, and *typed* communication channels. Evaluation is *data flow* oriented, despite the *on-demand* nature of Haskell functional processes. This feature makes Haskell$_{\#}$ different from related parallel functional languages based on process networks, such as *Caliban*[13, 28] and *Eden*[4], which makes use of a data demand evaluation strategy. These characteristics allows Haskell$_{\#}$ programs to be translated into Petri nets, for analysis of their important properties using automatic tools, such as INA (*Integrated Network analyzer*)[24].

The *first* version of Haskell$_{\#}$ [16] implemented explicit message passing primitives (*send* (`!`) and *receive* (`?=`)) on top of Haskell IO monad, operating over input and output ports that define the interface of functional processes. However, this approach made difficult the translation of programs into Petri Nets, due to breaking of *process hierarchy*. In its *second* version, the explicit communication primitives were eliminated from the functional code. Functional processes received values from other ones through communication ports mapped onto the arguments of their *main* functions. The semantic of functional processes was *strict*, despite of Haskell semantics being *non-strict*. Thus, process had to receive the values from all input ports before performing computation. This approach allows direct translation of programs into Petri Nets[15], but reduces the expressive power of the language for concurrency, because the impossibility to interleave computation and communication (*interleaving*). Recently, a new design of Haskell$_{\#}$ reconciled interleaving and process hierarchy[5], by the introduction of *lazy streams* and *non-strict* semantics to functional processes, taking advantage of the non-strict nature of *Haskell* functions and *lazy evaluation* provided by this language. From this point on, we refer to the three designs of Haskell$_{\#}$ as first, second and third ones, respectively.

This paper focuses on the translation of Haskell$_{\#}$ programs, in its third design, into Petri Nets, showing the efficacy of this approach to the formal analysis of properties of concurrent programs using automatic tools, such as INA. In what follows, Section 2 briefly introduces Petri Nets formalism. Section 3 formalizes the translation of Haskell$_{\#}$ programs into Petri nets. Section 4 presents examples of translation of some well-known concurrent programs implemented in Haskell$_{\#}$, providing examples of property analysis using INA tool. Conclusions and lines for further work are presented in Section 5.

## 2 Petri Nets

This section presents a brief introduction to *place-transition Petri nets*. Petri nets are a family of formal description techniques appropriate for modeling nondeterministic and concurrent systems[20, 23]. The mathematical foundation behind Petri nets allows for performing quantitative and qualitative analysis[19] of the specified systems.

### 2.1 Place-Transition Nets

A place-transition Petri net may be formalized as a tuple $(P, T, A, W, M_0)$, where: **(1)** $P$ is a finite set of places which can store an unlimited number of tokens. **(2)** $T$ is a finite set of transitions. **(3)** $P \cap T = \emptyset$. **(4)** $(P, T, A)$ is a bipartite graph with two types of vertices: places (in $P$) and transitions (in $T$), interconnected by directed arches (set $A$). **(5)** $A \subseteq P \times T \cup T \times P$. This indicates that an arch either goes from a place to a transition or from a transition to a place. **(6)** $W : A \rightarrow N$ is a function that maps each arch onto its *weight* (number of tokens removed/inserted from/into the source/target place). **(7)** $M_0 : P \rightarrow N$ is the *initial marking* which defines the initial number of tokens in each place, which presents the initial state of the network.

**Firing Rule.** A given transition is enabled to fire if, and only if, all input places have at least the number of tokens corresponding to the weight of the arch that links it to the transition. The new marking is obtained by removing this number of tokens from each input place and inserting them into each output place in number equal to the weight of the outgoing arch. The execution of actions is performed by firing transitions, modifying the token distribution.

### 2.2 A Simple Example

Figure 1 (a) presents a simple Place-Transition Petri net. The equivalent graphical representation is depicted on Figure 1 (b). One should note that weights are omitted as all arches have weight one.

### 2.3 Analysis of Petri Nets

One can divide the methods used to analyze Petri nets into three categories. The first method is based on the graph of reachability. The construction of the graph of reachability depends on the initial marking (the initial state of the system). Therefore, it may be used to analyze behavioral properties [12]. The main drawback of this method is the high computational complexity [9, 17]. The second method is based on the state equation [19, 25]. The main advantage of this method is the existence of simple linear algebraic equations which aid in

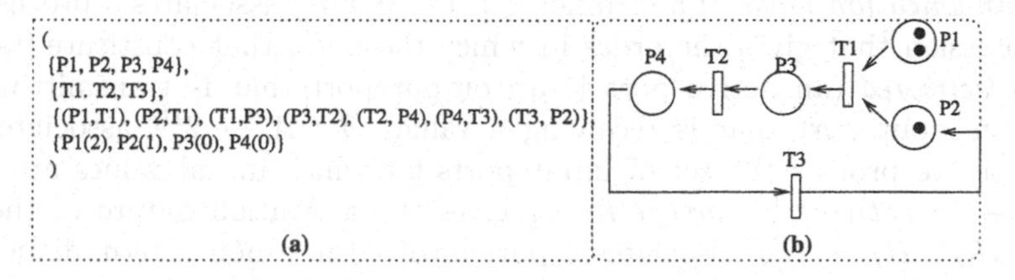


**Fig. 1.** Simple Place-Transition Petri net

determining properties of the nets. Unfortunately, it only allows for the analysis of properties when it is applied to general Petri nets. Necessary and sufficient conditions may be obtained by applying the matrix algebra for some subclasses of Petri nets[26]. The third method is based on reduction laws [19, 2]. It makes use of a set of transformation rules to reduce the size of the model, preserving the properties of the final net. However, it is possible that, for a given system and some set of rules, the reduction may not be complete. A more efficient and comprehensive analysis may be achieved by a cooperative use of these methods.

# 3 Translation Rules

Haskell$_\#$ programs abstract computational from concurrency concerns in two orthogonal layers. On the computational layer, functional processes are programmed in Haskell, while on the concurrent one, interaction between them is programmed in HCL (*Haskell$_\#$ Configuration Language*). This hierarchic view of concurrent programming, based on coordination approach [8], turns easier programming task and allows reasoning about communication topology of processes network apart from their implementation. In what follows, the translation of Haskell$_\#$ programs into Petri nets is detailed.

## 3.1 An Abstract Formal Description for Haskell$_\#$ Programs

To formalize translation rules, an algebraic representation for $Haskell_\#$ applications is now introduced. It captures the essence of the necessary HCL functionality, but abstracting from its formal syntax. Thus, a Haskell$_\#$ program $\Lambda_\#$ can be represented as a tuple:

$$\Lambda_\# = (\Lambda, \Pi_O, \Pi_I, \Pi_N, \Gamma, \pi, \tau, \gamma_O, \gamma_I, \beta, \sigma, \xi, \delta, \rho, \nu)$$

$\Lambda$ is a set of processes, $\Pi_O$ is a set of output ports, $\Pi_I$ is a set of input ports, $\Pi_N$ is a set of identifiers for non-deterministic choice of ports, $\Gamma$ is a set of channels, $\Lambda \cap \Pi_O \cap \Pi_I \cap \Pi_N \cap \Gamma = \emptyset$. The remaining components are functions. $\pi : \Pi_O \cup \Pi_I \rightarrow \Lambda$ associates a port with a process (the *interface* of a process $p$ is the set of ports $\{r \mid \pi(r) = p\}$), $\tau : \Pi_O \cup \Pi_I \rightarrow \{t \mid t \text{ is a Haskell type}\}$ associates each port with its type, $\gamma_O : \Gamma \rightarrow \Pi_O$ associates a channel with its *transmitter* port, $\gamma_I : \Gamma \rightarrow \Pi_I$ associates a channel with its *receiver* port, $\beta : \Gamma \rightarrow \{buffered \mid synchronous\}$ and $\sigma : \Gamma \rightarrow \{stream \mid single\}$ specify the *communication mode* of a channel, $\xi : \Lambda \rightarrow String$ associates a process with an expression that gives the order in which the ports that constitute its interface are *activated* (an active port is an output port that is transmitting a value or an input port that is receiving a value), $\delta : \Lambda \rightarrow 2^{\Pi_I}$ associates to each repetitive process the set of input ports for which initial values are given, $\rho : \Lambda \rightarrow \{repetitive \mid nonrepetitive\}$ gives the activation nature of the process, $\nu : \Pi_O \cup \Pi_I \rightarrow \Pi_N$ associates a port involved in a non-deterministic choice to the identifier of the non-determinisc set of ports. The following abstract grammar formally specifies the syntax of the $\xi$ expression:

$\xi_k ::= a \mid a \in \{x \mid x \in \Pi_I \cup \Pi_O \cup \Pi_N \wedge (x \in \Pi_I \cup \Pi_O \Rightarrow \nu(x) = \perp)\}$ *(port a)*
$\xi_k ::= \xi_1 > \xi_2 > \cdots > \xi_n$ *(sequence)*
$\xi_k ::= \xi_1 \mid \xi_2 \mid \cdots \mid \xi_n$ *(choice)*
$\xi_k ::= \xi^{*t}$ *(repeated activation. t is a synchronization label)*
$\xi_i ::= \xi^n$ *(bounded repeated activation)*

In $\xi$, ports are referenced only once. Ports inside the context of a *repeated activation* communicate in *stream* mode. Streams in Haskell$_{\#}$ are implemented by Haskell lazy lists. In $\xi$, they must be synchronized (balancing of communication) using *synchronization labels.* The functions $\gamma_O$ and $\gamma_I$ satisfies the condition in 1, which guarantees that channels are *unidirectional, point-to-point* and *typed.*

$$\forall c \in \Gamma,\ \gamma_O(c) = p_o \wedge \gamma_I(c) = p_i \Rightarrow \pi(p_o) \neq \pi(p_i) \wedge \tau(p_o) = \tau(p_i) \quad (1)$$

The Haskell$_{\#}$ application in Figure 2 is specified as below:

$$MergeApplication = (\Lambda, \Pi_O, \Pi_I, \Pi_N, \Gamma, \pi, \tau, \gamma_O, \gamma_I, \beta, \sigma, \xi, \delta, \rho, \nu)$$
$$\Lambda = \{a, b, m, c\},\ \Pi_O = \{a.l, b.l, m.c_l\},\ \Pi_I = \{m.a_l, m.b_l, c.l\},$$
$$\Pi_N = \{m.l\},\ \Gamma = \{c_1, c_2, c_3\}$$
$$\pi(a.l) = a,\ \pi(b.l) = b,\ \pi(m.a_l) = m,\ \pi(m.b_l) = m,\ \pi(m.c_l) = m, \pi(c.l) = c$$
$$\tau(*) = Int,\ \rho(*) = nonrepetitive,\ \beta(*) = synchronous,\ \sigma(*) = stream$$
$$\gamma_O(c_1) = a.l,\ \gamma_O(c_2) = b.l,\ \gamma_O(c_3) = m.c_l,\ \gamma_I(c_1) = m.a_l,\ \gamma_I(c_2) = m.b_l,\ \gamma_I(c_3) = c.l$$
$$\xi(a) = a.l^{*t},\ \xi(b) = b.l^{*t},\ \xi(m) = (m.l > m.c_l)^{*t},\ \xi(c) = c.l^{*t}$$
$$\nu(m.a_l) = m.l,\ \nu(m.b_l) = m.l,\ \delta = \emptyset$$

An application is *reactive* when all of its processes are repetitive, otherwise it is *transformational.* A reactive application never finishes, while a transformational one finishes when all of its nonrepetitive processes reach final state.

### 3.2 Translation Rules

The marking of the generated Petri net reflects the state of the original *Haskell$_{\#}$* application, defined as the set of states of its processes. *Places* correspond to *ports* and *transitions* model communication operations. A mark on a place specify that

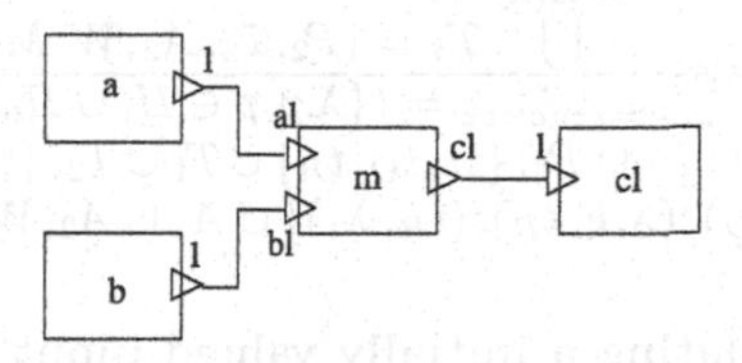


The process $m$ merges two lists (streams) of integers received non-deterministically from processes $a$ and $b$ and sends the merged list to process $c$.

**Fig. 2.** Merge Application

is corresponding port is active, representing the current state of a process. After completion of communication, modelled by the firing of a transition, the mark goes to the place corresponding to the next active port. The order in which ports are activated is given by $\xi$, which dictates how to arrange transitions and arches in the Petri net to allow its marking to reflect process state changing.

The translation can be performed in three phases. In the first one, Petri nets that reflects state transition of each process are generated. In the second one, the nets are synchronized using the information about *channels* in $\Gamma$. For each channel $c \in \Gamma$, the output transitions of the places corresponding to the ports $\gamma_O(c)$ and $\gamma_I(c)$ are unified. An exception occurs whenever one or both ports are involved in a non-deterministic choice, where the input transition is used. In the third phase, transitions that have the same *loop synchronization labels* are unified, ensuring that stream communication is properly balanced.

Following, the formal translation rules of Haskell$_\#$ programs into Petri nets are introduced. The translation mappings $\Upsilon_\Lambda$, $\Upsilon_\Gamma$ and $\Upsilon_\Sigma$ formalize respectively the three translation phases. The partial function $\mu$ gives the *synchronization label* of a transition. It is assumed that total functions $W$ and $M_0$ are one and zero, respectively, whenever not explicitly defined.

**$\Upsilon_{\Lambda_\#}$: Translating a general application $\Lambda_\#$**

$$\frac{\bigcup^{\lambda \in \Lambda} \Upsilon_\lambda = (P_\Lambda, T_\Lambda, A_\Lambda, W, M_0, \mu), \quad \bigcup^{\gamma \in \Gamma} \Upsilon_\gamma(T_\Lambda, A_\Lambda) = (P_\Gamma, T_\Gamma, A_\Gamma, W, M_0), \quad \Upsilon_\Sigma(T_\Gamma, (\bigcap_{\gamma \in \Gamma} A_\gamma) \cup (A_\Gamma - A_\Lambda)) = (T_\Sigma, A_\Sigma)}{\Upsilon_{\Lambda_\#} = (P_\Lambda \cup P_\Gamma, T_\Sigma, A_\Sigma, W, M_0)} \tag{2}$$

**$\Upsilon_\lambda$: Translating a transformational process (step 1)**

$$\frac{\Upsilon_\xi(\xi(\lambda), p_e, p_s) = (P_1, T_1, A_1, W, M_0, \mu), \quad \bigcup^{\delta \in \delta(\lambda)} \Upsilon_\delta = (P_2, T_2, A_2, W, M_0)}{\begin{array}{c}\Upsilon_{\lambda \mid \lambda \in \Lambda \wedge \rho(\lambda) = transformational} = (\{\lambda.p, p \in \Pi_i \cup \Pi_o \textit{ and } \pi(p) = \lambda\} \cup \\ \{\lambda.I, \lambda.F\} \cup P_1 \cup P_2, \{t_I, t_O\} \cup T_1 \cup T_2, \{(\lambda.I, t_I), \\ (t_O, \lambda.F), (t_I, p_e), (p_s, t_O)\} \cup A_1 \cup A_2, W, M_0, \mu), \; M_0(I) = 1\end{array}} \tag{3}$$

**$\Upsilon_\lambda$: Translating a reactive process (step 1)**

$$\frac{\Upsilon_\xi(\xi(\lambda), p_e, p_s) = (P_1, T_1, A_1, W, M_0, \mu), \quad \bigcup^{\delta \in \Delta(\lambda)} \Upsilon_\delta = (P_2, T_2, A_2, W, M_0)}{\begin{array}{c}\Upsilon_{\lambda \mid \lambda \in \Lambda \wedge \rho(\lambda) = reactive} = (\{\lambda.p, p \in \Pi_i \cup \Pi_o \textit{ and } \pi(p) = \lambda\} \cup \\ \{\lambda.I, \lambda.F\} \cup P_1 \cup P_2, \{t_I, t_O, t_R\} \cup T_1 \cup T_2, \{(\lambda.I, t_I), (t_O, \lambda.F), \\ (t_I, p_e), (p_s, t_O), (\lambda.F, t_R), (t_R, \lambda_I)\} \cup A_1 \cup A_2, W, M_0, \mu), \; M_0(I) = 1\end{array}} \tag{4}$$

**$\Upsilon_\delta$: Translating a initially valued input port (step 1)**

$$\frac{(\delta, t), (t, \delta') \in A}{\Upsilon_{\delta \mid \delta \in \Pi_I} = (\{p_1, p_2\}, \{t'\}, \{(p_1, t'), (t', p_2), (\delta, t'), (t, p_2), (p_2, t), (t', \delta')\}, W, M_0)} \tag{5}$$

**$\Upsilon_\xi$: Translating $\xi$ (port)**

$$\Upsilon_\xi(\pi, \pi, \pi, *) = (\pi, \emptyset, \emptyset, W, M_0, \mu), \pi \in \Pi_O \cup \Pi_I \tag{6}$$

**$\Upsilon_\xi$: Translating $\xi$ (non-determinisc ports)**

$$\frac{\{p \mid \nu(p) = \pi\} = \{\pi_1 \ldots \pi_n\}}{\begin{array}{c}\Upsilon_\xi(\pi, \pi_i, \pi_o, *) = (\{\pi_i, \pi_o, \pi_1, \ldots \pi_n\}, \{t^i{}_1, \ldots, t^i{}_n\} \cup \{t^o{}_1, \ldots, t^o{}_n\}, \\ \{\pi_i\} \times \{t^i{}_1, \ldots, t^i{}_n\} \cup \{(t^i{}_k, p^i{}_k), (p^i{}_k, t^o{}_k),\ i = 1 \ldots n\} \cup \\ \{t^o{}_1, \ldots, t^o{}_n\} \times \{\pi_o\}, W, M_0, \mu), \pi \in \Pi_N\end{array}} \tag{7}$$

**$\Upsilon_\xi$: Translating $\xi$ (choice)**

$$\frac{\Upsilon_\xi(\xi_k, p^i{}_k, p^o{}_k, t^o{}_k) = (P_k, T_k, A_k, W, M_0, \mu),\ k = 1 \ldots n,\ n \geq 2}{\begin{array}{c}\Upsilon_\xi(\xi_1|\ldots|\xi_n, p_i, p_o, *) = (\{p_i, p_o\} \cup (\bigcup^n P_k), \{t^i{}_1, \ldots, t^i{}_n\} \cup \{t^o{}_1, \ldots, t^o{}_n\} \cup \\ (\bigcup^n T_k)), \{p_i\} \times \{t^i{}_1, \ldots, t^i{}_n\} \cup \{t^o{}_1, \ldots, t^o{}_n\} \times \{p_o\} \cup \\ \{(t^i{}_1, p^i{}_1), \ldots, (t^i{}_n, p^i{}_n)\} \cup \{(p^o{}_1, t^o{}_1), \ldots, (p^o{}_n, t^o{}_n)\} \cup (\bigcup^n A_k), W, M_0, \mu)\end{array}} \tag{8}$$

**$\Upsilon_\xi$: Translating $\xi$ (sequence)**

$$\frac{\Upsilon_\xi(\xi_k, p^i{}_k, p^o{}_k, t_k) = (P_k, T_k, A_k, W, M_0, \mu),\ k = 1 \ldots n,\ n \geq 2}{\begin{array}{c}\Upsilon_\xi(\xi_1 > \ldots > \xi_n, p^i{}_1, p^o{}_n, *) = (\bigcup^n P_k, \{t_1, \ldots, t_{n-1}\} \cup (\bigcup^n T_k), \\ \bigcup^{n-1} \{(p^o{}_k, t_k), (t_k, p^i{}_{k+1})\} \cup (\bigcup^n A_k), W, M_0, \mu)\end{array}} \tag{9}$$

**$\Upsilon_\xi$: Translating $\xi$ (unbounded repetition)**

$$\frac{\Upsilon_\xi(\xi, p_i, p_o, t_o) = (P, T, A, W, M_0, \mu)}{\begin{array}{c}\Upsilon_\xi(\xi^{*t}, p, p, exit) = (P \cup \{p\}, T \cup \{t_o, t_i\}, A \cup \{(p, t_i), (t_i, p_i), (p_o, t_o), (t_o, p)\}, \\ W, M_0, \mu),\ \mu(exit) = t, \text{if } t \text{ is given}\end{array}} \tag{10}$$

**$\Upsilon_\xi$: Translating $\xi$ (bounded repetition)**

$$\frac{\Upsilon_\xi(\xi, p'_i, p'_o, t_4) = (P, T, A, W, M_0, \mu)}{\begin{array}{c}\Upsilon_\xi(\xi^n, p_i, p_o, *) = (P \cup \{p_i, p_o, p_1, p_2, p_3\}, T \cup \{t_1, t_2, t_3, t_4\}, A \cup \{(p_i, t_1), \\ (t_1, p_1), (p_1, t_3), (t_1, p_2), (p_2, t_3), (t_3, p'_i), (p'_0, t_4), (t_4, p_2), (t_4, p_3), (p_2, t_2), \\ (p_3, t_2), (t_2, p_0)\}, W, M_0, \mu), W(t_1, p_1) = W(p_3, t_2) = 2\end{array}} \tag{11}$$

**$\Upsilon_\gamma$: Synchronizing an individual buffered channel (step 2)**

$$\frac{\begin{cases}(\gamma_O(\gamma), t) \in A_\Lambda \\ (\gamma_I(\gamma), t') \in A_\Lambda\end{cases}, if\ \nu(\gamma_I(\gamma)) = \bot \begin{cases}(t', \gamma_I(\gamma)) \in A_\Lambda \\ (t, \gamma_O(\gamma)) \in A_\Lambda\end{cases}, if\ \nu(\gamma_O(\gamma)) \neq \bot}{\Upsilon_{\gamma \mid \gamma \in \Gamma \wedge \beta(\gamma) = buffered}(T_\Lambda, A_\Lambda) = (\{p\}, T_\Lambda, A_\Lambda \cup \{(t, p), (p, t')\}, W, M_0)} \tag{12}$$

**$\Upsilon_\gamma$: Synchronizing an individual synchronous channel (step 2)**

$$\frac{\begin{cases}(\gamma_O(\gamma), t) \in A_\Lambda \\ (\gamma_I(\gamma), t') \in A_\Lambda\end{cases}, if\ \nu(\gamma_I(\gamma)) = \bot \begin{cases}(t', \gamma_I(\gamma)) \in A_\Lambda \\ (t, \gamma_O(\gamma)) \in A_\Lambda\end{cases}, if\ \nu(\gamma_O(\gamma)) \neq \bot}{\Upsilon_{\gamma \mid \gamma \in \Gamma \wedge \beta(\gamma) = synchronous}(T_\Lambda, A_\Lambda) = (\emptyset, T_\Lambda - \{t'\}, A_\Lambda[t'/t], W, M_0)} \tag{13}$$

**$\Upsilon_\Sigma$: Synchronizing streams in unbounded repetitions (step 3)**

$$\frac{A_\Sigma = A_\Gamma[t_2/t_1], \forall t_1, t_2 \in T_\Gamma \mid \mu(t_1), \mu(t2) \neq \bot \wedge \mu(t_1) = \mu(t_2)}{\Upsilon_\Sigma(T_\Gamma, A_\Gamma) = (T_\Sigma, A_\Sigma)} \tag{14}$$

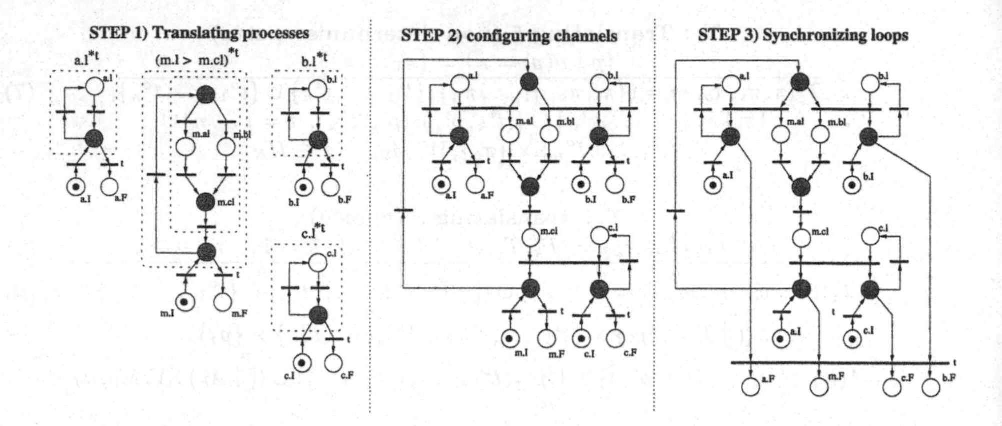


**Fig. 3.** Translation Process of Merge Application (Figure 2)

## 4 Examples

Three examples of Haskell$_\#$ applications that were translated to Petri nets using the rules above are now presented. A more detailed explanation on the implementation of these applications in Haskell$_\#$ may be found in reference [5].

### 4.1 Dining Philosophers (DP)

The *Dining Philosophers* is a well-known synchronization problem from concurrency theory proposed by Dijkstra[6]. It is used to model *deadlock*, a state where all processes belonging to a communicating group are waiting for another process in the group. The diagram on Figure 4 shows the network of functional processes of the solution to the dining philosophers problem in Haskell$_\#$. Applying the above translation rules, one may obtain the Petri net on Figure 5.

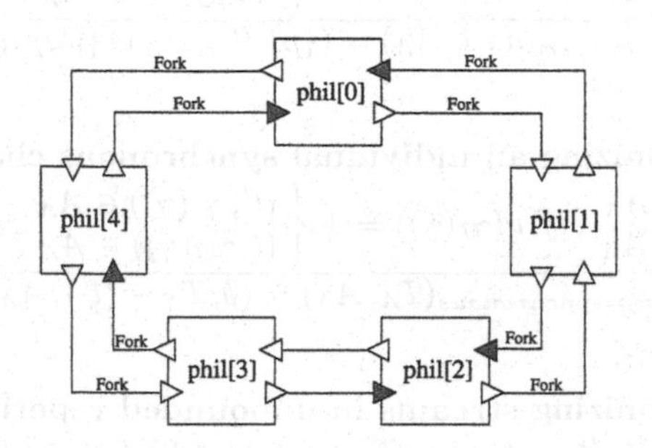


**Fig. 4.** Dinning Philosophers Haskell$_\#$ Process Network

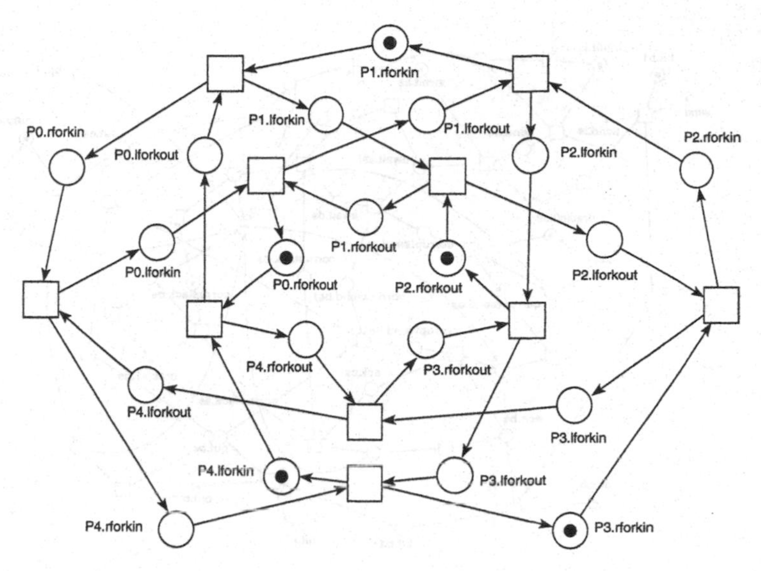


**Fig. 5.** Dinning Philosophers Petri Net

## 4.2 Alternating Bit Protocol (ABP)

The *Alternating Bit Protocol* is a simple yet effective protocol for managing retransmission of lost messages on low-level implementations of messaging-passing model[27]. Considering a receiver process A and a sender process B connected by two *stream channels*, the protocol ensures that whenever a message transmitted from B to A is lost, it is retransmitted. Figure 6 shows the process network of the ABP application specified in Haskell$_{\#}$. The translation into Petri nets is presented on Figure 7.

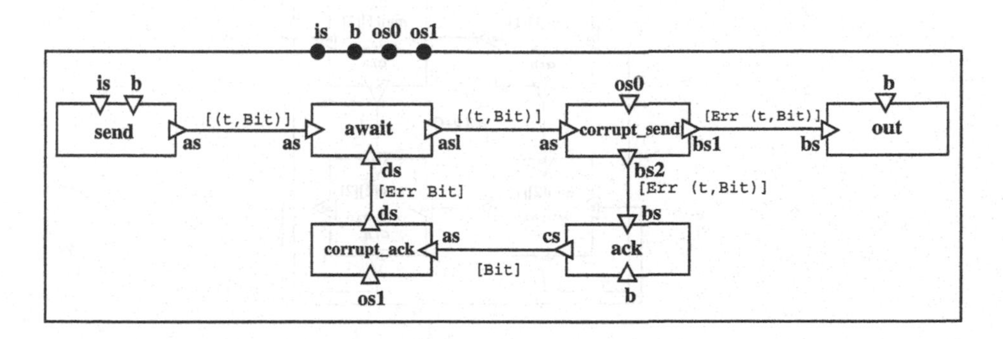


**Fig. 6.** Alternating Bit Protocol Haskell$_{\#}$ Process Network

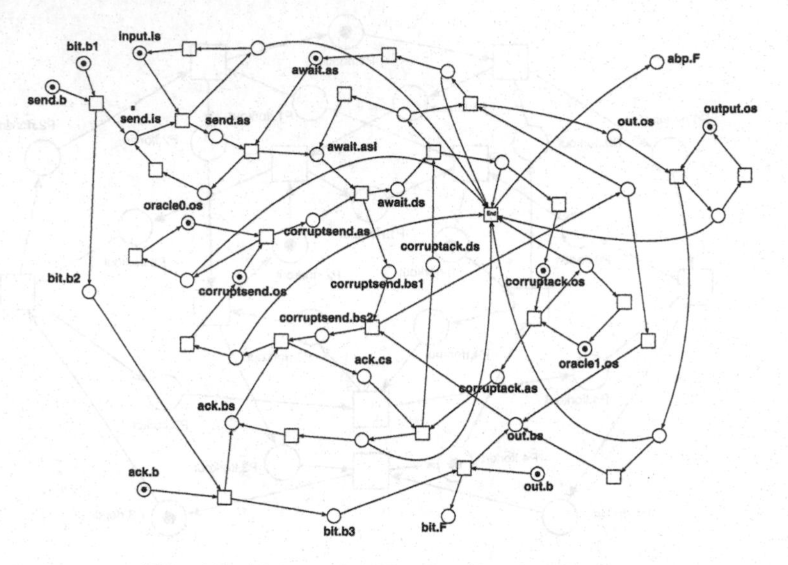


**Fig. 7.** Alternating Bit Protocol Petri Net

## 4.3 Matrix Multiplication on the Mesh (MMM)

The matrix multiplication on the mesh algorithm [18] uses a systolic approach to perform multiplication of two matrices of order $n$, where elements are displayed in a mesh of $n \times n$ processors. Figure 8 presents the network of functional processes for $2 \times 2$ matrixes in Haskell$_{\#}$. The generated Petri net is shown on Figure 9, but for $3 \times 3$ matrixes.

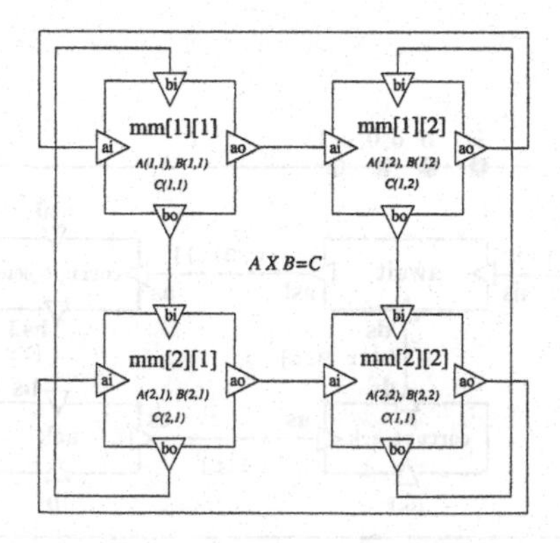


**Fig. 8.** Matrix Multiplication Haskell$_{\#}$ Process Network

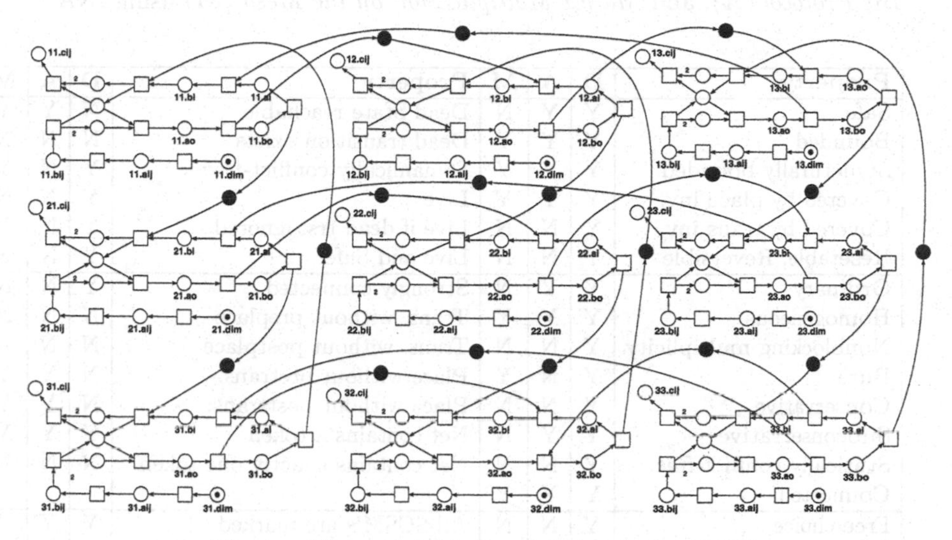


**Fig. 9.** Matrix Multiplication Petri Net (3X3)

### 4.4 Analysis of Formal Properties

Amongst many similar tools available for analysing properties of Petri nets, PEP(*Programming Environment based on Petri nets*)[3] was chosen. PEP is a comprehensive set of modeling, compilation, simulation, and verification components for Petri nets. PEP is integrated to INA (*Integrated System Analyzer*)[24], a powerful Petri net analyzer. INA was used to analyse properties of the applications presented. The effectiveness of the analysis depends on the programmer's ability to interpret the properties of Haskell$_\#$ programs in terms of the translated Petri net properties and vice-versa. We intend to design an interface for programmers to abstract Petri nets concerns on analysis, simplifying it. Next, we informally summarize some results of the analysis. The reader is invited to get PEP source files of the three applications presented from `www.cin.ufpe.br/~fhcj/`, and to reproduce the results obtained and also to prove some other properties that the reader may consider relevant. On Table 1, some general properties inferred about the corresponding Petri nets of **DP** (*Dining Philosophers*), **ABP** (*Alternating Bit Protocol*), and **MMM** (*Matrix Multiplication on the Mesh*) programs are summarized.

*Liveness* ensures absence of *deadlocks* in **DP**, proving its correctness. **ABP** and **MMM** are not *live*, because transformational applications always reach an end-state, where no transitions may be fired.

Using *reachability test* in INA, the reachability of end states for **ABP** and **MMM** programs (presence of a mark in each place of the sets {abp.F} and {11.F, 12.F, 13.F, 21.F, 22.F, 23.F, 31.F, 32.F, 33.F}, respectively) were proven. Also using this approach, it is possible to demonstrate that it is not possible

**Table 1.** Analysis of Static Properties for *Dining Philosophers* (**D**), *Alternating Bit Protocol* (**A**), and *Matrix Multiplication on the Mesh* (**M**) using INA

| Property | D | A | M | Property | D | A | M |
|---|---|---|---|---|---|---|---|
| Safe | Y | Y | N | Dead State reachable | N | Y | Y |
| Bounded | Y | Y | Y | Dead transition exists | N | N | N |
| Structurally bounded | Y | Y | Y | Dynamically conflict-free | Y | N | Y |
| Covered by place inv. | Y | Y | Y | Live | Y | N | N |
| Covered by trans inv. | Y | N | N | Live if dead trs. ignored | Y | N | N |
| Resetable, Reversible | Y | N | N | Live and Safe | Y | N | N |
| Ordinary | Y | Y | N | Strongly connected | Y | N | N |
| Homogeneous | Y | Y | Y | Trans. without preplace | N | N | N |
| Nonblocking multiplicity | Y | N | N | Trans. without postplace | N | N | N |
| Pure | Y | N | Y | Place without pretrans. | N | Y | Y |
| Conservative | Y | N | N | Place without posttrans. | N | Y | Y |
| Subconservative | Y | Y | N | Net contains a token | Y | Y | Y |
| Statically conflict-free | Y | N | N | Net contains exactly one token | N | N | N |
| Connected | Y | Y | Y | | | | |
| Free choice | Y | N | N | All SCSM's are marked | Y | Y | ? |
| Extended free choice | Y | N | N | Cov. by SCSM's cont. one token | Y | N | ? |
| Extended simple | Y | N | Y | Any minimal deadlock is SCSM | Y | N | ? |
| Marked graph | Y | N | N | State machine coverable | Y | Y | Y |
| State machine | N | N | N | State machine decomposable | Y | N | ? |
| Deadlock-trap property | Y | N | N | State machine allocatable | Y | N | N |
| All non-empty traps marked | Y | N | N | Rank less than NrOfPrevectors | Y | Y | ? |
| No deadlock | Y | N | N | | | | |

to occur a buffer overflow for **MMM**, because the number of tokens in *buffer* places (filled circles, in Figure 9) never exceeds the dimension of the matrix less one (two in the case of the example on Figure 9). Another result for **MMM** based on partial reachability analysis says that two buffers from channels that connect a process in common never have more than one token simultaneously.

Observe that the three applications are *state machine coverable*. This is intrinsic to $Haskell_{\#}$ programs. In this case, the sub-nets induced by each process are both a *state machine* and a *component*[1], ensuring the existence of at least one mark in the sub-net for all reachable marking (there is always an active port in a process). State-machine coverable Petri nets are an important class of Petri nets, with some special properties. To be *state-machine decomposable*, it is only necessary that the $Haskell_{\#}$ program be reactive, that is the case of **DP**.

All three applications are (structurally) bounded, ensuring that there is an upper bound for the number of marks in places, for every reachable marking.

[1] A component is a set of places that is a deadlock (siphon) and a trap simultaneously[24].

By the fact that the subnets induced by processes are *state-machine sufficiently marked components*, only places corresponding to buffers can be unbounded in a Haskell$_{\#}$ program. Thus, the possibility of buffer overflow may be predicted looking at the *boundness* property. This is another argument to prove absence of buffer overheads in **MMM**.

Non-trivial place invariants that say something about behavior of **DP**, **ABP** and **MMM** were found. For example, in **DP**, an invariant shows that there are always two philosophers sending their forks to their left neighbor, after eating. Due to *sub-conservativeness* of any Haskell$_{\#}$ program, all three Petri nets are *covered by place invariants* .

Invariants of transitions were also calculated. For **DP**, as expected, there is an invariant containing all transitions in the net. For **ABP**, there are two transition invariants. The first one contains the necessary transitions to be fired for `sender` process transmit a message correctly to the `receiver`, while the second one contains the necessary transitions to be fired for `await` process to retransmit a lost message. The transitions that constitute the latter invariant are a proper subset of the transitions contained in the former one, because `await` may try to send the message several times for each message transmission of the `sender`. For the **MMM** application, there is no transition invariant, as expected.

## 5 Conclusions and Lines for Further Work

The work presented herein is part of a task force on building a complete programming environment for high-performance concurrent computing for *clusters*[1] based on Haskell$_{\#}$. The aim of such environment is to allow concurrent programs to be built, simulated, benchmarked, executed, and having their properties to be formally analyzed. This paper focuses on the last item. Petri nets were chosen to be the formal model for Haskell$_{\#}$ programs because they are a valuable tool for modeling and analysis of concurrent systems, due to its simplicity, dissemination, expressive power, visual representation and simulation, and existence of automatic powerful tools for formal analysis, such as INA and PEP. Using the approach described in this paper, we intend to analyze properties of real complex large-scale concurrent programs. Structural analysis on the kind of Petri net generated by *Haskell$_{\#}$* programs is also being conducted.

## References

[1] Baker, M., Buyya, R., , & Hyde, D. (1999). Cluster Computing: A High Performance Contender. *IEEE Computer*, July, 79–83.

[2] Berthelot, G. (1986). Checking Properties of Nets Using Transformations. *Advanced in Petri Nets, LNCS 222*, 19–40.

[3] Best E. , Esparza J., Grahlmann B., Melzer S., Rómer S., Wallner F. (1997). The PEP Verification System. *Workshop on Formal Design of Safety Critical Embedded Systems (FEmSys'97) (tool presentation.*

[4] Breitinger, S., Loogen, R., Ortega Mallén, Y., & Peña, R. (1997). The Eden Coordination Model for Distributed Memory Systems. *High-Level Parallel Programming Models and Supportive Environments (HIPS)*.

[5] Carvalho Jr., F.H., Lima, R.M.F., & Lins, R.D. (2002). Coordinating Functional Processes with Haskell$_{\#}$. ACM Press (ed), *Proceedings of ACM Symposium on Applied Computing*, 393-400.

[6] Djkstra, E. W. (1968). The Structure of THE Programming System. *Communications of the ACM*, **11**, 341–346.

[7] Dongarra, J., Otto, S. W., Snir, M., & Walker, D. (1995). An Introduction to the MPI Standard. *Technical Report CS-95-274, University of Tennesee*, Jan. `http://www.netlib.org/tennessee/ut-cs-95-274.ps`.

[8] Gelernter, D., & Carriero, N. (1992). Coordination Languages and Their Significance. *Communications of the ACM*, **35**(2), 97–107.

[9] Hack, M. (1976). Decidability Questions for Petri Nets. *PhD Thesis, MIT*.

[10] Hoare, C. A. R. (1985). *Communicating Sequential Processes*. Prentice-Hall International Series in Computer Science.

[11] Hoare, C. A. R. (1988). *Occam 2 Reference Manual*. Prentice-Hall.

[12] Jensen, K. (1997). Colored Petri Nets. Basic Concepts, Analysis Methods and Practical Use. *Springer-Verlag. Three Volumes*.

[13] Kelly, P. (1989). Functional Programming for Loosely-coupled Multiprocessors. *Research Monographs in Parallel and Distributed Computing, MIT Press*.

[14] Lima, R. M. F., & Lins, R. D. (1998). *Haskell$_{\#}$*: A Functional Language with Explicit Parallelism. *LNCS 1573 (VECPAR'98 - International Meeting on Vector and Parallel Processing)*, June, 80–88.

[15] Lima, R. M. F., & Lins, R. D. (2000). Translating HCL Programs into Petri Nets. *Proceedings of the 14th Brazilian Symposium on Software Engineering*, João Pessoa, Brazil.

[16] Lima, R. M. F., Carvalho Jr., F. H., & Lins, R. D. (1999). *Haskell$_{\#}$*: A Message Passing Extension to Haskell. *CLAPF'99 - 3rd Latin American Conference on Functional Programming*, Mar., 93–108.

[17] Lipton, R. J. (1976). The Reachability Problem Requires Exponential Space. *New Haven, CT, Yale University, Department of Computer Science, Res. Rep. 62*, Jan.

[18] Manber, U. (1999). *Introduction to Algorithms: A Creative Approach*. Reading, Massachusetts: Addison-Wesley. chapter 12, pages 375–409.

[19] Murata, T. (1989). Petri Nets: Properties Analysis and Applications. *Proceedings of IEEE*, **77**(4), 541–580.

[20] Peterson, J. L. (1981). Petri Net Theory and the Modeling of Systems. *Prentice-Hall, Englewood, N. J.*

[21] Peyton Jones, S. L., & Hughes, J. (1999). Report on the Programming Language Haskell 98, A Non-strict, Purely Functional Language. Feb.

[22] Peyton Jones, S. L., Hall, C., Hammond, K., & Partain, W. (1993). The Glasgow Haskell Compiler: a Technical Overview. *Joint Framework for Information Technology Technical Conference*, 249–257.

[23] Reisig, W. (1985). Petri Nets. *Springer-Verlag, Berlin*.

[24] Roch, S., & Starke, P. (1999). Manual: Integrated Net Analyzer Version 2.2. *Humboldt-Universität zu Berlin, Institut für Informatik, Lehrstuhl für Automaten- und Systemtheorie*.

[25] Silva, M., & Teruel, E. (1995). Analysis of Autonomous Petri Nets with a Bulk Services and Arrivals. *Lecture Notes in Control and Information Science*, **199**, 131–143.

[26] Silva, M., & Teruel, E. (1996). Petri Nets for the Design and Operation of Manufacturing Systems. *CIMAT'96.*

[27] Tanenbaum, A. S. (1996). Computer Networks. *Prentice Hall.*

[28] Taylor, F. (1997). Parallel Functional Programming by Partitioning. *PhD Thesis, Department of Computing, Imperial College of Science, Technology and Medicine, University of London*, Jan.

# An Efficient Multi-processor Architecture for Parallel Cyclic Reference Counting

Rafael Dueire Lins

Departamento de Eletrônica e Sistemas
CTG, Universidade Federal de Pernambuco, Recife, PE, Brazil
rdl@ee.ufpe.br

**Abstract.** Multi-processor architectures are part of the technological reality of today. On the other hand, the software engineering community reached the consensus that memory management has to be performed automatically, without the interference of the programmer of applications. Reference counting is the memory management technique of most widespread use today. This paper presents a new architecture for parallel cyclic reference counting.

**Keywords:** Memory management, garbage collection, reference counting, cyclic graphs

## 1 Introduction

The concurrent programming language Java promoted a revolution to parallel programming by breaking with the client-server model, in which only data were allowed to be exchanged by programs. In Java, programs became "first-class" citizens in a distributed environment and may be passed between machines. One of the important points of Java is that it frees programmers from the burden and error prone task of performing memory management. The Java Virtual Machine (JVM) incorporates a garbage collector [12], an automatic dynamic memory management algorithm. Good performance figures for such environment is fundamental, thus Java ressurected the a critical need for high-performance, concurrent and incremental garbage collection.

The two most simple algorithms used for memory management are mark-scan and reference counting. The mark-scan garbage collection algorithm works in two phases. If a machine runs out of space, the computation process is suspended and the garbage collection algorithm is called. First, the algorithm traverses all the data structures (or cells) in use putting a mark in each cell visited. Then, the scan process takes place collecting all unmarked cells in a *free-list*. When the mark-scan process has finished, computation is resumed. The amount of time taken for garbage collection by the mark-scan algorithm is proportional to the size of the heap (the work space where cells are allocated). The copying algorithm is a modified version of the mark-scan algorithm which appeared with the advent of virtual memory operating systems.In the copying algorithm the heap is divided in two halves. The algorithm copies cells from one half to the other during collection. Its time complexity is proportional to the size of the graph in use. Mark-scan and copying algorithms generally traverse all the reachable data structures during garbage collection, which makes them unsuitable for real-time or large-virtual-memory applications.

In 1960, reference counting was developed by G.E.Collins [6] to avoid the long pauses in early LISP processors. The name of the scheme derives from the fact that each data structure or cell has an additional field which counts the number of references to it, i.e. the number of pointers to it. During computation, alterations to the data structure imply changes to the connectivity of the graph and, consequently, re-adjustment of the value of the count of the cells involved. Reference counting has the major advantage of being performed in small steps interleaved with computation, while other algorithms imply suspending computation for much longer. The disadvantage of the trivial algorithm for reference counting is the inability to reclaim cyclic structures, as remarked by J.H.McBeth in 1963 [20]. To solve this problem, a mixture of mark-scan and reference counting has already been used in the past. A detailed analysis of those algorithms may be found in [12]. The first general solution to cyclic reference counting only appeared in 1990 in reference [21]. Lins further improved that solution by himself and his colleagues by making the cycle analysis lazy [15] and introducing creation-time stamps to cells [16].

The first concurrent garbage collector was proposed by Steele [23]. In his architecture two processors share the same memory space. One of the processors is responsible for graph manipulation while the other performs garbage collection. In Steele's algorithm mark-scan and computation occur simultaneously.

Another parallel mark-scan algorithm is presented in [7], and was considered by Ben-Ari "one of the most difficult concurrent programs ever studied" [3]. This algorithm was implemented in hardware/software in the Intel iAPX-432 computer and iMAX operating system [22]. Kung and Song developed an improved mark-scan algorithm [9] based on the algorithm by Dijkstra and collaborators [7]. Based on the same algorithm Ben-Ari gave [3] several parallel mark-scan algorithms with a much simpler proof of correctness then the ones presented in [7,9].

All the algorithms mentioned above for parallel mark-scan seem to spend a lot of time colouring non-garbage cells and scanning the whole heap. As an alternative to mark-scan algorithms Wise proposed a on-board reference count architecture [24]. Kakuta-Nakamura-Iida [8] present a complex architecture based on reference counting, which is unable to deal with cyclic data structures. The first general concurrent architecture for reference counting was presented by Lins in [13], which worked with two processors: one in charge of graph rewritings, called *mutator*, and another dedicated to garbage collection, the *collector*. Reference [14] generalised the previous architecture *à la* Lamport [10] allowing any number of mutators and collectors to work concurrently. Recent work, developed at IBM T.J.Watson Research Center, aimed at the efficient implementation of concurrent garbage collection [1] in the context of the Jalapeño Java virtual machine [2]. That architecture is based on Lins' concurrent strongly-coupled algorithms [13,14], which on their turn are based on the cyclic reference counting algorithms presented in [21,15].

This paper presents a new concurrent algorithm for cyclic reference counting which parallelises Lins' new sequential algorithm [17], by far more efficient than his previous ones [21,15,16], reducing its complexity from 3Θ to 2Θ, where Θ is the size of the subgraph below the deleted pointer. The architecture introduces is generalised to work with any number of mutators and collectors. Besides that, the collaboration between collectors is increased removing a criticism made by Blelloch and Cheng [4] to Lins' previous concurrent architectures.

## 2 A Concurrent Architecture for Standard Reference Counting

In reference counting, each data structure or cell has an additional field which stores the number of references to it, i.e. the number of pointers to it. During computation, alterations to the data structure imply changes to the connectivity of the graph and, consequently, re-adjustment of the value of the count of the cells involved. It is assumed that cells are of constant fixed size. Unused cells are linked together in a structure called *free-list*. A cell ***B*** is *connected* to a cell ***A*** ($A \rightarrow B$), if and only if there is a pointer <*A*, *B*>. A cell ***B*** is *transitively connected* to a cell ***A*** ($A \overset{*}{\rightarrow} B$), if and only if there is a chain of pointers from ***A*** to ***B***. The initial point of the graph to which all cells is use are transitively connected is called *root*.

The starting milestone for the new architecture proposed herein is the architecture described in [13] for standard reference counting. There are two processors say $P_1$ and $P_2$, which will perform graph rewriting and garbage collection simultaneously. Both processors share the same memory area, the working space which is organised as a heap of cells. In case of simultaneous access from both processors to a given cell semaphores are used such as to guarantee that processor $P_1$ will have priority over processor $P_2$. There is also another shared data structure: the *Delete-queue*, which can be managed in different ways. For simplicity, one may assume it is organised as a FIFO. Processor $P_1$ is only allowed to push data onto the Delete-queue. Conversely, processor $P_2$ is only allowed to dequeue data from the Delete-queue. Processor $P_1$ has two registers called *top-free-list*, which stores a pointer to the top cell in the free-list, and *top-del-queue*, which stores a pointer to the top of the Delete-queue. Processor $P_2$ has two registers called *bot-free-list*, which stores a pointer to the last cell in the free-list, and *bot-del-queue*, which stores a pointer to the bottom of the Delete-queue. This architecture may be depicted as follows:

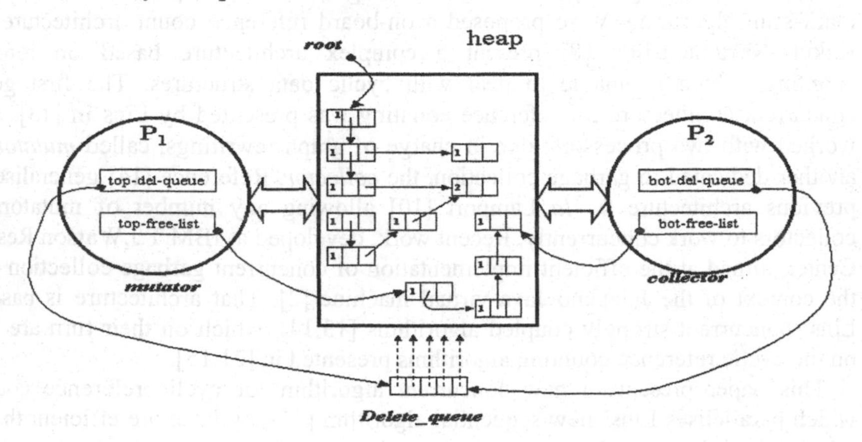

For the sake of simplicity the synchronisation that must be done when $P_1$ attempts to remove a node from an empty free-list or $P_2$ tries to get a reference from an empty Delete-queue is ignored. These situations should happen infrequently and any convenient synchronisation primitive can be used.

***Processor $P_1$ Instruction Set***

Processor $P_1$ is in charge of rewritings of the graph. Its instruction set comprises three basic operations: New, Copy, and Del. In order to keep our notation simple fields are not especified.

New tests if there are free cells on the free-list. If not empty it reads the information in register top-free-list and links it to the graph. New also gets the address of the new top of the free-list and saves it in register top-free-list. These operations are described as,

```
New (R) = if top-free-list not nil then
                    make pointer <R,top-free-list>
                    top-free-list := ^top-free-list
                else New(R)
```

where ^A means the information stored in A.

Copy copies information between cells. No special care is needed in order to keep the correct management of the data structures. If processor $P_1$ wants to copy some information, i.e. to make a pointer to a cell then this cell is transitively connected to root. Copy increments the reference count of the cell pointed at, thus:

```
Copy (R, <S,T>) = make pointer <R,T>
                  increment RC(T)
```

Del deletes pointers in the graph. It pushes a reference to the cell on top of the Delete-queue.

```
Del (<R,S>) = remove <R,S>
              ^top_Delete-queue := S
              increment top_Delete-queue
```

Processor $P_2$ will later perform the remaining operations for the effective re-adjustment of the graph.

***Processor $P_2$ Instruction Set***

Processor $P_2$ is the processor in charge of the deletion of pointers and feeding free cells onto the free-list. The main routine in $P_2$ is called Delete a routine which will run forever as the kernel of the operating system of processor $P_2$.

```
Delete = if Delete-queue not empty then
                S := bot_del-queue
                increment bot_del-queue
                Rec_del (S)
            else Delete
```

If the Delete-queue is not empty Delete calls Rec_del, as follows

```
Rec_del (S) = if RC (S) = 1 then
                  for T in Sons (S) do
                        Rec_del (T)
                        link_to_free-list (S)
              else Decrement RC (S)
```

where Sons(S) is the bag of all the cells T such that there is a pointer <S,T>.
The linking of a cell to the free-list is performed by the operations:

```
link_to_free-list (S) = ^S := ^bot-free-list
                        bot-free-list := S
```

As was mentioned before, standard reference counting is not able to recycle cyclic structures. The architecture presented above is upgraded to deal with cyclic graphs in the section below.

## 3 An Efficient Architecture for Cyclic Reference Counting

The architecture presented herein is based on the recently devised algorithm for cyclic reference counting [17] in uniprocessors. The general idea of the new algorithm is to perform a local mark-scan whenever a pointer to a shared structure is deleted. The algorithm works in two steps. In the first step, the sub-graph below the deleted pointer is scanned, rearranging counts due to internal references, marking nodes as possible garbage and also storing potential links to root. In step two, the cells pointed at by the links stored are visited directly. If the cell has reference count greater than one, the whole sub-graph below that point is in use and its cells should have their counts updated. In the final part of the second step, the algorithm collects garbage cells.

In addition to the information of number of references to a cell, an extra field is used to store the colour of cells. Two colours are used: green and red. Green is the stable colour of cells. All cells are in the free-list and are green to start with. To make the architecture proposed above work with this new algorithm, the instruction set of processor $P_2$ is exchanged by the routines presented below, keeping processor $P_1$ unaltered. Rec_del the routine recursively invoked by Delete, whenever a pointer is removed, now performs the following operations:

```
Rec_del (S) = if RC(S) = 1 then
                   Colour(S) := green
                   for T in Sons (S) do
                       Rec_del (T)
                 link_to_free-list (S)
              else
                   Decrement RC (S)
                   Mark_red (S)
                   Scan (S)
```

One can observe that the only difference to standard reference counting in the algorithm above rests in the last two lines of Rec_del, which are explained below. The algorithm makes use of a stack, called Jump_stack, to store references to nodes which can potentially link the sub-graph to root. Nodes are inserted in the Jump_stack by Mark_red, a routine which "simulates" the effect of detaching the subgraph from root, performing the following operations on the graph:

```
Mark_red(S) = If (Colour(S) == green) then
                  Colour(S) = red;
                  for T in Sons(S) do
                      Decrement_RC(T);
                      if (RC(T)>0 &&
                          T not in Jump_stack)
                      then Jump_stack = T;
                      if (Colour(T) == green)
                      then  Mark_red(T);
```

Scan(S) verifies whether the Jump_stack is empty. If so, the algorithm sends cells hanging from S to the free-list. If the jump-stack is not empty then there are nodes in the graph to be analysed. If their reference count is greater than one, there are external pointers linking the cell under observation to root and counts should be restored from that point on, by calling the ancillary function Scan_green(T).

```
Scan(S) =  If RC(S)>0 then Scan_green(S)
           else While (Jump_stack ≠ empty) do
                T = top_of_Jump_stack;
                    Pop_Jump_stack;
                    If (Colour(T) == red && RC(T)>0)
                    then
                        Scan_green(T);
                Collect(S);
```

Procedure Scan_green restores counts and paints green cells in a sub-graph in use, as follows,

```
Scan_green(S) = Colour(S) = green
                  for T in Sons(S) do
                      increment_RC(T);
                      if colour(T) is not green
                      then
                          Scan_green(T);
```

Collect(S) is the procedure in charge of returning garbage cells to the free-list, painting them green and setting their reference count to one, as follows:

```
Collect(S) =  If (Colour(S) == red) then
                for T in Sons(S) do
                    Remove(<S, T>);
                    RC(S) = 1;
                    Colour(S) = green;
                    free_list = S;
                    if (Colour(T) == red) then
                            Collect(T);
```

## 4 A Multi-mutator Architecture

This section generalises the architecture presented in the last section to work with any number of mutators. The mutators must be synchronised in some way so they do not interfere with one another. This synchronisation mechanism must enforce some partial ordering on mutator's operations, which are viewed as atomic actions. This means that if a processor $P_1^i$ has started an operation before a processor $P_1^j$ then operations will actually take place following this order.

This partial ordering must be enough to guarantee that the mutators correctly execute some sequential mutator algorithm. This avoids problems such as sending to the free-list cells still in use by performing the deletion of a pointer to a cell before a copy operation to the same cell. The implementation of this synchronisation is not of the concern of this paper, since it depends upon the details of the individual application. Synchronisation is also needed amongst mutators whenever removing nodes from a common free-list. The use of several separate free-lists associated with each mutator can reduce synchronisation delays. This can be implemented without any difficulty, but it is not considered further herein.

The picture below sketches the multi-mutator-one-collector architecture.

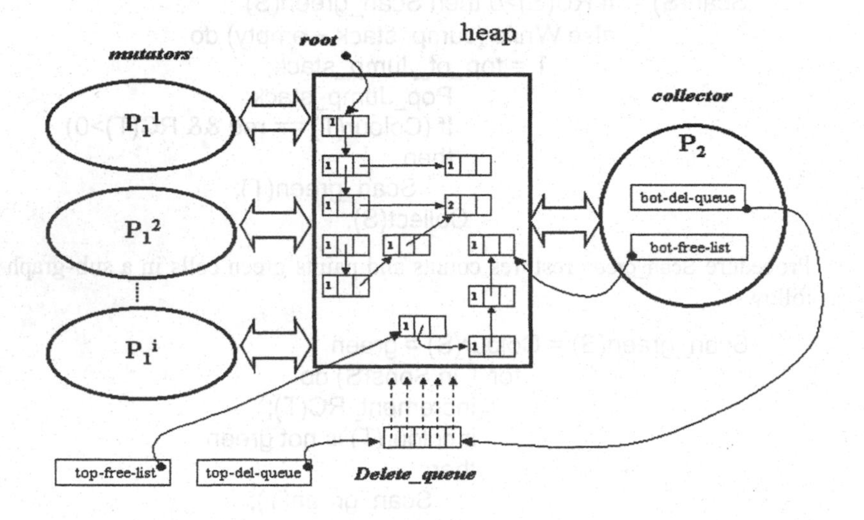


As one may observe in the picture above, instead of pointing directly to the top of the Delete-queue now each processor will keep a reference to an external register which points at the top of the Delete-queue. Similarly, for the top of the free-list.

The instruction set for the mutators is kept unchanged from the one before with only one mutator.

## 5 Using Multiple Collectors

There is a number of possible ways to extend the architecture presented in the last section to work with more than one collector. Our idea is to keep the philosophy of the one-mutator-one-collector architecture presented above, as much as possible, in which:

- mutators and collectors do not talk directly to each other.
- interfaces are simple and well defined.
- synchronisation between mutators and collectors when addressing interfaces is kept to a minimum.

Having the points above in mind the multi-mutator architecture presented in the last section can be pictured as:

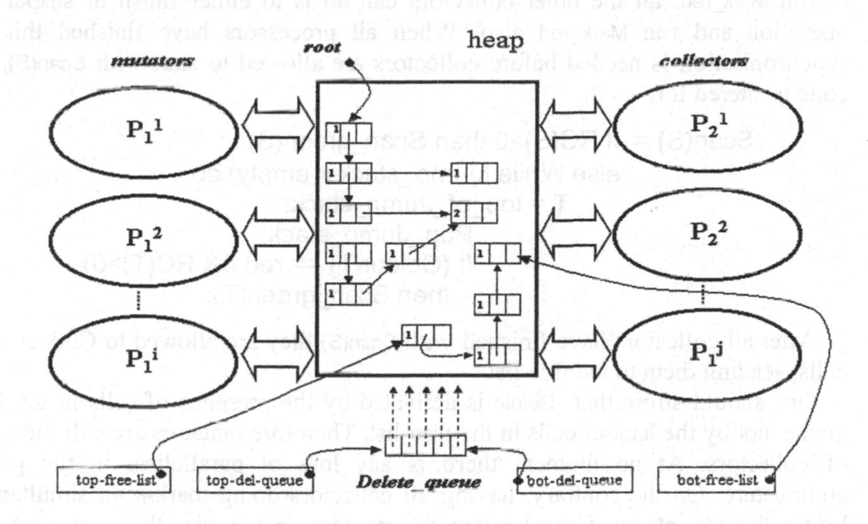


In a similar fashion to the multi-mutator architecture, instead of pointing directly to the bottom of the Delete-queue now each collector keeps a reference to external registers bot-del-queue and bot-free-list. The instruction set of each collector should be modified in order to avoid confusion during mark-scan. In the multi-processor shared memory architecture presented collectors are synchronised in such a way as to all of them to run the mark-scan simultaneously. The kernel routine which runs on the collectors $P_2^j$ is Delete.

```
Delete = if Delete-queue not empty then
              S := bot_del-queue
              increment bot_del-queue
              Rec_del(S)
         else Delete
```

If the Delete-queue is not empty Delete calls Rec_del, as follows,

```
Rec_del(S) = if  RC(S) = 1  then
                 Colour(S) := green
                 for T in Sons(S) do
                     Rec_del(T)
                 link_to_free-list(S)
               else Decrement RC(S)
                     Mark_red(S)
                     Synchronise_end_Mark_red
                     Scan(S)
                     Synchronise_end_Scan
                     Collect(S)
```

The control strategy for synchronisation is such as when one of the collectors start to run Mark_red, all the other collectors can do is to either finish or suspend their operation and run Mark_red also. When all processors have finished this phase synchronisation is needed before collectors are allowed to start with Scan(S), which code is altered for:

```
Scan(S) =  If RC(S)>0 then Scan_green(S)
           else While (Jump_stack ≠ empty) do
                 T = top_of_Jump_stack;
                     Pop_Jump_stack;
                     If (Colour(T) == red && RC(T)>0)
                        then Scan_green(T);
```

After all collectors have finished with Scan(S) they are allowed to Collect garbage cells, sending them to the free-list.

One should stress that Delete is activated by the presence of cells in the Delete-queue, not by the lack of cells in the free-list. Therefore mutators are still independent of collectors. At no moment there is any loss of parallelism in the proposed architecture. On the contrary, having all collectors doing mark-scan simultaneously brings the advantage of accelerating this process, in the case the mark-scan area is split up between collectors.

# 6 Increasing the Collaboration amongst Collectors

The paper by G.Blelloch and P.Cheng, in reference [4], describes a parallel architecture for garbage collection based on Replicating Coping Garbage Collection by S.M. Nettles and J.W. O'Toole [12]. They establish space and time bounds for operations performed by the garbage collector, making the architecture suitable for real-time applications. In their work [4], Blelloch and Cheng draw the following considerations about the multi-processor architecture presented in reference [14]:

> "Lins described a parallel cyclic reference counting collector for a shared memory architecture. Although his algorithm is concurrent....It is possible for one processor to perform all recursive deletes while the remaining processors do not participate. This can potentially result in violation of space bounds or in stalling of processors that are waiting for free space."

The scenario presented by Blelloch and Cheng is very unlikely to take place. One should observe that mutators do not depend on collectors at all, unless for feeding the free-list. The criticism presented only takes place in the architecture presented in this paper if:

1. the free-list is empty;
2. collectors start to Mark_red and the sub-graphs under analysis are largely different in size, allowing some or all but one of the collectors to stall, waiting for that phase to end and synchronise.

A number of solutions may be adopted to remove such an inconvenience in the new architecture. Possibly the simplest of them is to allow idle collectors to fetch new cells from the Delete-queue, in this very unlikely to happen scenario. Observe that the critical operation is Mark_red, which is critical for synchronisation, and may take longer to happen in the case of subgraphs of unbalanced sizes (Scan happens in "jumps" and Collect feeds the free-list). Notice that, empirical data have shown that for many different sorts of applications [12] the number of shared data structures is less than 10% of the total number of cells in use. Thus it is very likely that fetching new cells from the Delete-queue some of them will be recycled. If both the free-list and the Delete-queue are empty, during a Mark_red phase in which there are idle collectors, there is also a possibility of the busy collector "exporting" some of the suspended recursive calls of Mark_red to a global data structure, from which idle collectors may collaborate with busy ones.

One important feature of the architectures proposed herein is that in the case of an empty free-list mutators may exchange their rôle and become collectors, helping to feed the free-list. This is only possible because there is no direct communication neither amongst mutators, nor amongst collectors, nor between mutators and collectors.

## 7 Proof of Correctness

Formal proofs of the correctness of parallel algorithms is, in general, not simple [11,7]. Informal proofs of the correctness of the architectures presented are provided here.

The four architectures presented above are based on simple algorithms. It is general knowledge the correctness of the uniprocessor version of standard reference counting, as well as its inability to reclaim cyclic structures. An informal proof of the correctness of the cyclic reference count algorithm on which the architectures presented are based on appears in the original paper [17]. Therefore, our major concern is to analyse the interaction between mutator(s) and collector(s) for each of the architectures introduced.

The approach Lamport uses [10] for assuring the correctness of his multi-processor architecture is the *parallelization* of the sequential algorithm presented in [7] with the addition of some synchronisation elements. This is exactly the strategy adopted by the author used in the development of the architectures presented above, which was based on the sequential algorithms for uniprocessors presented in [17]. First, the original mutator algorithm is parallelised and then the same technique is applied to the collector algorithm. At last, collectors interact to allow to some move further faster to produce fresh cells to an empty free-list, yielding greater efficiency. Arguments are produced below in favour of the correctness of the architectures proposed.

***Standard Reference Counting***

In the standard reference counting architecture above there are only two interfaces between processors: the *free-list* and the *Delete-queue*. In general, there is no direct interaction between mutator and collector. If cells are claimed by processor $P_1$ and the free-list is empty $P_1$ must wait for $P_2$ to recycle cells. Conversely, if the Delete-queue is empty, processor $P_2$ waits for processor $P_1$ to push information onto the Delete-queue. This relationship between processors is producer/consumer coupling; this form of synchronisation can be achieved without any need for mutual exclusion.

If the two processors try to access the same cell, for instance if $P_1$ is copying a pointer to a cell and tries to increment its reference count while $P_2$ is in the process of decrementing this same count, then semaphores are used such as processor P1 to perform its operation first. Therefore the initial architecture proposed does not alter the basic mechanism of standard reference counting, keeping its correctness valid.

***Cyclic Reference Counting***

The only difference between the architecture for parallel cyclic reference counting above and the standard reference count one is that a local mark-scan is performed whenever one deletes a pointer which has the possibility of isolating a cycle.Altough this new architecture keeps the same interfaces of the standard reference count a new aspect must be analysed. The mark-scan phase of processor $P_2$ may occur simultaneously with rewritings of processor $P_1$. In case both processors try to access a given cell processor $P_1$ will have priority over processor $P_2$ as in the case of standard reference counting. The only piece of information in a cell both processors may try to modify simultaneously is the reference count of a given cell. Observe that the colour information of cells is irrelevant to processor $P_1$. Two points need to be stressed:

- One always makes a connection before breaking a previous one, i.e. one does copy before delete. One does not discard a pointer to a cell and try to copy the same pointer afterwards.
- All cells $P_1$ may copy a pointer to are transitively connected to root, i.e. are accessible from root.

The points above imply that if processor $P_2$ is mark-scanning a subgraph and processor $P_1$ makes a copy to one of the cells in the same subgraph this new pointer to the subgraph **is not** the only connection between this subgraph and root. The only effect this copy may have is to abbreviate the mark-scan phase of processor $P_2$, by finding this new external reference before the old ones. One may conclude that the architecture offers the correct interaction between processors $P_1$ and $P_2$.

***The Multi-mutator Architecture***

The existence of a partial ordering on the synchronisation of mutator operations is the key for the correctness of this architecture. This ordering must be such as to guarantee that mutators correctly execute some sequential algorithm, i.e. the sequence of operations is the same as the performed in the one-mutator architecture.

***The Multi-collector Architecture***

Similarly to the other architectures previously presented, synchronisation is used to avoid simultaneous access to a given cell. In the multi-collector architecture, if the Delete-queue is not empty then each collector will fetch a cell from the back of the delete-queue, by calling Delete. Rec_del is called on it. If the value of the count of this cell is one then it is painted green, it has its Sons analysed recursively and then it is sent to the free-list.

Otherwise, in the case of the reference count being greater than one, the local mark-scan takes place to check if an island of cell is not being isolated from the root of the graph. A new synchronisation mechanism is used to avoid confusion between phases of mark-scan amongst collectors. This synchronisation makes the work cooperative and increases the parallelism of the architecture, not allowing collectors to be at different phases of the mark-scan. Synchronisation after each phase of mark-scan assures that if there are $n$ collectors in the architecture and if the top $n$ cells on the Delete-queue point at $s$ cells with shared subgraph then after mark-scan the graph obtained is equal to $n\text{-}s\text{+}1$ sequential calls to in the one-mutator-one-collector architecture.

***Increasing the Colaboration amongst Collectors***

In this architecture, if the marking of one or more sub-graphs takes much longer than some other and the free-list is empty, then there may be idle collectors waiting tor synchonisation. The simple solution proposed of allowing collectors to fetch new cells for analysis from the Delete-queue does not alter the sysnchronisation milestone before Mark_red, causing no trouble for the correctness of the architecture. Permitting collectors to "export" suspended calls to Mark_red to a global data-structure, from which idle collectors may fetch entry pointers to subgraphs to be analysed, increases the collaboration between collectors, speeding-up the marking phase. This does not move the synchronisation barrier, keeping the sequentiality of the collectors, thus the correctness of the concurrent algorithm.

# 8 Conclusions

This paper presents a simple shared memory architecture for concurrent garbage collection based on reference counting, which is generalised in its final version to work with any number of mutators and collectors. The new architectures presented are far more efficient than Lins' architectures for parallel cyclic reference counting [13, 14], on which Bacon and Rajan based themselves to develop the concurrent garbage collector to the Jalapeño [1], a new Java virtual machine and compiler being developed at the IBM T.J.Watson Research Center. Reference [2] reports that using a set of eleven benchmark programs including the full SPEC benchmark suite, the Jalapeño garbage collector achieves maximum measured application pause times about two orders of magnitude shorter than the best previously published results and performance similar to a highly tuned non-concurrent but parallel mark-and-sweep garbage collector. It is reasonable to suppose that the implementation of the garbage collection algorithms presented herein in the Jalapeño machine may provide even better performance figures. This is still to be bourne out by experimental results to be presented together with a broad survey of the area in [18].

## Acknowledgements

This work was sponsored by CNPq and Recope-Finep, to whom the author is grateful.

## References


1. D.F. Bacon and V.T. Rajan. Concurrent Cycle Collection in Reference Counted Systems, Proceedings of European Conference on Object-Oriented Programming, June, 2001, Springer Verlag, LNCS vol 2072.
2. D.F. Bacon, C.R.Attanasio, H.B.Lee, R.T.Rajan and S.Smith. Java without the Coffee Breaks: A Nonintrusive Multiprocessor Garbage Collector, Proceedings of the SIGPLAN Conference on Programming Language Design and Implementation, June, 2001 (SIGPLAN Not. 36,5).
3. M.Ben-Ari. Algorithms for on-the-fly garbage collection. ACM Transactions on Programming Languages and Systems, 6(3):333--344, July 1984.
4. G.Blelloch and P.Cheng, On Bonding Time and Space for Multiprocessor Garbage Collection. In Proc. of ACM SIGPLAN Conference on Programming Languages Design and Architecture, March 1999.
5. D.G. Bobrow, Managing reentrant structures using reference counts, ACM Trans. On Prog. Languages and Systems **2** (3) (1980).
6. G.E. Collins, A method for overlapping and erasure of lists, Comm. of the ACM, 3(12):655—657, Dec.1960.
7. E.W.Dijkstra, L.Lamport, A.J.Martin, C.S.Scholten & E.M.F.Steffens. On-the-fly garbage collection: an exercise in cooperation. Communications of ACM, 21(11):966--975, November 1978.
8. Kakuta, Nakamura and Iida, Information Proc. Letters, 23(1):33-37, 1986.H.T.Kung and S.W.Song. An efficient parallel garbage collection system and its correctness proof. In IEEE Symposium on Foundations of Computer Science, pages 120--131. IEEE, 1977.
10. L.Lamport. Garbage collection with multiple processes: an exercise in parallelism. In Proc. of IEEE Conference on Parallel Processing, pages 50--54. IEEE, 1976.
11. D.Gries. An exercise in proving parallel programs correct. Communications of ACM, 20(12):921--930, December 1977.
12. R.E. Jones and R.D.Lins. Garbage Collection Algorithms for Dynamic Memory Management, John Wiley & Sons, 1996. (Revised edition in 1999).
13. R.D.Lins. A shared memory architecture for parallel cyclic reference counting, Microprocessing and microprogramming, 34:31—35, Sep. 1991.
14. R.D.Lins. A multi-processor shared memory architecture for parallel cyclic reference counting, Microprocessing and microprogramming, 35:563—568, Sep. 1992.
15. R.D.Lins. Cyclic Reference counting with lazy mark-scan, IPL 44(1992) 215—220, Dec. 1992.
16. R.D.Lins. Generational cyclic reference counting, IPL 46(1993) 19—20, 1993.
17. R.D.Lins. An Efficient Algorithm for Cyclic Reference Counting, Information Processing Letters, vol 83 (3), 145-150, North Holland, August 2002.

18. R.D.Lins. Garbage Collection in Shared Memory Architectures, **in preparation**.R.D.Lins and R.E.Jones, Cyclic weighted reference counting, in K. Boyanov (ed.), Proc. of Intern. Workshop on Parallel and Distributed Processing, NH, May 1993.
20. J.H. McBeth, On the reference counter method, Comm. of the ACM, 6(9):575, Sep. 1963.
21. A.D. Martinez, R. Wachenchauzer and R.D.Lins. Cyclic reference counting with local mark-scan, IPL 34(1990) 31—35, North Holland, 1990.
22. F.J.Pollack, G.W.Cox, D.W.Hammerstein, K.C.Kahn, K.K.Lai, and J.R.Rattner. Supporting Ada memory management in the iAPX-432, Proceedings of the Symposium on Architectural Support for Programming Languages and Operating Systems, pages 117--131. SIGPLAN Not. (ACM) 17,4, 1982.
23. G.L.Steele. Multiprocessing compactifying garbage collection. Communications of ACM, 18(09):495--508, September 1975.
24. D.S.Wise. Design for a multiprocessing heap with on-board reference counting, In J.P. Jouannaud, editor, Functional Programming Languages and Computer Architecture, volume LNCS 201, pages 289--304. Springer-Verlag, 1985.

# ROS: The Rollback-One-Step Method to Minimize the Waiting Time during Debugging Long-Running Parallel Programs

Nam Thoai, Dieter Kranzlmüller, and Jens Volkert

GUP Linz
Johannes Kepler University Linz
Altenbergerstraße 69, A-4040 Linz, Austria/Europe
nam.thoai@gup.jku.at

**Abstract.** Cyclic debugging is used to execute programs over and over again for tracking down and eliminating bugs. During re-execution, programmers may want to stop at breakpoints or apply step-by-step execution for inspecting the program's state and detecting errors. For long-running parallel programs, the biggest drawback is the cost associated with restarting the program's execution every time from the beginning. A solution is offered by combining checkpointing and debugging, which allows a program run to be initiated at any intermediate checkpoint. A problem is the selection of an appropriate recovery line for a given breakpoint. The temporal distance between these two points may be rather long if recovery lines are only chosen at consistent global checkpoints. The method described in this paper allows users to select an arbitrary checkpoint as a starting point for debugging and thus to shorten the temporal distance. In addition, a mechanism for reducing the amount of trace data (in terms of logged messages) is provided. The resulting technique is able to reduce the waiting time and the costs of cyclic debugging.

## 1 Introduction

Debugging identifies the tasks associated with detecting and eliminating errors in software in order to backtrack the errors from their place of occurrence to their origin. A popular technique for backtracking of errors is cyclic debugging, where a program is executed repeatedly to collect more information about its intermediate states. A necessary feature of a debugging system is breakpointing, which allows to halt and examine a program at interesting points during its execution [8].

The definition of breakpoints in parallel programs is more complex than in sequential programs because of the number of involved processes. Several approaches for parallel breakpoints have been proposed [3][10]. A commonly accepted technique is provided by causal distributed breakpoints [6]. These breakpoints allow to stop each process at the earliest possible state reflecting all events that happened before the selected place on one process, according to Lamport's partial order of events in a distributed system [12]. Assuming that the breakpoint affects all processes, this kind of distributed breakpoint is called a (strongly complete) global breakpoint set [2][10].

Besides the multiplicity of processes, parallel debugging has to address new kinds of errors established by process communication and synchronization. An example is nondeterministic behavior, where subsequent executions of a program may lead to different results, although the same input data is provided. This effect causes serious problems during debugging, because subsequent executions of the program may not reproduce the original bugs, and cyclic debugging is thus not possible. A solution to this irreproducibility problem is offered by record&replay mechanisms [11], which monitor a nondeterministic program's execution, and afterwards apply the obtained data to control arbitrary numbers of re-executions. These techniques ensure that the re-execution path is equivalent to the initially observed execution.

One of the biggest problems of re-executing parallel programs is the cost associated with restarting the program's execution at the beginning. This is especially unacceptable for long running programs, which are typical in the parallel computing community. In particular, it may take a substantial amount of waiting time to stop a program at a breakpoint, which is placed at an intermediate point of its execution. This prohibits using cyclic debugging for long-running parallel programs. Furthermore, it represents a serious challenge for building a parallel debugging tool, which must provide some degree of interactivity for the user's investigations.

An idea to reduce the waiting time is offered by combining checkpointing and debugging [14][27]. Checkpointing is the act of saving the state of a running program sometimes during execution, so that it may be reconstructed and restarted from this position later in time [18]. In concrete, checkpointing provides the possibility to restart a program at all the places, where a snapshot of the program has been saved [5]. This allows to decrease the total execution time of a program and thus the waiting time until reaching a breakpoint. Two important questions in this context are, whether this waiting time has an upper bound, and if it has, whether the upper bound can be determined by users. Answers to these questions are provided by the solution proposed in this paper.

The model used for parallel and distributed computation, and the terminologies of checkpointing as well as replay/rollback distance are described in Section 2. The reasons why debugging and checkpointing should be combined and how to guarantee the rollback distance has an upper bound are discussed in Section 3. The actual checkpointing method, the reduced message logging technique, and the construction of suitable recovery lines is shown in Section 4, before summarizing the paper with conclusions.

## 2 Distributed Computation and Checkpoints

### 2.1 Parallel and Distributed Computation

The technique described in this paper focuses on parallel and distributed programs, whose processes interact via message passing. For simplification, only point-to-point communication functions will be applied throughout the paper, although other means of communication and synchronization are equally applicable.

The execution of the program is modeled as an event graph, which represents events and their relations as observed during a program's execution [11]. According to our restriction, only sending and receipt of messages will trigger events. This limits the description to *send* and *receive events*, and a special kind of "artificial" events for local checkpoints. The partial ordering between the events is determined by Lamport's "happened before" relation, denoted as "$\rightarrow$" [12].

## 2.2 Checkpoints

Based on the event graph model of parallel and distributed computation, we define checkpoints, breakpoints, and the rollback/replay distance as the basic concepts of our combined checkpointing and debugging approach.

A *local checkpoint* is the local state of a process at a particular point in time. A *global checkpoint* is a set of local checkpoints, one from each process. When considering a global checkpoint *G*, two categories of messages are particularly important:

- *Orphan messages*, which are messages that have been delivered in *G*, although the corresponding send events occur only after the local checkpoints comprising *G*.
- *In-transit messages*, which are messages that have been sent but not delivered in *G*.

A global checkpoint is *consistent* if there is no orphan message with respect to it [9]. This condition can be described in the following way to check the consistency of a global checkpoint: if $(C_0, C_1, \ldots, C_n)$ $(n \geq 1)$ is a consistent global checkpoint, then $(\forall i, j)$ $(i \neq j, 0 \leq i, j \leq n)$ $\neg(C_i \rightarrow C_j)$. On contrary, an *inconsistent global checkpoint* is a global checkpoint that is not consistent.

In Fig. 1, message $m_1$ is an in-transit message of $(C_{0,0}, C_{1,1}, C_{2,0})$, while message $m_2$ is an orphan message of $(C_{0,1}, C_{1,2}, C_{2,0})$. $(C_{0,0}, C_{1,1}, C_{2,0})$ is a consistent global checkpoint but $(C_{0,1}, C_{1,2}, C_{2,0})$ are inconsistent global checkpoints due to $C_{2,0} \rightarrow C_{1,2}$ or orphan message $m_2$.

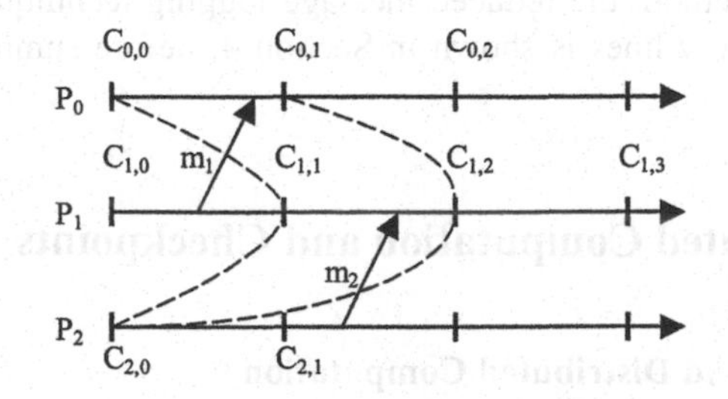


**Fig. 1.** Global checkpoints, in-transit messages and orphan messages

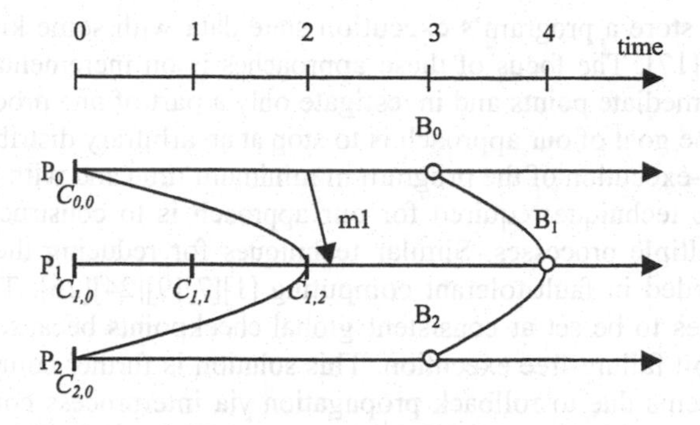


**Fig. 2.** Example for computing rollback/replay distance

Let $C_{p,i}$ denote the *i*-th checkpoint taken by process $P_p$ ($i \geq 0$, $0 \leq p \leq n$), where *n* is the total number of processes. We also assume, that process $P_p$ takes an initial checkpoint $C_{p,0}$ immediately before execution begins, and ends with a virtual checkpoint that represents the last state attained before termination. $I_{p,i}$ denotes the *i*-th checkpoint interval of process $P_p$ which includes all events occurring on process $P_p$ from $C_{p,i}$ to $C_{p,i+1}$ including $C_{p,i}$ but not $C_{p,i+1}$.

### 2.3 Rollback/Replay Distance

The running time from event $e_1$ to event $e_2$ on the same process is called the distance between $e_1$ and $e_2$, denoted $d(e_1, e_2)$. The distance between global states $G_1=(g_{1,0}, g_{1,1},\ldots, g_{1,n})$ and $G_2=(g_{2,0}, g_{2,1},\ldots, g_{2,n})$ is:

$$D = \max(d_i)\ (\forall i\colon 0 \leq i \leq n),\ \text{where } d_i = d(g_{1,i}, g_{2,i}).$$

Please, note that the definition of the distance between global states $G_1$ and $G_2$ is only valid if $g_{1,i} \rightarrow g_{2,i}$ or $g_{1,i} \equiv g_{2,i}$ ($\forall i\colon 0 \leq i \leq n$).

The distance between the recovery line and the corresponding distributed breakpoint is called *the rollback distance* or *the replay distance*.

An example for determining the replay distance is given in Fig. 2. The replay distance between ($C_{0,0}$, $C_{1,2}$, $C_{2,0}$) and ($B_0$, $B_1$, $B_2$) is 3 time units. Please note, that the replay time from the recovery line to the corresponding distributed breakpoint may differ from the replay distance, because the replay distance is based on the previous execution while the replay time is determined during re-execution. However, the replay distance offers a good estimate for the required replay time. In Fig. 2, if a programmer wants to stop at ($B_0$, $B_1$, $B_2$) while the program is recovered at ($C_{0,0}$, $C_{1,2}$, $C_{2,0}$), then the rollback/replay distance is 3 time units but the replay time is approximately 4 time units due to waiting of messages $m_1$.

## 3 Bounding Rollback/Replay Distance for Debugging

In order to reduce the waiting time when replaying a program from the beginning to a breakpoint, a variety of solutions has been proposed to restart the program from intermediate code positions [14][15][29][30]. Each of these ideas requires to periodically store a program's execution state data with some kind of checkpointing mechanism [17]. The focus of these approaches is on incremental replay in order to start at intermediate points and investigate only a part of one process at a time. As an extension, the goal of our approach is to stop at an arbitrary distributed breakpoint and to initiate re-execution of the program in minimum time and with a low overhead.

The basic technique required for our approach is to construct adequate recovery lines on multiple processes. Similar techniques for reducing the rollback execution time are needed in fault-tolerant computing [1][7][9][24][26]. These methods allow recovery lines to be set at consistent global checkpoints because inconsistent global states prohibit failure-free execution. This solution is further complicated in message-passing systems due to rollback propagation via interprocess communication [4]. In the worst case, cascading rollback propagation, which is called the *domino effect* [19], may force the system to restart from the initial state. Methods to avoid unbounded rollback are to enforce additional checkpoints [28] or to coordinate checkpoints [1][23]. However, limiting the rollback distance may still be impossible, or the associated overhead may be rather high, because many checkpoints are required and a lot of messages must be logged.

An example is given in Fig. 3. The most recent recovery line in Fig. 3a for the distributed breakpoint $(B_p, B_q)$ is $(C_{p,0}, C_{q,i})$. However, due to orphan message $m_1$, this recovery line is inconsistent and cannot be used by the available approaches. Instead, the nearest consistent recovery line is $(C_{p,0}, C_{q,0})$. Consequently, stopping at the distributed breakpoint $(B_p, B_q)$ requires replaying from checkpoint $C_{q,0}$ to $B_q$ on process $P_q$. Thus, $i+1$ intervals are required. If the value of $i$ is rather large, the corresponding distance between $C_{q,0}$ and $B_q$ is obviously long. This can be avoided by taking an additional checkpoint $C_{p,x}$ immediately after send event $e_{p,h}$ as shown in Fig. 3b. In this case, the recovery line $(C_{p,x}, C_{q,i})$ can be used for distributed breakpoint $(B_p, B_q)$ with a short replay time.

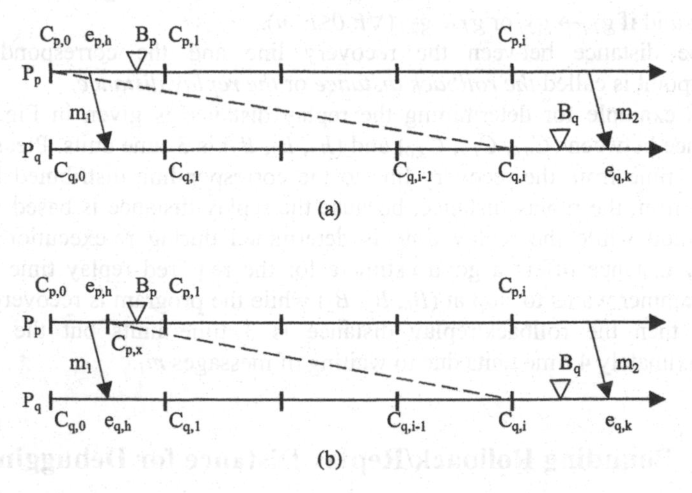


**Fig. 3.** Consistent and inconsistent recovery lines

The problem is that process $P_p$ does not know at send event $e_{p,h}$, where a breakpoint will be established and which send and receive events will happen on processes $P_p$, $P_q$. Thus it cannot decide whether to take an additional checkpoint or not. The traditional solution to limit the replay distance is to take a checkpoint immediately after every send event. Unfortunately, the overhead of this solution is rather high because many checkpoints are required.

The necessity of allowing only consistent global checkpoints is needed to avoid orphan messages, because they may substantially influence the final results. Assuming recovery line ($C_{p,0}$, $C_{q,i}$) of Fig. 3a, message $m_1$ will be an orphan message. By using this recovery line, receive event $e_{q,k}$ may accept message $m_1$ instead of message $m_2$, leading to incorrect final results.

This problem is implicitly solved by an integrated record&replay technique, where the re-execution enforces the same event ordering as observed during the record phase and orphan messages are skipped when need. With this *bypassing orphan messages* method, recovery lines can be established at both consistent and inconsistent global checkpoints. This establishes an upper bound of the replay distance, because any global checkpoint can be a recovery line.

For the recovery line ($C_{p,0}$, $C_{q,i}$) of Fig. 3a, all orphan messages - such as $m_1$ - are revealed by the data of the record phase. Therefore the send events of the orphan messages can be skipped during re-execution - message $m_1$ is not created - and the re-execution path of each process is equivalent to the initial execution. Furthermore, the distance between recovery line ($C_{p,0}$, $C_{q,i}$) and distributed breakpoint ($B_p$, $B_q$) is less than a checkpoint interval.

The replay distance has an upper bound $kT$, if the length of any checkpoint interval is equal or less than a constant value $T$ and the distance between any recovery line and the corresponding distributed breakpoint is not longer than $k$ intervals ($k$>0). Users can control the replay distance by changing the maximum length value $T$ of checkpoint intervals and the number of intervals $k$.

A remaining problem is the overhead due to logging messages. In fault tolerant computing, only actual and possible in-transit messages concerning the last recovery line need to be considered. Consequently, garbage collection algorithms are applied to reduce the number of logged messages [4][25]. However, these methods only support to set recovery lines at consistent global checkpoints [16][22]. For parallel program debugging, a different approach is required, because each message may affect a program's behavior and is therefore valuable for analysis activities. A bug may appear anywhere and thus breakpoints may be set at arbitrary places. Consequently, the way to choose a suitable recovery line for each distributed breakpoint with a low overhead of message logging and a small waiting time is necessary and difficult.

# 4 ROS: Rollback-One-Step Method

Based on the observation above, our combined checkpointing and debugging approach ROS is characterized by the following key issues:

(1) The core checkpointing technique.
(2) The reduced message logging method.
(3) The recovery line calculation based on the R2-graph.

### 4.1 Checkpointing Technique

For checkpointing, ROS uses an index-based communication-induced checkpointing [4], which works as follow:

The states of each process are saved periodically on stable storage. Whenever two processes communicate, the index of the current checkpoint interval is piggybacked on the transferred message. When a message is received, the receiver process knows, whether the message originated on the same or on a different checkpoint interval of the source process by comparing the piggybacked index with its current checkpoint interval. If the message originates from a higher checkpoint interval, a new checkpoint is immediately taken and the corresponding checkpoint interval is assigned the value of the piggybacked checkpoint interval. Furthermore, the checkpoint must be placed before the receive statement. For example, process $P_p$ receives message *m* with a piggybacked checkpoint interval index *j* in interval $I_{p,i}$. If $j > i$, then process $P_p$ takes checkpoint $C_{p,j}$ before processing message *m*. This checkpointing technique is comparable to Quasi-Synchronous Checkpointing as described in [13].

### 4.2 Message Logging

For the transferred messages, the following two rules are applied in order to reduce the overhead of message logging.

**Rule 1** Messages sent from i-th checkpoint interval ($i \geq 0$) and received in j-th checkpoint interval with $j > i$ must be logged.

**Rule 2** All messages sent from the (i-1)-th checkpoint interval of process $P_q$ ($i \geq 1$) and processed by process $P_p$ at the (i-1)-th checkpoint interval are not logged iff $(C_{p,i} \rightarrow C_{q,i+1}) \wedge ((\forall r (r \neq p,q))\ (C_{r,i} \rightarrow C_{p,i+1}) \Rightarrow (C_{r,i} \rightarrow C_{q,i+1}))$.

Based on these two rules, our method operates as follows: If process $P_q$ sends a message *m* to process $P_p$ in checkpoint interval $I_{q,j}$, the value of *j* is piggybacked on message *m*. When process $P_p$ receives message *m* in checkpoint interval $I_{p,i}$, the value of *j* is either equal or less than the value of *i* due to possible additional checkpoints, which may have been applied in them meantime by the checkpointing method. In case the value of *j* is less than *i*, message *m* is logged according to Rule 1. (An example for this case is given with message $m_1$ in Fig. 4.)

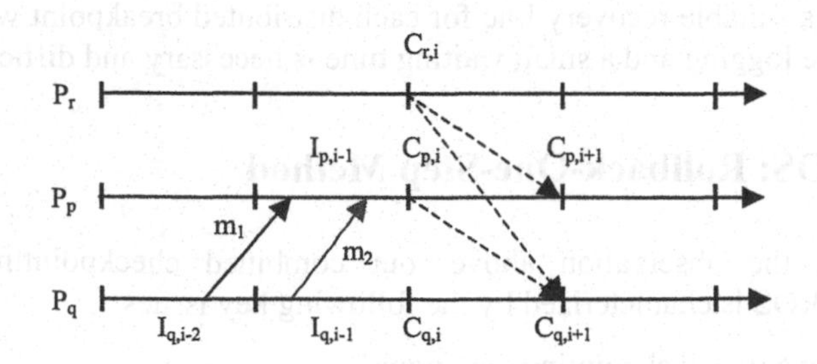


**Fig. 4.** Rollback-One-Step checkpointing

On contrary, in case the value of $j$ is equal to $i$, message $m$ may not need to be logged according to Rule 2. An example for this case is giving with message $m_2$ in Fig. 4. If the condition in Rule 2 does not hold, message $m$ must also be logged to stable storage. Please note, that the necessary information for this decision is only available at state of $(i+1)$-th checkpoint on each process. Thus, message $m$ is temporarily logged in this case. Whenever process $P_p$ arrives at checkpoint $C_{p,i+1}$, it will either store all messages received in $i$-th checkpoint interval on stable storage or remove them based on Rule 2.

With the checkpointing technique and the rules of message logging described above, the rollback/replay distance in the ROS method has an upper bound of two checkpoint intervals.

With an upper bound of the replay/rollback distance, users can lower the replay/rollback distance by reducing the value of the checkpoint interval length. However, this increases the number of checkpoints and thus the overhead of the approach. Consequently, a trade-off between the requirements of the rollback/replay distance and the overhead is necessary.

### 4.3 Logging Messages On-The-Fly

A problem of the technique described above is, that the necessary information for a decision of Rule 2 cannot be obtained at state $(i+1)$-th checkpoint on each process without additional messages. For example, in Fig. 4 process $P_q$ does not know all processes $P_r$s, which satisfy $C_{r,i} \rightarrow C_{p,i+1}$, at state $C_{q,i+1}$; and process $P_p$ does not know whether $C_{r,i} \rightarrow C_{q,i+1}$ is true or not at state $C_{p,i+1}$. As a solution, the condition can be released, so that all necessary information can be obtained at state of the $(i+1)$-th checkpoint without additional messages.

This new condition is based on two following observations:

(1) Only process $P_p$ can get accurate information about processes $P_r$s satisfying $C_{r,i} \rightarrow C_{p,i+1}$ at state $C_{p,i+1}$.

(2) Condition $C_{r,i} \rightarrow C_{q,i+1}$ may be examined based on knowledge of process $P_p$ at state $C_{p,i+1}$.

Observation (2) can be accepted because it only increases the number of logged messages, if the condition cannot be evaluated with the available information. For example, it may log a message with the knowledge available at process $P_p$, while that message would actually not be needed to log. Thus, the upper bound is not affected by this new condition. In addition, receiver-based message logging is suitable, because processes may log receipt messages based on Rule 2 and their available knowledge.

The knowledge of each process $P_p$ is stored in a corresponding *knowledge matrix* $MI_p$. This is an $nxn$ matrix of bits where $n$ is the total number of processes. Vector $MI_p[p]$ keeps knowledge of process $P_p$, where $MI_p[p][r]=1$ in interval $I_{p,i}$ describes $C_{r,i} \rightarrow C_{p,i+1}$. Process $P_p$ knows information about process $P_q$ through vector $MI_p[q]$, where $MI_p[q][r]=1$ in interval $I_{p,i}$ describes $C_{r,i} \rightarrow C_{q,i+1}$.

The knowledge matrix is comparable to matrix clocks, which are used to track down causality among events on processes [20][21]. Based on the same idea, each process learns about other processes through receipt of messages. However, in

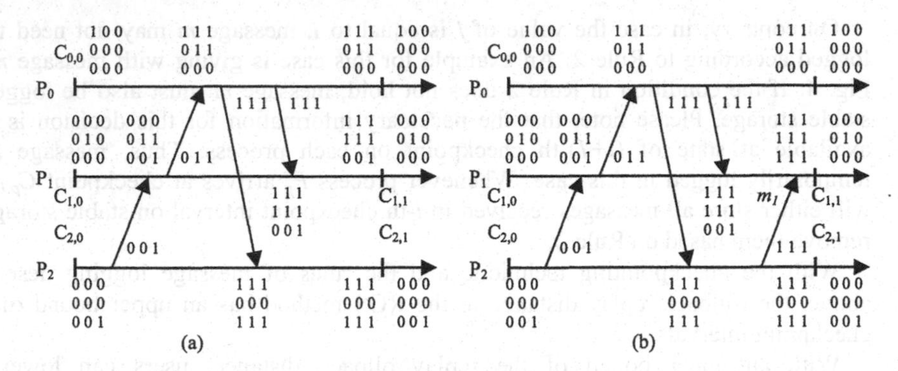


**Fig. 5.** Knowledge matrix

contrast to matrix clocks, each knowledge matrix is used to track down the relation among checkpoints related to the current interval.

The knowledge matrix on each process is computed as illustrated with the example of Fig. 5: Whenever process $P_p$ reaches a new checkpoint, it sets all elements of $MI_p$ to 0 except $MI_p[p][p]$ which is set to 1. When process $P_p$ sends a message to another process, the knowledge of process $P_p$ is piggybacked on the transmitted message. This means, that vector $MI_p[p]$ is piggybacked on message as a knowledge vector $VI$ in addition to the current checkpoint interval index. When process $P_p$ receives a message from process $P_q$ with the same checkpoint interval index, it learns the knowledge of process $P_q$ by updating its vector knowledge $MI_p[p]$ through bit-wise maximum operation and writes knowledge of process $P_q$ to $MI_p[q]$:

$$MI_p[p][i] = \max(MI_p[p][i], VI[i]) \ (\forall i: 0 \leq i \leq n-1)$$
$$MI_p[q][i] = VI[i] \ (\forall i: 0 \leq i \leq n-1)$$

When process $P_p$ arrives at state $C_{p,i+1}$, it possesses knowledge of all processes $P_r$s, which satisfy $C_{r,i} \rightarrow C_{p,i+1}$, based on vector $MI_p[p]$ (where $MI_p[p][r]=1$ implies $C_{r,i} \rightarrow C_{p,i+1}$). In addition, condition $C_{r,i} \rightarrow C_{q,i+1}$ is known on process $P_p$ through vector $MI_p[q]$ (where $MI_p[q][r]=1$ implies $C_{r,i} \rightarrow C_{q,i+1}$). As mentioned above, in some cases $C_{r,i} \rightarrow C_{q,i+1}$ may still be true although $MI_p[q][r]= 0$. For example, in Fig. 5a, process $P_1$ assumes $\neg(C_{0,0} \rightarrow C_{2,1})$ at state $C_{1,1}$ due to $MI_1[2][0]=0$ although $C_{0,0} \rightarrow C_{2,1}$. However, in Fig. 5b, process $P_1$ increases its knowledge through message $m_1$ so that $MI_1[2][0]=1$ and consequently decides that $C_{0,0} \rightarrow C_{2,1}$.

## 4.4 Overhead of Message Logging in ROS

The efficiency of above message logging method is demonstrated in Table 1 and Fig. 6 with the algorithms *Message Exchange, Poisson, FFT* and *Jacobi Iteration*. All programs have been implemented in MPI on a SGI Origin 3800 multiprocessor machine. *Message Exchange* is a simple program in which messages are sent and received from one process to others. *Poisson* program is a parallel solver for Poisson's equation. *FFT* program performs a Fast Fourier Transformation. *Jacobi Iteration* program is used to solve the system of linear equation. The results of the

**Table 1.** Message logging overhead

| Programs | Number of processes | Execution time / Checkpoint Interval (sec) | Total number of messages | Number of logged messages | Percentag e |
|---|---|---|---|---|---|
| Message Exchange | 4 | 16/2 | 120000 | 4629 | 3.86 |
| | 8 | 54/2 | 560000 | 10607 | 1.89 |
| | 16 | 115/2 | 1200000 | 14425 | 1.20 |
| Poisson | 4 | 16/2 | 149866 | 5230 | 3.49 |
| | 8 | 35/2 | 369084 | 4934 | 1.34 |
| | 16 | 83/2 | 864078 | 12521 | 1.45 |
| FFT | 4 | 27/2 | 270024 | 10747 | 3.98 |
| | 8 | 59/2 | 630056 | 2552 | 0.41 |
| | 16 | 130/2 | 1350120 | 16731 | 1.24 |
| Jacobi Iteration | 4 | 1730/2 | 49924 | 1952 | 3.91 |
| | 8 | 497/2 | 71928 | 2368 | 3.29 |
| | 16 | 259/2 | 151936 | 3115 | 2.05 |

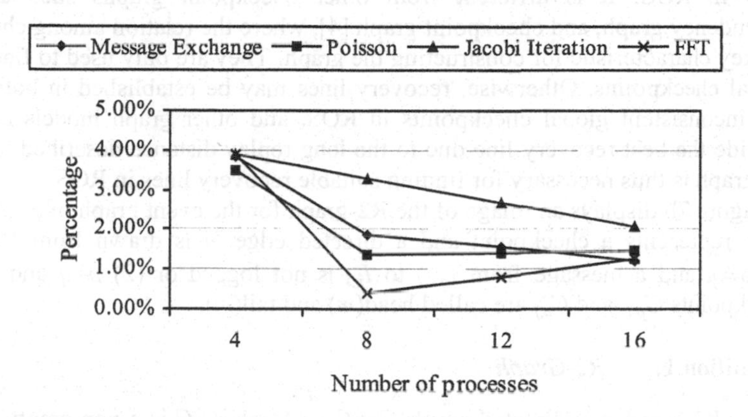


**Fig. 6.** Percentage of number of logged messages per number of total messages

reduced message logging method are shown in Table 1 and Fig. 6, where the vertical axis is the number of processes and the horizontal axis is the percentage of the number of logged messages per the number of total messages. Less than 4 percent of total number of messages need to be logged. This demonstrates the efficiency of Rule 1 and Rule 2 in reducing the overhead of message logging.

### 4.5 Recovery Line Calculation

In order to find a recovery line in ROS, the R2-graph has been developed. This graph is constructed based on checkpoints and in-transit messages, which are not logged.

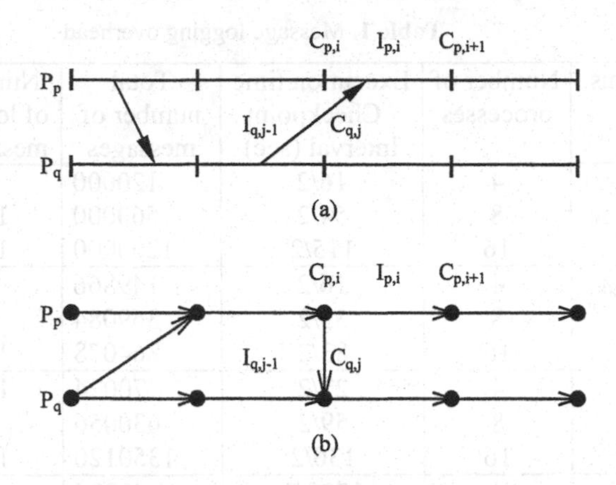


**Fig. 7.** (a) Checkpoints and not required messages; (b) R2-graph

Such a graph is required to find global checkpoints, which can be used as recovery lines in ROS. It is different from other checkpoint graphs such as rollback-dependency graph, and checkpoint graph [4], where the relation among checkpoints is the key characteristic for constructing the graph. They are only used to find consistent global checkpoints. Otherwise, recovery lines may be established in both consistent and inconsistent global checkpoints in ROS, and other graph models do often not provide the best recovery line due to the long replay distance described in Section 3. R2-graph is thus necessary for finding suitable recovery lines in ROS.

Figure 7b displays an image of the R2-graph for the event graph of Figure 7a. Each node represents a checkpoint and a directed edge $m$ is drawn from $C_{p,i}$ to $C_{q,j}$ if (1) p≠q, and a message from $I_{q,j-1}$ to $I_{p,i}$ is not logged or (2) $p=q$ and $j=i+1$. The checkpoints $C_{p,i}$ and $C_{q,j}$ are called head($m$) and tail($m$).

**Definition 1:** *R2-Graph*

An *R2-graph* is a directed graph $G=(C, \overset{d}{\rightarrow})$, where $C$ is a non-empty set of local checkpoints on all processes, while $\overset{d}{\rightarrow}$ is a relation connecting checkpoints, such that $C_{p,i} \overset{d}{\rightarrow} C_{q,j}$ means that there is an edge from $C_{p,i}$ to $C_{q,j}$ in $G$ with "head" at $C_{p,i}$ and "tail" at $C_{q,j}$.

The relation $C_{p,i} \overset{d}{\rightarrow} C_{q,j}$ only shows that in-transit messages of $(C_{p,i}, C_{q,j})$ from $I_{q,j-1}$ to $I_{p,i}$ are not logged. This relation can be opened to *direct relation* through the definition of *Direct R2-path* (*DR2-path*) below:

**Definition 2:** *DR2-path*

There is a *DR2-path* from checkpoint $C_{p,x}$ on process $P_p$ to checkpoint $C_{q,y}$ on process $P_q$ in R2-graph, denoted $C_{p,x} \overset{DR2}{\rightarrow} C_{q,y}$, iff a path $[m_1 \dots m_n]$ ($n \geq 1$) exists such that

1. $C_{p,x}$ is head($m_1$), and
2. for each $m_i$, $1 \le i < n$, tail($m_i$) and head($m_{i+1}$) are on $P_p$ or $P_q$, and tail($m_i$) is head($m_{i+1}$), and
3. $C_{q,y}$ is tail($m_n$).

This is called a direct relation because DR2-path is only constructed based on relation among checkpoints on a couple of processes. Direct relation is an extension of the relation in R2-graph by applying transitive characteristic on couple of processes. This means, that if $C_1$, $C_2$, $C_3$ are checkpoints on processes $P_p$ and $P_q$ and $(C_1 \overset{DR2}{\rightarrow} C_2) \wedge (C_2 \overset{DR2}{\rightarrow} C_3)$, then $C_1 \overset{DR2}{\rightarrow} C_3$.

Based on the R2-graph, an available recovery line can be computed by determining the DR2-path following description in theorem 1:

**Theorem 1** A global checkpoint $G=(C_0, C_1, ..., C_n)(n \ge 1)$ is a recovery line, if and only if there is no DR2-path between any two local checkpoints in $G$ $((\forall i,j \in [0,n],\ i \ne j)\ \neg(C_i \overset{DR2}{\rightarrow} C_j))$ . [1]

Theorem 1 can be applied to find the nearest recovery line for a distributed breakpoint with the following 3 steps approach:

1. The initial checkpoint set contains local checkpoints, where each checkpoint is *the nearest checkpoint before* the distributed breakpoint on each process.
2. If there is a DR2-path between any two checkpoints $C_{p,i}$, $C_{q,j}$ $(C_{p,i} \overset{DR2}{\rightarrow} C_{q,j})$, then $C_{q,j}$ is replaced by a checkpoint $C_{q,j-1}$ in the checkpoint set.
3. Continue Step 2 until there is no DR2-path between any two checkpoints in the set. The final checkpoint set is the recovery line.

If these steps are used to find a recovery line for a distributed breakpoint, Step 2 is at most applied $n$ times, because the rollback distance is equal or less than two checkpoint intervals.

## 5 Conclusions

This paper describes the problem of using consistent as well as inconsistent recovery lines when combining checkpointing and debugging. Although several comparable techniques have been proposed in the past, the distance between a recovery line and a corresponding distributed breakpoint has not been sufficiently investigated. The distance may be large, if recovery lines are only allowed at consistent global checkpoints.

In contrast to related work, the approach described in this paper permits recovery lines at both consistent and inconsistent global checkpoints. The problem of orphan messages can be solved by the bypassing orphan messages. In addition, the number of logged messages is small. The results have been implemented within the ROS

[1] A proof of Theorem 1 is not included in this paper due to the limited space.

method. This method guarantees, that the replay distance between the recovery line and the corresponding distributed breakpoint has an upper bound of two checkpoint intervals. Additionally, the paper introduces a general method to find the nearest recovery line from a distributed breakpoint based on R2-graph and DR2-path. The combination of all these techniques offers an appealing and efficient approach to reduce the large waiting times observed during cyclic debugging of parallel programs.

## References


[1] Chandy, K. M., and Lamport, L. "*Distributed Snapshots: Determining Global States of Distributed Systems*", ACM Transactions on Computer Systems 3 (1985), pp. 63-75.

[2] Cunha, J. C., and Lourenco, J. "*An Integrated Testing and Debugging Environment for Parallel and Distributed Programs*", Proc. of the 23rd EUROMICRO Conference, IEEE Computer Society Budapest, Hungary (1997), pp. 291-298.

[3] Dow, C. R., and Lin, C. M. "*Adaptive Distributed Breakpoint Detection and Checkpoint Space Reduction in Message Massing Programs*", Computers and Artificial Intelligence (2000), Vol. 19, pp. 547-568.

[4] Elnozahy, E. N., Johnson, D. B., and Wang, Y. M. "*A Survey of Rollback-Recovery Protocols in Message-Passing Systems*", Technical Report CMU-CS, Carnegie Mellon University, (October 1996), pp. 96-181.

[5] Feldman, S.I., Brown, Ch. B. "*Igor: A System for Program Debugging via Reversible Execution*", Proc. of the ACM SIGPLAN and SIGOPS Workshop on Parallel and Distributed Debugging (May 5-6, 1988), University of Wisconsin, Madison, Wisconsin, USA, SIGPLAN Notices (January 1989), Vol. 24, No. 1, pp. 112-123.

[6] Fowler, J., and Zwaenepoel, W. "*Causal Distributed Breakpoints*", Proc. of the 10th International Conference on Distributed Computing Systems (ICDCS) (1990), pp. 134-141.

[7] Garcia, I. C., and Buzato. L. E. "*Progressive Construction of Consistent Global Checkpoints*", In 19th IEEE International Conference on Distributed Computing Systems (ICDCS'99), Austin, Texas, EUA (June 1999).

[8] Haban, D., and Weigel, W. "*Global Events and Global Breakpoints in Distributed Systems*", Proc. of the 21st Annual Hawaii International Conference on System Sciences, Software Track, IEEE Computer Society (January 1988), Vol. 2, pp. 166-175.

[9] Hélary, J. M., Mostefaoui, A., and Raynal., M. "*Communication-Induced Determination of Consistent Snapshot*", IEEE Transaction on Parallel and Distributed Systems (September 1999), Vol. 10, No. 9.

[10] Kacsuk, P., "*Systematic Macrostep Debugging of Message Passing Parallel Programs*", In: Kacsuk, P., Kotsis, G., "Distributed and Parallel Systems (DAPSYS'98)", Future Generation Computer Systems, North-Holland (April 2000), Vol. 16, No. 6, pp. 597-607.

[11] Kranzlmüller, D. "*Event Graph Analysis for Debugging Massively Parallel Programs*", PhD Thesis, GUP Linz, Johannes Kepler University Linz, Austria (September 2000), http://www.gup.uni-linz.ac.at/~dk/thesis/thesis.php.

[12] Lamport, L. "*Time, Clocks, and the Ordering of Events in a Distributed System*", Communications of the ACM (July 1978), Vol. 21, No. 7, pp. 558-565.

[13] Manivannan, D. and Singhal, M. "*A Low Overhead Recovery Technique Using Quasi-Synchronous Checkpointing*", Proc. 16th IEEE International Conference on Distributed Computing Systems, Hong-Kong (1996), pp. 100-107.

[14] Netzer, R. H. B., and Xu, J. "*Adaptive Message Logging for Incremental Program Replay*", IEEE Parallel & Distributed Technology (November 1993), Vol. 1, No. 4, pp. 32-40.

[15] Netzer, R. H. B., Subramanian, S., and Xu, J. "*Critical-Path-Based Message Logging for Incremental Replay of Message-Passing Programs*", In 14th International Conference on Distributed Computing Systems, Poznan, Poland (June 1994).

[16] Netzer, R. H. B., and Xu, J. "*Sender-Based Message Logging for Reducing Rollback Propagation*", Proc. of the 7th IEEE Symposium on Parallel and Distributed Processing (SPDP '95).

[17] Pan, D.Z., and Linton, M.A. "*Supporting Reverse Execution of Parallel Programs*", Proc. of the ACM SIGPLAN and SIGOPS Workshop on Parallel and Distributed Debugging (May 5-6, 1988), University of Wisconsin, Madison, Wisconsin, USA, SIGPLAN Notices (January 1989), Vol. 24, No. 1, pp. 124-129.

[18] Plank, J. S. "An Overview of Checkpointing in Uniprocessor and Distributed Systems, Focusing on Implementation and Performance", Technical Report of University of Tennessee, UT-CS-97-372 (July 1997).

[19] Randel, B. "*System Structure for Software Fault Tolerance*", IEEE Transactions on Software Engineering TSE (June 1975), Vol. 1, No. 2, pp. 221-232.

[20] Raynal, M., and Singhal, M. "Logical Time: A Way to Capture Causality in Distributed Systems", IRISA (January 1995).

[21] Ruget, F. "*Cheaper Matrix Clocks*", Proc. of the 8th International Workshop on Distributed Algorithms, Springer-Verlag LNCS 857 (G. Tel and P. Vityani Eds) (1994), pp. 355-369.

[22] Wang, Y. M., and Fuchs, W. K. "*Optimistic Message Logging for Independent Checkpointing in Message Passing Systems*", Proc. of the 11th Symposium on Reliable Distributed Systems, (October 1992), pp. 147-154.

[23] Wang, Y. M., and Fuchs, W. K. "*Lazy Checkpoint Coordination for Bounding Rollback Propagation*", Proc. of the 12th Symposium on Reliable Distributed Systems (1993), pp. 78-85.

[24] Wang, Y. M. "*The Maximum and Minimum Consistent Global Checkpoints and Their Applications*", Proc. IEEE Symposium Reliable Distributed Systems (September 1995), pp. 86-95.

[25] Wang, Y. M., and Fuchs, W. K. "*Optimal Message Log Reclamation for Uncoordinated Checkpointing*", Fault-Tolerant Parallel and Distributed Systems, IEEE Computer Society Press (1995), pp. 24-29.

[26] Wang, Y. M. "*Consistent Global Checkpoints That Contains a Set of Local Checkpoints*", IEEE Transactions on Computers (1997), Vol. 46, No. 4, pp. 456-468.

[27] Yang, Z., and Marsland, T. "*Global Snapshots for Distributed Debugging*", Technical Report TR 92-03, Laboratory for Distributed and Parallel Computing, Computing Science Department, University of Alberta, Edmonton, Canada T6G 2H1 (1992).

[28] Zambonelli, F. "*On the Effectiveness of Distributed Checkpoint Algorithms for Domino-Free Recovery*", In 7th IEEE Symposium on High-Performance Distributed Computing (July 1998).

[29] Zambonelli, F., and Netzer, R. H. B. "*An Efficient Logging Algorithm for Incremental Replay of Message-Passing Applications*", Proc. of the 13th International Parallel Processing Symposium and 10th Symposium on Parallel and Distributed Processing (1999).

[30] Zambonelli, F., and Netzer, R. H. B. "*Deadlock-Free Incremental Replay of Message-Passing Programs*", Journal of Parallel and Distributed Computing 61 (2001), pp. 667-678.

# Distributed Paged Hash Tables

José Rufino[1] *, António Pina[2], Albano Alves[1], and José Exposto[1]

[1] Polytechnic Institute of Bragança, 5301-854 Bragança, Portugal
{rufino,albano,exp}@ipb.pt
[2] University of Minho, 4710-057 Braga, Portugal
pina@di.uminho.pt

**Abstract.** In this paper we present the design and implementation of DPH, a storage layer for cluster environments. DPH is a Distributed Data Structure (DDS) based on the distribution of a paged hash table. It combines main memory with file system resources across the cluster in order to implement a distributed *dictionary* that can be used for the storage of very large data sets with key based addressing techniques. The DPH storage layer is supported by a collection of cluster–aware utilities and services. Access to the DPH interface is provided by a user–level API. A preliminary performance evaluation shows promising results.

## 1 Introduction

Today commodity hardware and message passing standards such as PVM [1] and MPI [2] are making possible to assemble clusters that exploit distributed storage and computing power, allowing for the deployment of data-intensive computer applications at an affordable cost. These applications may deal with massive amounts of data both at the main and secondary memory levels. As such, traditional data structures and algorithms may no longer be able to cope with the new challenges specific to cluster computing.

Several techniques have thus been devised to distribute data among a set of nodes. Traditional data structures have evolved towards Distributed Data Structures (DDSs) [3, 4, 5, 6, 7, 8, 9] . At the file system level, cluster aware file systems [10, 11] already provide resilience to distributed applications. More recently a new research trend has emerged: online data structures for external memory that bypass the virtual memory system and explicitly manage their own I/O [12].

Distributed Paged Hashing (DPH[1]) is a cluster aware storage layer that implements a hash based Distributed Data Structure (DDS). DPH has been designed to support a Scalable Information Retrieval environment (SIRe), an ongoing research project with a primary focus on information retrieval and cataloging techniques suited to the World Wide Web.

* Supported by PRODEP III, through the grant 5.3/N/199.006/00, and SAPIENS, through the grant 41739/CHS/2001.

[1] A preliminary presentation of our work took place at the PADDA2001 workshop [13]; here we present a more in-depth and updated description of DPH.

J.M.L.M. Palma et al. (Eds.): VECPAR 2002, LNCS 2565, pp. 679–692, 2003.

The main idea behind DPH is the distribution of a paged hash table over a set of networked *page servers*. Pages are contiguous bucket sets[2], all with the same number of buckets. Because the amount of pages is initially set our strategy appears to be static. However, pages are created on–demand so the hash table grows dynamically.

A *page broker* is responsible for the mapping of pages to *page servers*. The mapping takes place just once for the lifetime of a page (page migration is not yet supported) and so the use of local caches at the service clients alleviates the *page broker*. In a typical scenario, the *page broker* is mainly active during the first requests to the DPH structure when pages are mapped to the available *page servers*. Because the local caches are incrementally updated the *page broker* will be relieved from further mapping requests.

The system doesn't rely only on the available main memory at each node. When performance is not the primary concern, a threshold based swap mechanism may also be used to take advantage of the file system. It is even possible to operate the system solely based on the file system, achieving the maximum level of resilience. The selection of the swap-out bucket victims is based on a Least–Recently–Used (LRU) policy.

The paper is organized as follows: section 2 covers related work, section 3 presents the system architecture, section 4 shows preliminary performance results and section 5 concludes and points directions for future work.

## 2 Related Work

Hash tables are well known data structures [14] mainly used as a fast key based addressing technique. Hashing has been intensively exploited because retrieval times are $O(1)$ when compared with $O(log\ n)$ for tree-structured schemes or $O(n)$ for sequential schemes. Hashing is classically *static* meaning that, once set, the bit–length of the hash index never changes and so the complete hash table must be initially allocated.

In dynamic environments, with no regular patterns of utilization, the use of static hash tables results on storage space waste if only a small bucket subset is used. Static hashing may not also be able to guarantee $O(1)$ retrieval times when buckets overflow. To counterwork these limitations several dynamic hashing [15] techniques have been proposed, such as Linear Hashing (LH) [16] and Extendible Hashing (EH) [17], along with some variants.

Meanwhile, with the advent of cluster computing, traditional data structures have evolved towards distributed versions. The issues involved aren't trivial because, in a distributed environment, scalability is a primary concern and new problems arise (consistency, timing, order, security, fault tolerance, hot–spots, etc.). In the hashing domain, LH* [3] extended LH [16] techniques for file and table addressing and coined the term *Scalable Distributed Data Structure* (SDDS). Distributed Dynamic Hashing (DDH) [4] offered an alternative

[2] In the DPH context, a bucket is a hash table entry where collisions are allowed and self–contained, that is, collisions don't overflow into other buckets.

approach to LH* while EH* [5] provided a distributed version of EH [17]. Although in a very specific application context, [18] have exploited a very similar concept to DPH, named *two–level hashing*. Distributed versions of several other classical data structures, such as trees [7, 8] and even hybrid structures, such as hash–trees [19], have also been designed. More recently, work has been done on hash based distributed data structures to support Internet services [9].

# 3 Distributed Paged Hashing

Our proposal shows that for certain classes of problems, an hybrid approach, that mixes static and dynamic techniques, may achieve good performance and scalability without the complexity of purely dynamic schemes.

When the dimension of the key space is unknown *a priori*, a pure dynamic hashing approach would incrementally use more bits from the hash index when buckets overflow and split. Only then storage consumption would expand to make room for the new buckets. Typically, the expansion takes place at another server, as distributed dynamic hashing schemes tend to move one of the splits to another server.

Although providing maximum flexibility, a dynamic approach increases the load on the network, not only during bucket splitting, but also when a server forwards requests from clients with an outdated view of the <bucket, server> mapping. Once we know in advance that the application domain (SIRe) will include a distributed web crawler, designed to extract and manage millions of URLs, then it doesn't make much sense not to start, from the beginning, using the maximum bit–length of the hash index. As such, DPH is a kind of hybrid approach that includes both static and dynamic features: it uses a fixed bit–length hash table, but pages (and buckets) are created on–demand and distributed across the cluster.

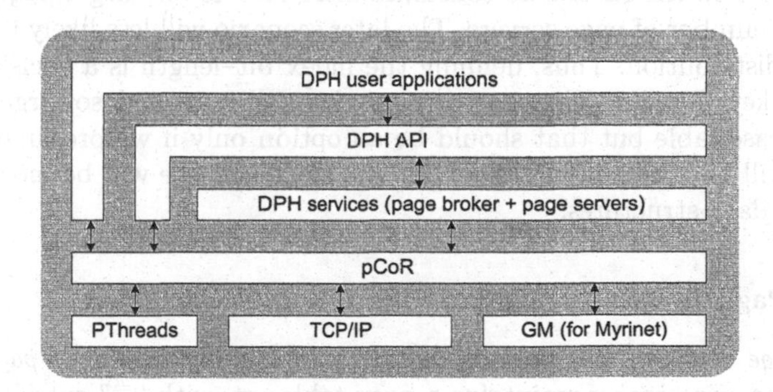


**Fig. 1.** The DPH architecture

### 3.1 Architecture

Figure 1 presents the architecture of DPH. User applications interface with the DPH core (the *page broker* and the *page servers*) through a proper API. The run-time system is provided by pCoR [20], a prototype of CoR [21]. CoR paradigm extends the process abstraction to achieve structured fine grained computation using a combination of message passing, shared memory and POSIX Threads. pCoR is both multithreaded and thread safe and already provides some very useful features, namely message passing (by using GM [22] over Myrinet) between threads across the cluster. This is fully exploited by the DPH API and services, which are also multithreaded and thread safe.

### 3.2 Addressing

The DPH addressing scheme is based on *one–level paging* of the hash table:

1. a static hash function $H$ is used to compute an index $i$ for a key $k$: $H(k) = i$;
2. the index $i$ may be split into a page field $p$ and an offset field $o$: $i =< p, o >$;
3. the hash table may be viewed as a set of $2^{\#p}$ pages, with $2^{\#o}$ buckets per page, where $\#p$ and $\#o$ are the (fixed) bit–length of the page and offset fields, respectively;
4. the page table $pt$ will have $2^{\#p}$ entries, such that $pt[j] = ps_j$, where $ps_j$ is a reference to the page server for page $j$.

$H$ is a 32 bit hash function[3], but smaller bit subsets from the hash index may be used, with the remaining bits being simply discarded. The definition of the page and offset bit–lengths are the main decisions to take prior to the usage of the DPH data structure. The more bits the page field uses, the more pages will be created, leading to a very sparse hash table (if enough page servers are provided), with a small number of buckets per page. Of course, the reverse will happen when the offset field consumes more bits: fewer, larger pages, handled by a small number of page servers. The later scenario will less likely take advantage of the distribution. Thus, defining the index bit–length is a decision dependent on the key domain. We want to minimize collisions and so large indexes may seem reasonable but that should be an option only if we presume that the key space will be uniformly used. Otherwise storage space will be mostly wasted on control data structures.

### 3.3 Page Broker

The *page broker* is responsible for the mapping of pages into *page servers*. As such, the *page broker* maintains a page table, *pt*, with $2^{\#p}$ entries, one for each page. When it receives a mapping request for page $p$, the *page broker* returns

[3] $H$ has been chosen from [23]. A comparison was made with other general hash functions from [24], [14] and [25], but no significant differences have been found, both in terms of performance and collision avoidance.

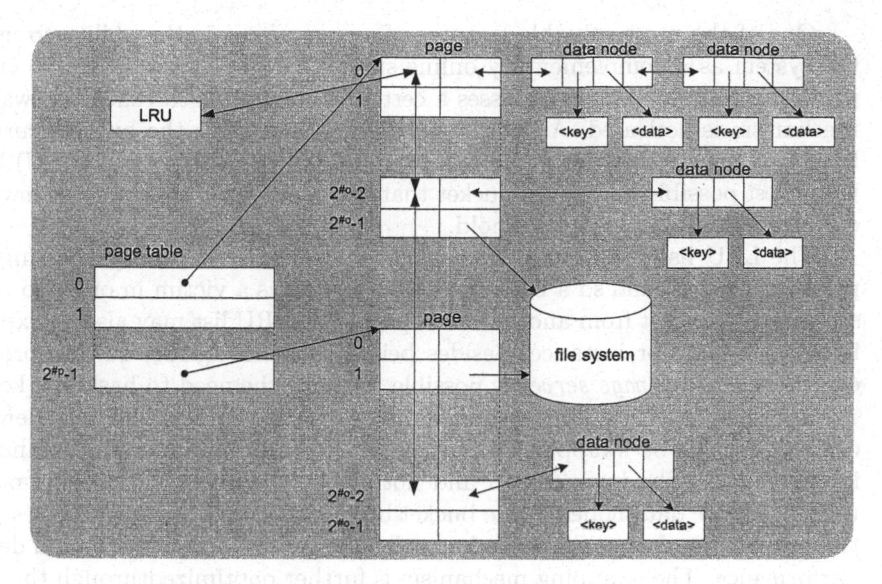


**Fig. 2.** Main data structures for a *page server*

*pt*[*p*], which is a reference to the *page server* responsible for the page *p*. It may happen, however, that this is the first mapping request for the page *p*. If so, the *page broker* will have to choose a *page server* to handle that page. A Round Robin (RR) policy is currently used over the available *page servers*, assuring that each handles an equal share of the hash table, but we plan to add the choice for more adaptive policies, such as weighted RR (proportional to the available node memory and/or current load, for instance) or others.

### 3.4 Page Servers

A *page server* hosts a page subset of the distributed hash table (as requested by the *page broker*, during the mapping process), and answers most of the DPH user level API requests (insertions, searches, deletions, etc.).

Figure 2 presents the main data structures for a *page server*. A *page table* with $2^{\#p}$ entries is used to keep track of the locally managed pages. A page is a bucket set with $2^{\#o}$ entries. A bucket is an entry point to a set of data nodes which are <key, data> pairs. Collisions are self contained in a bucket (*chaining*). Other techniques, like using available empty buckets on other pages (*probing*), wouldn't be compatible with the swapping mechanism[4].

Presently, buckets are doubly–linked lists. These rather inefficient data structures, with $O(n)$ access times, were used just to rapidly develop the prototype. In the future we plan to use more efficient structures, such as trees, skip–lists [26] or even dynamic hashing.

[4] This mechanism uses the bucket as the swap unit and depends on information kept therein to optimize the process.

One of the most valuable features of a *page server* is the ability to use the file system as a complementary online storage resource. Whenever the current user data memory usage surpasses a certain configurable threshold, a swapping mechanism is activated. A bucket victim is chosen, from the buckets currently hold in memory. The victim, chosen from a Least–Recently–Used (LRU) list, is the oldest possible referenced bucket that still frees enough memory to lower the current usage bellow the threshold.

The LRU list links every bucket currently in main memory, crossing local page boundaries, and so a bucket may be elected as a victim in order to release storage to a bucket from another local page. The LRU list may also be exploited in other ways. For instance, besides being a natural queue, quickly browsing every bucket in a *page server* is possible, without the need to hash any key.

Buckets that have been swapped-out to the file system are still viewed as online and will be swapped-in as they are needed. The swapping granularity is currently at the bucket level and not at the data node level. This may be unfair to some data nodes in the bucket but prevents too many small files (often thousands), one for each data node, at the file system level, which would degrade performance. The swapping mechanism is further ooptimizedthrough the use of a *dirty bit* per bucket, preventing unmodified buckets to be unnecessarily saved.

A *page server* may work with a zero threshold thus using the main memory just to keep control data structures and as an intermediate pool to perform the user request, after which the bucket is immediately saved to the file system and the temporary instance removed from main memory.

If a DPH instance has been terminated graciously (thus ssynchronizingits state with the file system), then it may be loaded again, on–demand: whenever a *page server* is asked to perform an operation on a bucket that is empty, it first tries to load a possible instance from the file system because an instance may be there from a previous shutdown of the DPH hash table. In fact, even after an unclean shutdown, partial recovery may be possible because uunsynchronized bucket instances are still loaded.

### 3.5 User Applications

User applications may be built on top of the DPH API and runtime environment. From a user application perspective, insertions, retrievals and removals are the main interactions with the DPH storage layer. These operations must have a key hashed and then mapped into the correct *page server*. This mapping is primarily done through a local cache of the *page broker* page table. A user application starts with an empty page table cache and so many cache misses will take place, forcing mapping requests to the *page broker*. This is done automatically, in a transparent way to the user application. Further mappings of the same page will benefit from a cache hit and so the user application will readily contact the relevant *page server*.

Presently, mapping never changes for the lifetime of a DPH instance[5] and so the cache will be valid during the execution of the user application. This way, a *page broker* will be a hot–spot (if ever) for a very limited amount of time. Our preliminary tests show no significant impact on performance during cache fills.

### 3.6 Client–Server Interactions

Our system operates with a relatively small number of exchanged messages[6]:

1. mapping a page into a *page server* may use zero, two, four or (very seldom) more messages: if the local cache gives a hit, zero messages were needed; otherwise the *page broker* must be contacted; if the page table gives a hit, only the reply to the user application is needed, summing up two messages; otherwise a *page server* must be contacted and so two more messages are needed (request and reply); of course, if the *page server* replied with a negative acknowledgement, the Round Robin search for another *page server* will add two more messages per *page server*;
2. insertions, retrievals and removals typically use two messages (provided a cache hit); however, insertions and retrievals may be asynchronous, using only one message (provided, once again, a cache hit); the later means that no acknowledge is requested from the *page server*, which translates into better performance, though the operation may have not be successfully performed and the user application won't be aware of it.

Once local caches become updated, and assuming the vast majority of the requests to be synchronous insertions, retrievals and deletions, we may set two messages as the upper bound for each interaction of a client with a DPH instance.

## 4 Performance Evaluation

### 4.1 Test–Bed

The performance evaluation took place in a cluster of five nodes, all running Linux Red Hat 7.2 with the stock kernel (2.4.7-10smp) and GM 1.5.1 [22]. The nodes were interconnected using a 1.28+1.28 Gbits/s Myrinet switch. Four of the nodes (A,B,C,D) have the following hardware specifics: two Pentium III processors at 733 Mhz, 512 Mb SDRAM/100 MHz, i840 chipset, 9Gb UDMA 66 hard disks, Myrinet SAN LANai 9 network adapter. The fifth node (E) has four Pentium III Xeon processors running at 700 Mhz, 1 Gb ECC SDRAM/100 MHz, ServerWorks HE chipset, 36 Gb Ultra SCSI 160 hard disk and a Myrinet SAN LANai 9 network adapter.

[5] We are referring to a live instance, on top of a DPH runtime system.

[6] We have restricted the following analysis to the most relevant interactions.

### 4.2 Hash Bit–Length

Because DPH uses static hashing, the hash bit–length must be preset. This should be done in such a way that overhead from control data structures and collisions are both minimized. However, those are conflicting requisites. For instance, to minimize collisions we should increase the bit–length, thus increasing the hash table height; in turn, a larger hash table will have more empty buckets and will consume more control data structures. We thus need a metric for the choice of the right hash bit–length.

**Metric Definition** Let $B_j$ be the number of buckets with $j$ data nodes, after the hash table has been built. If $k$ keys have been inserted, then $P_j = (B_j \times j)/k$ is the probability of any given key to have been inserted in a $B_j$ bucket. Also, let $N_j$ be the average number of nodes visited to find a key in a $B_j$ bucket. Once we have used linked lists to handle collisions, $N_j = (j + 1)/2$. Then, given an arbitrary key, the average number of nodes to be searched for the key is $N = \sum_j (P_j \times N_j)$. The overhead from control data structures is $O = C/(U + C)$, where $C$ is the storage consumed in control data structures and $U$ is the storage consumed in user data (keys and other possible attached data). Finally, our metric is defined by the ranking $R = nN \times oO$, where $n$ and $o$ are the percentual weights given to $N$ and $O$, respectively. For a specific scenario, the hash bit–length to choose will be the one that minimizes $R$.

**Application Scenario** The tests were performed, in a single cluster node (A), for a varying number of keys, using hash bit–lengths from 15 to 20. The page field of our addressing scheme used half of the hash; the other half was used as an offset in the page (for odd bit–lengths, the page field was favored). Keys were random unique sequences, 128 bytes wide; user data measured 256 bytes[7].

Figure 3 presents the rankings obtained. If an ideal general hash function (one that uniformly spspreads the hashes across the hash space, regardless of the randomness and nature of the keys) was used, we would expect the optimum hash bit–length to be approximately $log_2 k$, for each number of keys $k$. However, not only our general hash function [23] isn't ideal, but also the overhead factor must be taken into account. We thus observe that our metric is minimized when the bit–length is $log_2 k - 1$, regardless of $k$[8].

In order to determine if the variation of the key size would interfere with the optimum hash bit–length we ran another test, this time by varying the key size across {4, 128, 256}. Figure 4 shows the results for 125000 keys. It may be observed that $log_2 k - 1$ still is the recommended hash bit–length, independently of the key size[9]. The ranking is preserved because regardless of the key size, the hash function provides similar distributions of the keys; therefore, $N$ is approximately the same, while the overhead $O$ is the varying factor.

---

[7] Typical sizes used in the web crawler being developed under the SIRe project.

[8] For instance, 17 bits for the hash bit–length seems reasonable when dealing with a maximum of 125000 keys, but our metric gives 16 bits as the recommended value.

[9] This was also observed with 250000, 500000 and 1000000 keys.

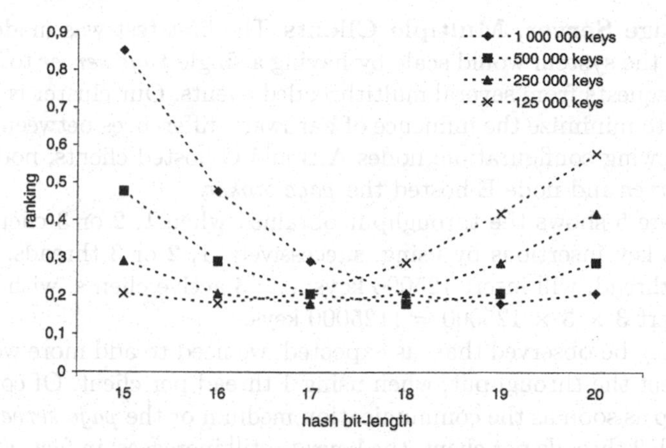


**Fig. 3.** $R$ for $n = 50\%$ and $o = 50\%$

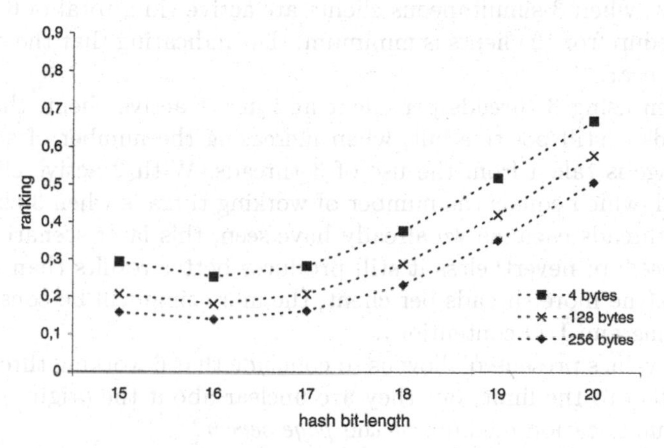


**Fig. 4.** Effect of the key size on $R$

### 4.3 Scalability

To evaluate the scalability of our system we produced another type of experiments using $k$=1500000 as the maximum number of keys. Accordingly with the metric $R$ defined in the previous experiment, the hash bit–length was set to $log_2 k - 1 = 19$ bits. Also, as previously, keys were random unique sequences, 128 bytes wide, with 256 bytes of attached user data. Each client thread was responsible for the handling of 125000 keys.

We measured insertions and retrievals. Insertions were done in newly created DPH instances and thus the measured times ("build times") accounted for cache misses and *page broker* mappings. The retrieval times and retrieval key rates are not presented, because they were observed to be only marginally better. The memory threshold was set high enough to prevent any DPH swapping.

**One Page Server, Multiple Clients** The first test was made to investigate how far the system would scale by having a single *page server* to attend simultaneous requests from several multithreaded clients. Our cluster is relatively small and so, to minimize the influence of hardware differences between nodes, we used the following configuration: nodes A,B and C hosted clients, node D hosted the *page server* and node E hosted the *page broker*.

Figure 5 shows the throughput obtained when 1, 2 or 3 clients make simultaneous key insertions by using, successively, 1, 2 or 3 threads: 1 active client, with 1 thread, will insert 125000 keys; ...; 3 active clients, with 3 threads each, will insert $3 \times 3 \times 125000 = 1125000$ keys.

It may be observed that, as expected, we need to add more working nodes to increment the throughput, when using 1 thread per client. Of course, this trend will stop as soon as the communication medium or the *page server* get saturated.

With 2 threads per client, the keyrate still increases; in fact, with just 1 client and 2 threads the throughput achieved is the same as with 2 clients with 1 thread each but, when 3 simultaneous clients are active (in a total of 6 client threads), the speedup from 2 clients is minimum, thus indicating that the saturation point may be near.

When using 3 threads per client and just 1 active client, the speedup from 2 threads is still positive but, when increasing the number of active clients, no advantage is taken from the use of 3 threads. With 2 active clients, 6 threads are used, which equals the number of working threads when 3 clients are active, with 2 threads each; as we already have seen, this later scenario produces very poor speedup; nevertheless it still produces better results than 2 clients with 3 threads (the more threads per client, the more time will be consumed in thread scheduling and I/O contention).

The values presented allow us to conclude that 6 working threads are pushing the system to the limit, but they are unclear about the origin of that behavior: the communication medium or the *page server*?

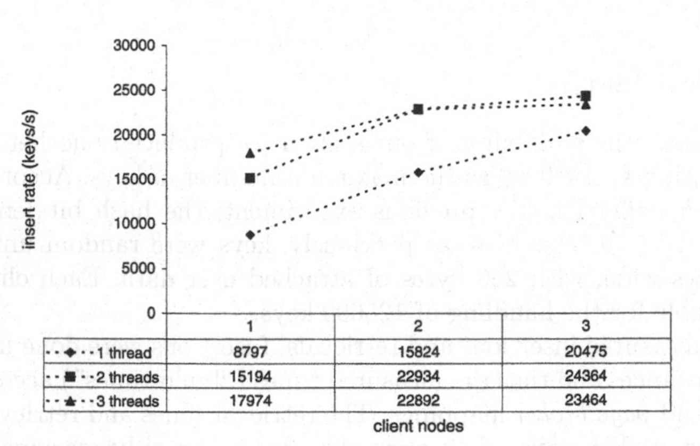


| | 1 | 2 | 3 |
|---|---|---|---|
| 1 thread | 8797 | 15824 | 20475 |
| 2 threads | 15194 | 22934 | 24364 |
| 3 threads | 17974 | 22892 | 23464 |

**Fig. 5.** Insert keyrate with one page server and multiple clients

**Two Page Servers, Multiple Clients** To answer the last question we added one more *page server* to the crew and repeated the tests. But, with just four nodes (the fifth hosted the *page broker* solely), we couldn't perform tests with more than 2 clients. Still, with a maximum of 3 threads per client, we were able to obtain results using a total of 6 threads.

Figure 6 sums up the test results by showing the improving on the insert rate when using one more *page server*. For 1 active client the gains are relatively modest. For 2 active clients the speedup is much more evident, specially when 3 threads per client are used, summing up 6 threads on overall.

The results presented allow us to conclude that by adding *page servers* to our system important performance gains may be obtained. However it remains to be done a quantitative study of the performance scaling in a cluster environment with much more nodes to assign both to clients and *page servers*.

**Multiple <Page Server, Client> Pairs** So far, we have decoupled clients and *page servers* on every scenario we have tested. It may happen, however, that both must share the same cluster node (as is the case for our small cluster). Thus, it is convenient to evaluate how the system scales in such circumstances.

As previously, the *page broker* was always kept at the node E and measurements were made with a different number of working threads in the client (1, 2 and 3). We started with a single node, hosting a client and a *page server*. We then increased the number of nodes, always pairing a client and a *page server*. The last scenario had four of these pairs, one per node, summing up to 12 active threads and accounting for a maximum of $12 \times 125000 = 1500000$ keys inserted.

Figure 7 shows the insert key rate. The 1–node scenario shows very low key rates with 2 and 3 threads. This is due to high I/O contention between the client threads and the *page server* threads. When the number of nodes is augmented, the key space, although larger, is also more scattered across the nodes, which

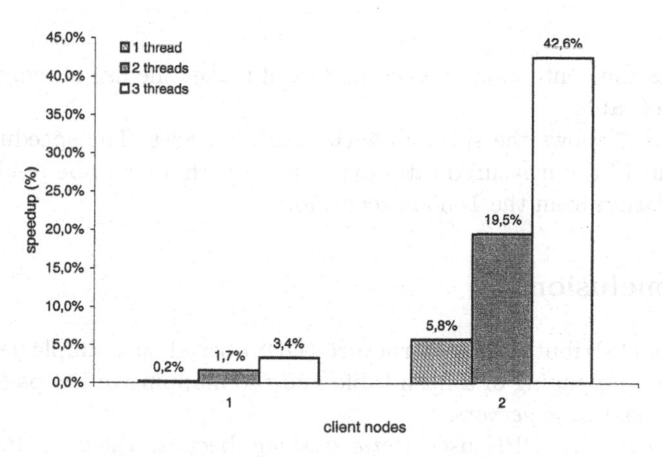


**Fig. 6.** Speedup with two *page servers*

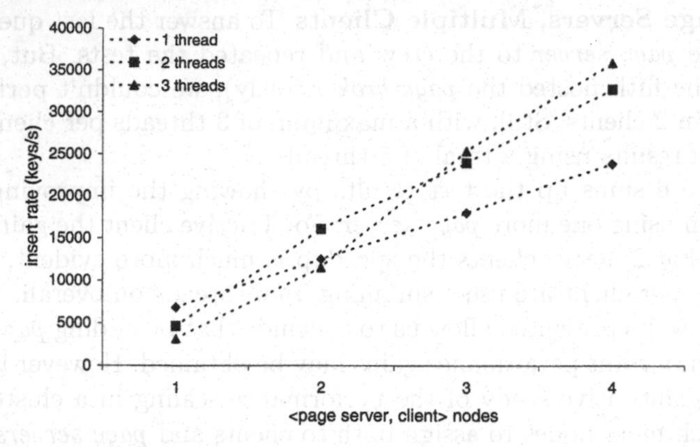


**Fig. 7.** Insert keyrate with multiple <page server, client> pairs

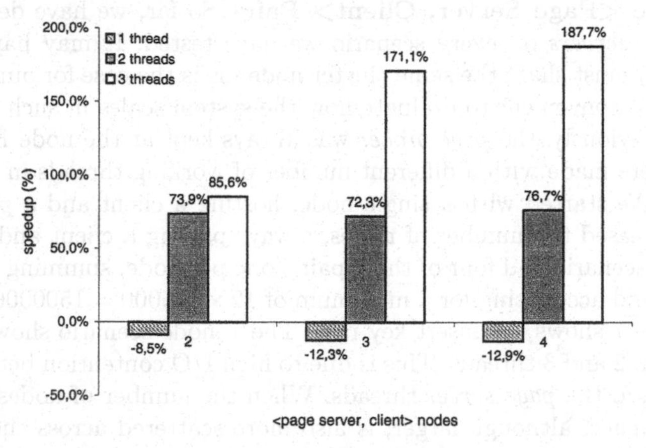


**Fig. 8.** Insert speedup with multiple <page server, client> pairs

alleviates the contention on each node and makes the use of more threads much more profitable.

Figure 8 shows the speedup with multiple nodes. The speedup refers to the increasing of the measured rates over the rates that could be predicted by linear extrapolation from the 1–node scenario.

## 5 Conclusions

DPH is a Distributed Data Structure (DDS) based on a simple yet very effective principle: the paging of a hash table and the mapping of the pages among a set of networked *page servers*.

Conceptually, DPH uses *static hashing*, because the hash index bit–length is set in advance. Also, the usage of a *page table* to preserve mappings between sections (pages) of the table and their locations (*page servers*) makes DPH a *directory* based [15] approach.

However, the hash table is not created at once, because it is virtually paged and pages are dynamically created, on–demand, being scattered across the cluster, thus achieving data balancing. Local caches at user applications prevent the *page broker* to become a *hot–spot* and provide some immunity to *page broker* failures (once established, mappings do not change and so the *page broker* can almost be dismissed).

Another important feature available in the DPH DDS is the capability to exploit the file system as a complementary on–line storage area, which is made possible through the use of a LRU/threshold based swapping mechanism. In this regard, DPH is very flexible in the way it consumes available storage resources, whether they are memory or disk based.

Finally, the performance evaluation we have presented shows that it is possible to define practical metrics to set the hash bit–length and that our selected hash function [23] preserves the (relative) rankings regardless of the key size. We have also investigated the scalability of our system and although we have observed promising results, further investigation is needed with many more nodes.

Much of the research work on hash based DDSs has been focused on dynamic hashing schemes. With this work we wanted to show that the increasing performance and storage capacity of modern clusters may also be exploited with great benefits using an hybrid approach.

In the future we plan to pursue our work in several directions: elimination of the *page broker* by using *directoryless* schemes, inspired by hash routing techniques, such as *consistent hashing* [27]; usage of efficient data structures to handle collisions and near zero–memory–copy techniques to improve performance; exploitation of cluster aware file systems (delayed due to the lack of choice on quality open–source implementations) and external memory techniques [12].

## References

[1] Al Geist, A. Beguelin, J. Dongarra, W. Jiang, R. Manchek, and V. Sunderam. *PVM: Parallel Virtual Machine. A User's Guide and Tutorial for Networked Parallel Computing.* Scientific and Engineering Computation. MIT Press, 1994.

[2] M. Snir, S. Otto, S. Huss-Lederman, David Walker, and J. Dongarra. *MPI - The Complete Reference.* Scientific and Engineering Computation. MIT Press, 1998.

[3] W. Litwin, M.-A. Neimat, and D. A. Schneider. LH*: Linear Hashing for Distributed Files. In *Proceedings of the ACM SIGMOD - International Conference on Management of Data*, pages 327–336, 1993.

[4] R. Devine. Design and implementation of DDH: a distributed dynamic hashing algorithm. In *Proceedings of the 4th Int. Conf. on Foundations of Data Organization and Algorithms*, pages 101–114, 1993.

[5] V. Hilford, F. B. Bastani, and B. Cukic. EH* – Extendible Hashing in a Distributed Environment. In *Proceedings of the COMPSAC '97 - 21st International Computer Software and Applications Conference*, 1997.

[6] R. Vingralek, Y. Breitbart, and G. Weikum. Distributed File Organization with Scalable Cost/Performance. In *Proceedings of the ACM SIGMOD - International Conference on Management of Data*, 1994.

[7] B. Kroll and P. Widmayer. Distributing a Search Tree Among a Growing Number of Processors. In *Proceedings of the ACM SIGMOD – International Conference on Management of Data*, pages 265–276, 1994.

[8] T. Johnson and A. Colbrook. A Distributed, Replicated, Data–Balanced Search Structure. Technical Report TR03-028, Dept. of CISE, University of Florida, 1995.

[9] S. D. Gribble, E. A. Brewer, J. M. Hellerstein, and D. Culler. Scalable, Distributed Data Structures for Internet Service Construction. In *Proceedings of the Fourth Symposium on Operating Systems Design and Implementation*, 2000.

[10] W. K. Preslan et all. A 64–bit, Shared Disk File System for Linux. In *Proceedings of the 7h NASA Goddard Conference on Mass Storage Systems and Tech. in cooperation with the Sixteenth IEEE Symposium on Mass Storage Systems*, 1999.

[11] P. H. Carns, W. B. Ligon, R. B. Ross, and R. Thakur. PVFS: A Parallel File System for Linux Clusters. In *Proceedings of the 4th Annual Linux Showcase and Conference*, pages 317–327. USENIX Association, 2000.

[12] J. S. Vitter. Online Data Structures in External Memory. In *Proceedings of the 26th Annual Intern. Colloquium on Automata, Languages, and Programming*, 1999.

[13] J. Rufino, A. Pina, A. Alves, and J. Exposto. Distributed Hash Tables. International Workshop on Performance-oriented Application Development for Distributed Architectures (PADDA 2001), 2001.

[14] D. E. Knuth. *The Art of Computer Programming – Volume 3: Sorting and Searching*. Addison-Wesley, 2nd edition, 1998.

[15] R. J. Enbody and H. C. Du. Dynamic Hashing Schemes. *ACM Computing Surveys*, (20):85–113, 1988.

[16] W. Litwin. Linear hashing: A new tool for file and table addressing. In *Proceedings of the 6th Conference on Very Large Databases*, pages 212–223, 1980.

[17] R. Fagin, J. Nievergelt, N. Pippenger, and H. R. Strong. Extendible hashing: a fast access method for dynamic files. *ACM Transactions on Database Systems*, (315-344), 1979.

[18] T. Stornetta and F. Brewer. Implementation of an Efficient Parallel BDD Package. In *Proceedings of the 33rd ACM/IEEE Design Automation Conference*, 1996.

[19] P. Bagwell. Ideal Hash Trees. Technical report, Computer Science Department, Ecole Polytechnique Federale de Lausanne, 2000.

[20] A. Pina, V. Oliveira, C. Moreira, and A. Alves. pCoR - a Prototype for Resource Oriented Computing. (to appear in HPC 2002), 2002.

[21] A. Pina. *$MC^2$ - Modelo de Computação Celular. Origem e Evolução*. PhD thesis, Dep. de Informática, Univ. do Minho, Braga, Portugal, 1997.

[22] Myricom. *The GM Message Passing System*, 2000.

[23] B. Jenkins. A Hash Function for Hash Table Lookup. *Dr. Doob's*, 1997.

[24] A. V. Aho, R. Sethi, and J. D. Ullman. *Compilers: Principles, Techniques and Tools*. Addison–Wesley, 1985.

[25] R. C. Uzgalis. General Hash Functions. Technical Report TR 91-01, University of Hong Kong, 1991.

[26] W. Pugh. Skip Lists: A Probabilistic Alternative to Balanced Trees. *Communications of the ACM*, 33(6):668–676, 1990.

[27] D. Kargeer, A. Sherman, A. Berkheimer, B. Bogstad, R. Dhanidina, K. Iwamoto, B. Kim, L. Matkins, and Y. Yerushalmi. Web Caching with Consistent Hashing. In *Proceedings of the 8th International WWW Conference*, 1999.

# A Meta-heuristic Approach to Parallel Code Generation

Barry McCollum, Pat H. Corr, and Peter Milligan

School of Computer Science, The Queen's University of Belfast
Belfast BT7 1NN, N. IRELAND
(b.mccollum,p.corr)@qub.ac.uk

**Abstract.** The efficient generation of parallel code for multi-processor environments, is a large and complicated issue. Attempts to address this problem have always resulted in significant input from users. Because of constraints on user knowledge and time, the automation of the process is a promising and practically important research area. In recent years heuristic approaches have been used to capture available knowledge and make it available for the parallelisation process. Here, the introduction of a novel approach of neural network techniques is combined with an expert system technique to enhance the availability of knowledge to aid in the automatic generation of parallel code.

## 1 Introduction

One of the principal limiting factor to the development of a fully automatic parallelisation environment is the automatic distribution of data to the available processors. Various aspects of this problem have proven to be intractable [1]. Recently heuristic techniques have re-invigorated research in this area. Indeed, this work complements and extends work presented at Vecpar 2000 on an environment for semi-automatic code transformation [2]. Systems such as expert systems, simulated annealing and genetic algorithms are currently being applied to the problem of data partitioning [3, 4]. These approaches could be deemed to be incomplete for the following reasons:

- They make generalising assumptions or tend to be biased towards particular problems [4].
- Most rely on intensive analysis of the possible distribution solutions which can be time consuming for anything but quite small programs. A lot of time can be spent tuning the parallel code to become as efficient as possible [4].
- They are machine specific and are not portable across a number of architectures [5].
- They rely on limited pattern matching thus not exploiting to the full the fact that similar structures exist within different sets of sequential code. Although some may use expert systems or knowledge based systems to advise on distribution strategies, this type of knowledge is not easily represented in such systems [6].

- In many cases the search space is pruned to make it sufficiently small to make the search for the solution fast [6].

However, while many of these heuristic approaches have been useful but limited, it is our contention that a combination of such approaches grounded in a consistent framework can offer a significant improvement.

Neural networks offer a method of extracting knowledge implicit in the code itself. This provides an alternative low-level, signal-based, view of the parallelisation problem in contrast to the high-level, symbolic view offered by an expert system. Combining both paradigms within a coherent knowledge hierarchy should improve the strategic intelligent guidance offered to users through access to a broader and deeper knowledge model. This knowledge model has been reported elsewhere [7].

This paper reports on this meta-heuristic approach for the purpose of data partitioning within a suite of existing knowledge based tools for automatic parallelisation, the KATT environment [3]. The architecture used consists a cluster of five Pentium processor workstations (COWs) with a UNIX operating system connected via a 10Mb bandwidth LAN switch in a star topology. Communication templates have been developed using C with embedded MPI statements to provide the necessary inter-processor communication. A SPMD programming model is used.

## 2 Existing Environments

In all areas, except an automatic approach, parallelisation is guided by data parallelism and based upon a user-provided specification of data distribution. Alternatively, the automatic approach invariably casts the problem as one of optimising processor usage while minimising communication. As an optimisation problem a range of techniques such as genetic algorithms, simulated annealing and formal mathematical notation have been brought to bare [4,5,7,8]. As yet, few of these approaches have been used within an overall parallelisation environment.

Research reported here is grounded within the framework of the on-going Knowledge Assisted Transformation Tools (KATT) project [3]. KATT provides an integrated software development and migration environment for sequential programmers. Central to the ethos of KATT is the belief that the user should have control over the extent of their involvement in the parallelisation process. For example, it should be possible for novice users to be freed from the responsibility of detecting parallelisable sections of code and distributing the code over the available processors. Alternatively, experienced users should have the facility to interact with all phases of the parallelisation and distribution of code. To enable a novice user to devolve responsibility entirely to the system implies that the system must have sufficient expert knowledge to accomplish the task. With the current KATT environment an expert system chooses either a column, row or block distribution depending on the assignment statements within the section of code under scrutiny. Cyclic and Block-Cyclic partitioning are not presently covered due to the complication of load balancing issues. Captured knowledge is currently held in two forms; explicitly as the facts and rules used by the existing expert system; and implicitly as hard-coded information about the target architecture for example, or inferred from within the code itself. The current expert system is limited due to its poor pattern matching capabilities and inability to generalise from the specific rules within its knowledge base.

The approach taken here enhances the existing expert system approach by the introduction of a second heuristic technique namely, neural networks.

## 3 Meta-heuristic Approach

An overview of the hybrid system is shown in figure 1. Here the neural network is responsible for suggesting an appropriate data partition given a set of characteristics derived from the code under consideration. The strategy expert system takes the suggested partition information together with the architectural information to determine a suitable data distribution strategy. This in turn is used to generate the actual parallel code for execution on the target architecture.

Distributions on the chosen architecture have been coded using MPI. These are, column block, column block cyclic, row block, row cyclic block and block block. Using Fortran with embedded MPI statements, code fragments have been distributed on the cluster of workstations (COWs) ensuring that the results before and after distribution are the same. The intention is to ascertain, through the experimentation process, how individual and particular combinations of characteristics are related to the choice of a particular data partition.

## 4 Data Partitioning Strategy

As data dependencies are a serious impediment to parallelism [9, 10], it is essential to remove and preferable eliminate them during the parallelisation process. Data dependency reduction can be achieved either by program transformations [10] or by internalisation [11]. program transformations allow the exploitation of low level parallelism and increase memory locality. These are cumbersome to do by hand but may have an important influence on overall performance. This method for data dependency reduction is currently used within the overall KATT project [3].

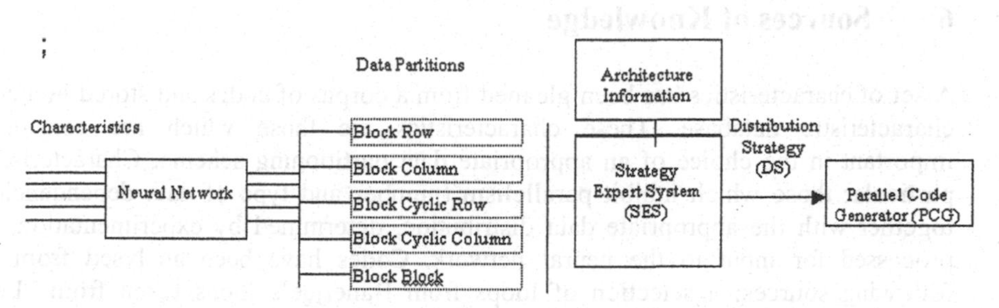


**Fig. 1.** A diagrammatic overview of the neural network, expert system based meta-heuristic environment for parallel code generation

Another effective way to reduce data dependencies is to internalise the data dependencies within each code partition so that all values required by computation local to a processor are available in its local domain. If this is not possible, the compiler must insert the appropriate communication statements to access non-local data. Due to the need to distribute data operated on by loops which have not necessarily gone through one of the established transformations [10, 12], the internalisation method is used and is the overriding factor when choosing how code should be partitioned and distributed.

## 5 The Neural Network Approach

Neural networks have achieved notable success in other areas where heuristic solutions are sought [14, 15]. In this work we intend to take advantage of their ability to generalise and extract patterns from a corpus of codes, one of the failings identified in alternative techniques. Two distinct approaches are currently being used:

- *an iterative data partitioning selection technique* which uses a multilayer perceptron model to recommend a particular partitioning, selected from a restricted set, to apply to an input loop structure. Training the neural network requires a representative selection of loops, each of which must be characterised and analysed to determine the appropriate transformation. This process has been carried out matching loop characteristics to the data partitioning which gives maximum speed up in loop execution.
- *a clustering technique* which uses a previously constructed feature map to determine an appropriate data partitioning for the code. The clusters formed are investigated to determine which data partitioning is applicable. The feature map may then be labelled such that a partitioning may be associated with an input.

The key issues are identifying sources of knowledge, deciding on the level of complexity to be dealt with, establishing a suitable characterisation scheme and acquiring training examples from a significant corpus of codes.

## 6 Sources of Knowledge

A set of characteristics has been gleaned from a corpus of codes and stored in a code characteristic database. These characteristics are those which are considered important in the choice of an appropriate data partitioning scheme. Characteristics, particular those which inhibit parallelism (number and type of data dependencies), together with the appropriate data distribution, determined by experimentation, are processed for input to the neural network. Codes have been analysed from the following sources: a selection of loops from Banerjee's loops taken from "Loop Transformations for Restructuring Compilers" [12] and Dongarra's parallel loops [13].

## 7 Definition and Coverage of Problem Space

To ensure that the completed neural network tool can deal appropriately with unseen code of arbitrary complexity it is essential that the training cases reflect the complexities found in real codes. These complexities are modeled by defining a problem space occupied by examples of real code. We define a number of dimensions, as shown in figure 2, which delineate this problem space, namely; number of dependencies, type of dependencies, degree of nesting, computational shape, presence of conditional statements, subroutine calls and symbolic bounds. Codes within this space may be labelled with a complexity level.

*Level 1*

This is the simplest level and includes that body of codes where it is clear parallelism is not appropriate, as communication will always dominate computation. Nevertheless, a number of examples are needed for completeness. The characteristics of this level are the following:

- Loops are perfectly nested i.e. only inner loop contains statements
- There are a maximum of two statements in the loop body
- Loops are normalised i.e. stride =1
- All loop bounds are known
- Arrays within loops may have a maximum of two dimensions.

*Level 2*

This level is defined by the introduction of data dependencies and an increased number of statements within loop bounds.

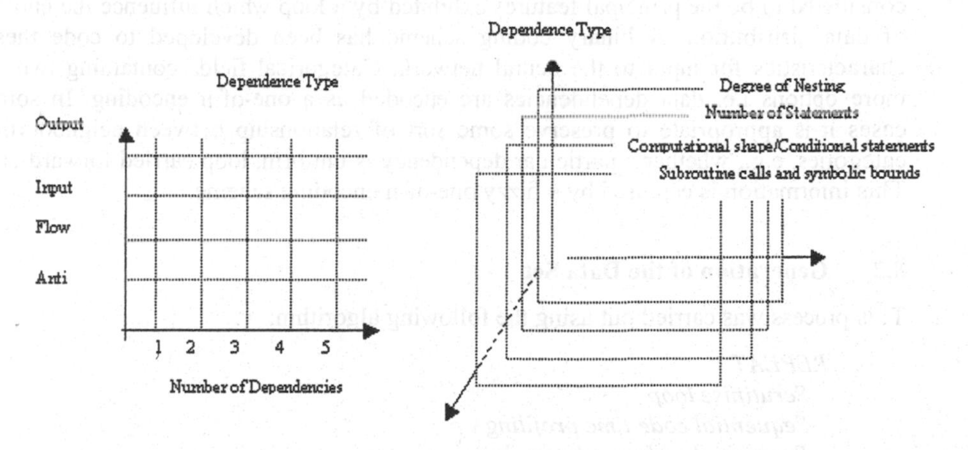


**Fig. 2.** Diagrammatic representation of the problem space

*Level 3*

This level sees the introduction of higher dimensions, statements between loops, conditional statements and varying shapes of computation space.

*Level 4*

At this level those factors that typically prevent data dependence analysis and data distribution are introduced. These include; linear expression being too complex, symbolic terms in subscript expressions and non-constant direction vectors.

Each piece of code studied may be mapped to a number of points in the defined problem space. To ensure significant coverage four hundred and fifty cases have been identified from the loop corpus and labelled with the appropriate level. These cases cover a number of discrete points within the problem space. However, once trained the neural network represents a continuous model of the problem space. As a result, any unseen input code, provided that it lies within the defined problem space, may be mapped to an appropriate data partition.

# 8 Training of Network

Example codes have been identified at levels 1, 2 and 3 within the loop corpus. These have been characterised and, by a process of experimentation, the partitioning strategy and most efficient data distribution identified for the target architecture.

## 8.1 Code Characterisation and Coding

The characterisation scheme used is shown in detail in Appendix A. These are considered to be the principal features exhibited by a loop which influence the choice of data distribution. A binary coding scheme has been developed to code these characteristics for input to the neural network. Categorical fields containing two or more options i.e. data dependencies are encoded as a one-of-n encoding. In some cases it is appropriate to preserve some sort of relationship between neighbouring categories, e.g., whether a particular dependency is uniform, loop carried forward etc. This information is captured by a fuzzy one-of-n encoding scheme.

## 8.2 Generation of the Data Set

This process was carried out using the following algorithm:

*REPEAT*
*Scrutinise loop*
*Sequential code time profiling*
*Sequential code results analysis*
*Characteristics manually extracted for use in production of parallel code*
*REPEAT*
*Chose Partitioning of Data*
*Production of Parallel Code*

*Parallel code results analysis*
*Parallel code time profiling*
*Store time*
*Store DPP if time is less than previous stored time*
*UNTIL All partitions have been used*
*Associate partition with code characteristics*
*UNTIL All loops analysed.*

After identification of an appropriate data partition the process of distribution requires access to further knowledge, e.g.,

- the number of processors available and the number to be used.
- distribution of loop bounds over the chosen processors.

This knowledge is provided by the expert system, which together with the neural model, draws from and exploits all available knowledge in advising on a suitable data distribution. Adding a communications harness using MPI then produces actual parallel code.

To prove the neural component we have, at this stage, fixed the architecture. All timings are taken on a four-processor workstation cluster. Once this process has been completed each code has been characterised and an appropriate partition identified. This data set is then used to train a multilayer perceptron capable of taking the characterisation of an unseen code and producing the appropriate data partition.

# 9 An Illustrative Example

To illustrate the process, consider the following code fragment:

```
DO 30, I = 1, rows
     DO 40, J = 1, cols
          A (I, J) = B (I +3, J)
          B (I, J) = A (I, J)
          M = I*J
          B (I, J) = B(I, J) *A(I, J)
     CONTINUE
CONTINUE
```

The relevant characteristics for partitioning are:

- 2 arrays
- Input dependence
- 2 Loop carried forward Anti dependence
- 2 Loop independent flow dependence
- Dependencies carried by outer loop

These characteristics are coded for input to the network which indicates that partitioning should be by column. Table 1 provides relative timings for both sequential and parallel partitioning for an array size of 800 by 800 (variables rows and cols in the code fragment above).

**Table 1.** Relative timings for various partitionings of an 800 by 800 array

| Sequential Time | 31.394252 |
|---|---|
| Column = 1 | 20.670667 |
| Column = 10 | 20.283152 |
| Column = 25 | 24.287525 |
| Column = 50 | 30.393759 |
| Column = 100 | 45.762204 |
| Column = 200 | 56.749110 |

The expert system, drawing on these facts together with the partitioning suggested by the neural network, indicates a particular distribution i.e. a column based distribution cyclically on four processors with a block size of 10.

## 10 Further Work

The system outlined above employs a multilayer perceptron trained using a supervised learning approach where both the input (the characterisations) and the output (the associated partition) must be known. This requirement represents a limitation in the system in that, in order to gather the training data, the target architecture must be fixed. The trained network then is applicable only to this fixed architecture. Work is ongoing on an architecture independent approach using an unsupervised mapping network - the Kohonen self-organising network. The Kohonen network will cause the input characteristics to cluster. These clusters may be labelled with a suitable data partition. This information will supplement the architectural details characterised within the expert system.

## 11 Conclusion

A meta-heuristic approach to data distribution has been developed and implemented. The approach involves the definition and modelling of the problem space, characterisation of the knowledge available and exploitation by appropriate paradigms, namely neural networks and expert systems. Results to date have proven the technique but are limited to a fixed architecture due to the nature of the neural model used. Work is progressing on an alternative approach using Kohonen networks which will result in an architectural independent, automatic data distribution technique.

## References


1. Ayguade, E., Garcia, J., Kremer, U., "Tools and techniques for automatic data layout", Parallel Computing 24 (1998) 557-578
2. P.H. Corr, P. Milligan and V. Purnell. "A Neural Network Based Tool for Semi-automatic Code Transformation." In VECPAR'2000. Selected Papers and Invited Talks from the 4th International Conference on Vector and Parallel Processing, Springer, Lecture Notes in Computer Science 1981, 2001, pp 142-153, ISBN 3-540-41999-3
3. P. J. P. McMullan, P. Milligan, P. P. Sage and P. H. Corr. A Knowledge Based Approach to the Parallelisation, Generation and Evaluation of Code for Execution on Parallel Architectures. IEEE Computer Society Press, ISBN 0-8186-7703-1, pp 58 - 63, 1997
4. Mansour, N., Fox, G., "Allocating data to distributed memory multiprocessors by genetic algorithms", Concurrency: Practice and Experience, Vol. 6(6), 485-504(September 1994)
5. Shenoy, U., Spikant, Y., Bhatkar, V., Kohli S., "Automatic Data Partitioning by Hierarchical Genetic Search", Parallel Algorithms and Applications, Vol. 14, pp 119-147.
6. Chrisochoides, N., Mansour, N., Fox, G., "A Comparison of optimisation heuristics for the data mapping problem", Concurrency: Practice and Experience, Vol. 9(5), 319-343 (May 1997)
7. McCollum B.G.C., Milligan P. and Corr P.H., "The Structure and Exploitation of Available Knowledge for Enhanced Data Distribution in a Parallel Environment", Software and Hardware Engineering for the 21st Century, Ed. N. E. Mastorakis, World Scientific and Engineering Society Press, 1999, pp139-145, ISBN 960-8052-06-8
8. K. Kennedy and U.Kremer," Automatic Data Alignment and Distribution for Loosely Synchronous Problems in an Interactive Programming Environment" Technical Report COMP TR91-155, Rice University, April 1991.
9. K. Knobe, J. Lukas and G. Steele, "Data Optimisation: Allocation of arrays to reduce communication on SIMD Machines", Journal of Parallel and Distributed Computing 8, 102-118 (1990)
10. Z. Shen, Z. Li and P. C. Yew, An Empirical Study of Fortran Programs for Parallelising Compilers", Technical Report 983, Centre for Supercomputing research and Development.
11. A. Dierstein, R. Hayer and T. Rauber, "Automatic Data Distribution and Parallelization" Paralel Programming 1995
12. U.Banerjee "Loop Transformations for Restructuring Compilers", Macmillan College Publishing Company, 1992
13. J. Dongara, "Atest Suite for Parallelising Compilers: Description and Example Results", Parallel Computing, 17, pp 1247-1255, 1991.
14. S Haykin "Neural Networks A Comprehensive Foundation" Macmillan College Publishing Company, Inc. 1994.
15. T. Kohonen, " The Self-Organising Map", Procedures of the IEEE, vol.78, pp 1464-1480, 1990

## Appendix A

The characterisation scheme used.

*N^o* *Characteristic*

1) Code section identification
2) Number of loops in code section
3) Number of statements before first Loop
4) Loop Identification
   a) Loop Lower Bound
   b) Loop Upper Bound
   c) Stride of loop
   d) Level of Nesting
   e) Degree of Nesting
   f) Number of statements In Loop Body
   g) Type of Loop Transformation underwent
   h) Sequential timing of loop
5) Number of Flow Dependence
6) Number of Anti Dependence
7) Number of Output Dependence
8) Number of Input Dependence
9) Number of Dependencies
   a) Data Dependence Type
   b) Within Single Statement
   c) Across Statements
   d) Loop Independent
   e) Loop carried forward i.e. Direction vector
   f) Loop carried backwards i.e. Direction vector
   g) Dependence Distance
   h) If Dependence is Uniform
   i) Which loop carries dependence
10) Sequential timing of code section

# Semidefinite Programming for Graph Partitioning with Preferences in Data Distribution

Suely Oliveira[1], David Stewart[2], and Takako Soma[1]

[1] The Department of Computer Science, The University of Iowa
Iowa City, IA 52242, USA
{oliveira,tsoma}@cs.uiowa.edu
[2] The Department of Mathematics, The University of Iowa
Iowa City, IA 52242, USA
dstewart@math.uiowa.edu

**Abstract.** Graph partitioning with preferences is one of the data distribution models for parallel computer, where partitioning and mapping are generated together. It improves the overall throughput of message traffic by having communication restricted to processors which are near each other, whenever possible. This model is obtained by associating to each vertex a value which reflects its net preference for being in one partition or another of the recursive bisection process. We have formulated a semidefinite programming relaxation for graph partitioning with preferences and implemented efficient subspace algorithm for this model. We numerically compared our new algorithm with a standard semidefinite programming algorithm and show that our subspace algorithm performs better.

## 1 The Graph Partitioning Problem and Parallel Data Distribution

Graph partitioning is universally employed in the parallelization of calculations on unstructured grids, such as finite element and finite difference calculations, whether using explicit or implicit methods. Once a graph model of a computation is constructed, graph partitioning can be used to determine how to divide the work and data for efficient parallel computation. The goal of the graph partitioning problem is to divide a graph into disjoint subgraphs subject to the constraint that each subgraph has roughly equal number of vertices, and with the objective of minimizing the number of edges that are cut by the partitionings. In many calculations the underlying computational structure can be conveniently modeled as a graph in which vertices correspond to computational tasks and edges reflect data dependencies. The objectives here are to evenly distribute the computations among the processors while minimizing interprocessor communication, so that the corresponding assignment of tasks to processors leads to efficient execution. Therefore, we wish to divide the graph into subgraphs with roughly equal

J.M.L.M. Palma et al. (Eds.): VECPAR 2002, LNCS 2565, pp. 703–716, 2003.

numbers of nodes with the minimum number of edges crossing between the subgraphs. Graph partitioning is an NP-hard problem [7]. Therefore, heuristics need to be used to get approximate solutions for these problems.

The graph partitioning problem for high performance scientific computing has been studied extensively over the past decades. The standard approach is to use Recursive Bisection [9, 21]. In Recursive Bisection, the graph is broken in half, the halves are halved independently, and so on, until there are as many pieces as desired.

The justification for traditional partitioning is that the number of edges cut in a partition typically corresponds to the volume of communication in the parallel application. Since communication is an expensive operation, minimizing this volume is extremely important in achieving high performance. In traditional recursive partitioning, after each step in a recursive decomposition the subgraphs are decoupled and interact no further. An edge crossing between two sets does not affect the later partitioning of either set. Consequently, there is nothing preventing the two adjacent vertices from being assigned to processors that are quite far from each other. A message between distant processors must traverse many wires, which are therefore rendered unavailable to transmit other messages. Conversely, if each message consumes only a small number of wires, more messages can be sent at once. In parallel computing, messages traveling between architecturally distant processors should be minimized by improving the data locality, since they tie up many communication links. Therefore, a good mapping is one that reduces message congestion and thereby preserves communication bandwidth. Many scientific computing applications of interest, for example those employing an iterative sparse solver kernel, have a structure in which many messages simultaneously compete for limited communication bandwidth. Good mappings are especially important in these cases.

Recently, Hendrickson et al. [8, 9, 11] pointed out problems with traditional models. Assume we have already partitioned the graph into left and right halves, and that we have similarly divided the left-half graph into top and bottom quadrants (see Figure 1). When partitioning the right-half graph between processors 3 and 4, we want the messages to travel short distances. The mapping shown in the left-hand of Figure 1 is better since the total message distance is less than that for the right-hand figure.

These models are obtained by associating to each vertex a value which reflects its net preference for being in one subgraph or another. Note that this preference is a function only of edges that connect the vertex to vertices which are not in the current subgraph. These preferences should be propagated through the recursive partitioning process. If the graph partitioning problem with preferences is relaxed as is done to obtain spectral graph partitioning, we obtain an extended eigenproblem: Find the minimum $\mu$ for which there is a $y \neq 0$ satisfying $Ay = \mu y + g$ with a specified norm [11].

In [18] we developed subspace methods to solve extended eigenproblems. In [19] we have developed a subspace algorithm for a SDP of the original graph partitioning. In this paper we will develop a semidefinite program for graph

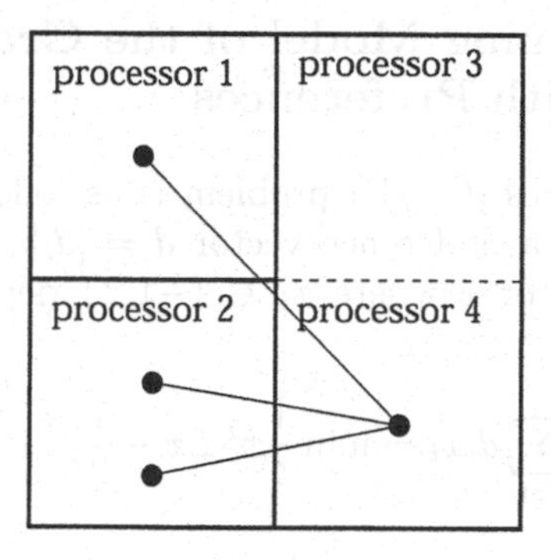


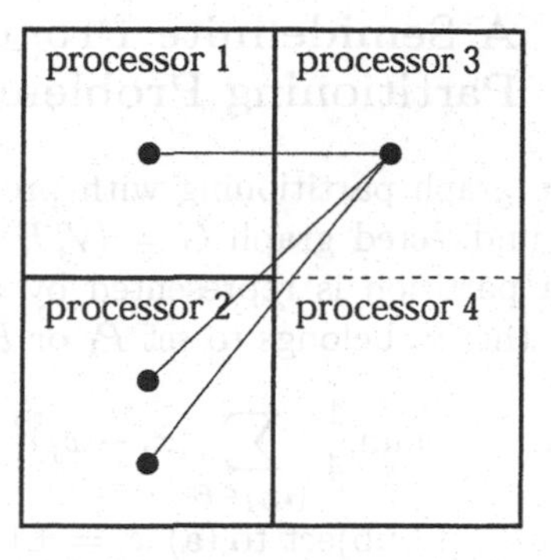


**Fig. 1.** Partition between the 4 processors

partitioning *with preferences* and present an efficient subspace algorithm for this model.

## 2 Spectral Graph Patitioning and Semidefinite Programming

Like combinatorial methods, spectral methods have proven to be effective for large graphs arising from FEM discretizations [21]. Software packages that use spectral methods combined with combinatorial algorithms include METIS [15], Chaco [10], JOSTLE [31], PARTY [22], and SCOTCH [20]. The spectral algorithms which are used in these packages are based on models that partition a graph by finding an eigenvector associated with the second smallest eigenvalue of its graph Laplacian matrix using an iterative method. This eigenvector model is used in [12, 13] for example.

Another approach to get better approximation of the graph partitioning problem is semidefinite programming. It gives tighter relaxations to the original graph partitioning problem than does spectral graph partitioning, leading to better partitionings.

The semidefinite programming problem [1, 24, 29] is the problem of minimizing a linear function over symmetric matrices with linear constraints and the constraint that the matrix is symmetric positive semi-definite. Semidefinite programming reduces to linear programming when all the matrices are diagonal. Currently there are various software packages for solving semidefinite programs available using interior-point algorithms, such as CSDP [4], SDPA [6], SDPpack [2], SDPT3 [28], SP [30], and a Matlab toolbox by Rendl [23]. Semidefinite programming relaxation technique for equal-partitioning problem has been developed in [14]. In the next section we develop an SDP relaxation for the GP problem with preferences. In Section 4 we present an efficient algorithm for the new model of Section 3

## 3 A Semidefinite Programming Model of the Graph Partitioning Problems with Preferences

The graph partitioning with preferences (GP/P) problem is as follows: Given an undirected graph $G = (V, E)$ and a preference vector $d = [d_i | i \in V]$, and if a partition is represented by a vector $x$ where $x_i \in \{-1, 1\}$ depending on whether $x_i$ belongs to set $P_1$ or $P_2$,

$$\min_x \frac{1}{4} \sum_{\{i,j\} \in E} (x_i - x_j)^2 - \frac{1}{2} \sum_{i \in V} d_i x_i = \min_x \frac{1}{4} x^T L x - \frac{1}{2} d^T x$$
$$\text{Subject to (a) } x_i = \pm 1, \ \forall i$$
$$\text{(b) } \sum_{i \in V} x_i \approx 0,$$

where $L$ is the graph Laplacian matrix. A straight forward way of approximating this problem is by using the following relaxation:

$$\min_x \frac{1}{4} x^T L x - \frac{1}{2} d^T x$$
$$\text{Subject to (a) } x^T x = n$$
$$\text{(b) } e^T x = 0,$$

where $e$ is the vector of all ones.

The above relaxation replaces constraint (a) $x_i = \pm 1, \ \forall i$ with $\sum_i x_i^2 = x^T x = n$, where $n = |V|$. A subspace algorithms for solving this extended eigenvalue relaxation was developed in [18]. In this paper we will develop a new semidefinite programming relaxation for the GP/P problem. This development is based on the following theorem which shows that the GP/P problem above can be rewritten as a semi-definite programming with a rank one constraint. An SDP relaxation can then be obtained by dropping the rank one constraint. In order to efficiently solve the SDP, we use a subspace approach.

**Theorem 1.** *The GP/P*

$$\min_x \frac{1}{4} x^T L x - \frac{1}{2} d^T x$$
$$\textit{Subject to (a) } x_i = \pm 1, \ \forall i \qquad (1)$$
$$\textit{(b) } \sum_{i \in V} x_i = 0$$

*is equivalent to*

$$\min_{\tilde{X}} \tilde{X} \bullet \frac{1}{4} \begin{pmatrix} 0 & -d^T \\ -d^T & L \end{pmatrix}$$
$$\textit{Subject to (a) } \tilde{X} \succeq 0$$
$$\textit{(b) } \xi = 1 \qquad (2)$$
$$\textit{(c) } \operatorname{diag}(X) = e$$
$$\textit{(d) } X \bullet (ee^T) = 0$$
$$\textit{(e) } \operatorname{rank}(X) = 1,$$

*where* $\tilde{X} = \begin{pmatrix} \xi & u^T \\ u & X \end{pmatrix}$, *and the solution to (2) is* $\tilde{X} = \begin{pmatrix} 1 & x^T \\ x & xx^T \end{pmatrix}$, *where* $x$ *is the solution of (1).*

**Proof:**

**(i)** If $x$ solves (1), then for $\tilde{X} = \begin{pmatrix} 1 & x^T \\ x & xx^T \end{pmatrix}$, the constraints are: $\tilde{X} \succeq 0$, $\text{diag}(xx^T) = [x_i^2 | i \in V] = e$, and $X \bullet (ee^T) = \text{tr}(Xee^T) = \text{tr}(xx^Tee^T) = \text{tr}(e^Txx^Te) = (e^Tx)^2 = 0$. Since

$$\begin{aligned}
\tilde{X} \bullet \frac{1}{4}\begin{pmatrix} 0 & -d^T \\ -d^T & L \end{pmatrix} &= \begin{pmatrix} 1 & x^T \\ x & xx^T \end{pmatrix} \bullet \frac{1}{4}\begin{pmatrix} 0 & -d^T \\ -d & L \end{pmatrix} \\
&= \frac{1}{4}\text{tr}\begin{pmatrix} -x^Td & -d^T + x^TL \\ -xx^Td & xx^TL - xd^T \end{pmatrix} \\
&= \frac{1}{4}(\text{tr}(xx^TL) - 2d^Tx) \\
&= \frac{1}{4}L \bullet xx^T - \frac{1}{2}d^Tx \\
&= \frac{1}{4}x^TLx - \frac{1}{2}d^Tx,
\end{aligned}$$

we have $\tilde{X} \bullet \frac{1}{4}\begin{pmatrix} 0 & -d^T \\ -d^T & L \end{pmatrix} = \frac{1}{4}x^TLx - \frac{1}{2}d^Tx$, so the value of (2) is less than or equal to the value of (1).

**(ii)** On the other hand, suppose $\tilde{X} = \begin{pmatrix} \xi & u^T \\ u & X \end{pmatrix}$ solves (2). Since $\text{rank}(X) = 1$, and $X \succeq 0$, $X = zz^T$ for some $z$. Also $\xi = 1$, so

$$\tilde{X} = \begin{pmatrix} 1 & u^T \\ u & zz^T \end{pmatrix} \succeq 0.$$

By taking one step of Cholesky factorization, we see that $\tilde{X}$ is positive semidefinite if and only if

$$zz^T - uu^T \succeq 0.$$

If we write $u = \beta z + w$, where $w \perp z$, then,

$$\begin{aligned}
w^T(zz^T - uu^T)w &= w^T(zz^T - (\beta z + w)(\beta z + w)^T)w \\
&= w^T(zz^T - (\beta^2 zz^T + \beta zw^T + \beta wz^T + ww^T))w \\
&= w^T(zz^T - \beta^2 zz^T - \beta zw^T - \beta wz^T - ww^T)w \\
&= w^T(-ww^T)w \\
&= -(w^Tw)^2 \geq 0.
\end{aligned}$$

So, $w = 0$. Therefore, $u = \beta z$. Then,

$$\begin{aligned}
zz^T - uu^T &= zz^T - (\beta z)(\beta z)^T \\
&= zz^T - \beta^2 zz^T \\
&= (1 - \beta^2)zz^T \succeq 0
\end{aligned}$$

is equivalent to $|\beta| \leq 1$, provided $z \neq 0$. If $z = 0$, then $zz^T - uu^T \succeq 0$, implies $u = 0$. In either case, $u = \beta z$, where $|\beta| \leq 1$.

Since $\tilde{X}$ minimizes $\tilde{X} \bullet \frac{1}{4}\begin{pmatrix} 0 & -d^T \\ -d^T & L \end{pmatrix}$ over the constraints in (2), and the only constraint in (2) involving $u$ is $\tilde{X} \succeq 0$ (which we just showed is equivalent to $u = \beta z$, where $|\beta| \leq 1$). We can then rewrite the problem in terms of variable $z$ in the following way

$$\begin{aligned}
\min_u \tilde{X} \bullet \frac{1}{4}\begin{pmatrix} 0 & -d^T \\ -d & L \end{pmatrix} &= \min_u \begin{pmatrix} 1 & u^T \\ u & zz^T \end{pmatrix} \bullet \frac{1}{4}\begin{pmatrix} 0 & -d^T \\ -d & L \end{pmatrix} \\
&= \min_u \frac{1}{4}\mathrm{tr}\begin{pmatrix} -u^T d & -d^T + u^T L \\ -zz^T d & -ud^T + zz^T L \end{pmatrix} \\
&= \min_u \frac{1}{4}(\mathrm{tr}(zz^T L) - 2d^T u) \\
&= \min_u \frac{1}{4} L \bullet zz^T - \frac{1}{2} d^T u \\
&= \min_u \frac{1}{4} z^T L z - \frac{1}{2} d^T u
\end{aligned}$$

Subject to (a) $u = \beta z$
(b) $|\beta| \leq 1$.

This minimum value is

$$\min_{-1 \leq \beta \leq +1} \frac{1}{4} z^T L z - \frac{1}{2}\beta d^T z = \frac{1}{4} z^T L z - \frac{1}{2}|d^T z|.$$

We can assume without loss of generality that $d^T z \geq 0$, since if we change the sign of $z$ then $X = zz^T$ is unchanged. So, this minimum value over $u$ is

$$\frac{1}{4} z^T L z - \frac{1}{2} d^T z.$$

Furthermore, the optimal choice of $u$ is $u = z$. So the value of (2) is

$$\begin{aligned}
&\min_z \frac{1}{4} z^T L z - \frac{1}{2} d^T z \\
&\text{Subject to (a) } \mathrm{diag}(zz^T) = e \\
&\qquad\qquad\text{(b) } (zz^T) \bullet (ee^T) = 0,
\end{aligned}$$

which is,

$$\begin{aligned}
&\min_z \frac{1}{4} z^T L z - \frac{1}{2} d^T z \\
&\text{Subject to (a) } z_i^2 = 1, \ \forall i \\
&\qquad\qquad\text{(b) } e^T z = 0.
\end{aligned}$$

Therefore, the value of (2) is

$$\begin{aligned}
&\min_z \frac{1}{4} z^T L z - \frac{1}{2} d^T z \\
&\text{Subject to (a) } z_i = \pm 1, \ \forall i \\
&\qquad\qquad\text{(b) } \sum_{i \in V} z_i = 0.
\end{aligned}$$

So, the minimum value of (2) is the minimum value of (1), and the solutions are related by $X = xx^T$ and $u = x$. □

The semidefinite programming relaxation removes the constraint $\text{rank}(X) = 1$:

$$\begin{aligned} &\min_{\tilde{X}} \tilde{X} \bullet \frac{1}{4}\begin{pmatrix} 0 & -d^T \\ -d^T & L \end{pmatrix} \\ &\text{Subject to (a) } \tilde{X} \succeq 0 \\ &\qquad\qquad\text{(b) } \xi = 1 \\ &\qquad\qquad\text{(c) } \text{diag}(X) = e \\ &\qquad\qquad\text{(d) } X \bullet (ee^T) = 0, \end{aligned}$$

where $\tilde{X} = \begin{pmatrix} \xi & u^T \\ u & X \end{pmatrix}$. To obtain the resulting partition, use the $u$ vector from $\tilde{X}$ just as the Fiedler vector is used in spectral graph partitioning.

Constraints (b) and (c) can be combined together as $\text{diag}(\tilde{X}) = e$, where $e$ is now a vector of length $n + 1$. Therefore,

$$\begin{aligned} &\min_{\tilde{X}} \tilde{X} \bullet \frac{1}{4}\begin{pmatrix} 0 & -d^T \\ -d^T & L \end{pmatrix} \\ &\text{Subject to (a) } \tilde{X} \succeq 0 \\ &\qquad\qquad\text{(b) } \text{diag}(\tilde{X}) = 1 \\ &\qquad\qquad\text{(c) } X \bullet (ee^T) = 0, \end{aligned}$$

It is easily shown that for the particular case when the graph partitioning problem is assumed without preferences (i.e. $d = 0$), the above SDP relaxation corresponds to the SDP relaxation of graph partitioning in [19].

## 4 Solving the New SDP Relaxation Efficiently

While solving semidefinite programs is believed to be a polynomial time problem, current algorithms typically take $\Omega(n^3)$ time and $\Omega(n^2)$ space, making them impractical for realistic data distribution problems. This makes the development of efficient algorithms important.

Subspace methods build sequences of subspaces that are used to compute approximate solutions of the original problem. They have been previously used to solve systems of linear equations [16] and to compute eigenvalues and eigenvectors [3]. Of the methods for eigenvalues, the Lanczos and Arnoldi methods use Krylov subspaces (generated using a single matrix and a starting vector), while Davidson-type methods (Generalized Davidson [25], Jacobi–Davidson [26]) use Rayleigh matrices. Theoretical studies of the Krylov subspace methods are well-known in the numerical analysis community, but they have only recently been developed for Davidson-type methods [5, 17]. Davidson-type subspace methods have been applied to spectral graph partitioning [12, 13].

A subspace algorithm for the SDP relaxation of the spectral partitioning model was developed in [17]. Below we describe a subspace algorithm for the semidefinite programming relaxation described above, in order to efficiently find high-quality solutions to the graph partitioning problem with preferences.

$$\min_{\tilde{X}} \tilde{X} \bullet \frac{1}{4}\begin{pmatrix} 0 & -d^T \\ -d & L \end{pmatrix}$$
$$\begin{aligned} \text{Subject to } & \text{(a) } \tilde{X} \succeq 0 \\ & \text{(b) } \xi = 1 \\ & \text{(c) } \mathrm{diag}(X) = e \\ & \text{(d) } X \bullet (ee^T) = 0, \end{aligned}$$

where $\tilde{X} = \begin{pmatrix} \xi & u^T \\ u & X \end{pmatrix}$. We define the operator $\mathcal{A}$ by $\mathcal{A}(\tilde{X})_i = \mathcal{A}_i \bullet \tilde{X} = \mathrm{trace}(\mathcal{A}_i \tilde{X})$ for $i = 1, 2, \dots, m$ where each $\mathcal{A}_i$ is an $(n+1) \times (n+1)$ matrix: $\mathcal{A}_1 = \begin{pmatrix} 0 & 0 \\ 0 & ee^T \end{pmatrix}$, $\mathcal{A}_2 = \mathrm{diag}(1, 0, 0, \dots, 0)$, $\mathcal{A}_3 = \mathrm{diag}(0, 1, 0, \dots, 0)$, ..., $\mathcal{A}_m = \mathrm{diag}(0, 0, 0, \dots, 1)$, where $m = n + 2$. Now consider the vector $a = (a_1, a_2, a_3, \dots, a_{m+1})^T = (0, 1, 1, \dots, 1)^T$ Note that $\mathcal{A}_1 \bullet \tilde{X} = a_1$ ensures that (d) $X \bullet (ee^T) = 0$; $\mathcal{A}_2 \bullet \tilde{X} = a_2$ ensures that (b) $\xi = 1$; and $\mathcal{A}_3 \bullet \tilde{X} = a_3$, ..., $\mathcal{A}_m \bullet \tilde{X} = a_m$ ensures that (c) $\mathrm{diag}(X) = e$.

Notice that the subspace framework here is similar to that of [17], the input data for the subspace SDP being the main difference. Below we present the main steps of our subspace semidefinite programming for extended eigenproblems.

**Algorithm 1 – Subspace Semidefinite Programming for SDP Relaxation of Graph Partitioning with Preferences**

1. Define $(n+1) \times (n+1)$ matrix $C = \frac{1}{4}\begin{pmatrix} 0 & -d^T \\ -d & L \end{pmatrix}$, $m \times 1$ vector $a$, and an operator $\mathcal{A}$ which is a set of $m$ number of $(n+1) \times (n+1)$ matrices.
2. Compute an interior feasible starting point $y$ for the original problem.
3. Define $(n+1) \times 3$ vector $V_1$ and $m \times 3$ vector $W_1$. The first column of $V_1$ should be the $e_1$, the first standard basis vector, and the the first two columns of $W_1$ should be $e_1$ and $e_2$ respectively.
4. Compute the semidefinite programming solution $(\hat{X}, \hat{y})$ on the subspace. The projected matrix, operator and vector on the subspace are $\hat{C} = V_j^T C V_j$, $\hat{\mathcal{A}}(\hat{X}) = W_j^T \hat{\mathcal{A}}(\hat{X}) = W_j^T \begin{pmatrix} V_j^T A_1 V_j \bullet \hat{X} \\ \vdots \\ V_j^T A_m V_j \bullet \hat{X} \end{pmatrix}$ and $\hat{a} = W_j^T a$, where $V_j$ and $W_j$ are the current orthogonal bases.
5. Estimate the "residual" matrix $\tilde{R} = C - \mathcal{A}^T(y)$ and compute $r$ which is an eigenvector corresponding to the minimum eigenvalue of $\tilde{R}$.
6. Orthonormalize $r$ against the current orthogonal basis $V_j$. Append the orthonormalized vector to $V_j$ to give $V_{j+1}$.
7. Estimate residual $p = \mathcal{A}(X) - a = \mathcal{A}(V_j \hat{X} V_j^T) - a$.

8. Orthonormalize $p$ against the current orthogonal basis $W_j$. Append the orthonormalized vector to $W_j$ to give $W_{j+1}$.
9. Compute $\tilde{X} = V_j \hat{X} V_j^T$ of original size, where $\tilde{X} = \begin{pmatrix} \xi & x^T \\ x & X \end{pmatrix}$.

## 5 Numerical Results

Here we describe how to reduce the number of constraints in original problem. The semidefinite programming formulation of extended eigenproblem is written as follows:

$$\begin{aligned} \min_{\tilde{X}} \ & \frac{1}{4} \begin{pmatrix} 0 & -d^T \\ -d & L \end{pmatrix} \bullet \tilde{X} \\ \text{Subject to } & \text{(a) } \tilde{X} \succeq 0 \\ & \text{(b) } \xi = 1 \\ & \text{(c) } \operatorname{diag}(X) = e \\ & \text{(d) } X \bullet (ee^T) = 0, \end{aligned}$$

where $\tilde{X} = \begin{pmatrix} \xi & x^T \\ x & X \end{pmatrix}$ and $X = xx^T$. The constraints (b) and (c) can be combined together.

$$\begin{aligned} \min_{\tilde{X}} \ & \frac{1}{4} \begin{pmatrix} 0 & -d^T \\ -d & L \end{pmatrix} \bullet \tilde{X} \\ \text{Subject to } & \text{(a) } \tilde{X} \succeq 0 \\ & \text{(b) } \mathcal{A}(\tilde{X}) = a \\ & \text{(c) } X \bullet (ee^T) = 0. \end{aligned}$$

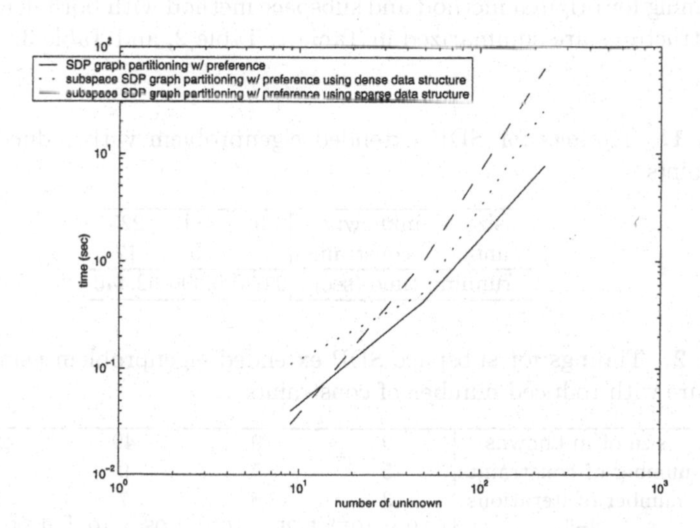


**Fig. 2.** Timings comparing subspace SDP extended eigenproblem against SDP extended eigenproblem

There are $n$ number of constrains from (c) in (3). We rewrite the problem so that we can reduce the number of constraints in the original problem. The operator $\mathcal{A}$ is defined in terms of $m$ diagonal matrices $A_i$ where $(A_i)_{jj}$ is one if $j \equiv i$ (mod m), and zero otherwise, $i = 1, \cdots, m$ and $m \leq n$:

$$\begin{aligned} &\min_{\tilde{X}} \frac{1}{4} \begin{pmatrix} 0 & -d^T \\ -d & L \end{pmatrix} \bullet \tilde{X} \\ &\text{Subject to (a) } \tilde{X} \succeq 0 \\ &\qquad\qquad\quad \text{(b) } \mathcal{A}(\tilde{X}) = (n/4m)/e \\ &\qquad\qquad\quad \text{(c) } X \bullet (ee^T) = 0, \end{aligned}$$

where $e$ is the vector of ones of the appropriate size.

The new algorithms have been implemented within the existing semidefinite software, CSDP [4] and matrix computation was utilized by Meschach [27]. The codes were run on a HP VISUALIZE Model C240 workstation, with a 326MHz PA-8200 processor and 512MB RAM.

The above subspace algorithm and CSDP were run with square matrices of various sizes. The subspace algorithm was implemented using both dense data structures and sparse data structures. Figure 2 compares the observed running timings for using CSDP directly, and the two implementations of the subspace algorithm. The vertical-axis shows the time in seconds. The horizontal-axis shows the number of unknowns: 9, 49, and 225. The number of constraints of the original problem for both methods was reduced to 5 for the matrix with 9 unknowns, 9 for the matrix with 49 unknowns, and 17 for the matrix with 225 unknowns, as described above. From this graph, we can see that the subspace algorithm takes less than the original algorithm as the number of unknowns increases. And implementation using sparse data structure perform much better. The timing for original method and subspace method with both dense and sparse data structures are summarized in Table 1, Table 2, and Table 3.

**Table 1.** Timings for SDP extended eigenproblem with reduced number of constraints

| size of unknowns | 9 | 49 | 225 |
|---|---|---|---|
| number of constraints | 5 | 9 | 17 |
| running time (sec) | 0.030 | 0.990 | 62.440 |

**Table 2.** Timings for subspace SDP extended eigenproblem using dense data structure with reduced number of constraints

| size of unknowns | 9 | 9 | 49 | 225 |
|---|---|---|---|---|
| number of constraints | 5 | 5 | 9 | 17 |
| number of iterations | 3 | 5 | 7 | 17 |
| $\|p\|_2$ | $1.19 \times 10^{-6}$ | $5.25 \times 10^{-7}$ | $1.08 \times 10^{-5}$ | $9.54 \times 10^{-5}$ |
| running time (sec) | 0.110 | 0.120 | 0.510 | 30.100 |

**Table 3.** Timings for subspace SDP extended eigenproblem using sparse data structure with reduced number of constraints

| size of unknowns | 9 | 9 | 49 | 225 |
|---|---|---|---|---|
| number of constraints | 5 | 5 | 9 | 17 |
| number of iterations | 3 | 5 | 7 | 15 |
| number of Arnoldi iterations | 9 | 9 | 30 | 30 |
| $\|p\|_2$ | $1.86 \times 10^{-6}$ | $6.24 \times 10^{-6}$ | $3.083 \times 10^{-5}$ | $4.10 \times 10^{-4}$ |
| running time (sec) | 0.040 | 0.130 | 0.420 | 7.690 |

The Table 4 shows the behavior of $\|p\|_2$ as the subspace was expanded in subspace SDP extended eigenproblem using dense data structure. Data was taken for the matrix with 9 unknowns and number of constraints is 5, the matrix with 49 unknowns and number of constraints is 9, and the matrix with 225 unknowns and number of constraints is 17. As you can see, all the cases converged at the end. The Table 5 is the result for subspace SDP extended eigenproblem using sparse data structure. The behavior of the $\|p\|_2$ is same as that for dense data structure.

**Table 4.** $\|p\|_2$ of subspace SDP extended eigenproblem using dense data structure with reduced number of constraints for matrices with number of unknown = 9, 49, and 225

| size of unknowns | 9 | 49 | 225 |
|---|---|---|---|
| number of constraints | 5 | 9 | 17 |
| number of iteration | | | |
| 1 | 68.7507477 | 12.4849939 | 15.341258 |
| 2 | 7.34615088 | 3.8352344 | 14.1316462 |
| 3 | 1.19470394e-06 | 4.51103926 | 1218.76562 |
| 4 | 5.49540289e-07 | 3.52121067 | 2978.45337 |
| 5 | 5.24935558e-07 | 4.89502668 | 16.0653 |
| 6 | | 0.531660855 | 3359.63013 |
| 7 | | 1.07650176e-05 | 17.5198021 |
| 8 | | 2.97533984e-06 | 5.35400295 |
| 9 | | 36342.1484 | 17.7656918 |
| 10 | | | 7.57762194 |
| 11 | | | 696.369324 |
| 12 | | | 5.29387665 |
| 13 | | | 737332.562 |
| 14 | | | 484.507904 |
| 15 | | | 16.7516613 |
| 16 | | | 0.000186197911 |
| 17 | | | 9.53856725e-05 |

**Table 5.** $\|p\|_2$ of subspace SDP extended eigenproblem using sparse data structure with reduced number of constraints for matrices with number of unknown = 9, 49, and 225

| size of unknowns | 9 | 49 | 225 |
|---|---|---|---|
| number of constraints | 5 | 9 | 17 |
| number of iteration | | | |
| 1 | 68.7507477 | 12.4849596 | 16.1067295 |
| 2 | 7.34615374 | 4.14506721 | 15.4141665 |
| 3 | 1.86033265e-06 | 5.64862299 | 16.7704239 |
| 4 | 7.83976702e-07 | 3.44969082 | 799.051331 |
| 5 | 6.24131007e-06 | 4.71650219 | 29.9862556 |
| 6 | | 1.86342728 | 7.08970928 |
| 7 | | 3.08254312e-05 | 11.917738 |
| 8 | | 33748.1367 | 844.419495 |
| 9 | | 36664.4961 | 14.5084686 |
| 10 | | | 447.563751 |
| 11 | | | 4.25445318 |
| 12 | | | 56.6501083 |
| 13 | | | 10.7289038 |
| 14 | | | 0.814486623 |
| 15 | | | 0.000409778644 |
| 16 | | | 11.1356792 |
| 17 | | | 201.696182 |

## References


[1] F. Alizadeh. Interior-point methods in semidefinite programming with applications to combinatorial optimization. *SIAM Journal on Optimization,* 5(1):13-51, 1995.

[2] F. Alizadeh, J. P. A. Haeberly, M. V. Nayakkankuppam, M. L. Overton, and S. Schmieta. SDPpack user's guide - version 0.9 beta for Matlab 5.0. Technical Report TR1997-737, Computer Science Department, New York University, New York, NY, June 1997.

[3] W. E. Arnoldi. The principle of minimized iteration in the solution of the matrix eigenproblem. *Quarterly Applied Mathematics,* 9:17-29, 1951.

[4] B. Borchers. CSDP, 2.3 User's Guide. *Optimization Methods and Software,* 11(1):597-611, 1999.

[5] M. Crouzeix, B. Philippe, and M. Sadkane. The Davidson method. *SIAM J. Sci. Comput.,* 15(1):62-76, 1994.

[6] K. Fujisawa, M. Kojima, and K. Nakata. SDPA User's Manual - Version 4.50. Technical Report B, Department of Mathematical and Computing Science, Tokyo Institute of Technology, Tokyo, Japan, July 1999.

[7] M. R. Carey, D. S. Johnson, and L. Stockmeyer. Some simplified NP-complete problems. *Theoretical Computer Science,* 1:237-267, 1976.

[8] B. Hendrickson. Graph partitioning and parallel solvers: Has the emperor no clothes? (extended abstract). In *Lecture Notes in Computer Science,* volume 1457, 1998.

[9] B. Hendrickson and T. G. Kolda. Graph partitioning models for parallel computing. *Parallel Comput.*, 26(12):1519-1534, 2000.

[10] B. Hendrickson and R. Leland. The Chaco user's guide, version 2.0. Technical Report SAND-95-2344, Sandia National Laboratories, Albuquerque, NM, July 1995.

[11] B. Hendrickson, R. Leland, and R. Van Driessche. Enhancing data locality by using terminal propagation. In *Proc. 29th Hawaii Intl. Conf. System Science,* volume 16, 1996.

[12] M. Holzrichter and S. Oliveira. A graph based Davidson algorithm for the graph partitioning problem. *International Journal of Foundations of Computer Science,* 10:225-246, 1999.

[13] M. Holzrichter and S. Oliveira. A graph based method for generating the Fiedler vector of irregular problems. In *Lecture Notes in Computer Science,* volume 1586, pages 978-985. Springer, 1999. Proceedings of the 11th IPPS/SPDP'99 workshops.

[14] S. E. Karisch and F. Rendl. Semidefinite programming and graph equipartition. In P. M. Pardalos and H. Wolkowicz, editors, *Topics in Semidefinite and InteriorPoint Methods,* volume 18, pages 77-95. AMS, 1998.

[15] G. Karypis and V. Kumar. METIS: Unstructured graph partitioning and sparse matrix ordering system Version 2.0. Technical report, Department of Computer Science, University of Minnesota, Minneapolis, MN, August 1995.

[16] C. Lanczos. Solution of systems of linear equations by minimized iterations. *J. Research Nat'l Bureau of Standards,* 49:33-53, 1952.

[17] S. Oliveira. On the convergence rate of a preconditioned subspace eigensolver. *Computing,* 63(2):219-231, December 1999.

[18] S. Oliveira and T. Soma. A multilevel algorithm for spectral partitioning with extended eigen-models. In *Lecture Notes in Computer Science,* volume 1800, pages 477-484. Springer, 2000. Proceedings of the 15th IPDPS 2000 workshops.

[19] S. Oliveira, D. Stewart, and T. Soma. A subspace semidefinite programming for spectral graph partit ioning. In P.M.A Sloot, C.K.K. Tan, J.J. Dongarra, and A. G. Hoekstra, editors, *Lecture Notes in Computer Science,* volume 2329, pages 10581067. Springer, 2002. Proceedings of International Conference on Computational Science-ICCS 2002, Part 1, Amsterdam, The Netherlands.

[20] F. Pellegrini. SCOTCH 3.1 user's guide. Technical Report 1137-96, Laboratoire Bordelais de Recherche en Informatique, Universite Bordeaux, France, 1996.

[21] A. Pothen, H. D. Simon, and Kang-Pu K. Liou. Partitioning sparse matrices with eigenvectors of graphs. *SIAM J. Matrix Anal. Appl.,* 11(3):430-452, 1990.

[22] R. Preis and R. Diekmann. The PARTY Partitioning-Library, User Guide - Version 1.1. Technical Report tr-rsfb-96-024, University of Paderborn, Germany, 1996.

[23] F. Rendl. A Matlab toolbox for semidefinite programming. Technical report, Technische Universitdt Graz, Institut fiir Mathematik, Kopernikusgasse 24, A-8010 Graz, Austria, 1994.

[24] F. Rendl, R. J. Vanderbei, and H. Wolkowicz. primal-dual interior point algorithms, and trust region subproblems. *Optimization Methods and Software,* 5:1-16, 1995.

[25] Y. Saad. *Numerical Methods for Large Eigenvalue Problems.* Manchester University Press, Oxford Road, Manchester M13 9PL, UK, 1992.

[26] G. L. G. Sleijpen and H. A. Van der Vorst. A Jacobi-Davidson iteration method for linear eigenvalue problems. *SIAM J. Matrix Anal. Appl.,* 17(2):401-425, 1996. Max-min eigenvalue problems,

[27] D. E. Stewart and Z. Leyk. *Meschach: Matrix Computations in C.* Australian National University, Canberra, 1994. Proceedings of the CMA, # 32.

[28] K. C. Toh, M. J. Todd, and P. H. Tiitüncii. SDPT3 - a Matlab software package for semidefinite programming, version 2.1. Technical report, School of Operations Research and Industrial Engineering, Cornell University, Ithaca, NY, September 1999.

[29] L. Vandenberghe and S. Boyd. Semidefinite programming. *SIAM Review,* 38:49-95, 1996.

[30] L. Vandenberghe and S. Boyd. SP Software for semidefinite programming User's guide, version 1.0. Technical report, Information System Laboratory, Stanford University, Stanford, CA, November 1998.

[31] C. Walshaw, M. Cross, and M. Everett. Mesh partitioning and load-balancing for distributed memory parallel systems. In B. Topping, editor, Proc. *Parallel & Distributed Computing for Computational Mechanics, Lochinver, Scotland,* 1998.

# A Development Environment for Multilayer Neural Network Applications Mapped onto DSPs with Multiprocessing Capabilities

Eugenio S. Cáner, José M. Seixas, and Rodrigo C. Torres

COPPE/EE/Federal University of Rio de Janeiro (UFRJ)
CP 68504, Rio de Janeiro 21945-970, Brazil
{eugenio,seixas,torres}@lps.ufrj.br
http://www.lps.ufrj.br

**Abstract.** This paper concerns the development of artificial neural network applications in digital signal processors (DSPs) with multiprocessing capabilities. For defining an efficient partition of processing tasks within target DSP boards, a user-friendly development environment evaluates different parallelism approaches for the network design according to specific device features. The development environment supports two different levels of experience (novice, expert) on both parallel system and neural network designs, enabling novice users to access system resources easily and shorten design cycle time. Pre-processing methods based on topological mapping and principal component analysis are also supported by the system, so that compact and efficient neural network system designs can be implemented.

## 1 Introduction

Artificial neural networks (ANNs) have been attracting a considerable attention as problem solvers of a wide range of real-world problems. One important feature that has been explored quite successfully on these networks is their inherent parallelism, which allows nonlinear adaptive processing to be performed at high speeds. Today, instead of an universal algorithm that is supposed to solve almost any problem at hand, ANNs are being increasingly considered as powerful processing blocks that interface with other nonneural processing blocks for building an overall hybrid processing system [1]. This tendency of integrating ANNs into a set of techniques that attacks the problem translates many times into pre-processing methods, which are applied to raw data. Whenever accumulated knowledge or exact methods are available they should be used on raw data, so that pre-processed data can be fed into the input nodes of the network. Pre-processing methods may also reduce ANN's complexity, so that the overall processing speed can be increased, which is very attractive for real-time applications [2, 3].

In this paper, we focus on feedforward multilayer artificial neural networks that are trained by the backpropagation algorithm [4]. This supervised learning

J.M.L.M. Palma et al. (Eds.): VECPAR 2002, LNCS 2565, pp. 717–730, 2003.

algorithm for multilayer topologies deserves attention, as the majority of the neural applications involve backpropagation training steps. On the other hand, as one crucial factor of such applications is the long time required by the learning process, especially when the ANN models become large, parallel processing can be considered in a practical design [5]. Thus, the aim of this work is to present an efficient development framework for neural network applications (working in conjunction with classical - or neural - pre-processing schemes), which is also capable to exploit the natural parallelism of the neural processing. For this, digital signal processor (DSP) technology with multiprocessing support is used. DSPs have become an effective choice for parallel processing implementations, both in terms of performance and cost. Moreover, neural processing is based on computing inner products and accessing look-up-tables, and DSPs are defined as processors with optimized topology (Harvard) for such computation tasks, which makes them more suitable for our application, with respect to general-purpose processors.

A multiprocessor board based on (four) ADSP-21062 DSPs [6] was selected as target platform: ASP-P14 [7]. Different parallelization approaches for both training and production phases of the neural network designs were evaluated according to main features (processing resources, connectivity, communication overheads) of the multiprocessing system. In this way, both novice and expert users can explore efficiently the potentialities of the processing platform using valuable implementation advices furnished by the proposed system.

The main contributions of the proposed neural developing system comes from its conceptual design. The system is coded in C language and it is based on simple and efficient modeling of the target multiprocessing environment, so that it is not restricted to the specific DSP family (SHARC) chosen in this work. Thus, the main concepts developed here can be used to build similar development systems for others DSP platforms. In addition, based on such modeling, the system furnishes advices for choosing the best parallel approach for the target problem, taking also into account the preprocessing phase, which is supported by a number of implemented techniques. Novice and expert users can fully exploit such features from different development environments available from the system.

The paper is organized as follows. Section 2 focuses on the main features of the DSP system. Section 3 describes different parallelization schemes for the backpropagation algorithm, and Section 4 shows how to map them optimally onto the ASP-P14 system. Section 5 describes the main features of the development environment proposed. Finally, in Section 6, some conclusions are derived.

## 2 System Hardware

One efficient approach for implementing fast and cost effective multiprocessing applications is to explore DSP technology, which has been attracting increasing attention as processors become faster and more powerful in computation capabilities.

The ASP PCI [7] family of processing boards can be installed into a PC and provides a high performance PCI interface to the SHARC (Super Harvard Architecture) family of DSPs [6]. The ASP-P14 integrates four SHARCs with 12 megabits of DRAM. A block diagram of the system is shown in Fig. 1.

The ADSP-21062 is part of the SHARC family. It is a 40 MHz, 32-bit DSP that has the ADSP-21020 as its core processor. The SHARC forms a complete system-on-a-chip, adding a dual-ported on-chip SRAM (2 Mbit deep) and integrated I/O peripherals supported by a dedicated I/O bus. With its on-chip instruction cache, the processor can execute every instruction in a single cycle (25 nanoseconds).

The ADSP-21062 includes functionality and features that allow the design of multiprocessing DSP systems. The unified address space allows direct interprocessor accesses of each ADSP-21062's internal memory. Distributed bus arbitration logic is included on-chip for simple, glueless connection of systems containing up to six ADSP-21062 and a host processor. Maximum throughput for interprocessor data transfer is claimed to be 240 Mbytes/sec over an external port.

The ADSP-21062 features six 4-bit link ports that provide additional I/O capabilities. Link port I/O is especially useful for point-to-point interprocessor communication in multiprocessing systems. The link ports can operate independently and simultaneously, with a maximum data throughput of 40 Mbytes/sec.

The development software includes an ANSI C Compiler. The compiler includes numerical C extensions for array selection, vector math operations, complex data types, circular pointers, and variably-dimensioned arrays. Other components of the development software include a C Runtime Library with custom DSP functions, C and Assembly language Debugger, Assembler, Assembly Library/Librarian, Linker, and Simulator.

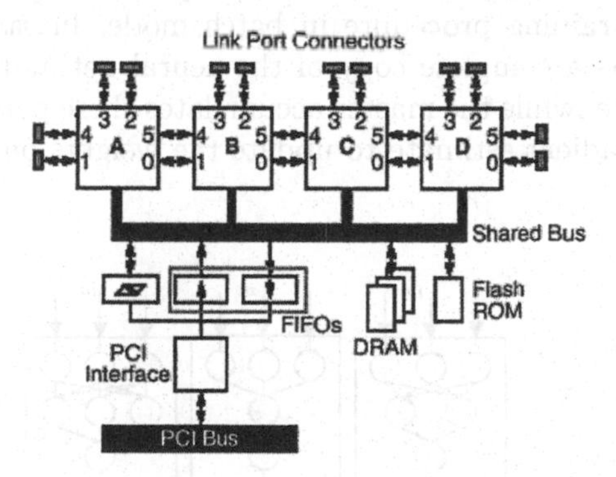


**Fig. 1.** ASP-P14 block diagram (quad-DSP version)

## 3 Source of Parallelism for Neural Networks

Different approaches to the problem of parallelizing the training process for neural networks have been developed, considering specific features of supervised learning algorithms. Here, multilayer feedforward topologies trained by means of the backpropagation algorithm are considered. The models of parallelism to be discussed are also employed in the validation and production phases, although adapted to the fact that the computation effort is considerably smaller for such phases.

The back-propagation algorithm reveals basically four different kinds of parallelism [8, 9], which are described next.

*Session Parallelism*: As it is well known, there is a considerable number of parameters to be selected in a standard ANN design. This difficulty combined with the absence of a general constructive method for determining the optimal network topology, and the corresponding freedom the designer has on choosing such topology, makes the ANN design rather involving for complex applications. To decrease the design cycle time, the whole training sessions may be performed in parallel. For this, each training session runs on a processing element (PE) and consists of a network, the training data, and a training algorithm, but using its own version of the network and a specific choice of parameters, e.g. momentum, learning rate, type of activation function, etc. All training sessions operate on the same set of training data in parallel (Fig. 2). Thus, this approach offers a fast way for an optimal tuning of the training parameters and also for evaluating the best network topology for a given application. Session parallelism is a rather trivial type of parallelism, but on the other hand it is the most efficient one for large-scale parallel computers, due to its low percentage of communication activities.

*Sample Parallelism*: The required gradient vector estimation is a result of adding its partial components, which correspond to the patterns visited in an epoch, from a training procedure in batch mode. In sample parallelism (Fig. 3) each slave has a complete copy of the neural net and computes part of the gradient estimate, while the master accumulates these partial estimates and uses the complete gradient estimate to update the weights on the networks running

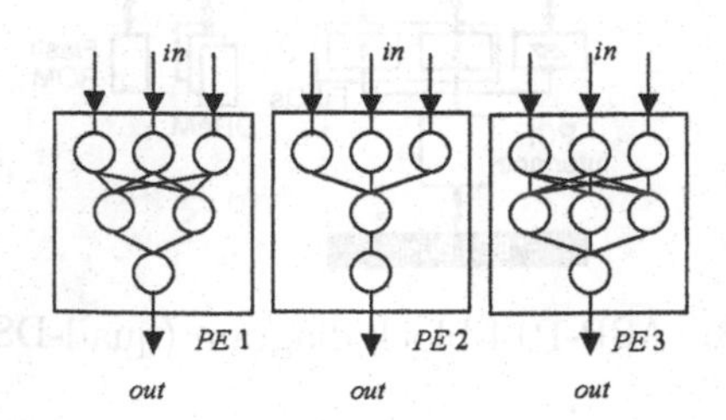


**Fig. 2.** Session Parallelism. Three processing elements (PEs) are shown, for illustration purpose

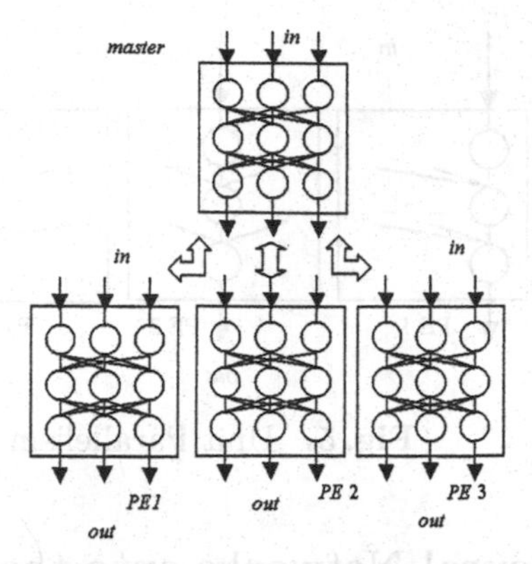


**Fig. 3.** Sample Parallelism

in the slaves. Sample parallelism is more adequate for training procedures based on large batch sizes.

*Pipelining*: Allows the training patterns to be pipelined between the layers, that is, the hidden and output layers are computed on different processors. While the output layer processor computes output and error values for the present training pattern, the hidden layer processor processes the next training pattern. The forward and backward phase may also be parallelized in a pipeline. Pipelining (Fig. 4) requires a training procedure in batch mode.

*Unit Parallelism*: This fine-grain approach partitions the neural net so that each processing node gets about the same quantity of synaptic weights. That is, each processor computes part of the intermediate values during forward and backward phases of the learning procedure for a given training set. Fig. 5 sketches this approach. Unit parallelism is rather communication intensive compared to sample parallelism and is more effective for large networks.

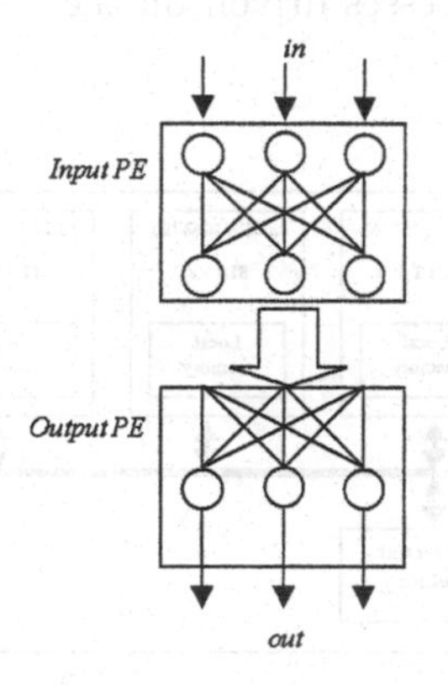


**Fig. 4.** Layer Pipelining

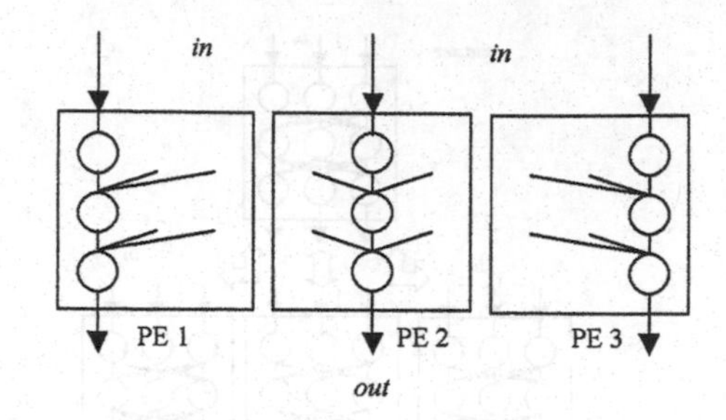


**Fig. 5.** Unit Parallelism

## 4 Mapping Neural Networks onto the ASP-P14 System

Cluster multiprocessing is best suited for applications where a fair amount of flexibility is required. This is especially true when a system must be able to support a variety of different tasks, some of which may be running concurrently. The cluster multiprocessing configuration using master-slave paradigm for the ASP-P14 system is shown in Fig. 6.

The target application can be loaded onto the network of SHARC processors in a number of ways [7]. We use the *tops* utility, a program that loads the application onto the SHARC network and services input/output requests from the executable running on the DSP labeled as $A$, which acts as master.

The master processor can communicate with the slave processors in a variety of ways: direct writing and reading; single-word transfers through buffers; communication via DMA and via link ports.

*Direct W/R*: The master node may access both the external memory of the system and the internal memories of the slave processors. The master may directly access the internal memory and IOP registers (control, status, or data buffer registers) of slave processors by reading or writing to the appropriate address locations in the specific multiprocessor memory space of the SHARCs. Each slave monitors addresses driven on the external bus and responds to any

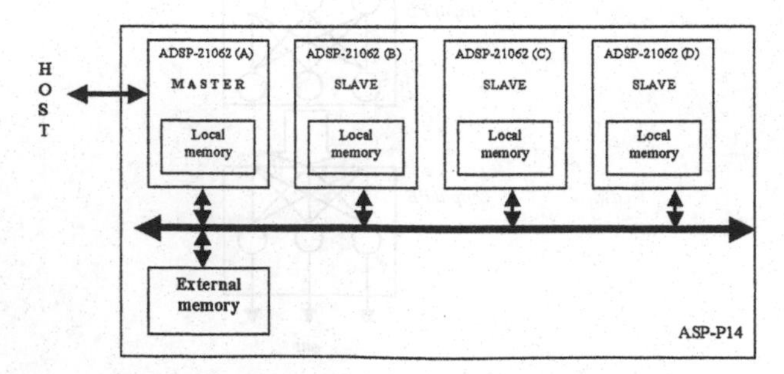


**Fig. 6.** Parallel bus and point-to-point communication for ASP-14 system

that fall within its multiprocessor memory space range [6]. These accesses are invisible to the slave core processor, because they are performed through the external port and via the on-chip I/O bus - not the DM bus or PM bus. This is an important distinction, because it allows uninterrupted program execution through the slave's core processor. When a Direct Write to a slave occurs, address and data are both latched on-chip by the I/O processor. The I/O processor buffers the address and data in a special set of FIFO buffers, so that up to six Direct Writes can be performed before another is delayed. Direct Reads cannot be pipelined like Direct Writes.

*Single-Word Transfers Through Buffers*: In addition to direct reading and writing, the master DSP bus can transfer data to and from the slaves through the external port FIFO buffers, EPB0-3. Each of these buffers, which are part of the IOP register set, is a six-location FIFO. Both single-word transfers and DMA transfers can be performed through the EPBx buffers. The ADSP-21062 DMA controller handles DMA transfers internally, but the ADSP-21062 core must handle single-word transfers. Because the EPBx buffers are six-deep FIFOs (in both directions), the master and the slave's core are able to read data efficiently. Continuous single-word transfers can, thus, be performed in real-time, with low latency and without using DMA.

*DMA*: The master can set up external port DMA channels to transfer blocks of data to and from a slave's internal memory or between a slave's internal memory and the external memory of the board using two modes. To set up the DMA transfer, the master must initialize the slave's control and parameter registers for the particular DMA channel to be used. Once the DMA channel is set up, the master may simply read from (or write to) the corresponding EPBx buffer on the slave, in order to perform the required data transfer. This is referred to as MODE 1. Another way is to set up its own DMA controller (MODE 2) to perform the transfers. If the slave's buffer is empty (or full), the access is extended until data are available (or stored). This method allows faster and more efficient data transfers.

*Link Port*: A bottleneck might exist within the cluster, as only two processors can communicate over the shared bus during each cycle and, consequently, other processors are held off until the bus is released. Since the SHARC can also perform point-to-point link port transfers within a cluster, this bottleneck can easily be eliminated. Data links between processors can dynamically be set up and initiated over the common bus. All six link ports can operate simultaneously on each processor. A disadvantage of the link ports is that individual transfers occur at a maximum of 40 Mbytes/sec, a lower rate than that one of the shared parallel bus. Since the link ports' 4-bit data path is smaller than the processor's native word size, the transfer of each word requires multiple clock cycles. Link ports may also require more software overhead and complexity because they must be set up on both sides of the transfers before they can occur.

Figures 7 and 8 summarize the measured communication performance of the ASP-P14 system, when the data size to be transmitted is varied. The performance is measured for different communication methods. This measurement

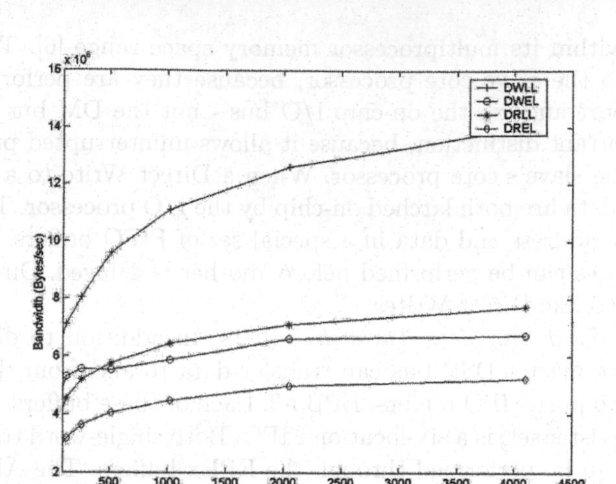


**Fig. 7.** Communication performance of the ASP-P14 using Direct W/R. See text

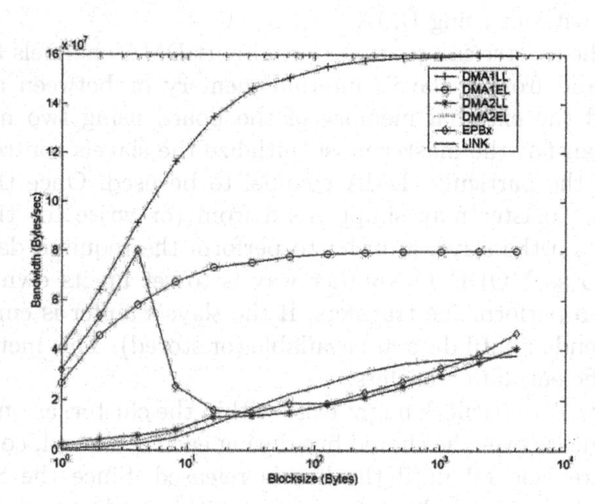


**Fig. 8.** Communication performance of the ASP-P14. See text

helps to define the best communication algorithm to be used in each ANN model and to estimate the time required for training, testing and production phases. Fig. 7 shows the communication performance for Direct W/R between local memories (DWLL/DRLL) and between the external memory and any local memory (DWEL/DREL). Fig. 8 shows the achieved performance for the two DMA modes: for MODE1, DMA1LL (communication between local memories) and DMA1EL (between the external memory and any local memory) and, in a similar way, for Mode 2, DMA2LL and DMA2EL. Fig. 8 also shows the achieved communication performance for single-word transfers through EPBx buffers (EPBx) and for Link Port (LINK).

During ANN simulation, initially the master reads design parameters from the configuration file (including number of layers, number of neurons, learning rate, etc), stored on the host PC, and retransmits them to all slaves using EPBx buffers. Each slave processor allocates the required memory space for the weighting vectors, output and error vectors, etc., and retransmits its address location to the master. In this way, future communications will be able to be accomplished via direct access or via DMA channel. For the sample parallelism, the master sends different training vectors to slaves, using DMA channel to transfer data. In case of unit parallelism, the master transfers training vectors using a combination of DMA and the broadcast-writes, which allow simultaneous data transmission to all DSPs in a multiprocessing system.

## 5 The Development Environment

The goal of the proposed development environment is to exploit models of parallelism for ANN applications in target machines, providing accurate performance predictions. For this purpose, a collection of subroutines written in C language implements the models discussed on Section 3. Using performance measurements of the communication channels of the target machine, simple parallel models can be emulated. Based on such models, users can get design advices from the development environment, both on how many processors are really required and what is the kind of parallelism that best suites the user's needs. Thus, even novices can come close to optimal designs without much effort.

The development environment provides an user-friendly interface for the training kernels with external data and network descriptions. This interface provides a comfortable way to configure the ANN and analyze the design performance. In addition, pre-processing of input data is supported by means of built-in functions and specific tools that can be used for the development of the required preprocessing algorithm of the target application. Topological mapping (rings, grouping neighbor data channels, etc) [10] and principal component analysis [4] are both supported as preprocessing schemes on raw data.

Fig. 9 sketches the basic structure stressing the interconnection among modules. A Graphic User Interface (*GUI*) was developed to work together with ASP-P14, in our specific case, but it can be adapted to different machines.

Using *GUI*, the user defines the configuration of the neural network (type, architecture, training parameters, type of parallelism, etc.) in a comfortable and efficient way, generating a Configuration File that holds the main design parameters. Depending on the chosen neural network, *GUI* requests one of the simulation programs of the "Neural Networks Package" (*NNP*) to be executed. *NNP* simulates the neural network defined in the Configuration File using the Input Data, and generates the corresponding Output File with the results from the simulation. *GUI* reads the Output File and generates different plots for allowing visualization of results (weights, errors, etc). Finally, it also generates an application report.

The user is able to choose or customize different preprocessing methods (principal component analysis, topological mapping), and the appropriate set of parameters. In other words, *GUI* accomplishes all the communication between the user and *NNP*, serving as a link among them.

Once started, the *GUI* asks the user about his skills in parallel programming and neural network design, splitting the users into two categories: beginners and advanced. For beginners, the environment provides a step-by-step configuration of the system, with helping facilities for each step. With this option, a novice user can exercise the environment purely at the *GUI* level.

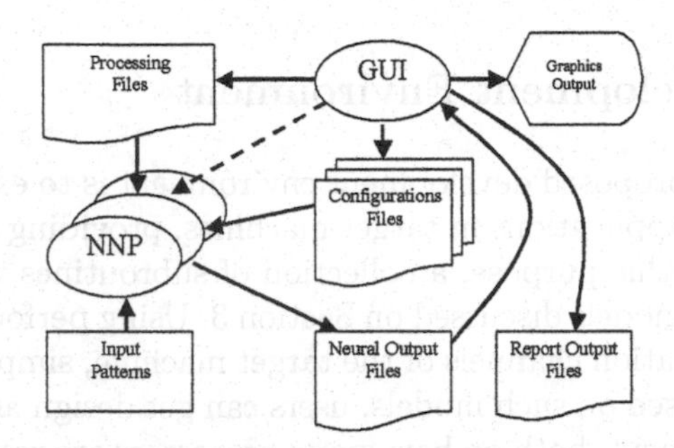


**Fig. 9.** Program building model

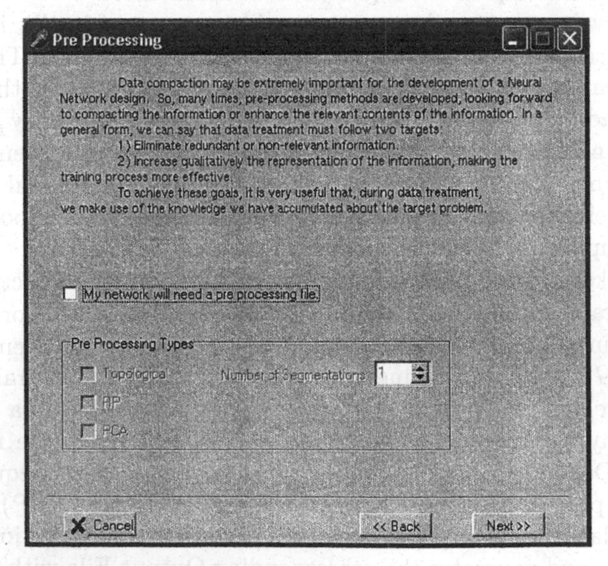


**Fig. 10.** Pre-processing for beginners

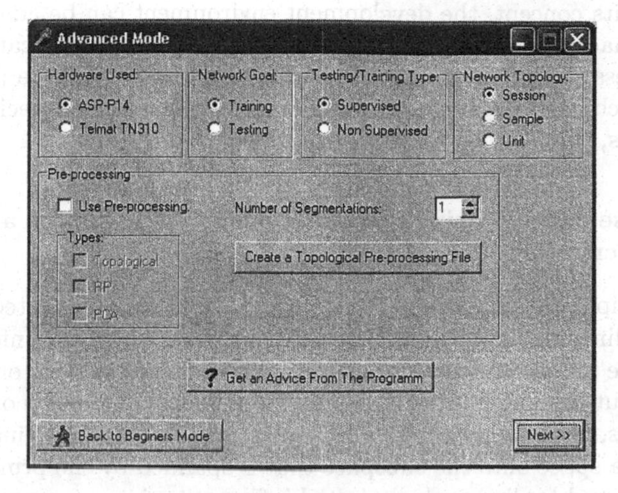


**Fig. 11.** Screen used for advanced users configuration

Fig. 10 illustrates how pre-processing may be accomplished through beginners mode operation. As it can be seen, an explanatory text is presented to help user's choices, requiring more time in this configuration phase, but providing the user with useful information about parallelism, neural networks and the system itself. A typical sequence of system screens presented to novice users is as follows:

1. Presentation screen, where the user can define what will be his operational mode (in this case, "Beginners").
2. Execution mode (Network Training, Testing or Production).
3. If training is chosen, the training method[1].
4. Pre-processing method (PCA, topological, specific or customized) and possible data segmentation (see Fig. 10).
5. Parallelism to be used (Session, Sample or Unit).
6. General configuration of the network (number of layers, number of processors, momentum, etc).
7. Node distribution in each layer and/or in each processor.
8. Transfer function running in each layer.
9. Finally, a screen with the results obtained during the execution is displayed.

For the experts, a fast and direct way to configure the system is made available (see Fig. 11). In this case, the options appear in groups, making the configuration process much more efficient for those who are already used to the system.

All options for the system can be accessed through buttons or menus. A Demo mode operation is also available, which allows the user to have a brief understanding on how the *GUI* works.

[1] Although not discussed here, non-supervised training is also supported by the system.

From its concept, the development environment can be adapted to different parallel machines with minor effort. Basically, the communication performance and processing speed for each processing element have to be measured for the target machine. Also, in case the applications require more specific preprocessing algorithms, they can be easily incorporated to the system.

### 5.1 Case Study: Principle Component Analysis for a Particle Discriminator

The principal component analysis (PCA) is well known as a technique that can perform dimensionality reduction with preservation of the information spread around the full data input space [4]. It aims at finding a set of orthogonal vectors in the input data space that accounts for as much as possible of the data's variance. Thus, one may expect that by projecting data of the original N-dimensional input data space onto the subspace that is spanned by the principal component directions and feeding such projected information into a neural classifier will be an efficient way to identify classes of events originally distributed in data input space.

ANNs have been used to perform principal component extraction [11]. In particular, the topology shown in Fig. 12, which shows a set of networks with a single hidden layer, N inputs and N outputs and exhibiting linear neurons, can extract components through backpropagation training. The first component is extracted by using one single unit in the hidden layer (Fig. 12a) and training the network in such way that the output vector is made as close as possible to the input vector. At the end of the training phase, the hidden unit reveals the first principal component in its weight vector. The other components are extracted in a similar way, by fixing the weights of the network used to extract the previous components, adding one more unit in the hidden layer (Fig. 12b) and training the free weights of the resulting network. The mean squared error (MSE) is the figure of merit used during the whole training phase. After M components have been extracted (typically, M << N), and observing that MSE does not significantly reduces by adding more components, one may consider that the set of M extracted components describes with adequate accuracy the event distribution in the original input space.

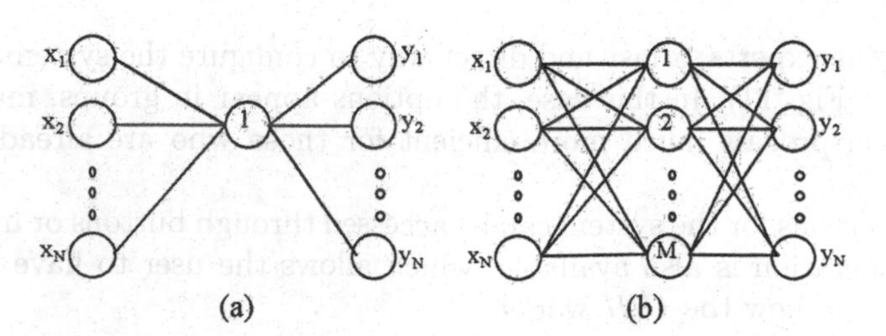


**Fig. 12.** Neural network topology for PCA

This neural principal component analysis was successfully applied to a particle discrimination problem in high-energy physics [12]. For a collider experiment that will be operational by the year 2007, the Large Hadron Collider (LHC)[13], an online validation system is being designed for triggering on very rare events, which are of interest to the experiment but are masked by the huge background noise generated by the colliding machine. This online system searches for discriminating (among other physics channels) electrons, which represent messengers of interesting events, from jets of particles, which fake the electron signal and represent the background noise to be removed. This electron/jet discriminator may be designed using ANNs and PCA can reduce significantly the dimensionality of the original high-dimensional input data space (121 components in this analysis). In this case, only seven components proved to be enough for input data reconstruction.

Such neural principal component analysis was implemented through the proposed development environment, as parallel implementation is required to cope with the high event rate (LHC will be colliding particles at periods of 25 nanoseconds). Both sequential and unit parallelism methods were evaluated, according to system advice.

## 6 Conclusion

A development environment for implementing artificial neural network applications in digital signal processor (DSP) systems with multiprocessing capabilities has been presented. Using simple models of the most efficient paradigms for parallel implementation of both training and testing phases of the neural network design, users can have a good estimation of the overall performance that can be achieved in practice. The system supports two levels of users, so that both novice and expert users, in terms of parallel processing applications or neural network designs, can exploit system features. In addition, built-in preprocessing schemes help users to integrate to the neural processing task their accumulated knowledge on the specific application, so that compact and efficient designs can be accomplished.

As modern DSPs match well the main computational requirements of artificial neural networks, and the proposed development environment can be adapted to different DSP systems, by measuring the communication performance of the system and using these performance curves for adjusting the schemes of parallel processing to be supported by the DSP system, a variety of applications can efficiently be developed.

## Acknowledgements

We are thankful for the financial support that has been provided by CNPq, FAPERJ, FUJB (Brazil), and CEADEN (Cuba) to this work.

## References

[1] Littmann, E.: Trends in Neural Network Research and an Application to Computer Vision. New Computing Techniques in Physics Research III. World Scientific (1994) 253–268
[2] Bock, R. K., Carter, J., Legrand, I. C.: What can Artificial Neural Networks do for the Global Second Level Trigger, ATLAS/DAQ Note, CERN (March 1994)
[3] Kiesling, C. et al: MPI: The H1 Neural Network Trigger. Sixth International Workshop on Software Engineering, Artificial Intelligence and Expert Systems on High-Energy and Nuclear Physics, Crete, Greece (1999)
[4] Haykin, S.: Neural Networks, a Comprehensive Foundation, 2nd. Edition. Prentice-Hall (1999)
[5] Zomaya, A. Y. H. (Editor): Parallel & Distributed Computing Handbook. McGraw-Hill (1996)
[6] Analog Device, Inc, User Manual ADSP-21060/21062. USA (1995)
[7] Transtech Parallel Systems, ASP SHARC Handbook. USA (1997)
[8] Singer, A.: Implementation of artificial neural networks on the Connection Machine, Parallel Computing Vol. 14 (1990) 305–315
[9] Nordstrom, T., Svensson, B.: Using And Designing Massively Parallel Computers for Artificial Neural Networks, Journal Of Parallel And Distributed Computing, 14 (1992) 260–285
[10] Seixas, J. M., Caloba, L. P., Souza, M. N., Braga, A. L., Rodrigues, A. P.: Neural Second-Level Trigger System Based on Calorimetry. Computer Physics Communications, Vol. 95 (1996) 143–157
[11] Simpson, P.: Artificial Neural Systems: Foundations, Paradigms, Applications and Implementations. Pergamon Press (1990)
[12] Seixas, J. M., Caloba, L. P., Kastrup, B.: A Neural Second-Level Trigger System Based on Calorimetry and Principal Component Analysis. New Computing Techniques in Physics Research IV. World Scientific (1995)
[13] CERN: `http://www.cern.ch`

# Author Index

# Lecture Notes in Computer Science

For information about Vols. 1–2532

please contact your bookseller or Springer-Verlag

Vol. 2533: N. Cesa-Bianchi, M. Numao, R. Reischuk (Eds.), Algorithmic Learning Theory. Proceedings, 2002. XI, 415 pages. 2002. (Subseries LNAI).

Vol. 2534: S. Lange, K. Satoh, C.H. Smith (Ed.), Discovery Science. Proceedings, 2002. XIII, 464 pages. 2002.

Vol. 2535: N. Suri (Ed.), Mobile Agents. Proceedings, 2002. X, 203 pages. 2002.

Vol. 2536: M. Parashar (Ed.), Grid Computing – GRID 2002. Proceedings, 2002. XI, 318 pages. 2002.

Vol. 2537: D.G. Feitelson, L. Rudolph, U. Schwiegelshohn (Eds.), Job Scheduling Strategies for Parallel Processing. Proceedings, 2002. VII, 237 pages. 2002.

Vol. 2538: B. König-Ries, K. Makki, S.A.M. Makki, N. Pissinou, P. Scheuermann (Eds.), Developing an Infrastructure for Mobile and Wireless Systems. Proceedings 2001. X, 183 pages. 2002.

Vol. 2539: K. Börner, C. Chen (Eds.), Visual Interfaces to Digital Libraries. X, 233 pages. 2002.

Vol. 2540: W.I. Grosky, F. Plášil (Eds.), SOFSEM 2002: Theory and Practice of Informatics. Proceedings, 2002. X, 289 pages. 2002.

Vol. 2541: T. Barkowsky, Mental Representation and Processing of Geographic Knowledge. X, 174 pages. 2002. (Subseries LNAI).

Vol. 2542: I. Dimov, I. Lirkov, S. Margenov, Z. Zlatev (Eds.), Numerical Methods and Applications. Proceedings, 2002. X, 174 pages. 2003.

Vol. 2543: O. Bartenstein, U. Geske, M. Hannebauer, O. Yoshie (Eds.), Web Knowledge Management and Decision Support. Proceedings, 2001. X, 307 pages. 2003. (Subseries LNAI).

Vol. 2544: S. Bhalla (Ed.), Databases in Networked Information Systems. Proceedings 2002. X, 285 pages. 2002.

Vol. 2545: P. Forbrig, Q. Limbourg, B. Urban, J. Vanderdonckt (Eds.), Interactive Systems. Proceedings 2002. XII, 574 pages. 2002.

Vol. 2546: J. Sterbenz, O. Takada, C. Tschudin, B. Plattner (Eds.), Active Networks. Proceedings, 2002. XIV, 267 pages. 2002.

Vol. 2547: R. Fleischer, B. Moret, E. Meineche Schmidt (Eds.), Experimental Algorithmics. XVII, 279 pages. 2002.

Vol. 2548: J. Hernández, Ana Moreira (Eds.), Object-Oriented Technology. Proceedings, 2002. VIII, 223 pages. 2002.

Vol. 2549: J. Cortadella, A. Yakovlev, G. Rozenberg (Eds.), Concurrency and Hardware Design. XI, 345 pages. 2002.

Vol. 2550: A. Jean-Marie (Ed.), Advances in Computing Science – ASIAN 2002. Proceedings, 2002. X, 233 pages. 2002.

Vol. 2551: A. Menezes, P. Sarkar (Eds.), Progress in Cryptology – INDOCRYPT 2002. Proceedings, 2002. XI, 437 pages. 2002.

Vol. 2552: S. Sahni, V.K. Prasanna, U. Shukla (Eds.), High Performance Computing – HiPC 2002. Proceedings, 2002. XXI, 735 pages. 2002.

Vol. 2553: B. Andersson, M. Bergholtz, P. Johannesson (Eds.), Natural Language Processing and Information Systems. Proceedings, 2002. X, 241 pages. 2002.

Vol. 2554: M. Beetz, Plan-Based Control of Robotic Agents. XI, 191 pages. 2002. (Subseries LNAI).

Vol. 2555: E.-P. Lim, S. Foo, C. Khoo, H. Chen, E. Fox, S. Urs, T. Costantino (Eds.), Digital Libraries: People, Knowledge, and Technology. Proceedings, 2002. XVII, 535 pages. 2002.

Vol. 2556: M. Agrawal, A. Seth (Eds.), FST TCS 2002: Foundations of Software Technology and Theoretical Computer Science. Proceedings, 2002. XI, 361 pages. 2002.

Vol. 2557: B. McKay, J. Slaney (Eds.), AI 2002: Advances in Artificial Intelligence. Proceedings, 2002. XV, 730 pages. 2002. (Subseries LNAI).

Vol. 2558: P. Perner, Data Mining on Multimedia Data. X, 131 pages. 2002.

Vol. 2559: M. Oivo, S. Komi-Sirviö (Eds.), Product Focused Software Process Improvement. Proceedings, 2002. XV, 646 pages. 2002.

Vol. 2560: S. Goronzy, Robust Adaptation to Non-Native Accents in Automatic Speech Recognition. Proceedings, 2002. XI, 144 pages. 2002. (Subseries LNAI).

Vol. 2561: H.C.M. de Swart (Ed.), Relational Methods in Computer Science. Proceedings, 2001. X, 315 pages. 2002.

Vol. 2562: V. Dahl, P. Wadler (Eds.), Practical Aspects of Declarative Languages. Proceedings, 2003. X, 315 pages. 2002.

Vol. 2565: J.M.L.M. Palma, J. Dongarra, V. Hernández, A. Augusto Sousa (Eds.), High Performance Computing for Computational Science – VECPAR 2002. Proceedings, 2002. XVII, 732 pages. 2003.

Vol. 2566: T.Æ. Mogensen, D.A. Schmidt, I.H. Sudborough (Eds.), The Essence of Computation. XIV, 473 pages. 2002.

Vol. 2567: Y.G. Desmedt (Ed.), Public Key Cryptography – PKC 2003. Proceedings, 2003. XI, 365 pages. 2002.

Vol. 2568: M. Hagiya, A. Ohuchi (Eds.), DNA Computing. Proceedings, 2002. XI, 338 pages. 2003.

Vol. 2569: D. Gollmann, G. Karjoth, M. Waidner (Eds.), Computer Security – ESORICS 2002. Proceedings, 2002. XIII, 648 pages. 2002. (Subseries LNAI).

Vol. 2570: M. Jünger, G. Reinelt, G. Rinaldi (Eds.), Combinatorial Optimization – Eureka, You Shrink!. Proceedings, 2001. X, 209 pages. 2003.

Vol. 2571: S.K. Das, S. Bhattacharya (Eds.), Distributed Computing. Proceedings, 2002. XIV, 354 pages. 2002.

Vol. 2572: D. Calvanese, M. Lenzerini, R. Motwani (Eds.), Database Theory – ICDT 2003. Proceedings, 2003. XI, 455 pages. 2002.

Vol. 2574: M.-S. Chen, P.K. Chrysanthis, M. Sloman, A. Zaslavsky (Eds.), Mobile Data Management. Proceedings, 2003. XII, 414 pages. 2003.

Vol. 2575: L.D. Zuck, P.C. Attie, A. Cortesi, S. Mukhopadhyay (Eds.), Verification, Model Checking, and Abstract Interpretation. Proceedings, 2003. XI, 325 pages. 2003.

Vol. 2576: S. Cimato, C. Galdi, G. Persiano (Eds.), Security in Communication Networks. Proceedings, 2002. IX, 365 pages. 2003.

Vol. 2578: F.A.P. Petitcolas (Ed.), Information Hiding. Proceedings, 2002. IX, 427 pages. 2003.

Vol. 2580: H. Erdogmus, T. Weng (Eds.), COTS-Based Software Systems. Proceedings, 2003. XVIII, 261 pages. 2003.

Vol. 2581: J.S. Sichman, F. Bousquet, P. Davidsson (Eds.), Multi-Agent-Based Simulation II. Proceedings, 2002. X, 195 pages. 2003. (Subseries LNAI).

Vol. 2583: S. Matwin, C. Sammut (Eds.), Inductive Logic Programming. Proceedings, 2002. X, 351 pages. 2003. (Subseries LNAI).

Vol. 2585: F. Giunchiglia, J. Odell, G. Weiß (Eds.), Agent-Oriented Software Engineering III. Proceedings, 2002. X, 229 pages. 2003.

Vol. 2586: M. Klusch, S. Bergamaschi, P. Edwards, P. Petta (Eds.), Intelligent Information Agents. VI, 275 pages. 2003. (Subseries LNAI).

Vol. 2587: P.J. Lee, C.H. Lim (Eds.), Information Security and Cryptology – ICISC 2002. Proceedings, 2002. XI, 536 pages. 2003.

Vol. 2588: A. Gelbukh (Ed.), Computational Linguistics and Intelligent Text Processing. Proceedings, 2003. XV, 648 pages. 2003.

Vol. 2589: E. Börger, A. Gargantini, E. Riccobene (Eds.), Abstract State Machines 2003. Proceedings, 2003. XI, 427 pages. 2003.

Vol. 2590: S. Bressan, A.B. Chaudhri, M.L. Lee, J.X. Yu, Z. Lacroix (Eds.), Efficiency and Effectiveness of XML Tools and Techniques and Data Integration over the Web. Proceedings, 2002. X, 259 pages. 2003.

Vol. 2591: M. Aksit, M. Mezini, R. Unland (Eds.), Objects, Components, Architectures, Services, and Applications for a Networked World. Proceedings, 2002. XI, 431 pages. 2003.

Vol. 2592: R. Kowalczyk, J.P. Müller, H. Tianfield, R. Unland (Eds.), Agent Technologies, Infrastructures, Tools, and Applications for E-Services. Proceedings, 2002. XVII, 371 pages. 2003. (Subseries LNAI).

Vol. 2593: A.B. Chaudhri, M. Jeckle, E. Rahm, R. Unland (Eds.), Web, Web-Services, and Database Systems. Proceedings, 2002. XI, 311 pages. 2003.

Vol. 2594: A. Asperti, B. Buchberger, J.H. Davenport (Eds.), Mathematical Knowledge Management. Proceedings, 2003. X, 225 pages. 2003.

Vol. 2595: K. Nyberg, H. Heys (Eds.), Selected Areas in Cryptography. Proceedings, 2002. XI, 405 pages. 2003.

Vol. 2597: G. Păun, G. Rozenberg, A. Salomaa, C. Zandron (Eds.), Membrane Computing. Proceedings, 2002. VIII, 423 pages. 2003.

Vol. 2598: R. Klein, H.-W. Six, L. Wegner (Eds.), Computer Science in Perspective. X, 357 pages. 2003.

Vol. 2599: E. Sherratt (Ed.), Telecommunications and beyond: The Broader Applicability of SDL and MSC. Proceedings, 2002. X, 253 pages. 2003.

Vol. 2600: S. Mendelson, A.J. Smola, Advanced Lectures on Machine Learning. Proceedings, 2002. IX, 259 pages. 2003. (Subseries LNAI).

Vol. 2601: M. Ajmone Marsan, G. Corazza, M. Listanti, A. Roveri (Eds.) Quality of Service in Multiservice IP Networks. Proceedings, 2003. XV, 759 pages. 2003.

Vol. 2602: C. Priami (Ed.), Computational Methods in Systems Biology. Proceedings, 2003. IX, 214 pages. 2003.

Vol. 2604: N. Guelfi, E. Astesiano, G. Reggio (Eds.), Scientific Engineering for Distributed Java Applications. Proceedings, 2002. X, 205 pages. 2003.

Vol. 2606: A.M. Tyrrell, P.C. Haddow, J. Torresen (Eds.), Evolvable Systems: From Biology to Hardware. Proceedings, 2003. XIV, 468 pages. 2003.

Vol. 2607: H. Alt, M. Habib (Eds.), STACS 2003. Proceedings, 2003. XVII, 700 pages. 2003.

Vol. 2609: M. Okada, B. Pierce, A. Scedrov, H. Tokuda, A. Yonezawa (Eds.), Software Security – Theories and Systems. Proceedings, 2002. XI, 471 pages. 2003.

Vol. 2611: S. Cagnoni, J.J. Romero Cardalda, D.W. Corne, J. Gottlieb, A. Guillot, E. Hart, C.G. Johnson, E. Marchiori, J.-A. Meyer, M. Middendorf, G.R. Raidl (Eds.), Applications of Evolutionary Computing. Proceedings, 2003. XXI, 708 pages. 2003.

Vol. 2612: M. Joye (Ed.), Topics in Cryptology – CT-RSA 2003. Proceedings, 2003. XI, 417 pages. 2003.

Vol. 2614: R. Laddaga, P. Robertson, H. Shrobe (Eds.), Self-Adaptive Software: Applications. Proceedings, 2001. VIII, 291 pages. 2003.

Vol. 2615: N. Carbonell, C. Stephanidis (Eds.), Universal Access. Proceedings, 2002. XIV, 534 pages. 2003.

Vol. 2616: T. Asano, R. Klette, C. Ronse (Eds.), Geometry, Morphology, and Computational Imaging. Proceedings, 2002. X, 437 pages. 2003.

Vol. 2618: P. Degano (Ed.), Programming Languages and Systems. Proceedings, 2003. XV, 415 pages. 2003.

Vol. 2619: H. Garavel, J. Hatcliff (Eds.), Tools and Algorithms for the Construction and Analysis of Systems. Proceedings, 2003. XVI, 604 pages. 2003.

Vol. 2620: A.D. Gordon (Ed.), Foundations of Software Science and Computational Structures. Proceedings, 2003. XII, 441 pages. 2003.

Vol. 2621: M. Pezzè (Ed.), Fundamental Approaches to Software Engineering. Proceedings, 2003. XIV, 403 pages. 2003.

Vol. 2622: G. Hedin (Ed.), Compiler Construction. Proceedings, 2003. XII, 335 pages. 2003.

Vol. 2623: O. Maler, A. Pnueli (Eds.), Hybrid Systems: Computation and Control. Proceedings, 2003. XII, 558 pages. 2003.

Vol. 2626: J.L. Crowley, J.H. Piater, M. Vincze, L. Paletta (Eds.), Computer Vision Systems. Proceedings, 2003. XIII, 546 pages. 2003.